# HUMAN GROWTH

## 1
## Principles and Prenatal Growth

HUMAN GROWTH

# HUMAN GROWTH

## 1
## Principles and Prenatal Growth

**Edited by**

## Frank Falkner
*The Fels Research Institute*
*Wright State University School of Medicine*
*Yellow Springs, Ohio*

**and**

## J. M. Tanner
*Institute of Child Health*
*London, England*

*BAILLIÈRE TINDALL · London*

A BAILLIÈRE TINDALL book published by Cassell Ltd.
35 Red Lion Square, London WC1R 4SG
and at Sydney, Auckland, Toronto, Johannesburg
an affiliate of Macmillan Publishing Co. Inc., New York

HUMAN GROWTH

Volume 1   Principles and Prenatal Growth      ISBN 0 7020 0731 5
Volume 2   Postnatal Growth                    ISBN 0 7020 0732 3
Volume 3   Neurobiology and Nutrition          ISBN 0 7020 0733 1

© 1978 Plenum Press, New York
A Division of Plenum Publishing Corporation
227 West 17th Street, New York, N.Y. 10011

First published 1978

# Contributors

**PETER A. J. ADAM**
Professor of Pediatrics
Case Western Reserve University
    School of Medicine
Cleveland Metropolitan General
    Hospital
Cleveland, Ohio

**KARLIS ADAMSONS**
Director and Chairman
Department of Obstetrics and
    Gynecology
Brown University Program in
    Medicine
Providence, Rhode Island

**MICHEL L. AUBERT**
Département de Pédiatrie et Génétique
Clinique Universitaire de Pédiatrie
Université de Genève
Genève, Switzerland

**STEPHEN M. BAILEY**
Center for Human Growth and
    Development
University of Michigan
Ann Arbor, Michigan

**CATHERINE BONAÏTI-PELLIÉ**
Clinique et Unite de Rechèrches de
    Génétique Médicale
Hôpital des Enfants-Malades
Paris, France

**C. O. CARTER**
Director, MRC Clinical Genetics Unit
Institute of Child Health
London, England

**CHARLOTTE CATZ**
Pregnancy and Infancy Branch
National Institute of Child Health and
    Human Development
National Institutes of Health
Bethesda, Maryland

**JOSEPH DANCIS**
Professor and Chairman
Department of Pediatrics
New York University School of
    Medicine
New York, New York

**FRANK FALKNER**
Scientific Director, Fels Research
    Institute
Fels Professor of Pediatrics
Wright State University School of
    Medicine
Yellow Springs, Ohio
Visiting Professor of Pediatrics, and
    Obstetrics and Gynecology
University of Cincinnati School of
    Medicine

**PHILIP FELIG**
Professor and Vice Chairman
Department of Internal Medicine
Chief, Section of Endocrinology
Yale University School of Medicine
New Haven, Connecticut

**JEAN FRÉZAL**
Professeur de la Clinique de Génétique
  Médicale
Clinique et Unité de Recherches de
  Génétique Médicale, INSERM
Hôpital des Enfants-Malades
Paris, France

**STANLEY M. GARN**
Professor of Human Development
Center for Human Growth and
  Development
The University of Michigan
Ann Arbor, Michigan

**GERALD E. GAULL**
New York State Institute for Basic
  Research in Mental Retardation
Staten Island, New York
Mount Sinai School of Medicine of the
  City University of New York, New
  York

**HARVEY GOLDSTEIN**
Professor and Head, Department of
  Statistics and Computing
University of London
Institute of Education
London, England

**RICHARD J. GOSS**
Professor of Biology
Division of Biological and Medical
  Sciences
Brown University
Providence, Rhode Island

**A. R. HAYWARD**
Department of Immunology
Institute of Child Health
London, England

**M. J. R. HEALY**
Professor of Computing and Statistics
London School of Hygiene and Tropi-
  cal Medicine
London, England

**FRITZ A. HOMMES**
Laboratory of Developmental
  Biochemistry
Department of Pediatrics
University of Groningen School of
  Medicine
Groningen, The Netherlands

**W. A. MARSHALL**
Professor of Human Biology
University of Technology
Loughborough, Leicestershire,
  England

**ETTORE MARUBINI**
Professor of Biometry
Department of Biometry and Medical
  Statistics
University of Milan
Milan, Italy

**R. A. McCANCE**
Sidney Sussex College
Cambridge, England

**JACK METCOFF**
Professor of Pediatrics
Professor of Biochemistry and Molecular Biology
Health Sciences Center
University of Oklahoma
Oklahoma City, Oklahoma

**D. F. ROBERTS**
Department of Human Genetics
The University of Newcastle upon Tyne
Newcastle upon Tyne, England

**ELIZABETH B. ROBSON**
Assistant Director
MRC Human Biochemical Genetics Unit
Galton Laboratory
University College
London, England

**JACQUES F. ROUX**
Professor of Obstetrics and Gynecology
University of Montréal
Director, Department of Obstetrics and Gynecology
Hôtel Dieu/Ste.-Justine Hospitals
Montréal, Quebec, Canada

**HENNING SCHNEIDER**
Assistant Professor
Department of Obstetrics and Gynecology
New York University School of Medicine
New York, New York

**DOUGLAS R. SHANKLIN**
Professor of Obstetrics and Gynecology
Professor of Pathology
University of Chicago
Pathologist-in-Chief
Chicago Lying-in Hospital
Chicago, Illinois

**PIERRE C. SIZONENKO**
Professeur-Assistant
Département de Pédiatrie et Génétique
Clinique Universitaire de Pédiatrie
Université de Genève
Genève, Switzerland

**D. A. T. SOUTHGATE**
Dunn Nutritional Laboratory
Medical Research Council
University of Cambridge
Cambridge, England

**ELSIE M. WIDDOWSON**
Addenbrooke's Hospital
Cambridge, England

**SUMNER J. YAFFE**
Chief, Division of Clinical Pharmacology
Children's Hospital of Philadelphia
Professor of Pediatrics and Pharmacology
University of Pennsylvania
Philadelphia, Pennsylvania

# *Preface*

Growth, as we conceive it, is the study of change in an organism not yet mature. Differential growth creates form: external form through growth rates which vary from one part of the body to another and one tissue to another; and internal form through the series of time-entrained events which build up in each cell the specialized complexity of its particular function. We make no distinction, then, between growth and development, and if we have not included accounts of differentiation it is simply because we had to draw a quite arbitrary line somewhere.

It is only rather recently that those involved in pediatrics and child health have come to realize that growth is the basic science peculiar to their art. It is a science which uses and incorporates the traditional disciplines of anatomy, physiology, biophysics, biochemistry, and biology. It is indeed a part of biology, and the study of human growth is a part of the curriculum of the rejuvenated science of Human Biology. What growth is not is a series of charts of height and weight. Growth standards are useful and necessary, and their construction is by no means void of intellectual challenge. They are a basic instrument in pediatric epidemiology. But they do not appear in this book, any more than clinical accounts of growth disorders.

This appears to be the first large handbook—in three volumes—devoted to Human Growth. Smaller textbooks on the subject began to appear in the late nineteenth century, some written by pediatricians and some by anthropologists. There have been magnificent mavericks like D'Arcy Thompson's *Growth and Form*. In the last five years, indeed, more texts on growth and its disorders have appeared than in all the preceeding fifty (or five hundred). But our treatise sets out to cover the subject with greater breadth than earlier works.

We have refrained from dictating too closely the form of the contributions; some contributors have discussed important general issues in relatively short chapters (for example, Richard Goss, our opener, and Michael Healy); others have provided comprehensive and authoritative surveys of the current state of their fields of work (for example, Robert Balázs and his co-authors). Most contributions deal with the human, but where important advances are being made although data from the human are still lacking, we have included some basic experimental work on animals.

Inevitably, there are gaps in our coverage, reflecting our private scotomata, doubtless, and sometimes our judgment that no suitable contributor in a particular field existed, or could be persuaded to write for us (the latter only in a couple of instances, however). Two chapters died on the hoof, as it were. Every reader will notice the lack of a chapter on ultrasonic studies of the growth of the fetus; the manuscript, repeatedly promised, simply failed to arrive. We had hoped, also, to include a chapter on the very rapidly evolving field of the development of the visual processes, but here also events conspired against us. We hope to repair these omissions in a second edition if one should be called for; and we solicit correspondence, too, on suggestions for other subjects.

We hope the book will be useful to pediatricians, human biologists, and all concerned with child health, and to biometrists, physiologists, and biochemists working in the field of growth. We thank heartily the contributors for their labors and their collective, and remarkable, good temper in the face of often bluntish editorial comment. No words of praise suffice for our secretaries, on whom very much of the burden has fallen. Karen Phelps, at Fels, handled all the administrative arrangements regarding what increasingly seemed like innumerable manuscripts and rumors of manuscripts, retyped huge chunks of text, and maintained an unruffled and humorous calm through the whole three years. Jan Baines, at the Institute of Child Health, somehow found time to keep track of the interactions of editors and manuscripts, and applied a gentle but insistent persuasion when any pair seemed inclined to go their separate ways. We wish to thank also the publishers for being so uniformly helpful, and above all the contributors for the time and care they have given to making this book.

Frank Falkner
James Tanner

*Yellow Springs and London*

# Contents

## III Genetics

*Chapter 8*

### Introduction to Genetic Analysis

*Jean Frézal and Catherine Bonaïti-Pellié*

*Chapter 9*

### The Genetics of Human Fetal Growth

*D. F. Roberts*

*Chapter 10*

### The Genetics of Birth Weight

*Elizabeth B. Robson*

## Chapter 15

## Fetal Measurements

D. A. T. Southgate

## Chapter 16

## Implications for Growth in Human Twins

Frank Falkner

## Chapter 17

## Association of Fetal Growth with Maternal Nutrition

Jack Metcoff

## Chapter 18

## Carbohydrate, Fat, and Amino Acid Metabolism in the Pregnant Woman and Fetus

Peter A. J. Adam and Philip Felig

*Chapter 19*

**Pre- and Perinatal Endocrinology**

*Pierre C. Sizonenko and Michel L. Aubert*

*Chapter 20*

**Development of Immune Responsiveness**

*A. R. Hayward*

*Chapter 21*

**Fetal Growth: Obstetric Implications**

*Karlis Adamsons*

# Contents of Volumes 2 and 3

# I

# *Developmental Biology*

# 1

# *Adaptive Mechanisms of Growth Control*

## *RICHARD J. GOSS*

## *1. Mechanisms of Organ and Tissue Growth*

At a meeting in Rome in 1894, Professor Giulio Bizzozero (1894) speculated before his fellow pathologists that the various tissues and organs of the body might be classified according to their mitotic potentials. He referred to those tissues which are in a constant state of proliferation as *elementi labili*. In others, the cells multiply during maturation, but become mitotically stable in the adult. These he classified *elementi stabili*. Finally, *elementi perenni* included the nervous and muscular tissues, the cells of which are incapable of division beyond early stages of development. These categories have since come to be known as mitotically renewing, expanding, and static tissues, respectively (Figure 1).

### *1.1. Mitotic Potentials*

Renewing tissues, such as the epidermis, mucosal epithelium of the gut, seminiferous epithelium, and blood cells, are in a continual state of multiplication as cellular losses are balanced by replacements. The germinative zone, spatially distinct from the differentiated compartment, is composed of relatively undifferentiated cells in a constant state of proliferation. Some of their descendants then undertake a course of differentiation which precludes further mitosis and usually leads to their demise.

In expanding tissues there is no incompatibility between differentiation and proliferation. The cells of glands never lose the potential for mitosis, although in the adult organism there is no need for proliferation except in cases of injury or increased work load. These tissues differ from renewing ones in that when growth occurs the distribution of dividing cells is diffuse. Since all cells are capable of division, there is no need for a germinative compartment distinct from the differentiated one.

---

*RICHARD J. GOSS* • Brown University, Providence, Rhode Island.

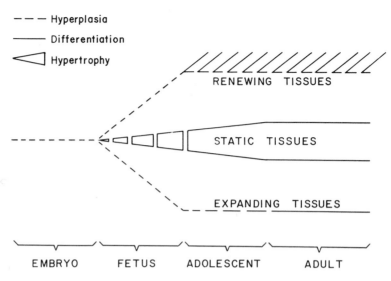

Fig. 1. Alternate pathways of growth and differentiation in tissues and organs. Renewing tissues continue to proliferate throughout life, giving off descendants that differentiate into nonmitotic cells. Static tissues permanently lose the capacity to divide as they differentiate early in life; thereafter they grow by hypertrophy. Expanding tissues do not stop dividing until body growth ceases, despite their fully differentiated condition. (From Goss, 1967.)

Mitotically static tissues, namely, the nervous system and striated muscle, do not proliferate beyond early stages of development when neuroblasts and myoblasts begin to differentiate into neurons and muscle fibers. These cells, like those in expanding organs, are capable of living as long as the organism as a whole survives. In point of fact, their longevity accounts for the chief difference between static and renewing tissues. In both cases the differentiated cells are incapable of division. In static tissues the germinative compartment is separated from the differentiated one in time. In renewing tissues the separation is spatial.

There are some tissues that defy classification. Skeletal elements, for example, possess growth zones (perichondrium, periosteum, epiphyseal plate) and therefore have something in common with renewing tissues. However, since there is little if any cell loss, they are in a sense expanding tissues. The adrenal cortex is also hard to classify. It too has a growth zone at the interface between the zona glomerulosa and the zona fasciculata. It might therefore be thought of as a renewing tissue, the more so for the centripetal migration of its cells and their disappearance from the zona reticularis. Yet the adrenal cortex resembles expanding tissues in that its differentiated cells are still capable of proliferation. A final exception to the rule is the lens, which shares attributes with all three types of tissues. It possesses a growth zone at the equator where lens epithelial cells proliferate and give rise to new lens fibers. In this sense it is a renewing tissue, and like other renewing tissues the differentiated lens fibers are themselves mitotically incompetent. Indeed, they lose all of their cytological organelles as well as their nuclei, yet their lifelong persistence testifies to the fact that they are in a living limbo. Because the cells of the lens are sequestered rather than lost, and because the lens ceases to grow when the body as a whole stops enlarging, there is reason to classify the lens as an expanding organ. It also resembles mitotically static tissues in that the lens fibers are nonmitotic.

### 1.1.1. Mitosis versus Differentiation

It is generally held that differentiation and proliferation are mutually exclusive (Figure 2). This would seem to be the case with respect to static and renewing tissues. Neurons and muscle fibers do not divide; they grow by cellular hypertrophy. However, skeletal muscle fibers, being multinucleate, are capable of augmenting their nuclear populations by fusion with mitotically competent mononucleate myoblasts. The alternative mode of growth would have been to permit nuclear division in differentiated multinucleate muscle fibers, but if this were the case all nuclei in a given fiber would be expected to divide synchronously. Such cells would have to double their nuclear complement every time they needed to enlarge, rather than add new nuclei one at a time.

The nonmitotic condition of highly specialized cells in static and renewing tissues would appear to be correlated with their intracellular retention of specific products. Differentiation is characterized by the production of specific proteins, and in nonproliferative cells these are not secreted. Hence, it may be the presence of hemoglobin in erythrocytes, keratin in epidermal cells, crystallins in lens fibers, or myosin in muscle fibers that makes it impossible for such cells to divide.

This is in contrast to the ease with which fully differentiated cells in expanding tissues can divide when called upon to do so. Virtually all cells in exocrine and endocrine glands, as well as the kidney, are able to multiply throughout life, whether it be during normal ontogeny or in response to injury or overwork. Unlike their counterparts in renewing and static tissues, the differentiated cells of expanding organs make their products for export in the form of hormones or the wide variety of materials secreted by exocrine glands. Yet even when these products are still present in the cytoplasm prior to release, they do not prevent cell division.

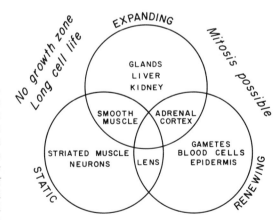

Fig. 2. Relationships between static, renewing, and expanding tissues. Static tissues differ from the others in their lack of mitotic potential. Renewing tissues have growth zones spatially distinct from the differentiated compartment, and the life spans of their cells are relatively short. Expanding tissues are capable of cell division in the fully differentiated state; their specific products are secreted from the cells and their functions are, in general, chemical in nature. As indicated, some tissues and organs of the body share attributes in common with more than one of these categories. (From Goss, 1967.)

In general, the cells of expanding organs perform chemical functions, while those of renewing and static tissues are more physical in their activities (e.g., contraction, conduction, oxygen transport). In the case of skeletal tissues the cell-specific products are cartilaginous and bony matrices. While the function of these tissues is clearly mechanical, their products are extracellular; yet fully differentiated chondrocytes and osteocytes seldom divide. Perhaps it is the proximity of these cells to their products within which they are imprisoned which discourages division. Only when released from their matrices do they regain their proliferative potentials.

## 1.2. Proliferation of Functional Units

Bizzozero's classification of tissue growth in terms of their mitotic profiles has been of great value in our interpretation of histology, both normal and pathological. To understand how growth is regulated, however, it is necessary to ask to what extent organs enhance their physiological efficiencies when they increase in size. In this respect cellular proliferation is not always sufficient to promote functional competence most efficiently. It can do so only when the functional units of a tissue are the cells themselves.

A functional unit may be defined as the smallest irreducible structure still capable of performing the physiological activities specific to the tissue or organ. In many cases, the functional unit may be represented by cells, as in the blood cells or in many of the endocrine glands. In other tissues, however, the functional unit may be subcellular, as in the case of the sarcomeres and myofibrils in muscle fibers, or the synapses of neurons. Growth in these tissues involves the production of new functional units, but the extent to which they may be augmented is limited by the maximum dimensions beyond which cells cannot grow.

In some organs the functional units are multicellular. They cannot be increased simply by the multiplication of their constituent cells, although this does result in their enlargement and therefore a limited increase in functional efficiency. It would be more to the advantage of the organism, however, were these organs capable of proliferating their functional units at the histological level of organization. In some cases this is possible, as in follicle production in ovaries and thyroid glands, or in the differentiation of new secretory acini in exocrine glands. Because of the apparently unlimited capacity for augmenting their populations of functional units at the histological level of organization, these organs are capable of remarkable feats of growth and regeneration.

Not every organ whose functional units are multicellular is capable of increasing their numbers. In the kidney, for example, the number of nephrons is fixed early in life and cannot be increased after maturation. Thus, compensatory growth of the kidney following unilateral nephrectomy (Nowinski and Goss, 1969) is achieved by cellular hyperplasia and hypertrophy. These processes account for the enlargement of preexisting nephrons, but do not provide for the *de novo* formation of new ones. The same situation obtains in the pulmonary alveoli of lungs, the villi of the small intestine, and the seminiferous tubules of the testes. Growth is possible in these adult organs, but it is achieved only by increasing the sizes of the functional units, not their numbers. Hypertrophy of functional units at the histological level of organization does in fact increase the functional efficiency of the organs in question, but not nearly so much as would have been the case had new functional units been created.

The aforementioned examples lead to the conclusion that if an organ can multiply its functional units, it is capable of unlimited growth and regeneration without sacrificing its functional efficiency. Conversely, those which cannot are able to compensate for deficiencies only to a limited extent. The ultimate sizes to which they can grow are strictly limited by the numbers of functional units that formed in early developmental stages. Thus, it is the level of organization at which hyperplasia takes place that determines the growth potentials of the organs in the body. For reasons that are difficult to understand, some organs are endowed with unlimited powers of growth while others are handicapped by built-in restrictions. It is even more curious when one realizes that some of the most vitally essential organs of the body are the very ones which are least capable of compensatory growth and regeneration. The heart and the brain cannot multiply their cells, nor can the kidneys and lungs augment their populations of nephrons and alveoli. On the other hand, many organs which are not indispensable for survival, such as the thyroid, mammary glands, exocrine pancreas, and salivary glands, readily increase their sizes or compensate for deficiencies by compensatory growth to prodigious extents.

### 1.2.1. Indeterminate Growth in Lower Vertebrates

Why are some organs so proficient at growing while others are so limited? The answer must be sought in their phylogenetic histories. The restrictions described above apply to warm-blooded vertebrates, but our cold-blooded ancestors do not necessarily live by the same rules that obtain for birds and mammals (Goss, 1974). Certain kinds of fishes, amphibians, and reptiles continue to grow throughout life, albeit at a decelerating pace. Studies have shown that in some of these forms the organs which in mammals are so limited in their growth can continue to enlarge as long as the body as a whole grows. The kidneys of fishes, for example, do not lose the ability to differentiate new nephrons, as a result of which the lifelong growth of the kidney is paralleled by proportional increases in functional competence (Nash, 1931). Likewise, the heart of a large fish contains myofibers which are approximately the same dimensions as those in small ones. This suggests that there must be an increase in their numbers, although the source of these new cardiac muscle fibers remains to be discovered (Goss, 1971). Even the brains of fishes appear to be capable of increasing their complements of neurons by the proliferation and differentiation of cells (Luquet and Hannequart, 1974). In the retina, new photoreceptors are added by the division of cells peripherally followed by their differentiation into new rods and cones (Blaxter and Jones, 1967). Thus, where mammals are so restricted in their growth, lower vertebrates seem to have unlimited potential for enlargement.

## 2. Regulation of Organ and Tissue Growth

Whatever may be their modes of growth, every organ and tissue in the body must be subject to regulatory mechanisms that control the dimensions to which they grow. Beyond early developmental stages, the so-called ''normal'' size of an organ represents a compromise between influences tending to promote either atrophy or hypertrophy. In most cases thus far explored, these influences are closely related to the physiological factors which control functional activity. In this way, organs of the body tend to enlarge when overworked and atrophy from disuse, thus ensuring that the mass of an organ is optimally suited to the work it must perform (Goss, 1964, 1966, 1967, 1972).

The methods of investigation by which these mechanisms of growth regulation have been explored are aimed at upsetting the normal balance between structure and function. Compensatory growth can be brought about by increasing physiological demands on the amount of a tissue available. This can be achieved either by decreasing the mass of an organ, so that the remaining portions must work harder to carry out their normal work loads, or by heightening functional demands upon an otherwise normal mass of tissue. Conversely, atrophy is brought about by transplanting extra organs without increasing the work load or by reducing the demands for physiological activity by various experimental interventions, the nature of which differs from organ to organ.

The various organs in the body can be categorized according to whether their growth is regulated by factors that are systemically distributed or restricted to the local environment. In the former case, the tissue performs a function which serves the body as a whole, and its growth and physiology are subject to control mechanisms emanating from outside its own confines. In the latter case, the domain of the tissue's function is spatially restricted, and the control of its growth is similarly limited.

### 2.1. Visceral Organs: Humoral Growth Control

Not surprisingly, many of the organs which serve a body-wide function are subject to hormonal controls. The classic situation in which target organs are regulated by trophic hormones serves to remind us of how intimately correlated are the function and the growth of such structures. Typically, if the hormones produced by the target glands are reduced, the output of trophic hormones is correspondingly increased. Thus, interruption of thyroxine secretion results in the production of a goitre due to chronic stimulation by TSH. On the other hand, if both ovaries are removed but part of one is transplanted to the spleen, its output of estrogen is counteracted by steroid degradation in the liver and in due course the ovarian transplant may become so overstimulated by gonadotrophic hormones that granulomas may develop (Guthrie, 1954). In the case of the adrenal cortex, unilateral ablation results in compensatory hypertrophy of the opposite gland, unless, of course, adrenal cortical steroids are administered as replacement therapy (Ganong and Hume, 1955), in which case atrophy may supervene owing to the reduced secretion of ACTH. In the case of the testis the population of interstitial cells is subject to the influences of ICSH, yet there is evidence that the seminiferous tubules with which they are so intimately associated may normally hold their numbers in abeyance. If the testis is destroyed by ischemia induced by cadmium chloride, the few interstitial cells which may survive beneath the capsule will proliferate into a mass of endocrine cells in excess of the original population and may in due course become neoplastic (Reddy *et al.*, 1973). A similar hyperplasia of interstitial tissue is also seen in cryptorchid testes in which the seminiferous tubules are rendered inactive (Iturriza and Irusta, 1969). The corpora lutea of ovaries are also subject to both stimulatory and inhibitory influences, their growth being promoted by LH while their demise is triggered by luteolysin from the nongravid uterus.

Other endocrine glands are not regulated by trophic hormones but fall under the control of circulating materials whose concentration they control. Such direct feedbacks are illustrated by the parathyroid gland which becomes hyperplastic when the serum calcium declines. In like fashion, the zona glomerulosa of the adrenal cortex enlarges under conditions of sodium deficiency, while the cells of the islets of Langerhans are responsive to blood glucose levels. It would seem that if what a gland does is to control the level of a substance in the blood, then it is this substance to which the activity and growth of the gland is directly responsive.

The renewal of blood cells, whereby the losses from the circulating blood are counterbalanced by the production of new ones in the marrow and lymphoid organs, is carefully regulated by poietins. Erythrocyte production, for example, is triggered in the marrow by erythropoietin produced by the kidney in response to decreased availability of oxygen or increased demands for it (Krantz and Jacobson, 1970). The control of erythropoiesis is paralleled by comparable mechanisms for the regulation of other blood cells. Presumably each kind of leukocyte is subject to its own kind of poietin, but uncertainties as to the exact functions of the granulocytes are accompanied by equal ignorance concerning the nature of the granulopoietins which may regulate their production. Lymphocytes may be controlled by lympho-poietins, yet they are also very responsive to the presence of antigens which trigger their proliferation, if not their production of antibodies. Platelet production is a special case, for in mammals, unlike all other vertebrates which possess nucleated thrombocytes, they are not cellular entities. Nevertheless, their numbers in the circulation are carefully regulated by a serum factor (thrombopoietin) which may in turn control the production of megakaryocytes (Rolovic *et al.*, 1970). Again, the source and stimulation for thrombopoietin production remains unknown (Odell, 1972).

Even less well understood are the circulating factors which are hypothetically responsible for regulating the growth of such physiologically complex organs as the liver (Bucher and Malt, 1971), kidney (Nowinski and Goss, 1969), and lungs (Tartter and Goss, 1973). While each of these organs readily undergoes compensatory growth in response to reductions in mass, or increases in work load, the intensity with which such problems have been investigated in these organs is surpassed only by the lack of satisfactory explanations of how their growth is stimulated on the one hand and then inhibited when a functionally adequate mass is finally restored.

It is perhaps no coincidence that organs such as those mentioned above which serve systemic needs are generally classified as visceral organs. Accordingly, the most visceral of them all, namely, the gut, is perhaps also subject to growth regulations which are not altogether local (Dowling, 1970; Tilson and Wright, 1972). Although experiments have shown that the growth of the stomach or intestines can be stimulated by local surgical interventions, such as partial occlusion (Crean *et al.*, 1969), there is still reason to believe that the digestive tract may adjust its dimensions to the overall size of the body for which it must absorb nutrients. For example, atrophy of the intestines may occur in animals chronically fed intravenously (Levine *et al.*, 1973). Conversely, hyperphagia typically leads to growth of the gut, provided it is accompanied by a corresponding increase in the nutritional content of the diet (Jervis and Levin, 1966).

The digestive tract is also affected by certain internal controls. Thus, the parietal cells of gastric glands become hyperplastic under the influence of the hormone gastrin (Crean *et al.*, 1968). In the small intestines, resection of the anterior half typically leads to hypertrophy of the ileum, while loss of the posterior half brings about only a modest growth response in the jejunum (Young and Weser, 1974). Here there is reason to suspect that some intraluminal factor may be operating in a downstream direction to enhance the functional competence of the remaining segments of the intestine. This is accompanied by elongation of the ileal villi, thereby increasing their surface areas for absorption. Experiments have shown that this increased villus height may not be a direct response to the resection of more anterior segments of the gut, but to the attending proximity to the pancreas (Altmann, 1971). Hence, if the duodenal segment into which the pancreatic duct empties is transposed to the ileum, the villi in the latter segment still grow longer even though nothing has been removed.

## 2.2. Nonvisceral Organs: Mechanical Control of Growth

Surrounding the viscera is an envelope of supporting tissues which are under a rather different set of controls, both functionally and structurally. These tissues, including skin, bone, muscle, and blood vessels, are not primarily governed by factors which are systemically distributed. Thus, a reduction in the mass of any one of these tissues is not accompanied by compensatory growth in all homologous tissues wherever they may be located. Instead, their growth responses are local ones, and this presumably reflects the fact that the jobs they do are limited to the local neighborhood.

The skeletal system is a case in point. The length to which a bone grows is apparently autonomous and predetermined, but its lateral growth as well as its density are subject to various mechanical forces impinging upon them. An appendicular bone forced to support extra weight will not grow longer but it will become considerably thicker and denser than normal (Tulloh and Romberg, 1963). Unused bones, such as those in the stumps of amputated legs, typically develop osteoporosis (Jenkins and Cochran, 1969). Even the mode of fracture healing is subject to the mechanical conditions prevailing in the vicinity of the injury. The callus formed at the site of a broken bone first differentiates into cartilage, and secondarily this is replaced by ossification. However, the healing of a drill hole in the shaft of a bone involves little if any chondrification; the defect tends to be repaired by direct ossification (Radden and Fullmer, 1969). It would appear that the degree of mobility in the fracture as opposed to the drill hole may be responsible for the production of a cartilaginous callus. Indeed, the repeated fracture of a bone day after day will promote the differentiation of excess cartilage until such time as the bulk of the chondrogenic callus precludes further refracturing whereupon ossification and remodeling supervene (Lindholm *et al.*, 1970).

Like the bones to which they are attached, tendons adjust their dimensions to the forces to which they are subjected. They are also capable of remarkable feats of repair, for a severed tendon is able to bridge the gap by the prompt development of a callus composed of fibroblasts which then produce new collagen to knit together the two stumps. Indeed, the efficiency with which tendons adhere to surrounding tissues is an inconvenience in experiments designed to study the effect of tenotomy on muscle growth and maintenance.

Few tissues have been studied so intensively in recent years as the muscular system. As a result, our understanding of the factors responsible for its hypertrophy under some circumstances or atrophy under others has been deepened by a series of fascinating breakthroughs.

The uniqueness of skeletal muscle has been responsible for much of the research attention it has received. It is, first and foremost, made up of multinucleate cells which are among the largest in the body. The highly specialized nature of the contractile elements provides a convenient model system for the analysis of cellular differentiation. Finally, the intimate association of striated muscle with the skeletal system on the one hand and the nervous system on the other afford opportunities for experimental interventions unparalleled in any other tissue of the body.

Basic to our understanding of muscle is its dependence upon the nervous system. Denervation typically leads to the atrophy of the affected muscle. As the diameters of the muscle fibers decrease, the myofibrils they contain are reduced. This in turn is brought about by the loss of myofilaments from their surfaces. This atrophy is readily reversible upon reinnervation. Atrophy may also be achieved by immobilization of a limb or by cutting the tendon. It would appear that no matter how disuse is achieved, the morphological integrity of the muscle is diminished in the absence of the stimulus of a normal work load.

It is well known that neuromuscular transmission is mediated via acetylcholine and that the resulting physiological stimulation is important in maintaining the structural integrity of the fiber. However, acetylcholine is not to be regarded as the neurotrophic factor which is directly responsible for muscle maintenance, for there are some exceptional instances in which muscle growth can occur in the absence of innervation. One such case involves the anterior latissimus dorsi muscle of the bird which, when denervated, hypertrophies instead of atrophies (Feng *et al.*, 1963). The explanation for this unexpected result is very simple. This muscle normally is used to hold the wing up, and when denervated the wing droops (Sola *et al.*, 1973). The muscle is therefore stretched and this leads to its hypertrophy. If the wings of such birds are artificially kept from drooping, then the anterior latissimus dorsi atrophies (Stewart *et al.*, 1972).

A somewhat comparable situation obtains in the case of the diaphragm. Each half of the diaphragm is innervated by its own phrenic nerve, and if one of these is cut the corresponding hemidiaphragm undergoes hypertrophy instead of the expected atrophy (Stewart *et al.*, 1972). This phenomenon would appear to be due to the rhythmic stretching caused by the periodic contractions of the contralateral hemidiaphragm. Indeed, when both phrenic nerves are cut neither side hypertrophies (Feng and Lu, 1965).

These exceptions to the rule that nerves are necessary for muscle maintenance require that the rule be changed. Clearly, the important factor is that a muscle requires tension in order to avoid atrophy (Stewart, 1972). Even in immobilized limbs, atrophy occurs only in muscles that are held in a relaxed position. Those that are immobilized in the stretched position tend to grow despite their inability to move (Williams and Goldspink, 1973).

Perhaps the most compelling evidence for the important role of tension in muscle maintenance is found in the growth of infant muscles. When denervated, growth of such muscles goes on, but at a reduced rate compared with normally innervated ones (Stewart, 1968). Thus, although there is an absolute gain in weight, there is a relative atrophy during the course of maturation. This reduced rate of growth in denervated infant muscles is to be explained by the tension exerted upon them by the continued elongation of the bones to which they are attached. If denervated infant muscles are also tenotomized, they undergo little or no further growth (Stewart, 1968).

It now seems clear that the quest for the elusive neurotrophic factor in muscle growth has been unsuccessful for the very good reason that there is no factor as such. Instead, there are *conditions* upon which the integrity of the muscle depends, not the least of which is the tension exerted by or upon the muscle which appears to be of utmost importance in regulating its growth. Nevertheless, the fact that muscle can be induced to grow in the absence of nerves does not prove categorically that innervation *per se* is not without its influence. To test this it would be necessary to demonstrate the opposite of the experiment showing that denervated muscles can hypertrophy, namely, that innervated muscles can be made to atrophy. This is exactly what happens following tenotomy, but it is noteworthy that additive atrophy ensues tenotomy plus denervation (Binkhorst, 1972).

In contrast to skeletal muscle, the striated muscle of the heart consists of small fibers, incapable of anaerobic respiration, usually with one or at the most two nuclei, without motor innervation, and lacking satellite cells. Its inability to regenerate is made up for in part by its capacity for hypertrophy. The diameters of the ventricular myocardial fibers may double during maturation from infancy to adulthood and can nearly double again under pathological conditions leading to hypertrophy. Cardiac hypertrophy occurs in response to increased work load brought about by hypertension. This can be stimulated experimentally by partial constriction of the aorta (Fanburg and Posner, 1968) or pulmonary artery (Laks *et al.*, 1974) causing the ventricles to pump against an increased head of pressure. It is not uncommon for one chamber to hypertrophy in the absence of the other, as for example in the case of left ventricular hypertrophy following aortic constriction. On the other hand, the right ventricle will enlarge not only after pulmonary arterial constriction, but also following unilateral pneumonectomy (Cohn, 1939) or hypoxia (Gibson and Harris, 1972). These experimental interventions, all of which interfere with respiration, also lead to polycythemia. For reasons yet to be explained, it has been found that hypertransfusion polycythemia alone, in the absence of hypoxia, will also contribute to right ventricular hypertrophy (Swigart, 1965).

The natural inequity between the right and left sides of the ventricles is correlated with the differences in their work loads. This uneven shape of the heart, however, is not genetically determined. It develops solely in response to the greater functional demands on the left ventricle after birth. Prenatally the two sides of the ventricle are equal in size as well as work load (Latimer, 1965). Only when the ductus arteriosus closes off after birth does the function of the left ventricle exceed that of the right. Indeed, for a short period postnatally the reduction in the function of the right ventricle leads to a cessation in its growth (Emery and Mithal, 1961).

Blood vessels, like the heart, adjust their dimensions to the work to be performed. Increased pressure leads to an enlargement of the lumen and a thickening of the wall, particularly in arteries (Rodbard *et al.*, 1967). If a segment of vein is grafted into an artery, its normally thin walls thicken considerably (Jones and Dale, 1958), although it has never been shown to develop true arterial histology. Perhaps the response of arteries to increased intraluminal pressure is best illustrated by the development of collateral circulation. If one lung is removed, for example, the opposite pulmonary artery enlarges to dimensions commensurate with the main one (Edwards, 1939). Meanwhile, the bronchial artery on the ligated side undergoes marked enlargement in compensation for the unused pulmonary artery (Liebow *et al.*, 1950), and collateral anastomoses may develop between the bronchial arterial supply and the pulmonary veins (Vidone and Liebow, 1957). Elsewhere in the body, wherever an artery may be occluded, the blood is detoured through formerly minor vessels which subsequently attain the proportions of the one that was blocked. One gets the unmistakable impression that the smallest artery has the latent capacity to grow to aortic dimensions.

Other hollow organs lined with smooth muscle are also stimulated to hypertrophy when stretched by internal pressure. The prodigious growth of the pregnant uterus, while responsive to hormonal stimuli, is primarily attributable to the growth of the fetus and its accumulated amniotic fluids. Artificial distension of the uterus by inflation will also promote its hypertrophy (DeMattos *et al.*, 1967).

The urinary bladder exhibits a similar reaction, for its size is adjusted according to the hydrostatic pressure of the urine it contains. Micturition is normally triggered by the buildup of internal pressure which, via stretch receptors, brings about a reflex contraction of the smooth muscle in its wall thus voiding the contained urine. If the ureters are diverted elsewhere, disuse atrophy of the bladder ensues (Schmaelzle *et al.*, 1969). Conversely, chronic distension of the bladder causes hypertrophy (Peterson *et al.*, 1973). This is most effectively achieved by cutting both pudendal nerves thus interrupting the reflex arc responsible for emptying the bladder. Under these circumstances, the bladder is not prevented from filling with urine, but it is unable to empty itself. The hydrostatic pressure builds up until overflow incontinence occurs. Such a stretched bladder becomes hypertrophied (Jacobson, 1945). Curiously, following unilateral denervation of the dog or cat bladder, despite the fact that the urine can be voided by contraction of the innervated side, hypertrophy of the operated side still takes place (Langworthy and Kolb, 1938; Jacobson, 1945). However, the explanation of this intriguing phenomenon would appear to involve hydrostatic pressure, since bilaterally denervated bladders do not hypertrophy when their ureters have been diverted (Veenema *et al.*, 1952).

## 2.3. Nervous Control of Growth

Most organs whose function depends upon their innervation tend to atrophy following denervation. Chief among these are the salivary glands, the secretory activity of which is regulated by the autonomic nervous system. Hence, either sympathetic or parasympathetic denervation leads to atrophy of the salivary glands (Hall and Schneyer, 1973), while hypertrophy is promoted by a variety of stimuli, not the least of which is treatment with isoproterenol, a sympathomimetic drug (Schneyer, 1973). If a rat is fed a bulk diet, its salivary glands will also enlarge (Wells and Peronance, 1967). Conversely, a liquid diet causes their atrophy. This would suggest that demands for increased salivation promote the growth of the salivary glands, a possibility supported by experiments in which rats have been held under excessively warm conditions. In the absence of sweat glands, they salivate profusely and lick their fur in order to keep cool, and this in turn is accompanied by enlargement of the salivary glands (Elmér and Ohlin, 1970). Still another approach to this has involved the repeated amputations of the lower incisors which have been found also to promote salivary gland hypertrophy (Wells *et al.*, 1959). Moreover, if just one lower incisor is amputated, the glands on that side enlarge, but not if the nerves are cut (Wells and Peronace, 1964). One can only conclude that the intake of food is monitored by innervation of the lower incisors and that the size of the salivary glands is adjusted accordingly.

Equally interesting are the effects of feeding proteolytic enzymes in the diet (Ershoff and Bajwa, 1963). This also promotes growth of the salivary glands, but not if the enzymes are given by stomach tube. It turns out that if the glossopharyngeal nerves are cut, the salivary glands do not grow in response to proteolytic enzymes (Wells *et al.*, 1965). Their influence is therefore mediated via the taste buds which, in the absence of innervation, degenerate within a week. Clearly, the salivary glands are subject to many influences which are channeled through their innervation. Not surprisingly, all of these influences are associated with the oral cavity and the physiological activity of the salivary glands.

### 2.4. Intrinsic versus Systemic Control of Organ Growth

Central to the problem of growth regulation is the question of whether each organ controls its own growth or is subservient to influences from the body as a whole. To test these alternatives, it is necessary to graft organs from animals of one size into those of another, an approach best carried out between young and old individuals. Such heterochronic grafts have been studied in several systems. The long bones of infant mice and rats, for example, have been transplanted subcutaneously and found to continue their elongation to at least 80% of their normal lengths regardless of host age (Felts, 1959). In a study of heart growth, Dittmer *et al.* (1974) transplanted hearts from infant rats as auxiliary grafts into other infant or adult hosts. In both cases, the transplants continued to grow to normal adult dimensions. Metcalf (1964) transplanted one-day-old spleens into adult mice and noted that subsequent growth occurred, but only if the hosts had been splenectomized. In general, there would appear to be some tissues and organs of the body whose growth may be autonomous, and others in which enlargement is closely bound up with systemic factors responsible for maintaining a fixed proportion of a given kind of tissue in the body.

In the cases cited above, the transplanted organs were not entirely satisfactory inasmuch as they were not provided with normal revascularization from the outset. In casting about for the ideal system in which to study heterochronic grafts, it is important above all to utilize sufficiently inbred strains of animals to avoid undue immunological rejections and obviate the necessity for immunosuppressive therapy. The largest animal which fulfills these requirements is the rat. Another requirement relates to the minimal sizes of blood vessels which can be successfully anastomosed by microvascular surgery. In the rat, this is of necessity limited to those organs close enough to the aorta and vena cava to permit surgical union of these vessels to those of the recipient. Of all such organs (heart, lungs, liver, gut, kidneys) the kidney is the one best adapted to sharing, without replacing, the functions of the host's homologous organs. Figure 3 illustrates the various combinations of grafts and hosts which have been explored thus far.

The transplantation of extra kidneys was first successfully attempted by Dittmer and Goss (1969), who found that subsequent atrophy was confined to the extra transplanted kidney (Figure 3). Klein and Gittes (1973) confirmed these findings, although Silber and Crudup (1973, 1974) and Silber and Malvin (1974) reported that the sizes of extra transplanted kidneys remained normal, as did those of the host's own kidneys. The discrepancy in these results may be attributable to the fact that the latter authors estimated the dimensions of their kidneys by intravenous pyelograms while the former sacrificed their animals and weighed the kidneys.

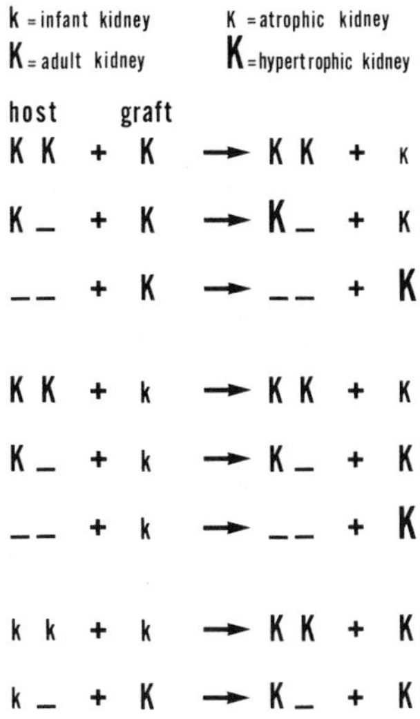

k = infant kidney    K = atrophic kidney

**K** = adult kidney    **K** = hypertrophic kidney

Fig. 3. Growth of heterochronic kidney grafts in inbred strains of rats. The top three examples illustrate the reactions of extra adult kidneys grafted to intact or nephrectomized adult hosts. The middle three are cases of auxiliary infant kidneys transplanted to adult hosts. At the bottom are infant or adult kidneys in infant hosts. See text for fuller explanation and reference sources.

| host | graft | | |
|------|-------|----|----|
| K K | + K | → K K | + K |
| K _ | + K | → K _ | + K |
| _ _ | + K | → _ _ | + **K** |
| K K | + k | → K K | + K |
| K _ | + k | → K _ | + **K** |
| _ _ | + k | → _ _ | + **K** |
| k k | + k | → K K | + K |
| k _ | + K | → K _ | + K |

In the experiments by Dittmer and Goss (1969) and Klein and Gittes (1973), it is curious that only the transplanted kidney should undergo atrophy (Figure 3). This cannot be ascribed solely to the fact that the transplantation operation might have predisposed this kidney to atrophy, for when one of the host's own kidneys was removed the degree of atrophy in the graft was considerably reduced while the host's own remaining kidney hypertrophied (Klein and Gittes, 1973). If both host kidneys were removed, the transplant then underwent hypertrophy. The explanation for this differential reaction may be found in terms of "renal counterbalance" as proposed by Hinman (1943). This concept grew out of his studies of recovery in hydronephrotic kidneys. If the duration of experimental hydronephrosis was not so long as to produce irreversible damage, the recovery of the affected kidney was accelerated when the opposite intact organ had been removed. Conversely, the hydronephrotic kidney failed to recover, even after removal of the ureteral obstruction, if in the meantime the opposite kidney had been allowed to undergo compensatory hypertrophy. Thus, when one kidney compensates for its partner, the differential in their division of labor (as well as size) tends to persist even after the original experimental intervention has been discontinued. The preferential atrophy of transplanted extra kidneys, therefore, is consistent with this interpretation, even though it does not explain it.

The reactions of infant kidneys to transplantation into adult hosts are even more interesting. Investigations in both dogs (Claman *et al.*, 1963) and humans (Andersen *et al.*, 1974; Silber, 1974*a*) have proven that adults deprived of their own kidneys can survive on one or two donor infant kidneys grafted *en bloc*. Ross *et al.* (1970), Baden *et al.* (1973), and Silber (1974*a*) have studied the growth of such transplants, showing that when infant kidneys were grafted into adults whose own kidneys had been removed, the infant kidneys not only took over the renal functions of the host but underwent considerable enlargement during the several weeks or months they were followed. This confirms the capacity of transplanted kidneys to continue their development, but does not tell us whether the growth of such kidneys represents compensation on the part of the grafts for the missing kidneys of the hosts or is an expression of innate potentials for growth in maturing organs. Since none of the aforementioned authors transplanted infant kidneys into intact hosts, it was not possible to distinguish in these experiments between normal vs. compensatory growth.

This gap in the story has recently been filled by a brilliant series of experiments carried out by Silber (1974*b*, 1976). When single infant rat kidneys were grafted into intact adult hosts the transplants continued to grow but did not quite achieve adult proportions. Infant kidneys in unilaterally nephrectomized adult hosts, however, grew to normal adult dimensions, while infant kidneys in bilaterally nephrectomized adult hosts not only fulfilled their normal growth potential but became hypertrophic. If an infant kidney is transplanted into an intact infant host, all three kidneys grow to normal adult sizes, the rat eventually acquiring a superabundance of renal mass. Finally, when an adult kidney was grafted into a unilaterally nephrectomized infant rat, the host's remaining kidney continued to grow to normal adult size while the transplanted adult kidney remained unchanged. These transplantations are diagrammatically illustrated in Figure 3.

Collectively considered, these experiments suggest that the infant kidney has a capacity for growth which will express itself even in the presence of a normal complement of renal tissue, and presumably in the absence of excess physiological demands. This has led Silber to propose that compensatory renal hypertrophy is qualitatively different from the "obligatory growth" expressed during normal maturation. This is an important interpretation in that it implies for the first time that compensatory growth is not merely a continuation of the normal processes of ontogeny. Such a view favors the possibility that each organ of the body may be programmed to fulfill a certain potential for enlargement irrespective of somatic size or physiological demands. Not all parts of the body may follow this rule, but if certain ones do so then they might serve as pace setters for the growth of less autonomous parts.

## 3. Discussion

From the foregoing examples, it is evident that no two tissues grow in the same way. Some enlarge by cellular hypertrophy, others by hyperplasia. In some the functional units are readily augmented, while in others they are fixed early in development. There are some organs which have a potential for unlimited growth, while others are unaccountably restricted in their regenerative potential. Yet out of this bewildering profusion of examples, the student of growth is obliged to seek common denominators which may represent those attributes of growth which are so important as to be virtually ubiquitous.

It is axiomatic that the components of living tissues are in a constant state of renewal (Goss, 1970). Different tissues, however, renew themselves at different levels of organization. Those whose cells can divide do so at the cellular level. Accordingly, their capacity for turnover at the subcellular level is considerably abridged, as in the inability of red blood cells to undergo hemoglobin turnover. The cells of static tissues, incapable of mitosis, must renew themselves at the subcellular and molecular levels. Myosin is constantly degraded and resynthesized in skeletal muscle fibers, although the exact mechanism of this replacement remains to be explained. In nerve fibers there is a ceaseless flow of axoplasm from the nerve cell body. In the retina, the photoreceptors are also in the state of constant renewal. The rod outer segment, for example, is synthesized proximally as previously formed material is pushed distally toward the pigment epithelium. Here, fragments of the apical ends are from time to time phagocytized by the pigment epithelium cells (Young and Bok, 1969).

Whatever may be the mechanism by which the components of a tissue turn over, or the level of organization at which renewal occurs, there is a need for regulatory mechanisms to assure that the rate of replacement keeps pace with the rate of depletion. These same control mechanisms are presumably responsible for stimulating repair and compensatory hypertrophy and for turning off growth when the organ mass is restored to normal. Much attention is focused on the nature of stimulators of synthesis, but the control of degradative processes is equally important.

Despite vast differences in the mechanisms of growth control in different tissues of the body, the relationship between size and function cannot be denied. In those tissues which are best understood, growth is governed by influences closely related to physiological demands, a logical arrangement which enables the organs of the body to adjust their dimensions to the functional loads they are called upon to perform. Nevertheless, not all facets of growth are this easily explained. For example, the organs and tissues of an embryo grow considerably before they become functional, yet their rates of growth are carefully controlled. In the adult organism, most organs are represented by several times more tissue than is required for their subsistence. This provides a comfortable margin of safety in cases of emergency, but is not easily reconciled with the functional demand theory of growth regulation. Indeed, when the normal physiological activities of an organ are bypassed, the resultant disuse atrophy seldom leads to the complete disappearance of the tissue concerned. A minimal mass is maintained providing a reserve from which regrowth can occur if necessary. Nevertheless, the retention of mophological structure in the absence of function is difficult to explain in terms of physiological regulation.

Aside from pondering the problems of growth of individual organs and tissues, one must not neglect the question of what determines the size of the body as a whole (Goss, 1975). Each organ is a predictable proportion of the body, but it is not clear whether their sizes are determined by the dimensions to which the organism grows or *vice versa*. If the body is merely the sum of its parts, then one or more of them could be the limiting factors to which all others adjust their rates of growth. The rates at which these organs enlarge may be predetermined early in development, if not set down in the genome of the zygote. It is tempting to predict that future exploration of organ growth potentials, using the technique of heterochronic grafting, will reveal that those organs with fixed numbers of functional units (Goss, 1966, 1967) enlarge to determinate sizes while others, capable of potentially unlimited multiplication of their functional units, grow to whatever dimensions may be permitted or required by systemic demands for functional activity.

Embryonic and fetal growth are themselves the least understood of all. It is dangerous to assume that such obvious attributes of organisms are under genetic control, for this does not explain how such genes might operate at the physiological level, nor does it take into account the possible role played by maternal influences that are not genetic. Prenatal growth is correlated with the enlargement of the placenta. Since removal of a fetus does not prevent continued placental growth (Petropoulos, 1973), it is tempting to conclude that the placenta may in some way affect the growth and size of its fetus. Yet the placenta itself is profoundly affected by the intrauterine environment (e.g., litter size and crowding) as well as the ovarian hormones themselves. The loss of corpora lutea by ovariectomy brings about placental hypertrophy, presumably owing to its attempts to compensate for the missing corpora (Csapo, 1969). Not until experimental biologists find a way to grow ova from a species of one size in the uterus of an animal of a different size will it be possible categorically to prove whether or not bodily dimensions and growth rate are inherently determined.

## 4. Summary

The mechanisms by which organs and tissues grow have profound implications for the ultimate sizes to which they can enlarge, their capacities for regeneration, and the efficiency with which they function. In nonmitotic tissues growth is achieved by the multiplication of cytological organelles, but it is limited by the ultimate sizes beyond which their cells cannot enlarge. Organs capable of mitosis can theoretically grow to indeterminate sizes if, in doing so, their functional units are increased proportionately as in the cases of most endocrine and exocrine glands. When physiological activity depends upon functional units which are organized at the histological level but which cannot be increased in number in the adult animal, functional efficiency cannot keep pace with increases in organ mass because structures such as renal nephrons and pulmonary alveoli can only grow larger by cell division at the expense of vital surface–volume relationships. In all, some organs of the body can grow efficiently and theoretically without limit, while others have serious restrictions imposed upon their growth potentials. It is a curious thing that in higher vertebrates some of the most vital organs seem to be endowed with the least capacity for growth and regeneration while less important organs have the greatest potential for growth.

Organ sizes are adjusted to the functions they are called upon to perform. Accordingly, growth-regulating mechanisms are closely associated with those physiological factors which govern the functional activities of organs and tissues. Most visceral organs, which serve the body as a whole, do business with the circulatory system. As such, their functions, as well as their growth, are regulated by humoral factors such as trophic hormones and circulating electrolytes. Other parts of the body, including supporting tissues as well as certain hollow organs, are subject to growth control by local mechanical factors. These may take the form of stresses and strains on bones and tendons, tension in skeletal muscle, and hydrostatic pressure in the heart, blood vessels, uterus, and bladder. Local growth control may also be under the influence of innervation, as in the taste buds and salivary glands.

Finally, there are some organs whose growth is still not understood. In some cases, such as the liver, kidneys and lungs, it is generally agreed that growth is controlled by functional demands. The physiological complexities of these organs, however, have made it difficult thus far to pinpoint the exact influences which may be responsible for turning their growth on and off. In other parts of the body there is little reason to believe that growth is necessarily correlated with physiological demands. Included among these are such interesting and intriguing structures as integumentary appendages (hairs, feathers, teeth), the sometimes cyclic growth of which is under intrinsic control, appendicular bones with inherent rates of elongation, and the embryo or fetus itself, the growth of which either controls or is controlled by the parts of which it is composed.

# 5. References

Altmann, G., 1971, Influence of bile and pancreatic secretions on the size of the intestinal villi in the rat, *Am. J. Anat.* **132**:167.

Andersen, O. S., Jonasson, O., and Merkel, F. K., 1974, En bloc transplantation of pediatric kidneys into adult patients, *Arch. Surg.* **108**:35.

Baden, J. P., Wolf, G. M., and Sellers, R. D., 1973, The growth and development of allotransplanted neonatal canine kidneys, *J. Surg. Res.* **14**:213.

Binkhorst, R. A., 1972, Hypertrophy and atrophy of rat plantar flexors, *Experientia* **28**:268.

Bizzozero, G., 1894, An address on the growth and regeneration of the organism, *Br. Med. J.* **1**:728.

Blaxter, J. H. S., and Jones, M. P., 1967, The development of the retina and retinomotor responses in the herring, *J. Mar. Biol. Assoc. U.K.* **47**:677.

Bucher, N. L. R., and Malt, R. A., 1971, *Regeneration of Liver and Kidney,* Little, Brown, Boston.

Claman, M., Balfour, J., and Forbes, D., 1963, Survival of homologous transplants of neonatal kidneys in dogs, *Surg. Gynecol. Obstet.* **124**:227.

Cohn, R., 1939, Factors affecting the postnatal growth of the lung, *Anat. Rec.* **75**:195.

Crean, G. P., Rumsey, R. D. E., Hogg, D. F., and Marshall, M. W., 1968, Experimental hyperplasia of the gastric mucosa, in: *The Physiology of Gastric Secretion* (L. S. Semb and J. Myren, eds.), pp. 82–85, Universitetsforlaget, Oslo.

Crean, G. P., Hogg, D. F., and Rumsey, R. D. E., 1969, Hyperplasia of the gastric mucosa produced by duodenal obstruction, *Gastroenterology* **56**:193.

Csapo, A., 1969, The luteo-placental shift, the guardian of pre-natal life, *Postgrad. Med. J.* **45**:57.

DeMattos, C. E. R., Kempson, R. L., Erdos, T., and Csapo, A., 1967, Stretch-induced myometrial hypertrophy, *J. Fertil. Steril.* **18**:545.

Dittmer, J., and Goss, R. J., 1969, Regulation of renal mass after transplantation of an extra kidney, *Am. Zool.* **9**:607.

Dittmer, J. E., Goss, R. J., and Dinsmore, C. E., 1974, The growth of infant hearts grafted to young and adult rats, *Am. J. Anat.* **141**:155.

Dowling, R. H., 1970, Small bowel resection and bypass—recent developments and effects, in: *Modern Trends in Gastro-enterology* (W. I. Card and B. Creamer, eds.), Vol. 4, pp. 73–104, Appleton-Centruy-Crofts, New York.

Edwards, F. R., 1939, Studies in pneumonectomy and the development of a two-stage operation for the removal of a whole lung, *Br. J. Surg.* **27**:392.

Elmér, M., and Ohlin, P., 1970, Salivary glands of the rat in a hot environment, *Acta Physiol. Scand.* **79**:129.

Emery, J. L., and Mithal, A., 1961, Weights of cardiac ventricles at and after birth, *Br. Heart J.* **23**:313.

Ershoff, B. H., and Bajwa, G. S., 1963, Submaxillary gland hypertrophy in rats fed proteolytic enzymes, *Proc. Soc. Exp. Biol. Med.* **113**:879.

Fanburg, B. L., and Posner, I., 1968, Ribonucleic acid synthesis in experimental cardiac hypertrophy in rats, *Circ. Res.* **23**:123.

Felts, W. J. L., 1959, Transplantation studies of factors in skeletal organogenesis. 1. The subcutaneously implanted immature long-bone of the rat and mouse, *Am. J. Phys. Anthropol.* **17**:201.

Feng, T.-P., and Lu, D.-X., 1965, New lights on the phenomenon of transient hypertrophy in the denervated hemidiaphragm of the rat, *Sci. Sin.* **14**:1772.

Feng, T. P., Jung, H. W., and Wu, W. Y., 1963, The contrasting trophic changes of the anterior and posterior latissimus dorsi of the chick following denervation, in: *The Effect of Use and Disuse on Neuromuscular Functions* (E. Gutmann and P. Hnik, eds.), pp. 431–441, Elsevier Publishing Company, Amsterdam.

Ganong, W. F., and Hume, D. M., 1955, Effect of hypothalamic lesions on steroid-induced atrophy of adrenal cortex in the dog, *Proc. Soc. Exp. Biol. Med.* **88**:528.

Gibson, K., and Harris, P., 1972, Effects of hypobaric oxygenation, hypertrophy and diet on some myocardial cytoplasmic factors concerned with protein synthesis, *J. Mol. Cell. Cardiol.* **4**:651.

Goss, R. J., 1964, *Adaptive Growth,* Academic Press, New York.

Goss, R. J., 1966, Hypertrophy versus hyperplasia, *Science* **153**:1615.

Goss, R. J., 1967, The strategy of growth, in: *Control of Cellular Growth in Adult Organisms* (H. Teir and T. Rytömaa, eds.), pp. 3–27, Academic Press, New York.

Goss, R. J., 1970, Turnover in cells and tissues, in: *Advances in Cell Biology* (D. M. Prescott, L. Goldstein, and E. McConkey, eds.), pp. 233–296, Appleton-Centruy-Crofts, New York.

Goss, R. J., 1971, Adaptive growth of the heart, in: *Cardiac Hypertrophy* (N. R. Alpert, ed.), pp. 1–10, Academic Press, New York.

Goss, R. J. (ed.), 1972, *Regulation of Organ and Tissue Growth,* Academic Press, New York.

Goss, R. J., 1974, Aging versus growth, *Perspect. Biol. Med.* **17:**485.

Goss, R. J., 1975, The right size, in: *Growth and Productivity of Meat Producing Animals* (D. Lister, D. N. Rhodes, V. R. Fowler, and M. F. Fuller, eds.), pp. 237–254, Plenum Press, New York.

Guthrie, M. J., 1954, The structure of intrasplenic ovaries in mice, *Anat. Rec.* **118:**305.

Hall, H. D., and Schneyer, C. A., 1973, Role of autonomic pathways in disuse atrophy of rat parotid, *Proc. Soc. Exp. Biol. Med.* **143:**19.

Hinman, F., 1943, The condition of renal counterbalance and the theory of renal atrophy of disuse, *J. Urol.* **49:**392.

Iturriza, F. C., and Irusta, O., 1969, Hyperplasia of the interstitial cells of the testis in experimental cryptorchidism, *Acta Physiol. Lat. Am.* **19:**236.

Jacobson, C. E., Jr., 1945, Neurogenic vesical dysfunction. An experimental study, *J. Urol.* **53:**670.

Jenkins, D. P., and Cochran, T. H., 1969, Osteoporosis: The dramatic effect of disuse of an extremity, *Clin. Orthop. Relat. Res.* **64:**128.

Jervis, E. L., and Levin, R. J., 1966, Anatomic adaptation of the alimentary tract of the rat to the hyperphagia of chronic alloxan-diabetes, *Nature* **210:**391.

Jones, T. I., and Dale, W. A., 1958, Study of peripheral autogenous vein grafts, *Arch. Surg.* **76:**294.

Klein, T. W., and Gittes, R. F., 1973, Three-kidney rat: Renal isografts and renal counterbalance, *J. Urol.* **109:**19.

Krantz, S. B., and Jacobson, L. O., 1970, *Erythropoietin and the Regulation of Erythropoiesis,* University of Chicago Press, Chicago.

Laks, M. M., Morady, F., Garner, D., and Swan, H. J. C., 1974, Temporal changes in canine right ventricular volume, mass, cell size, and sarcomere length after banding the pulmonary artery, *Cardiovasc. Res.* **8:**106.

Langworthy, O. R., and Kolb, L. C., 1938, Histologic changes in the vesicle muscle following injury of the peripheral innervation, *Anat. Rec.* **71:**249.

Latimer, H. B., 1965, The weight and thickness of the two ventricular walls in the newborn dog heart, *Anat. Rec.* **152:**225.

Levine, G. M., Steiger, E., and Deren, J. J., 1973, The importance of oral intake in maintenance of rat small intestinal mass and disaccharidase activity, *Clin. Res.* **21:**828.

Liebow, A. A., Hales, M. R., Bloomer, W. E., Harrison, W., and Lindskog, G. E., 1950, Studies on the lung after ligation of the pulmonary artery. II. Anatomical changes, *Am. J. Pathol.* **26:**177.

Lindholm, R. V., Lindholm, T. S., Toikkanen, S., and Leino, R., 1970, Effect of forced interfragmental movements on the healing of tibial fractures in rats, *Acta Orthop. Scand.* **40:**721.

Luquet, P., and Hannequart, M.-H., 1974, Relations entre la longueur du poisson, le poids du cerveau et sa teneur en acides nucléiques chez la Carpe royale, *C.R. Acad. Sci.* **278:**3371.

Metcalf, D. 1964, Restricted growth capacities of multiple spleen grafts, *Transplantation* **2:**387.

Moss, F. P., and Leblond, C. P., 1970, Nature of dividing nuclei in skeletal muscle of growing rats, *J. Cell Biol.* **44:**459.

Nash, J., 1931, The number and size of glomeruli in the kidneys of fishes and observations on the morphology of the renal tubules of fishes, *Am. J. Anat.* **47:**425.

Nowinksi, W. W., and Goss, R. J. (eds.), 1969, *Compensatory Renal Hypertrophy,* Academic Press, New York.

Odell, T. T., Jr., 1972, Regulation of the megakaryocyte-platelet system, in: *Regulation of Organ and Tissue Growth* (R. J. Goss, ed.), pp. 187–195, Academic Press, New York.

Peterson, C. M., Goss, R. J., and Atryzek, V., 1973, Hypertrophy of the rat urinary bladder following reduction of its functional volume, *J. Exp. Zool.* **187:**121.

Petropoulos, E. A., 1973, Maternal and fetal factors affecting the growth and function of the rat placenta, *Acta Endocrinol. (Suppl. 176)* **73:**9.

Radden, B. G., and Fullmer, H. M., 1969, Morphological and histochemical studies of bone repair in the rat, *Arch. Oral Biol.* **14:**1243.

Reddy, J., Svoboda, D., Azarnoff, D., and Dawar, R., 1973, Cadmium-induced Leydig cell tumors of rat testis: Morphologic and cytochemical study, *J. Natl. Cancer Inst.* **51:**891.

Rodbard, S., Ikeda, K., and Montes, M., 1967, An analysis of mechanisms of post stenotic dilatation, *Angiology* **18:**349.

Rolovic, Z., Baldini, M., and Dameshek, W., 1970, Megakaryocytopoiesis in experimentally induced immune thrombocytopenia, *Blood* **35:**173.

Ross, G., Jr., Cosgrove, M. D., Dragan, P., Mowat, P., Battenberg, J., and Goodwin, W. E., 1970, Growth of homotransplanted puppy kidneys, *Surg. Forum* **21**:526.

Schmaelzle, J. F., Cass, A. S., and Hinman, F., Jr., 1969, Effect of disuse and restoration of function on vesical capacity, *J. Urol.* **101**:700.

Schneyer, C. A., 1973, A growth-suppressive influence of L-isoproterenol on postnatally developing parotid gland of rat, *Proc. Soc. Exp. Biol. Med.* **143**:899.

Silber, S. J., 1974*a*, Renal transplantation between adults and children: Differences in renal growth, *J. Am. Med. Assoc.* **228**:1143.

Silber, S. J., 1974*b*, Compensatory and obligatory renal growth in babies and adults, *Aust. N.Z.J. Surg.* **44**:421.

Silber, S. J., 1976, Growth of baby kidneys transplanted into adults, *Arch. Surg.* **111**:75.

Silber, S. J., and Crudup, J., 1973, Kidney transplantation in inbred rats, *Can. J. Surg.* **16**:551.

Silber, S. J., and Crudup, J., 1974, The three-kidney rat model, *Invest. Urol.* **11**:466.

Silber, S., and Malvin, R. L., 1974, Compensatory and obligatory renal growth in rats, *Am. J. Physiol.* **226**:114.

Sola, O. M., Christensen, D. L., and Martin, A. W., 1973, Hypertrophy and hyperplasia of adult chicken anterior latissimus dorsi muscles following stretch with and without denervation, *Exp. Neurol.* **41**:76.

Stewart, D. M., 1972, The role of tension in muscle growth, in: *Regulation of Organ and Tissue Growth* (R. J. Goss, ed.), pp. 77–100, Academic Press, New York.

Stewart, D. M., Sola, O. M., and Martin, A. W., 1972, Hypertrophy as a response to denervation in skeletal muscle, *Z. Vgl. Physiol.* **76**:146.

Swigart, R. H., 1965, Polycythemia and right ventricular hypertrophy, *Circ. Res.* **17**:30.

Tartter, P. I., and Goss, R. J., 1973, Compensatory pulmonary hypertrophy after incapacitation of one lung in the rat, *J. Thorac. Cardiovasc. Surg.* **66**:147.

Tilson, M. D., and Wright, H. K., 1972, Adaptational changes in the ileum following jejunectomy, in: *Regulation of Organ and Tissue Growth* (R. J. Goss, ed.), pp. 257–270, Academic Press, New York.

Tulloh, N. M., and Romberg, B., 1963, An effect of gravity on bone development in lambs, *Nature* **200**:438.

Veenema, R. J., Carpenter, F. G., and Root, W. S., 1952, Residual urine, an important factor in interpretation of cystometrograms, an experimental study, *J. Urol.* **68**:237.

Vidone, R. A., and Liebow, A. A., 1957, Anatomical and functional studies of the lung deprived of pulmonary arteries and veins, with an application in the therapy of transposition of the great vessels, *Am. J. Pathol.* **33**:539.

Wells, H., and Peronace, A. A. V., 1964, Synergistic autonomic nervous regulation of accelerated salivary gland growth in rats, *Am. J. Physiol.* **207**:313.

Wells, H., and Peronace, A. A. V., 1967, Functional hypertrophy and atrophy of the salivary glands of rats, *Am. J. Physiol.* **212**:247.

Wells, H., Zackin, S. J., Goldhaber, P., and Munson, P. L., 1959, Increase in weight of the submandibular salivary glands of rats following periodic amputation of the erupted portion of the incisor teeth, *Am. J. Physiol.* **196**:827.

Wells, H., Peronace, A. A. V., and Stark, L. W., 1965, Taste receptors and sialadenotrophic action of proteolytic enzymes in rats, *Am. J. Physiol.* **208**:877.

Williams, P. E., and Goldspink, G., 1973, The effect of immobilization on the longitudinal growth of striated muscle fibres, *J. Anat.* **116**:45.

Young, E. A., and Weser, E., 1974, Nutritional adaptation after small bowel resection in rats, *J. Nutr.* **104**:994.

Young, R. W., and Bok, D., 1969, Participation of the retinal pigment epithelium in the rod outer segment renewal process, *J. Cell Biol.* **42**:392.

# 2

# Human Biochemical Development

## GERALD E. GAULL, FRITZ A. HOMMES, and JACQUES F. ROUX

## 1. Introduction

There seems to be a relationship between the metabolic requirements for rapid growth and the development of enzyme systems. For example, the enzymes involved in anabolism are particularly active in early development, whereas those involved in catabolism are less active or even virtually absent (Sereni and Principi, 1965). The differentiation of tissue function, i.e., enzymatic differentiation, is the result of the synthesis of new enzymes rather than of the activation of enzymes already present (Knox *et al.*, 1956). It proceeds in a stepwise manner consisting of the simultaneous addition of clusters of new enzymes (Greengard, 1971), after tissue morphology has become relatively distinctive. Furthermore, the schedule of enzymatic differentiation is alterable, at least in part, in that premature or postmature birth can advance or delay the appearance of some enzymes (Nemeth, 1959; Dawkins, 1961).

The work of Greengard has been of particular importance in demonstrating that enzymatic differentiation seems to follow fairly regular patterns among species, especially as regards the enzymes which are relatively central in metabolism. She has put forward the concept of the activity quotient ($AQ$) (Greengard, 1976)

$$AQ = \frac{\text{units of activity/g immature liver}}{\text{units of activity/g mature liver}}$$

The use of this quotient facilitates comparison of data from different laboratories. Collection of such data of human liver from her own and other laboratories shows that during the second trimester of gestation (the period to which studies of human

GERALD E. GAULL • New York State Institute for Basic Research in Mental Retardation, Staten Island, New York, and Mount Sinai School of Medicine of the City University of New York, New York.    FRITZ A. HOMMES • University of Groningen, Groningen, The Netherlands. JACQUES F. ROUX • Hotel Dieu/Sainte-Justine Hospitals, University of Montreal, Montreal, Quebec, Canada.

fetal liver are necessarily limited), enzymatic activity changes little with time, although this is not true for brain. It is clear from these data that the $AQ$ (Table I) is appreciably different from 1.0 in virtually every case. This difference suggests that the functional need for various enzymatic activities changes during development. The majority of enzymes studied increase in activity during development, although some decrease in activity with age. In addition, the subcellular distribution of some enzymes changes during development, e.g., in fetal liver a greater proportion of glutamate dehydrogenase is present in the particulate fraction than is present in the soluble fraction. In the few studies available, the $AQ$ of various enzymes in neonatal liver is the same or only slightly greater than in fetal (2nd trimester) liver.

Use of the $AQ$ allows some useful comparisons between hepatic enzyme differentiation in man and in the rat. Enzymes denoted by A in Table I decrease in rat liver during late fetal or early postnatal development; these are the enzymes with an $AQ < 1.0$ in 2nd trimester human fetal liver. Enzymes denoted by B, C, or D all increase during development (Figure 1): the B group increases about day 17 of the

Table I.  Developmental Changes in Hepatic Enzyme Activities[a]

| Enzyme[b] | Fetal age (weeks) | Human liver Units/g Fetal | Human liver Units/g Adult | $AQ^c$ | Rat liver Adult[d] (units/g) | Rat liver Developmental behavior[e] |
|---|---|---|---|---|---|---|
| Thymidine kinase | 10.5–19 | 16.9 ± 15.1 | 1.8 ± 1.06 | 9.40 | 0.59 ± 0.1 | A |
| Phosphoserine phosphatase | 17–19 | 1.44 ± 0.21 | 0.225 | 6.40 | 0.56 ± 0.04 | A |
| Ornithine decarboxylase | 15–28 | 19.7 | 3.2 | 6.20 | | A |
| Hexokinase | | | | | | |
| S | 10.5–19 | 0.5 + 0.21 | 0.29 ± 0.05 | 2.04 } 2.60 | 0.32 ± 0.1 | A |
| P | | 0.51 ± 0.21 | <0.1 | 5.10 | <0.1 | A |
| Glucose-6-phosphate dehydrogenase | 17–19 | 2.14 ± 0.46 | 1.0; 0.93 | 2.10 | 1.2 ± 0.2 | A |
| Ornithine aminotransferase | 12–22 | 0.51 ± 0.3 | 0.42 ± 0.28 | 1.30 | 3.6 ± 0.8 | D |
| Pyruvate carboxylase | 9–18 | 2.1 | 2.6 | 0.67 | 7.3 | B |
| Glutamate dehydrogenase | | | | | | |
| S | 10.5–19 | 0.28 ± 0.4 | 18.5 ± 5.5 | 0.15 } 0.63 | 46.8 ± 10.7 | |
| P | | 29.3 ± 15 | 27.7 ± 8.6 | 1.06 | 2.4 ± 1.6 | B |
| Phenylalanine hydroxylase | 10.5–22 | 162 ± 20 | 272 ± 58 | 0.59 | 1055 ± 119 | B |
| Carbamyl phosphate synthetase | 8–28 | 2.2 | 3.8 | 0.57 | | B |
| Fructose-1,6-diphosphatase | 9–18 | 1.0 | 2.6 | 0.54 | 7.3 | B |
| Arginase | 16–22 | 642 ± 127 | 1380 ± 140 | 0.46 | 2735 ± 419 | B |
| Aspartate aminotransferase | | | | | | |
| S | 10.5–22 | 13.3 ± 3.9 | 48.1 ± 11.8 | 0.28 } 0.43 | 44 ± 0.79 | B |
| P | | 19 ± 4.6 | 27.4 ± 7.4 | 0.69 | 96 ± 12 | B |
| Glucose-6-phosphatase | 17–19 | 1.0 ± 0.5 | 2.3; 2.0 | 0.38 | 5.0 ± 0.5 | B |
| Ornithine transcarbamylase | 8–28 | 32 | 103 | 0.31 | | B |
| Methionine adenosyl transferase | 9–28 | 0.037 | 0.123 | 0.30 | | B |
| Cystathionine synthase | 9–28 | 3.5 | 16.3 | 0.21 | | B |
| Argininosuccinase | 8–25 | 0.85 | 4.9 | 0.17 | | B |
| Malate dehydrogenase (NADP⁺) | 19 | 0.03 ± 0.02 | 0.19 ± 0.02 | 0.16 | 0.9 ± 0.1 | D |
| Phosphoenolpyruvate carboxykinase | 9–18 | 0.75 | 9.1 | 0.082 | 2.2 | C |
| Tyrosine aminotransferase | 10.5–22 | 0.02 ± 0.02 | 0.27 ± 0.1 | 0.074 | 1270 ± 150 | C |
| Alcohol dehydrogenase | 9–27 | 227 | 3391 | 0.067 | | D |
| Phenylalanine pyruvate aminotransferase | 18–19 | 0.03 ± 0.03 | 0.58; 0.38 | 0.062 | 1.3 ± 0.1 | C |
| Alanine aminotransferase | 10.5–19 | 2.0 ± 1 | 41.3 ± 18.4 | 0.043 | 57.5 ± 8 | D |
| Glucokinase | 18 | 0.01; 0.01 | 0.53; 0.2 | 0.027 | 2.55 ± 0.8 | D |
| Cystathionase | 9–28 | 0 | 0.181 | 0.0 | | B |
| Protein | | | | | | |
| S | 10.5–19 | 68 ± 17 | 97.8 ± 11.7 | | 97 ± 14 | |
| P | | 74.5 ± 9.9 | 88.3 ± 14.9 | | 123 ± 17 | |

[a]Adapted from Greengard, 1976.
[b]Data for the enzymes are from Greengard's laboratory or taken by her from the literature. Where necessary activities were converted to standard units (μmol/min) per g.
[c]$AQ$ was calculated by Greengard as described in the text; for enzymes measured separately in the soluble (S) and particulate (P) fractions of liver homogenates AQ is given for each and for the sum.
[d]Values for adult rat liver are shown if reported in the same publication.
[e]The development of rat liver enzymes is denoted by letters explained in the text (Section 1) and in Figure 1. For references to the original data cf. Greengard (1976).

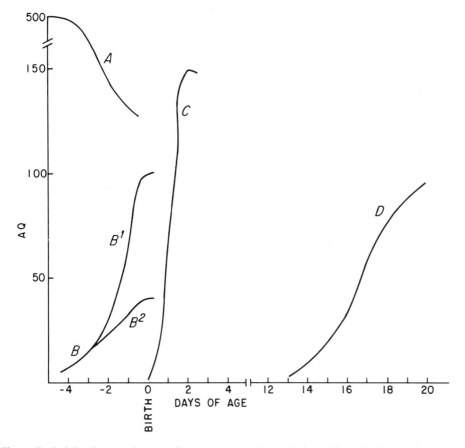

Fig. 1. Typical developmental curves of enzyme clusters in developing rat liver. The late fetal cluster (B) may attain their mature values at birth (B[1]) or later (B[2]). The neonatal cluster (C) increases immediately after birth and may have a temporary "overshoot." The enzymes of the late suckling cluster (D) begin their increase just prior to weaning.

21-day gestation of the rat, the C group increases on the first postnatal day, and the D group increases just before weaning. The important principle is that the amounts of enzyme activities in developing human liver usually change in the same direction as they do in rat liver. Furthermore, enzymes which have an *AQ* of around 0.2–0.6 tend to belong the the B group ("late fetal cluster"), whereas those with an *AQ* of <0.1 belong to the C ("postnatal cluster") and D ("late suckling cluster") group. Greengard emphasizes the species similarity in the order of appearance.

The schedule of enzyme development in liver has a certain teleorganic wisdom. Glycogenesis, glycogenolysis, and urea synthesis are unnecessary *in utero,* but the requisite enzymes are present before birth in anticipation of the need immediately after birth. Gluconeogenesis and detoxification mechanisms are less imperative and emerge sometime after birth. Studies in the developing rat show that the stepwise emergence of enzyme clusters is related to sequential changes in endocrine secretions (Greengard, 1970).

The study of human biochemical development is, at best, fraught with difficulties. For all practical purposes it is only 2nd trimester fetuses which are available for study. Fetuses from the first trimester are either stillborn or, if aborted, often have been aborted by curettage. In either case, they are unsuitable for biochemical analysis. Fetuses from the third trimester are seldom available because, if born spontaneously, they are entering the viable stage or are severely abnormal. Even as this chapter is being written, the availability of 2nd trimester fetuses is decreasing. The use of prostaglandins as abortifacients is replacing hysterotomy (and hypertonic saline) for 2nd trimester abortions. Furthermore, in the United States, the "abortion issue" is one of the most bitterly fought political issues, and the antiabortionists have attacked all fetal research. The irony, of course, is that the enormous clinical value of fetal studies to all of our offspring is unappreciated by those who would stop these studies. Well-meaning, but poorly informed, apologists within the biomedical research establishment have confused the issue further by their acquiescence to the view that we have no need to study the human fetus, but in its stead can study standard laboratory animals—especially subhuman primates. Although some biochemical phenomena are the same in subhuman primates as they are in man, many are different: we bear about the same relationship to the monkey as a log cabin does to a Corbusier building. Thus, what follows is an attempt to summarize and to interpret biochemical development in man from a combination of what is known to be true as general principles in lower animals and primates and of what can be pieced together from fragmentary evidence in human material.

## 2. Development of Carbohydrate Metabolism and Energy Transformation

### 2.1. Analytical and Interpretative Problems in the Study of Carbohydrate Metabolism during Development

The problems associated with the study of the development of carbohydrate metabolism in man are by no means easy ones, because, for obvious reasons, the degrees of freedom for experiments with man are limited. Studies of carbohydrate metabolism on autopsy material suffer especially from the drawback that autopsies are usually performed several hours after death. Thus, a new variant of unknown magnitude is introduced, namely deterioration of the tissue after death and changes in the tissue before death by insufficient circulation or by the disease process itself.

### 2.2. Changes in Glycogen Synthesis and Breakdown during Development

Shelley (1961) wrote in her classical review on "Glycogen Reserves and Their Changes at Birth and in Anoxia" that "for obvious reasons there is no reliable information on the glycogen reserves of new-born human infants." That situation remains virtually unchanged. Frequent references to laboratory animal studies have to be made and the data extrapolated to man. Species differences may obscure the real issues not only because of differences in the regulation of metabolic pathways, enzyme compartmentation, etc., but also because of the degree of development relative to birth, which is known to vary widely among species. The species differences emphasize the urgent need for developmental biochemical studies in man.

Tissue culture techniques have been of some help but are of limited value, since the cultured tissue tends to dedifferentiate when the tissue is repeatedly subcultured. An exception to this general rule is, perhaps, fibroblasts. These cells are of limited value, however, because many enzymes or functions occurring in other tissues are insufficiently expressed or not expressed in fibroblasts. Furthermore, it has been shown that cultured explants of fetal liver, for example, may lose up to 60% of their total soluble protein into the medium, which certainly limits the application of this technique for developmental studies (MacDonnell *et al.,* 1975).

A vast literature exists on the assessment of carbohydrate metabolism in the human newborn and infant. The reason is the frequent occurrence of neonatal hypoglycemia which necessitated the development of methods of diagnosis for this life-threatening or otherwise disabling condition. Glycogen mobilization can be tested by the measurement of blood glucose at different time intervals after intravenous administration of glucagon or adrenalin. Glucagon mobilizes glycogen from the liver, and adrenalin mobilizes glycogen from muscle as well as from liver. For a sound interpretation of the results, simultaneous measurements of insulin, growth hormone, lactate, and pyruvate as well as of ketone bodies are essential.

There is also a vast literature on the assessment of glycolysis, which can be tested by an oral or, better, an intravenous glucose-loading test. The functional capacity for gluconeogenesis can be tested at several steps of the metabolic pathway. Intravenous alanine loading (Fernandes and Blom, 1974) has been proposed to test the complete pathway. A rise in blood glucose after intravenous alanine administration indicates that all enzymes of gluconeogenesis are functional, provided the patient has been fasted for a long enough time to lower the blood sugar to a sufficiently low level at the start of the loading test. Such a loading test, however, is not without danger and should be carried out under carefully controlled conditions. A long fast may be harmful to a patient, especially in cases of impaired gluconeogenesis. A dihydroxyacetone- or glycerol-loading test may be used to test for deficiencies at the level of fructose-1,6-diphosphatase (Green *et al.,* 1971; Steinmann *et al.,* 1975). Disturbances in fructose utilization likewise can be tested by a fructose-loading test (Steinmann *et al.,* 1975).

For reasons which are unclear, the glucogenic amino acid aspartate does not give rise to an increase in blood glucose in normal fasted adults. Therefore this can not be used to test for deficiencies at the level of phosphoenolpyruvate carboxykinase (Hommes *et al.,* 1976). Disturbances in pyruvate utilization usually give rise to lactacidemia. This can be tested very easily, since it has been shown that lactacidemia is associated with lactaciduria (Daalmans-DeLange and Hommes, 1974; Fernandes and Blom, 1976). A preliminary screening for lactaciduria followed by a blood lactate determination in cases positive in the urine screening, therefore, will suffice. It should be emphasized, however, that definite conclusions can be drawn only after diagnosis at the molecular level, i.e., by direct measurement of the enzymes in biopsy material.

Glycogen synthesis and breakdown is a complicated process. Both glycogen synthetase and phosphorylase exist in interconvertible forms (Figure 2), and there are at least seven other enzymes involved in the regulation of this interconversion. Virtually nothing is known about the development of this enzyme system in human tissues. Conflicting reports are available as to the time when glycogen first appears in the liver. Histochemical techniques demonstrated its presence as early as 6 weeks of gestation (Shapovalov, 1961, 1962) but chemical methods could not demonstrate its presence before 8 weeks (Villee, 1954) or 13 weeks (Bourne *et al.,* 1966). The reason for this discrepancy may be the different extraction techniques that were used in these studies. Villee (1954) used boiling 30% KOH, whereas Bourne (1966) employed cold 10% trichloroacetic acid. The former method is

GERALD E. GAULL
ET AL.

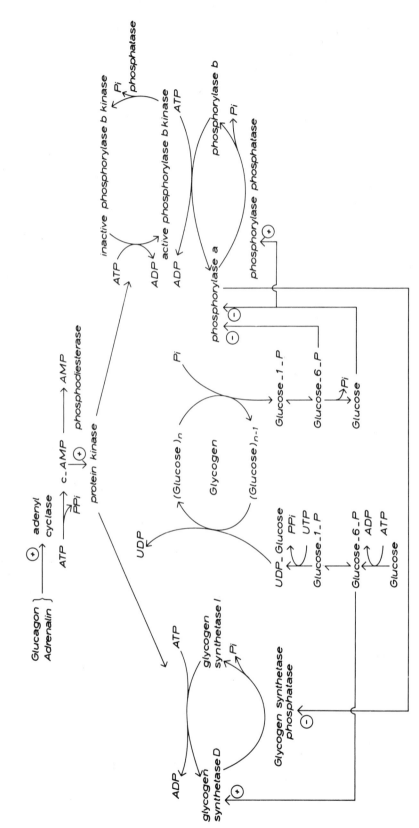

Fig. 2. Metabolic pathways of glycogen synthesis and breakdown. The circled plus and minus signs mean activation and inhibition, respectively, of enzymes.

generally considered to give the highest yield. It can be concluded, therefore, that glycogen is present in human fetal liver at least as early as the 8th week of gestation, and, consequently, glycogen synthetase must be present. The structure of glycogen of fetal liver does not seem to differ from that of adult liver (Bourne *et al.*, 1966; Manners *et al.*, 1971). The glycogen content of human fetal liver increases with fetal length (10–30 cm crown–heel length) to about 4% (Gennser *et al.*, 1971; Capkova and Jirasch, 1968). Data on older fetuses are not available. No difference could be observed between male and female fetuses (Gennser *et al.*, 1971). Cardiac muscle, lung, brain, kidney, and diaphragm contain glycogen at the earliest age tested, about 9 weeks of gestation (Villee, 1954).

The developmental pattern of glycogen deposition varies with the tissue. The glycogen content of cardiac muscle and kidney remained essentially constant from week 9 to 21; the brain glycogen content decreased; while the glycogen content of liver, lung, and diaphragm increased during that period (Villee, 1954). Increases in the glycogen content of fetal skeletal muscle to values well above those found in adult skeletal muscle have been reported in fetuses of 30–40 weeks gestation (Manners *et al.*, 1971).

Experiments with rat fetuses have yielded evidence that glucocorticoids control the regulation of glycogen storage in the liver. Decapitation *in utero* or adrenalectomy *in utero* before or at 3 days before birth prevented in part the normal increase in liver glycogen content. ACTH or cortisol given to decapitated fetuses restored glycogen deposition. The increase in liver glycogen observed in the unoperated fetus runs parallel to an increase in glycogen synthetase activity, which likewise is prevented by decapitation or adrenalectomy. Similarly, administration of cortisol to operated fetuses gives an increase in glycogen synthetase parallel to the increase in glycogen content (Jacquot, 1959). These *in vivo* studies have been confirmed by *in vitro* studies of cultured hepatocytes. Primary cultures of liver of 7–4 days before birth (that is before the *in vivo* accumulation of glycogen starts) responded to cortisol or dexamethasone administration by a rapid increase in glycogen content and an increase in glycogen synthetase activity (Plas *et al.*, 1973, 1976). The synthesis of glycogen synthetase in the fetal liver, like that of the adult liver (De Wulf and Hers, 1967) seems to be under the control of glucocorticoids.

Evidence has been presented that insulin effects the conversion of glycogen synthetase to its active form. Eisen *et al.*, (1973) confirming the studies of Plas *et al.* (1973,1976) on the induction of glycogen synthetase by hydrocortisone, found an increase in the active form of glycogen synthetase after incubation of explants of fetal rat liver in culture with insulin. The total synthetase activity remained unchanged. Therefore, insulin can stimulate glycogen deposition by increasing the percentage of active glycogen synthetase, but only in fetal liver near term. Explants of liver obtained 16 days prior to birth failed to show such a response.

Similar studies with the fetal rat heart ventricle (Jost, 1966) showed that decapitation of the fetus had no effect on heart glycogen content, unless the mother was adrenalectomized. ACTH or cortisol acetate administration restored normal glycogen deposition. Although fetal heart tissue reacts somewhat differently from fetal liver, it is evident that glucocorticoids play an important role in glycogen deposition in fetal heart as well.

Thus two other factors have to be considered in order to understand the *in vivo* development of glycogen accumulation: the development of glucocorticoid secretion and the development of glucocorticoid receptors. Plasma corticosterone concentrations in the perinatal rat have been measured by Holt and Oliver (1968) and in the postnatal rat by Tigner and Barnes (1975). A maximum in corticoid hormone was observed 2 days before birth, decreasing at birth and increasing immediately after birth. A second maximum was observed 15–20 days after birth. Glucocorticoids enter the cell and are tightly bound to highly specific receptor proteins in the cytoplasm. These complexes are then modified in the cytoplasm in such a way that they can be taken up by the nucleus, again at specific nuclear receptors. They can then interact with the gene to give specific enzyme induction (Tomkins, 1972).

The development of cytosolic glucocorticoid-binding proteins in rat liver has been studied by Singer and Litwack (1971) and by Cake *et al.* (1973). Three cortisol-metabolite-binding proteins have been identified in rat liver cytosol. The total capacity for steroid binding in the cytosol increases 10-fold from 1 day before birth to the adult level, which is reached 24 hr after birth (Singer and Litwack, 1971). The relative proportions of the binders change as well. Binder I is constant, regardless of age, but binders II and III in fetal liver cytosol are reversed from the adult pattern, binder II being high in the fetal liver. This fraction starts to decrease after the 16th postnatal day, and binder III starts to increase. The levels of the various binder proteins are similar to the adult at about the 40th postnatal day. The developmental pattern of the dexamethasone binder, as observed by Cake *et al.* (1973), closely resembles that of binder III of Singer and Litwack (1971). The data available for nuclear receptors seem to indicate that the nuclei of fetal rat liver have as many receptors as those of the 2-day-old animal (Cake *et al.*, 1973). Evidence has been presented (Beato *et al.*, 1973) that only binder III moves to the nucleus, binders I and II being of the transcortin type, the proteins responsible for transport of glucocorticoids in plasma.

Figure 3 summarizes these data. The changes in circulating glucocorticoids run parallel to the changes in liver glycogen content and glycogen synthetase activity. The proposed role of glucocorticoids for the induction of glycogen synthetase is consistent with the developmental pattern of plasma glucocorticoid-binding proteins and cytosolic glucocorticoid-binding protein. The question arises whether or not

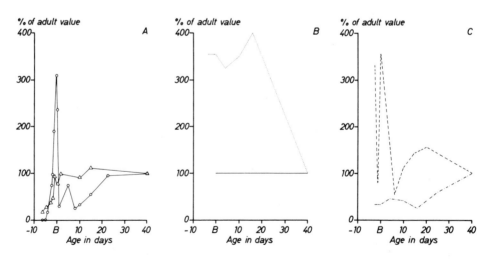

Fig. 3. Changes in rat liver glycogen (O———O), glycogen synthetase (Δ———Δ) (Figure 3A); glucocorticoid binder I (———) and II ( . . . . . .) (Figure 3B); plasma corticosterone (-----) and glucocorticoid binder III (-·-··-) (Figure 3C) as a function of age. The data are given as percentage of the values found in the adult animal and calculated from the data given by Holt and Oliver (1968), Tigner and Barnes (1975), Ballard and Oliver (1972), and Singer and Litwack (1971).

glucocorticoids also are responsible for glycogen deposition in man. Abramovich and Wade (1969b) have measured 17-hydroxycorticosteroid concentrations in amniotic fluid from 11 weeks of gestation to term and found appreciable concentrations at the earliest age studied. Although it has been shown that corticoids can be transferred from the maternal to the fetal circulation, there are reasons to believe that these steroids are of fetal origin (Abramovich and Wade, 1969a). A mechanism of glycogen deposition in human liver similar to that found in experimental animals, therefore, is consistent with currently available data.

It is generally believed—and numerous animal studies are available to support the concept (for review see Shelley, 1961)—that glycogen accumulated in liver during the later stages of pregnancy serves as a reserve energy supply to bridge the gaps between birth, the first feeding, and the onset of gluconeogenesis. Such a reserve supply of energy might be needed especially to provide the brain with glucose, even though the brain of the human newborn does not rely entirely on glucose to fulfill its energy demands. The utilization of ketone bodies can replace glucose to a significant extent (Kraus et al., 1974). Other tissues have a lower capacity for ketone body utilization, however, and are dependent, therefore, on glucose as an energy source, at least in part. This dependence implies that the machinery for glycogenolysis must be functional at birth and that glucose-6-phosphatase must be present as well (cf. Figure 2).

The activity of glucose-6-phosphatase in liver at midterm is about one third that of adult liver (Villee, 1953; Auricchio and Rigillo, 1960). Boxer et al. (1974) have investigated fetuses of younger ages and found that the activity of this enzyme starts to increase at 12 weeks of gestation. Further data on the activity of this enzyme at younger ages are lacking, but Villee (1954) demonstrated a considerable glucose production by human fetal liver slices at 10 weeks of gestation. Between 12 and 15 weeks of gestation the enzyme can be induced by glucagon and dibutyryl cyclic AMP. These compounds also can induce this enzyme in fetal rat liver, although only during later stages of pregnancy (Boxer et al., 1974). A considerable increase in glucose-6-phosphatase activity is observed during the last 10 weeks of pregnancy (Manners et al., 1971) to values observed in the adult. Many species have an overshoot in glucose-6-phosphatase activity during the neonatal period. Data on this effect in man, however, are not available. Liver phosphorylase activity starts to increase during the last 10 weeks of pregnancy (Manners et al., 1971) to values slightly above those found in adult liver. These data were obtained from livers of anencephalic fetuses, which may have suffered from a disturbed pituitary–adrenal axis. This axis has been proven to be of prime importance for glycogen deposition (see above), but it also has been shown that decapitation of rat fetuses did not modify the normal development of phosphorylase (Jacquot and Kretchmer, 1964). Similarly studies of in vitro cultures of fetal rat hepatocytes have shown that the phosphorylase activity is independent of the presence of glucocorticoids (Plas et al., 1973, 1976). It is likely, therefore, that the data obtained from the liver of the anencephalic fetuses represent the true developmental pattern of phosphorylase. Data on the development in human liver of adenyl cyclase, protein kinase, phosphorylase kinase, and phosphorylase phosphatase are not available.

In addition to the phosphorylase system, the complete breakdown of glycogen requires the debranching enzyme to hydrolyze the 1,6-glucosidic linkages. Amylo-1,6-glucosidase is present in human fetal liver, at least by the 30th week of gestation, at activity levels comparable to those found in the adult liver (Manners et al., 1971). Another route for glycogen breakdown is provided by the acid maltase, amylo-1,4-glucosidase. The physiological role of this lysosomal enzyme is not known with certainty; however, it does serve an indispensable purpose because congenital absence of this enzyme (Pompe's disease) is incompatible with long life. Rosenfeld *et al.* (1970) have shown that administration of adrenalin elicits acid maltase activity in the rat, and Gennser *et al.* (1971) demonstrated an increase in the activity of this enzyme in human fetal liver after 10 min of asphyxia. These observations emphasize the importance of the phosphorylase-independent pathway of glycogen breakdown; however, neither pathway can take over the other's role. Gennser *et al.* (1971) found a linear increase in human liver acid maltase activity from 13 to 23 weeks gestation to values which are about threefold those observed in adult liver. The activity decreases to slightly above the value observed in the adult during the last 10 weeks of gestation (Manners *et al.*, 1971).

Experiments with newborn rats have shown that glycogenolysis can be stimulated by glucagon or dibutyryl cyclic AMP in the third hour after birth, but not at earlier times after birth (Snell and Walker, 1973), although phosphorylase and the auxiliary enzymes necessary for activation of phosphorylase (cf. Figure 2) are present at birth (Novak *et al.*, 1972; Holmes and Beere, 1971). The reason for this delay in glycogen breakdown in the early postnatal period is unknown. This phenomenon does not seem to be restricted to rats because Gennser *et al.* (1971) found a similar situation during acute intrauterine asphyxia in the human fetus at midterm. Not only did the glycogen content fail to decrease, but there was no increase in the activity of phosphorylase either. The measurements of phosphorylase activity, however, are inconclusive. These measurements have been carried out in the presence of fluoride and AMP. Both compounds are known to stimulate phosphorylase *b* (Stalmans and Hers, 1975). A discrimination between the activity of phosphorylase *a* and *b*, therefore, has not been made. Furthermore, preincubation of the liver homogenate in the presence of ATP and $MgCl_2$ did not increase the total phosphorylase activity. The technique of sampling of the tissue used allowed for a considerable activation of phosphorylase (Stalmans *et al.*, 1974). It is likely, therefore, that all phosphorylase was in the *a* form in the experiments described by Gennser *et al.* (1971). Nevertheless, fetal liver does accumulate glycogen although glycogenolysis starts only after birth.

It is not clear why phosphorylase is inactive in fetal liver. Studies with isolated fetal rat hepatocytes have shown that a high glucose concentration in the medium prevents glycogen breakdown (Hommes, 1975), presumably by binding glucose to phosphorylase *a*. A similar observation has been made by Sparks *et al.* (1976) using isolated perfused fetal monkey liver obtained near term. A glucose concentration in the perfusate of 5.5 mM decreased the phosphorylase activity to very low levels. Pines *et al.* (1976) observed similar effects in fetal rat liver, but only in fetuses of 21 days gestation. In rat liver from fetuses of younger gestational ages there was no decrease in phosphorylase *a* after administration of glucose, although glycogen synthetase was activated. The phosphorylase *a*–glucose complex is less active than phosphorylase *a* without bound glucose; furthermore, it is more susceptible to the action of phosphorylase *a* phosphatase (Stalmans, *et al.*, 1972) (cf. Figure 2). This enzyme of the adult liver is stimulated by glucose (Stalmans *et al.*, 1972) but is less stimulated by glucose in fetal liver (De Vos and Hers, 1974).

It has been inferred that the inactivation of phosphorylase *a* is a prerequisite for the activation of glycogen synthetase, because phosphorylase *a* inhibits glycogen synthetase phosphatase (Stalmans *et al.*, 1974). The level of phosphorylase *a* activity at younger gestational ages, however, is below the threshold necessary for glycogen synthetase activation, which may contribute to glycogen accumulation (Pines *et al.*, 1976). Another possibility is that fetal phosphorylase is different from adult phosphorylase. Isoelectric focusing experiments, indeed, have shown that fetal rat liver contains a phosphorylase isoenzyme not found in adult liver (Sato *et al.*, 1972). Similar observations, however, are not available for human liver.

The presence of glucose-6-phosphatase in the liver allows glycogen to be degraded to glucose which then can be used by other organs of the body. Muscle does not contain glucose-6-phosphatase; nevertheless, it does show glycogen accumulation during the last stages of pregnancy and a decrease in glycogen content within 1–3 days after birth to the level observed in the adult animal (cf. Shelley, 1961). During the last 10 weeks of gestation, human fetal mucle contains twice as much glycogen as human adult muscle (Manners *et al.*, 1971). The activities of phosphorylase, amylo-1,4-glucosidase, and amylo-1,6-glucosidase were essentially the same as the adult activities during that period (Manners *et al.*, 1971).

Studies with skeletal muscle of the fetal rhesus monkey (Bocek *et al.*, 1973) have shown that at 60% of gestation epinephrine can stimulate glycogenolysis (as measured by increased lactate production and decreased glycogen content after addition of epinephrine to muscle fiber groups *in vitro*). The glycogen content of fetal muscle at that stage of pregnancy is about the same as in muscle of the adult animal (Shelley, 1961). A threefold increase in adenylate cyclase activity was observed, while similar experiments with skeletal muscle of the adult animal showed a fourfold increase in adenylate cyclase activity. The complete machinery for glycogen breakdown, therefore, must be present at this stage of pregnancy. If it is permissible to extrapolate these data to man, the implication would be that the human fetus at 24 weeks of gestation is virtually completely equipped for glycogen breakdown in skeletal muscle.

Further evidence for early development of phosphorylase in muscle has been provided by studies on cultures of myogenic cells from breast muscle of the 11-day chick embryo. After the onset of fusion of mononucleated myoblasts to multinucleated myotubes, phosphorylase starts to increase (Schmidt *et al.*, 1975). A number of other enzymes of energy transformation, e.g., phosphofructokinase (Schmidt *et al.*, 1975), creatine phosphokinase, and aldolase (Turner *et al.*, 1976), start to increase as well. Although experimental evidence has been presented that fusion and enzyme synthesis are not necessarily coupled (Holtzer *et al.*, 1972; Schudt *et al.*, 1975), both phenomena occur simultaneously during normal development. In the human embryo, myotubes can be observed first around the 5th week and increase rapidly thereafter (Fenichel, 1966). This pinpoints the start of the development of phosphorylase in human skeletal muscle at around the 5th week of gestation. Glycogen synthetase reaches its maximum activity earlier than phosphorylase during *in vitro* myoblast differentiation (Schudt *et al.*, 1975). This also is the case *in vivo*, as exemplified by developmental studies of these two enzymes in skeletal muscle of the rat hind legs (Margreth *et al.*, 1970). The development of phosphorylase is delayed until 10 days after birth. This developmental pattern is similar to that observed for other parameters of muscle differentiation and contrasts with the development of glycogen synthetase, which starts immediately after birth.

Very little is known about the glycogen content of human tissues in the period following birth. Novak and Monkus (1972) analyzed the glycogen content of subcutaneous adipose tissue—a relatively easily accessible tissue—during the newborn period and found that the glycogen content at less than 4 hr of age was inversely proportional to the length of labor. This finding emphasizes the importance of glycogen stores to the organism during periods of impaired oxgenation of the tissue. The glycogen content of adipose tissue in the first 4 hr of life was about 2.5 times that of older infants. Since adipose tissue shows a very low activity of glucose-6-phosphatase, if any (Weber *et al.,* 1965), the glucose released from glycogen must have been used within the cells. This again illustrates the importance of glycogen reserves during the period following birth before feeding has begun.

## 2.3. Changes in Glycolysis during Development

Glycolysis is the anaerobic degradation of glucose of lactic acid. For each mole of glucose broken down to lactate, 2 moles of ATP are generated (Figure 4). This pathway serves as an important emergency mechanism for short periods of oxygen deprivation as well as a major fuel supply for the tricarboxylic acid cycle during aerobic fermentation of glucose.

Developmental changes in the rate of glycolysis and in the control of this process have been reported, albeit mostly for laboratory animals. These changes are different for many tissues and may be the result of altered enzyme activities or of a different isoenzyme pattern. Isoenzymes of hexokinase, aldolase, pyruvate kinase, and lactate dehydrogenase are known.

Brown *et al.* (1967) have shown in adult human liver (also observed in adult rat liver) an isoenzyme migrating to the cathode as opposed to the four isoenzymes that phosphorylate glucose which migrate to the anode in starch-gel electrophoresis. The isoenzyme is a low $K_m$ hexokinase. A high $K_m$ glucokinase is present also in human liver. Morrison (1967) demonstrated that this glucokinase is located in the hepatic parenchymal cells and not in the bile duct epithelial cells, which contain hexokinases exclusively. The glucokinase of human liver is an inducible enzyme (Borrebaek *et al.,* 1970) as is that of rat liver.

Developmental studies on hexokinase isoenzymes in human tissues have not been reported, except for the study of Holmes *et al.* (1967), who showed that erythrocytes of newborn infants contain predominantly the type II hexokinase, whereas erythrocytes of adults contain type I and type III hexokinases. The absence of type II hexokinase in the erythrocytes of adults is not related to the age of the red blood cells. Developmental studies of hepatic glucokinase in the neonatal rat (Walker and Holland, 1965) have shown that glucokinase activity is virtually absent until 16 days after birth. Adult activities are reached 10–12 days later.

Starvation or alloxan-induced diabetes retarded the normal development, suggesting that both glucose and insulin are necessary for induction of glucokinase. This has been confirmed by the studies of Jamdar and Greengard (1970), who showed that this enzyme can be induced prematurely by hydrocortisone at 9 days after birth (but not earlier) when an injection of this hormone is followed by an injection of glucose. This premature induction of glucokinase activity could be prevented by estradiol and mannoheptulose, an inhibitor of insulin secretion. Adrenalectomy partially prevented the normal development of glucokinase. Premature induction of glucokinase by hydrocortisone was not confirmed by Partridge *et al.* (1975) using essentially the same injection scheme as that of Jamdar and Greengard. The latter authors, however, started the first injection of the hormone at 6 or 7 days after birth, while the former started the first injection at 10 days after birth. This difference may have been critical, since it is conceivable that the tissue had lost the competence to react to the hormone (cf. Greengard, 1975). Indeed, it has been shown that in the normal adult rat neither hydrocortisone nor glucose, nor a combination of the two, can induce glucokinase (Jamdar and Greengard, 1970).

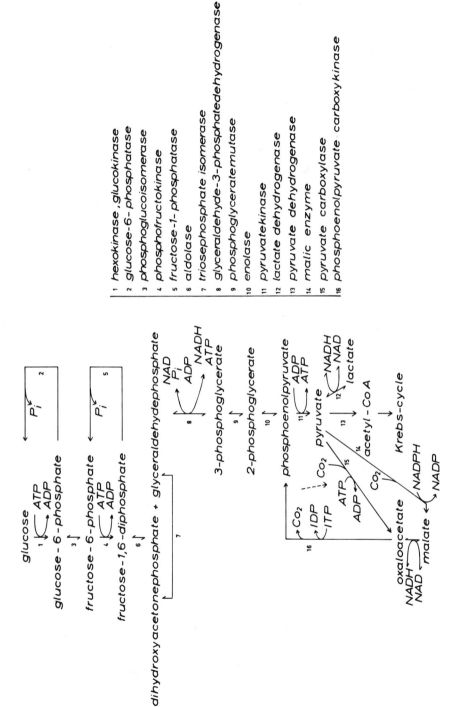

Fig. 4. Metabolic pathways of glycolysis and gluconeogenesis.

1 hexokinase , glucokinase
2 glucose-6- phosphatase
3 phosphoglucoisomerase
4 phosphofructokinase
5 fructose-1- phosphatase
6 aldolase
7 triosephosphate isomerase
8 glyceraldehyde-3-phosphatedehydrogenase
9 phosphoglyceratemutase
10 enolase
11 pyruvatekinase
12 lactate dehydrogenase
13 pyruvate dehydrogenase
14 malic enzyme
15 pyruvate carboxylase
16 phosphoenolpyruvate carboxykinase

Partridge *et al.* (1975) showed that glucokinase can be induced prematurely by thyroxine, as early as the 2nd postnatal day, when administration of thyroxine was followed by injection of glucose. Furthermore, the start of the increase in liver glucokinase in normal development coincided with a peak in the concentration of plasma thyroid hormone at about 16 days after birth. Administration of *n*-propyl-thiouracil as a thyroid inhibitor prevented the normal development of glucokinase, which could be reversed by simultaneous administration of triiodothyronine. It seems, therefore, that thyroid hormone plays an essential role as a normal physio-logical stimulus for induction of glucokinase, next to glucocorticoids, glucose, and insulin. It is likely that these factors also play a role in the induction of glucokinase in human liver, but no data are available on whether induction takes place during or after gestation.

Three different aldolases are known, each consisting of four subunits of molecular weight of about 40,000 (Morse and Horecker, 1968). Aldolase A is present in muscle and has a ratio of activities towards fructose-1,6-diphosphate and fructose-1-phosphate of 50 (Tsunematzu and Shiraishi, 1969); aldolase B is present in liver and shows a ratio of activities towards these substrates of 1.0 (Tsunematzu and Shiraishi, 1969); and aldolase C can be isolated from brain (Pennhoet *et al.*, 1966). The development of aldolase B in liver is of particular interest because this enzyme preferentially cleaves fructose-1-phosphate to dihydroxyacetone phosphate and glyceraldehyde, an obligatory reaction in fructose utilization. Fetal rat liver contains both the A and B type aldolases. At 9 days before birth about 90% of the total activity is the result of the A type. The B type increases gradually in activity, until at birth it reaches a level of about 40% of total activity (Hommes and Draisma, 1970). It has been suggested by Rutter *et al.* (1963) that the A type aldolase is involved primarily in glycolysis, whereas the B type aldolase participates in fructose metabolism and gluconeogenesis. No data are available for the development of human liver aldolase, but the fact that normal newborns, in contrast to infants suffering from hereditary fructose intolerance (cf. Van den Berghe, 1975), have no difficulties in the utilization of fructose suggests that in man type B aldolase has developed to a sufficient level at birth.

Pyruvate kinase plays an important role in the regulation of the dynamic balance between glycolysis and gluconeogenesis in the liver. During glycolysis the enzyme must be active, while during gluconeogenesis the activity must be reduced in order to conserve energy. The liver contains two types of pyruvate kinase: The L type shows an allosteric response to its substrate phosphoenolpyruvate, is inhibited by ATP and by alanine, and is activated by fructose-1,6-diphosphate. The other pyruvate kinase, the M type, is not an allosteric enzyme. The L type pyruvate kinase is confined to the parenchymal cells, whereas the M type isoenzyme is located exclusively in the Kupfer cells (Crisp and Pogson, 1972: Van Berkel *et al.*, 1972). The development of these isoenzymes in rat liver has been studied by Middleton and Walker (1972) and by Osterman *et al.* (1973). The fetal liver contains predominantly the M type, although the amount may vary with the strain of rat. The activity of the M isoenzyme decreases gradually after birth, whereas the L isoen-zyme starts to increase to the adult level around 15 days after birth. This decrease in M type pyruvate kinase may be the result of the decrease in nonparenchymal cells (Greengard, *et al.,* 1972). The possibility that even the fetal hepatocyte contains 0exclusively the L type pyruvate kinase, therefore, must be considered seriously. The kidney, another tissue capable of gluconeogenesis, contains almost exclusively the M isoenzyme. Only 10% of the L isoenzyme is found in the adult rat kidney and virtually none is present at birth (Osterman *et al.,* 1973). Pyruvate kinase isoenzymes in fetal human tissues of 11–16 weeks gestation have been studied by Foulkner and Jones (1975). Brain, heart, kidney and lung showed the M type only, whereas a substantial amount of the L type was observed in the liver. The M type isoenzyme was observed in the liver also, but the question remains whether or not this isoenzyme is located in the hepatocytes.

Lactate dehydrogenase, the last enzyme in anaerobic glycolysis, consists of four subunits. The enzyme exists in five molecular forms, which are the five possible combinations in tetramer form of two genetically different chains. Pfleiderer and Wachsmuth (1961) found that all human fetal tissues have a rather similar LDH isoenzyme pattern, provided the tissues are from fetuses younger than 14 weeks of gestation (cf. Tahasu and Hughes, 1969). Maturation to the patterns observed in the adult tissues starts around 15 weeks of gestation and has been virtually completed at 30 weeks of gestation; however, this maturation does not occur simultaneously in all tissues (Francesconi and Villee, 1969; Werthamer *et al.,* 1973). The physiological significance of these isoenzyme changes is unclear. One of the subunits is more sensitive to inhibition by pyruvate. Whether or not this difference is physiologically significant is questionable, because the reaction catalyzed by lactate dehydrogenase has been found to be near equilibrium in a variety of tissues under many different conditions (Bucher, 1970). Certain species including man, rat, mouse, hamster, and horse contain in their livers mainly the LDH subunit less inhibited by pyruvate, as do others such as pig, cow, sheep, deer, and goat (Fine *et al.,* 1963). The last group of animals, however, maintain the reaction catalyzed by lactate dehydrogenase near equilibrium.

In addition to glucose, fructose and galactose may serve as carbohydrate sources after birth. Data on the development of ketohexokinase and galactokinase in human liver are not available, but Villee *et al.* (1958) have shown that slices of midterm human fetal liver utilize fructose, suggesting that ketohexokinase of human liver has developed earlier than is usually found in animals (Walker, 1963; Ballard and Oliver, 1965). Galactokinase activity of human erythrocytes decreases with age, the adult level being reached at the age of about 5 years (Ng *et al.,* 1965). Since it has been demonstrated that the rate of removal of intravenously administered galactose (Vink and Kroese, 1959) and the rate of glucose formation after galactose administration (Haworth and Ford, 1963) is higher in infants and young children than it is in adults, the change in activity of galactokinase in erythrocytes may reflect that in the liver.

Several enzymes have been pinpointed as contributing to the control of the glycolytic flux: the glucose phosphorylating system, phosphofructokinase, and pyruvate kinase. Accurate data on the development of these enzymes in human tissues are not available, but tentative conclusions may be drawn from studies on laboratory animals. The development of glycolytic enzymes has been studied in liver, brain, kidney, and muscle (Burch *et al.,* 1963; Lowry, 1971; Stave, 1964; Hommes and Wilmink, 1968; Schaub *et al.,* 1971). Complex patterns of development of glycolytic enzymes have been observed which are difficult to correlate with control properties of this metabolic pathway. Analyses of intermediates of aerobic and anaerobic glycolysis suggested that the glucose phosphorylating step contributed most to the control of glycolysis, with phosphofructokinase exerting only fine control (Hommes, 1971). Such a conclusion was confirmed by studies on isolated fetal rat liver cells (Hommes and Berger, 1974) applying the theory of control strength as developed by Heinrich and Rapoport (1974).

The regulation of glycolysis in brain poses the additionally complicating factor of changing proportions of glycolytic enzymes in the particle-bound and in the soluble forms. The fact that most of the hexokinase activity in brain is bound to mitochondria has been well documented. MacDonnell and Greengard (1974) and Baquer *et al.* (1975) observed that the increase in hexokinase activity in rat brain is predominantly the result of an increase in the particle-bound form of the enzyme. Hexokinase, located outside the inner mitochondrial membrane and external to the ATP barrier, would confer biological advantage (Newsholme *et al.,* 1968) in that the bound form has a lower $K_m$ for ATP and a higher $K_i$ for glucose-6-phosphate. It has been inferred, furthermore, that bound hexokinase might preferentially use mitochondrial ATP, thus lowering the intramitochondrial ATP/ADP ratio, which in turn would maintain pyruvate dehydrogenase in the dephosphorylated, active form (Baquer *et al.,* 1975). The flux through hexokinase, thus, is linked to the pyruvate dehydrogenase flux. Indeed, the development of pyruvate dehydrogenase in brain closely parallels the development of the particle-bound hexokinase (Wilbur and Patel, 1974; Land and Clark, 1975). Several other glycolytic enzymes have been reported to be associated with particulate fractions of brain homogenates, either with the $100,000g$ precipitate (Baquer *et al.,* 1975) or with the $P_2$ fraction of the Gray-Whittaker procedure (MacDonnell and Greengard, 1974). This does not necessarily mean, however, that these enzymes are bound to cell organelles, because soluble forms may have been occluded within synaptosomes formed during homogenization of the tissue. This problem requires further investigation.

The glycolytic capacity of several human tissues has been determined by Villee (1954). Liver, kidney, lung, and heart show a virtually constant rate of pyruvate production from glucose from 7–10 weeks gestation to term. Glycolysis of the cerebral cortex increased fivefold from 9 to 21 weeks. When slices of these tissues were incubated under aerobic conditions, a stimulation of glycolysis was observed (Villee and Hagerman, 1958). A functional significance has been ascribed to this effect as an adaptive mechanism to provide energy during birth when the supply of oxygen may be less effective.

### 2.4. Changes in Gluconeogenesis during Development

During intrauterine life the fetus receives glucose continuously from the mother, and the fetal plasma glucose concentration is only slightly lower than the maternal plasma glucose concentration. Glucose is transported easily across the placenta (King *et al.,* 1971), but this supply ceases abruptly at birth. When the glycogen reserves are exhausted and feeding has not yet started, the newborn infant must be able to synthesize glucose from noncarbohydrate precursors, such as glycerol, lactate, or glucogenic amino acids, a process called gluconeogenesis. When lactate is the substrate for gluconeogenesis, the enzymes necessary for the conversion of lactate to glucose are essentially the same as for the conversion of glucose to lactate, except for the steps of this process which are irreversible under physiological conditions. At these steps other enzymes are necessary (Figure 4). These are glucose-6-phosphatase, fructose-1,6-diphosphatase, and the enzymes necessary for the conversion of pyruvate to phosphoenolpyruvate: pyruvate carboxylase and phosphoenolpyruvate carboxykinase.

Animal studies have shown that the ability of liver and kidney to carry out gluconeogenesis is linked to the appearance of phosphoenolpyyruvate carboxykinase (Yeung and Oliver, 1968; Thorndike, 1972; Arinze, 1975), the development of pyruvate carboxylase being slightly ahead of that of phosphoenolpyruvate carboxy-kinase (Ballard and Hanson, 1967). Such a situation seems to prevail in man as well, since Lindros and Räihä (1968) demonstrated that the activity of pyruvate carboxyl-ase in the liver of 3- to 5-month-old fetuses was only slightly lower than in adult liver, whereas the activity of phosphoenolpyruvate carboxykinase at these ages was only one tenth that of the adult liver. The activity of the latter enzyme at birth is the same as that found in the adult (Hommes, 1974). Younger gestational ages were studied by Kirby and Hahn (1973), who found that at 9 weeks of gestation phosphoenolpyruvate carboxykinase activity of liver was about three times higher than at 12 weeks. This decrease was mainly the result of a decrease in the cytosolic enzyme. Phosphoenolpyruvate carboxykinase has been induced prematurely in fetal rat liver by glucagon, epinephrine, and norepinephrine as adenylcyclase-activating compounds (Yeung and Oliver, 1968). Dibutyryl cyclic AMP also induced phosphoenolpyruvate carboxykinase in explants of human liver of 7–21 weeks of gestation. The addition of oleic acid and carnitine to the explants also resulted in an increase in phosphoenolpyruvate carboxykinase activity. The latter treatment had no effect on tyrosine aminotransferase activity, an enzyme which can be induced by dibutyryl cyclic AMP (Kirby and Hahn, 1973). These increases in activity are presumably the result of induction of the cytosolic enzyme, since it is generally accepted that it is the cytosolic enzyme which can be induced (Shrago *et al.*, 1963; Garber and Hanson, 1971). Glucocorticoids also may have an effect on fetal phosphoenolpyruvate carboxykinase, as in the adult (Shrago *et al.*, 1963; Hennig *et al.*, 1968). Kirby and Hahn (1973) observed a tenfold increase in fetal liver phos-phoenolpyruvate carboxykinase activity (12.5 weeks gestation) after treatment of the mother for 8 days with prednisolone (20 mg/day).

Fructose-1,6-diphosphatase of human liver develops between the 12th and 15th weeks of gestation (Boxer *et al.*, 1974) to values found in adult liver. The data of Villee (1954), who observed production of [$^{14}$C]glucose from [$^{14}$C]pyruvate in slices of human fetal liver around the 12th week, are in agreement with this observation. Fructose-1,6-diphosphatase could not be induced by glucagon or dibutyryl cyclic AMP when explants of fetal liver were cultured; however, adult liver reacts with a rapid increase of fructose-1,6-diphosphatase activity after intravenous glucagon administration (Greene *et al.*, 1974). It seems, therefore, that the human fetal liver at 12–15 weeks of gestation is not yet competent to react with an increase in fructose-1,6-diphosphatase activity after stimulation by glucagon (cf. Greengard, 1975), a capacity which is apparently acquired at a later age only.

The subcellular distribution of phosphoenolpyruvate carboxykinase in liver is different in many species, and this seems to affect the control of gluconeogenesis. Whereas in adult rat liver about 90% of this enzyme is located in the cytosol, in adult human liver about 80% is found in mitochondria (Söling *et al.*, 1971; Wieland *et al.*, 1968). Man resembles the guinea pig in this respect (Söling *et al.*, 1973). Evidence has been presented that in human liver the mitochondrial enzyme is different from the cytosolic enzyme (Diesterhaft *et al.*, 1971), as it is in rat liver (Ballard and Hanson, 1969). The question of the relative importance of the mito-chondrial and cytosolic enzyme in the control of gluconeogenesis arises in those species in which most of the activity is located in the mitochondria. Under condi-tions of increased demand for gluconeogenesis, the cytosolic enzyme is induced (Shrago *et al.*, 1963; Garber and Hanson, 1971). This suggests that under these conditions the main carbon flow passes the cytosolic enzyme. Evidence for this has been given by Peng *et al.* (1973), who found that addition of tryptophan or quinolinic acid to liver slices from starved guinea pigs inhibited glucose production 60%, whereas these inhibitors of phosphoenolpyruvate carboxykinase inhibited gluconeogenesis in liver slices of starved rats 90%. Tryptophan had virtually no effect on gluconeogenesis in liver slices of the fed guinea pig. Quinolinic acid is not taken up by guinea pig liver mitochondria (Söling *et al.*, 1971).

The experiments described by Arinze (1975), carried out with guinea pig fetuses of about 2–5 days prior to normal delivery, may have some relevance to man. The activity of the mitochondrial enzyme was virtually the same in the fetus, the fasted newborn, and the fed and fasted adult guinea pig liver. In contrast, activity of the cytosolic enzyme was low in fetal liver and high in liver of the fasted newborn and of the fasted adult animal. Perfusion of fetal liver with 2 mM lactate or glycerol yielded a rate of glucose production of about 0.7 $\mu$mol/min/g. Söling *et al.* (1973) have reported rates of 1.5 and 0.7 $\mu$mol/min/g with 20 mM lactate and 10 mM glycerol, respectively, as substrate in the perfused liver of the adult starved guinea pig. The rates observed in the fetal animal are very close to those observed in the adult animal. Therefore, a considerable glucose production is possible, although activity of cytosolic phosphoenolpyruvate carboxykinase is very low. If these findings can be extrapolated to man, it is likely that the normal human fetus near the end of gestation has the capacity for gluconeogenesis. This does not necessarily mean that the human fetus actually uses that capacity.

It is well known that normal newborns, and especially small-for-gestational-age infants and premature infants, develop hypoglycemia very easily. This suggests that the mechanisms for triggering and for control of gluconeogenesis in the perinatal period are more complicated than can be understood by simple observation of the developmental pattern of the enzymes participating in this process. Evidence for this also has been given by Swiatek *et al.* (1970) and by Tildon *et al.* (1971), who observed hypoglycemia in the starved newborn pig, despite the fact that a considerable cytosolic phosphoenolpyruvate carboxykinase activity was present.

Although part of the hypoglycemia in the newborn period may be caused by enzyme limitation, the inability of the newborn pig to maintain high levels of plasma free fatty acids may contribute as well (Swiatek, 1971; Tildon *et al.*, 1971). Fatty acids enhance gluconeogenesis by establishing a high intramitochondrial acetyl-CoA concentration, which is necessary for activation of pyruvate carboxylase (Seufert *et al.*, 1971; Rudermann *et al.*, 1971), at least in rat liver. Guinea pig liver, however, fails to respond with an increased rate of gluconeogenesis upon addition of fatty acids (Willms *et al.*, 1971). The reason for this species difference is unclear. Furthermore, glucagon may stimulate gluconeogenesis (Exton *et al.*, 1966, 1971) over and above that by fatty acids. A maximal effect of glucagon is dependent upon the concomitant oxidation of fatty acids (Frölich and Wieland, 1971).

Evidence has been presented that the small-for-gestational-age infant suffers from a functional delay in the development of a rate-controlling factor in hepatic gluconeogenesis. Mestyan *et al.* (1974) observed a smaller increment in blood glucose to an intravenous alanine load in small-for-gestational-age infants than that observed in normal term infants. Intravenously administered alanine is known to stimulate glucagon secretion (Muller *et al.*, 1917; Wise *et al.*, 1973a), and it does so in the normal human fetus at term (Wise *et al.*, 1973b). Similar data for the small-for-gestational-age infant are not available. Haymond *et al.* (1974) observed a significant inverse correlation of plasma glucose and of glucose and alanine in hypoglycemic small-for-gestational-age infants, suggesting a causative relation between plasma glucose concentration and gluconeogenic substrates.

Analyses of plasma amino acids showed that at birth amino acids entering gluconeogenesis at the level of pyruvate, $\alpha$-ketoglutarate, and oxaloacetate were well above control values, while at 2 hr of age only those entering at the level of pyruvate remained significantly increased. No differences were observed at 24 hr of age. This is an interesting observation, since it has been shown that glucagon stimulates gluconeogenesis from lactate, pyruvate, alanine, and serine (Ross *et al.*, 1967; Lardy *et al.*, 1974), i.e., substrates entering gluconeogenesis at the level of pyruvate. It has been shown recently that glucagon stimulates the uptake of pyruvate via the mitochondrial pyruvate carrier by increasing the $V_{max}$ of this transport system without affecting the $K_m$ for pyruvate (Titheradge and Coore, 1976). A delayed glucagon secretion or a delayed response to glucagon rather than delayed enzyme maturation, therefore, could be causative in neonatal hypoglycemia, since plasma cortisol levels were significantly higher in the small-for-gestational-age infants than they were in control infants (Haymond *et al.*, 1974). A similar conclusion can be drawn from the studies of Sperling *et al.* (1974), who observed a sluggish increase in plasma glucagon in the first 2 hr after birth in normal born infants, while plasma glucose decreased rapidly within 2 hr after birth. The plasma insulin concentration remained essentially constant. It is well known that administration of glucagon to the newborn (Cornblath *et al.*, 1961) during the first hours after birth gives a rise in blood sugar.

Experiments with newborn rats also indicate a lag in the time of onset of gluconeogenesis. Surgically delivered rats had very low liver phosphoenolpyruvate carboxykinase activities. The activity increased markedly during the first 2 hr after birth. Induction *in utero* of phosphoenolpyruvate carboxykinase by treatment of the dam with progesterone or by reduction in litter size decreased the lag in the time of onset of gluconeogenesis. It could not completely prevent it, however, despite the fact that the activity of phosphoenolpyruvate carboxykinase had been increased to the adult level (Pearce *et al.*, 1974). These studies show that immediately after birth a rate-controlling factor for gluconeogenesis exists other than limitation by enzyme activity. Ballard (1971*a,b*) showed that gluconeogenesis did not occur in neonatal rats until the liver was sufficiently oxygenated and a sufficiently high NAD/NADH ratio and energy-charge ratio was obtained. However, Ross *et al.* (1967) provided evidence that the rate of gluconeogenesis is virtually constant over a wide range of cytosolic NAD/NADH ratios, at least in adult rat livers which were sufficiently oxygenated. The data of Ballard (1971*a*) suggest that the energy-charge ratio has not yet decreased at all under hypoxic conditions when the mitochondrial NAD/NADH ratio and the rate of gluconeogenesis have decreased considerably. The mitochondrial NAD/NADH ratio, therefore, seems to be of more importance than the energy-charge ratio or cytosolic NAD/NADH ratio. Inadequate oxygenation may be a contributing factor in the delay in the onset of gluconeogensis in human liver, since marked changes in the circulation of the liver take place after birth (cf. Walsh and Lind, 1970). The data of Villee (1954),who observed production of labeled glucose from labeled pyruvate in slices of human fetal liver of 12 weeks gestation, do not contradict this hypothesis, since slices may be better oxygenated in an experimental setting then in the liver *in vivo*.

## 2.5. Changes in Mitochondrial Functions during Development

Mitochondria serve an important role in metabolism, not only as the site of oxidative phosphorylation but also because of the compartmentation of metabolites and ions. This compartmentation is made possible by the properties of the inner membrane of the mitochondria, which serves as an additional regulatory mechanism available to the cell.

### 2.5.1. Oxidative Phosphorylation

A considerable literature has accumulated recently on the question whether or not mitochondria of the fetus are as tightly coupled as those of the adult (Pollak and Duck-Chang, 1973). Using a variety of substrates, carefully prepared mitochondria show the same P/O ratios as mitochondria of the adult tissue, whether the mitochondria are isolated from whole rat embryos or from fetal liver at 17 days gestation (Mackler *et al.*, 1973). This observation contrasts with those of Holtzman and Moore (1973). They found that the P/O ratios of mitochondria prepared from different brain regions of rats younger than 2 weeks of age with glutamate and malate as substrates were lower than the expected value of 3. Normal values were observed with succinate as substrate. Mitochondria prepared from liver did not show this difference. The method used for the isolation of the brain mitochondria, however, required a step with proteinase, in contrast to the method used for the isolation of liver mitochondria (Holtzman and Moore, 1971). The possibility that brain mitochondria of younger rats are more susceptible to proteolytic attack with concomitant loss of coupling cannot be excluded. Mitochondria isolated from whole human fetuses of 6–10 weeks gestation demonstrate P/O ratios similar to those of mitochondria isolated from adult tissues (Cammer and Moore, 1972). Therefore, it cannot be stated definitely that the P/O ratio of fetal mitochondria is different from that of adults.

Several studies, however, have shown that the respiratory control ratio (RCR), as defined by Chance and Williams (1956) is lower in mitochondria isolated from fetal tissues. A serious problem with these studies is the degree of contamination of the isolated mitochondria with other cell organelles, which is difficult to assess because the marker enzymes used to determine that degree of contamination in many cases had not been developed yet. These contaminants may contribute to ATPase activity, thus lowering the RCR. A different mechanism has been proposed by Pollak (1975). Preincubation of fetal rat liver mitochondria with ATP increased the RCR twofold. GTP or low concentrations of ADP did not give this effect. The RCR of liver mitochondria of prematurely delivered rats, subsequently exposed to a 90-min maturation period, was found to be higher than it was without the maturation period. This ATP effect was explained by a specific steric interaction of ATP with the inner mitochondrial membrane, resulting in enhanced oxidative phosphorylation by an autocatalytic mechanism. This process of mitochondrial maturation requires nonfunctional mitochondria of the liver *in utero* and a triggering mechanism, possibly oxygen. Such changes in RCR have not been observed in mitochondria of neonatal swine liver (Mersmann *et al.*, 1972), developing rabbit heart (Sordahl *et al.*, 1972), developing bovine heart (Warshaw, 1969), or rabbit heart (Sordahl *et al.*, 1972). It should be pointed out, however, that it is possible to prepare fetal rat liver mitochondria, provided sufficient care is taken in the preparation procedure, with RCR values comparable to those observed in mitochondria from the adult liver (Berger, personal communication). The evidence for decreased respiratory control ratios in fetal mitochondria, therefore, is not conclusive.

The oxidative capacity of many tissues increases during development. This may be the result of an increase in the oxidative capacity per mitochondrion, an increase in the number of mitochondria, or both. The cytochrome content of liver mitochondria is independent of age (DeVos *et al.,* 1968; Jakovcic *et al.,* 1971). The specific activities of the respiratory enzyme, however, are lower in fetal liver mitochondria (Jakovcic *et al.,* 1971; Hallman 1971; Mackler *et al.,* 1971). The reason for this may be related to the phospholipid composition of the inner mitochondrial membrane. It is well known that the activities of the carriers of the respiratory chain depend on the phospholipid composition of the membrane. Developmental changes in the phospholipid and fatty acid composition of mitochondrial membranes have been demonstrated (Jakovcic *et al.,* 1971; Luit *et al.,* 1975).

The number of mitochondria per hepatocyte increases fourfold during development in the rat (Hommes, 1975). The increased oxidative capacity of the liver during development, therefore, is the result of an increased number of mitochondria as well as of an increased functional specific activity of the respiratory carriers. This increase takes place immediately before and after birth. It is not known whether or when similar changes take place in human liver, but the data of Villee (1954) on the oxidation of $[2\text{-}^{14}\text{C}]$pyruvate by human fetal liver slices showed no difference between 7–10 weeks gestation and term, indicating that, at least functionally, the human fetal liver has matured at 7–10 weeks gestation.

The developmental pattern of cytochromes in cardiac muscle seems to be different from that in liver. Hallman (1971) showed for rat heart that the cytochrome content per mitochondrion doubled from 2 days before birth to 2 days after birth. This increase in cytochrome content per mitochondrion also can be deduced from spectra reported by Sordahl *et al.* (1972) for developing rabbit heart. These developmental changes, however, do not take place simultaneously in every tissue, since similar changes take place in brain mitochondria, but in the period from 15 to 20 days after birth (Chepelinsky and De Lores Arnaiz, 1969). The arterial oxygen tension, furthermore, has an effect on mitochondrial respiratory capacity of the newborn, as illustrated by the studies of Mela *et al.* (1975) on mitochondria of the newborn dog heart. Hypoxia increases the respiratory capacity, whereas reoxygenation decreases it. This effect may have some physiological singificance, since the arterial $Po_2$ increases considerably after birth. A cytoplasmically mediated *in vivo* control mechanism for this effect seems likely, since isolated mitochondria fail to show such an effect.

Two other enzymes, although not belonging to the group of cytochromes, need special consideration because they catalyze reactions of prime importance for metabolism. The mitochondrial FAD-linked α-glycerophosphate oxidase plays an important role in the α-glycerophosphate shuttle, a mechanism for the transfer of reducing equivalents from cytosol to the mitochondrion. In order for this shuttle to be operative, both the cytoplasmic α-glycerophosphate dehydrogenase and the mitochondrial α-glycerophosphate oxidase are required. The former enzyme is present in fetal rat liver (De Vos *et al.,* 1968; Ward and Walker, 1973), but the latter does not develop until after birth (De Vos *et al.,* 1968; Hemon, 1967, 1968; Hemon and Berbey, 1968). This implies that an α-glycerophosphate shuttle cannot be operative in the fetal rat liver. Therefore, the transfer of reducing equivalents across the inner mitochondrial membrane in fetal rat liver must occur via the malate–

aspartate shuttle, because that is the only mechanism for which all the enzymes are present in the fetal liver (Hommes and Richters, 1969).

The other enzyme which needs special consideration is pyruvate dehydrogenase, which occurs in two forms, an active, dephosphorylated form and an inactive, phosphorylated form. A specific kinase and a specific phosphatase are present in mitochondria to catalyze the interconversion of the two forms (Linn *et al.*, 1969*a,b*). The development of pyruvate dehydrogenase in liver has been studied by Knowles and Ballard (1974). Its activity increases fourfold from fetal to adult life, with progressively less of the pyruvate dehydrogenase in the active form. Experiments with isolated fetal hepatocytes have provided evidence that the activity of the pyruvate dehydrogenase complex is determined mainly by the phosphorylation–dephosphorylation mechanism, whereas regulation by the mitochondrial NAD/NADH and CoA-SH/acetyl-CoA ratio is of minor importance (Berger and Hommes, 1975). It was observed that towards the end of gestation progressively less glycolytically generated pyruvate is oxidized by the mitochondrion. This was mainly the result of a higher rate of glycolysis, whereas the flux through pyruvate dehydrogenase remained fairly constant. About 80% of the pyruvate dehydrogenase complex was in the active form. This activity could account for the observed rate of pyruvate oxidation. Toward the end of gestation proportionally more of the energy demand of the liver is provided by glycolytically generated ATP. The pyruvate not oxidized by the mitochondria is converted to lactate. The NAD generated in this reaction thus becomes available for glycolysis. The lactate is transported to the mother. The size of this lactate transport from the fetus to the mother was found to be sufficient to account for the overproduction of lactate in fetal glycolysis (Kraan and Dias, 1975).

Pyruvate dehydrogenase activity in brain follows a different developmental pattern. The activity increases about eightfold from birth to maturity (Land and Clark, 1975; Berger and Hommes, 1975; Cremer and Teal, 1974; Wilbur and Patel, 1974). Up to the age of 20 days most of the pyruvate dehydrogenase is in the active form, the proportion present in the inactive form being greater in older animals. These changes may be related to the changing pattern of energy supply of the brain and lipogenesis, i.e., a change from a contribution to these processes by ketone bodies during the newborn period to predominantly glucose in older animals (Cremer and Heath, 1975). It is not known when similar changes take place in man. It is likely, however, that the general developmental pattern will be the same.

### 2.5.3. Krebs Citric Acid Cycle Enzymes

The development of most of the enzymes of the tricarboxylic acid cycle has not been studied in great detail, but some data are available for rat liver (Hommes *et al.*, 1971) and rat brain (Wilbur and Patel, 1974). Maturation in liver mitochondria takes place at an earlier stage than in brain mitochondria. It should be mentioned that the activity of ATP–citrate lyase is considerably higher in fetal liver than it is in newborn liver (Greengard and Jamdar, 1971). This higher activity may be related to the higher rate of fatty acid synthesis in fetal liver, directing the flux from pyruvate to fatty acid synthesis. In adult liver the complete tricarboxylic acid participates in the removal of pyruvate (Hommes, 1970). A special case is succinic dehydrogenase, which is low in fetal liver, possibly by limitation of the flavin part of this enzyme (DeVos *et al.*, 1968). Since carbon is withdrawn from the cycle for synthetic reactions, an additional supply of C-4 units is required to guarantee an efficient operation of the cycle. Such C-4 units may be supplied by the pyruvate carboxylase catalyzed reaction. The development of this enzyme in liver has already been discussed. Pyruvate carboxylase development in brain has been studied by Wilbur and Patel (1974) and by Land and Clark (1975). During the first three weeks of life the ratio of pyruvate carboxylase to pyruvate dehydrogenase activity is higher than in the adult. This directs the flux of pyruvate to oxaloacetate potentiated by the generation of acetyl-CoA from the utilization of ketone bodies (acetyl-CoA being an activator of pyruvate carboxylase) in the immature brain, thus facilitating the synthesis of fatty acids and steroids (cf. Land and Clark, 1975).

### 2.5.4. Translocations

The outer and inner membranes of the mitochondrion separate the content of this cell organelle from the cytosol. This physical separation of cellular compartments creates the possibility for control of metabolism other than by alteration of enzyme activity. Although the outer mitochondrial membrane is permeable to most compounds of low molecular weight, the inner membrane of the mitochondrion is not. Specific carriers of translocators are present in the inner membrane to mediate exchange-diffusion processes, sometimes energy linked (cf. Klingenberg, 1970). These translocations are species specific and genetically determined. The activity of a given translocator may vary widely among mitochondria of different tissues within the same species.

Developmental aspects of translocator activity have received only limited investigation, but some studies have been reported which stress the importance of developmental changes in mitochondrial transport processes. The delay in the onset of gluconeogenesis after birth, which may be related to a maturation of the pyruvate transporting system has already been mentioned. Mela *et al.* (1975) reported a decrease in $Ca^{2+}$ transport after birth in dog heart mitochondria, possibly induced by a higher arterial $Po_2$. Fatty acyl group translocation affords another example. The carnitine long-chain fatty acyl transferase forms an integral part of the uptake mechanism of fatty acids by mitochondria for fatty acid oxidation (Fritz *et al.*, 1974). Several studies have shown that this transferase activity is low in fetal and neonatal heart mitochondria (Warshaw, 1970, 1972; Warshaw and Terry, 1970; McMillin Wood, 1975). Although these changes in rat tissues take place mostly after birth, similar developmental changes in human tissues appear to take place much earlier (Hahn and Skala, 1973). The activity of a given translocator is determined not only by the amount of the translocator but by the phospholipid environment in the membrane as well. Changes in the phospholipid composition of the mitochondrial membrane (Jakovcic *et al.*, 1971) as well as in the fatty acid composition of the inner mitochondrial membrane (Luit *et al.*, 1975) have been reported. Considerably more detailed investigations will be required before definite conclusions can be drawn as to the importance of translocations in metabolic regulation during development.

## 3. Development of Lipid Metabolism

### 3.1. Lipid Biochemistry

The development and differentiation of fetal organs demands the synthesis of new cell structures, especially membranes, which contain phospholipids and sterols. The energy needs after separation of the fetus from the mother are met by the catabolism of free fatty acids (FFA) released from adipose tissues and derived from glucose. Adipose tissue also plays an important role in insulating the newborn against the outside cold. For these reasons, the fetus accumulates 5–10 g/day of fat in adipose tissue during the last six weeks of gestation, which would correspond to an 18-kg weight gain in a 70-kg man.

Section 3 will concern itself only with the lipid metabolism of the human embryo and fetus between the 8th and 40th weeks of gestation. (The metabolism of steroids and sterols is presented in another chapter of this book.) The reader interested in obtaining comparative animal data about the metabolism of fetal lipids is encouraged to read other reviews (Biezenski, 1975; Hahn and Novak, 1975; Harding, 1970; Hull, 1975; Myant, 1970; Robertson and Sprecher, 1968; Roux and Yashioka, 1970). Animal data and postnatal studies will be mentioned only if they are necessary to the understanding of human fetal lipid metabolism.

The importance of lipids to fetal growth and development cannot be understood without some basic knowledge of lipid biochemistry. These compounds are water-insoluble organic substances which are extractable by nonpolar solvents such as chloroform, ether, and benzene. The most important members of these heterogeneous substances may be grouped as follows:

Fatty acids
Neutral fats (acylglycerols)
    1. Mono-, di-, and triglycerides
    2. Glyceryl ethers
    3. Glycosyl glycerides
Phospholipids
    1. Phosphatides
    2. Phosphoglycerides and phosphoinositides
Sphingolipids
    1. Ceramides
    2. Sphingomyelins
    3. Glycosphingolipids
Aliphatic alcohols and waxes
Terpenes and steroids
Lipids combined with other types of compounds
    1. Lipoproteins
    2. Lipopolysaccharides
    3. Proteolipids

### 3.1.1. Biological Function of Lipids

In normal adult man, 16% of body weight is lipid, which serves three general functions: (1) as structural components of membranes (phospholipids and cholesterol), (2) as intracellular storage depots of metablolic fuel (triglycerides), (3) as a transport form of metabolic fuel (free fatty acids, lipoproteins, chylomicrons).

Phospholipids (Figure 5) and cholesterol are the components of cell membranes, endoplasmic reticulum and mitochondria, as they have both polar (hydrophilic) and nonpolar (hydrophobic) groups (Figure 5). These are found also in neurons and in myelin of brain tissue (O'Brien and Sampson, 1965).

Fig. 5. Space-filling model of phosphatidylcholine showing the compactness of the molecule (From Lehninger, *Biochemistry,* Worth Publishers, New York, 1970, p. 197).

Neutral fats are stored in liver, white adipose tissue, and brown fat. They represent the energy reserve of the body and insulate it against trauma and cold. The compactness of the triglyceride molecule (Figure 6) permits its storage in a minimum of space. One gram of lipid has a potential energy of 9.3 kcal/g, which is more than twice that of 1 g of carbohydrate or of protein (4.1 kcal/g).

Fatty acids (Figure 7), mostly in the nonesterified form (FFA), are present in serum associated with albumin. They are released from adipose tissue triglyceride and their rapid turnover and metabolism accounts for a large portion of the energy utilized daily. Lipids are digested in the intestinal tract and transported in the form of chylomicrons. The role of chylomicrons in the fetus has not been defined yet.

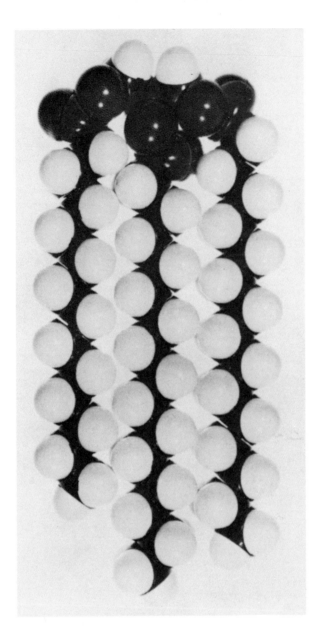

Fig. 6. Space-filling model of a tri-glyceride (tripalmitin) (From Lehninger, *Biochemistry,* Worth Publishers, New York, 1970, p. 193).

### 3.1.2. Metabolism of Lipids

Adipose tissue, mammary glands, liver, brain, heart, muscle, and endocrine tissue metabolize lipids actively. Three stages of catabolism and anabolism of the major nutrients can be distinguished (Figure 8). They serve to emphasize the close relationship between lipid metabolism and that of carbohydrate, and, to a lesser extent, that of protein.

Acetyl-CoA plays a major role in the synthesis of fatty acids and ketone bodies (Figure 9). In insulin deficiency or in fasting, the acetyl-CoA formed by the oxidation of fatty acids cannot be metabolized completely. The animal will tend then to convert more acetyl-CoA into ketone bodies (acetoacetate and β-hydroxybutyrate).

### 3.1.3. Synthesis of Lipids

The *de novo* synthesis of fatty acids is accomplished by a complex of enzymes called fatty acid synthetase. This complex of seven enzymes is located in the cytosol and catalyzes the following overall reaction:

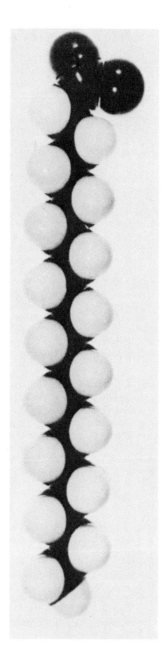

Fig. 7. Space-filling model of a free fatty acid (stearic acid) (From Lehninger, *Biochemistry*, Worth Publishers, New York, 1970).

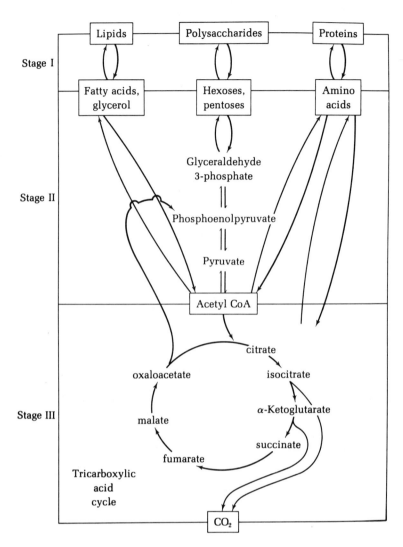

Fig. 8. The three stages of anabolism and catabolism are shown in this figure as well as the central position of acetyl-CoA. The interrelationships among lipids, polysaccharides, and protein are also clearly identified (From Lehninger, *Biochemistry,* Worth Publishers, New York, 1970).

$$\text{Acetyl-CoA} + 7 \text{ malonyl-CoA} + 14\text{NADPH} + 14\text{H} \rightarrow$$
$$CH_3(CH_2)_{14}COOH + 7CO_2 + 8CoA + 14 \text{ NADP} + 6H_2O$$
palmitic acid

Malonyl-CoA is derived from acetyl-CoA after $CO_2$ fixation under the catalytic action of acetyl-CoA carboxylase. This enzyme is a rate-limiting step in fatty acid synthesis and is influenced by the addition of citric acid, $\alpha$-ketoglutaric acid, and fatty acyl carnitine derivatives. Cytoplasmic acetyl-CoA is not obtained directly from mitochondrial acetyl-CoA. The latter may be converted to citrate, which crosses the mitochondrial membrane, or it may cross the membrane in the form of acetylcarnitine. The mitochondria elongate both saturated and unsaturated fatty

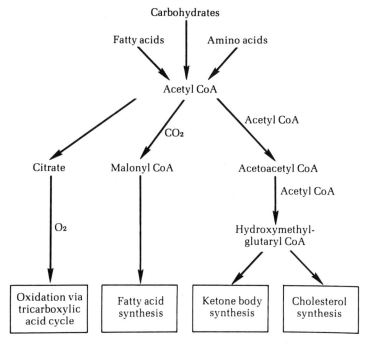

Fig. 9. Importance of acetyl-CoA and hydroxymethylglutaryl-CoA in the integration of metabolism (From Lehninger, *Biochemistry,* Worth Publishers, New York, 1970, p. 274).

acids by the addition of acetyl-CoA. Microsomes perform the same function but use malonyl-CoA.

The mitochondrion is the site of fatty acid breakdown via a $\beta\omega$ oxidation. Beta-oxidation is the most common pathway for palmitate and oleate degradation in mammalian tissues. The fatty acid enters the mitochondria with the help of carnitine, which acts as an acyl carrier. The acylation is catalyzed by two transferases, one acting on short-chain fatty acids and the other on palmitic acid. The enzymes are probably located within the mitochondrial barrier (Bremer, 1963). The oxidation of 1 mole of palmityl-CoA to $CO_2$ will yield 131 moles of ATP.

Triglycerides (triacylglycerols) or neutral lipids are synthesized in liver, adipose tissue, and intestinal mucosa from L-glycerol-3-phosphate and fatty acyl-CoA. L-Glycerol-3-phosphate is produced during glycolysis by the action of the cytoplasmic NAD-linked glycerol phosphate dehydrogenase:

Dihydroxyacetone phosphate + NADH
$$+ H^+ \rightarrow \text{L-glycerol-3-phosphate} + NAD^+$$

It can be formed also by the action of glycerol kinase:

$$ATP + glycerol \rightarrow \text{L-glycerol-3-phosphate} + ADP$$

Two molecules of fatty acyl-CoA are then acylated to the hydroxyl group of the L-glycerol-3-phosphate to form L-phosphatidic acid. A phosphatase removes the phosphoric acid group. The diacylglycerol then reacts with a molecule of a fatty acyl-CoA to yield the triglyceride:

*GERALD E. GAULL*
*ET AL.*

$$2R-C-S-CoA \qquad \text{Fatty acyl-CoA}$$
$$\overset{\|}{O}$$

$$+$$

$$CH_2OH$$
$$\overset{|}{H}COH$$
$$CH_2-O-PO_3H_2 \qquad \text{L-Glycerol-3-phosphate}$$

$$\longrightarrow 2CoA-SH$$

$$H_2C-O-C-R$$
$$\overset{\|}{O}$$

$$HC-O-C-R' \qquad \propto\text{-Phosphatidic acid*}$$
$$\overset{\|}{O}$$

$$H_2C \quad O-PO_3H_2$$
$$\qquad\qquad\qquad\qquad \text{Phosphatase}$$

$$\longrightarrow P_i$$

$$H_2C-O-C-R$$
$$HC-O-C-R'$$
$$H_2COH \quad O \qquad \text{Diacylglycerol}$$

1 mole diacylglycerol + 1 mole of fatty acyl-CoA → 1 mole of triglyceride

The phosphoglycerides (phospholipids) are formed from phosphatidic acid (see * in previous example) and cytidine nucleotide (CTP). The transformation of phosphatidylethanolamine to phosphatidylcholine requires three methyl groups which are provided by *S*-adenosylmethionine. Phosphatidylcholine can be made also from preformed choline ("salvage" pathway). This pathway is probably the major one used to synthesize lung surfactant.

Choline + ATP → phosphorylcholine + ADP

Phosphorylcholine with CTP and 1,2-diacylglycerol will yield phosphatidycholine (lecithin). Choline kinase, cytidyltransferase, and choline phosphotransferase are the enzymes involved.

### 3.1.4. Regulation of Lipid Metabolism

It is evident from the previous figures that lipid metabolism depends in part on carbohydrate and protein metabolism. However, the activity of the tricarboxylic acid cycle, the synthesis of NAD and NADPH, and the presence of amino acids increase the complexity of the regulation. There is a direct relationship between fatty acids and glucose metabolism. The uptake and metabolism of glucose by skeletal and heart muscle is inhibited by FFA. When the plasma FFA concentration increases, glucose is diverted to adipose tissue and more triglyceride is synthesized. The FFA output then diminishes, and glucose is diverted from adipose tissue to muscle, resulting in an increased rate of FFA release.

Certain hormones influence fatty acid metabolism: Insulin and estrogen stimulate glucose uptake, glycogenesis, fatty acid synthesis, and glycerol formation by the cells. Thyroxine, by its uncoupling action on oxidative phosphorylation, regulates lipid synthesis; therefore, hypophysectomy and thyroidectomy decrease lipogenesis.

Adipose tissue is a target organ for hormones which stimulate lipolysis. Epinephrine, ACTH in large amounts, TSH, and glucagon stimulate "a hormone-sensitive lipase" via cyclic AMP. Cortisol potentiates the effects of epinephrine. Growth hormone elicits a slow, prolonged response in adipose tissue with resulting FFA release. Prostaglandin E decreases the lipolysis produced by epinephrine, TSH, ACTH, and glucagon on adipose tissue. The effect on lipid metabolism (Maugh, 1975) of somatostatin (which inhibits glucagon and insulin release) has not been studied.

It is now easily understood that fasting, obesity, diabetes, overeating, exposure to cold, and mental stress may alter the metabolism of blood and tissue lipids. Glucose metabolism is different in each of these conditions, and the supply of acetyl-CoA and L-glycerol-3-phosphate for lipid synthesis is affected. Furthermore, during fasting, fatty acid synthetase, citrate cleavage enzyme, and desaturase activities diminish. Excess fatty acid storage will result in adipose tissue deposition, and the catabolism of lipids will result in adipose tissue depletion and in an excess supply of acetyl-CoA with the production of ketone bodies.

It should be noted that the cardivascular system and the activity of the autonomic nervous system, the hypothalamus, and the peripheral nerve endings will indirectly control lipid metabolism by altering the rate of blood flow and the delivery of substrates to tissues.

### 3.1.5. Analytical Techniques and Evaluation of Lipid Metabolism

The following techniques have facilitated the study of lipid metabolism in the last decade (Lehninger, 1970; White *et al.,* 1968; Stumpf, 1969).

*Thin-Layer Chromatography.* The procedure of thin-layer chromatography permits the separation of complex lipid mixtures on silica gel. The technique is excellent if the appropriate standard lipids are used. However, every effort should be made to identify the lipid class by specific chemical techniques.

*Gas Chromatography.* Gas–liquid chromatography makes use of the partition coefficient of fatty acids mixed in a carrier gas flowing on the top of a stationary phase coating a glass or steel column. It is possible to separate complex mixtures of fatty acids by this technique (Karmen, 1963).

*Computers.* The use of computers to study rates of lipid metabolism has much improved the collection and interpretation of data (Baker, 1969).

*Techniques of Estimating Rates of Irreversible Disposal of Blood-Borne Substrates in Vivo.* The disappearance rate of labeled lipids administered in one single intravenous injection or after continuous infusion permits the calculation of the disposal rate of lipids (Heath and Barton, 1973).

### 3.2. Lipid Concentration and Fatty Acid Composition during Development

Nine to sixteen percent of the total body wet weight of the human fetus at term is composed of lipid (Fehling, 1877; Widdowson, 1950). Fat deposition in early gestation takes place at a rate of 28–35 mg fat/day and reaches its maximum of 4.9 g/

Table II.  Changes in Fetal Lipid Content of the Whole Body at Various
States of Gestation[a]

| Fetal age (weeks) | Fetal weight (g) | Fetal lipids (g) | Percent of lipids |
|---|---|---|---|
| 12–14 | 20 | 0.102 | 0.5 |
| 18–20 | 200 | 1.000 | 0.5 |
| 24–25 | 635 | 16.9 | 2.6 |
| 36–27 | 2240 | 156.8 | 7.0 |
| 40–42 | 3240 | 295.8 | 9.0 |

[a]From Fehling, 1877.

day near term (Fehling, 1877). At the tenth week of gestation, 4.8 g lipid/100 g fetus (dry weight) can be extracted. This lipid concentration increases to 28.2 g/100 g fetus (dry weight) at term (Watanabe, 1967). The rapid increase in lipid deposition is reflected in part by the fetal adipose tissue which is grossly visible by the 24th week of gestation and continues to grow until term. The lipid content of fetuses at various ages of gestation is shown in Table II (Fehling, 1877).

Each organ of the human fetus demonstrates an increase in lipids during development (Figures 10,11,12, and 13). This increase is proportional to the weight of the organ, whether expressed as total protein concentration, as wet weight, or as dry weight (Roux et al., 1971). After 32 weeks of gestation, the same trend continues, with the liver triglycerides increasing more strikingly (Watanabe, 1967). The predominant lipid fractions of all tissues studied are phospholipids and sterol (Figure 13). Brain sterol deposition begins to increase after the 23rd–40th week of gestation, whereas brain phospholipids are detected from the 16th week on. Brain lipid deposition continues postnatally (Clausen et al., 1965; Cumings et al., 1958; Hansen and Clausen, 1968).

Lung phospholipids increase steadily from the 12th week of gestation until

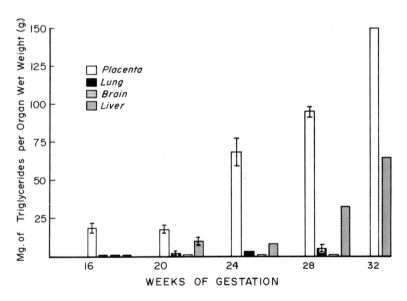

Fig. 10.  Triglycerides content of major human fetal organs during development. Results are the average of the tissues of 3–5 fetus ± the standard error. (From Roux et al., 1971, with permission.)

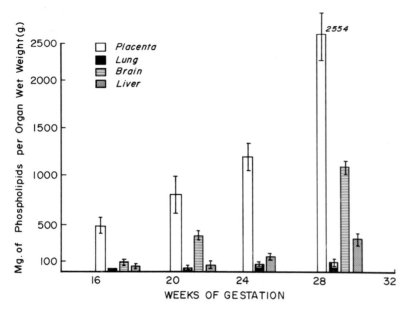

Fig. 11. Phospholipids content of major human fetal organs during development. Results are the average of the tissues of 3–5 fetus ± the standard error. (From Roux *et al.*, 1971, with permission.)

term, which is in agreement with the histological findings in the human fetal lung. The lung changes from a canalicular to an alveolar structure and starts to synthesize surfactant (Avery and Said, 1965; Felt, 1965; Fujiwara *et al.*, 1968). The placenta shows the most regular lipid deposition of all the tissues studied, suggesting that it may have an important role in fetal lipid metabolism, perhaps in the synthesis and release of fatty acids or in their transport. A search of the literature has not revealed any data on the lipid composition of the heart, muscle, or kidney in the human fetus.

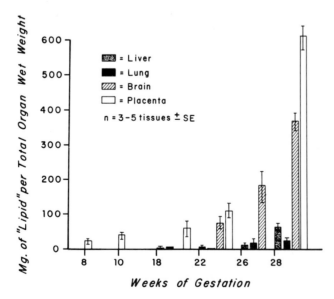

Fig. 12. "Sterol" content of major human fetal organs during development. Results are the average of the tissues of 3–5 fetus ± the standard error. (From Roux *et al.*, 1971, with permission.)

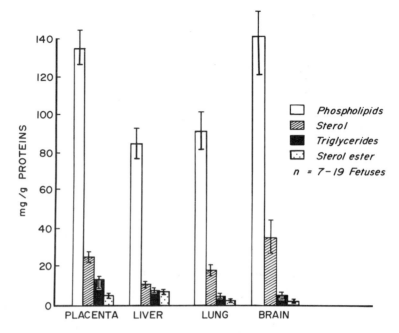

Fig. 13. Concentration of the lipid fractions in each tissue of human fetus to demonstrate the predominance of phospholipids and sterols. (From Roux *et al.,* 1971, with permission.)

When the various lipid fractions are methylated and their fatty acid compositions determined by gas chromatography, differences among fetal and adult tissue are demonstrable (Table III). Fetal tissue phospholipids contain more palmitate and stearate and less oleate and linoleate than adult tissue. Lung phospholipids, however, contain as much oleate as adult adipose tissue. Hirsch (1960) made similar observations on the adipose tissue of the fetus and newborn. He showed that profound differences exist in fatty acid composition between maternal and fetal adipose tissue at term (Hirsch, 1960). This difference is confirmed in the plasma free fatty acid composition of the newborn which, 2 hr postnatally, closely resembles the composition of its own adipose tissue triglycerides (King *et al.,* 1971*a*). The data of Roux *et al.* (1971) confirm these observations (Table IV).

The higher percentage of palmitic acid in adipose tissue triglycerides of the fetus than in those of the mother indicates a very active *de novo* fat synthesis from

Table III.   Percentage of Fatty Acid Composition of Fetal Phospholipids between the 14th and 32nd Week of Gestation[a]

| | Fatty acid[b] | | | | | | |
|---|---|---|---|---|---|---|---|
| | 14C | 16C | 16C[1] | 18C | 18C[1] | 18C[2] | 18C[3] |
| Placenta | 0.7 ± 0.1 | 50 ± 2.1 | 2.1 ± 0.4 | 23.0 ± 1.1 | 16.7 ± 1.4 | 4.5 ± 1.4 | 0 |
| Fetal liver | 0.5 ± 0.1 | 41 ± 1.1 | 4.4 ± 0.3 | 24.4 ± 1.6 | 23.0 ± 0.9 | 1.7 ± 0.4 | 0.6 ± 0.3 |
| Fetal lung | 1.0 ± 0.2 | 41 ± 1.5 | 3.6 ± 0.6 | 18.9 ± 1.4 | 50.3 ± 1.4 | 2.7 ± 0.6 | 1.2 ± 0.5 |
| Fetal brain | 3.3 ± 0.1 | 40 ± 1.9 | 6.2 ± 0.4 | 22.0 ± 0.7 | 20.5 ± 0.6 | 0 | 0 |
| Adult adipose tissue | 0 | 31 ± 1.3* | 5.7 ± 0.4 | 10.8 ± 3.0* | 38.5 ± 5.6* | 10.8± 5.4* | 0 |

[a]From Roux and Yoshioka, 1970.
[b]The results are the average of 3–6 determinations ± the standard error. They represent the percent of the total mass injected in the column. C denotes number of carbon atoms in fatty acids. 1, 2, 3 denote number of double bonds in fatty acids.(*) Denotes significantly different ($P < 0.01$) when compared with placenta, liver, lung, and brain.

Table IV.   Fatty Acid Composition of Maternal and Fetal Adipose Tissue between the 24th and 28th Week of Gestation[a,b]

|  | Mother (four determinations) | Fetus (ten determinations) |
|---|---|---|
| $C_{12}$ | 0 | 0 |
| $C_{14}$ | 2.10 ± 0.82 | 2.40 ± 0.20 |
| $C^1_{14}$ | 0.13 ± 0.08 | 0.15 ± 0.05 |
| $C_{16}$ | 29.10 ± 0.64 | 46.20* ± 2.10 |
| $C^1_{16}$ | 6.50 ± 0.80 | 7.00 ± 0.80 |
| $C_{18}$ | 5.00 ± 1.20 | 7.00 ± 0.70 |
| $C^1_{18}$ | 38.40 ± 4.90 | 31.00* ± 1.40 |
| $C^2_{18}$ | 12.80 ± 4.10 | 1.50* ± 0.40 |
| $C^3_{18}$ | 0.40 ± 0.20 | 0 |

[a]From Roux *et al.*, 1971, with permission of the author and *Pediatrics*. Copyright American Academy of Pediatrics, 1971.
[b]The results are expressed as the mean percent fatty acid composition per wet weight ± SE. Determinations carried out in duplicate. (*) $P < 0.05$ when compared with the mother.

acetyl-CoA and glucose near term. The small percentage in $C^1_{18}$, $C^2_{18}$, and $C^3_{18}$ fatty acids (oleate, linoleate, and linolenate) present in fetal adipose tissue demonstrates that the desaturase and elongating enzymes are not very active *in utero*.

One of the implications of these observations is that the origin of embryonic and fetal lipids in early gestation could be from maternal fatty acids crossing the placenta. With advancing gestation there is a gradual shift to *de novo* lipid synthesis from glucose in fetal tissue.

There is a striking increase in human brain growth (wet weight) starting at the 30th week of gestation, which continues for the next two postnatal years (Dobbing, 1974) (Figure 14). This increase requires continuous sterol and phospholipid deposition in the various parts of the brain, as demonstrated elegantly by Svennerholm and Vanier (1972) (Figures 15,16,17,18, and 19). The distribution of fatty acids in

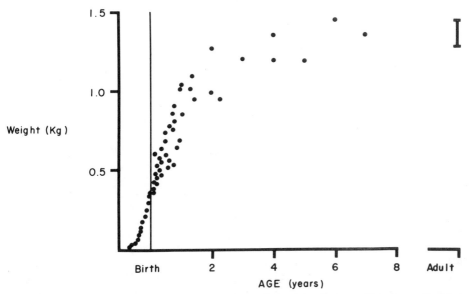

Fig. 14. Weight of the human whole brain at various periods of development. Note the continuation of growth up to two years of age. (From Dobbing, 1974, with permission of author and Saunders Publishers.)

GERALD E. GAULL
ET AL.

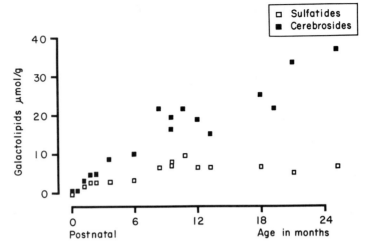

Fig. 15. Concentration of cerebrosides and sulfatides in cerebral white matter during development. (From Svennerholm and Vanier, 1972, with permission.)

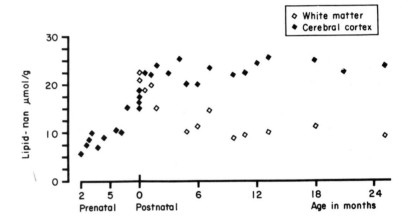

Fig. 16. Concentration of gangliosides, expressed as lipid *N*-acetylneuraminic acid, in cerebrum during development. (From Svennerholm and Vanier, 1972, with permission.)

different regions of developing fetal brain is illustrated in Table V (Srinivasa Rao and Subba Rao, 1974). This change is of a higher magnitude in the cerebellum than in other regions studied and correlates well with the overall growth spurt of the brain a few weeks before birth. Similar data have been compiled for *Macaca mulatta* (Kerr *et al.*, 1973). In this species, the brain reaches 70% of its adult weight before birth, whereas the human brain reaches only 25–30% of the adult weight before birth. This difference in brain development has important consequences for the scientist who wants to extrapolate monkey data to man.

Nikolasev *et al.* (1975) have shown that the proportion of phosphatidylcholine in lung is the same in the early fetus and the adult. This correlates well with the observations made at the other stages of gestation (Table III). The evidence for the presence of the methylation pathway in the synthesis of phosphatidylcholine is the synthesis of lecithin with myristic acid on $C_2$. Since myristic ($C_{14}$) acid is not found in increasing proportion with fetal age in the lung, the methylation pathway for phosphatidylcholine synthesis does not appear to play an important role in surfactant synthesis. Similar data have been reported in the fetal rhesus monkey (Kerr and Helmuth, 1975).

The lipid components of placental tissue at term and of umbilical cord blood have been compared with those of the maternal blood (Table VI) (Kerr and Helmuth, 1975; Robertson *et al.*, 1969). The phospholipid content of early and late human placenta can be found in the paper of Nikolasev *et al.* (1973). In fetal blood the FFA concentration (Table VI) as well as the concentration of other lipids is lower than in maternal blood (with the exception of lysolecithin which is higher in

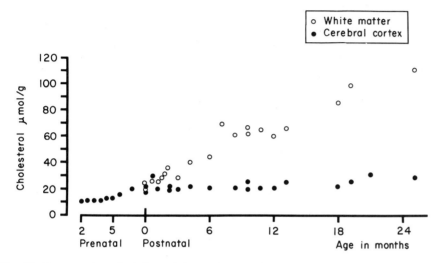

Fig. 17. Concentration of cholesterol in cerebrum during development. (From Svennerholm and Vanier, 1972, with permission.)

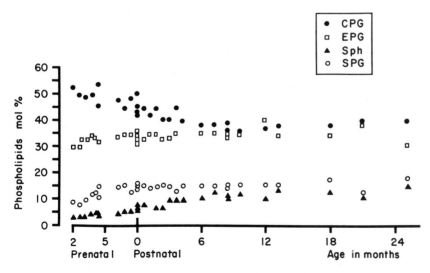

Fig. 18. Percentage distribution of major phospholipids in cerebral cortex during development. CPG: choline phosphoglyceride; EPG: ethanolamine phosphoglyceride; Sph: sphingomyelin; SPG: serine phosphoglyceride. (From Svennerholm and Vanier, 1972, with permission.)

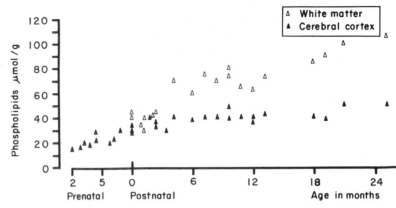

Fig. 19. Concentration of total phospholipids in cerebrum during development. (From Svennerholm and Vanier, 1972, with permission.)

*Table V.    Distribution of Fatty Acid Groups in Different Regions of Developing Fetal Brain*[a]

| Gestational age (weeks) | Percentage of total methyl esters[b] and nature of fatty acid group | | | | | | | | | | | |
|---|---|---|---|---|---|---|---|---|---|---|---|---|
| | Saturated | | | Unsaturated | | | Monoenes | | | Polyenes[c] | | |
| | CB | CL | MO | CB | CL | MO | CB | CL | MO | CB | CL | MO |
| 11–13 (whole brain) | | 63.3 | | | 36.7 | | | 31.3 | | | 4.7 | |
| 20–22 | 56.7 | 52.8 | 58.0 | 43.3 | 47.2 | 42.1 | 24.9 | 18.1 | 32.3 | 16.3 | 21.1 | 9.4 |
| 24–26 | 56.5 | 50.3 | 49.7 | 43.5 | 49.7 | 47.1 | 28.7 | 33.1 | 29.5 | 12.4 | 11.1 | 14.1 |
| 30–32 | 60.5 | 52.2 | 50.5 | 39.4 | 47.8 | 49.5 | 23.2 | 32.7 | 31.4 | 11.6 | 12.0 | 15.1 |
| 33–35 | 64.9 | 60.8 | 63.1 | 35.1 | 39.2 | 36.9 | 22.2 | 30.7 | 29.7 | 12.8 | 9.8 | 8.8 |
| Term | 52.2 | 48.9 | 48.0 | 47.8 | 51.1 | 52.8 | 24.6 | 23.0 | 25.6 | 18.9 | 26.2 | 18.2 |

[a]From Srinivasa Rao and Subba Rao, 1974.
[b]CB = cerebrum, CL = cerebellum, and MO = medulla oblongata.
[c]Includes tetra-, penta-, and hexaenoic acids of 20 and 22 carbon units only.

*Table VI.    Lipid Components of Placental Tissue at Term, Umbilical Cord Blood, and Maternal Blood*[a]

| Component | Mother's blood | Placental tissue | Umbilical cord blood |
|---|---|---|---|
| Microequivalents per liter or per kilogram of tissue (net weight) | | | |
| Free fatty acids | 1250 + 78 | 4922 + 664 | 586 + 78 |
| Milligrams per 100 ml or per 100 g of tissue (wet weight) | | | |
| Free cholesterol | 93 + 7 | 221 + 11 | 29 + 3 |
| Cholesterol ester | 224 + 9 | 64 + 8 | 76 + 5 |
| Triglyceride | 159 + 14 | 89 + 7 | 43 + 5 |
| Phospholipid | 316 + 14 | 886 + 39 | 111 + 7 |
| Milligrams per 100 ml or per 100 g of tissue (wet weight) of phospholipid phosphorus | | | |
| Lecithin | 6.94 + 1.27 | 18.73 | 2.75 + 0.47 |
| Cephalin | 0.75 + 0.17 | | 0.25 + 0.04 |
| Phosphotidylethanolamine | | 11.44 | |
| Phosphatidylserine | | 3.45 | |
| Lysolecithin | 0.21 + 0.09 | 0.71 | 0.30 + 0.07 |
| Sphingomyelin | 1.59 + 0.47 | 5.01 | 0.84 + 0.18 |
| Milligrams per 100 ml of lipoprotein lipid | | | |
| Alpha lipoproteins | 257 + 71 | | 147 + 40 |
| Beta lipoproteins | 847 + 176 | | 224 + 61 |

[a]From Robertson and Sprecher, 1968.

fetal blood). This indicates that some lipids in fetal blood may not be transferred directly from the maternal to the fetal circulation. The placenta acts as a selective barrier, and its role in FFA transport will be discussed later.

FFAs are released at birth in large quantities into the newborn circulation. This release is secondary to the environmental changes taking place after birth, because changes in the environmental temperature do alter the release of fetal FFA (Roux, 1966; Roux and Romney, 1967; Van Duyne, 1965).

Differences exist among various species in total body or organ lipid concentration and composition. Like man, rabbits, guinea pigs, and rhesus monkeys increase their tissue lipid content during intrauterine life. Serum free fatty acid concentration of fetal guinea pigs, however, is higher than that of the mother (Roux, 1966; Hershfield and Nemeth, 1968; Kayden et al., 1969). The body of the lamb fetus, in contrast to that of the human fetus, contains less lipids than its mother (Van Duyne, 1965); Body et al., 1966). Lipid concentration in rat and mouse liver shows a remarkable increase immediately after birth, whereas the reverse is true in rabbits (Roux, 1966; Morikawa et al., 1965; Morikawa and Eguchi, 1966).

Such differences should caution the investigator not to extrapolate to man lipid metabolic data obtained in other species. Depending on the research objective to be reached, an appropriate animal model should be selected and supporting data should be obtained in the human fetus.

## 3.3. Changes in Lipid Metabolism in Vitro and in Vivo during Development and Origin of Fetal Lipids

Popjak demonstrated in rabbits and guinea pigs that fetal tissue was capable of synthesizing fatty acids, cholesterol, and phospholipids *in vitro* and *in vivo* from acetate, deuterated water, and phosphate (Popjack, 1954). He suggested that lipid deposition in the fetus is not so much characterized by an increased rate of synthesis as it is by a decreased rate or absence of the degradative process. These observations were extended and corroborated in other species such as lamb, rat, mouse, rabbit, monkey, and man (Roux, 1966; Dhopeshwarkar et al., 1973, Coltart, 1972; Coutts and MacNaughton, 1969; Fain and Scow, 1966; Hummel and Wagner, 1975; Plotz et al., 1968; Roux et al., 1967; Smith and Abraham, 1970; Van Duyne et al., 1960, 1962; Villee, 1969).

Villee showed that human tissue, as early as the 12–20th week of gestation, is capable of incorporating labeled glucose, fructose, acetate, citrate, and amino acids into lipids *in vitro*. He calculated that the net rate of fetal lipid synthesis is greater than that of the adult (Villee and Loring, 1961). This was confirmed *in vivo* by Coltart in the human fetus and extended to the synthesis of cholesterol esters in the adrenals and liver of previable human fetus (Coltart, 1972).

It has been observed also that cell-free supernatant fluid and microsomes of liver, lung, brain, and placenta of the human fetus, studied between the 8th and 28th week of gestation, incorporate $[1\text{-}^{14}C]$acetate into sterol and phospholipids (Figures 20 and 21). This incorporation is marked in brain, liver, and placenta. At 22 weeks of gestation, citrate and acetate are converted into lipids by slices of liver, brain, lung, and placenta (Roux et al., 1971; Sabata et al., 1968). When palmitate is used as substrate (Figure 22) in the presence of placental slices, the synthesis of phospho-

*GERALD E. GAULL*
*ET AL.*

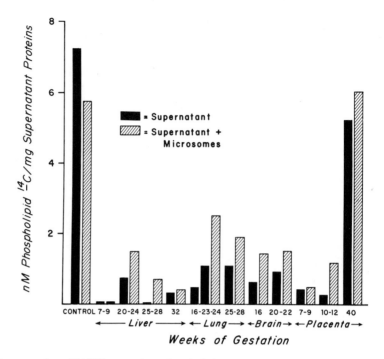

Fig. 20. Incorporation of [1-¹⁴C]acetate into phospholipids by human fetal tissue during development. Results are the average of two experiments in quadruplicate. The data are compared with those obtained in cell fraction of rat mammary gland (control) incubated in the same system. The results are expressed in mμM of [¹⁴C]phospholipid derived from [1-¹⁴C]acetate. (From Roux *et al.*, 1967; Roux and Yoshioka, 1970.)

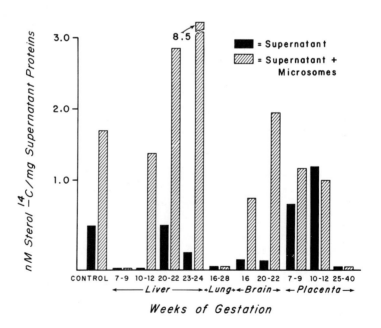

Fig. 21. Incorporation of [1-¹⁴C] into sterol by human fetal tissue during development. Data are expressed in mμM of [¹⁴C]sterol. The experimental design is the same as in Figure 20. (From Roux *et al.*, 1971.)

lipids and triglycerides is rapid. The synthesis of sterol and sterol ester is small (Yoshioka and Roux, 1972). These data indicate that from an early embryonic stage, human tissues contain the fatty acid synthetase, citrate-cleavage enzyme, glycerol phosphate dehydrogenase, phosphatase, NADH, $NAD^+$, and cytidine nucleotide necessary to synthesize lipids from carbohydrate.

These observations made *in vitro* suggest that fetal tissue may use carbohydrates and fatty acids for lipid synthesis. In a study of the enzymes of glucose and fatty acid metabolism in early and term placenta, Diamant *et al.* (1975) noted that, although the activity of the enzymes involved in fatty acid synthesis was low, the activity of the pentose phosphate shunt was relatively high. This pathway provides NADPH and glycerol for lipid synthesis.

It is known that in maternal plasma, free fatty acids are transported to other tissues to be catabolized to $CO_2$ or converted into lipids (see above). It was postulated that maternal FFA and cholesterol could be the precursors of fetal lipids. It was demonstrated later *in vivo* in sheep, rabbit, man, guinea pig, monkey, and rat and by perfusion of human and guinea pig placentas that palmitate, linoleate, and oleate can cross the placental barrier and possibly be incorporated into fetal lipids (Van Duyne *et al.*, 1960; Portman *et al.*, 1969). The rate of transfer is extremely rapid (2–3 min) and takes place against a concentration gradient. Cholesterol also was shown to be incorporated *in vivo* into the myelin of the fetal human brain and cholesterol esters of the liver and the adrenals (Poltz *et al.*, 1968; Cekan *et al.*, 1973).

From these data numerous calculations have been made to determine how much maternal glucose and fatty acid is incorporated into fetal lipids. Maternal glucose provides approximately 40 kcal/kg/day to the human fetus after crossing the placenta (Sabata *et al.*, 1968). This corresponds to approximately 4–5 g fat/kg body weight manufactured in one day. If all the glucose transferred to the fetus is converted to lipid, then the fetus would synthesize 120 g fat/kg body weight in 30 days. Actually the human fetus accumulates this amount of fat in four weeks

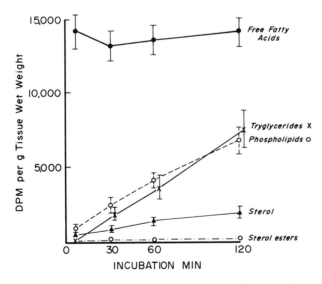

Fig. 22. *In vitro* [1-$^{14}$C]palmitate incorporation into lipids by term placental slices at 37°C. Two (2.0) millimol of [1-$^{14}$C]palmitate has been incubated with the placental slices in the presence of 10 millimol of glucose (Roux, 1974). The slices were removed at the time of incubation indicated on the abscissa. The methods have been described. (From Yoshioka and Roux, 1972.)

(Fehling, 1877; Sabata *et al.*, 1968). Glucose alone, then, is sufficient to account for the synthesis of lipid during fetal development. Since 6.8 mmol of free fatty acid per day is transferred into the fetal compartment of the perfused human placenta, Dancis *et al.* (1973) have calculated that this daily supply of FFA would enable synthesis of 70 g of lipid a month, or approximately 20% of the 350 g lipid accumulated in the last month of gestation. By extrapolation from sheep, Van Duyne *et al.* (1960, 1962), have calculated that the transport of maternal FFA to the fetus can account for all the fetal lipid deposition. Indeed, 1.6–2.1% of [1-$^{14}$C]palmitate injected intravenously to a pregnant ewe crossed the placenta (Van Duyne *et al.*, 1960, 1962; Elphick *et al.*, 1975). The turnover rate calculated from this experiment theoretically would account fot the total accumulation of fatty acids in the rabbit fetus provided none was oxidized (Van Duyne *et al.*, 1960, 1962; Elphick *et al.*, 1975). In the guinea pig and rhesus monkey, the administration of essential and nonessential fatty acids to the mother was accompanied by the appearance within minutes of these acids in the fetal compartment and their incorporation into fetal tissue lipids (Hershfield and Nemeth, 1968; Kayden *et al.*, 1969; Portman *et al.*, 1969). The transfer rate of palmitate and linoleate from mother to fetus was found to be 2.2 mg/24 hr, respectively. This would account for all the linoleate stored in the fetus and half of the palmitate. The other half could have been synthesized from glucose (Hershfield and Nemeth, 1968).

From these data, it appears that both glucose and fatty acids of maternal origin can account for fetal lipid deposition. Furthermore, the human fetus uses FFA because the umbilical artery concentration of FFA is lower than that of the umbilical vein (Sheath *et al.*, 1972). Dancis' evaluation in the human fetus seems to be the most realistic estimate, since fetal tissue shows *in situ* lipid synthesis from carbohydrate and fatty acids. Although FFA cross the placenta, the transfer of phospholipids and triglycerides from mother to fetus is minimal (Smith and Abraham, 1970; Biezenski, 1969; Biezenski and Kimmel, 1969).

Cholesterol crosses the placenta of guinea pigs, rabbits, and man quite readily. Labeled cholesterol administered to the mother can be found in fetal tissues such as liver and adrenal, but brain shows the lowest concentration (Coutts and Mac-Naughton, 1969; Plotz *et al.*, 1968; Connor and Lin, 1967).

It is important to realize that fetal nutrition is dependent not only on maternal glucose and fatty acids, but on other precursors as well. This allows the fetus flexibility to draw its necessary energy for growth and development from different nutrients.

An extensive study of the *in vitro* metabolism of palmitic acid and glucose in developing tissues of the rhesus monkey demonstrated that these two compounds are precursors of fetal lipids (Tables VII and VIII) and $CO_2$ (Roux and Myers, 1974). In a similar study done in man (Figure 23 and Table IX), it can be seen that each tissue uses palmitic acid. The incorporation of this substrate into lipids or $CO_2$ is a direct function of its concentration and duration of incubation (Figure 23). Beatty and Bocek (1970) reported similar data using rhesus monkey muscle and demonstrated an active palmitate uptake into the fetal muscle and rapid conversion of the substrate to $CO_2$. Other investigators reported similar findings in different species (Fain and Scow, 1966; Roux *et al.*, 1967; Smith and Abraham, 1970; Villee and Loring, 1961; Yoshioka and Roux 1972).

The production of carbon dioxide *in vitro* by fetal tissue (Tables VII and VIII and Figure 23) is a clear indication that fatty acids are catabolized and, therefore, a source of energy for the fetal cell. The enzymes of $\beta$-oxidation must be present in fetal tissues. It should be noted that placenta and fetal liver catabolize palmitate

Table VII.   *[1-¹⁴C]Palmitate Conversion to Lipids or ¹⁴CO₂ and Glucose Utilization by Tissue Slices from Rhesus Monkeys during Development*[a,b]

| | Young fetus | Term fetus | Newborn monkey | Adult |
|---|---|---|---|---|
| Glucose utilization | | | | |
| Brain | | 2.2 ± 5.1 (14) | 15.1 ± 2.2 (8) | 20.7 ± 2.2 (8) |
| Lung | | −14.9 ± 7.5 (14) | 6.3 ± 1.4 (8) | 0.2 ± 4.4 (8) |
| Liver | | −26.7 ± 4.4 (14) | −28.3 ± 6.7 (8) | −23.6 ± 6.0 (8) |
| Phospholipids | | | | |
| Brain | 1729.0 ± 404.0 (9) | 710.0 ± 54.0 (4) | 873.0 ± 95.0 (7) | 265.0 ± 110.0 (6) |
| Lung | 2296.0 ± 298.0 (9) | 2621.0 ± 638 (3) | 2759.0 ± 349.0 (8) | 3876.0 ± 307.0 (6) |
| Liver | 1876.0 ± 220.0 (10) | 873.0 ± 80.0 (4) | 833.0 ± 216 (8) | 2277.0 ± 294.0 (5) |
| Sterols | | | | |
| Brain | 443.0 ± 85.0 (10) | 314.0 ± 35.0 (4) | 288.0 ± 43.0 (7) | 157.0 ± 54.0 (8) |
| Lung | 279.0 ± 58.0 (10) | 194.0 ± 31.0 (4) | 470.0 ± 76.0 (8) | 524.0 ± 69.0 (8) |
| Liver | 624.0 ± 98.0 (9) | 243.0 ± 68.0 (4) | 294.0 ± 38.0 (8) | 1148.0 ± 198.0 (8) |
| Triglycerides | | | | |
| Brain | 1124.0 ± 137.0 (9) | 483.0 ± 26.0 (4) | 318.0 ± 47.0 (7) | 160.0 ± 100.0 (6) |
| Lung | 1436.0 ± 391.0 (9) | 1073.0 ± 174.0 (3) | 1743.0 ± 256.0 (8) | 2266.0 ± 714.0 (6) |
| Liver | 6293.0 ± 1032.0 (10) | 3282.0 ± 631.0 (4) | 1571.0 ± 240.0 (8) | 3248.0 ± 710.0 (5) |
| Sterol esters | | | | |
| Brain | 12.1 ± 1.4 (8) | 13.2 ± 2.0 (4) | 11.8 ± 2.0 (7) | 17.0 ± 5.0 (8) |
| Lung | 18.6 ± 2.3 (9) | 14.0 ± 6.0 (3) | 28.0 ± 5.0 (8) | 64.0 ± 14.0 (8) |
| Liver | 89.2 ± 11.2 (10) | 54.0 ± 10.0 (4) | 40.0 ± 4.3 (8) | 20.0 ± 5.4 (7) |
| ¹⁴CO₂ | | | | |
| Brain | 40.8 ± 6.3 (9) | 65.0 ± 5.0 (3) | 60.0 ± 7.0 (8) | 54.0 ± 7.0 (8) |
| Lung | 127.3 ± 13.6 (9) | 94.0 ± 12.0 (4) | 349.0 ± 81.0 (8) | 430.0 ± 31.0 (8) |
| Liver | 276.2 ± 51.5 (10) | 202.0 ± 34.0 (4) | 233.0 ± 26.0 (8) | 273.0 ± 86.0 (8) |

[a]From Roux and Myers, 1974.
[b]The results are the mean disintegrations per minute ± SE of the number of determinations shown in parentheses. Glucose utilization is expressed in micromoles per gram of wet weight per hour of incubation; − means glucose production. The conversion of [1-¹⁴C]palmitate to lipids and carbon dioxide is expressed in disintegrations per minute per gram of wet weight per hour of incubation.

faster than other fetal tissues; they are also the most active metabolically during development.

The breakdown of fatty acids and their transport in and out of the mitochondria is under the control of palmityltransferase, which is present in term placenta (Karp *et al.,* 1971) and in fetal tissue (Hahn and Skala, 1972*a*; Novak *et al.,* 1973*a,b*). Although fatty acid breakdown takes place in fetal tissue *in vitro,* it does not mean that this happens *in vivo* to the same extent. The data available suggest, however, that glucose and fatty acids are metabolized by the fetus according to the needs of growth and differentiation.

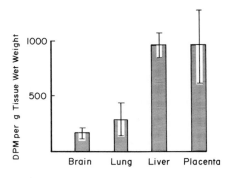

Fig. 23. Respiratory ¹⁴CO₂ production from [1-¹⁴C] palmitate by tissue of 12- to 13-week-old human fetus. The experimental design is the same as in Figure 22. Incubation time was 1 hr. The results are the average of four experiments carried out in duplicate ± the standard error. (From Yoshioka and Roux, 1972.)

Table VIII.   [U-[14]C]Glucose Utilization and Conversion into Lipids and Carbon Dioxide by Tissue
Slices of Rhesus Monkey during Development[a,b]

|  | Fetus | Newborn monkey | Adult |
|---|---|---|---|
| Glucose utilization |  |  |  |
| Brain | 33.6 ± 2.1 (14) | 34.5 ± 2.1 (4) | 42.9 ± 0.4 (4) |
| Lung | 15.3 ± 1.5 (12) | 20.8 ± 3.4 (4) | 24.4 ± 1.1 (4) |
| Liver | −25.5 ± 7.4 (14) | −25.3 ± 5.8 (4) | −20.1 ± 13.6 (4) |
| Phospholipids |  |  |  |
| Brain | 249.2 ± 43.1 (14) | 123.0 ± 5.0 (4) | 61.0 ± 2.9 (4) |
| Lung | 221.3 ± 17.3 (12) | 229.0 ± 54.0 (4) | 356.0 ± 12.8 (4) |
| Liver | 176.2 ± 30.8 (14) | 77.0 ± 13.0 (4) | 110.0 ± 8.0 (4) |
| Sterols |  |  |  |
| Brain | 80.8 ± 25.9 (14) | 32.0 ± 2.0 (4) | 24.0 ± 5.0 (4) |
| Lung | 26.9 ± 4.3 (7) | 45.0 ± 8.0 (4) | 28.0 ± 0.0 (1) |
| Liver | 94.6 ± 32.6 (14) | 35.0 ±0.0 (1) | 23.0 ± 4.0 (4) |
| Triglycerides |  |  |  |
| Brain | 98.8 ± 13.7 (11) | 13.0 ± 0.0 (1) | 20.0 ± 0.0 (1) |
| Lung | 48.2 ± 12.2 (8) | 73.0 ± 37.0 (4) | 52.0 ± 15.0 (4) |
| Liver | 113.1 ± 20.2 (14) | 145.0 ± 54.0 (4) | 37.0 ± 5.0 (4) |
| Sterol esters |  |  |  |
| Brain | 17.1 ± 1.6 (6) | ND | ND |
| Lung | 16.0 ± 0.0 (2) | ND | ND |
| Liver | 30.5 ± 10.2 (6) | ND | ND |
| [14]CO$_2$ |  |  |  |
| Brain | 18,460.0 ± 2017.0 (12) | 25,858.0 ± 2997.0 (4) | 33,189.0 ± 9206.0 (4) |
| Lung | 3819.0 ± 462.0 (12) | 5973.0 ± 1135.0 (4) | 9136.0 ± 1040.0 (4) |
| Liver | 3218.0 ± 1006.0 (12) | 2694.0 ± 409.0 (2) | 3783.0 ± 279.0 (4) |

[a]From Roux and Myers, 1974.

[b]The results are the means ± SE of the number of determinations shown in parentheses. Glucose utilization is in micromoles per gram of wet weight per hour of incubation. All other data are in disintegrations per minute per gram of wet weight per hour of incubation. Production, ND = not determined.

The brain has a rapid rate of growth before birth and anabolizes lipids for myelin synthesis (Tables VII and IX). The brain produces less $CO_2$ from palmitate than from other fetal tissues (Figure 23). Since the blood–brain barrier of the fetus is more permeable to biological substances than it is in the adult (Stumpf, 1969; Sperry, 1962) and since palmitic acid injected intraperitoneally to young rats is incorporated into brain lipids (Dhopeshwarkar *et al.*, 1973), the fetal brain must rely on lipid precursors during development. Lipid metabolism in the lung will be reviewed in more detail below.

Table IX.   [1-[14]C]Palmitic Acid Incorporation into Tissue Lipids and Conversion to
Respiratory CO$_2$ by Human Fetal Tissue[a]

|  | Brain | Lung | Liver |
|---|---|---|---|
| Phospholipids | 4111 ± 1421 | 3720 ± 598 | 6050 ± 931 |
| Sterols | 704 ± 458 | 940 ± 518 | 1245 ± 641 |
| Free fatty acids | 7578 ± 1232 | 7646 ± 1450 | 4407 ± 1314 |
| Triglycerides | 5503 ± 1803 | 2656 ± 388 | 8897 ± 2129 |
| Sterol esters | 37 ± 15 | 86 ± 23 | 754 ± 172 |
| Respiratory $CO_2$ | 316 ± 174 | 505 ± 205 | 1062 ± 199 |

[a]Tissue was incubated at 37.0° in 95% $O_2$–5% $CO_2$ for 1 hr in 2.2 mmol [1-[14]C]palmitic acid (specific activity, 15 $\mu$ Ci/mmol) and 10 mmol glucose. Results are expressed as dpm/g wet weight and represent mean ± SE of 5 experiments.

Fig. 24. The distribution of white (WAT) and brown (BAT) adipose tissue in the newborn infant (BAT redrawn from Dawkins and Hull, *Sci. Am.* **213**:62, 1965).

## 3.4. Adipose Tissue

The storage of fat in the adipose tissue of the fetus represents an important part of the total energy reserves accumulated in anticipation of the metabolic demands of the neonatal period. In the previous section deposition and synthesis of lipids were discussed, and a discussion of the metabolism of lipids in fetal adipose tissue was omitted. The metabolism of adipose tissue during fetal development and the neonatal period has been reviewed extensively (Hahn and Novak, 1975; Harding, 1970; Novak *et al.*, 1973*a,b*; Sperry, 1962). Because of the availability of these reviews, only the important features of the adipose tissue lipid metabolism during development will be outlined.

Two kinds of adipose cells have been described in mammals, the white fat cell and the brown fat cell. The two cells differ histologically, and in man, the distribution of these two different fat cells is quite characteristic (Figure 24). In addition, brown fat is found around the heart and the great vessels. The purpose of brown fat is to produce heat by oxidation of the triglyceride fatty acids in case of an abrupt change in the outside temperature. The white fat and brown fat also have an important role to play in insulating the body against heat loss and cold.

The complexity of the release of fatty acid in brown fat is shown below (Hahn and Novak, 1975):

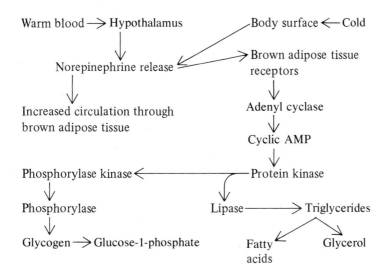

It is evident that this cascade of metabolic events is activated after birth. But *in utero,* the hypothalamus, the receptors of norepinephrine in the adipose tissue, and the enzymes of lipid catabolism have to be ready to play their part around the time of birth. This sequence is made possible by the chronological appearance of the various metabolic regulators with development.

The substrates necessary for triglyceride synthesis have been discussed previously. They are carried by the bloodstream in the form of glucose and of fatty acid bound to albumin and to lipoprotein triglyceride. Insulin promotes the transport of glucose across the cell membrane, and it regulates the intracellular glucose concentration. A lipolytic enzyme, lipoprotein lipase (LPL) is synthesized in the vicinity of the adipose cell, within the endothelium of adipose tissue capillaries, and is released into the plasma. LPL promotes the hydrolysis of lipoprotein-bound triglycerides in plasma and presents to the adipose tissue cell a constant flow of fatty acid substrates. LPL acitivity increases in fetal fat at the time that fetal adipose tissue deposition takes place (Harding, 1970; Harding and Ralph, 1970) (Figure 25).

Another interesting aspect of the human fetal adipose cell is that, as measured by glycerol release, added glucose stimulates lipolysis *in vitro,* whereas it does not have this effect on the adult adipocyte (Harding, 1970, Harding and Ralph, 1970). Furthermore, Novak *et al.* (1975) and Hahn and Skala (1972*b*) demonstrated that the mechanisms involved in fatty acid oxidation in mitochondria of developing white adipocytes of the human newborn are similar to those described for mammalian brown fat (i.e., the oxidation of fatty acids was found to be both carnitine-dependent and carnitine-independent in human newborn subcutaneous fat mitochondria). This characteristic disappears with age (Novak *et al.,* 1975).

The fact that the white fat may serve some of the functions of the brown fat

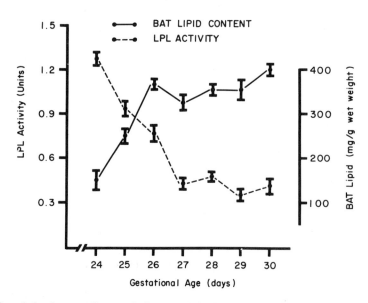

Fig. 25. The relation between lipoprotein lipase activity in brown adipose tissue of the fetus and the content of fat in the brown adipose tissue as gestation progresses to term. Units LPL activity represents μmol of free fatty acid liberated after 1-hr incubation *in vitro* per gram brown adipose tissue. (From Harding, 1970, with permission of author and *Clin. Obstet. Gynecol.*)

may explain why brown fat is hardly visible in the human fetus. The ability of the white adipose tissue to function as brown fat at birth would confer a considerable survival advantage to the human newborn, with the action of norepinephrine and glucose, which complement each other, regulating the degree of lipolysis after birth.

### 3.5. Endocrine Control of Lipid Metabolism

There is evidence that during development, the endocrine organs of the fetus control lipid metabolism. Rabbit or rat fetuses have been decapitated surgically at various periods of gestation (Bearn *et al.,* 1967; Picou, 1968) and despite decapitation, the body continued to grow until term. By beheading the fetus through the mouth, or through the neck, the thyroid is left in place, or removed, with the pituitary (Beard *et al.,* 1967; Picou, 1968). More lipid is accumulated at term in the absence of the pituitary and thyroid if the surgical procedure is performed at 24 days of gestation. It is reported also that hypophysectomized–thyroidectomized fetuses at term have a higher serum triglyceride and cholesterol content than the control fetuses at term (Beard *et al.,* 1967; Picou, 1968). The fat content of rabbit fetal liver and the ratio of total body lipid to proteins in the rat fetus is increased after decapitation at the base of the neck. Endocrine control could be mediated by the known activities of pituitary and thyroid hormones or by some lipolytic agent being produced by the pituitary. Jack and Milner (1975) have reported that decapitation retards growth in body weight of the fetus. This retardation can be corrected by ACTH administration to the fetus. Interestingly enough, injecting ACTH into a fetus on the 24th day of gestation produces a greater growth retardation than decapitation alone. This experiment indicates that the pituitary and adrenal control fetal growth and lipid deposition and that an overproduction of steroids by the fetoplacental unit has a deleterious effect.

The content of phospholipids in the brain of the rabbit fetus is not influenced by the administration of methylthiouracil to the mother, although this substance crosses the placenta and inhibits fetal thyroxin secretion (Cuaron *et al.,* 1963; Webster and Young 1948). Thyroxin, therefore, does not affect the synthesis of phospholipids in fetal brain cells. In the human fetus, there are indications that the pituitary and adrenal glands control the deposition of adipose tissue because anencephalic newborn infants have been shown to have a larger amount of subcutaneous adipose tissue than normal infants (Angevine, 1938).

Since lipid synthesis is dependent on glucose and fatty acid availability to tissue, the action of insulin, growth hormone, and glucagon are important to consider. Indeed, in the adult these hormones modulate lipid metabolism (see Section 3.1.4). In the first trimester of gestation, insulin has been shown to be present in human fetal pancreas (Lehninger, 1970; Bjoerkman *et al.,* 1966; Espinoza *et al.,* 1970; Kling and Kotas, 1975). Growth hormone (HGH) has been found in cord blood at term. The concentration of HGH in fetal venous blood is higher than it is in arterial blood, and as HGH does not cross the placenta, it is likely that it is secreted by the fetal pituitary. Its role in the fetus is still unknown (Yen *et al.,* 1965). Human chorionic somatomammotropin (HCS) has been measured in trace amounts in the term human fetus (Grumbach *et al.,* 1968). Since HCS favors FFA formation and has a glucose-sparing effect, its role in fetal metabolism must be minimal. The parenteral administration of insulin and glucagon to the rhesus monkey fetus has been studied (Bocek and Beatty, 1969; Chez *et al.,* 1970). Fetal glucose, fatty acids,

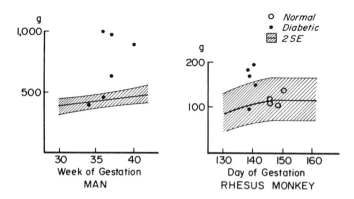

Fig. 26. Placental weight increases in diabetic gestation of man and monkey during development. It shows the similar change in placental weight in the experimental model (i.e., rhesus monkey).

growth hormone, and placental prolactin showed minimal changes in concentration after direct injection of insulin into the fetus, although adult monkey tissue responds to insulin. Fetal hypoglycemia did not alter the concentration of growth hormone in the plasma of fetal monkeys. Fetal plasma insulin level, however, was enhanced both by fetal and by maternal glucagon as well as by tolbutamide administration (Mintz et al., 1969). In man, a single administration of glucose to the mother does not release fetal insulin in large amounts (Tobin et al., 1969). In contrast, when glucose is infused at a constant rate, fetal insulin is released (Obenshain et al., 1970). This release has been shown also in the rhesus monkey (Reynolds and Pitkin, 1976). It appears, thus, that insulin and growth hormone are released in the fetus in response to a prolonged administration of glucose.

The bulk of these data shows that the human fetus controls its lipid metabolism during development with the help of the known adult-type hormones. The action and release of these hormones during gestation must be different quantitatively from those of the adult, because the endocrine system of the fetus has its own particular metabolic needs. The metabolic actions of somatomedin and somatostatin on lipid in the growing fetus still require investigation. The development of steroid hormone production in the human placenta and fetus, although related to sterol production, will not be discussed here. An exhaustive review of this subject was published in 1969 and remains up to date (Villee, 1969).

Prostaglandins $E_2$ and $F_2$ have been found to be derived from arachidonic acid, a long-chain fatty acid, present in the fetal membranes (Schwartz et al., 1975); a decrease in arachidonic acid was measurable during active labor. This last observation has stimulated investigators to postulate that progesterone near term is bound to receptors in the amiotic membranes and stimulates the destruction of lysosomes with activation of phospholipase. The activity of these receptors, therefore, must be different at term than before term. Phospholipase, thus, releases arachidonic acid from the amniotic membrane and prostaglandin is synthesized (Sisson and Plotz, 1967). If these speculations and observations are correct, the lipid metabolism of the fetus is directly related to the onset of labor through the hypothalamopituitary and adrenal axis (Sisson and Plotz, 1967). This concept is summarized in the following scheme and suggests that the action of hormones in fetal tissues may be modulated by a change in receptor activity during development:

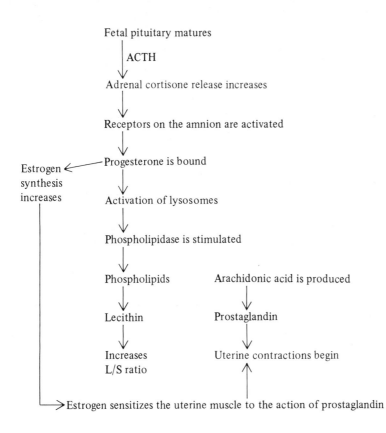

Fetal pituitary matures

ACTH

Adrenal cortisone release increases

Receptors on the amnion are activated

Progesterone is bound → Estrogen synthesis increases

Activation of lysosomes

Phospholipidase is stimulated

Phospholipids          Arachidonic acid is produced

Lecithin               Prostaglandin

Increases L/S ratio    Uterine contractions begin

Estrogen sensitizes the uterine muscle to the action of prostaglandin

## 3.6. Infants of Diabetic Mothers

The infant of the diabetic mother shows many deviations from normal, among which are high perinatal mortality and high fetal or birth weight. The overmaturation of the diabetic newborn is in part the result of a surplus of fat acquired during intrauterine life (Baird, 1969; Joassin *et al.*, 1967; Osler and Pederson, 1960). Thus, hyperglycemia and hyperinsulinism in these fetuses causes additional synthesis of body fat. These infants have an hyperplasia of the pancreatic islets and an increase in insulin response to glucose loading (Espinoza *et al.*, 1970; Baird, 1969; Joassin *et al.*, 1967; Schwartz, 1968). Recent studies show that careful management of the diabetic patient can result in lower perinatal mortality and a normal-size fetus (Beard *et al.*, 1971; Reynolds *et al.*, 1975). Induction of experimental diabetes in the rhesus monkey by the administration of streptozotocin (SZ) has reproduced the clinical picture found in the human fetus and infant of the diabetic mother (Reynolds *et al.*, 1975). For instance, the placental weight is increased above normal in this diabetic model as it is in man (Figure 26). It was found that brain and hepatic phosphatidylethanolamine as well as brain monopalmitin were increased in the fetus of the diabetic monkey (Reynolds *et al.*, 1975). The incorporation of [1-$^{14}$C]palmitate was greater in fetal tissues of diabetic mothers than in those of normal mothers (Reynolds *et al.*, 1975). Similar observations were made with [U-$^{14}$C]glucose in the placenta and $^{14}CO_2$ production was increased in fetal brain and heart. Such findings correlated well with the increased concentration of triglycerides and $C_{16}$ fatty acids (palmitic acid) found in phospholipids of fetal placenta, liver, and lung from the diabetic mothers (Reynolds *et al.,* 1975).

It was concluded that in the diabetic-like state, there is an increase in $\beta$-oxidation and Krebs-cycle activity. Furthermore, the fetal tissues of the diabetic mother were more sensitive to the added insulin than were those of the normal mother (Reynolds *et al.,* 1975). Although small numbers of animals were studied, the findings in the monkey model are consistent with those in the fetus of the human diabetic mother. In the human fetus, the availability of excess glucose and FFA will result in an excess of fat storage with hyperinsulinemia and catabolism of glucose and fatty acids. This process is an attempt to burn the excess fuel. Thus, in the clinical management of the diabetic mother, every attempt must be made to control her hyperglycemia. Good control reduces perinatal wastage, but does not prevent it completely. Even in a well-controlled diabetic, an excess reesterification of FFA within the fetal adipose tissue and a greater glucose utilization after birth is measurable (Persson *et al.,* 1973). Other metabolic factors must play a role during fetal development, even though the maternal glycemia is well controlled. It is possible also that a perfect homeostatic control of the maternal glycemia is impossible to obtain. It would be of great interest to study the changes in proteins, amino acids, and fatty acids taking place in diabetic mothers because fetal glucose is only one of the compounds which participates in fetal growth and development.

### 3.7. Effects of Maternal Dietary Alterations on Fetal Lipids

The fetus responds to maternal dietary alterations. For instance, the fetal lamb during maternal starvation catabolizes amino acids after the first two days and up to the seventh day (Simmons *et al.,* 1974). The glucose uptake by the fetus during starvation is less than 40% that of the normal fed state (Simmons *et al.,* 1974). In pregnant women fasted for 84–90 hr, at the 16th to the 22nd week of gestation, maternal hypoglycemia, hyperketonemia, and an increase in glycerol levels were observed both in maternal and in amniotic compartments (Kim and Felig, 1972). Furthermore, in the fetal lamb 50% of the metabolic requirement is provided by glucose (Tsoulos *et al.,* 1971). This indicates that proteins, fatty acids, and other unknown substrates may have a role to play in providing the fetus with the necessary fuel for growth. The bulk of these data indicate that in the lamb fetus and the human fetus metabolic adjustments take place with drastic dietary alterations.

Diet influences lipid metabolism in the adult (Noble *et al.,* 1971; Abraham, 1970). Since maternal fatty acids, cholesterol, and glucose cross the placenta, maternal diet should influence the concentration and composition of fetal lipids. There are, however, few well-controlled experiments to support this. The term fetuses of pregnant rabbits fed diets rich in cholesterol showed no alterations in serum cholesterol, phospholipids, or triglyceride concentrations (Sisson and Plotz, 1967, 1968). In another study, pregnant rabbits were fed for eight days, starting at the 20th day of gestation, with a diet rich in fat and cholesterol, or a standard diet. Maternal and fetal plasma lipids were studied. Using the former diet, there was a decrease of free cholesterol and triglyceride palmitoleate and an increase in oleyl triglycerides in fetal blood. These observations correlated well with the composition of the fat–cholesterol-rich diet (Sisson and Plotz, 1968; Shephard *et al.,* 1974). Other changes in fetal lipid took place, such as an increase in free fatty acids and cholesterol palmitate, and a decrease in cholesterol linoleate and phospholipid arachidonate which did not correlate with the diet.

Marked variations in the quality of tissue lipid took place in pregnant rats fed a diet containing up to 40% by weight of fat with varying degrees of unsaturation (Chaikoff and Robinson, 1933). The fat iodine number varied from 36.3 to 145 in the mothers, but only from 61.5 to 103 in the fetal rat. Modifications of fetal fat within certain limits appear to be possible by altering the composition of maternal diet. The chronic feeding of cholesterol to pregnant guinea pigs was found to influence the uptake of cholesterol by the fetus (Hirsch *et al.,* 1960).

In man, feeding unsaturated fatty acids (hexadecanoate and pentadecanoate) in the first 20 weeks of pregnancy is accompanied by an increase in their concentration in fetal cord blood (Soderhjelm, 1953). In another study, pregnant women in early gestation were fed diets composed of various fatty acids (Watanabe, 1950). At the time of delivery or therapeutic abortion, the fatty acid composition of fetal tissues was determined. Alterations in the lipid composition of fetal tissues and of placenta were found. The limited data, however, are difficult to interpret. In a study which was controlled better, 30 pregnant women had an accurate estimate of their dietary intake and food composition, starting in the first or second trimester of gestation (Roux *et al.,* 1971). They ingested 35–101 g of fat per day representing 36.6% of their caloric intake. Fat composition in the diet was the same in all groups. No correlation was found between the mothers' nutritional intake and the composition of a wide range of nutrients in the plasma of the neonates.

In pathological conditions such as diabetes, the fetus of the rabbit shows an increase in fetal plasma cholesterol, linoleate, oleate, free fatty acids (linoleate), phospholipid, and palmitate as well as a decrease in cholesterol palmitate (Sisson and Plotz, 1968). The mothers were fed a constant peanut oil cholesterol diet. In the diabetic rabbit, placental transport and fetal lipid metabolism, at least, seem to be grossly different from the normal. In man, however, the total lipid fatty acid patterns of umbilical cord blood were not altered in the group of newborns with intrauterine growth failure when the data were compared with those of normal infants (Robertson *et al.,* 1969).

All the animal data suggest that alterations in maternal diets influence the composition and concentration of fetal lipids. The changes are small, however, when compared to those observed in the mother, and large alterations in dietary intake must occur before any changes are observed. The rate of transfer of glucose and of fatty acids across the placenta, the rate of synthesis of lipids by the fetus and their degradation, utilization, and resynthesis, may account for this. All these factors will modify the response of fetal metabolism to alterations in maternal diets.

## 3.8. Postmaturity, Placental Insufficiency, Hypoxia, and Lipid Metabolism

Hypoxia, placental insufficiency, and postmaturity are associated factors which produce infants having loss of vernix caseosum, and a thin physical appearance associated with loose, wrinkled parchment-like skin (Clifford, 1954). It is interesting to note that the small-for-date infant sometimes resembles the postmature fetus (Andrews, 1960). They both probably suffer from malnutrition secondary to an insufficient placenta.

The postmature fetal rabbit has been studied extensively (Roux, 1966; Roux and Romney, 1967; Harding and Ralph, 1970; Roux *et al.,* 1964). Harding has shown that there is an increase in the rate of lipolysis of brown adipose tissue in the postmature fetus (Harding and Ralph, 1970). The lipid content of the liver is diminished, probably for the same reason (Roux *et al.,* 1964). Plasma FFA concentration, however, is the same as that of the term fetus, despite an increase in the turnover rate of palmitate (Roux and Romney, 1967). Since brown adipose tissue is important in the temperature control of the newborn rabbit, an increased metabolism of FFA in postmaturity may affect neonatal survival (Hahn and Novak, 1975; Hull, 1969). It should be noted that the metabolism of pyruvate is not altered in liver and heart of the postmature fetus. In the postmature brain incubated under nitrogen, lipids cannot be synthesized, whereas the liver still has a significant rate of lipogenesis (Roux *et al.*, 1964). This difference suggests that lipid synthesis in fetal brain is more sensitive to anoxia or hypoxia than that in fetal liver. The metabolic changes taking place in the postmature animal are of a discrete nature, but may have long-term consequences in the immediate and late neonatal periods.

It seems that the postmature placenta is capable of supplying the rabbit fetus with enough nutrients to continue growth until the 34th day of pregnancy. This is two days beyond the normal calculated day of confinement. Fetal death in the postmature rabbit, however, takes place suddenly between the 34th and 35th day of gestation. It is possible that the oxygen supply crossing the placenta is insufficient to compensate for the demands of growth after term.

Since lipogenesis is curtailed by anoxia or hypoxia in some fetal tissues, including the lung, it is easy to understand the work of Finley *et al.* (1964), who showed that a reduction in blood flow to a segment of adult dog lung is associated with a disappearance of surfactant in the impaired segment.

It is interesting that in human pregnancy, the concentration of placental triglycerides is increased in eclamptic women (Nelson, 1966). Since the placenta plays an essential role in the synthesis of complex lipids from fatty acids and in the transfer of these lipids to the fetus, it is conceivable that in eclampsia one or both of these functions is affected.

### 3.9. Lipids in Amniotic Fluid

The rate of exchange between the maternal and the amniotic compartments is extremely rapid. Because the fetus sheds amniotic cells and vernix caseosa, produces urine, and swallows amniotic fluid, amniotic fluid contains steroids and a large range of substances which are in dynamic equilibrium with the fetus, the placenta, and the maternal compartment. It is understandable, therefore, that analysis of amniotic fluid has been useful in fetal evaluation, both in genetic diagnosis and in the evaluation of fetal maturation.

Scarpelli, Biezenski, and Nelson have been the primary investigators of the structure of amniotic fluid lipids (Biezenski *et al.*, 1968; Nelson, 1969; Roux *et al.*, 1973*a*). Scarpelli demonstrated (Figure 27) that amniotic fluid had surface tension activity similar to the lung and related that to the amniotic phospholipid content (Roux *et al.*, 1973*a*). Biezensky defined the concentration and structure of amniotic fluid lipids in normal and abnormal gestation (Table X). Nelson (1969) observed that the total lipid concentration of human amniotic fluid increases from 10–20 mg/100 ml before 28 weeks of gestation to 40–60 mg/100 ml at 36 weeks. He found altered phospholipid concentration in certain abnormal gestations (respiratory distress syndrome and anencephaly with polyhydramnios).

Gluck *et al.* (1971) and Gluck and Kulovich (1973) were the first investigators to publish a simple technique to determine the ratio of lecithin to sphingomyelin in amniotic fluid and to relate it to the clinical condition of the baby at birth. They

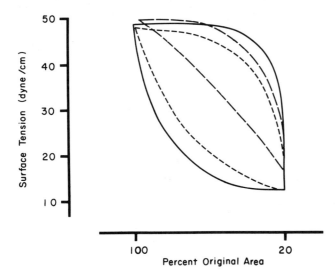

Fig. 27. Surface tension–area diagrams of minced fetal lung (solid line), tracheal fluid (short dashed line), and amniotic fluid (long dashed line).

found that lecithin increased linearly between the 34th and 38th weeks of gestation (Figure 28). The lecithin was precipitated from amniotic fluid with acetone, and the increase in acetone-precipitable lecithin suggested to these authors that this phospholipid might be surfactant material originating from the fetal lung. Their speculation was confirmed when they noticed that fetuses with an L/S ratio greater than 2 rarely developed hyalin membrane disease (Gluck *et al.*, 1971). They were quick to point out, however, that the L/S ratio can be unreliable in high-risk pregnancies because of a possible stress to the fetus in obtaining amniotic fluid (Gluck and Kulovich, 1973). In a similar study, Roux *et al.* (1972) corroborated these observations except for the fact that many antepartum low ratios were not always accompanied by respiratory distress syndrome or hyalin membrane disease (Figure 29).

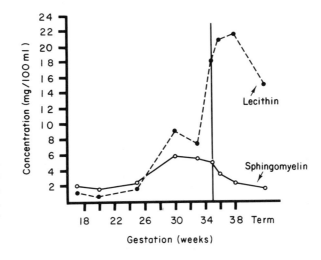

Fig. 28. Concentration of amniotic fluid lechithin and sphingomyelin during development. Note the sharp increase in lecithin between weeks 33 and 37 of pregnancy. (From Gluck *et al.*, 1971, with permission of the author and *Am. J. Obstet. Gynecol.*)

GERALD E. GAULL
ET AL.

Table X. Concentrations of Amniotic Fluid Lipids in Normal and Abnormal Gestation[a,b]

| Class of lipid | Normal (53)[c] (24 wk-labor) | Hemolytic disease of newborn (8) | Toxemia of pregnancy (8) | Diabetes (6) | IUD > 2 weeks[d] (8) | IUD < 2 weeks[e] (7) |
|---|---|---|---|---|---|---|
| Monoglycerides | $0.23 \pm 0.02$ | $0.20 \pm 0.007$ | $0.32 \pm 0.12$ | $0.35 \pm 0.10$ | $0.67 \pm 0.14^f$ | $0.12 \pm 0.03^f$ |
| Diglycerides | $0.89 \pm 0.11$ | $0.68 \pm 0.15$ | $0.53 \pm 0.13$ | $0.78 \pm 0.20$ | $1.00 \pm 0.30$ | $0.44 \pm 0.10^f$ |
| Triglycerides | $1.21 \pm 0.12$ | $0.96 \pm 0.30$ | $0.94 \pm 0.08$ | $1.29 \pm 0.32$ | $1.41 \pm 0.53$ | $0.99 \pm 0.29$ |
| Free fatty acids | $2.17 \pm 0.13$ | $1.24 \pm 0.20$ | $1.47 \pm 0.21$ | $1.92 \pm 0.57$ | $4.72 \pm 0.92^f$ | $1.86 \pm 0.39$ |
| Free cholesterol | $1.40 \pm 0.10$ | $1.03 \pm 0.20$ | $0.94 \pm 0.23$ | $1.55 \pm 0.26$ | $6.50 \pm 1.51^f$ | $1.87 \pm 0.47$ |
| Cholesterol esters | $2.00 \pm 0.15$ | $0.99 \pm 0.17^f$ | $1.24 \pm 0.25^f$ | $1.12 \pm 0.17$ | $2.78 \pm 0.37$ | $2.05 \pm 0.45$ |
| Hydrocarbons | $1.75 \pm 0.12$ | $1.43 \pm 0.38$ | $2.16 \pm 0.14$ | $2.12 \pm 0.67$ | $4.89 \pm 1.42$ | $2.15 \pm 0.60$ |
| Total nonpolar lipids | $9.65 \pm 0.48$ | $6.56 \pm 1.05^f$ | $7.60 \pm 0.75$ | $9.16 \pm 1.10$ | $22.66 \pm 3.97^f$ | $9.50 \pm 1.59$ |
| Phospholipids | $3.99 \pm 0.38$ | $3.67 \pm 0.92$ | $2.82 \pm 0.58$ | $4.95 \pm 2.01$ | $5.10 \pm 0.71$ | $4.77 \pm 1.66$ |
| Total lipids | $13.64 \pm 1.02$ | $10.23 \pm 1.73$ | $10.33 \pm 0.91$ | $14.11 \pm 3.64$ | $24.89 \pm 3.69^f$ | $13.09 \pm 3.24$ |

[a] From Biezenski. 1973.
[b] Values in mg/100 ml ± SE.
[c] Number of cases in parenthesis.
[d] Intrauterine death more than two weeks before collection of fluid.
[e] Intrauterine death less than two weeks before collection of fluid.
[f] Statistically significant ($P \leq 0.01$).

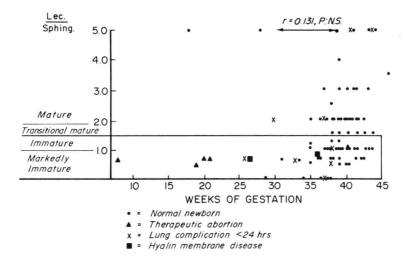

Fig. 29. L/S ratios of individual clinical cases obtained 24 hr before birth as they relate to the neonatal outcome. The increase in L/S with gestational age is shown, but note that many low L/S (markedly immature ratios) do not result in sick neonates. (From Roux *et al.*, 1973*a*, with permission of the author and *Am. J. Obstet. Gynecol.*)

Experiments in which Condorelli *et al.* (1974) injected [1-$^{14}$C]palmitate into fetal lambs in which the tracheas were ligated indicated that the lecithin of the amniotic fluid originates from many different sources, with the fetal bladder being one of them. Biezenski and Abramovich (Biezenski, 1973; Abramovich *et al.*, 1975; Roux and Frosolono, 1977; Russell *et al.*, 1974) were of the same opinion, and Roux demonstrated that a L/S ratio can be obtained from amniotic membranes, pharyngeal aspirate, and amniotic fluid, but not from the vernix caseosa (Roux *et al.*, 1973*a*) (Figure 30). Furthermore, it was shown that the precipitation of lecithin and sphingomyelin with acetone was not requisite for the demonstration of an increase in the L/S ratio with gestational age. Moreover, the clinical accuracy of the test was better without acetone precipitation (Table XI) (Roux *et al.*, 1974). In the fetus of the monkey and the baboon which had its trachea ligated, Gluck *et al.* (1974) showed that the L/S ratio determined after acetone precipitation was lower. They found also that the synthesis of amniotic fluid lecithin via the methylation pathways (see background information on phospholipids) was possible in the rhesus monkey in early gestation, but not in the baboon. It was found also that lung surfactant isolated by sucrose-gradient fractionation made up 15% of the total amniotic fluid lecithin. This ratio does not change in the last two trimesters of gestation (Roux and Frosolono, 1977). When the percent fatty acid composition of surface-active lecithin is measured in amniotic fluid, lung, and amnion, the fatty acid composition profile of amniotic fluid resembles that of lung (Roux and Frosolono, 1978). At no time during development, however, is myristic acid ($C_{14}$) increasing. This indicates that the methylation pathway in human lung and amniotic fluid is not predominant (Table XII). In that sense, the human fetus is more comparable to the baboon fetus (Gluck *et al.*, 1974). Furthermore, a study of the surface-tension activity of the amniotic fluid surface-active lecithin, carried out at term, shows an hysteresis which is similar to that of the lung. Therefore, term amniotic fluid must contain some lung surface-active material (Figure 31).

The bulk of these somewhat controversial data can be reconciled if we accept the idea that amniotic fluid lecithin originates from different sources, with one of

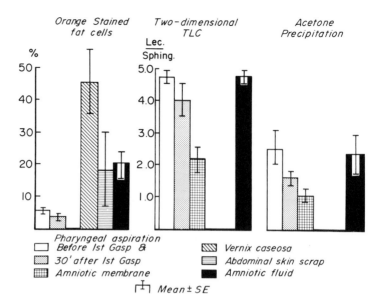

Fig. 30. Comparison of three different techniques to determine fetal maturity. Orange stained fat cells are found mainly in vernix caseosa and abdominal skin scrape, whereas lecithin and sphingomyelin are measurable in amniotic fluid, amniotic membranes, and pharyngeal aspirations. The differences between the acetone precipitation and the two-dimensional thin-layer chromatography are only quantitative. (From Nakamura *et al.*, 1972, with permission of the author and *Am. J. Obstet. Gynecol.*)

them being the lung. This may explain why a low L/S ratio does not always precede RDS at birth (Figure 29). It is consistent also with the fact that the L/S ratio correlates with gestational age, the degree of fetal maturation, and the metabolic state. Pathological pregnancies may alter lecithin formation in lung, liver, amnion, placenta, and this may affect the amniotic fluid lecithin concentration curve. Indeed, the lecithin coming from the amniotic–tracheal fluid mixes with a pool of lecithin originating from other sources. It is also of interest to find that amniotic fluid

Table XI.   Fatty Acid Composition of Amniotic Fluid Lecithin in the Acetone-Precipitated and Soluble Extract[a]

| Fatty acid chain length | Weeks of gestation | Acetone-soluble lecithin | Acetone-precipitable lecithin |
|---|---|---|---|
| $C_{12}$ | 18–35 | $2.6 \pm 1.6$ (13)[b] | N.D.[e] (13) |
|  | 36–43 | $0.9 \pm 0.5$ (8) | N.D. (8) |
| $C_{14}$ | 18–35 | $3.2 \pm 2.2$ (13) | $3.8 \pm 2.1$ (13) |
|  | 36–43 | $5.3 \pm 0.7$ (8) | $6.3 \pm 1.7$ (8) |
| $C_{16}{}^{c}$ | 18–35 | $57.5 \pm 4.4$ (13) | $60.9 \pm 4.9$ (13) |
|  | 36–43 | $75.7 \pm 3.4^{d}$ (8) | $80.9 \pm 2.6^{d}$ (8) |
| $C_{18}$ | 18–35 | $16.9 \pm 1.2$ (13) | $21.3 \pm 2.1$ (13) |
|  | 36–43 | $6.4 \pm 0.8^{d}$ (8) | $6.7 \pm 1.0^{d}$ (8) |
| $C_{18}^{1}{}^{c}$ | 18–35 | $18.1 \pm 3.5$ (13) | $14.0 \pm 0.8$ (13) |
|  | 36–43 | $10.0 \pm 2.2^{d}$ (8) | $6.1 \pm 0.5^{d}$ (8) |
| $C_{20}$ | 18–35 | $1.7 \pm 0.9$ (13) | N.D. |
|  | 36–43 | $2.0 \pm 0.9$ (8) | N.D. |

[a]From Roux *et al.*, 1974.
[b]The numbers are the average ± SE of the number of determinations indicated in parentheses.
[c]$C_{16}$—palmitate; $C_{18}^{1}$—oleate.
[d]This average is significantly different from that obtained at 18 to 35 weeks.
[e]N. D.—not detectable.

*Table XII. Percent Fatty Acid Composition of Surfactant (IB-Lecithin) from Lung, Amnion, and Amniotic Fluid[a]*

| Fatty acids | Lung | | | | Amniotic fluid | | | Amnion | |
|---|---|---|---|---|---|---|---|---|---|
| | 10–24[b] (3)[c] | 28–36 (3) | 4-Day-old newborn (1) | Adult (1) | 14–21 (2) | 33 (6) | 37–42 (6) | 10–24 (6) | 37–42 (6) |
| $C_{14}$ | 2.4 ± 0.5[d] | 5.3 ± 1.0 | 3.1 | 6.1 | 6.3 | 2.7 ± 0.4 | 5.8 ± 0.8 | 5.4 ± 1.6 | 4.0 ± 0.5 |
| $C_{16}$ | 69.7 ± 1.8 | 72.0 ± 9.6 | 69.3 | 65.7 | 62.2 | 77.2 ± 3.8 | 82.3 ± 2.6 | 60.0 ± 6.2 | 55.0 ± 1.6 |
| $C_{16}^1$ | 1.5 ± 0.8 | 3.4 ± 0.3 | 4.3 | 5.1 | 2.1 | 2.9 ± 0.5 | 2.5 ± 0.5 | 3.0 ± 1.1 | 2.5 ± 0.6 |
| $C_{18}$ | 7.1 ± 1.9 | 3.3 ± 1.1 | 7.4 | 3.0 | 4.0[d] | 5.9 ± 1.6 | 2.6 ± 0.5 | 5.4 ± 1.0[d] | 12.1 ± 1.4 |
| $C_{18}^1$ | 12.6 ± 2.7 | 10.0 ± 5.6 | 10.7 | 8.1 | 5.2[d] | 5.0 ± 0.9 | 2.3 ± 0.5 | 13.5 ± 2.8[d] | 9.8 ± 0.8 |

[a] Values are mean ± SE.
[b] Weeks of gestation.
[c] The numbers in parentheses represent the number of duplicate determinations.
[d] $P < 0.05$ significantly lower or higher than the mean obtained in the next higher gestational age.

phosphatidic acid phosphohydrolase increases with the concentration of lecithin. This enzyme is of importance in the regulation of phospholipid biosynthesis in fetal lung (Jimenez *et al.*, 1975). This provides further evidence that some of the amniotic fluid lecithin is of pulmonary origin. Irrespective of the source of amniotic fluid phospholipids, the L/S ratio is an excellent clinical test and, with the foam test

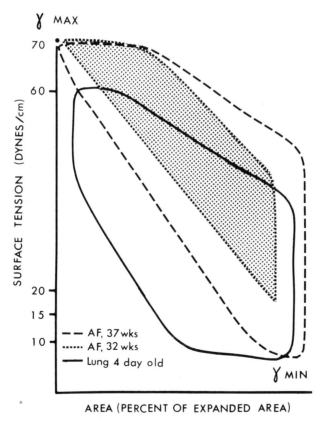

Fig. 31. Surface-tension curve of lung surface active material in the newborn lung and human amniotic fluid (AF). Note the smaller hysteresis of the preterm amniotic fluid (dotted line).

(Shephard *et al.*, 1974; Clements *et al.*, 1972; Merola *et al.*, 1974; Roux *et al.*, 1973), a good index of amniotic surface-active lecithin. These tests help the perinatologist to decide when to interrupt a pregnancy at risk. Finally, the increase in concentration of lecithin during gestation is not unique. Triglyceride also increases in amniotic fluid and possibly could be an index of fat deposition (Nelson, 1971). For the reader interested in evaluating the L/S ratio, the following articles are suggested (Russell *et al.*, 1974; Badham and Worth, 1975; Caspi *et al.*, 1975; Ekelund *et al.*, 1973; Nakamura *et al.*, 1972; Olson and Graven, 1974; Pomerance *et al.*, 1971; Roux, 1974; Schirar *et al.*, 1975; Schreyer *et al.*, 1974; Singh and Zuzpan, 1973; Freeman *et al.*, 1974; Bayer *et al.*, 1973; Harding *et al.*, 1973; Nelson, 1972).

A discussion of amniotic fluid lecithin is incomplete without a few remarks on the metabolism of lecithin in the lung. For a complete discussion of this subject, the reader is referred to the book by Scarpelli *et al.* (1968) and to the articles by Farrell and Morgan (1975) and Scarpelli (1967). Suffice it to say that, as in the other fetal tissues, carbohydrates and fatty acids are the precursors of lung lipids. In addition to phospholipids found in other tissues, the lung synthesizes surfactant (e.g., dipalmytoyl lecithin) which prevents the newborn and adult lung from collapsing during expiration. Surfactant is synthesized like other phospholipids using choline kinase, cytidyltransferase, and choline phosphotransferase. These enzymes modulate surfactant synthesis, their activity in fetal lung increasing with gestation and the production of lecithin. Cortisol or betamethazone accelerates surfactant synthesis. The production of surfactant is so stimulated that it may prevent lung tissue development (Gross *et al.*, 1975; Torday *et al.*, 1975). This action is mediated via the cortisol receptor present in lung (Watanabe, 1967; Villee, 1969; Ballard and Ballard, 1974; Giannopoulos, 1973; Giannopoulos *et al.*, 1973). Brain phospholipid synthesis does not seem to be altered by cortisol administration, in contrast to that of the liver which increases (Gross *et al.*, 1975). The reader may now understand why the increase in lecithin production in amniotic fluid was related to lung surfactant production and maturation by Gluck and Kulovich. Furthermore, the stimulation of lung surfactant production by cortisol or betamethazone gives support to Liggins' findings (Villee and Loring, 1961). He observed that corticoids administered parenterally to patients in premature labor at 32 weeks of pregnancy or less diminished the incidence of hyalin membrane disease in the premature infant. The suppression of fatty acid synthetase activity in the fetal lung by cortisone and the multiple target organs of this hormone, however, should serve to caution the clinician in its use (Gross *et al.*, 1975). Indeed, we may permit the premature fetus to breathe, but at the expense of lung or liver metabolism and maturation. More studies are needed to clarify its therapeutic value.

The effect of cortisone can be seen in amniotic fluid. The L/S ratio increases in man and baboon, but the administration of metapirone, which blocks 11-betahydroxylation, and hence cortisone synthesis, diminishes the L/S ratio without altering the development of the surface activity of the lung (Kling and Kotas, 1975). This finding suggests that amniotic fluid lecithin is not solely an index of lung surface-active material.

Gluck has speculated also on the stress effect of labor on lung surfactant synthesis. Cortisone would be released and surfactant production augmented (Gluck and Kulovich, 1973). It was suggested then to wait at least 48 hr after premature rupture of membranes in order to permit the fetus to prepare its lungs. This suggestion has been challenged by Freeman *et al.* (1974), who showed that the

L/S ratio of oxytocin-stressed fetuses did not increase, and by Jones *et al.* (Kling and Kotas, 1975), who in a large series of premature babies did not demonstrate a stress effect of a long labor on the human fetus. In view of the complexity of fetal lipid metabolism in each tissue, its multiple hormonal regulations, its interaction with protein and carbohydrate metabolism, it is unlikely that amniotic fluid lipid metabolism can be interpreted adequately solely as a reflection of fetal lung surfactant production and cortisol-receptor activity or stress.

### 3.10. Clinical Significance of Lipid Metabolism

It is apparent that fetal lipid synthesis and deposition in brown and white fat directly influences the neonate's tolerance to cold and controls its fuel supply. The premature and the small-for-date fetuses or neonates are at a disadvantage because they do not have adequate fat storage. Therefore, thermal neutrality should be maintained after birth, and feeding with glucose, proteins, and fatty acids should start early in order to supply the necessary nutrients.

The use of L/S ratio has been discussed extensively, and it has offered the perinatologist another test of fetal maturation which relates directly to lipid metabolism in the lung and other organs. It has opened large avenues of investigation, especially that of the fetal pituitary–adrenal axis. Data in this area will be forthcoming and welcome to perinatologists.

The great interest in lung surfactant production has helped to explain some of the pathogenesis of neonatal respiratory distress and hyalin membrane disease. If it were possible to stimulate lung surfactant production without inhibiting lung lipid synthesis, we might have an important therapeutic tool to help the premature infant breathe and survive.

It seems that hydrolysis of phospholipids in the amniotic membrane with the production of arachidonic acid and prostaglandin may well trigger the onset of normal labor.

Further investigations are required in pregnancy at risk to determine the importance of an altered maternal environment on fetal lipid metabolism. Such investigations would supply important information on the relationship of complex enzyme systems to disease, may show differences in the proportion of nutrients reaching the fetus at risk and, in general, are necessary in order to understand the relationship of maternal disease to fetal nutrition.

Finally, the sensitivity of the fetal organs to hormones as well as the presence of steroid receptors must be investigated in order to understand fetal metabolic development and growth. Such data may well shed light on cell growth in normal and abnormal conditions.

## 4. Development of Amino Acid and Protein Metabolism

A central process in the growth and differentiation of cells and tissues is protein synthesis, and this process requires the optimal availability of all the amino acids contained in protein. Many of the amino acids needed for synthesis of proteins are transported via the placenta from the maternal circulation, although some can be synthesized by the fetus. Protein synthesis, even in mature tissues which are not growing, is a dynamic process in which proteins are being continuously synthesized and resynthesized. In growing tissues the rate of synthesis exceeds the rate of degradation, i.e., there is net protein synthesis. In tissues which are not growing, the rate of synthesis just equals the rate of degradation, unless protein is secreted by the tissue (as in the pancreas, which has a considerable net protein synthesis even in the adult). When proteins are broken down some amino acids are reutilized and some are degraded or lost by excretion and epidermal desquamation. The fetus accumulates considerable protein of many types, with numerous functions, and it does this in an intricately programmed manner, at a precisely regulated rate.

The section on protein and amino acid metabolism in this chapter will attempt to summarize what is known about the ways in which these aspects of metabolism are adapted to human development. The major means by which nitrogen reaches the fetus is by transplacental transfer of free amino acids, although other types of nitrogenous compounds as well as some proteins are transferred via the placenta. General aspects of the regulation of this process of placental transfer of amino acids have been reviewed extensively (Szabo and Grimaldi, 1970) and will be touched upon here only as they are somehow different in the human; however, quantitative data on synthesis of proteins by the human fetus are meager. This section will not cover general aspects of protein synthesis, but rather what is known about the ways in which the metabolism of amino acids and proteins is adapted and regulated in human development. Furthermore, it will not discuss in detail the synthesis of particular proteins found in plasma. An excellent summary of the development of plasma proteins has been published recently (Gitlin and Gitlin, 1975).

The enzymatic and metabolic adaptations for amino acid and protein metabolism during fetal development and the perinatal period will be emphasized. The clinical and nutritional implications of these enzymatic adaptations are numerous (see below). Recent evidence suggests that we are beginning to be able to alter this enzymatic development in ways which may ease the adaptation to extrauterine life of the prematurely born infant. Since knowledge in the area of amino acid and protein metabolism in human development is fragmentary, a comprehensive treatment is impossible. What follows is a summary of what is emerging as the most important and characteristic aspects of this area of metabolism. Although we shall not limit discussion to liver, that organ will get major emphasis because most is known about it. Amino acid and protein metabolism in other organs, especially brain, will be discussed when interesting differences and contrasts are known and are relevant.

## 4.1. Amino Acid Synthesis during Development

### 4.1.1. Synthesis of Tyrosine

In the human adult about 60% of tyrosine is derived from phenylalanine (Rose and Wixom, 1955b). The enzyme system which hydroxylates phenylalanine, an essential amino acid, to tyrosine is a complex which includes phenylalanine hydroxylase and a tetrahydropterin cofactor. Adequate amounts of cofactor depend, in turn, on a system to regenerate it: dihydropterin reductase and NADPH (Kaufman, 1969; Friedman and Kaufman, 1973). It had been thought that phenylalanine hydroxylase belonged to the enzyme cluster that emerges in the postnatal period (Kenney *et al.*, 1958), and Snyderman (1971) has proposed that tyrosine is an essential amino acid in the preterm infant. Ryan and Orr (1966), however, demonstrated that [$^{14}$C]phenylalanine injected into the umbilical vein of previable human fetuses (22nd–24th week of gestation) is converted to [$^{14}$C]tyrosine. Jakubovic (1971) showed that phenylalanine dihydropteridine reductase and tetrahydrobiopterin are present in liver of human fetuses between the 11th and the 20th weeks of gestation. Friedman and Kaufman (1971) found phenylalanine hydroxylase in human liver at 23 weeks gestation. Räihä (1973) has shown unequivocally that phenylalanine hydroxylase activity is present in human fetal liver as early as the 8th week of gestation and that activities similar to those found in mature human liver are attained by the 13th week of gestation. The presence of phenylalanine hydroxylase activity in human fetal liver has been shown also by Cartwright *et al.* (1973).

Greengard has shown that human hepatic phenylalanine hydroxylase has an *AQ* of 0.59 (see Section 1) during the second trimester of pregnancy; furthermore, premature birth does not immediately stimulate its synthesis, as it does with tyrosine aminotransferase (Del Valle and Greengard, 1976) (see below). Similar results were found by McLean, Merwick, and Clayton (1973). Thus, in man, as in the rat (McGee *et al.,* 1972; Tourian *et al.,* 1972), phenylalanine hydroxylase approaches functional levels of activity at an earlier stage of development than do the enzymes of tyrosine degradation (Del Valle and Greengard, 1976) (see below). The concept that tyrosine is essential in the preterm infant requires reexamination. On the one hand, the enzymatic machinery for the conversion to phenylalanine to tyrosine is present in the preterm infant. On the other hand, it is still possible that with the rapid rate of growth of the preterm infant the activity of this complex enzyme is not adequate and that there is a dietary requirement for tyrosine during development.

## 4.1.2. The Development of Transsulfuration

Cystathionase, the last enzyme on the pathway of transsulfuration of methionine to cysteine (Figure 32), cleaves cystathionase with production of cysteine and $\alpha$-ketobutyrate. In adult man, 90% of methionine sulfur goes down this pathway (Rose and Wixom, 1955a). Activity of cystathionase is virtually absent in extracts of human fetal liver and brain during the 2nd trimester, although there is considerable activity in human fetal kidney (Sturman *et al.,* 1970; Gaull *et al.,* 1972). Radiochemical perfusion studies (Gaull *et al.,* 1972) and immunochemical studies (Pascal *et al.,* 1972) indicated that trace amounts of cystathionase, at most, are present in 2nd trimester human fetal liver. Heinonen and Räihä (1974) have shown that cystathionase can be induced in human liver explants obtained from 2nd trimester human fetuses by dexamethasone, glucagon, or dibutyryl cycle AMP plus theophylline. Failure to induce cystathionase in the three smallest fetuses suggests that a certain developmental stage must be reached before such competence is acquired.

Activities of methionine adenosyltransferase and of cystathionine synthase present in fetal human liver were less than those present in mature human liver (Sturman *et al.,* 1970; Gaull *et al.,* 1972) but were considerable. Furthermore, the placenta had no transsulfuration enzyme activity (Sturman *et al.,* 1970; Gaull *et al.,* 1972). Therefore, it was proposed that cystine might be an essential amino acid in the human fetus, at least until sometime after birth when hepatic cystathionase activity appeared (Sturman *et al.,* 1970; Gaull *et al.,* 1972, 1973b). It is impossible to be certain exactly when full activity of cystathionase appears in term neonates; however, changes in the concentrations of cystathionine and cystine in plasma and urine of preterm infants fed pooled human milk or various artificial formulas suggest that cystathionase may no longer be limiting for cystine synthesis by about the 2nd postnatal week (Gaull *et al.,* 1977). Based on nitrogen balance studies in infants on a defined mixture of amino acids, Snyderman found at least one preterm infant who still had a need for cystine when restudied at 5 months of age (1971).

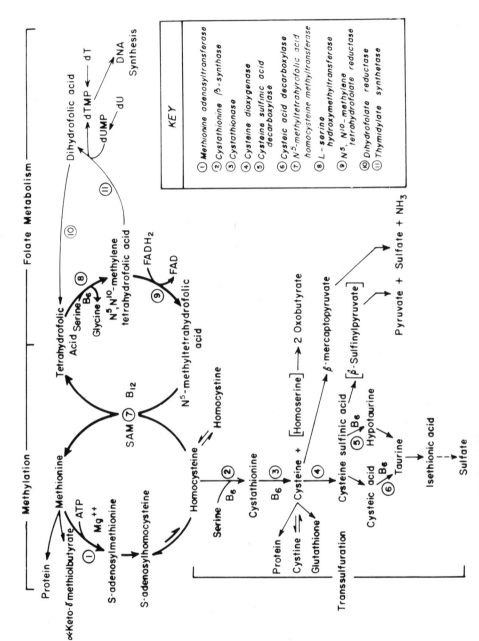

Fig. 32. Metabolism of sulfur-containing amino acids and related compounds.

Because the third trimester of gestation is not available for systematic study in man, animal models have been studied: In fetal rat liver, cystathionase is present at about 10% of the activity present in adult rat until about the 19th day (of a 22-day gestation) when cystathionase activity increases sharply (Heinonen, 1973). The monkey is closer to man in many aspects of development of sulfur metabolism than the rat, although there are some differences (Sturman *et al.*, 1973, 1976*a,b*). Activity of hepatic cystathionase in the monkey is virtually absent until near the end of gestation when it gradually increases to reach the activity found in mature liver at about 5 months after birth (Sturman *et al.*, 1976*b*) (Figure 33A). The concentration of cystathionine, the substrate of cystathionase, is higher in fetal monkey liver than it is in the liver of the adult monkey (Figure 33B). The activity of cystathionine synthase, the enzyme which synthesizes cystathionine from homocysteine and serine, is higher in late fetal monkey liver than it is in adult monkey liver (Figure 33C). This may represent a difference from man or may be because of the different relative periods of gestation studied. The situation in monkey brain is the obverse: The concentration of cystathionine is lower in fetal monkey brain than it is in adult monkey brain (occipital lobe) (Figure 34A). The activity of cystathionine synthase reflects closely the concentration of its product cystathionine (Figure 34B). Cystathionase activity in brain is low throughout development (Sturman *et al.*, 1976*b*). In summary, in the monkey, and by analogy in man, cystathionase is limiting to transsulfuration in liver only prior to birth, whereas in brain, it is limiting to transsulfuration throughout development.

The concentration of most amino acids is about twofold higher in fetal plasma than it is in maternal plasma (see also, for example, Szabo and Grimaldi, 1970; Glendening *et al.*, 1961; Young and Prenton, 1969; Young, 1971). A notable exception is cystine which in cord blood at term is only about 1.1 (Lindblad, 1971) to 1.3 (Ghadimi and Pecora, 1964). Actual determinations of the fetal/maternal ratio of cystine during the 2nd trimester, however, show a ratio of 1.0 or less; therefore, the transplacental transfer of cystine and cysteine were examined systematically (Gaull *et al.*, 1973*a*). In the case of most amino acids, the transfer from maternal blood to fetal blood must overcome a considerable difference in concentration. Despite structural differences in the placentas of various animals, the transport systems are essentially the same. They are energy-dependent (''active''), saturable, and stereospecific, and amino acids of similar structure compete for transport by their particular system. The rate of transfer is related in part to blood flow especially on the maternal side (Young, 1971). During the 2nd trimester, intravenous loads of L-methionine (Figure 35), L-leucine, and L-ornithine are transferred from maternal to fetal plasma against a twofold difference in concentration. After a similar infusion of L-cystine or L-cysteine to the mother, there was a slower and smaller increment in the plasma concentrations of these amino acids in the fetus (Figures 36 and 37). There was no transfer of D-cystine from mother to fetus (Figure 38) (Gaull *et al.*, 1973*a*). It was concluded that the transfer of cystine and cysteine is carrier-mediated and is not simple diffusion; however, the concentration in fetal plasma is not higher than in maternal plasma. Since the *in vitro* uptake of [$^{35}$S]cystine by slices of human fetal liver was less than that of [$^{35}$S]methionine (Gaull *et al.*, 1972), it seemed unlikely that the failure to maintain a plasma concentration which is higher in the fetus than in the mother is a result of avid uptake of cystine by fetal tissues. It is of interest that the acidic amino acids glutamate and aspartate are also transferred across the rat placenta slowly (Wapnir and Dierks-Ventling, 1971).

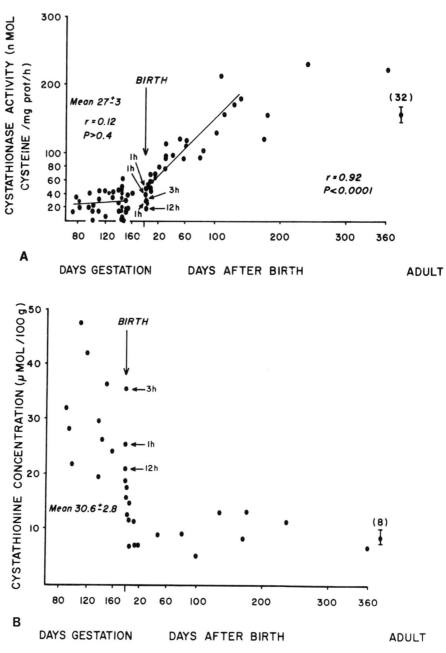

Fig. 33. (A) Activity of hepatic cystathionase in monkey liver during development. (B) Concentration of cystathionine in monkey liver during development. (C) Activity of hepatic cystathionine synthase in monkey liver during development. (From Sturman *et al.*, 1976*b*, with permission.)

Perhaps it is important during fetal development to protect the brain from high concentrations of cystine because of its acidic metabolites cysteine sulfinic acid and cysteic acid (cf. Gaull *et al.*, 1975, for a more complete discussion of this rather speculative point). The slow placental transfer and the limited ability to synthesize cystine both would be acting to protect the brain in this regard.

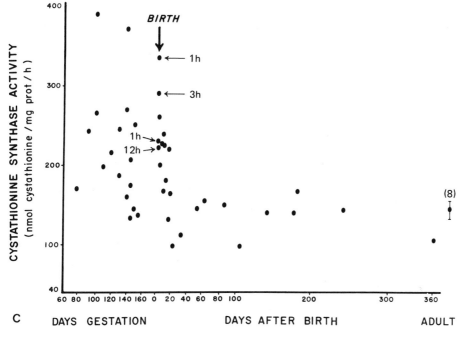

Fig. 33 (*Continued*)

## 4.1.3. The Role of the Methionine Cycle in Development

It was striking in the studies on the development of the transsulfuration pathway that the pathway was shut off at cystathionase, which is distal to cystathionine β-synthase (Figure 32). Cystathionine accumulated, but homocyst(e)ine did not (Sturman *et al.*, 1970; Gaull *et al.*, 1972). Activity of 5-methyltetrahydrofolate (MTHF) homocysteine methyltransferase, the $B_{12}$-linked enzyme which remethylates homocysteine back to methionine, is much greater in fetal human liver and brain than it is in mature human liver and brain (Gaull *et al.*, 1973*b*). Activity of the alternative enzyme for remethylation of homocysteine to methionine, betaine-homocysteine methyltransferase, is much lower in fetal liver than it is in mature liver (Gaull *et al.*, 1973*b*); its activity in brain is low both in fetal brain and in mature brain. The apparent $K_m$ for MTHF methyltransferase is of the order of $10^{-5}$ M, whereas that both of cystathionine synthase and of *S*-admosyl-homocysteine hydrolyase is $10^{-3}$ M. Furthermore, activity of the last two enzymes is relatively low in fetal tissues. The MTHF methyltransferase, therefore, would be expected to compete effectively for the available homocysteine, thus completing an active cycle of demethylation and remethylation of the sulfur of methionine.

The "methionine cycle," the existence of which had been well recognized in mature tissues, was postulated to be especially active in human fetal liver and brain (Gaull *et al.*, 1973*b*; Gaull, 1973*a,b*). An active methionine cycle, thus, is interpreted as an enzymatic adaptation which, during development, would act to conserve the sulfur of methionine (homocysteine) at the price of making cyst(e)ine essential and would facilitate the biosynthetic reactions related to this cycle. The absence of cystathionase in the human fetal liver, according to this hypothesis, would maximize the conservation of methionine sulfur, since none would be

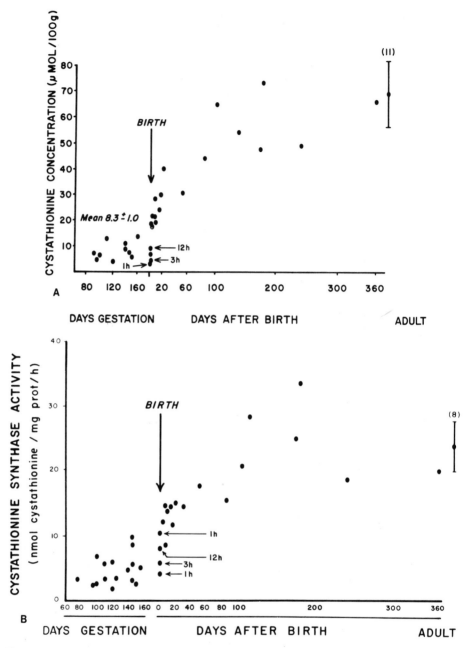

Fig. 34. (A) Concentration of cystathionine in monkey brain during development. (B) Activity of cystathionine synthase in monkey brain during development. (From Sturman *et al.*, 1976*b*, with permission.)

converted to cysteine. This cycle, in turn, interlocks with a cycle for the generation and regeneration of the key folate, tetrahydrofolate (Figure 32).

There is considerable ancillary evidence in support of this teleological hypothesis (Figure 39): (1) There is greater incorporation of [$^{35}$S]methionine into proteins of fetal human liver than there is into proteins of adult human liver (Gaull *et al.*, 1972). (2) There is virtually no conversion of [$^{35}$S]methionine to [$^{35}$S]cystine by perfused human fetal liver (Gaull *et al.*, 1972). (3) In human fetal liver, at the time when activity of MTHF methyltransferase is high, there is more rapid transfer of the β-carbon of serine into the *de novo* synthesis of DNA via thymidylate (Sturman *et al.*, 1975) than there is in adult human liver. (4) There is greater synthesis of the polyamines, spermidine and spermine, from *S*-adenosylmethionine and ornithine, in fetal human liver than there is in mature human liver (Sturman and Gaull, 1974). (5) Methionine adenosyltransferase activity and MTHF methyltransferase activities are highest in the lung of man, monkey, and rabbit around the time when each

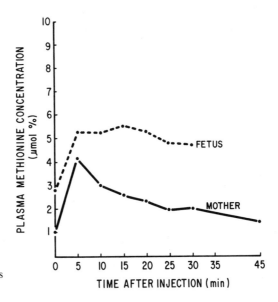

Fig. 35. Transfer of L-methionine across human placenta (Gaull *et al.*, 1973*a*).

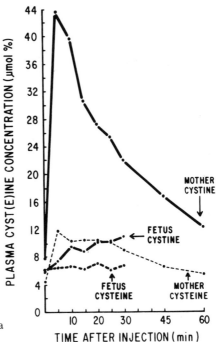

Fig. 36. Transfer of L-cystine across human placenta (Gaull *et al.*, 1973*a*).

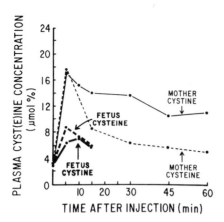

Fig. 37. Transfer of L-cysteine across human placenta (Gaull *et al.*, 1973*a*).

species acquires the respiratory capacity to survive (Sternowsky *et al.*, 1976). These high enzymatic activities could be important in the lung for detoxification reactions after birth; however, even if the "methylation pathway" plays no role in the synthesis of surface-active lecithins (see above), it is still possible that *S*-adenosylmethionine could be a methyl donor via the choline pathway (Salerno and Beeler, 1973).

Thus, the increased enzymatic capacity of this cycle is an enzymatic adaptation which could provide the potential for: (1) more methionine for protein synthesis at times of rapid net protein synthesis; (2) the facilitation of the *de novo* pathway of synthesis of DNA during times of cellular proliferation; (3) more hepatic synthesis of spermidine for stabilization of RNA, assuming it has functions in mammals similar to those it has in bacteria, and more synthesis of spermine for its role in DNA metabolism at times of cellular growth and proliferation (cf. Ingoglia and

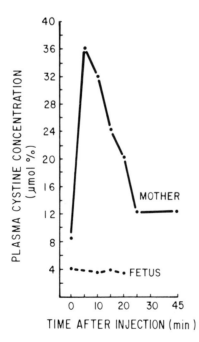

Fig. 38. Transfer of D-cystine across human placenta (Gaull *et al.*, 1973*a*).

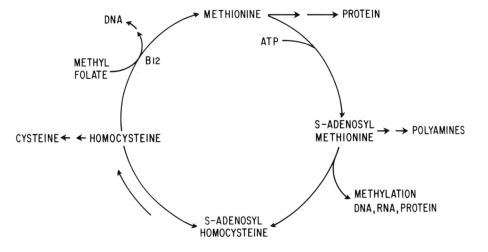

Fig. 39. The "methionine cycle."

Sturman [1977] for a more complete discussion of this point); and (4) the facilitation of methylation reactions associated with biosynthesis or detoxification.

### 4.1.4. Taurine in Development

Taurine is one of the most ubiquitous and abundant ninhydrin-positive compounds in the body, but it takes part in few known biochemical reactions (cf. Jacobsen and Smith, 1968; and Huxtable and Barbeau, 1976, for excellent general reviews). It is conjugated with bile acids in the liver, and limited deamination to isethionic acid apparently takes place in a few organs. Taurine can be converted to inorganic sulfate by bacteria but not by any tissues of mammals; it has been thought of mainly as an end product of metabolism. It is being discussed in the section on synthesis because of recent evidence that it has a number of important physiological roles and that it may be essential in some growing animals and not a mere "end product." Taurine plays a role in the regulation of membrane excitability in cardiac and skeletal muscle and may function as a neurotransmitter or neuromodulator in brain and retina. Finally, it may be involved in some endocrine and reproductive functions.

Although the greatest amount of information about taurine in development is available for brain and retina, we shall continue to focus mainly on liver as the central organ in metabolism. Brain and retina will be discussed in order to highlight and to explain the metabolic behavior of taurine in developing liver. Moreover, taurine metabolism in these three organs is interrelated. Details of taurine in developing brain and retina have been reviewed recently (Sturman *et al.*, 1977*a*).

Fetal human liver has a higher concentration of taurine than mature human liver, and there is no significant correlation between this concentration and increasing crown–rump length (Sturman and Gaull, 1975). Analysis of liver obtained at autopsy of two human neonates suggested that the concentration of taurine in liver decreases rapidly after birth. In the monkey, the concentration of taurine in fetal liver also is higher than it is in mature liver, and there is no significant correlation between taurine concentration and gestational age (Sturman and Gaull, 1975). After birth, the taurine concentration of monkey liver decreases rapidly, and concentrations similar to those found in mature monkey liver are attained by 1–2 weeks of age (Figure 40) (Sturman and Gaull, 1975).

The taurine concentration also is considerably higher in fetal human brain than it is in mature human occipital lobe. The difference in taurine concentration between fetus and adult is greater in brain than it is in liver. Moreover, during the 2nd trimester, there is a significant negative correlation between brain taurine concentration and increasing crown–rump length. Monkey brain is similar to human brain in this regard except that only the period of late gestation was studied and no change in taurine concentration could be demonstrated until the decrease which takes place in the postnatal period (Figure 41) (Sturman and Gaull, 1975).

The most striking difference between liver and brain is that the postnatal decrease in taurine concentration is very rapid in liver, whereas that in brain occurs slowly. When this decrease in brain taurine concentration of monkey, rabbit, and rat is plotted as a function of percent of time to weaning, it can be demonstrated that this concentration decreases in a linear fashion to reach concentrations at the time of weaning equal to those found in the respective adults (Figure 42).

The plasma taurine concentration of the human fetus is higher than that of either the pregnant or the nonpregnant adult human female (Glendening *et al.*, 1961; Ghadimi and Pecora, 1964). It appears to decrease during gestation, since the taurine concentration of plasma taken from the umbilical vein of infants at various times during gestation decreases with length of gestation (Ghadimi and Pecora, 1964; Bickel and Souchon, 1955). Furthermore, there is a relatively rapid decrease in urine and plasma taurine concentration during the early postnatal period (Lindblad, 1971; Bickel and Souchon, 1955). The taurine concentration of amniotic fluid decreases with increasing crown–rump length during the second trimester of pregnancy (Reid *et al.*, 1971). The monkey differs somewhat from man with regard to the changes in taurine concentration in plasma and amniotic fluid during development (Sturman and Gaull, 1975).

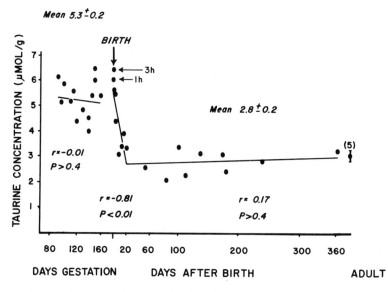

Fig. 40. Concentration of taurine in rhesus monkey liver during development. (From Sturman and Gaull, 1975, with permission.)

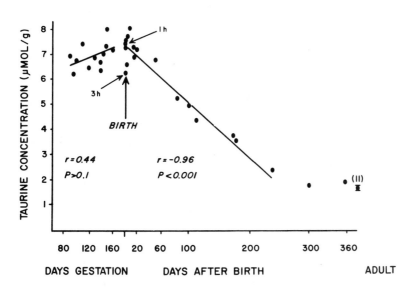

Fig. 41. Concentration of taurine in rhesus monkey occipital lobe during development. (From Sturman and Gaull, 1975, with permission.)

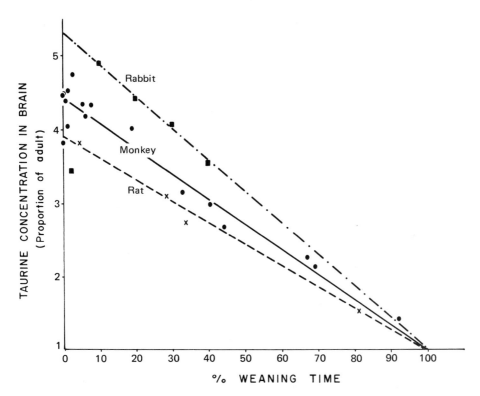

Fig. 42. Concentrations of taurine in brain of infant monkey (●), rabbits (■), and rats (×) plotted as proportion of the concentration in brain of the appropriate adult as a function of weaning time for each species. (From Sturman and Gaull, 1975, with permission.)

The time course of the decrease from the plasma taurine concentration of the fetus at term (20–39 $\mu$mol/100 ml) to that found in postnatal man (about 6 $\mu$mol/100 ml) has not been determined precisely, because the plasma and urine taurine concentrations of the neonate is a function of diet. Preterm infants fed synthetic formulas derived from bovine milk, especially casein-predominant formulas, had a progressive decrease in plasma taurine concentration and an immediate decrease in urine taurine concentration which was greater than that of preterm infants fed pooled human milk (Gaull et al., 1977). The activity of hepatic cysteinesulfinic acid decarboxylase had been known to be lower in adult man than it is in the adult rat. Recent evidence (Gaull et al., 1977) shows that fetal hepatic cysteinesulfinic acid decarboxylase is also lower in man than it is in the rat or even the monkey and cat. Thus, the low activity of this enzyme in man, both fetal and mature, suggests that there is a dietary requirement for taurine, especially in the rapidly growing human infant. Human milk, unlike bovine milk and artificial formulas derived from it, contains a considerable amount of taurine (Gaull et al., 1977); it is second in free amino acid concentration only to glutamate. The taurine concentration of milk is highest early in lactation, although after the initial decrease, the total amount of taurine ingested increases progressively because of the larger volume of milk consumed (Rassin et al., 1977).

Taurine transfer from mother to infant via milk has been studied in the rat (Sturman et al., 1977b). A single injection of [$^{35}$S]taurine was given to the dam 6–8 hr after birth. The labeled taurine so administered was secreted in the milk and accumulated in liver and brain of the pup. The uptake of labeled taurine by the brain of the pup was slower than that of the liver of the pup, and there was a slower turnover in the brain of the pup so that at 5 days of age there was considerably more [$^{35}$S]taurine in brain than there was in liver. Thus, even in the rat, an animal with a considerable ability to synthesize taurine via cysteinesulfinic acid decarboxylase, dietary taurine is a significant source of tissue taurine during development, especially in brain.

Although the taurine concentration of whole brain and that in the soluble fraction of brain decrease during development, there is a relative enrichment of the synaptosomal fraction by taurine (Rassin et al., 1976c). There seems to be a large soluble pool of taurine in brain which decreases during development and whose function has not been delineated. There is also a smaller pool associated with nerve ending particles, which may have a neurotransmitter or neuromodulator function and which is constant during development and, thereby, becomes relatively enriched during development.

The evidence for the involvement of taurine in retinal function is even more compelling. In contrast to brain, the concentration of taurine in retina increases during development of the rat (Macaione et al., 1974) and cat (Knopf et al., 1978), both of which are born with their eyes closed. There is no change in taurine concentration in monkey retina during the last part of gestation (Sturman et al., unpubl.); what happens in human retina is unknown (cf. Sturman et al., 1976c, for a discussion of this). It is of interest, in regard to taurine in retina, that kittens fed a taurine-deficient diet develop changes in the electroretinogram and retinal degeneration (Hayes et al., 1975a,b; Schmidt et al., 1976; Berson et al., 1976). Blindness ensues unless their diets are supplemented with taurine. Amounts of methionine, cystine, or $SO_4^{2-}$ equimolar to taurine will not reverse the retinal changes. The human neonate conjugates bile acids chiefly with taurine during the first week or two of postnatal life and gradually converts to predominantly glycine-conjugated bile acids (Encrantz and Sjovall, 1959; Poley et al., 1964). In contrast, the cat can conjugate bile acids with taurine only (Knopf *et al.,* 1978; Schmidt *et al.,* 1976). This conversion from predominantly taurine-conjugated bile acids to predominantly glycine-conjugated bile acids may reduce the taurine requirement of the human neonate enough to avoid the overt clinical blindness found in taurine-deficient kittens. More subtle or ephemeral ocular changes have not been noted, but they should be looked for, especially in the rapidly growing preterm infant fed artificial formulas deficient in taurine. Retinal blindness in the taurine-deficient cat may be the result of at least five factors: (1) rapid increase in retinal taurine concentration after birth, (2) high taurine concentration in cat retina, (3) low rate of taurine synthesis in the cat, (4) rapid rate of growth of the kitten, (5) inability of the cat to convert from bile acid conjugation with taurine to conjugation with glycine.

### 4.1.5. Synthesis of Histidine

There is no dietary requirement for histidine demonstrable in the normal human adult. A dietary requirement in the infant has been shown by Snyderman and co-workers (1959*b*), who demonstrated a lower rate of weight gain and decreased nitrogen retention when histidine was eliminated from the diet of a 9-kg infant for 3 weeks. Although histidine biosynthesis has not been studied in human fetal liver, Snyderman's study suggests that it is probably also essential in the human fetus.

### 4.1.6. Synthesis of Glycine and Serine

The presence of serine hydroxymethyltransferase in human fetal liver (Gaull *et al.,* 1973*b*) suggests that it is able to synthesize glycine readily. Like many other enzymes of liver, no change in activity of this enzyme is found during the second trimester of gestation. In brain, however, there is a decrease in activity with increasing crown–rump length during the same period (Gaull *et al.,* 1973*b*).

The presence of a high *AQ* (cf. Section 1) for hepatic phosphoserine phosphatase (6) suggests that the ability of the human fetus to synthesize serine is considerable. Cheung *et al.* (1968) have presented data that in fetal liver of a variety of animals, including man, serine biosynthesis occurs primarily via D-3-phosphoglycerate rather than D-glycerate. The activity of phosphoglycerate dehydrogenase is greater than that of D-glycerate dehydrogenase in fetal liver, and the relative overall conversion of each substrate to serine was comparable to the relative activity of each respective enzyme. This pathway from glycolytic intermediates would seem of particular importance in the fetus (see Section 2). Indeed, there is a larger transfer of the $\beta$-carbon of serine to thymidylate (the pathway of *de novo* synthesis of DNA) (Sturman *et al.,* 1975) in fetal human liver than there is in adult human liver, as might be expected in a rapidly proliferating tissue. Thus, not only does human fetal liver show considerable early competence for glycine biosynthesis from serine derived from glycolytic intermediates, but this ability is linked to the generation of one-carbon fragments for DNA synthesis, yet another demonstration of Nature's teleorganic good sense.

## 4.2. Amino Acid Metabolism and Urea Synthesis (Figure 43)

### 4.2.1. Urea Synthesis

About 90% of the $\alpha$-amino nitrogen from amino acids released during catabolism of proteins and amines is converted to ammonia (Colombo, 1971). The latter

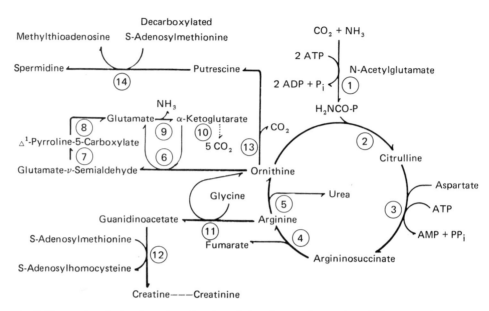

Fig. 43. Urea cycle and related reactions: 1, carbamyl phosphate synthetase; 2, ornithine transcarbamylase; 3, argininosuccinate synthetase; 4, argininosuccinase; 5, arginase; 6, ornithine ketoacid aminotransferase; 7, spontaneous; 8, $\Delta^1$-pyrroline 5-carboxylate dehydrogenase; 9, glutamate dehydrogenase; 10, tricarboxylic acid cycle; 11, glycine amidinotransferase; 12, guanidinoacetate methyltransferase; 13, ornithine decarboxylase; 14, putrescine propyltransferase.

compound is extremely toxic (Bessman and Bessman, 1955) and requires a large amount of water for its effective elimination. Man is a ureotelic animal in whom urea, easily soluble in water, is the major mode of detoxication and excretion of ammonia nitrogen. In the fetus, it is excreted, via the placenta, by the maternal kidney or by the fetal kidney into amniotic fluid. Exchange of fluid between mother and fetus is extremely rapid. After birth the infant is in an environment relatively poor in water; therefore, ammonia detoxication and excretion rely largely on urea synthesis.

Man's ureotelic existence depends upon the development of the enzymes of urea synthesis during fetal development. Needham (1963) cites observations dating from the early 19th century of the presence of urea in the bladder of the human embryo. Enzymes for synthesis of urea were measured first by Kennan and Cohen (1959): arginase, carbamyl phosphate synthetase, ornithine transcarbamylase, and argininosuccinic acid synthetase were found as early as the 12th week of human gestation. Räihä and Suihkonen (1968a,b) showed that all enzymes of the urea cycle were present in 10- and 12-week human fetuses in activities similar to those of newborn rats and that liver slices of 16-week fetuses were able to perform overall synthesis of urea. The activity of the arginine synthetase system, or argininosuccinate cleaving enzyme, is the lowest of the enzymes of the urea cycle and probably limits urea synthesis. The activity of the arginine synthetase system is of the same order of magnitude in the perinatal period as it is in later life. Studies in isolated perfused human fetal liver (Kekomäki et al., 1971) demonstrated that urea was released into the perfusion medium at a steady rate. Addition of high concentrations of alanine (10 mmol) to the perfusion medium failed to increase the rate of urea synthesis in perfused liver obtained during the first trimester. It seems that the capacity for urea synthesis in human liver has a limited capability for expansion.

Thus, its quantitative significance as a means of removal of excess nitrogen is probably small during gestation. After birth, urea synthesis, as reflected by plasma urea nitrogen concentration, parallels directly the amount of protein ingested (Fomon, 1967; Omans *et al.*, 1961). This parallelism is true also in the smallest preterm infants, in whom differences of plasma urea nitrogen concentrations with increased protein loads can be demonstrated as early as the first postnatal week (Räihä *et al.*, 1976).

Colombo (1968) found that carbamyl phosphate synthetase, the arginine synthetase system (argininosuccinate cleaving enzyme), and arginase show two early peaks of activity at about 60 and 90 days of gestation and that both then increase towards the adult range towards the end of pregnancy. The activity of ornithine transcarbamylase remains constant throughout gestation until towards the end when a distinct increase to the lower end of the adult range is observed. In the first postnatal year carbamyl phosphate synthetase and ornithine transcarbamylase are within the same range as adult values. The values for arginase and argininosuccinate cleaving enzyme remain relatively low in comparison to adults. Argininosuccinate synthetase, the third enzyme in the cycle was not measured.

From a teleological point of view, there seems to be much less of a need for urea synthesis prenatally than there is postnatally. Nevertheless, Kennan and Cohen (1959) calculated from their data that, towards the end of gestation, the human fetus produces 3 g of urea per day, which is eliminated via the maternal kidney. Further direct evidence of the prenatal functioning of urea synthesis is the accumulation of argininosuccinic acid in amniotic fluid of fetuses identified *in utero* (Colombo and Richterich, 1968; Shih and Littlefield, 1970; Goodman *et al.*, 1973; Billmeier *et al.*, 1974; Fleisher *et al.*, 1977). In these cases argininosuccinic acid excretion by the mother occurs during pregnancy. Recent determinations of argininosuccinic acid in the tissues of an affected fetus with a deficiency of arginine synthetase system gives further indication of the ability of the human fetus to synthesize urea (Fleisher *et al.*, 1977). The synthesis of urea in the fetus is a function which is difficult to detect, but becomes more evident in the presence of an inherited enzymatic deficiency of the urea cycle. Experiments in primates by Hutchinson *et al.* (1962) demonstrated an exchange rate of 1–2 $\mu$mol urea/h/kg fetus. The preponderance of evidence suggests that urea synthesis occurs in the human fetus, but a reduced rate compared to that which takes place postnatally, and with a limited capability for expansion of this synthetic ability (Hutchinson *et al.*, 1962). As a corollary, it seems likely that any brain damage that occurs in inborn errors of urea synthesis probably starts antenatally.

Since ammonia toxicity manifests its effects especially in brain, Colombo (1971, pp. 40–47) measured a number of enzymes of ammonia metabolism in fetal brain. From the 2nd month of gestation the brain has enzymes which facilitate ammonia detoxication: glutamate dehydrogenase, glutaminase 1, glutamine synthetase, $\alpha$-glutamyl transpeptidase, glutamate decarboxylase, and argininosuccinate cleaving enzyme. The latter enzyme may be present in brain for synthesis of arginine rather than for urea. Arginine forms $\gamma$-guanidobutyric acid in brain by transamidation with $\gamma$-aminobutyric acid, the product of glutamate decarboxylase (Pisano *et al.*, 1957, 1963). Indeed, the synthesis of arginine may be looked upon as a function of the urea cycle in addition to that of excretion of ammonia. In this regard there appears to be no dietary requirement for arginine in the human infant (Snyderman *et al.*, 1959*a*).

In addition to the synthesis of arginine, the urea cycle regenerates a mole of ornithine for each mole of ornithine it utilizes to form citrulline. The control of the metabolism of this compound is of considerable metabolic importance. It is degraded to glutamate by a γ-transamination between ornithine and α-ketoglutarate to form glutamic-γ-semialdehyde, which spontaneously cyclizes to Δ'-pyrroline-5-carboxylate, which, in turn, is enzymatically dehydrogenated to glutamate. The studies of Kekomäki *et al.* (1969) suggest that the activity of ornithine ketoacid aminotransferase is rate limiting and that it is the factor which determines the rate of catabolism of ornithine in human liver. The reverse reaction, i.e., the synthesis of glutamate, is thought to be essentially irreversible, although it has been suggested that it exists in liver and kidney in two forms, each functioning in different directions (Volpe *et al.*, 1969). The specific activity of this enzyme is considerably higher in fetal human liver than it is in mature human liver (Kekomäki *et al.*, 1969), the maximal activity being attained prior to the 5th month of gestation. At birth, the activity of ornithine ketoacid aminotransferase still exceeds that of normal adults. The *AQ* of this enzyme is about 7 (Greengard, 1970), which is the opposite of that in rat liver. This complete difference, like that in the development of betaine-homocysteine methyltransferase (Gaull *et al.*, 1973*b*), remains unexplained. Ornithine decarboxylase, another pathway of metabolism of ornithine, also has an *AQ* of about 6 (Greengard, 1970), which is similar to the rat.

### 4.2.2. Oxidation of Phenolic Amino Acids

The catabolism of tyrosine (Figure 44) has been studied extensively in fetal and perinatal liver (Kretchmer *et al.*, 1956, 1957; Räihä *et al.*, 1971). This process begins by transamination of tyrosine with α-ketoglutarate by tyrosine aminotransferase. Activity of this enzyme increases slowly during gestation. In the liver of the preterm infant it has about 10% of the activity found in adult human liver. The activity of tyrosine aminotransferase in human fetal liver during the 2nd trimester is approximately twice that found in fetal rat liver. There apparently is an immediate postnatal increase, similar to that found in the newborn rat, since the activity in newborn infants dying in the first 12–24 hr after birth is considerably higher than that found in the fetus (Del Valle and Greengard, 1976).

Tyrosine aminotransferase is highly responsive to hormonal influences (cf. for example: Sereni and Principi, 1965; Greengard, 1971, 1970). In fetal human liver, as in fetal rat liver, it can be induced by corticosteroids in explants at about the end of the 2nd trimester (Räihä *et al.*, 1971). Thus, the liver can respond to hormone-induced enzyme synthesis only after a certain developmental stage is reached.

The *p*-hydroxyphenylpyruvic acid (pHPP) which is produced by transamination of tyrosine is then oxidized by pHPP oxidase, and the resultant homogenistic acid is oxidized by a series of reactions to form fumarate and acetoacetate. Because the low activity of pHPP oxidase of fetal liver can be increased several-fold by reducing agents (Goswami and Knox, 1961), this enzyme is thought to be present in fetal rat liver in a partially inactive form. Glucagon but not hydrocortisone is ineffective in the induction of tyrosine aminotransferase *in vivo* (Greengard, 1970, 1971), but both are effective *in vitro* (Wicks, 1968, 1969), giving further evidence at least for glucagon inhibition *in vivo* in fetal liver. Rutter (1969) suggests that growth hormone, which represses tyrosine aminotransferase in adult liver, might repress it *in utero* and that the hypoglycemia that develops in the perinatal period leads to glucagon release and induction of this enzyme. Kretchmer and co-workers (1957; Räihä *et al.*, 1971) reported that the low activity of pHPP oxidase in the liver of the preterm infant can be increased *in vitro* by ascorbic acid, suggesting that in

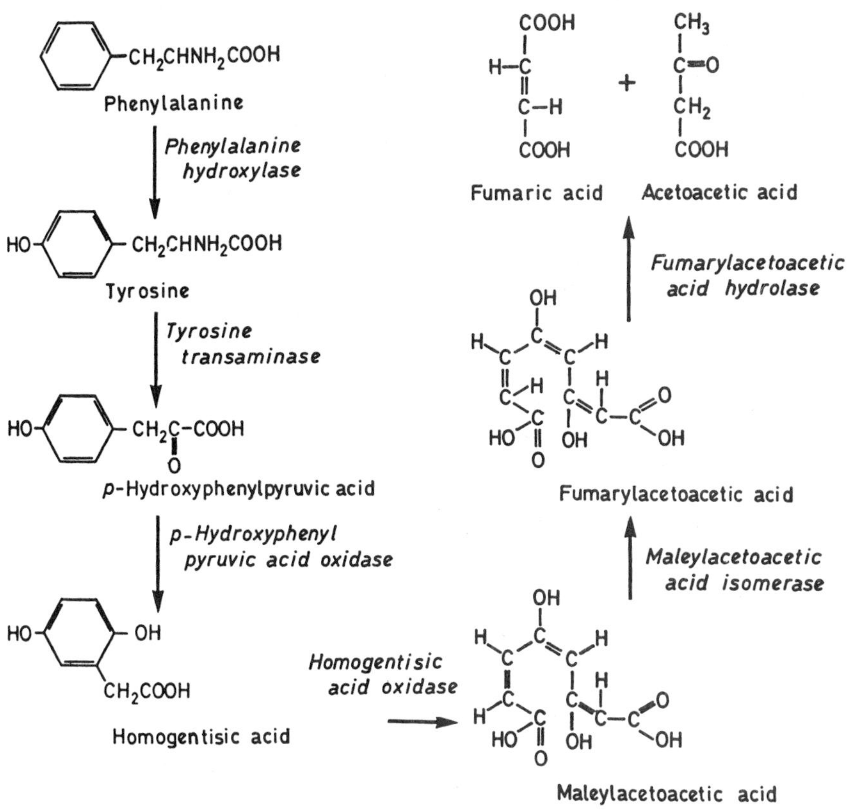

Fig. 44. Oxidation of phenylalanine and tyrosine. (Reproduced with permission from *The Principles of Human Biochemical Genetics* by H. Harris, North Holland/American Elsevier.)

immature human liver, as well, the pHPP oxidase present is in a partially inactive form which can be activated. This enzyme appears one week later than tyrosine aminotransferase, the enzyme which immediately precedes it on the pathway. The developmental relationship of these two enzymes suggests that substrate activation could play a role in the development of pHPP oxidase.

Understanding the development of the enzymes of tyrosine oxidation is of considerable clinical importance. It has been known for some time that preterm infants fed high-protein formulas develop some hyperphenylalaninemia and considerable tyrosinemia and excrete large amounts of *p*-hydroxyphenylpyruvate, *p*-hydroxyphenyllactate, and *p*-hydroxyphenylacetate in the urine (tyrosyluria) (cf., for example: Levine *et al.*, 1941; Mathews and Partington, 1967; Avery *et al.*, 1967; LaDu and Gjessing, 1972). The hyperphenylalaninemia, tyrosinemia, and tyrosyluria may persist for 4–6 weeks. It has been shown more recently that the tyrosinemia is greater in casein-predominant formulas (bovine milk protein) that in whey-predominant formulas (Rassin *et al.*, 1976). Early studies suggested that supplementation with ascorbate and/or folate would alleviate these biochemical abnormalities. Systematically controlled studies performed more recently, (Bakker *et al.*, 1975; Partington and Mathews, 1967; Mohanram and Kumar, 1975), however, indicate that the tyrosinemia and tyrosyluria found in preterm infants fed high-protein diets is not corrected by massive doses of either vitamin. Plasma concentrations of phenylalanine as high as those found in phenylketonuria were observed in a few infants fed a formula providing 4.5 protein/kg/day of a casein-predominant formula (Rassin *et al.*, 1976). Individual infants fed this same formula had plasma tyrosine concentrations which were 30 times the mean found in similar infants fed pooled human milk. The fact that the concentrations of tyrosine were higher than those of phenylalanine is compatible with the presence of phenylalanine hydroxylase in immature human liver (see above).

A number of studies have attempted to relate later intellectual or neurological impairment of children who were born prematurely to the tyrosinemia resulting from feeding high-protein diets in early infancy (Menkes *et al.*, 1972; Partington, 1968; Goldman *et al.*, 1974; Mamunes *et al.*, 1976). Most of these studies presume that any such impairment would be mediated by tyrosine *per se*, both because of the high tyrosine concentrations found and because of the structural relationship of tyrosine and its acidic metabolites to phenylalanine and its acidic metabolites. In light of the extent and severity of the other imbalances in amino acid metabolism that accompany the tyrosinemia and tyrosyluria (Gaull *et al.*, 1977; Omans *et al.*, 1961; Zuckerman *et al.*, 1970), however, it seems advisable to consider all metabolic parameters possible rather than just those most readily measured, i.e., tyrosine and phenylalanine. The multiplicity of pathogenetic possibilities to explain the brain dysfunction associated with disordered amino acid metabolism have been reviewed critically (Gaull *et al.*, 1975).

Tryptophan pyrrolase, the enzyme which converts tryptophan to formylkynurenine also has been studied extensively because it was found to be absent in fetal rabbit and guinea pig liver, to develop in the early postnatal period, and to be subject to experimental manipulation by various inducers and inhibitors of protein synthesis. In the rat it appears in the late suckling period (Greengard, 1970). Corticosteroids act by activation of messenger RNA synthesis, whereas substrate acts by stabilizing tryptophan oxidase (Nemeth, 1963). Tryptophan pyrrolase can be induced in adult man as well as in adult rat, and it can be induced in the rat fetus by sequential treatment with hydrocortisone and tryptophan (Greengard, 1971). Substrate induction can occur only after the normal developmental increase begins and must be considered to be the result, rather than the cause, of differentiation of this enzyme. The technical difficulties in the determination of tryptophan in plasma have dampened enthusiasm for research in the metabolism of this essential amino acid. This is a pity because it is important not only in protein metabolism, but in brain metabolism, in particular because it is precursor for the neutrotransmitter serotonin (5-hydroxytryptamine). Although studied extensively in liver of animals, especially by Nemeth (1963) and Greengard (1970), it seems not to have been studied in human liver.

### 4.2.3. Catabolism of Other Amino Acids

Amino acids, in general, are converted to an intermediate which can be oxidized through the final common pathway, the tricarboxylic acid cycle. Oxidative deamination and transamination allow immediate entry into this cycle. Glutamate is oxidized more rapidly than the other amino acids as a result of oxidative deamination with a pyridine nucleotide. This enzyme has been studied by Herzfeld *et al.* (1976). In adult human liver it is equally divided between the soluble and particulate fractions. The concentration of only the soluble enzyme increased after birth. The *AQ* of the soluble fraction is 0.15, that of the particulate fraction is 1.06, and that of the homogenate is 0.63. In rat liver, glutamate dehydrogenase is entirely particulate and emerges in the neonatal cluster. The further study of the late-emerging soluble

fraction in man would be of interest in light of the large amounts of glutamate in
currently popular intravenous solutions.

Alanine aminotransferase also was studied by Herzfeld *et al.* (1976). The
enzyme is mostly soluble, both in man and in rat. It is virtually absent from fetal and
neonatal liver in both species and appears in the D, late suckling, cluster in rat liver.
The small fraction of this enzyme, which is particulate, persists in man, whereas it
decreases significantly in the rat.

Aspartate aminotransferase was studied in detail in rat liver by Herzfeld and
Greengard (1971) and more recently by Greengard (1976) in man. In man, the *AQ* of
the soluble enzyme is 0.28, the particulate is 0.69, and the whole homogenate is
0.43. In the rat, it appears with the B, late fetal, cluster.

Phenylalanine pyruvate aminotransferase also has been studied by Greengard
(1976). It has an *AQ* of 0.062 and appears in the neonatal (C) cluster.

## 4.3. Protein Metabolism

The process of protein synthesis, in a general sense, is understood in great
detail, and there is a considerable literature on protein synthesis during development in animals. The literature on protein synthesis in developing man, however, is
scant. What follows is an attempt to draw the rough outlines of what is known about
this process in the human fetus. It is presented, of necessity, in rather general
terms.

Protein synthesis in the fetus differs considerably from that in infants and
adults: a larger proportion of protein synthesis is channeled into structural protein,
since the placenta and maternal organs can perform many necessary metabolic
functions during earlier stages of development. As the individual organs assume
their ultimate form and function, the emphasis on synthesis of structural protein
gradually shifts to that of functional proteins (e.g., enzymes, hormones, antibodies).

Fetal proteins originate from the free amino acids transferred from mother to
fetus. Dancis, Braverman, and Lind (1957) showed that the major precursor of fetal
protein is neither protein synthesized in the placenta nor the actual transfer of intact
maternal protein. They demonstrated that by 3 months of gestation human fetal
liver is capable of synthesizing plasma proteins, with the exception of γ-globulin
which is derived largely from the mother.

Although human infants gain much more weight after birth during the suckling
period than before birth, the *rate* at which they increase their weight is far greater in
the antenatal period because the weight of the ovum is so small relative to the birth
weight. Fetal growth is exponential at first, but later slows up and is the resultant of
the specific growth rates of individual organs. Thus, net protein synthesis proceeds
at a faster rate during fetal development than at any other time in life. It is likely that
the turnover of some proteins is no greater than in the adult and that a considerable
factor in the greater net protein synthesis is a slow rate of degradation (see below).

The placenta grows most rapidly during early pregnancy. A study of minced
human placenta by Mori (1965) demonstrated that protein synthesis, as measured
by the incorporation of a mixture of amino acids uniformly labeled with $^{14}C$, was
higher at 3 months gestation than it was at 5 and 10 months and that it paralleled the
RNA concentration of placenta.

In the fetus itself, protein deposition occurs mainly during the last part of
pregnancy. Nitrogen analyses of fetuses that did not survive to term suggest that
almost two thirds of the total body nitrogen of the term infant is acquired during the

last 2 months (Swanson and Iob, 1939). Lichtenstein (1931) showed that the fetal $\alpha$-amino nitrogen, as measured in cord blood, is highest in the smallest premature infants and decreases as development progresses. This decrease was confirmed for individual free amino acids some years later by Ghadimi and Pecora (1964). Thus, as gestation progresses, the amino acid requirements for fetal protein synthesis are apparently satisfied by lower blood concentrations. As pointed out by Widdowson (1969), however, the nutrient supply of a fetus depends more upon the quantity of blood reaching the fetus than it does upon the concentration of nutrient in the blood. This caveat notwithstanding, amino acids are transferred from mother to fetus via the placenta by a process of active transport, i.e., carrier-mediated, saturable, energy-dependent, sterospecific and operating against a chemical gradient (Szabo and Grimaldi, 1970; Young, 1971). This process results in the maintenance of higher concentrations of amino acids in fetal blood than in maternal blood, with the notable exception of cystine and cysteine. These sulfur-containing amino acids are transferred, to the human fetus at least, by a carrier-mediated system, but without maintaining a concentration difference (see above). What is true in human fetal blood is true also for human fetal liver, i.e., most amino acids are present in higher concentrations than in mature human liver (Ryan and Carver, 1966).

The synthesis of a protein (Figure 45) is an enormous task of assembling a complex molecule containing scores of amino acids in the same sequence each time the molecule is produced. It employs a precise coding system which programs the insertion of a specific amino acid into a specific position in the protein chain and programs the initiation as well as the termination of that chain. By determining the primary structure (the amino acid composition in proper sequence), it also establishes the secondary structure (a helix which is stabilized by hydrogen bonding between carbonyl and imino groups of the peptide bonds) as well as the tertiary structure (the spatial configuration which condenses and stabilizes these complex molecules). Some proteins also have a quaternary structure in which polypeptides are polymerized into dimers, tetramers, etc.

Genetic information is carried as a sequence of bases in DNA, which is in turn programmed on RNA for the orderly synthesis of a new protein. In a general way, the genetic information in DNA transcribes a complementary sequence of bases on a new RNA chain (messenger RNA). The process by which the genetic information in a mRNA molecule directs the order of insertion of the specific amino acid during protein synthesis is called translation. The translational process involves several steps which include: activation of the 20 amino acids and their subsequent covalent linkage with the 20 different specific molecules referred to as transfer RNA to form acyl-tRNAs. The latter complex carriers diffuse to specific sites on the mRNA–ribosomal complex on which surface the orderly condensation takes place, followed by a discharge of the aminoacyl moieties to form the primary structure of a new protein molecule.

The DNA in a cell is confined almost entirely to the nucleus, and in a diploid human nucleus is rather constant at about 6.2 pg (Winick and Rosso, 1973). Thus, in an organ with mononuclear cells having diploid nuclei only, one can measure the cell number of an organ by measuring the DNA of that organ. Once cell number is determined, the average weight per cell, protein/cell, etc., can be determined by analyzing the total amount of the constituent and dividing by the cell number. Thus, the protein/DNA, RNA/DNA, and lipid/DNA ratios may be calculated. The general concept is that increase in DNA content of an organ is a measure of increase in cell number, whereas weight/DNA or protein/DNA is a measure of increase in cell mass. Early phases of growth proceed entirely by an increase in cell division. DNA, weight, and protein all increase proportionally, resulting in a constant cell size as expressed either by weight/DNA ratio or by protein/DNA ratio. The rate of cell division decreases at different rates in different organs. Protein content continues its increase, resulting in an increased protein/DNA ratio. Increase in cell size, therefore, occurs as a consequence of the decreased rate of cell division in the presence of a constant rate of increase in protein. With maturity the rate of protein synthesis and that of protein breakdown reach a steady state. Later phases of growth consist of an increase in cell number and cell size, but with the increase in cell size predominant. Finally, there is a third phase consisting of an increase in cell size as DNA synthesis decreases and protein synthesis continues to increase. Growth ceases when protein synthesis and breakdown reach a steady state. It is important to realize that these phases are not discrete, but rather merge continuously with each other. Thus, protein synthetic activity during fetal development follows a general sequence, the exact schedule of which varies from organ to organ. First, there is an increase in DNA content parallel to hyperplastic growth or increase in cell number. This is followed by an increase in RNA content during hypertrophic growth or increase in cell size. Finally, protein synthetic activity increases, abetted by the high concentration of free amino acids available. Substrates and hormonal influences play a significant regulatory role.

Kapeller-Adler and Hammad (1972) made such measurements in various human fetal tissues at different stages of development during the first half of gestation (6–20 weeks). The variations in nucleic acid and protein concentrations during this period are characteristic for each organ. Although the RNA/DNA ratio does not vary much with the stage of pregnancy, the protein/DNA ratios of fetal liver, brain, and heart increase progressively. Among the fetal organs examined, liver and heart had the highest RNA/DNA and protein/DNA ratios. The maximal rate of cell division of liver occurred during the 3rd–4th month with a progressive increase in protein/DNA ratio (cell size) during this period and continuing on. During the same time DNA content in brain and in heart was decreasing, i.e., the rate of cell division, whereas the protein/DNA ratios in these organs was continuously increasing.

Nucleic acid synthesis in human fetal tissues (8–15 weeks gestation) has been studied by autoradioagraphy in explants of various human fetal tissues using [$^3$H]uridine as a RNA precursor and [$^3$H]thymidine as a DNA precursor (Mukherjee *et al.*, 1973). Unique curves for DNA and RNA synthesis so derived were found for each organ studied. In each organ, however, when the rate of DNA synthesis decreased, the rate of RNA synthesis increased, and vice versa, indicating a periodicity in nucleic acid synthesis. These waves of DNA and RNA synthesis were interpreted as cell division followed by differentiation and metabolic activity.

[$^{35}$S]-L-Methionine and [$^{35}$S]-L-cysteine are incorporated into protein in human fetal liver, brain, kidney, and pancreas; cystine does not seem to be rate limiting to protein synthesis in liver (Gaull *et al.*, 1972). [$^{35}$S]-L-Methionine was incorporated more rapidly into fetal human liver protein than it was into postnatal human liver protein (Heinonen, 1974). It was found, further that $^{35}$S derived from methionine was demonstrated in the cystine in the acid-insoluble fraction of a slice of human liver, whereas [$^{35}$S]inorganic sulfate did not serve as a precursor for cystine in liver protein. Since cystathionase is absent from these livers, it was suggested (Heino-

GERALD E. GAULL
ET AL.

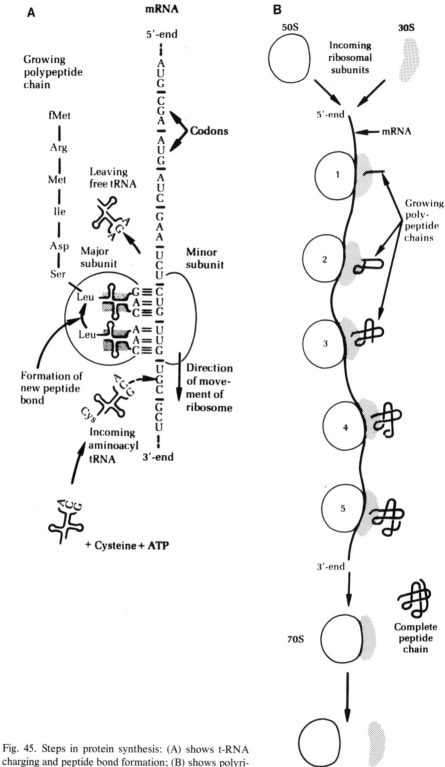

Fig. 45. Steps in protein synthesis: (A) shows t-RNA charging and peptide bond formation; (B) shows polyribosome formation. (From Lehninger, *Biochemistry*, Worth Publishers, New York, 1970.)

nen, 1974) that the homocysteine sulfur (derived from methionine) was incorporated into cystine via sulfide and serine. This would require both homocysteine desulfurase, yielding sulfide, and a serine sulfhydrase, yielding cysteine from serine and sulfide.

Incorporation of [$^{35}$S]methionine and [$^{35}$S]cysteine into tissue protein was measured also in the fetal monkey (Sturman *et al.*, 1973), which is similar to the human fetus in some, albeit not in all, aspects of sulfur metabolism. When these compounds were injected into the monkey fetus, methionine was more avidly incorporated into all fetal tissue proteins than was cysteine. A notable exception was fetal plasma proteins in which specific activities were similar after injection of [$^{35}$S]methionine and of [$^{35}$S]cysteine. The lower specific activities of fetal proteins when the mother is injected with [$^{35}$S]cysteine compared to [$^{35}$S]methionine may be a function of the slower transfer of cystine across the placenta (Glendening *et al.*, 1961; Gaull *et al.*, 1973*a*). When the mother was injected with [$^{35}$S]methionine, the specific activity of protein isolated from brain, heart, and gastrocnemius muscle was considerably higher in the fetus than in the mother. When [$^{35}$S]cysteine was injected into the mother, however, the specific activity of protein isolated from maternal organs was always higher than in all fetal organs. The pancreas of the mother, when she was injected, and the pancreas of the fetus, when it was injected, had protein of the highest specific activity. In experiments in which [$^{35}$S]methionine was injected, [$^{35}$S]cysteine could be detected in protein hydrolysates only of fetal plasma and of a few maternal organs. The specific activity of protein purified from the plasma both of mother and of fetus increased more rapidly with time after injection when methionine was injected than when cysteine was injected.

Picou and Taylor-Roberts (1969), using methods which do not depend upon administration of a single dose of labeled amino acid, estimated the mean rate of total body protein synthesis in older infants to be about 6 g/kg/day. Nicholson (1970), using the less satisfactory single-dose method of administration of San Pietro and Rittenberg (1953), found rates of whole body protein synthesis in 3 preterm infants (29–68 days of age) to be 11–15 g/kg/day. Young *et al.* (1975) and Pencharz *et al.* (1977) studied one term and 6 preterm infants fed a commercial cow milk formula. The infants were given [$^{15}$N]glycine orally over a 30- to 36-hr period and the $^{15}$N enrichment of urinary urea N was determined. Mean whole body protein synthesis was calculated to be $17.8 \pm 7.1$ g/kg/day and breakdown to be $15.5 \pm 7.3$ g protein/kg/day.

Picou and Taylor-Roberts (1969), in older infants, and the MIT group (Young *et al.*, 1975; Pencharz *et al.*, 1977), in preterm infants, found the intensity of nitrogen metabolism per unit body weight to be considerably higher than that of adults. Nicholson (1970) found rates of protein synthesis, when based on surface area, to be comparable to those in adults as measured in similar studies. The MIT group observed that differences in intensity of protein metabolism among groups is eliminated when total body protein synthesis is related to the energy expenditure for different age groups. These workers found that only 17% of the protein synthesized by the preterm infant appears as a net gain in body protein. They concluded, therefore, that the high energy and protein intakes required for normal growth are more a function of high rates of protein turnover than of growth as such.

The data on the synthesis of specific proteins, especially those of plasma, are perhaps the best available on protein synthesis *per se,* since sensitive specific immunochemical methods are available for measurement. A recent, comprehensive review (Gitlin and Gitlin, 1975) obviates the necessity for detailed coverage here.

Nevertheless, certain highlights are worthy of mention from the point of view of protein synthesis, as opposed to metabolism, transfer, etc. By 3–4 weeks of gestation, the liver of the human fetus already can synthesize prealbumin, $\alpha$-fetoprotein, $\alpha_1$-antitrypsin, transferrin, $\beta$-lipoprotein, $\alpha_2$-macroglobulin, and $C_1$-esterase inhibitor. Within the next 2 weeks $\alpha_1$-acid glycoprotein, ceruloplasmin, and then fibrinogen can be synthesized. These proteins, therefore, can be synthesized when hepatic cells are emerging and beginning to differentiate. Plasma proteins also can be synthesized for some time by the yolk sac and by lung and gastrointestinal tract.

Most proteins in the fetus are present at lower concentrations earlier in gestation. $\alpha$-Fetoprotein is a notable exception, occurring in cord plasma in concentrations 20,000 times those in the adult (Gitlin *et al.,* 1972). From about 14 weeks of gestation to approximately 30 weeks, its concentration in fetal plasma decreases exponentially because of the rapid growth of the fetus and its relatively lower rate of synthesis (Gitlin *et al.,* 1972). Synthesis of $\alpha$-fetoprotein later in life by diseased liver or carcinomas may represent a derepression phenomenon (Gitlin *et al.,* 1972).

In the fetus, the plasma proteins in amniotic fluid increase during development. Proteins can leave the amniotic fluid by diffusion, in part, but mostly by fetal swallowing. The protein is then hydrolyzed in the fetal gastrointestinal tract. There is considerable maternal–fetal exchange of protein, mostly via the placenta. The contribution of the maternal plasma proteins to the fetal plasma will differ according to the protein, and some of the mechanisms are complex. For most proteins, the rate at which it is transferred from mother to fetus is small in comparison to the rate of fetal synthesis. IgG is a notable exception (Gitlin *et al.,* 1972).

Another specific protein which has been studied in the human fetus is the brain-specific S-100 protein. The S-100 protein appears first in the phylogenetically older brain areas (caudally), with delay in appearance in the frontal cortex until the 30th week of gestation. By the 20th week there was about 50% of the maximum value found at term (Zuckerman *et al.,* 1970).

Over the past decade there has been an explosion of information on all details of protein synthesis. Application of this detailed knowledge to the study of protein synthesis in the human fetus, however, has been limited to study of specific placental proteins. This limitation is the result, in part, of the amounts of tissues available, as well as the necessity to have fresh tissue. The clinical interest in placental hormones and the availability of antibodies to these hormones have abetted this important work.

These placental hormones are glycoproteins, i.e., they are composed of a protein core with branched carbohydrate side chains, which usually terminate with sialic acid. Some of them share common $\alpha$-subunits (Vaitukaitis *et al.,* 1976).

Benagiano *et al.* (1972) demonstrated that fresh human placental tissue from first-trimester pregnancies incubated with [$^{14}$C]leucine incorporated it into soluble protein both in soluble and particulate fractions. The highest radioactive incorporation correlated with the highest concentrations of human chorionic gonadotrophin (HGC) as assayed (immunochemically). Patrito *et al.* (1973) also incubated first-trimester human placenta with [$^{3}$H]leucine and isolated labeled HCG from tissue and incubation medium by use of antiserum to HCG. The amount of radioactive HCG released into the medium was approximately 85% of the total HCG synthesized. Masuo *et al.* (1974), using [$^{3}$H]proline and cultured chorion from first-trimester pregnancies, showed that HCG as well as human chonionic follicle-stimulating hormone (HCFSH) could be synthesized *in vitro*. There was some evidence that biosynthesized HCG had some properties different from those of native HCG.

Synthesis of the $\alpha$-subunit of HCG was demonstrated by Landefeld *et al.* (1976) in a cell-free system made up of polysomes derived from first-trimester placenta and soluble fraction from ascites tumor cells. The level of radioactivity in four tryptic peptides which comigrated with tryptic peptides from authentic HCG was fivefold greater with polysomes from first-trimester placentas than it was with those from term placentas. Since the rate of *total* protein synthesis between first trimester and term placenta was approximately equal, the decrease in HCG blood levels after the first trimester is probably the result of a selective decrease in the rate of synthesis of HCG.

The placenta synthesizes a lactogenic hormone (HPL) as well, which reaches maximal blood levels at about 30 weeks gestation. In a system employing polysomes derived from placenta and cell sap from ascites cell tumor, Boime and Boguslawski (1974) demonstrated synthesis of HPL immunoreactive material specific for placental ribosomes, suggesting that this material is indeed placental HPL, synthesis of which was directed by the corresponding mRNA. The amounts of HPL synthesized in the cell-free system containing term polysomes was fourfold that with the same system containing first-trimester polysomes. The smaller amounts of HPL synthesized in the first-trimester system were not the result of a lower rate of total protein synthesis. Chatterjee *et al.* (1976) also studied the synthesis of HCG and HPL in *in vitro* systems. Membrane-bound polysomes from term placentas were more active in synthesis of both peptide hormones than were the free polysomes. About 8% of nascent peptide chains released by incubation of polysomes from term placentas was accounted for by HPL and about 2% by HCG. In first-trimester human placentas 11% was accounted for by HCG, whereas none was accounted for by HPL. When mRNA was prepared from polysomes obtained from 20- and 40-week gestation placentas, HPL accounted for 0.4 and 2%, respectively, of total protein synthesis, whereas HCG accounted for 8 and 2%, respectively. Thus, the lower blood levels of HCG and the higher blood levels of HPL as term approaches are reflected in the relative *in vitro* rates of synthesis of them by placental polysomes. The abundance of the polysomes is determined by the availability of their respective mRNAs at various times in gestation.

## *5. References*

Abraham, S., 1970, Effect of diet on hepatic fatty acid synthesis, *Am. J. Clin. Nutr.* **23**:1120.

Abramovich, D. R., and Wade, A. P., 1969*a,* Transplacental passage of steroids: The presence of corticosteroids in amniotic fluid, *J. Obstet. Gynaecol. Br. Commonw.* **76**:610.

Abramovich, D. R., and Wade, A. P., 1969*b,* Levels and significance of 17-oxosteroids and 17-hydroxycorticosteroids in amniotic fluid throughout pregnancy, *J. Obstet. Gynaecol. Br. Commonw.* **76**:893.

Abramovich, D. R., Keeping, J. D., and Thom, H., 1975, The origin of amniotic fluid lecithin, *Br. J. Obstet. Gynaecol.* **82**:204.

Andrews, B. F., 1970, Small-for-date infant, *Pediatr. Clin. N. Am.* **17**:30.

Angevine, D. M., 1938, Pathologic anatomy of hypophysis and adrenals in anencephaly, *Arch. Pathol.* **26**:507.

Arinze, I. J., 1975, On the development of phospoenolpyruvate carboxykinase and gluconeogenesis in guinea pig liver, *Biochem. Biophys. Res. Commun.* **65**:184.

Auricchio, S., and Rigillo, N., 1960, Glucose-6-phosphatase activity of the human foetal liver, *Biol. Neonat.* **2**:146.

Avery, M. E., and Said, S., 1965, Surface phenomena in lungs in health and disease, *Medicine* **33**:503.

Avery, M. E., Clow, C. L., Menkes, J. H., Ramos, A., Scriver, C. R., Stern, L., and Wasserman, B. P., 1967, Transient tyrosinemia of the newborn, dietary and clinical aspects, *Pediatrics* **39**:378.

Badham, L. P., and Worth, G. J., 1975, Critical assessment of phospholipid measurement in amniotic fluid, *Clin. Chem.* **21**:1441.

Baird, J. D., 1969, Some aspects of carbohydrate metabolism in pregnancy with special reference to the energy metabolism and hormonal status of the infant of the diabetic woman and the diabetogenic effect of pregnancy, *J. Endocrinol.* **44**:139.

Baker, N., 1969, The use of computers to study rates of lipid metabolism, *J. Lipid Res.* **19**:1.

Bakker, H. D., Wadman, S. K., vanSprang, F. J., van Der Heiden, C., Ketting, D., and DeBree, P. K., 1975, Tyrosinemia and tyrosyluria in healthy prematures: Time courses not vitamin C-dependent, *Clin. Chim. Acta.* **61**:73.

Ballard, F. J., 1971*a*, Regulation of gluconeogenesis during exposure of young rats to hypoxic conditions, *Biochem. J.* **121**:169.

Ballard, F. J., 1971*b*, The development of gluconeogenesis in rat liver, Controlling factors in the newborn, *Biochem. J.* **121**:265.

Ballard, F. J., and Hanson, R. W., 1967, Phosphoenolpyruvate carboxykinase and pyruvate carboxylase in developing rat liver, *Biochem. J.* **104**:866.

Ballard, F. J., and Hanson, R. W., 1969, Purification of phosphoenolpyruvate carboxykinase from cytosol fraction of rat liver and the immunochemical demonstration of differences between this enzyme and the mitochondrial phosphoenolpyruvate carboxykinase, *J. Biol. Chem.* **244**:5623.

Ballard, F. J., and Oliver, I. T., 1963, Glycogen metabolism in embryonic chick and neonatal rat liver, *Biochim. Biophys. Acta* **71**:578.

Ballard, F. J., and Oliver, I. T., 1965, Carbohydrate metabolism in liver from fetal and neonatal sheep, *Biochem. J.* **95**:191.

Ballard, P. L., and Ballard, R. A., 1974, Cytoplasmic receptor glucocorticoids in lung of the human fetus and neonate, *J. Clin. Invest.* **53**:477.

Baquer, N. Z., McLean, P., and Greenbaum, A. L., 1975, Systems relationships and the control of metabolic pathways in developing brain, in: *Normal and Pathological Development of Energy Metabolism* (F. A. Hommes and C. J. van den Berg, eds.), p. 109, Academic Press, London.

Bayer, H., Bonnar, J., Phizackerley, P. J. P., Moore, R. A., and Wylie, F., 1973, Amniotic fluid phospholipids in normal and abnormal pregnancies, *J. Obstet. Gynecol. Brit. Comm.* **80**:333.

Beard, R. W., Turner, R. C., and Oakley, N. W., 1971, An investigation into the control of blood glucose in fetuses of normal and diabetic mothers, in: *Perinatal Medicine* (P. J. Huntingford, R. W. Beard, F. E. Hytten, and J. W. Scopes, eds.), p. 114, S. Karger, New York.

Bearn, J. G., Antonis, A., and Pilkington, T. R. E., 1967, Foetal and maternal plasma triglyceride and cholesterol levels in the rabbit after foetal hypophysectomy by decapitation *in utero*, *J. Endocrinol.* **37**:479.

Beato, M., Kalimi, M., Konetamo, M., and Feigelson, P., 1973, Interaction of glucocorticoids with rat liver nuclei. II. Studies on the nature of the cytosol transfer and the nuclear acceptor site, *Biochemistry* **12**:3372.

Beatty, C. I., and Bocek, R. M., 1970, Metabolism of palmitate by fetal, neonatal, and adult muscle of the rhesus monkey, *Am. J. Physiol.* **219**:1311.

Benagiano, G., Pala, A., Meiriaho, M., and Ermini, M., 1972, Biosynthesis of human chorionic gonadotrophin *in vitro*: Incorporation of [$^{14}$C]-L-leucine, *J. Endocrinol.* **55**:387.

Berger, R., and Hommes, F. A., 1975, Regulation of pyruvate metabolism in fetal rat liver, in: *Normal and Pathological Development of Energy Metabolism* (F. A. Hommes and C. J. van den Berg, eds.), pp. 97–107, Academic Press, London.

Berson, E. L., Hayes, K. C., Rabin, A. R., Schmidt, S. Y., and Watson, G., 1976, Retinal degeneration in cats fed casein. II. Supplementation with methionine, cysteine or taurine, *Invest. Ophthalmol.* **15**:52.

Bessman, S. P., and Bessman, A. N., 1955, The cerebral and peripheral uptake of ammonia in liver disease with an hypothesis for the mechanism of hepatic coma, *J. Clin. Invest.* **32**:622.

Bickel, H., and Souchon, F., 1955, Die Papierchromatographie in der Kinderheilkunde, *Arch. Kinderheilk* **31**(Suppl. 31):42.

Biezenski, J. J., 1969, Role of placenta in fetal lipid metabolism, *Am. J. Obstet. Gynecol.* **104**:1177.

Biezenski, J. J., 1973, Origin of amniotic fluid lipids. III. Fatty acids, *Proc. Soc. Exp. Biol. Med.* **142**:1326.

Biezenski, J. J., 1975, Fetal lipid metabolism, in: *Obstetrics–Gynecology Annual Review* (R. M. Wynn, ed.), Vol. 4, p. 39, Appleton-Century-Croft, New York.

Biezenski, J. J., and Kimmel, B., 1969, Inability of fetal liver to synthesize plasma phospholipids, *Proc. Soc. Exp. Biol. Med.* **130**:1238.

Biezenski, J. J., Pomerance, W., and Goodman, J., 1968, Studies on the origin of amniotic fluid lipids. I. Normal composition, *Am. J. Obstet. Gynecol.* **102**:853.

Billmeier, G. J., Molinary, S. V., Wilroy, R. S., Duenas, D. A., and Brannon, M. E., 1974, Argininosuccinic aciduria: Investigation of an affected family. *J. Pediatr.* **84**:85.

Bjoerkman, N., Hellerstroem, C., and Hellman, B., 1966, Cell types in the endocrine pancreas of the human fetus, *Z. Zellforsch.* **72**:425.

Bocek, R. M., and Beatty, C. H., 1969, Effect of insulin on the carbohydrate metabolism of fetal rhesus monkey muscle, *Endocrinology* **85**:615.

Bocek, R. M., Young, M. K., and Beatty, C. H., 1973, Effect of insulin and epinephrine on the carbohydrate metabolism and adenylate cyclase activity of rhesus fetal muscle, *Pediatr. Res.* **7**:787.

Body, D. R., Shorland, F. B., and Gass, J. P., 1966, Fetal and maternal lipids of Romney sheep, *Biochim. Biophys. Acta* **125**:207.

Boime, I., and Boguslowski, S., 1974, Radioimmunoassay of human placental lactogen synthesized on ribosomes isolated from first trimester and third trimester placenta, *FEBS Lett.* **45**:104.

Borrebaek, B., Hultman. L., Nillson, L. H., Roch-Norlund, A. E., and Spydervold, O., 1970, Adaptable glucokinase activity of human liver, *Biochem. Med.* **4**:469.

Bourne, E. J., McLean, A., and Prodham, J. B., 1966, The structure and deposition of human liver glycogen, *Biochem. J.* **98**:678.

Boxer, J., Kirby, L. T., and Hahn, P., 1974, The response of glucose-6-phosphatase in human and rat liver cultures to dibutyryl-cyclic AMP, *Proc. Soc. Exp. Biol. Med.* **145**:901.

Bremer, J., 1963, Carnitine in intermediary metabolism. The biosynthesis of palmityl carnitine by cell subfractions, *J. Biol. Chem.* **238**:2774.

Brown, J., Miller, D. M., Holloway, M. T., and Lewe, G. D., 1967, Hexokinase isoenzymes in liver and adipose tissue of man and dog, *Science* **155**:205.

Bucher, Th., 1970, The state of the DPN system in liver. An analysis of pyridine nucleotide levels; surface fluorescence and redox potentials of indicator metabolite couples in the hemoglobin-free perfused rat liver, in: *Pyridine Nucleotide Dependent Dehydrogenases* (H. Sund, ed.), p. 439, Springer Verlag, Berlin.

Burch, H. B., Lowry, O. H., Kuhlmans, A. M., Skerjance, J., Diamont, E. J., Lowry, S. R., and Van Dippe, J., 1963, Changes in patterns of enzyme of carbohydrate metabolism in the developing rat liver, *J. Biol. Chem.* **238**:2267.

Cake, M. H., Ghisalbert, A. V., and Oliver, I. T., 1973, Cytoplasmic binding of dexamethason and induction of tyrosine aminotransferase in neonatal rat liver, *Biochem. Biophys. Res. Commun.* **54**:983.

Cammer, W., and Moore, C. L., 1972, Biochemical properties of human fetal mitochondria, *Biol. Neonat.* **21**:259.

Capkova, A., and Jirasch, J. E., 1968, Glycogen reserves in organs of human foetuses in the first half of pregnancy, *Biol. Neonat.* **13**:129.

Cartwright, E. C., Connellan, J. M., and Danks, D. M., 1973, Some properties of phenylalanine hydroxylase in human foetal liver, *Aust. J. Exp. Biol. Med. Sci.* **51**:559.

Caspi, E., Schreyer, I., Schreyer, P., Weinraub, Z., and Tamir, I., 1975, Amniotic fluid volume, total phospholipid concentration, and L/S ratio in term pregnancies, *Obstet. Gynecol.* **46**:584.

Cekan, Z., Juneja, H., and Diczfalusy, E., 1973, De novo synthesis of cholesteryl esters in previable human fetuses, *Biochim. Biophys. Acta* **296**:196.

Chaikoff, I. L., and Robinson, A., 1933, Studies in fetal fat, *J. Biol. Chem.* **100**:13.

Chance, B., and Williams, C. R., 1956, The respiratory chain and oxidative phosphorylation, *Adv. Enzymol.* **17**:65.

Chatterjee, M., Baliga, B. S., and Munro, H. N., 1976, Synthesis of human placental mitogen and human chorionic gonadotrophin by polyribosomes and messenger RNA's from early and full term placentas, *J. Biol. Chem.* **251**:2945.

Chepelinsky, A. B., and De Lores Arnaiz, G. R., 1969, Levels of cytochromes in rat brain mitochondria during post-natal development, *Biochim. Biophys.* **197**:321.

Cheung, G. P., Cotropia, J. P., and Sallach, H. J., 1968, Comparative studies of enzymes related to serine metabolism in fetal and adult liver, *Biochim. Biophys. Acta* **170**:334.

Chez, R. A., Mintz, D. H., and Horger, E. O., 1970, Factors affecting the response to insulin in the normal subhuman pregnant primate, *J. Clin. Invest.* **49**:1517.

Clausen, L., Lou, H., and Anderson, H., 1965, Phospholipids and glycolipid patterns of infant and fetal brain, *J. Neurochem.* **12**:595.

Clements, J. A., Platzker, A. C. G., Tierney, D. F., Hobel, C. J., Creasy, R. K., Margolis, A. J., Thibeault, D. W., Tooley, W. A., and Oh, W., 1972, Assessment of the risk of the respiratory distress syndrome by a rapid test for surfactant in amniotic fluid, *N. Engl. J. Med.* **286**:1077.

Clifford, S. H., 1954, Postmaturity with placental dysfunction: Clinical syndrome and pathologic findings, *J. Pediatr.* **44**:1.

Colombo, J. P., 1971, Congenital disorders of the urea cycle and ammonia detoxication, *Monogr. Paediatr.*

Colombo, J. P., and Richterich, R., 1968, Urea cycle enzymes in the developing human fetus, *Enzym. Biol. Clin.* **9**:68.

Coltart, T. M., 1972, Effect on fetal liver lipids of $^{14}$C glucose administered intravenously to the mother, *J. Obstet. Gynaecol. Br. Commonw.* **79**:639.

Condorelli, S., Cosmi, E. V., and Scarpelli, E. M., 1974, Extrapulmonary source of amniotic fluid phospholipids, *Am. J. Obstet. Gynecol.* **118**:842.

Connor, W. E., and Lin, D. S., 1967, Placental transfer of cholesterol-$^{14}$C into rabbit and guinea pig fetus, *J. Lipid Res.* **8**:558.

Cornblath, M., Gonson, A. F., Nicopoulos, D., Beans, G. S., Hollander, R. J., and Gordon, M. H., 1961, Studies of carbohydrate metabolism in the newborn infant. III. Some factors influencing the capillary blood sugar and the response to glucagon during the first hours of life, *Pediatrics* **27**:378.

Coutts, J. R. T., and MacNaughton, M. C., 1969, Metabolism of [4-$^{14}$C]cholesterol in the previable human fetus, *J. Endocrinol.* **44**:481.

Cremer, J. E., and Teal, H. M., 1974, The activity of pyruvate dehydrogenase in rat brain during postnatal development, *FEBS Lett.* **39**:17.

Cremer, J. E., Teal, H. M., and Heath, D. F., 1975, Regulatory factors in glucose and ketone body utilization by the developing brain, in: *Normal and Pathological Development of Energy Metabolism* (F. A. Hommes and C. J. van den Berg, eds.), p. 133, Academic Press, London.

Crisp, D. M., and Pogson, C. J., 1972, Glycolytic and gluconeogenic enzyme activities in parenchymal and non-parenchymal cells from mouse liver, *Biochem. J.* **126**:1009.

Cuaron, A., Gamble, J., and Myant, N. B., 1963, Effect of thyroid deficiency on the growth of the brain and on the deposition of brain phospholipids in foetal and newborn rabbits, *J. Physiol. (London)* **168**:613.

Cumings, J. N., Goodwin, H., and Woodward, E. M., 1958, Lipids in the brains of infants and children, *J. Neurochem.* **2**:289.

Daalmans-De Lange, M. M., and Hommes, F. A., 1974, The urinary lactate excretion in children, *Helv. Paediatr. Acta* **29**:599.

Dancis, J., Braverman, N., and Lind, J., 1957, Plasma protein synthesis in the human fetus and placenta, *J. Clin. Invest.* **36**:398.

Dancis, J., Jansen, V., Kayden, H. J., Schneider, H., and Levitz, M., 1973, Transfer across perfused human placenta. II. Free fatty acids, *Pediatr. Res.* **7**:192.

Dawkins, M. J. R., 1961, Changes in glucose-6-phosphatase activity in liver and kidney at birth, *Nature* **191**:72.

Del Valle, J. A., and Greengard, O., 1976, Phenylalanine hydroxylase and tyrosine aminotransferase in human fetal and adult liver, *Pediatr. Res.* **11**:2.

De Vos, M. A., Wilmink, C. W., and Hommes, F. A., 1968, Development of some mitochondrial oxidase systems of rat liver, *Biol. Neonat.* **13**:83.

De Vos, P., and Hers, H. G., 1974, Glycogen metabolism in the foetal rat, *Biochem. J.* **140**:331.

De Wulf, H., and Hers, H. G., 1967, The stimulation of glycogen synthesis and of glycogen synthetase in the liver by glucocorticoids, *Eur. J. Biochem.* **2**:57.

Dhopeshwarkar, G. A., Subramanian, C., and Mead, J. F., 1973, Metabolism of 1-$^{14}$C palmitic acid in the developing brain: Persistence of radioactivity in the carboxylcarbon, *Biochim. Biophys. Acta* **296**:257.

Diamant, Y. Z., Mayorek, N., Neuman, S., and Shafrir, E., 1975, Enzymes of glucose and fatty acid metabolism in early and term human placenta, *Am. J. Obstet. Gynecol.* **121**:58.

Diesterhaft, M. D., Shrago, E., and Sallach, H. J., 1971, Human liver phosphoenolpyruvate carboxykinase. Evidence for a separate mitochondrial and cytosol enzyme, *Biochem. Med.* **5**:297.

Dobbing, J., 1974, The later development of the brain and its vulnerability, in: *Scientific Foundations of Pediatrics* (J. A. Davis and J. Dobbing, eds.), p. 565, Saunders, Philadelphia.

Ekelund, L., Arvidson, G., and Astedt, B., 1973, Amniotic fluid lecithin and its fatty acid composition in respiratory distress syndrome, *J. Obstet. Gynaecol. Brit. Commonw.* **80**:912.

Elphick, M. C., Hudson, D. G., and Hull, D., 1975, Transfer of fatty acids across the rabbit placenta, *J. Physiol.* **252**:29.

Eisen, H. J., Goldfine, J. D., and Glinsmann, W. H., 1973, Regulation of hepatic glycogen synthesis during fetal development: Roles of hydrocortisone, insulin receptors, *Proc. Natl. Acad. Sci. U.S.A.* **70**:3454.

Encrantz, J. C., and Sjovall, J., 1959, On the bile acids in duodenal contents of infants and children. Bile acids and steroids, *Clin. Chim. Acta* **4**:793.

Espinoza, A., Driscoll, S. G., and Steinke, J., 1970, Insulin release from isolated human fetal pancreatic islets, *Science* **168**:1111.

Exton, J. H., Corbin, J. G., and Park, C. R., 1966, Control of gluconeogenesis in liver. II. Differential effects of fatty acids and glucagon on ketogenesis and gluconeogenesis in the perfused rat liver, *J. Biol. Chem.* **244**:4095.

Exton, J. H., Ui, M., Lewis, S. B., and Park, C. R., 1971, Mechanism of glucagon activation of gluconeogenesis, in: *Regulation of Gluconeogenesis* (H. D. Söling and B. Willms, eds.), p. 160, Thieme Verlag, Stuttgart.

Fain, J. N., and Scow, R. O., 1966, Fatty acid synthesis *in vivo* in maternal and fetal tissue of the rat, *Am. J. Physiol.* **210**:19.

Farrell, P. M., and Morgan, T. E., 1975, Lecithin biosynthesis in the developing lung, in: *The Development of the Lung* (W. A., Hodson, ed.), p. 309, Dekker, New York.

Fehling, I. H., 1877, Beiträge zur Physiologie des placentären Stoffverkehrs, *Arch. Gynäkol.* **11**:523.

Felt, J. M., 1965, Carbohydrate and lipid metabolism of lung tissue *in vitro, Med. Thorac.* **22**:89.

Fenichel, G. M., 1966, A histochemical study of developing skeletal muscle, *Neurology* **16**:741.

Fernandes, J., and Blom, W., 1974, The intravenous L-alanine tolerance test as a means for investigating gluconeogenesis, *Metabolism* **23**:1149.

Fernandes, J., and Blom, W., 1976, Urinary lactate excretion in normal children and in children with enzyme defects of carbohydrate metabolism, *Clin. Chim. Acta* **66**:345.

Fine, J. H., Kaplan, N. O., and Kuftinec, D., 1963, Developmental changes of mammalian lactic dehydrogenases, *Biochemistry* **2**:116.

Finley, T. N., Tooley, W. H., and Swenson, W. E., 1964, Pulmonary surface tension in experimental atelectasis, *Am. Rev. Respir. Dis.* **89**:372.

Fleisher, L. D., Rassin, D. K., Rogers, P., Desnick, R. J., and Gaull, G. E., 1977, Argininosuccinic aciduria: Prenatal diagnosis and studies of an affected fetus, *Pediatr. Res.* **11**:455.

Fomon, S. J., 1967, *Infant Nutrition,* Saunders, Philadelphia.

Foulkner, A., and Jones, C. F., 1975, Pyruvate kinase isoenzymes of the human fetus, *FEBS Lett.* **53**:167.

Francesconi, R. P., and Villee, A. C., 1969, Lactate dehydrogenase isoenzymes in human and rat fetal liver and lung, *Life Sci.* **8**:33.

Freeman, R. K., Bateman, B. G., Goebelsman, U., Arce, J. J., and James, J., 1974, Clinical experience with the amniotic fluid lecithin/sphingomyelin ratio, *Am. J. Obstet. Gynecol.* **119**:239.

Friedman, P. A., and Kaufman, S., 1971, A study of the development of phenylalanine hydroxylase in fetuses of several mammalian species, *Arch. Biochem. Biophys.* **146**:321.

Friedman, P. A., and Kaufman, S., 1973, Some characteristics of partially purified human liver phenylalanine hydroxylase, *Biochim. Biophys. Acta* **293**:56.

Fritz, I. B., Kapec, B., and Brosnan, J. T., 1974, Localization of carnitine palmitoyl transferases on inner membrane of mitochondria and their possible role in the regulation of fatty acyl group translocation, in: *Regulation of Hepatic Metabolism* (F. Lundquist and N. Tygstrup, eds.), p. 482, Munksgaard, Copenhagen.

Frölich, J., and Wieland, O., 1971, Dissociation of gluconeogenic and ketogenic action of glucagon in the perfused rat liver, in: *Regulation of Gluconeogenesis* (H. D. Söling and B. Willms, eds.), p. 179, Thieme Verlag, Stuttgart.

Fujiwara, T., Adams, F. H., and Sipos, S., 1968, Alveolar and whole lung phospholipids of the developing fetal lamb lung, *Am. J. Physiol.* **215**:375.

Garber, A. J., and Hanson, R. W., 1971, The interrelationships of the various pathways forming gluconeogenic precursors in guinea pig liver in mitochondria, *J. Biol. Chem.* **246**:589.

Gaull, G. E., 1973*a*, Sulfur amino acids, folate, and DNA: Metabolic interrelationships during fetal development, *Proceedings of the Sir Joseph Barcroft Centenary Symposium,* Cambridge Univ. Press, p. 339.

Gaull, G. E., 1973*b*, Interrelationships of sulfur amino acids, folate and DNA in human brain development, in: *Symposium on Developmental Biochemistry—Inborn Errors of Metabolism* (F. Hommes and C. J. van den Berg, eds.), p. 131, Academic Press, New York.

Gaull, G. E., 1975, Methionine adenosyltransferase: Development and deficiency in the human, in: *Symposium on Normal Development & Pathology of Energy Metabolism in Inborn Errors of Metabolism* (F. Hommes and C. J. van den Berg, eds.), p. 11, Academic Press, New York.

Gaull, G. E., Sturman, J. A., and Räihä, N. C. R., 1972, Development of mammalian sulphur metabolism: Absence of cystathionase in human fetal tissues, *Pediatr. Res.* **6**:538.

Gaull, G. E., Räihä, N. C. R., Saarikoski, S., and Sturman, J. A., 1973*a*, Transfer of cyst(e)ine and methionine across the human placenta, *Pediatr. Res.* **7**:908.

Gaull, G. E., Van Berg, W., Räihä, N. C. R., and Sturman, J. A., 1973*b*, Development of methyltransferase activities of human fetal tissues, *Pediatr. Res.* **12**:527.

Gaull, G. E., Tallan, H. H., Rassin, D. K., and Lajtha, A., 1975, Pathogenesis of brain dysfunction in disorders of amino acid metabolism, in: *Biology of Brain Dysfunction* (G. E. Gaull, ed.), Vol. III, p. 47, Plenum Press, New York.

Gaull, G. E., Rassin, D. K., Räihä, N. C. R., and Heinonen, K., 1977, Milk protein quantity and quality in low-birth-weight infants: III. Effects on sulfur-containing amino acids in plasma and urine, *J. Pediatr.* **90**:348.

Gennser, G., Lundquist, I., and Nillson, E., 1971, Glycogenolytic activity in the liver of the human foetus, *Biol. Neonate* **19**:1.

Ghadimi, H., and Pecora, P., 1964, Free amino acids of cord blood as compared with maternal plasma during pregnancy, *Pediatrics* **33**:500.

Giannopoulos, G., 1973, Glucocorticoid receptors in lung. I. Specific binding of glucocorticoid to cytoplasmic components of rabbit fetal lung, *J. Biol. Chem.* **248**:3876.

Giannopoulos, G., Mulay, S., and Solomon, S., 1973, I. Glucocorticoid receptors in lung, II. Specific binding of corticoids to nuclear components of rabbit fetal lung, *J. Biol. Chem.* **248**:5016.

Gitlin, D., and Gitlin, J. D., 1975, Fetal and neonatal development of human plasma proteins, in: *The Plasma Proteins; Structure, Function, and Genetic Control,* 2nd. ed. (F. W. Putnam, ed.), Vol. II, p. 264, Academic Press, New York.

Gitlin, D., Perricelli, A., and Gitlin, G. M., 1972, Synthesis of α-fetoprotein by liver, yolk sac, and gastrointestinal tract of the human conceptus, *Cancer Res.* **32**:979.

Glendening, M. D., Margolis, A. J., and Page, E. W., 1961, Amino acid concentrations in fetal and maternal plasma, *Am. J. Obstet. Gynecol.* **81**:591.

Gluck, L., and Kulovich, M. V., 1973, Lecithin/Sphingomyelin ratios in amniotic fluid in normal and abnormal pregnancy, *Am. J. Obstet. Gynecol.* **113**:539.

Gluck, L., Kulovich, M. V., Borer, R. C. Brenner, P. H., Anderson, G. G., and Spellacy, W. N., 1971, Diagnosis of the respiratory distress syndrome by amniocentesis, *Am. J. Obstet. Gynecol.* **109**:440.

Gluck, L., Chez, R. H., Kulovich, M. V., Hutchison, D. L., and Niemann, W. H., 1974, Comparison of phospholipid indicators of fetal lung maturity in the amniotic fluid of the monkey and baboon, *Am. J. Obstet. Gynecol.* **120**:524.

Goldman, H. I., Goldman, J. S., Kaufman, I., and Liebman, O. B., 1974, Late effects of early dietary protein intake on low-birth-weight infants, *J. Pediatr.* **85**:764.

Goodman, S. I., Mace, J. W., Turner, B., and Garrett, W. J., 1973, Antenatal diagnosis of argininosuccinic aciduria, *Clin. Genet.* **4**:236.

Goswami, M. N., and Knox, W. E., 1961, Developmental changes of *p*-hydroxypyruvate-oxidase activity in mammalian liver, *Biochim. Biophys. Acta* **50**:35.

Greene, H. L., Stifel, F. B., and Herman, R. H., 1972, Ketotic hypoglycemia due to hepatic fructose-1,6-diphosphatase deficiency, *Am. J. Dis. Child.* **124**:415.

Greene, H. L., Taunton, O. D., Stifel, F. B., and Herman, R. N., 1974, The rapid changes of hepatic glycolytic enzymes and fructose-1,6-disphosphatase activities after intravenous glucagon in humans, *J. Clin. Invest.* **53**:44.

Greengard, O., 1970, The developmental formation of enzymes in rat liver, in: *Biochemical Actions of Hormones* (G. Litwack, ed.), p. 53, Academic Press, New York.

Greengard, O., 1971, Enzymic differentiation in mammalian tissues, *Essays Biochem.* **7**:159.

Greengard, O., 1975, Hormonal regulation of enzyme synthesis in differentiating mammalian tissues, in: *Normal and Pathological Development of Energy Matabolism* (F. A. Hommes and C. J. van den Berg, eds.), p. 55, Academic Press, London.

Greengard, O., 1977, Enzymic differentiation of human liver: Comparison with the rat model, *Pediatr. Res.* **11**:669.

Greengard, O., and Jamdar, S. C., 1971, The prematurely promoted formations of liver enzymes in suckling rats, *Biochim. Biophys. Acta* **237**:476.

Greengard, O., Federman, M., and Knox, W. E., 1972, Cytomorphometry of developing rat liver and its application to enzymic differentiation, *J. Cell Biol.* **52**:261.

Gross, I., Rooney, S. A., and Warshaw, J. B., 1975, The influence of cortisol on the enzymes of fatty acid synthesis in developing mammalian lung and brain, *Pediatr. Res.* **9**:752.

Grumbach, M. M., Kaplan, S. L., and Sciarra, J. J., 1968, Chorionic growth hormone prolactin (CGP) secretion, disposition, biologic activity in man and postulated function as the growth hormone of the second half of pregnancy, *Ann. N.Y. Acad. Sci.* **148**:501.

Hahn, P., and Novak, M., 1975, Development of brown and white adipose tissue, *J. Lipid Res.* **16**:79.

Hahn, P., and Skala, J., 1972*a*, Carnitine and heat production in developing rats, *Fed. Proc.* **31**:224.

Hahn, P., and Skala, J., 1972*b*, Carnitine and brown adipose tissue metabolism in the rat during development, *Biochem, J.* **127**:107.

Hahn, P., and Skala, J., 1973, Carnitine transferase in human fetal tissues, *Biol. Neonat.* **22**:9.

Hallman, M., 1971, Changes in mitochondrial respiratory chain proteins during perinatal development. Evidence of the importance of environmental oxygen tension, *Biochim. Biophys. Acta* **253**:360.

Hansen, I. B., and Clausen, J., 1968, Fatty acids of the human fetal brain, *Scand. J. Clin. Lab. Invest.* **22**:231.

Harding, P., Possmayer, F., Milne, K., Jaco, N. T., and Walters J. H., 1973, Amniotic fluid phospholipids and fetal maturity, *Am. J. Obstet. Gynecol.* **115**:298.

Harding, P. G. R., 1970, The metabolism of brown and white adipose tissue in the fetus and newborn, *Clin. Obstet Gynecol.* **13**:685.

Harding, P. G. R. and Ralph, E. D., 1970, Effects of chronic hypoxia on lipolysis in brown adipose tissue in the fetal rabbit, *Am. J. Obstet. Gynecol.* **106**:907.

Haworth, J. C., and Ford, J. D., 1963, Variation of the oral galactose test with age, *J. Pediatr.* **63**:276.

Hayes, K. C., Carey, R. E., and Schmidt, S. Y., 1975*a*, Retinal degeneration associated with taurine deficiency in the cat, *Science* **188**:949.

Hayes, K. C., Rabin, A. R., and Berson, E. L., 1975*b*, An ultrastructural study of nutritionally induced and reversed retinal degeneration in cats, *Am. J. Pathol.* **78**:505.

Haymond, M. W., Karl, I. E., and Pagliari, A. S., 1974, Increased gluconeogenesis substrates in the small-for-gestational-age infant, *N. Engl. J. Med.* **291**:322.

Heath, D. F., and Barton, R. N., 1973, The design of experiments using isotopes for the determination of the rates of disposal of blood-borne substrates, *in vivo* with special reference to glucose, ketone bodies, free fatty acids and proteins, *Biochem. J.* **136**:503.

Heinonen, K., 1973, Studies on cystathionase activity in rat liver and brain during development, *Biochem. J.,* **136**:1011.

Heinonen, K., 1974, The uptake of labelled sulphur by the proteins of human liver slices incubated with L-$^{35}$S-methionine or $^{35}$S-sulfate, *Life Sci.* **15**:463.

Heinonen, K., and Räihä, N. C. R., 1974, Induction of cystathionase in human fetal liver, *Biochem. J.* **144**:607.

Heinrich, R., and Rapoport, T. A., 1974, A linear steady state treatment of enzymatic chains. General properties, control and effector strength, *Eur. J. Biochem.* **42**:89.

Hemon, P., 1967, L'activité α-glycérophosphate oxidase du foie de rat au cours du développement, *Biochim. Biophys. Acta* **132**:175.

Hemon, P., 1968, Malate dehydrogenase (decarboxylating NADP) and α-glycerophosphate oxidase in the developing rat, *Biochim. Biophys. Acta* **151**:681.

Hemon, P., and Berbey, P., 1968, Changes of enzyme activities with diet and thyroxine during postnatal development of the rat, *Biochim. Biophys. Acta* **170**:235.

Hennig, H. V., Stumpf, B., Okly, B., and Seubert, W., 1968, On the mechanism of gluconeogenesis. III. The glucogenic capacity and the activities of pyruvate carboxylase and PEP-carboxylase of rat kidney and rat liver after cortisol treatment and starvation, *Biochem. Z.* **344**:274.

Hershfield, M. S., and Nemeth, A. M., 1968, Placental transport of free palmitic and linoleic acids in the guinea pig, *J. Lipid Res.* **9**:461.

Herzfeld, A., and Greengard, O., 1971, Aspartate aminotransferase in rat tissues: Changes with growth and hormones, *Biochim. Biophys. Acta* **237**:88.

Herzfeld, A., Rosenoer, V. M., and Raper, S. M., 1976, Glutamine dehydrogenase, alanine aminotransferase, thymidine kinase and arginase in fetal and adult human and rat liver, *Pediatr. Res.* **10**:960.

Hirsch, J., Farquhar, J. W., and Ahrens, E. M., 1960, Studies of adipose tissue in man, *Am. J. Clin. Nutr.* **8**:499.

Holmes, E. W., Jr., Malone, J. J., Winegard, A. I., and Oski, F. A., 1967, Hexokinase isoenzymes in human erythrocytes: Association of type II with fetal hemoglobin, *Science* **156**:646.

Holt, P. G., and Oliver, I. T., 1968, Plasma corticosterone concentration in the perinatal rat, *Biochem. J.* **108**:339.

Holtzer, H., Weintraub, H., Mayne, R., and Mochan, B., 1972, The cell cycle, cell lineages and cell differentiation, *Curr. Top. Dev. Biol.* **7**:229.

Holtzman, D., and Moore, C., 1971, A micromethod for the study of oxidative phosphorylation, *Biochim. Biophys. Acta* **234**:1.

Holtzman, D., and Moore, C., 1973, Oxidative phosphorylation in mature rat brain mitochondria, *Biol. Neonate* **22**:230.

Hommes, F. A., 1970, Development of Krebs cycle enzyme in rat liver mitochondria, in: *Metabolic Pathways in Mammalian Embryos During Organogenesis and its Modification by Drugs* (R. Bass, F. Beck, H. J. Merker, D. Neubert, and R. Randhahn, eds.), p. 325, Freie Universität, Berlin.

Hommes, F. A., 1971, Development of enzyme systems in glycolysis, in: *Metabolic Processes in the Foetus and Newborn Infant* (J. H. P. Jonxis H. K. Visser, and J. A. Troelstra, eds.), p. 3, Stenfert Kroese, Leiden.

Hommes, F. A., 1974, unpublished observations.

Hommes, F. A., 1975, Energetic aspects of late fetal and neonatal metabolism, in *Normal and Pathological Development of Energy Metabolism* (F. A. Homes and C. J. van den Berg, eds.), p. 1, Academic Press, London.

Hommes, F. A., and Beere, A., 1971, The development of adenyl cyclase in rat liver, kidney, brain and skeletal muscle, *Biochim. Biophys. Acta* **237**:296.

Hommes, F. A., and Berger, R., 1974, The control of glycolysis in isolated rat liver cells, *Z. Physiol. Chem.* **355**:1209.

Hommes, F. A., and Draisma, M. I., 1970, The development of L- and M-type aldolase in rat liver, *Biochim. Biophys. Acta* **222**:251.

Hommes, F. A., and Richters, A. R., 1969, Mechanism of oxidation of cytoplasmic reduced nicotinamide adrenine dinucleotide in the developing rat liver, *Biol. Neonate* **14**:359.

Hommes, F. A., and Wilmink, C. W., 1968, Developmental changes of glycolytic enzymes in rat brain, liver and skeletal muscle, *Biol. Neonat.* **12**:181.

Hommes, F. A., Luit-De Haan, G., and Richters, A. R., 1971, The development of some Krebs cycle enzymes in rat liver mitochondria, *Biol. Neonat.* **17**:15.

Hommes, F. A., Bendien, K., Elema, J., Bremer, H. J., and Lombeck, I., 1976, Two cases of phosphoenolpyruvate carboxykinase deficiency, *Acta Pediatr. Scand.* **65**:233.

Hull, D., 1969, Nutrition and temperature control in the newborn baby, *Proc. Nutr. Soc.* **28**:56.

Hull, D., 1975, Storage and supply of fatty acids before and after birth, *Br. Med. Bull.* **31**:32.

Hummel, L., Schirrmeister, W., and Wagner, H., 1975, Quantitative evaluation of the maternal–fetal transfer of free fatty acids in the rat, *Biol. Neonat.* **26**:263.

Hutchinson, D. L., Kelly, W. T., Friedmann, E. A., and Plentl, A. A., 1962, The distribution and metabolism of carbon labelled urea in pregnant primates, *J. Clin. Invest.* **41**:1745.

Huxtable, R., and Barbeau, A., eds., 1976, *Taurine,* Raven Press, New York.

Ingoglia, N. A., and Sturman, J. A., 1978, The axonal transport of putrescine, spermidine and spermine in the goldfish visual system: Speculation on the association of axonally transported spermidine and tRNA in regenerating optic nerves, in: *Advances in Polyamines* (R. A. Campbell, ed.), Raven Press, New York, **2**:169.

Jack, P. M. B., and Milner, R. D. G., 1975, Effect of decapitation and ACTH on somatic development of the rabbit fetus, *Biol. Neonate* **26**:195.

Jacobsen, J. G., and Smith, L. H., Jr., 1968, Biochemistry and physiology of taurine and taurine derivatives, *Physiol. Rev.* **48**:429.

Jacquot, R. L., 1959, Rechèrches sur le contrôle endocrinien de l'accumulation du glycogène dans le foie chez le foetus de rat, *J. Physiol. (Paris)* **51**:655.

Jacquot, R. L., and Kretchmer, N., 1964, Effect of decapitation on enzymes of glycogen metabolism, *J. Biol. Chem.* **239**:1301.

Jakovcic, S., Haddock, J., Getz, G. S., Rabinowitz, M., and Swift, H., Mitochondrial development in liver of foetal and newborn rats, *Biochem. J.* **121**:341.

Jakubovic, A., 1971, Phenylalanine-hydroxylating system in the human fetus at different developmental ages, *Biochim. Biophys. Acta* **237**:469.

Jamdar, S. C., and Greengard, O., 1970, Premature formation of glucokinase in developing rat liver, *J. Biol. Chem.* **245**:2779.

Jimenez, J. M., Schultz, M. F., and Johnston, J. M., 1975, III. Amniotic fluid phosphatidic acid phosphohydrolase (papase) and its relation to the lecithin/sphingomyelin ratio. *Obstet. Gynocol.* **46**:588.

Joassin, G., Parker, M. C., and Pildes, R. S., 1967, Infants of diabetic mothers, *Diabetes* **16**:306.

Jost, A., 1966, Problems of fetal endocrinology: The adrenal glands, *Rec. Progr. Horm. Res.* **22**:541.

Kapeller-Adler, R., and Hammad, W. A., 1972, A biochemical study on nucleic acids and protein synthesis in the human fetus and its correlation with relevant embryological data, *J. Obstet. Gynaecd. Br. Commonw.* **79**:924.

Karmen, A., 1963, Biological implications of gas chromatography, *Science* **142**:163.

Karp, W., Sprecher, H., and Robertson, A., 1971, Palmityl-CoA; Carnitine palmityl transferase identification in human placental tissue, *Biol. Neonate* **18**:341.

Kaufman, S., 1969, Phenylalanine hydroxylase of human liver: Assay and some properties, *Arch. Biochem. Biophys.* **134**:249.

Kayden, H. J., Dancis, J., and Money, W. L., 1969, Transfer of lipids across the guinea pig placenta, *Am. J. Obstet. Gynecol.* **104**:564.

Kekomäki, M. P., Räihä, N. C. R., and Bickel, H., 1969, Ornithine ketoacid aminotransferase in human liver with reference to patients with hyperornithinemia and familial protein intolerance, *Clinc. Chim. Acta* **23**:203.

Kekomäki, M., Seppala, M., Ehnholm, C., Schwartz, A. L., and Raivio, K., 1971, Perfusion of isolated human fetal liver: Synthesis and release of α-fetoprotein and albumin, *Int. J. Cancer* **8**:250.

Kannan, A. L., and Cohen, P. P., 1959, Biochemical studies of the developing mammalian fetus. I. Urea cycle enzymes, *Dev. Biol.* **1**:511.

Kenney, F. T., Reem, G. H., and Kretchmer, N., 1958, Development of phenylalanine hydroxylase in mammalian liver, *Science* **127**:85.

Kerr, G. R., and Helmuth, A. C., 1975, Growth and development of the fetal rhesus monkey, *Biol. Neonate* **25**:10.

Kerr, G. R., Helmuth, A. C., and Waisman, H. A., 1973, Growth and development of the fetal rhesus monkey. IV. Fractional lipid analyses of the developing brain, *Growth* **37**:41.

Kim, Y. J., and Felig, P., 1972, Maternal and amniotic fluid substrate levels during caloric deprivation in human pregnancy, *Metabolism* **21**:507.

King, K. C., Adams, P. A., and Laskowski, D. E., 1971*a*, Sources of fatty acids in the newborn, *Pediatrics* **47**(suppl. 2):192.

King, K. C., Butt, E., Jr., Raivio, K., Räihä, N., Roux, J., Teramo, K., Wamaguchi, K., and Schwartz, R., 1971*b*, Human maternal and fetal insulin response to arginine, *N. Engl. J. Med.* **285**:607.

Kirby, L., and Hahn, P., 1973, Enzyme induction in human fetal liver, *Pediatr. Res.* **7**:75.

Kling, R. O., and Kotas, R. V., 1975, Endocrine influences on pulmonary maturation and the lecithin/sphingomyelin ratio in the fetal baboon, *Am. J. Obstet. Gynecol.* **121**:664.

Klingenberg, M., 1970, Metabolite transport in mitochondria: An example for intracellular membrane function, *Essays Biochem.* **6**:119.

Knopf, K., Sturman, J. A., Armstrong, M., and Hayes, K. C., 1978, Taurine: A new essential amino acid for the cat, *J. Nutr.*, in press.

Knowles, S. E., and Ballard, F. J., 1974, Pyruvate dehydrogenase activity in rat liver during development, *Biol. Neonat.* **24**:41.

Knox, W. E., Auerbach, V. H., and Lin, E. C. C., 1956, Enzymatic and metabolic adaptation in animals, *Physiol. Rev.* **36**:164.

Kraan, G. P. B., and Dias, T., 1975, Size of ʟ-lactate transport from the fetal rat to the mother animal, *Biol. Neonat.* **26**:9.

Kraus, H., Schlenker, S., and Schwedesky, D., 1974, Developmental changes of cerebral ketone body utilization in human infants, *Z. Physiol. Chem.* **355**:164.

Kretchmer, N., Levine, S. A., and McNamara, H., 1957, The *in vitro* metabolism of tyrosine and its intermediates in the liver of the premature infant, *Am. J. Dis. Child.* **93**:19.

Kretchmer, N., Levine, S. Z., McNamara, H., and Barnett, H. L., 1956, Certain aspects of tyrosine metabolism in the young. I. The development of the tyrosine oxidizing system in human liver, *J. Clin. Invest.* **35**:236.

LaDu, B. N., and Gjessing, L. R., 1972, Tyrosinosis and Tyrosinemia, in: *The Metabolic Basis of Inherited Disease,* 3rd ed. (J. B. Stanbury, J. B. Wyngaarden, and D. S. Fredrickson, eds.), p. 296, McGraw-Hill, New York.

Land, J. M., and Clark, J. B., 1975. The changing pattern of brain mitochondrial substrate utilization during development, in: *Normal and Pathological Development of Energy Metabolism* (F. A. Hommes and C. J. van den Berg, eds.), p. 155, Academic Press, London.

Landefeld, T., Boguslowski, S., Corash, L., and Boime, I., 1976, The cell-free synthesis of the alpha subunit of human chorionic gonadotrophin, *Endocrinology* **98**:1220.

Lardy, H. A., Zahlten, R. N., Stratman, F. W., and Cook, D. E., 1974, Regulation of hepatic gluconeogenesis by glucagon, in: *Regulation of Hepatic Metabolism* (F. Lundquist and N. Tygstrupp, eds.), p. 19, Munksgaard, Copenhagen.

Lehninger, A. L., 1970, *Biochemistry,* p. 189, Worth Publishers, New York.

Levine, S., Marples, E., and Gordon, H., 1941, A defect in the metabolism of tyrosine and phenylalanine in premature infants: 1. Identification and assay of intermediary products, *J. Clin. Invest.* **20**:199.

Lichtenstein, A., 1931, Untersuchungen an Nabelschnurblut bei frühgeborenen und ausgetragenen Kindern mit besonderer Berücksichtigung der Amino-säuren, *Z. Kinderheilkd* **51**:748.

Lindblad, B. A., 1971, The plasma aminogram in "small for dates" newborn infants, in: *Metabolic*

*Processes in the Foetus and Newborn Infant* (J. H. P. Jonxis, H. K. A. Visser, and J. A. Travoelstra, eds.), p. 109, H. E. Stenfert Knoese, N. V., Leiden.

Lindros, K. O., and Räihä, N. C. R., 1968, Activity of enzymes involved in gluconeogenesis in developing human liver, *Scand. J. Clin. Lab. Invest.* **21**(Suppl. 101):6.

Linn, T. C., Pettit, F. N., and Reed, J. J., 1969*a*, α-Keto acid dehydrogenase complexes, X. Comparative studies of regulatory complex from kidney, heart and liver mitochondria, *Proc. Natl. Acad. Sci. U.S.A.* **64**:227.

Linn, T. C., Pettit, F. H., and Reed, J. J., 1969*b*, α-Keto acid dehydrogenase complexes. XI. Regulation of the activity of the pyruvate dehydrogenease complex from beef kidney mitochondria by phosphorylation and dephosphorylation, *Proc. Natl. Acad. Sci. U.S.A.* **64**:234.

Lowry, O. H., 1971, Changes in patterns of enzymes of carbohydrate metabolism in the developing rat kidney, *Pediatrics* **47**:199.

Luit, H., Berger, R., and Hommes, F. A., 1975, The fatty acid composition of some cellular membranes of fetal rat liver, *Biol. Neonate* **26**:1.

Macaione, S., Ruggeri, P., De Luca, F., and Tucci, G., 1974, Free amino acids in developing rat retina, *J. Neurochem* **22**:887.

MacDonnell, P. C., and Greengard, O., 1974, Enzymes in intracellular organelles of adult and developing rat brain, *Arch. Biochem. Biophys.* **163**:644.

MacDonnell, P. C., Ryder, E., Del Valle, J. A., and Greengard, O., 1975, Biochemical changes in cultured foetal rat liver explants, *Biochem. J.* **159**:269.

Mackler, B., Grace, R., and Duncan, H. M., 1971, Studies of mitochondrial development during embryogenesis in the rat, *Arch. Biochem. Biophys.* **144**:603.

Mackler, B., Grace, R., Haynes, B., Bargman, G. J., and Shepard, T.H., 1973, Studies of mitochondrial energy systems during embryogenesis in the rat, *Arch. Biochem. Biophys.* **158**:662.

Mamunes, P., Prince, P. E., Thornton, N. H., Hunt, P. S., and Hitchcock, E. S., 1976, Intellectual deficits after transient tyrosinemia in the term neonate, *Pediatrics* **57**:675.

Manners, D. J. M., Schutt, W. H., Stark, J. R., and Thambyraja, V., 1971, Studies on the structure and metabolism of glycogen in some abnormal human infants, *Biochem. J.* **124**:461.

Margreth, A., Di Mauro, S., Tortorini, A., and Salviati, G., 1970, Glycogen synthetase in developing and adult skeletal muscle, *Biochem. J.* **122**:597.

Masuo, T., Ashitaka, Y., Mochizuki, M., and Tojo, S., 1974, Chorionic gonadotrophin synthesized in cultured trophoblast, *Endocrinol. J.* **21**:499.

Mathews, J., and Partington, N. W., 1964, The plasma tyrosine levels of premature babies, *Arch. Dis. Child.* **39**:271.

Maugh, T. H., II, Diabetes. III. New hormones promise more effective therapy, *Science* **188**:920.

McGee, M. M., Greengard, O., and Knox, W. E., 1972, The quantitative determination of phenylalanine hydroxylase in rat tissues, *Biochem. J.* **127**:669.

McLean, A., Marwick, M. J., and Clayton, B. E., 1973, Enzymes involved in phenylalanine metabolism in the human foetus and child, *J. Clin. Patho.* **26**:678.

McMillin Wood, J., 1975, Carnitine palmityl transferase in neonatal and adult heart and liver mitochondria. Effect of phospholipase C treatment, *J. Biol. Chem.* **250**:3062.

Mela, L., Goodwin, C. W., and Miller, L. D., 1975, Correlation of mitochondrial cytochrome concentration and activity to oxygen availability in the newborn, *Biochem. Biophys. Res. Commun.* **64**:384.

Menkes, J. H., Welcher, D. W., Leir, H. S., Dallas, J., and Gretsky, N. E., 1972, Relationship of elevated blood tyrosine to the ultimate intellectual performance of premature infants, *Pediatrics* **49**:218.

Merola, J. G. L., Johnson, L. M., Bolognese, R. J., and Corson, S. L., 1974, Determination of fetal pulmonary maturity by amniotic fluid lecithin/sphingomyelin ratio and rapid shake test, *Am. J. Obstet. Gynecol.* **119**:243.

Mersmann, H. J., Goodman, J., Hook, J. M., and Anderson, S., 1972, Studies on the biochemistry of mitochondria and cell morphology in the neonatal swine hepatocyte, *J. Cell Biol.* **53**:335.

Mestyan, J., Schultz, K., and Horvath, M., 1974, Comparative glycemic response to alanine in normal term and small-for-gestational-age infants, *J. Pediatr.* **85**:276.

Middleton, A. C., and Walker, D. G., 1972, Comparison of the properties of two forms of pyruvate kinase in rat liver and determination of their separate activities during development, *Biochem. J.* **127**:721.

Mintz, D. H., Chez, R. A., and Horger, E. O., 1969, III. Fetal insulin and growth hormone metabolism in the subhuman primate, *J. Clin. Invest.* **48**:176.

Mohanram, M., and Kumar, A., 1975, Ascorbic acid and tyrosine metabolism in preterm and small-for-dates infants, *Arch. Dis. Child.* **50**:235.

Mori, M., Study of protein biosynthesis in fetus and placenta. I. Incorporation of [14]C-amino acids into the human placenta, *Am. J. Obstet. Gynecol.* **93**:1164.

Morikawa, Y., and Eguchi, Y., 1966, Perinatal changes of the amount of liver fat in the rat, *Endocrinol. Jpn.* **13**:184.

Morikawa, Y., Eguchi, Y., and Hashimoto, Y., 1965, Sudden increase of fat in the liver of mice at birth, *Nature* **206**:1368.

Morrison, G. R., 1967, Hexokinase and glucokinase activities in bile duct epithelial cells and hepatic cells from normal rat and human liver, *Arch. Biochem. Biophys.* **122**:569.

Morse, D. E., and Horecker, B. L., 1968, The mechanism of action of aldolases, *Adv. Enzymol.* **31**: 125.

Mukherjee, A. B., Hastings, C., and Cohen, M. M., 1973, Nucleic acid synthesis in various organs of developing human fetuses, *Pediatr. Res.* **7**:696.

Muller, W. A., Falcone, G. R., and Unger, R. H., 1917, The effect of alanine on glucagon secretion, *J. Clin. Invest.* **50**:2215.

Myant, N. B., 1970, Lipid metabolism, in: *Scientific Foundations of Obstetrics and Gynaecology,* p. 354, Heinemann, London.

Nakamura, J., Roux, J. F., Brown, E. G., and Sweet, A. Y., 1972, Total lipids and the lecithin/ sphingomyelin ratio of amniotic fluid: An antenatal test of lung immaturity, *Am. J. Obstet. Gynecol.* **113**:363.

Needham, J., 1963, Protein utilization in mammalian embryonic life, in: *Chemical Embryology,* Vol. 2, Hafner, New York.

Nelson, G. H., 1966, Lipid metabolism in toxemia of pregnancy, *Clin. Obstet. Gynecol.* **9**:882.

Nelson, G. H., 1969, Amniotic fluid phospholipid patterns in normal and abnormal pregnancies at term, *Am. J. Obstet. Gynecol.* **105**:1072.

Nelson, G. H., 1971, Relationship between amniotic fluid triglyceride level and fetal maturity, *Am. J. Obstet. Gynecol.* **111**:930.

Nelson, G. H., 1972, Relationship between amniotic fluid lecithin concentration and respiratory distress syndrome, *Am. J. Obstet Gynecol.* **112**:827.

Nemeth, A. M., 1959, Mechanisms controlling change of tryptophan peroxidase activity in developing mammalian liver, *J. Biol. Chem.* **234**:2921.

Nemeth, A. M., 1963, Biochemical events underlying the development and adaptive increase in trypto- phan pyrrolase activity, in: *Advances in Enzyme Regulation* (G. Weber, ed.), Vol. 1, p. 57, Permagon Press, New York.

Newsholme, E. A., Ralleston, F. S., and Taylor, K., 1968, Factors affecting the glucose-6-phosphate inhibition of hexokinase from cerebral cortex tissue of the guinea pig, *Biochem. J.* **106**:193.

Ng, W. G., Donnell, G. N., and Bergren, W. R., 1965, Galactokinase activity in human erythrocytes of individuals at different ages, *J. Lab. Clin. Med.* **66**:115.

Nicholson, J. F., 1970, Rate of protein synthesis in premature infants, *Pediatr. Res.* **4**:389.

Nikolasev, V., 1975, Individual phospholipids contents and their fatty acid-compositions in early normal human embryonic lung tissue, *Lipids* **9**:827.

Nikolasev, V., Resh, B. A., Meszaro, J., Szontagh, F. E., and Karady, I., 1973, Phospholipid content of early human placenta and fatty acid composition of the individual phospholipids, *Steroid Lipid Res.* **4**:76.

Noble, R. C., Christie, W. W., and Moore, J. H., 1971, Diet and the lipid composition of adipose tissue in the young lamb, *J. Sci. Food Agric.* **22**:616.

Novak, E., Drummond, G. I., Skala, J., and Hahn, P., 1972, Developmental changes in cyclic AMP, protein kinase, phosphorylase kinase and phosphorylase in liver, heart and skeletal muscle of the rat, *Arch. Biochem. Biophys.* **150**:511.

Novak, E., Hahn, P., Penn, D., Monkus, E., and Skala, J., 1973. The role of carnitine in subcutaneous white adipose tissue from newborn infants, *Biol. Neonate* **23**:11.

Novak, M., and Monkus, E., 1972, Metabolism of subcutaneous adipose tissue in the immediate postnatal period in human newborns. I. Developmental changes in lipolysis and glycogen content, *Pediatr. Res.* **6**:73.

Novak, M., Penn, D., and Monkus, E., 1973, Regulation of lipolysis in human neonatal adipose tissue, *Biol. Neonate* **22**:451.

Novak, M., Penn-Walker, D., and Monkus, E. F., 1975, Oxidation of fatty acids by mitochondria obtained from newborn subcutaneous (white) adipose tissue, *Biol. Neonate* **25**:95.

Obenshain, S. S., Adam, P., King, K., Teramo, K., Raivo, K., Räihä, N., and Schwartz, R., 1970, Human fetal insulin response to sustained maternal hyperglycemia, *N. Engl. J. Med.* **283**:566.

O'Brien, J. S., and Sampson, E. L., 1965, Lipid composition of the normal human brain, gray matter, white matter and myelin, *J. Lipid Res.* **6**:537.

Olson, E. B., and Graven, S. N., 1974, Comparison of visualization methods to measure the lecithin/ sphingomyelin ratio in amniotic fluid, *Clin. Chem.* **20**:1408.

Omans, W. B., Barness, L. A., Rose, C. S., and György, P., 1961, Prolonged feeding studies in premature infants, *J. Pediatr.* **59**:951.

Osler, M., and Pedersen, J., 1960, Body composition of the newborn infants of diabetic mothers, *Pediatrics* **26**:985.

Osterman, J., Fritz, P. J., and Wuntek, T., 1973, Pyruvate kinase isoenzyme from rat tissues. Developmental studies, *J. Biol. Chem.* **248**:1011.

Partington, M. W., 1968, Neonatal tyrosinemia, *Biol. Neonate* **12**:316.

Partington, M. W., and Mathews, J., 1967, The prophylactic use of folic acid in neonatal hypertyrosinemia, *Pediatrics* **39**:776.

Partridge, N. C., Hoh, C. H., Weaver, P. K., and Oliver, I. T., 1975, Premature induction of glucokinase in the neonatal rat by thyroid hormones, *Eur. J. Biochem.* **51**:49.

Pascal, T. A., Gillam, B. M., and Gaull, G. E., 1972, Cystathionase: Immunochemical evidence for absence from human fetal liver, *Pediatr. Res.* **6**:773.

Patrito, L. C., Flury, A., Rosato, J., and Martino, A., 1973, Biosynthesis *in vitro* of chorionic gonadotrophin from human placenta, *Hoppe-Seyler's Z. Physiol. Chem.* **354**:1129.

Pearce, P. H., Buischell, B. J., Weaver, P. K., and Oliver, I. T., 1974, The development of phosphopyruvate carboxylase and gluconeogenesis in neonatal rats, *Biol. Neonate* **24**:320.

Pencharz, P. B., Steffee, W. P., Rand, W. M., Cochran, W., Scrimshaw, N. S., and Young, V. R., 1977, Protein metabolism in human neonates: Nitrogen balance studies, estimated obligatory loss and whole body turnover, *Clin. Sci. Mol. Med.* **52**:485.

Peng, Y. S., Brooks, M., Elson, C., and Shrago, E., 1973, Contribution of the cytosol and mitochondrial pathways to phosphoenolpyruvate formation during gluconeogenesis, *J. Nutr.* **103**:1489.

Penhoet, E., Rajkumar, T., and Rutter, W. J., 1966, Multiple forms of fructose diphosphate aldolase in mammalian tissues, *Proc. Natl. Acad. Sci. U.S.A.* **56**:1275.

Persson, B., Gentz, J., and Kellum, M., 1973, Metabolic observations in infants of strictly controlled diabetic mothers, *Acta Paediatr. Scand.* **62**:465.

Pfleiderer, G., and Wachsmuth, E. D., 1961, Alters—und funktionsabhängige Differentierung der Laktatdehydrogenase menschlicher Organe, *Biochem. Z.* **334**:185.

Picou, L., 1968, Teneur en lipides du foetus du rat et du rat nouveau-né, action des glandes endocrines, *J. Physiol. (Paris)* **60**:275.

Picou, D., and Taylor-Roberts, T., 1969, The measurement of total protein synthesis and catabolism and nitrogen turnover in infants of different nutritional states and receiving different amounts of dietary protein, *Clin. Sci.* **36**:283.

Pines, M., Bashan, N., and Moses, S. W., 1976, Glucose effect on glycogen synthetase and phosphorylase in fetal rat liver, *FEBS Lett.* **62**:301.

Pisano, J. J., Mitoma, C., and Udenfriend, S., 1957, Biosynthesis of $\gamma$-guanidinobutyric acid from $\gamma$-aminobutyric acid and arginine, *Nature (London)* **180**:1125.

Pisano, J. J., Abraham, D., and Udenfriend, S., 1963, Biosynthesis and disposition of $\gamma$-guanidinobutyric acid in mammalian tissue, *Arch. Biochem. Biochem.* **100**:323.

Plas, C., Chapeville, F. and Jacquot, R. L., 1973, Development of glycogen storage ability under cortisol control in primary cultures of rat foetal hepatocytes, *Dev. Biol.* **32**:82.

Plas, C., Chapeville, F., and Jacquot, R. L., 1976, Rechèrches sur la différentation functionelle du foie chez foetus de rat. Influence de l'hydrocortisone sur synthèse du glycogène par les hépaticytes foetaux en culture, *C.R. Acad. Sci.* **270**:2846.

Plotz, E. J., Kabara, J. J., and Davis, M. E., 1968, Studies on the synthesis of cholesterol in the brain of the human fetus, *Am. J. Obstet. Gynecol.* **101**:534.

Poley, J. R., Dower, J. C., Owen, C. A., and Stickler, G. B., 1964, Bile acids in infants and children, *J. Lab. Clin. Med.* **63**:838.

Pollak, J. K., 1975, The maturation of the inner membrane of foetal rat liver mitochondria, an example of a positive-feedback mechanism, *Biol. J.* **150**:477.

Pollak, J. K., and Duck-Chang, C. G., 1973, Changes in rat liver mitochondria and endoplasmic reticulum during development and differentiation, *Enzyme* **15**:139.

Pomerance, W., Biezenski, J. J., Moltz, A., and Goodman, J., 1971, Origin of amniotic fluid lipids. II. Abnormal pregnancy, *Obstet Gynecol.* **38**:379.

Popjack, G., 1954, The origin of fetal lipids, *Cold Spring Harbor Symp. Quant. Biol.* **19**:200.

Portman, O. W., Behrman, R. E., and Soltys, P., 1969, Transfer of free fatty acids across the primate placenta, *Am. J. Physiol.* **216**:143.

Räihä, N. C. R., 1973, Phenylalanine hydroxylase in human liver during development, *Pediatr. Res.* **7**:1.

Räihä, N. C. R., and Suihkonen, J., 1968*a*, Development of urea-synthesizing enzymes in human liver, *Acta Paediatr. Scand.* **57**:121.

Räihä, N. C. R., and Suihkonen, J., 1968*b*, Factors influencing the development of urea-synthesizing enzymes in rat liver, *Biochem. J.* **107**:793.

Räihä, N. C. R., Schwartz, A. L., and Lindroos, M. C., 1971, Induction of tyrosine-α-ketoglutarate transaminase in fetal rat and fetal human liver in organ culture, *Pediatr. Res.* **5**:70.

Räihä, N. C. R., Heinonen, K., Rassin, D. K., and Gaull, G. E., 1976, Milk protein quantity and quality in low-birth-weight infants: I. Metabolic responses and effects on growth, *Pediatrics* **57**:659.

Rassin, D. K., Gaull, G. E., Heinonen, K., and Räihä, N. C. R., 1977*a*, Milk protein quantity and quality in low-birth-weight infants. II. Effects on selected essential and non-essential amino acids in plasma and urine, *Pediatrics* **59**:407.

Rassin, D. K., Gaull, G. E., Räihä, N. C. R., and Heinonen, K., 1977*b*, Milk protein quantity and quality in low-birth-weight infants, IV. Effects on tyrosine and phenylalanine in plasma and urine, *J. Pediatr.* **90**:356.

Rassin, D. K., Sturman, J. A., and Gaull, G. E., 1977*c*, Taurine in developing rat brain: Subcellular distribution and association with synaptic vesicles of [$^{35}$S]taurine in maternal, fetal and neonatal brain, *J. Neurochem.* **28**:41.

Rassin, D. K., Sturman, J. A., and Gaull, G. E., 1977*d*, Taurine in milk: Species variation, *Pediatr. Res.* **11**:28.

Reid, D. W. J., Campbell, D. J., and Yakyniyshyn, L., 1971, Quantative amino acids in amniotic fluid and maternal plasma in early and late pregnancy, *Am. J. Obstet. Gynecol.* **111**:251.

Reynolds, W. A., and Pitkin, R. M., 1976, Fetal insulin response to glucose: A reexamination, *Gynecol. Invest.* **7**:59.

Reynolds, W. A., Chez, R. A., Roux, J. F., and Yoshioka, T., 1975, Observations on the tissue lipids of infants born of diabetic monkeys, in: *Fetal and Postnatal Cellular Growth, Hormones and Nutrition* (D. B. Cheek ed.), p. 323, Wiley, New York.

Robertson, A. F., and Sprecher, H., 1968, A review of human placental lipid metabolism and transport, *Acta Paediatr. Scand.* **183**:1.

Robertson, A. F., Sprecher, H., and Wilcox, J. P., 1969, Total lipid fatty acid patterns of umbilical cord blood in intrauterine growth failure, *Biol. Neonate* **14**:28.

Rose, W. C., and Wixom, R. L., 1955*a*, Amino acid requirements of man: XIII, Sparing effect of cystine on the methionine requirement, *J. Biol. Chem.* **216**:95.

Rose, W. C., and Wixom, R. L., 1955*b*; The amino acid requirements of man: XIV The sparing effect of tyrosine on the phenylalanine requirement, *J. Biol. Chem.* **217**:95.

Rosenfeld, E. L., Papovo, I. A., and Orlova, V. S., 1970, Influence of adrenaline on the splitting of glycogen and maltose by α-amylase of liver, heart and skeletal muscle of the rat, *Bull. Soc. Chim. Biol.* **52**:1111.

Ross, B. D., Hems, R., and Krebs, H. A., 1967, The rate of gluconeogenesis from various precursors in the perfused rat liver, *Biochem. J.* **102**:342.

Roux, J. F., 1966, Lipid metabolism in the fetal and neonatal rabbit, *Metabolism* **15**:856.

Roux, J. F., 1974, Monitoring of labor in high-risk centers, in: *Clinical Perinatology* (S. Aladjen and A. K. Brown, eds.), p. 336, C. V. Mosby, St. Louis.

Roux, J. F., and Frosolono, M., 1978, The surfactant activity of amniotic fluid, *Am. J. Obstet. Gynec.* **130**:562.

Roux, J. F., and Myers, R. E., 1974, *In vitro* metabolism of palmitic acid and glucose in the developing tissue of the rhesus monkey, *Am. J. Obstet. Gynecol.* **118**:385.

Roux, J. F., and Romney, S. L., 1967, Plasma free fatty acids and glucose concentrations in the human fetus and newborn exposed to various environmental conditions, *Am. J. Obstet. Gynecol.* **97**:268.

Roux, J. F., and Yoshioka, T., 1970, Lipid metabolism in the fetus during development, *Clin. Obstet. Gynecol.* **13**:595.

Roux, J. F., Romney, S. L., and Dinnerstein, A., 1964, Environmental and aging effects of postmaturity on fetal development and carbyhydrate metabolism, *Am. J. Obstet. Gynecol.* **90**:546.

Roux, J. F., Grigorian, A., and Takeda, Y., 1967, *In vitro* lipid metabolism in the developing human fetus, *Nature* **216**:819.

Roux, J. F., Takeda, Y., and Grigorian, A., 1971, Lipid concentration and composition in human fetal tissue during development, *Pediatrics* **48**:540.

Roux, J. F., Nakamura, J., Brown, E., and Sweet, A. Y., 1972, The lecithin–sphingomyelin ratio of amniotic fluid: An index of fetal lung maturity, *Pediatrics* **49**:464.

Roux, J. F., Nakamura, J., and Brown, E., 1973a, Further observations on the determination of gestational age by amniotic fluid analysis, *Am. J. Obstet. Gynecol.* **116**:633.

Roux, J. F., Nakamura, J., and Brown, E., 1976b, Assessment of fetal maturation by the foam test, *Am. J. Obstet. Gynecol.* **117**:280.

Roux, J. F., Nakamura, J., and Frosolono, M., 1974, Fatty acid composition and concentration of lecithin in the acetone fraction of amniotic fluid phospholipids, *Am. J. Obstet. Gynecol.* **119**:838.

Rudermann, N. B., Toews, C. J., Lowy, C., and Shafrir, E., 1971, Inhibition of hepatic gluconeogenesis by pent-4-enoic acid: Role of redox state, acetyl-CoA and ATP, in: *Regulation of Gluconeogenesis* (H. D. Söling and B. Willms, eds.), p. 194, Thieme Verlag, Stuttgart.

Russell, P. T., Miller, W. J., and McLain, C. R., 1974, Palmitic acid content of amniotic fluid lecithin as an index to fetal lung maturity, *Clin. Chem.* **20**:1431.

Rutter, W. J., 1969, Independently regulated synthetic transitions in foetal tissues, in: *Foetal Autonomy* (G. E. W. Wolstenholme and M. O'Connor, eds.), p. 59, J. A. Churchill, London.

Rutter, W. J., Blostein, R. E., Woodfin, B. M., and Weber, C. S., 1963, Enzyme variants and metabolic diversification, *Adv. Enzym. Reg.* **1**:39.

Ryan, W. L., and Carver, M. J., 1966, Free amino acids of human foetal and adult liver, *Nature* **212**:292.

Ryan, W. L., and Orr, W., 1966, Phenylalanine conversion to tyrosine by the human fetal liver, *Arch. Biochem. Biophys.* **113**:684.

Sabata, V., Wolf, H., and Lausmann, S., 1968, Role of free fatty acids, glycerol, ketone bodies and glucose in the energy metabolism of the mothers and fetus during delivery, *Biol. Neonate* **13**:7.

Salerno, D. M., and Beeler, D. A., 1973, The biosynthesis of phospholipids and their precursors in rat liver involving *de novo* methylation, and base-exchange pathways, *in vivo, Biochim. Biophys. Acta* **326**:325.

San Pietro, A., and Rittenberg, D., 1953, A study of the rate of protein synthesis in humans. II. Measurement of the metabolic pool and the rate of protein synthesis, *J. Biol. Chem.* **201**:457.

Sato, K., Morris, H. P., and Weinhouse, S., 1972, Phosphorylase: A new isoenzyme in rat hepatic tumors and fetal liver, *Science* **178**:879.

Scarpelli, E. M., 1967, The lung, tracheal fluid, and lipid metabolism of the fetus, *Pediatrics* **40**:951.

Scarpelli, E. M., 1968, *The Surfactant System of the Lung,* Lea and Febiger, Philadelphia.

Schaub, J., Gutman, I., and Lippert, H., 1971, Developmental changes of glycolytic and gluconeogenic enzymes in fetal and neonatal rat liver, *Horm. Metal. Res.* **4**:110.

Schirar, A., Vielh, J. P., Alcindor, A., and Gautray, J. P., 1975, Amniotic fluid phospholipids and fatty acids in normal pregnancies, *Am. J. Obstet. Gynecol.* **121**:653.

Schreyer, P., Tamir, I., Bukovsky, I., Weinraub, Z., and Caspi, E., 1974, Amniotic fluid total phospholopids versus lecithin/sphingomyelin ratio in the evaluation of fetal lung maturity, *Am. J. Obstet. Gynecol.* **120**:909.

Schudt, C., Gaertner, U., Dohlen, G., and Pette, D., 1975, Calcium related changes of enzyme activities in energy metabolism of cultured embryonic chick myoblasts end myotubes, *Eur. J. Biochem.* **60**:579.

Schmidt, S. Y., Berson, E. L., and Hayes, K. C., 1976, Retinal degeneration in cats fed casein. I. Taurine deficiency, *Invest. Ophthalmol.* **15**:47.

Schwartz, B. E., Schultz, M. F., MacDonald, P. C., and Johnston, J. M., 1975, Initiation of human parturition. III. Fetal membrane content of prostaglandin $E_2$ and $F_{2a}$ precursor, *Obstet. Gynecol.* **46**:564.

Schwartz, R., 1968, Life before birth: Metabolic fuels in the fetus, *Proc. R. Soc. Med. (Suppl.)* **61**:123.

Seufert, C. D., Herlemann, E., Albrecht, E., and Seubert, W., 1971, Purification and properties of pyruvate carboxylase from rat liver, in: *Regulation of Gluconeogenesis* (H. D. Söling and B. Willms, eds.), p. 11, Thieme Verlag, Stuttgart.

Sereni, F., and Principi, N., 1965, The development of enzyme systems, *Pediatr. Clin. North Am.* **12**:515.

Shapovalov, U. N., 1961, *Arkh. Anat. Gistol. Embriol.* **40**:34; *ibid.* **42**:46, 1962.

Sheath, J., Grimwade, J., Waldron, K., Bickley, M., Taft, P., and Wood, C., 1973, Arteriovenous nonesterified fatty acids and glycerol differences in the umbilical cord at term and their relationships to fetal metabolism. *Am. J. Obstet. Gynecol.* **113**:358.

Shelley, H. J., 1961, Glycogen reserves and their changes at birth and anoxia, *Br. Med. Bull.* **17**:137.

Shephard, B., Buhi, W., and Spellacy, W. N., 1974, Critical analysis of the amniotic fluid shake test, *Obstet. Gynecol.* **43**:558.

Shih, V. E., and Littlefield, J. W., 1970, Argininosuccinase activity in amniotic-fluid cells, *Lancet* **1**:45.

Shrago, E., Lardy, H. A., Nordlie, R. C., and Forster, D. O., 1963, Metabolic and hormonal control of phosphoenolpyruvate carboxylase and malic enzyme in rat liver, *J. Biol. Chem.* **238**:3188.

Simmons, M. A., Meschia, G., Makowski, E. L., and Battaglia, F. C., 1974, Fetal metabolic response to material starvation, *Pediatr. Res.* **8**:830.

Singer, S., and Litwack, G., 1971, Effects of age and sex on ³H-cortisol uptake, binding and metabolism in liver and on enzyme induction capacity, *Endocrinology* **88**:1448.

Singh, E. J., and Zuzpan, F. P., 1973, Amniotic fluid lipids in normal human pregnancy, *Am. J. Obstet. Gynecol.* **117**:919.

Sisson, J. A., and Plotz, E. J., 1967, Plasma lipid in maternal and fetal rabbits fed stock and peanut oil-cholesterol diets, *J. Nutr.* **92**:435.

Sisson, J. A., and Plotz, E. J., 1968, Maternal and fetal plasma lipids in alloxon diabetic and non-diabetic rabbits fed a peanut oil-cholesterol diet, *Exp. Mol. Pathol.* **9**:197.

Smith, S. T., and Abraham, S., 1970, Fatty acid synthesis in developing mouse liver, *Arch. Biochem.* **136**:112.

Snell, K., and Walker, D. G., 1973, Glucose metabolism in the newborn rat. Hormonal effects *in vivo*, *Biochem. J.* **134**:899.

Snyderman, S. E., 1971, The protein and amino acid requirements of the premature infant, in: *Metabolic Processes in the Foetus and Newborn Infant* (J. H. P. Jonxis, H. K. A. Visser, and J. A. Travoelstra, eds.), p. 128, H. E. Stenfert Knoese, N. V., Leiden.

Snyderman, S. E., Boyer, A., and Holt, L. E., Jr., 1959*a*, The arginine requirement of the infant, *Am. J. Dis. Child.* **97**:192.

Snyderman, S. E., Prose, E. H., and Holt, L. E., Jr., 1959*b*, Histidine, an essential amino acid for the infant, *Am. J. Dis. Child.* **98**:459.

Soderhjelm, L., 1953, Fat absorption studies. VI. Passage of polyunsaturated fatty acids through the placenta, *Acta Soc. Med. Upsalien* **58**:239.

Söling, H. D., Willms, B., and Kleineke, J., 1971, Regulation of gluconeogenesis in rat and guinea pig, in: *Regulation of Gluconeogenesis* (H. D. Söling and B. Willms, Eds.), p. 210, Thieme Verlag, Stuttgart.

Söling, H. D., Kleineke, J., Willms, B., Janson, G., and Kuhn, A., 1973, Relationship between intracellular distribution of phosphoenolpyruvate carboxykinase, regulation of gluconeogenesis and energy cost of glucose formation, *Eur. J. Biochem.* **37**:233.

Sordahl, L. A., Crow, C. A., Kroft, G. H., and Schwartz, A., 1972, Some ultrastructural aspects of heart mitochondria associated with development: Fetal and cardiomyopathic tissue, *J. Mol. Cell. Cardiol.* **4**:1.

Sparks, J. W., Lyncj, A., Chez, R. A., and Glinnsman, W. H., 1976, Glycogen regulation in isolated perfused near term monkey liver, *Pediatr. Res.* **10**:51.

Sperling, M. A., DeLameter, P. V., Philips, D., Fiser, R. N., Oh, W., and Fisher, D. A., 1974, Spontaneous and amino acid stimulated glucagon secretion in the immediate postnatal period. Relation to glucose and insulin, *J. Clin. Invest.* **53**:1159.

Sperry, W. M., 1962, Biochemistry of brain during early development, in: *Neurochemistry* (K. A. C. Elliott, I. H. Page, and J. H. Quastel, eds.), p. 55, Charles C. Thomas, Springfield, Illinois.

Srinivasa Rao, P., and Subba Rao, K., 1974, Fatty acid composition of phospholipids in different regions of developing human fetal brain, *Lipids* **8**:374.

Stalmans, W., and Hers, H. G., 1975, The stimulation of liver phosphorylase b by AMP, fluoride and sulfate, *Eur. J. Biochem.* **54**:341.

Stalmans, W., De Barsy, Th., Laloux, M., De Wulf, H., and Hers, H. G., 1972, Phosphorylase a as a glucose receptor, in: *Metabolic Interconversion of Enzymes* (O. Wieland, E. Helmrich, and H. Holzer, eds.), p. 121, Springer Verlag, Berlin.

Stalmans, W., De Wulf, H., Hue, L., and Hers, H. G., 1974, The sequential inactivation of glycogen in liver after administration of glucose to mice and rats, *Eur. J. Biochem.* **41**:127.

Stave, U., 1964, Age dependent changes of metabolism. I. Studies of enzyme patterns of rabbit organs, *Biol. Neonate* **6**:128.

Steinmann, B., Baerlocher, B., and Gitzelmann, R., 1975, Heriditäre Störungen des Fruktose Stoffwechsels. Belastungsproben mit Fruktose, Sorbitol und Dihydroxyaceton, *Nutr. Metab.* **18** (Suppl. 1):115.

Sternowsky, H. J., Räihä, N. C. R., and Gaull, G. E., 1976, Methionine adenosyltransferase and transmethylation in fetal and neonatal lung of the human, monkey and rabbit, *Pediatr. Res.* **10**:545.

Stumpf, P. K., 1969, Metabolism of fatty acids, *Annu. Rev. Biochem.* **38**:159.

Sturman, J. A., and Gaull, G. E., 1974, Polyamine biosynthesis in human fetal liver and brain, *Pediatr. Res.* **8**:231.

Sturman, J. A., and Gaull, G. E., 1975, Taurine in brain and liver of the developing human and monkey, *J. Neurochem* **25**:831.

Sturman, J. A., Gaull, G., and Räihä, N. C. R., 1970, Absence of cystathionase in human fetal liver: Is cystine essential? *Science* **169**:74.

Sturman, J. A., Niemann, W. H., and Gaull, G. E., 1973, Metabolism of $^{35}$S-methionine and $^{35}$S-cystine in the pregnant rhesus monkey, *Biol. Neonate* **21**:16.

Sturman, J. A., Gaull, G. E., and Räihä, N. C. R., 1975, DNA synthesis from the $\beta$-carbon of serine by fetal and mature human liver, *Biol. Neonate* **27**:17.

Sturman, J. A., Gaull, G. E., and Niemann, W. H., 1976a, Activities of some enzymes involved in homocysteine methylation in brain, liver and kidney of the developing rhesus monkey, *J. Neurochem* **27**:425.

Sturman, J. A., Gaull, G. E., and Niemann, W. H., 1976b, Cystathionine synthesis and degradation in brain, liver and kidney of the developing monkey, *J. Neurochem.* **26**:457.

Sturman, J. A., Rassin, D. K., and Gaull, G. E., 1977a, Taurine in development, *Life Sci.* **21**:1.

Sturman, J. A., Rassin, D. K., and Gaull, G. E., 1977b, Taurine in developing rat brain: Transfer of [$^{35}$]taurine to pups via milk, *Pediatr. Res.* **11**:28.

Svennerholm, L., and Vanier, M. T., 1972, The distribution of lipids in the human nervous system. II. Lipid composition of human fetal and infant brain, *Brain Res.* **47**:457.

Swanson, W. W., and Iob, V., 1939, The growth of fetus and infant as related to mineral intake during pregnancy, *Am. J. Obstet. Gynecol.* **38**:382.

Swiatek, K. R., 1971, Development of gluconeogenesis in pig liver slices. *Biochim. Biophys. Acta* **252**:274.

Swiatek, K. R., Choo, K. L., Choo, H. L., Cornblath, M., and Tildon, J. T., 1970, Enzymatic adaptations in newborn pig liver, *Biochim. Biophys. Acta* **222**:145.

Szabo, A. J., and Grimaldi, R. D., 1970, The metabolism of the placenta, *Adv. Metab. Disord.* **5**:185.

Tahasu, T., and Hughes, B. P., 1969, Lactate dehydrogenase isoenzyme patterns in human skeletal muscle, *J. Neurol. Neurosurg. Psychiatr.* **32**:175.

Thorndike, J., 1972, Comparison of the levels of three enzymes in developing livers of rats and mice, *Enzyme* **13**:52.

Tigner, J. C., and Barnes, R. H., 1975, Effect of postnatal malnutrition on corticosteroid levels in male albino rats, *Proc. Soc. Exp. Biol. Med.* **149**:80.

Tildon, J. T., Swiatek, K. R., and Cornblath, M., 1971, Phosphoenolpyruvate carboxykinase in the developing pig liver, *Biol. Neonate* **17**:436.

Titheradge, M. A., and Coore, H. G., 1976, The mitochondrial pyruvate carrier, its exchange properties and its regulation by glucagon, *FEBS Let.* **63**:45.

Tobin, J. D., Roux, J. F., and Soeldner, J. S., 1969, Human fetal insulin response after acute maternal gluclose administration during labor, *Pediatrics* **44**:668.

Tomkins, G. M., 1972, Regulation of gene expression in mammalian cells, *Harvey Lect.* **68**:37.

Torday, J. S., Smith, B. T., and Giround, G. J. P., 1975, The rabbit fetal lung as a glucocorticoid target tissue, *Endocrinology* **96**:1462.

Tourian, A., Treiman, D. M., and Carr, J. S., 1972, Developmental biology of hepatic phenylalanine hydroxylase activity in fetal and neonatal rats synchronized by conception, *Biochim. Biophys. Acta* **279**:484.

Tsoulos, N. G., Colwill, J. R., Battaglia, F. C., Makowski, E. L., and Mescia, G., 1971, Comparison of glucose, Fructose, and O$_2$, uptakes by fetuses of fed and starved ewes, *Am. J. Physiol.* **221**:234.

Tsunematzu, K., and Shiraishi, T., 1969, Aldolase isozymes in human tissue and serum. With special reference to cancer diagnosis, *Cancer* **24**:637.

Turner, D. C., Gmür, R., Siegrist, M., Burckhardt, E., and Eppenberger, H. M., 1976, Differentiation in cultures derived from embryonic chicken muscle. I. Muscle specific enzyme changes before fusion in EGTA-synchronized cultures, *Dev. Biol.* **48**:258.

Vaitukaitis, J. L., Ross, G. T., Braunstein, G. D., and Rayford, P. L., 1976, Gonadotropins and their subunits: Basic and clinical studies, *Rec. Prog. Horm. Res.* **32**:289.

Van Berkel, Th. J. C., Koster, J. F., and Hulsman, W. C., 1972, Distribution of L- and M-type pyruvate kinase between parenchymal and Kupfer cells of rat liver, *Biochim. Biophys. Acta* **276**:425.

Van den Berghe G., 1975, Biochemical aspects of hereditary fructose intolerance, in: *Normal and Pathological Development of Energy Metabolism* (F. A. Hommes and C. J. van den Berg, eds.), p. 211, Academic Press, London.

Van Duyne, C. M., Parker, H. R., Havel, R. J., and Holm, L. W., 1960, Free fatty acid metabolism in fetal and newborn sheep, *Am. J. Physiol.* **199**:987.

Van Duyne, C. M., Havel, R. J., and Felts, J. M., 1962, Placental transfer of palmitic acid-1-$^{14}$C in rabbits, *Am. J. Obstet. Gynecol.* **84**:1070.

Van Duyne, C. M., 1965, Free fatty acid metabolism during perinatal life, *Biol. Neonate* **9**:115.

Villee, C. A., 1953, Regulation of blood glucose in the human fetus, *J. Appl. Physiol.* **5**:437.

Villee, C. A., 1954, The intermediary metabolism of human fetal tissues, *Cold Spring Harbor Symp. Quant. Biol.* **19**:186.

Villee, C. A., and Hagerman, D. D., 1958, Effect of oxygen deprivation on the metabolism of fetal and adult tissues, *Am. J. Physiol.* **194**:457.

Villee, C. A., and Loring, J. M., 1961, Alternative pathways of carbohydrate metabolism in fetal and adult tissues, *Biochem. J.* **81**:488.

Villee, C. A., Hagerman, D. D., Holmberg, N., Lind, J., and Villee, D. B., 1958, The effects of anoxia on the metabolism of fetal tissues, *Pediatrics* **22**:953.

Villee, D. B., 1969, Development of endocrine function in the human placenta and fetus, *N. Engl. J. Med.* **281**:473.

Vink, C. L. J., and Kroese, A. A., 1959, Liver function and age, *Clin. Chim. Acta* **4**:674.

Volpe, P., Sawamura, R., and Strecker, H. J., 1969, Control of ornithine-γ-transaminase in rat liver and kidney, *J. Biol. Chem.* **244**:719.

Walker, D. G., 1963, Development of hepatic enzymes for the phosphorylation of glucose and fructose, *Adv. Enzyme Reg.* **3**:163.

Walker, D. G., and Holland, G., 1965, The development of hepatic glucokinase in the neonatal rat, *Biochem. J.* **97**:845.

Walsh, S. Z., and Lind, J., 1970, The dynamics of the fetal heart and circulation and its alternation at birth, in: *Physiology of the Perinatal Period* (U. Stave, ed.). Vol. 1. p. 141, Appleton-Century-Crofts, New York.

Wapnir, R. A., and Dierks-Ventling, C., Placental transfer of amino acids, *Biol. Neonate,* **17**:373, 1971.

Ward, C. J., and Walker, D. G., 1973, Regulation of enzyme development for glycerol utilization by neonatal rat liver, *Biol. Neonate* **23**:403.

Warshaw, J. B., 1969, Cellular energy metabolism during fetal development. I. Oxidative phosphorylation in the fetal heart, *J. Cell Biol.* **41**:651.

Warshaw, J. B., 1970, Cellular energy metabolism during fetal development. III. Deficient acetyl-CoA synthetase, acetyl carnitine transferase and oxidation of acetate in the fetal bovine heart, *Biochim. Biophys. Acta* **223**:409.

Warshaw, J. B., 1972, Cellular energy metabolism during fetal development. IV. Fatty acids activation, acyl transfer and fatty acid oxidation during development of the chick and rat, *Dev. Biol.* **28**:537.

Warshaw, J. B., and Terry, M. L., 1970, Cellular energy metabolism during fetal development. II. Fatty acid oxidation by the developing heart, *J. Cell Biol.* **44**:354.

Watanabe, Y., 1967, Study on fetal lipid metabolism, *J. Jpn. Obstet. Gynecol. Soc.* **19**:1187.

Weber, G., Hird, H. J., Stamm, N. B., and Wagle, D. S., 1965, Enzymes involved in carbohydrate metabolism in adipose tissue, in: *Handbook of Physiology* (A. E. Renold and G. F. Cahill, eds.), p. 225, American Physiological Society, Washington.

Webster, R. C., and Young, W. C., 1948, Thiouracil treatment of the female guinea pig, effect on gestation and the offspring, *Anat. Rec.* **101**:722.

Werthamer, S., Freiberg, S., and Amaral, L., 1973, Quantitation of lactate dehydrogenase isoenzyme pattern of the developing human fetus, *Clin. Chim. Acta* **45**:5.

White, A. Handler, P., and Smith, E. L., 1968, *Principles of Biochemistry,* p. 57, McGraw-Hill, New York.

Wicks, W. D., 1969, Induction of hepatic enzymes by adenosine 3′, 5′-monophosphate in organ culture, *J. Biol. Chem.* **244**:3941.

Wicks, W. D., 1968, Induction of tyrosine α-ketoglutarate transaminase in fetal rat liver, *J. Biol. Chem.* **243**:900.

Widdowson, E. M., 1950, Chemical composition of newly born mammals, *Nature* **166**:626.

Widdowson, E., 1969, How the foetus is fed, *Proc. Nutr. Soc.* **28**:17.

Wieland. O., Evertz-Prusse, E., and Stukowski, B., 1968, Distribution of pyruvate carboxylase and phosphoenolpyruvate carboxylase in human liver, *FEBS Lett.* **2**:26.

Wilbur, D. O., and Patel, M. S., 1974, Development of mitochondrial pyruvate metabolism in rat brain, *J. Neurochem.* **22**:709.

Willms, B., Kleineke, J., and Soling, H. D., 1971, On the redox state of NAD$^+$/NADH systems in guinea pig liver under different experimental conditions, in: *Regulation of Gluconeogenesis* (H. D. Söling and B. Willms, eds.), p. 103, Thieme Verlag, Stuttgart.

Winick, M., and Rosso, P., 1973, Effects of malnutrition on brain development, in: *Biology of Brain Dysfunction* (G. E. Gaull, ed.), Vol. 1, p. 301, Plenum Press, New York.

Wise, J. K., Hendler, R., and Felig, R., 1973*a*, Evaluation of alpha-cell function by infusion of alanine in normal, diabetic and obese subjects, *N. Engl. J. Med.* **287**:288.

Wise, J. K., Lyall, S. S., and Felig, R., 1973*b*, Evidence of stimulation of glucagon secretion by alanine in the human fetus at term, *J. Clin. Endocrinol. Metab.* **37**:345.

Yen, S. S. C., Pearson, O. H., and Stratman, S. T., 1965, Growth hormone levels in maternal and cord blood, *J. Clin. Endocrinol.* **25**:655.

Yeung, D., and Oliver, I. T., 1968, Factors effecting the premature induction of phosphopyruvate carboxykinase in neonatal rat liver, *Biochem. J.* **108**:325.

Yoshioka, T., and Roux, J. F., 1972, *In vitro* metabolism of palmitic acid in human fetal tissue, *Pediatr. Res.* **6**:675.

Young, M., 1971, Placental transport of free amino acids, in: *Metabolic Processes in the Foetus and Newborn Infant* (J. H. P. Jonxis, H. K. A. Visser, and J. A. Travoelstra, eds.), p. 97, H. E. Stenfert Knoese, N. V., Leiden.

Young, M., and Prenton, M. A., 1969, Maternal and fetal plasma amino acid concentrations during gestation and in retarded fetal growth, *J. Obstet. Gynaecol. Br. Commonw.* **76**:333.

Young, V. R. Steffee, W. P., Pencharz, P. B., Winterer, J. C., and Scrimshaw, N. S., 1975, Total body protein synthesis in relation to protein requirements at various ages, *Nature* **253**:192.

Zuckerman, J. E., Herschman, H. R., and Levine, L., 1970, Appearance of a brain specific antigen (the 100-S protein) during human foetal development, *J. Neurochem.* **17**:247.

# 3

# *Developmental Pharmacology*

## *CHARLOTTE CATZ and*
## *SUMNER J. YAFFE*


## *1. Introduction*

An organism reaches maturity only after successfully completing a continuous series of intricate and interlocking events. The sequence begins at conception, and the processes of growth and development advance in predictable fashion. The sequence is also associated with a wide variation in physiologic functions, many of which vary in proportion to body surface and/or body weight. These have been the accepted basis for adjusting drug dosage for individuals of different sizes and ages, but extrapolation of adult dosages to the developing organism has resulted in some unexpected toxic effects. Foreign compounds administered to pregnant women have given rise to concern, and their safety for embryos and fetuses has been increasingly questioned since the therapeutic catastrophe of thalidomide. Surveys have shown that a significant number of undesirable and unnecessary chemicals (drugs, pollutants) reach the intrauterine guest and may modify its normal development. Some effects are noticeable at birth, while others are not so evident or do not become manifest until later in development.

The possibility of treating fetal disease by drugs administered to the mother is a relatively new approach and will be achieved only when the pharmacokinetics of different compounds is precisely characterized in the maternal–placental–fetal unit.

At all stages of development the solution to the safe and effective administration of drugs is to base their prescription upon scientific data obtained for the particular age group under consideration. This is applicable to the newborn (especially in the usage of newer potent medications in intensive-care nurseries) as well as to infants, children, and adolescents in whom significant physiological changes are taking place.

---

*CHARLOTTE CATZ* • National Institute of Child Health and Human Development, National Institutes of Health, Bethesda, Maryland.    *SUMNER J. YAFFE* • University of Pennsylvania, Philadelphia, Pennsylvania.

This chapter focuses on age-related physiologic factors which may influence the disposition of drugs in the organism without ignoring the possible effects that compounds may have on the growth and development of the same individual.

## 2. Teratology

Classical teratology refers to the study of congenital malformations grossly visible at birth, which are induced during the organogenetic period. Today the concepts of teratology have been widened and include any morphological, biochemical, or behavioral defect induced at any stage of gestation and detected either at birth or later in life (World Health Organization, 1967).

The causes of congenital malformations are usually multiple and result from an interaction between genetic and environmental factors. Etiologic factors are identifiable in only one third of human anomalies, reflecting the difficult task of establishing cause-and-effect relationships. The production and occurrence of an anatomic malformation is dependent upon the interplay of four teratological principles: timing, nature of the foreign compound and its accessibility to the embryo, concentration attained and duration of contact, and genetic makeup of the host.

### 2.1. Timing

The timing of action of a teratogen determines the type of malformation produced. Immediately after fertilization and during preimplantation the embryo is considered relatively resistant to environmental effects. Damage at this stage would be severe, causing death of the embryo and abortion. The transport of a number of drugs into tubular fluid and into the blastocyst has been shown in pregnant rabbits (Fabro and Sieber, 1969), and similar events in humans are quite possible. During this period of embryologic development all cells are functionally equivalent in terms of their totipotentiality; if a great number are damaged or killed the blastocyst will die, whereas if the number is small the remaining cells may replace them and normal development ensue.

The embryonic or organogenetic period (15–56 days postfertilization) is characterized by extreme sensitivity to teratogenic agents. The type of anomaly produced will relate closely to the stage of development at which intervention occurs. Tables have been developed in regard to periods of greater sensitivity for different organs which may or may not be maturing simultaneously. As a consequence it is possible from a specific cluster of anomalies to predict when the teratogenic effect took place.

The third period in human embryology (56th day to parturition) begins after the first trimester. By this time most organs are formed, although with the important exceptions of genital apparatus, teeth, and maturation of the central nervous system. During this period the fetus is relatively resistant to teratogens as far as organ formation is concerned. However, teratogens may interfere with the pattern of growth and development or produce pharmacological effects similar to those observed in postnatal life: for example, the fetal hemorrhage produced by maternal treatment with dicumarol (Hirsh *et al.*, 1970). Teratologic effects induced *in utero* may not be manifest until long after birth, occasionally even not until adulthood. Following maternal treatment with tranquilizing drugs or with sex hormones various types of behavioral changes have been reported in the offspring (Desmond *et al.*, 1969; Money, 1971). A tragic example of late or delayed teratogenic effect relates to the transplacental carcinogenesis caused by diethylstilbestrol (DES) (Herbst *et al.*, 1971). Although the mechanism of action is not yet clear, the observation stands most likely as an example of molecular memory at work. A prospective study of pregnant women vaccinated against polio disclosed a tenfold increase in the incidence of neural tube tumors in their offspring, with a higher risk when immunization took place in the early months of gestation (Heinonen *et al.*, 1973). The mechanisms responsible are not clear, and to further complicate the issue it was found that some batches of vaccine were contaminated with an oncogenic virus.

## 2.2. Nature of Compound

Several agents can induce the same anatomic malformation, and one agent may cause more than one type of malformation. In humans, only a few teratogens have been proven beyond a doubt, although many are suspect. Some of the proven teratogens are androgenic steroid hormones, producing virilization of the female infant; thalidomide, producing limb defects; rubella, producing hearing loss, heart defects, and cataracts; irradiation, producing microcephaly; and mercury poisoning, producing multiple anomalies including mental retardation.

## 2.3. Dosage

The teratogenic dose is situated between the minimum concentration producing a temporary effect and the maximum causing fetal death. In most cases the teratogenic zone is narrow and the dose–effect curve has a steep slope. Two proven human teratogens, thalidomide and rubella virus, do not produce permanent damage to the maternal organism. This is in contrast to polychlorinated biphenyl intoxication or maternal Yusho disease (Kuratsune *et al.*, 1976).

## 2.4. Duration of Contact

The duration of teratogenic exposure is also of importance since this may lead to a cumulative effect. Chronic administration of a therapeutic agent may also cause enzyme induction or inhibition which in turn may alter the concentration of the drug and/or metabolites at critical moments during intrauterine development.

## 2.5. Genetic Makeup

Genetic makeup plays an important role in determining the individual response to similar environmental factors and explains differences noted among species, strains, and individuals. In humans, for instance, not all mothers who took thalidomide during the susceptible period gave birth to malformed infants. This consideration imposes limitations on extrapolation of data obtained in animal models and emphasizes the fact that definitive answers can only be secured in human beings.

## 2.6. Further Considerations

Present knowledge of embryogenesis does not permit a clear delineation of mechanisms responsible for abnormal development. Noxious agents may act at many periods where interference with regulatory processes, such as organization of genetic material in the nucleus, transmission of information to the cytoplasm,

*CHARLOTTE CATZ and*
*SUMNER J. YAFFE*

intercellular communication, spatial arrangements, specialization of somatic cells, etc., will lead to alterations and aberrations in the biochemistry, physiology, or morphology of the embryo. Eradication of birth defects will not be achieved by a no-drug policy during pregnancy or by surveillance programs alone. These steps are useful, but only as temporary and rather inefficient measures. The solution will become available only when a thorough understanding of the mechanisms of embryonic growth and maturation will allow the institution of preventive measures based upon rationality.

## 3. Pharmacology

### 3.1. General Principles

Therapeutic agents administered to living organisms undergo a series of steps which eventually lead to their interaction with specific receptors and the production of a pharmacologic effect. Following absorption the drug enters the vascular system, circulates either in free form or bound to proteins and is distributed throughout the body. The effect of the drug ceases when it is eliminated, either by excretion or by biotransformation followed by excretion. Route of administration changes absorption and absorption regulates the rate of drug appearance into the circulation. For the fetus these routes are transplacental (reflecting maternal treatment), gastrointestinal and cutaneous (both because of fetal immersion and swallowing of amniotic fluid), and intraperitoneal (e.g., intrauterine transfusion). Specific drug characteristics, such as rate of drug dissolution, water solubility, and degree of ionization, modify absorption after oral administration, but in age-related phenomena it is necessary to consider the host and its peculiarities. Absorption will be influenced among other things by gastrointestinal content, pH, motility, and mesenteric blood flow.

Drug distribution is governed by other elements including lipid solubility, degree of ionization, and binding properties to proteins in plasma and specific tissues, as well as by regional blood flow.

Biotransformation of drugs occurs mainly in the liver, with other organs participating to a lesser degree. Drug excretion is mainly via the kidney, with minor pathways also available such as saliva, bile, feces, sweat, etc.

Although the above-mentioned steps occur at all stages of development, there are many quantitative and qualitative differences. Maturation implies continuous change, first in a very complex unit constituted by mother, placenta, and fetus, and then in a single individual, the newborn, who is adapting to extrauterine life. The data regarding all these factors will be reviewed under fetal, neonatal, and pediatric pharmacology. The multiple factors governing drug disposition are continually operative. It must be stressed that even if the pharmacologic principles were to remain unchanged, the age change in the host and the resulting maturational events introduce variables; these will be reviewed in the following sections.

### 3.2. The Mother

Pregnancy produces many physiological changes in the organism which in turn influence drug disposition and effect. Some reports pinpoint these differences, which are important not only for maternal therapy but also to understand unplanned fetal exposure to the drug. Several sources have documented the high intake of foreign compounds throughout gestation (Hill, 1973; Nelson and Forfar 1971; Medalie *et al.*, 1972).

Although absorption of certain nutrients seems enhanced during pregnancy, no information is available for drugs. An indirect evaluation suggests a decreased absorption of orally administered salicylamide in 11 parturients (Rauramo *et al.*, 1963). Drug distribution is changed as a result of a lower oncotic pressure (20% in the last trimester) and a gradual increase in total water content and blood volume. All of this results in hemodynamic adaptations characteristic of the pregnant state. The relative hypoalbuminemia, especially in the third trimester, is responsible for the decrease in drug-binding capacity as measured for sulfisoxazole and Congo red (Csögor *et al.*, 1968). The rate of drug excretion also reflects changes in renal function such as increase in glomerular filtration rate and modifications in tubular reabsorptive capacity. In late pregnancy, changes in body position will most likely affect renal excretion of foreign compounds as shown by the decrease in urine volume and clearances of inulin and PAH in the supine position. The metabolic capability to modify drugs also undergoes some changes during pregnancy, and some studies in humans demonstrate this alteration. Glucuronidation studied in parturients for salicylamide by measuring free and conjugated drug in blood suggested inhibition of glucuronide formation (Rauramo *et al.*, 1963). In a similar study demethylation of pethidine and hydroxylation of promazine were studied by measuring the urinary excretion of metabolized and unchanged drugs (Crawford and Redofsky, 1966). The data pointed towards a decreased metabolism of pethidine but no differences for promazine. The mechanisms behind these changes during pregnancy are not clear but may be related to hormonal changes such as increased amounts of reduced progesterone derivatives. These may exert an inhibitory action on the activities of certain drug-metabolizing enzymes. Furthermore, complications of pregnancy introduce modifications in drug disposition at a time when medications are needed. For example, toxemia is associated with a further increase in the amount of body water and abnormal liver function tests are frequently reported.

### 3.3. The Placenta

The transplacental route is the only one available by which compounds taken by the pregnant woman reach the embryo or fetus. The placenta is a complex organ and the interplay of several factors influences the net transfer which occurs across it from the maternal arterial supply by way of the intravillous space into the umbilical veins. Maternal factors, and the process of labor influence the amount of drug offered to the fetus. In active labor the supply is reduced during contractions as a consequence of decreased blood flow. Drug transfer studies have been made by measuring the concentration of a drug in cord blood after its administration to the mother at specified times before delivery or prior to therapeutic abortions. Information about rate of equilibration has been pieced together from a number of separate investigations in different individuals. Some designs have used sequential fetal blood sampling during delivery. During the equilibration process the highest drug levels will be found in venous cord blood, the next highest in fetal arterial blood (including arteriolar cord blood), and the lowest in the fetal tissue as a whole, especially in fetuses with a poor blood supply.

Several mechanisms are involved in the transfer of both exogenous and endogenous substances across the placenta: simple diffusion, facilitated diffusion,

*CHARLOTTE CATZ and
SUMNER J. YAFFE*

active transport, and special processes. In simple diffusion, substances cross the placenta from a region of higher to one of lower concentration according to molecular size, ionic dissociation, and lipid solubility. Therefore, drugs with a high lipid solubility and a low ionic strength will diffuse rapidly, (e.g., barbituric acid derivatives). Facilitated diffusion occurs in proportion to the concentration gradient but at an accelerated speed by using a carrier system (e.g., glucose and other sugars). Some substances are transferred only after metabolic conversion by the placenta. Transport of riboflavin, for example, from maternal to fetal plasma appears to involve the uptake of flavin adenine dinucleotide by the placenta and its enzymatic cleavage into free riboflavin prior to its entry into fetal circulation. Basically, any substance present in sufficient concentration on the maternal side will eventually reach the fetus, and most drugs appear to cross by passive diffusion (Ginsburg, 1971).

Drug transfer is influenced by the metabolic and functional changes of the placenta that occur during the course of pregnancy. Tissue layers interposed between the fetal capillaries and maternal blood become progressively thinner, and surface area increases. Transport of sodium ion increases up to 36 weeks of gestation and then declines to parturition. The specific metabolic patterns of the placenta are influenced by aging, for example, a decline in the rate of anaerobic glycogenolysis occurs as term approaches. The pattern of intermediary metabolism can be altered by hypoxia and by the influences of hormones such as estrogen and cortisone. Moreover, certain disease states such as diabetes mellitus and toxemia produce morphological alterations and possibly functional changes of the placenta.

Drug metabolism has been demonstrated in human placenta homogenates and in trophoblast cells grown in tissue cultures (Juchau and Dyer, 1972). Oxidation reactions occur at term for certain compounds such as aniline and benzpyrene. The *in vitro* studies are not easily interpretable for the functional intact placenta, as homogenates contain parts from different types of cells, contaminants such as erythrocytes and plasma, and nonenzymatic catalysts for reduction reactions. Conjugation pathways are minimally active in general, and glucuronidation in particular has not been detected. In summary, it is impossible to assess at the present time the role played by the intact placenta in drug biotransformation. However, if the above-mentioned enzymic pathways are functional, they would modify the distribution of drugs between mother and fetus by altering their lipid solubility.

Simultaneously drugs and their metabolites can compete for enzymes catalyzing the conversion of endogenous substances, such as hormones, and thus disturb the internal environment of the fetus. Finally, they can inhibit the placenta's own metabolism and modify transport mechanisms requiring energy expenditure.

### 3.4. Placental Circulation

Placental circulation may be affected by drugs, and this in turn produces changes in the transfer of oxygen and nutrients as well as of the drug itself. Uterine blood flow when expressed per kilogram of total tissue (uterus, placenta, fetus) increases toward term, but when related to weight of fetus it decreases at the end of pregnancy. During labor, contractions may cause diminution of uterine blood flow, and this may protect the fetus from receiving large amounts of anesthetics and analgesic drugs. But in certain circumstances it may produce the opposite effect by keeping drugs within the fetus.

*In vitro* studies have shown that human umbilical arteries and veins respond only to alpha stimulation, and lack, or have nonfunctioning, beta receptors as demonstrated by a lack of vascular response to either isoproterenol or propanolol. Narcotic compounds, such as meperidine, morphine, and codeine, and hallucinogenic substances, such as LSD, affect placental circulation by causing vasoconstriction (Eltherington *et al.*, 1968). Hypoxia in the fetus is accompanied by hemodynamic changes in uterine blood flow, probably modifying drug transfer as well.

### 3.5. The Fetus

After crossing the placenta, substances administered to the mother reach the fetus. The drug exposure of fetal tissues is dependent on the kinetics of equilibration between maternal and fetal blood and on the drug's distribution within the fetus. The usage of drugs, such as anesthetics and analgesics, during parturition may or may not have a prolonged specific influence but certainly may affect the neonatal adaptation to the new external environment. There is no absorption into the fetal organism in the accepted classical sense. In general drugs are transferred from maternal to fetal circulation as described above. A foreign compound injected into the peritoneum crosses membranes according to the already stated basic pharmacological principles. However, intestinal absorption is operative in regard to components of amniotic fluid. An example is the presence of maternal albumin which is swallowed by the fetus and then digested in the gastrointestinal tract. The fetus swallows between 5 and 70 ml of amniotic fluid per hour, and the amount of drug presented to the intestine will depend upon its concentration. In theory, for some conjugated drugs excreted by the kidney, the active enzyme $\beta$-glucuronidase present in the intestinal mucosa would allow for the reabsorption and therefore recirculation of the parent drug. Another theoretical absorptive route is through the fetal skin, which is known to be permeable to water and probably other substances, including drugs.

### 3.5.1. Drug Distribution

The differences in pre- and postnatal circulation and their relationship to drug distribution are not known. In the fetus the major part of the blood in the umbilical vein passes directly through the liver rather than into the vena cava through the ductus venosus. The fetal brain receives a much greater proportion of the cardiac output than at later stages; consequently specific tissue concentrations may vary from those observed in the newborn infant. During hypoxia the circulation through the fetal brain may be increased. The total mass of red cells of the fetus explains, for instance, why trichloroethylene is taken up in greater amounts than by the maternal red cells. Body composition varies with fetal growth. Total body water decreases from about 94% of body weight at 16 weeks of gestational age to about 76% at term, which reflects a diminution in the amount of extracellular fluids. Thus water-soluble drugs have a greater volume of distribution at earlier stages of development. Fat is minimal in very young fetuses and is deposited mainly during the final trimester of intrauterine life. As a consequence, lipid-soluble drugs (e.g., thiopental, diazepam) vary in their distribution in accordance with fat content. For different organs, and therefore for specific effect of drugs, these variables are extremely important. An example would be the higher water content of the fetal brain together with a small degree of myelinization. This would limit the distribution of lipid-soluble compounds in the brain and influence their distribution to some other tissues away from the specific receptors. An example is diphenylhydantoin which is present in greater amounts in the adrenal cortex, myocardium, and corporea lutea than in the central nervous system (Mirkin, 1971).

Quantitative and qualitative differences exist between immature and mature individuals in the binding of drugs to plasma proteins. Some studies have compared neonate to adult and very few have considered the fetus. However, in some cases drug binding to fetal plasma proteins appears less efficient than in the newborn. Endogenous compounds may interfere with the process as well, and the implications for maternofetal drug distribution are obvious.

### 3.5.2. Receptors

Autonomic receptors are functional in the ileum of early human fetuses (12–24 weeks old). A cholinergic agent, acetylcholine, produces contractions and adrenergic substances, such as epinephrine, norepinephrine and isoprenaline, relaxation of the intestinal muscles. Dose–effect curves do not change with age, but the strength of response increases. Beta-blocking agents inhibit intestinal relaxation and alpha blockers are inactive. These results differ considerably from studies in adult humans where both alpha and beta receptors participate in the relaxation of intestinal muscles (McMurphy and Boreus, 1968).

The above refers to the only study considering developmental pattern in human fetuses. Receptors in different organs appear at various times of intrauterine life, as seen by measurable responses to external stimuli. Most likely their maturation proceeds at different rates in different organs, and this information, if acquired, would permit the planning of maternal and/or fetal therapy according to specific receptor sensitivity.

### 3.5.3. Metabolism

The fetus utilizes the placenta as the major excretory route involving crossing membranes. Foreign compounds must be lipid soluble. Therefore it appears that enzymic transformation into water-soluble compounds to facilitate renal excretion would be superfluous. Drug oxidation, a major pathway, involves an electron-transport system utilizing cytochrome $P_{450}$ as the final step. The components of this chain have been measured in human fetal liver and when expressed per gram of tissue present in liver microsomal preparations was found to be similar in amount in the fetus and adult (Yaffe *et al.*, 1970; Pelkonen *et al.*, 1973). Early experimental studies indicated that oxidizing activity was evident for endogenous substrates and not for xenobiotic compounds. However, more recently significant activity of the oxidizing pathway utilizing drug substrates has been shown (Rane and Gustafsson, 1973). Studies with some frequently utilized drugs have demonstrated that diphenylhydantoin, tricyclic antidepressants, and diazepam are biotransformed and that accumulation of some metabolites occurs in the fetus. Drug-metabolizing activity is not only found in fetal liver, but in a significant proportion in fetal adrenal tissues, which exhibit a high concentration of cytochrome $P_{450}$. Another observation developing over the last decade is that these enzyme activities can be significantly augmented by the administration of certain drugs, constituting the process known as drug induction. The possibility of utilizing this mechanism as a clinical tool for "fetal therapy" has been debated in the medical literature. No conclusions can be drawn at this time, and caution must be exerted until more fundamental data are obtained. The opposite phenomenon of drug inhibition must be viewed in a similar manner.

### 3.5.4. Excretion

At this stage of development the fetus utilizes the maternal organism as the chief excretory route for foreign compounds. Drugs are transferred once more across the placenta but in the opposite direction. The fetal kidney produces urine during the last trimester of pregnancy and contributes largely to the production of amniotic fluid in which drugs and metabolites of either maternal or fetal origin can be measured. The fetus swallows amniotic fluid, removing a limited amount of foreign compounds which in certain cases may be reabsorbed by the fetal intestine and therefore recirculated. The consequence of this process would be prolonged fetal exposure to drugs with the possibility of continuing effects which might or might not be deleterious to the developing organism.

## 3.6. The Newborn

The newborn infant handles medications in a different fashion than at later ages. The picture is complicated by continuous anatomical and physiological changes needed for adaptation to the external environment. As in any age group drug action will depend on several factors, including absorption, volume of distribution, plasma protein and tissue binding, receptor sensitivity, metabolism, and excretion.

### 3.6.1. Absorption

The rate and pattern of drug absorption are modified by changes in the morphology and function of the gastrointestinal tract. Some modifications such as bacterial colonization and rapid alterations in splanchnic circulation begin at birth. The surface area available will influence the rate and extent of absorption, and the gastrointestinal tract represents a larger portion of the body in the newborn than in later life. In addition, intestinal permeability appears to be augmented, and the passage of intact protein molecules has been observed in early life. Gastrointestinal absorption of drugs is regulated by two major factors: pH-dependent diffusion and emptying time, both of which undergo continuous changes after birth. Gastric pH is within the neutral range (6.5–8.0) at birth and falls during the first 24 hr to values of 1.0–3.0. After gastric acidity reaches its maximum, little acid secretion occurs for a week. Therefore relative achlorhydria is found for the next 9–10 days. Histological studies of the gastric mucosa demonstrate a close correlation with acid secretion. Premature infants have less gastric acid and a corresponding lack of mucosal development. Adult values for gastric pH are attained by 3 years of age. The low gastric acidity explains the greater serum concentrations of some drugs (penicillin G, ampicillin, and nafcillin) in the newborn than in older infants and children (Huang and High, 1953; Silverio and Poole, 1973; O'Connor *et al.,* 1965). This phenomenon is also observed at older ages in conjunction with hypochlorhydria and is due to less inactivation of penicillin.

*CHARLOTTE CATZ and
SUMNER J. YAFFE*

Transit time also varies with age. Gastric emptying is slower in the newborn and continues to be less than in adults until 6 months of age. Moreover, movements are irregular and unpredictable and can be modified by diet and feeding schedules. This has been observed in studies with riboflavin, which is absorbed in the proximal small intestine by a saturable transport process. Infants less than one week of age have a slow and prolonged absorption, and maximum concentrations achieved in serum were only one fifth of that in older individuals. However, the total amount of riboflavin absorbed was similar in both age groups since in the young absorption extended over a longer period (Jusko *et al.*, 1970). The results can be interpreted as the consequence of a slower transit rate, although this factor may be coupled to age-related alterations in the spatial distribution of transport processes along the gastrointestinal tract. A small but continuing increase in riboflavin absorption occurs from 3 months to adulthood. Other studies have indicated an augmentation in D-xylose absorption with age in a similar pattern to that recorded for riboflavin.

Following the intramuscular or subcutaneous administration of drugs, the rate of absorption will depend on the regional blood flow, which changes during the first 2 weeks of life. Variations exist in the relative blood flow to different muscular groups, and this may be accentuated under hypoxic conditions. A comparison of absorption of drugs during the neonatal period points to a less efficient mechanism for the oral absorption of phenobarbital (Wallin *et al.*, 1974), nalidixic acid (Rohwedder *et al.*, 1970), and phenytoin (Jalling *et al.*, 1970). Efficient oral absorption has been reported for various sulfonamides (Sereni *et al.*, 1968) and digoxin (Morselli *et al.*, 1975). Slow parenteral absorption has been reported for digoxin (Steiness *et al.*, 1974). Diazepam is absorbed efficiently following either intramuscular or oral administration (Morselli *et al.*, 1973).

### 3.6.2. Drug Distribution

Total body water (TBW), extracellular water (ECW), and intracellular water (ICW) are functional compartments where different types of drugs are distributed. Total body water is much greater in this age group and varies from 85% of body weight in small premature infants, to 70–75% in full-term infants, reaching about 60% at 1 year. The extracellular volume is approximately 40–45% of the body weight in comparison to 27% in the 1-year-old and 15–20% in the adult. This change reflects mainly interstitial volume as the plasma compartment remains relatively constant (4–5% of body weight) throughout life. Intracellular water increases from 34% at birth to 43% at 3 months of age and then decreases to a value equal to that of the newborn. Between years 1 and 3 all these compartments increase and then gradually decline to adult values. Fat content is lower in the premature (1%) than in the normal full-term infant (15%).

These variations are for the whole body, within a single organ or tissue the pattern may be different. In general drugs must be distributed in the ECW to reach their receptors, and the size of this compartment will influence drug concentration. With these variations in body composition, changes in drug distribution can be anticipated.

In plasma, drugs bind mostly to albumin, and this depends upon the molecular structure of the drug and the nature of the bond formed with the protein. It can be affected by qualitative or quantitative changes of the plasma proteins as well as by competition for binding sites with endogenous or exogenous compounds. Although full-term infants have plasma albumin in an equivalent value to that found in adults, total proteins and plasma protein binding reach adult values by 1 year of age (Miyoshi *et al.,* 1966). Salicylate protein binding in the plasma of newborns is reduced, therefore the fraction of free circulating drug increases (Krasner *et al.,* 1973). Simultaneously the newborn has a higher extracellular volume, which signifies that the apparent volume of distribution of salicylate is higher than in older children for similar concentrations (Levy and Garrettson, 1974). On the contrary, diazepam distribution appears to be lower as binding is equal at both ages, and this combines with a high lipid solubility of the compound and a reduced fat tissue in the newborn (Kanto *et al.,* 1974). Decreased binding resulting in a greater volume of distribution is reported for phenobarbital, which achieves high concentration in plasma for free drug (60–64%) and reaches even higher levels in infants with hyperbilirubinemia (Ehrnebo *et al.,* 1971). Diphenylhydantoin has an unbound fraction of 11% in infants compared to 7% in adults. Sulfaphenazole, sulphamethoxypyrazine, ampicillin, nafcillin, and imipramine have less affinity for neonatal plasma albumin.

The differences in binding between neonatal and adult plasmas may be due not only to a lower concentration of plasma protein but also to a qualitative difference in the binding properties. Endogenous substances during the first few days of life (hormones, free fatty acids) may occupy binding sites, thus reducing binding capacity. Indeed, Rane *et al.* (1971) showed that bilirubin competes with diphenylhydantoin for binding sites on the albumin molecule, and the unbound fraction doubled with increased bilirubin concentration.

Ignorance of differences in plasma protein binding in the neonate may have tragic consequences, as was noted in a controlled study of two prophylactic antibacterial regimes (Silverman *et al.,* 1956). Two groups of premature infants received either penicillin and sulfisoxazole or oxytetracycline, and a higher mortality rate with an increased incidence of kernicterus was reported for the sulfisoxazole treated group. These infants had a lower concentration of bilirubin in serum. The displacement of this endogenous compound from the albumin-binding site permitted its transfer into the brain. Sodium benzoate contained in injectable preparations of diazepam and caffeine causes a similar displacing phenomenon.

In summary, the decreased binding, the increased apparent volume of distribution of drugs, and the interactions with other substrates play an important role in the therapeutic and toxic effects of compounds administered to the immature organism.

### 3.6.3. Receptors

Drugs exert their effect at receptor sites. It is therefore important to determine whether receptors function during the perinatal period. The reactivity of alpha-adrenergic receptors in the iris, studied in premature and full-term infants, showed that they are effective and do respond to direct- or indirect-acting sympathomimetic drugs. Indeed, phenylephrine produces mydriasis of varying degree in all infants without any correlation to birth weight or gestational age. When tyramine or hydroxyamphetamine (which act by releasing norepinephrine) are instilled, there is a correlation between the extent of mydriasis produced and both gestational age and birth weight, mydriasis occurring only in the more mature babies. This data may signify that the immature infant is unable to release norepinephrine or that he has a shortage of stored norepinephrine.

3.6.4. Metabolism

CHARLOTTE CATZ and
SUMNER J. YAFFE

The newborn infant appears to have limited capabilities to biotransform drugs. The low activities of the enzymes responsible for mixed-function oxidation and glucuronide conjugation may be partially compensated by an adequate sulfation mechanism. Since fetal liver is able to produce sulfate conjugates of steroids, it is accepted that this function is mature at birth. The rate of maturation for different pathways appears to be dissimilar, as demonstrated by evaluation of several drugs administered to infants.

Conjugating capacity was studied for sulfobromophtalein following its administration to healthy full-term and premature infants. Half-lives were longer in premature infants than in mature ones, but the differences disappeared by 3 months of age. Bilirubin behaves similarly, and following intravenous administration the elimination half-life in newborns is much longer than in older children. The difference disappears by 8 weeks of age. The underlying mechanism for the increased retention of sulfobromophtalein was suggested to be hepatic immaturity and insufficient secretion into the bile, although other contributing factors, such as low concentration of hepatic carrier protein for bilirubin and BSP, must be considered. Conjugating capacity was studied for p-aminobenzoic acid, which is coupled to glucuronic acid and glycine, and p-aminophenol (the oxidized metabolite of acetanilide), which is glucuronidated before excretion (Vest and Rossier, 1963). Both were eliminated at a slower rate in newborn and premature infants. The decreased activity for oxidative pathways has been demonstrated in studies involving several drugs. p-Aminophenol, the oxidized metabolite of acetanilide, achieves a peak concentration in serum of newborn infants later than in older children.

Tolbutamide administered orally or intravenously to normal full-term babies has a prolonged plasma retention during the first two days of life (Nitowsky et al., 1966). The plasma disappearance of the drug shows an inverse correlation to the appearance of the oxidized metabolite, carboxytolbutamine, in the urine. Further proof of decreased oxidative ability was gathered when aminopyrine half-life was shown to decrease from the first to the eighth day of life in normal full-term infants (Reinicke et al., 1970). A functional deficiency in reduction mechanisms was suggested by observing that a decreased ratio of conjugated/free compounds follows the exogenous administration of cortisol to newborns. A decreased capability to form hydroxylated metabolites of diazepam and mepivacaine exists in the newborn. In the neonate the apparent plasma half-life is prolonged when compared to that of older children, and differences exist as well between full-term and premature infants. Two slower pathways seem to explain those results: low N-demethylation and almost absent hydroxylation. Hydroxylated metabolites of diazepam present in the urine of newborns are of maternal origin and not formed by the infant (Mandelli, 1975). Environmental factors, such as other drugs, modify drug-metabolizing activities. The possibility of increasing deficient enzyme systems in the newborn is an obvious therapeutic goal in certain clinical situations. It was observed retrospectively that children of epileptic mothers on phenobarbital treatment throughout pregnancy had a low serum bilirubin.

Phenobarbital given to the mother or directly to the newborn results in a significantly lower bilirubin concentration in the infant when compared to untreated control groups. The underlying mechanisms are complex, probably involving enhanced hepatic uptake, increased glucuronidation, and excretion into the bile. The increase in glucuronide formation was demonstrated by administering salicylamide and measuring the excretion of its conjugated metabolites in urine. The apparent plasma half-life of diazepam is decreased as well when infants are exposed to phenobarbital or other inducing agents either during intrauterine life or during the first few postnatal days. In these infants, an increase in hydroxylation leads to the presence of specific metabolites, changing the ratio between hydroxy and nonhydroxy derivatives in urine (Sereni, 1973). Diphenylhydantoin is eliminated rather slowly by the newborn, although the rate can achieve adult values following induction. For phenobarbital itself the formation of hydroxylated metabolites is limited during the first 24 hr of life, and then it increases progressively.

For other drugs, such as bupivacaine and carbamazepine, no differences were observed between plasma half-lives in newborn and adult individuals. This can be explained by the occurrence of self-induction as these drugs are administered chronically to pregnant women for specific indications (Rane *et al.,* 1975). Salicylate is a drug commonly used, and its disposition in the newborn has been studied in detail. The infants eliminate salicylate slower than mature individuals and thus have an increased exposure to the drug. Following a single dose of aspirin shortly before delivery, the qualitative pattern of metabolites excreted, in contrast to the quantitative, is similar at all ages. Chronic maternal ingestion of the drug produced a rapid maturation of conjugation with glycine (formation of salicylurate), although the total elimination of the drug was still slow. A plausible explanation is that the glucuronidation and the rate of renal excretion are immature (Levy, 1975).

### 3.6.5. Excretion

The majority of drugs, whether unchanged or metabolized, are removed from the body by renal excretion. In the newborn, kidney function has not yet reached its full development. The anatomy and function of the kidney evolve with age. All glomeruli are present by the end of 36 weeks of gestation, and the postnatal changes include elongation and maturation of the tubules of some of the nephrons. This would indicate that the full-term human infant has a structural integrity in contrast to the prematurely born who may have a deficiency in glomeruli and short, immature tubules.

After placental separation, a rapid increase in the renal circulation occurs in the first 2–3 days. Glomerular filtration studied by mannitol clearance may reach adult values at 10–20 weeks postpartum. Evaluation was also carried out by inulin-clearance technique. In infants varying in gestational age from 25 to 42 weeks, clearance was directly proportional to gestational age when measured at 2–3 days postpartum, and when measured at 1–9 weeks of age it was proportional to postconceptional age. It was similar to that of infants of similar gestational age studied at 2–3 days of age. Thiosulfate, a test substance undergoing more than 80% glomerular filtration, is eliminated three times slower in newborns. The differences from older children disappear by 3 weeks of age. The importance of mature glomerular filtration for drug elimination is best examplified by antibiotic agents of the aminoglycoside group (kanamycin, streptomycin, gentamicin) which are excreted unchanged. Their half-lives in premature infants are longer than in full-term newborns, and both are longer than in older children (Axline and Simon, 1964). In order to prevent an accumulation of these drugs in the body and the development of toxic effects, alterations in dosage and schedules of administration have been recommended. Digoxin, also dependent on glomerular filtration for its excretion, is cleared very slowly in the newborn (Morselli *et al.,* 1975). The glomerular filtrate

CHARLOTTE CATZ and
SUMNER J. YAFFE

will be altered by tubular function, which involves reabsorption and secretion. The tubular transport maximum for $p$-aminohippuric acid is markedly lower in the newborn infant, reaching adult values by 30 weeks of age. The dye phenol red is secreted up to 94% by the proximal tubules. The elimination half-life after intravenous administration to newborns is four times longer than in more mature individuals. Maturation appears to be completed within the first 6 months of life. Induction of the tubular secretion may occur by repeated drug exposure due to an increased synthesis of the carrier protein. Studies of tubular reabsorption, using the sulfonamide sulfamethoxypyracine, demonstrated that this drug, which undergoes high tubular reabsorption, has a prolonged elimination in the first days of life (Dost and Gladtke, 1969). Taken as an isolated observation, this would seem a contradiction if the postulate of a tubular reabsorptive maturation process is accepted. A plausible explanation is that it is extremely difficult to detect changes in reabsorption by measuring elimination half-life because glomerular filtration is also deficient at this stage of development. Drugs depending upon tubular secretion, such as penicillin G and a number of the semisynthetic penicillins, are eliminated slowly by newborns. Their rates of excretion are directly related to postnatal age, and they achieve adult levels by 3–4 weeks of age (Axline et al., 1967). Comparison of young premature (30–33 weeks gestational age) and small-for-gestational-age infants showed similar curves for excretion of ampicillin, but they showed a difference from that of older prematures (34–35 weeks). The latter have a peak excretion significantly higher in the first 2 hr, but the total drug excreted over 12 hr was identical in all three groups (Gamboa, 1976). The regulation of acid–base balance is not as efficient as at later stages of development, and this may affect excretion of drugs by the kidney. In mature individuals the urine pH changes rhythmically during the 24-hr day, according to periods of sleep and wakefulness. An acid pH occurs during sleep, and this will modify tubular reabsorption of drugs, depending upon their stage of ionization. These changes in acid secretion are not observed in infants less than 1 year of age, and the rhythmic urinary drug excretion pattern is absent. Indeed, evaluation of renal excretion of two sulfonamides confirmed this phenomenon in infants (Krauer et al., 1968).

### 3.7. Children and Adolescents

Only occasionally has a population from infancy until the end of adolescence been studied from the pharmacological point of view. The presumption is that no further changes occur in the processes of drug disposition and action after infancy. However, an investigation of imipramine binding in children 7–10 years showed less variability than in adults. Also the percentage of drug binding was not as high, and adult values were attained at puberty (Winsberg et al., 1974). Drug metabolism was assessed by determining the half-life of theophylline, which undergoes demethylation and mixed-function oxidation. A shorter half-life of the drug was obtained in the adolescent than in the adult. Children between 1 and 8 years of age metabolize antipyrine and phenylbutazone much faster than adult patients (Alvares et al., 1975). The same phenomenon is documented for acetomenophen. These fragmentary and sporadic observations indicate that for a variety of drug substrates the child has an increased metabolic capability. There has been no systemic evaluation of the interaction between the active process of growth that extends from the neonatal period to adulthood and the specific pharmacologic principles discussed in the previous sections. It is also fundamental to examine the influence that disease may have on normal processes. An example is the low serum concentrations attained with dicloxacillin in adolescent patients with cystic fibrosis when compared to normal individuals of the same age. The patients exhibited a high renal clearance of the drug, as well as of creatinine (Jusko *et al.*, 1975).

## 4. Drug Survey

A review of developmental pharmacology is incomplete without a short enumeration of published deleterious effects attributable to specific compounds. It is important to realize that favorable influences do occur as well but are often not reported. Absence of effects is not necessarily good, if fetal or newborn treatment is intended or if the foreign compound present in small amounts interacts with other drugs administered simultaneously.

A myriad of minor or major congenital malformations have been attributed to commonly used medications such as aspirin (Richards, 1969; Nelson and Forfar, 1971) and diazepam (Safra and Oakley, 1975). Other evaluations have pronounced aspirin safe during pregnancy with no increase in congenital anomalies in the fetuses exposed (Turner and Collins, 1975; Slone *et al.*, 1976). An increased number of birth defects are reported in diabetic women in general, and in many cases those have been specifically attributed to the use of sulfonylurea compounds (Campbell, 1963) and tolbutamide (Campbell, 1961; Schiff *et al.*, 1970). Skeletal deformities have been reported in one infant exposed to anticoagulants *in utero* (Petifor and Benson, 1975). It is extremely difficult to prove cause-and-effect relationships in the majority of cases, but the reports support either epidemiological conclusions or a strong suspicion by an educated observer.

The relationship of streptomycin to deafness and 8th nerve damage is well established (Conway and Birt, 1965; Robinson and Canibon, 1964). A similar situation applies to the masculinization of the female fetus by testosterone and 17-substituted steroid hormones. Several reports have appeared linking the usage of oral contraceptives in early pregnancy to an increase in congenital heart defects and transposition of the great vessels (Levy *et al.*, 1973; Nora and Nora, 1973). A prospective study suggests an association between the rate of limb reduction anomalies in male infants exposed during early development to progesterone/estrogen combinations (Janerich *et al.*, 1974; Nora and Nora, 1974). Presently, diphenylhydantoin is considered as a possible cause of cleft lip and palate and digital hypoplasia in infants receiving the drug during critical periods of development (Zellweger, 1974; Hanson *et al.*, 1976). On the other hand, Shapiro *et al.* (1976) consider the epileptic environment as the culprit and not necessarily the medication.

Functional changes in the fetus and newborn occur also as a consequence of maternal medication. Cigarette smoking by a pregnant woman reduces the proportion of fetal breathing movements (Manning *et al.*, 1975), and nicotine appears to be partly responsible for this observation (Genner *et al.*, 1975). Antithyroid compounds and iodine have produced goiters (Burrow, 1965; Jancu *et al.*, 1974). Antidiabetic agents resulted in severe hypoglycemia in the newborn (Zucker and Simon, 1968; Kemball *et al.*, 1970). Psychopharmacologic agents, and specifically chlorpromazine, caused extrapyramidal dysfunction (Tamer *et al.*, 1969). Progesterone therapy was responsible for tomboyish behavior and higher IQ in girls exposed *in utero* (Ehrhardt and Money, 1967). Increased fetal heart rate has been noted in response to maternal diazepam (Sereni *et al.*, 1973). Neonatal depression has been described after either general (Cohen and Olson, 1970) or local anesthesia during delivery (Rogers, 1970), magnesium sulfate therapy (Lipsitz, 1971), or the administration of hypnotics and sedatives (Ploman and Persson, 1957). The depression is described as a temporary phenomenon, but no specific follow-up of these infants has been carried to determine long-term effects (Brazelton, 1970; Scanlon *et al.*, 1974). Low birth weight and growth retardation are described in the offspring of addicted mothers (Zelson, 1973), of smoking women (Meyer *et al.*, 1976), and in those who had lead poisoning during the third trimester of pregnancy (Angle and McIntire, 1964). Maternal alcoholism causes intrauterine growth failure, including the brain (Jones *et al.*, 1973), and tetracycline has been implicated in a depression of long-bone growth which reverses when drug exposure ceases (Cohlan *et al.*, 1963). Early aborted fetuses exposed to oral contraceptives demonstrated a lethal growth disorganization (Poland and Ash, 1973). Increased growth, on the other hand, has been observed following prolonged therapy with $\beta$-agonist catecholamine derivatives (Unkehaun, 1974). The mechanism of action seems to derive from changes caused by the medication in maternal intermediary metabolism which in turn modifies fetal homeostasis. Extrauterine adaptive ability is not free from the influence of maternal medication. Neonatal depression mentioned previously may interfere with the sequence of changes needed to achieve homeostasis in the newborn. Indeed, decreased body temperature, low Apgar scores, and hypotonia may occur following intake of diazepam by the mother (Rosanelli, 1970; Cree *et al.*, 1973). *In utero* exposure to diethylstilbestrol has not been recognized as deleterious for the fetus or newborn; however, it is now unequivocally accepted as the etiological factor in the increased incidence of adenocarcinoma of the vagina in young adult women (Herbst *et al.*, 1971) and epididymal cysts with probable infertility in young adult men (Bibbo, 1975). The latter has been proven experimentally, and it may well apply to the clinical situation as well (McLachlan *et al.*, 1975).

## 5. Conclusions

The administration of a drug to the growing and developing infant or child presents a unique problem to the physician. Not only must he be constantly aware of the changes in drug dosage which are determined by alterations in processes of disposition at different ages, but he also must be cognizant of the fact that the drug may affect the developmental process itself and that this alteration may not be apparent for many years. The problem is made even more difficult when drugs are administered to the pregnant woman because she is then interposed between the drug actions and effects upon the intrauterine guest, which is more likely than not the unintended drug recipient. It is quite clear, therefore, that the task of establishing a basis for rational prescribing is extremely difficult in the developing organism. In addition, ethical considerations in evaluating drugs in pregnant women and children add to the constraints placed upon the drug prescriber. Immaturity and the general process of growth and development increase the risk potential, thereby demanding more substantial evidence of benefit. Nonetheless this data base must be obtained, and the information gap which currently exists in pediatric pharmacology must be narrowed and eliminated. The observation of drug effects must extend beyond traditional anatomic congenital malformations and consider influences on behavioral development, learning ability, and modifications in the processes of growth and maturation. Indeed we should shift our focus from the identification of adverse effects to a determination of optimal dosage. This goal can only be achieved after adequate study of drug disposition and action in the developing sick infant and child.

In the preceding sections we have attempted to highlight our current state of knowledge concerning drug action and disposition in the fetus, neonate, and older child. It is quite clear that information is far from adequate. In contrast, application of newer methodologies and laboratory techniques to human patients have generated a considerable body of data on drug disposition in adults during the past decade. Some of these techniques have been applied to infants and children, but most attention has been given to the neonatal period in which several serious adverse reactions have motivated additional studies. Data on drug disposition in other age groups, particularly in diseased children where drugs are prescribed for therapeutic purposes, are very scanty. The effects of illness on pharmacologic reactivity in the developing infant and child may be extremely great. The problem of furnishing the required evidence of safety and effectiveness in this important segment of the population can be resolved only by greatly intensified investigative efforts in the area of developmental pharmacology.

## 6. References

Alvares, A. P., Kapelner, S., Sassa, S. and Kappas, A., 1975, Drug metabolism in normal children, lead poisoned children and normal adults, *Clin. Pharmacol. Ther.* **17**:179–183.

Angle, C. R., and McIntire, M. S., 1964, Lead poisoning during pregnancy: Fetal tolerance of calcium disodium edetate, *Am. J. Dis. Child.* **108**:436–439.

Axline, S. G., and Simon, H. J., 1964, Clinical pharmacology of antimicrobial agents in premature infants. I. Kanamycin, streptomycin and neomycin, *Antimicrob. Agents Chemother.* **135**:138.

Axline, S. G., Yaffe, S. J., and Simon, H. J., 1967, Clinical pharmacology of antimicrobials in premature infants. I. Ampicillin, methicillin, oxacillin, neomycin and colistin, *Pediatrics* **39**:97–107.

Bibbo, M., 1975, Follow-up study of male and female offspring of DES-treated mothers, *J. Reprod. Med.* **15**:29–32.

Brazelton, T. B., 1970, Effect of prenatal drugs on the behavior of the neonate, *Am. J. Psychiatry* **126**:1261–1266.

Burrow, G. N., 1965, Neonatal goiter after maternal propylthiouracil therapy, *J. Clin. Endocrinol. Metab.* **25**:103–108.

Campbell, G. D., 1961, Possible teratogenic effect of tolbutamide in pregnancy, *Lancet* **1**:891–892.

Campbell, G. D., 1963, Chlorpropamide and foetal damage, *Br. Med. J.* **1**:59–60.

Ciancio, S. G., Yaffe, S. J., and Catz, C. S., 1972, Gingival hyperplasia and diphenylhydantoin, *J. Period.* **43**:411–415.

Cohen, S. N., and Olson, W. A., 1970, Drugs that depress the newborn infant, *Pediatr. Clin. N. Am.* **17**:835–850.

Cohlan, S. Q., Bevelander, G., and Tiamsic, T., 1963, Growth inhibition of prematures receiving tetracycline: A clinical and laboratory investigation of tetracycline-induced bone fluorescence, *Am. J. Dis. Child.* **105**:453–461.

Conway, N., and Birt, B. D., 1965, Streptomycin in pregnancy: Effect on the foetal ear, *Br. Med. J.* **2**:260–263.

Crawford, J. S., and Redofsky, S., 1966, Some alterations in the pattern of drug metabolism associated with pregnancy, oral contraceptives and the newly born, *Br. J. Anaesth.* **38**:116–131.

Cree, J. E., Meyer, J., and Hailey, D. M., 1973, Diazepam in labour: Its metabolism and effect on the clinical condition and thermogenesis of the newborn, *Br. Med. J.* **4**:251–255.

Csögor, S., Csutak, J., and Pressler, A., 1968, Modifications of albumin transport capacity in pregnant women and newborn infants, *Biol. Neonate* **13**:211–218.

Desmond, M. M., Rudolph, A., Hill, R. M., Claghorn, J. L., Dreesen, P. R., and Burgdorff, I., 1969, Behavioral alterations in infants born to mothers on psychoactive medication during pregnancy.

**CHARLOTTE CATZ and
SUMNER J. YAFFE**

Symposium on Mental Retardation, Texas Institute of Mental Sciences, Austin Texas, University of Texas Press, p. 235.

Dost, F. H., and Gladtke, E., 1969, Pharmacokinetik des 2-sulfanilamido-3-methoxypyrazin bein kind, *Arzneim. Forsch.* **19:**1304–1307.

Ehrhardt, A. A., and Money, J., 1967, Progestin-induced hermaphroditism: I.Q. and psychosexual identity in a study of ten girls, *J. Sex. Res.* **3:**83–87.

Ehrnebo, M., Agurell, S., Jalling, B., and Boreus, L. O., 1971, Age differences in drug binding by plasma proteins: Studies on human foetuses, neonates and adults, *Eur. J. Clin. Pharmacol.* **3:**189–193.

Eltherington, L. G., Stoff, J., Hughes, T., and Melmon, K. L., 1968, Constriction of human umbilical arteries: Interaction between oxygen and bradykinin, *Circ. Res.* **22:**747–752.

Fabro, S., and Sieber, S. M., 1969, Caffeine and nicotine penetrate the preimplantation blastocyst, *Nature (London)* **223:**410–411.

Gamboa, R., 1976, personal communication.

Genner, G., Marsal, K., and Brantmark, B., 1975, Maternal smoking and fetal breathing movements, *Am. J. Obstet. Gynecol.* **123:**861–867.

Ginsburg, J., 1971, Placental drug transfer, *Annu. Rev. Pharmacol.* **11:**387–408.

Hanson, J. W., Myrianthopoulos, N. C., Harvey, M. A. S., and Smith, D. W., 1976, Risks to the offspring of women treated with hydantoin anticonvulsants, with emphasis on the fetal hydantoin syndrome, *J. Pediatr.* **89:**662–668.

Heinonen, O. P., Shapiro, S., Monson, R. P., Hartz, S. C., Rosenberg, L., and Slone, D., 1973, Immunization during pregnancy against poliomyelitis and influenza in relation to childhood malignancy, *Int. J. Epidemiol.* **2:**229.

Herbst, A. L., Ulfelder, H., and Roskanzer, D. C., 1971, Adenocarcinoma of the vagina. Association of maternal stillborn therapy with tumor appearance in young women, *N. Engl. J. Med.* **284:**878–881.

Hill, R. M., 1973, Drugs ingested by pregnant women, *Clin. Pharmacol. Ther.* **14:**654.

Hirsh, J., Cade, J. F., and O'Sullivan, E. F., 1970, Clinical experience with anticoagulant therapy during pregnancy, *Br. Med. J.* **1:**270–273.

Huang, N. N., and High, R. H., 1953, Comparison of serum levels following the administration of oral and parenteral preparations of penicillin to infants and children of various age groups, *J. Pediatr.* **42:**657–668.

Jancu, T., Boyanower, Y., and Laurian, N., 1974, Congenital goiter due to maternal ingestion of iodide, *Am. J. Dis. Child.* **128:**528–530.

Jalling, B., Boreus, L. O., Rane, A., and Sjoqvist, F., 1970, Plasma concentrations of diphenylhydantoin in young infants, *Pharmacol. Clin.* **2:**200–202.

Janerich, D. T., Piper, J. M., and Glebatis, D. M., 1974, Oral contraceptives and congenital limb-reduction defects, *N. Engl. J. Med.* **291:**697–700.

Jones, K. L., Smith, D. W., Ulleland, C. N., and Streissguth, A. P., 1973, Pattern of malformation in offspring of alcoholic mothers, *Lancet* **1:**1267–1271.

Juchau, M. R., and Dyer, D. C., 1972, Pharmacology of the placenta, *Pediatr. Clin. N. Am.* **19:**65–79.

Jusko, W. J., Khanna, N., Levy, G., Stern, L., and Yaffe, S., 1970, Riboflavin absorption and excretion in the neonate, *Pediatrics* **45:**945–951.

Jusko, W. J., Mosovich, L. L., Gerbracht, L. M., Mattar, M. E., and Yaffe, S. J., 1975, Enhanced renal excretion of dicloxacillin in patients with cystic fibrosis, *Pediatrics* **56:**1038–1044.

Kanto, J., Erkkola, R., and Sellman, R., 1974, Perinatal metabolism of diazepam, *Br. Med. J.* **1:**641–642.

Kemball, M. L., McIver, C., Milner, R. D., Nourse, C. H., Schiff, D., and Tiernan, S. R., 1970, Neonatal hypoglycaemia in infants of diabetic mothers given sulphonylurea drugs in pregnancy, *Arch. Dis. Child.* **45:**696–701.

Krasner, J., Giacoia, G. P., and Yaffe, S. J., 1973, Drug-protein binding in the newborn infant, *Ann. N.Y. Acad. Sci.* **126:**101–114.

Krauer, B., Spring, P., and Dettli, L., 1968, Zur Pharmakokinetik der Sulfonamide im ersten Lebensjahr. *Pharmacol. Clin.* **1:**47–53.

Kuratsune, M., Masuda, Y., and Nagayama, J., 1976, Some of the recent findings concerning Yusho. Conference Proceedings National Conference on Polychlorinated Byphenyls, Environmental Protection Agency, Office of Toxic Substances, pp. 14–29.

Levy, G., 1975, Salicylate pharmacokinetics in the human neonate, in: *Basic and Therapeutic Aspects of Perinatal Pharmacology*, (P. L. Morselli, S. Garattini, and F. Sereni, eds.), pp. 319–330, Raven Press, New York.

Levy, G., and Garrettson, L. K., 1974, Kinetics of salicylate elimination by newborn infants of mothers who ingested aspirin before delivery, *Pediatrics* **53:**201–210.

Levy, E. P., Cohen, A., and Fraser, F. C., 1973, Hormone treatment during pregnancy and congenital heart defects, *Lancet* **1:**61.

Lipsitz, P. J., 1971, The clinical and biochemical effects of excess magnesium in the newborn, *Pediatrics* **47:**501.

Mandelli, M., Morselli, P. L., Nerdio, S., Pardi, G., Principi, N., Sereni, F., and Tognoni, G., 1975, Placental transfer of diazepam and its disposition in the newborn, *Clin. Pharmacol. Ther.* **17:**564–572.

Manning, F., Wyn Pugh, E., and Boddy, K., 1975, Effect of cigarette smoking on fetal breathing movements in normal pregnancies, *Br. Med. J.* **1:**552–553.

McLachlan, J. A., Newbold, R. R., and Bullock, B., 1975, Reproductive tract lesions in male mice exposed prenatally to diethylstilbestrol, *Science* **190:**991–992.

McMurphy, D. M., and Boreus, L. O., 1968, Pharmacology of the human fetus; adrenergic receptor function in the small intestine, *Biol. Neonate* **13:**325–339.

Medalie, J. H., Serr, D., Neufeld, H. N., Brown, M., Berandt, N., Sternberg, M., Sive, P., Schoenfeld, S., Fuchs, Z., and Karo, S., 1972, The use of medicines before and during pregnancy: Preliminary results from an epidemiological study of congenital defects, *Adv. Exp. Med. Biol.* **27:**481–487.

Meyer, M. B., Jonas, B. S., and Tonascia, J. A., 1976, Perinatal events associated with maternal smoking during pregnancy, *Am. J. Epidemiol.* **103:**464–476.

Mirkin, B. L., 1971, Diphenylhydantoin placental transport, fetal localization, neonatal metabolism and possible teratogenic effects, *J. Pediatr.* **78:**329–337.

Money, J. W., 1971, Prenatal hormones and intelligence: A possible relationship, *Impact Sci. Soc.* **21:**285–290.

Morselli, P. L., Principi, N., Tognoni, G., Reali, E., Belevedere, G., Standen, S. M., and Sreni, F., 1973, Diazepam elimination in premature and full term infants and children, *J. Perinat. Med.* **1:**133–141.

Morselli, P. L., Assael, B. M., Gomeni, R., Mandelli, M., Marini, A., Reali, E., Visconti, U., and Sereni, F., 1975, Digoxin pharmacokinetics during human development, in: *Basic and Therapeutic Aspects of the Perinatal Pharmacology* (P. L. Morselli, S. Garattini, and F. Sereni, eds.), pp. 377–392, Raven Press, New York.

Miyoshi, K., Saijo, K., Kotani, Y., Kashiwagi, T., and Kawai, H., 1966, Chaacteristic properties of fetal human albumin (Alb.F) in isomerization equilibrium, *Tokushima J. Exp. Med.* **13:**121–132.

Nelson, M. M., and Forfar, J. O., 1971, Associations between drugs administered during pregnancy and congenital abnormalities of the fetus, *Br. Med. J.* **1:**523–527.

Nitowsky, H. M., Matz, L., and Berzofsky, J. A., 1966, Studies on oxidative drug metabolism in the full-term newborn infant, *Pediatr. Pharmacol. Ther.* **69:**1139–1142.

Nora, J. J., and Nora, A. H., 1973, Birth defects and oral contraceptives, *Lancet* **1:**941–942.

Nora, J. J., and Nora, A. H., 1974, Can the pill cause birth defects?, *N. Engl. J. Med.* **291:**731–732.

O'Connor, W. J., Warren, G. H., Edrada, L. S., Mandala, P. S., and Rosenman, S. B., 1965, Serum concentrations of sodium nafcillin in infants during the perinatal period, *Antimicrob. Agents Chemother.* **5:**220–222.

Pelkonen, O., Kaltiala, E. H., Larmi, T. K. I., and Karki, N. T., 1973, Comparison of activities of drug-metabolizing enzymes in human fetal and adult livers, *Clin. Pharmacol. Ther.* **14:**840–846.

Petifor, J. M., and Benson, R., 1975, Congenital malformations associated with the administration of oral anticoagulants during pregnancy, *J. Pediatr.* **86:**459–462.

Ploman, L., and Persson, B. H., 1957, On the transfer of barbiturates to the human foetus and their accumulation in some of its vital organs, *J. Obstet. Gynecol. Br. Commonw.* **64:**706–708.

Poland, B. J., and Ash, K. A., 1973, The influence of recent use of an oral contraceptive on early uterine development, *Am. J. Obstet. Gynecol.* **116:**1138–1142.

Rane, A., and Gustafsson, J. A., 1973, Formation of a 16,17-transglycolic metabolite from a 16-dehydro-androgen in human fetal liver microsomes, *Clin. Pharmacol. Ther.* **14:**833–839.

Rane, A., Lunde, P. K. M., Jalling, B., Yaffe, S. J., and Sjoqvist, F., 1971, Plasma protein binding of diphenylhydantoin in normal and hyperbilirubinemic infants, *J. Pediatr.* **78:**877–882.

Rane, A., Bertilsson, L., and Palmer, L., 1975, Disposition of placentally transferred carbamazepine Tegretol[R] in the newborn, *Eur. J. Clin. Pharmacol.* **8:**283–284.

Rauramo, L., Pulkkinen, M., and Hartiala, K., 1963, Glucuronide formation in partureints, *Ann. Med. Exp. Biol. Fenn.* **41:**32–37.

Reinicke, C., Rogner, G., Frenzel, J., Maak, B., and Klinger, W., 1970, Die Wirkung von Phenylbuta-zone und Phenobarbital auf die Amidopyrin-Elimination, die Bilirubin-Gesamtkozentration in Serum und einige Blutgerinnungsfaktoren beim neugeborenen Kindern, *Pharmacol. Clin.* **2:**167–172.

Richards, I. D. G., 1969, Congenital malformations and environmental influences in pregnancy, *Br. J. Prev. Soc. Med.* **23:**218–225.

**CHARLOTTE CATZ and
SUMNER J. YAFFE**

Robinson, G. C., and Canibon, K. G., 1964, Hearing loss in infants of tuberculosis mothers treated with streptomycin during pregnancy, *N. Engl. J. Med.* **271:**949–951.

Rogers, R. E., 1970, Fetal bradycardia associated with paracervical block anesthesia in labor, *Am. J. Obstet. Gynecol.* **106:**913–916.

Rohwedder, H. J., Simon, C., Kubler, W., and Hohfnauer, M., 1970, Untersuchungen über die Pharmakokinetik von Nalidixinsaure bei Kindern verschiedenen Alters, *Z. Kinderheilkd.* **109:**124–134.

Rosanelli, K., 1970, Über die Wirkung von pranatal verabreichtem Diazepman auf das Fruhgeborene, *Geburtsh. Fraunheilk* **30:**713–724.

Safra, M. J., and Oakley, G. P., Jr., 1975, Association between cleft lip with or without cleft palate and prenatal exposure to diazepam, *Lancet* **2:**478–480.

Sardemann, H., Madsen, K. S., and Frus-Hansen, B., 1976, Follow-up of children of drug-addicted mothers, *Arch. Dis. Child.* **51:**131–134.

Scanlon, J. W., Brown, W. V., Jr., Weiss, J. B., and Alper, M. H., 1974, Neurobehavioral responses of newborn infants after maternal epidural anesthesia, *Anesthesiology* **40:**121–128.

Schiff, D., Aranda, J. V., and Stern, L., 1970, Neonatal thrombocytopenia and congenital malformations associated with administration of tolbutamide to the mother, *J. Pediatr.* **77:**457–458.

Sereni, F., 1973, The need for further data, *Clin. Pharmacol. Ther.* **14:**662–665.

Sereni, F., Perletti, L., Marubini, E., and Mars, G., 1968, Pharmacokinetic studies with a long-acting sulfonamide in subjects of different ages, *Pediatr. Res.* **2:**29–39.

Sereni, F., Mandelli, M., Principi, N., Tognoni, G., Pardi, G., and Morselli, P. L., 1973, Induction of drug metabolizing enzyme activities in the human fetus and in the newborn infant, *Enzyme* **15:**318–329.

Shapiro, S., Slone, D., Hartz, S. C., Rosenberg, L., Siskind, V., Monson, R. R., Mitchell, A. A., Heindnen, O. P., Idanpaan-Heikkila, J., Haro, S., and Saxen, L., 1976, Anticonvulsants and parental epilepsy in the development of birth defects, *Lancet* **1:**272–274.

Silverio, J., and Poole, J. W., 1973, Serum concentrations of ampicillin in newborn infants after oral administration. *Pediatrics* **51:**578–580.

Silverman, W. A., Anderson, D. H., Blanc, W. A., and Crozier, D. N., 1956, A difference in mortality rate and incidence of kernicterus among premature infants allotted to two prophylactic antibacterial regimens, *Pediatrics* **18:**614–621.

Slone, D., Heinonen, O. P., Kaufman, D. W., Siskind, V., Monson, R. R., and Shapiro, S., 1976, Aspirin and congenital malformations, *Lancet* **1:**1373–1375.

Steiness, E., Svendsen, O., and Rasmussen, F., 1974, Plasma digoxin after parenteral administration. Local reaction after intramuscular injection, *Clin. Pharmacol. Ther.* **16:**430–434.

Tamer, A., McMey, R., Arias, D., Worley, L., and Fogel. B. J., 1969, Phenothiazine-induced extrapyramidal dysfunction in the neonate, *J. Pediatr.* **75:**479–480.

Turner, J., and Collins, E., 1975, Fetal effects of regular salicylate ingestion in pregnancy, *Lancet* **2:**338–339.

Unkehaun, V., 1974, Effects of sympathomimetic tocolytic agents on the fetus, *J. Perinat. Med.* **2:**17–29.

Vest, M. F., and Rossier, R., 1963, Detoxification in the newborn: The ability of the newborn infant to form conjugates with glucuronic acid, glycine, acetate and glutathione, *Ann. N.Y. Acad. Sci.* **111:**183–197.

Wallin, A., Jalling, B., and Boreus, L. O., 1974, Plasma concentrations of phenobarbital in the neonate during prophylaxis for neonatal hyperbilirubinemia, *J. Pediatr.* **85:**392–398.

Winsberg, B. G., Perel, J. M., Hurwic, M. J., and Klutch, A., 1974, Imipramine protein binding and pharmacokinetics in children, in: *Phenothiazines and Structurally Related Drugs* (I. S. Forrest, C. J. Carr, and E. Usdin, eds.), pp. 425–431, Raven Press, New York.

World Health Organization, 1967, Principles for the testing of drugs for teratogenicity: Report of a WHO scientific group, *WHO Tech. Rep. Ser.* **364:**1–18.

Yaffe, S. J., Rane, A., Sjoqvist, F., Boreus, L. O. and Orrenius, S., 1970, The presence of a monooxygenase system in human fetal liver microsomes, *Life Sci. (II)* **9:**1189–1200.

Zellweger, H., 1974, Anticonvulsants during pregnancy: A danger to the developing fetus?, *Clin. Pediatr.* **13:**338–346.

Zelson, C., 1973, Infant of the addicted mother, *N. Engl. J. Med.* **288:**1393–1395.

Zucker, P., and Simon, G., 1968, Prolonged symptomatic neonatal hypoglycemia associated with maternal chlorpropamide therapy, *Pediatrics* **42:**824–825.

# 4

# *Glimpses of Comparative Growth and Development*

## R. A. McCANCE and
## ELSIE M. WIDDOWSON

### 1. Introduction

Minot (1891) pointed out that from conception to maturity guinea pigs grew at an average rate of 1.82 g/day, rabbits at one of 6.3 g/day, and man at one of 6.69 g/day. Men therefore grew larger than rabbits because they grew longer but rabbits grew larger than guinea pigs because they grew faster. This was the first time there had been numerical demonstration of the two ways by which one could attain a large size, and Minot concluded that it would be well worthwhile to make a study of this. We agree, and this chapter is an attempt to show what has been done.

### 2. Growth, Rates of Growth, and Age

#### 2.1. Conception to Birth

All mammalian species start life as a single cell, the fertilized ovum, and this cell does not differ much in size whatever the ultimate mass of the organism is going to be. Growth in the early stages of gestation in all species is brought about entirely by cell division, with little or no increase in cell size, so at first the fetus can only grow as fast as its cells can divide. Cell division, however, goes on more rapidly in the first weeks after conception in some species than in others. The rat, for example, increases from a single cell to two to three thousand million during its first three weeks of growth (Winick and Noble, 1965), at which time it leaves the uterus weighing 5 g. Adolph (1972) has gathered together information about the ages from

---

***R. A. McCANCE*** • Sidney Sussex College, Cambridge, England.    ***ELSIE M. WIDDOWSON*** • Addenbrooke's Hospital, Cambridge, England.

R. A. McCANCE and
ELSIE M. WIDDOWSON

conception at which various species reach a weight of 1 g. These are shown in Table I. The hamster, rat, and rabbit grow to a weight of 1 g in about 2 weeks, the pig and cat in 4–5 weeks, while the human fetus takes 8 weeks to reach this weight. The rate of growth soon after conception sets the pace for the subsequent growth *in utero,* hence the weights at birth for similar lengths of gestation. This is illustrated by the growth of three species shown in Figure 1. This figure shows the weight of the human fetus at various ages of gestation (Lubchenco *et al.,* 1963) compared with those of the fetal lamb (Joubert, 1956) and foal (Meyer and Ahlswede, 1976). The scale used for weight is a continuously doubling one since this seemed the best way of expressing growth in the early stages of development. The lamb is born at about the same body weight as the human infant, but the gestation period is only half as long. The sheep fetus grows more rapidly than the human initially, and the differences in weight at each gestational age thereafter depend upon the rates of growth and hence the weights achieved during the first few weeks after conception. The fetus of the horse has a gestation period longer than that of man; at 160 days it weighs almost the same as the lamb at term (146 days).

Figure 2 extends this to other species. It shows the weights at birth of 17 species plotted against their length of gestation. The points appear to fall on three curves. The small rodents, mouse, rat, and rabbit, grow extremely rapidly—more rapidly than the sheep or horse, and it looks as though this must also be true of the fetal whale. The cat and pig, and presumably also dog, guinea pig, goat, sheep, porpoise, and hippopotamus grow less rapidly after conception, but they still grow very fast. Man, the ox, the macaque monkey, and probably also the elephant, grow more slowly in the early stages, and this results in lower weights after similar times in the uterus. Minot's (1891) generalization therefore applies from conception to birth as well as from conception to maturity. The newborn hippopotamus is larger than the newborn human baby because it grows faster. The human fetus grows much larger than the rat before it is born because it stays in the uterus longer. Its cells have much more time to double and redouble their number, and, moreover, the cells of the organs of the human fetus begin to increase in size by about 25 weeks gestation (Widdowson *et al.,* 1972*a*), whereas growth of the organs of the rat by cell hypertrophy does not begin until after birth (Winick and Noble, 1965).

Not only do species grow in size at different rates before birth, but their progress towards maturity is faster in some species than others. Animals born after a short gestation period tend to be less mature at birth than those born after a longer time in the uterus. Thus the mouse, rat, rabbit, kitten, and puppy are born with eyes

Table I. Time from Conception at which Six Species Reach a Weight of 1 Gram[a]

|  | Days |
|---|---|
| Hamster | 14.5 |
| Rabbit | 17.0 |
| Rat | 17.5 |
| Pig | 26 |
| Cat | 36 |
| Man | 55 |

[a]From Adolph, 1972.

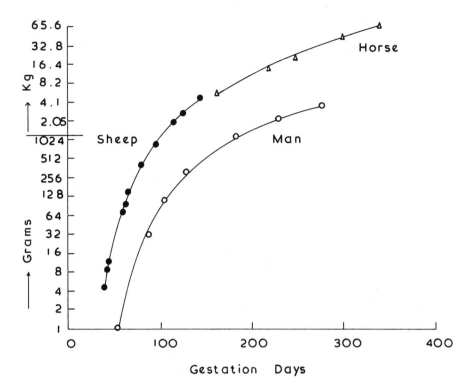

Fig. 1. Growth before birth of the sheep, horse, and man.

and ears closed, although the newborn rabbit has sharp well-developed incisor teeth. However, although these species are less mature than the human infant at birth they are much more mature than the human fetus after the same time from conception. The guinea pig is even more remarkable in that after about 10 weeks gestation it is born so mature that it runs about and even begins to eat solid food within the next few days.

Progress towards maturity is characterized by fundamental changes in the chemical composition of the body. One of these is a fall in the percentage of water coupled with a rise in the percentage of solids as the cells make protein and the bones become calcified. Figure 3 shows the percentage of solids in the lean body tissue of the developing human fetus compared with percentages for other species at their time of birth. Not only do these other species grow in size more rapidly than the human fetus, but the percentage of solids in their bodies increases more rapidly too. Some have less solids per 100 g of lean body tissue at the time of birth than the human infant, but at the same time from conception they all have considerably more.

## 2.2. Birth to Maturity

Brody and Ragsdale (1922) reported that the growth of animals could be represented as taking place in three waves or phases—an infantile, a juvenile, and an adolescent one. These phases were identified by plotting the rates of growth in weight in each species in terms of g/unit of time. Each phase was symmetrical about

R. A. McCANCE and
ELSIE M. WIDDOWSON

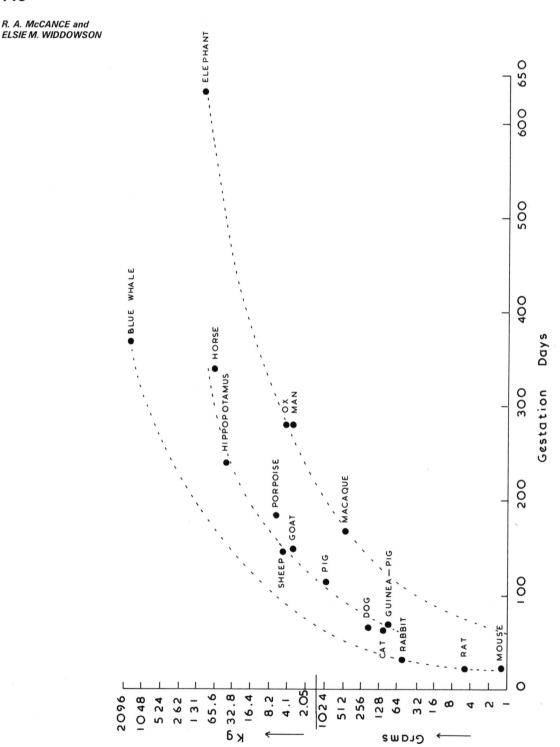

Fig. 2. Birth weights of 17 species and their lengths of gestation. The dotted curves show that the species can be classified roughly into those that grow (1) very fast from conception whether they are destined to become small or large animals; (2) fast; and (3) comparatively slowly.

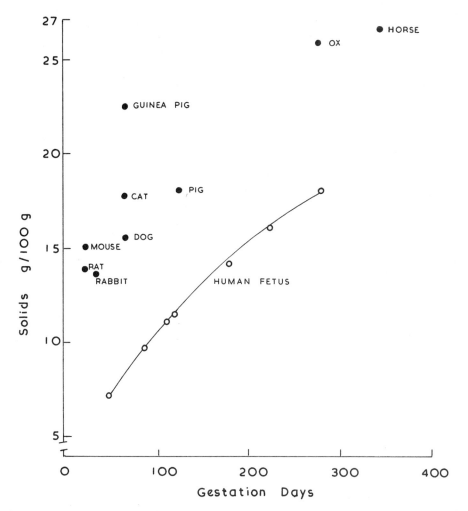

Fig. 3. Percentage of water in the lean body tissue of nine species at birth and of the human fetus at various ages from conception to birth.

its center, at which time the rate of growth for that particular phase was at its peak. The whole of the infantile phase took place before birth in the calf and reached a maximum before or soon after in the majority of other animals they studied. The juvenile phase was easily identified in the mouse, rat, guinea pig, rabbit, lamb, and calf, but not in man, in whom it was replaced by a long period of steady growth. The peak of the adolescent phase they found to coincide with puberty. Brody (1945) later pointed out that if the mature weight of each species was taken to be 100 and a suitable time scale was adopted for the growth curve of each species, the curves for all species except man became virtually superimposable from birth, but that the peak rate of the adolescent growth phase (i.e., sexual maturity) fell at very different points along the curve. The idea of a suitable time scale for comparing the growth curve of two different species was not a new one, for Donaldson (1908) had employed the same technique some time before to show how the growth of the rat's brain could be made a suitable "model," so far as it went, for that of man. This division of growth into three phases seems to have been adopted by those interested in the growth of primates (Napier and Napier, 1967; Schultz, 1969), without their appreciating the central part puberty and sexual maturity played in Brody's thinking, or how his curves were derived. Consequently the same terms have been applied to different periods on the time scales of growth.

Brody *et al.* (1926) considered also the linear dimensions of growth in some farm animals, and Brody (1945) requoted those for the cow, but made his generalizations on the basis of weight. This was perhaps natural in one interested mainly in domesticated farm animals whose market value depends upon weight rather than length or height, and weight is what matters commercially. Weight is the correct parameter to use for some comparisons, but it is now generally conceded to be a poor index of maturity, and poorest of all perhaps in man. The trouble about weight is the dominating part played by the preceding nutrition of the animal in determining what it shall be and the difficulty of measuring it in some species.

There were some very fine students of growth in the first half of this century, D'Arcy Thompson, Minot, Donaldson, and Jackson among them, and Brody probably thought as much about growth as any man alive today, but he had his limitations. He lacked a sufficiently wide experience of human growth, and much of the modern evidence about it (and about the growth of wild species) was not available to him. This led him to suggest that the absence of the juvenile phase of accelerated growth in man was due to undernutrition. He would not, moreover, have placed so much emphasis on the late development of puberty in man had he realized that the elephant, with nearly the same life span, also takes a long time to reach puberty, more than 10 years sometimes, or that other animals as well as man do not become sexually mature till they are fully grown or nearly so. The blue whale, for example, becomes sexually mature at 5 or 6 years of age at a length of about 23.5 m and may not grow to more than 25.5 m. This would make puberty of a young female whale occur when she was 93% of her mother's mature length. The average girl of 14 at menarche is 98% of her mature height. A boy of 15 is 96% of his. Similar figures are available for the see whale and the fin whale, which become sexually mature at 87 and 90% of their mature length (Alpers, 1963; Asdell, 1965; Slijper, 1962; Small, 1971). The female northern fur seal becomes sexually mature at 3–4 years of age at a length of about 110 cm and may not grow much longer, although she may only be 50% of her mature weight if she lives another 5 or 10 years (Scheffer and Wilke, 1953). The male becomes sexually mature not long after the female at a weight of 50 kg and a length of about 120 cm, and only then do his weight and length begin to increase in a postpubertal growth spurt which, by 9 or 10 years of age, leaves his weight upwards of 200 kg and length 200 cm. The elephant seal grows in a similar way (Laws, 1953, 1956*a,b*), and in this species the males are sexually and functionally mature enough by 4 years of age to copulate in the water with the virgin females 2 or 3 years old. Tanner (1962) considered that in the case of the fur seals we were dealing with a remarkable postpubertal growth spurt, confined entirely to the male. But are we? The female northern fur seals do go on increasing in length to some extent after puberty and may double their weight before they are fully mature. The female elephant seals grow from a length of 390 cm at puberty to 450 cm after it. It is true that the males grow much more, but so do they in other species that become sexually mature long before they have stopped growing, e.g., the rat and pig. There is one thing more to consider: The male elephant seal is in rut for a relatively short season each year. After this, his gonads regress in size and their production of hormones is greatly reduced for the next 9 months or so. Could this help to promote the growth of the body?

The large sea animals are not the only mammals other than man to become sexually mature when they are nearly or quite full grown. The beagle is said to be fully grown before it reaches puberty (Anderson, 1970) which may be unusual. Other breeds of dog, however, seem to be not far off full size at sexual maturity. It is impossible to be exact about a species, the females of which only come on heat at six monthly intervals, but the evidence available about boxers indicates that they are about 90% of their mature height at the shoulder at their first estrus and are 87% of their mature weight (Ireland, 1977). Hayden and Gambino (1966) reported that the small pocket mouse, whose adult weight is only 8–10 g, attained adult dimensions before it would breed, but here again one must be cautious in assessing the age of "puberty" in short-lived rodents with the capacity of bearing several litters a year, and seasonal periodicity in breeding.

Figure 4 shows diagrammatically the height and weight of a number of mam-

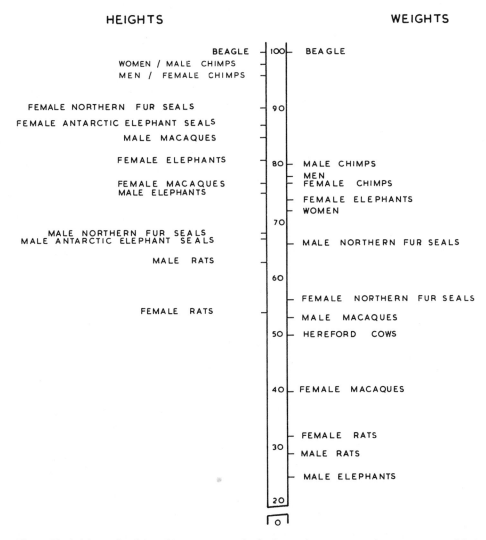

Fig. 4. The heights and weights of an assortment of animals at puberty expressed as a percentage of their heights and weights at maturity. Sources: Altman and Dittmer (1972); Cupps *et al.* (1969); Foote (1969); Gavan (1953); Gregory *et al.* (1963); Grether and Yerkes (1940); Hafez (1962); Hanks (1969); King (1964); Laws (1966); Laws *et al.* (1975); Perry (1954); Quirang (1939); Thompson (1942).

mals at puberty as a percentage of their mature height and weight. By height is meant length for the marine animals, shoulder height for the elephant and bovines, and crown–rump length for rats and the primates other than man. The figures can only be approximate, for the exact age of puberty is not all that easy to define. Menarche in the primates for example, or the first estrus of an animal, cannot be equated with sexual maturity as defined by the capacity to reproduce. Some facts, however, are clear: (1) All species included in Figure 4, except perhaps the beagle, are nearer their mature height at puberty than their mature weight. No figures could be found for the height of Hereford cattle, but from data given by Brody, particularly Brody *et al.* (1926), the generalization appears to be true of Holstein and Jersey cattle. The cows of both breeds seem to be about 42% and 31–32% of their mature heights and weights, respectively, at first estrus. (2) In a number of species the males continue to grow for a considerable time in height or weight or both after puberty. The males of the northern fur seal, the Antarctic elephant seal, the African elephant, and the rat do so, and there are others. (3) Females may also have a long period of growth in height and weight after puberty, for example cattle, northern fur seals, and rats. (4) In the light of the data in this figure, generalizations are probably more likely to be misleading than valuable.

The one thing which stands out about human growth is not that puberty falls so late on the physical growth curve, but the exceptional acceleration of growth that precedes and accompanies it. It takes man a long time after puberty to become mentally mature, but it may take an elephant nearly as long. Probably not, however, for nature has not endowed the elephant with the potential possibilities of man. His trunk is useful, but a man's hands are much more so, and man has a larynx which can make accurate articulation, and therefore speech, possible. He also has a wide range of binocular color vision. Puberty in man may be at an age of 15 or so, but who is to say when a man's intellectual powers or physical skills are at their prime? Roger Bannister was 25 before he ran a mile in less than four mintues, and Picasso continued to paint highly original pictures till he was an old man. Donaldson (1895) wrote thoughtfully about this, and so did Tanner (1962); one can only wish that they had committed a few more of their thoughts to paper.

## 3. The Chemistry of Growth

Without knowing any biochemistry, early students of growth realized that the process they were studying implied change. They could see that no two parts of the body grew at the same rate, and when histology and biochemistry began to make their contributions to the subject, it became clear that the composition of the body must be changing as well as its shape. "La fixité du milieu intérieur" was a great thought, but Moulton (1923) made another stride forward by pointing out that the great variable in an animal was the amount of fat in it. Apart from fat, the composition of the lean body mass of all adult mammals was fundamentally the same in terms of water, protein, and ash. Furthermore, although the proportions of these change during growth, they change in a consistent manner. Order had been produced out of chaos, and everything that has been discovered since that time has supported this. It is now well established that there is an age of chemical maturity just as there is of sexual maturity. Each species reaches its age of chemical maturity at a chronological time from conception peculiar to itself and in keeping with its lifespan. Before this age the percentage of water and concentration of sodium in the fat free body tissue fall while the concentration of protein, potassium, and calcium rise (Widdowson and Dickerson, 1964). In all species the soft tissues reach chemical maturity before the bones.

Widdowson and Dickerson (1964) produced an historical and factual review of the mineral composition of the body and its organs, both of animals and man. Table II shows that at 80 days from conception the human fetus still had more sodium and chloride and less nitrogen, potassium, calcium, phosphorus, and magnesium per kg of lean body tissue than a newborn mouse, rat, rabbit, guinea pig, kitten, and puppy, although these are born after considerably less than 80 days in the uterus. This is further evidence that the small species develop chemically, as well as grow in size, considerably faster than man. Even after another 200 days gestation the human fetus at term is not as mature chemically as the newborn guinea pig with a gestation period of 67 days. The human fetus begins, however, to lay down fat in its body at a much earlier stage of chemical development of its lean tissues than most other species (Widdowson, 1950). Even the newborn calf and foal, which are far more highly developed in every other way than the human infant, have very little fat in their adipose tissue at term (Meyer and Ahlswede, 1976).

Thus, although chemical development follows fundamentally the same pathways before and after birth, the rate at which the changes take place varies a great deal from one species to another. It would be interesting to know whether the blue whale, which from Figure 2 seems as though it must grow very rapidly in the early weeks after conception, but which spends a year in the uterus, is a fast or slow chemical developer before it is born.

The mineral composition of the body provides many examples of species, sex, and age differences. While the ratio of calcium to phosphorus in bone is similar in all species of mammals, the degree of calcification of the matrix is not. Man has less highly calcified bone than the pig, cat, rabbit, and rat (Widdowson and Dickerson, 1964). Nevertheless, the concentration of calcium in the lean part of the adult human body is nearly twice that of other animals whose bodies have been analyzed. Large animals must have thick bones to support their weight, and the weight of the skeleton increases at a faster rate than that of the body as one moves from small species to large ones (Schmidt-Nielsen, 1972). An elephant probably has a larger percentage of calcium in its body than man.

The red cells of cats, dogs, cattle, goats, and some breeds of sheep contain more sodium than potassium, in contrast to all other species examined, which have

*Table II.  Chemical Composition of Newly Born Mammals Compared with that of a Human Fetus*

| Species | Gestation (days) | Weight (g) | Fat (g/kg) | Fat-free body (g/kg) | | | | Fat-free body (meq / kg) | | |
|---|---|---|---|---|---|---|---|---|---|---|
| | | | | N | Ca | P | Mg | Na | K | Cl |
| Rat | 21 | 5 | 11 | 16 | 3.1 | 3.6 | 0.25 | 84 | 65 | 67 |
| Rabbit | 31 | 60 | 40 | 18 | 4.8 | 3.6 | 0.23 | 78 | 53 | 56 |
| Cat | 63 | 118 | 18 | 24 | 6.6 | 4.4 | 0.26 | 92 | 60 | 66 |
| Dog | 63 | 200 | 14 | 21 | 4.9 | 3.9 | 0.17 | 81 | 58 | 60 |
| Guinea pig | 67 | 95 | 100 | 29 | 12.3 | 7.5 | 0.46 | 71 | 69 | — |
| Man | 80 | 11 | 5 | 9 | 1.7 | 1.8 | 0.08 | 108 | 49 | 84 |
| | 280 | 3500 | 160 | 23 | 9.6 | 5.6 | 0.26 | 82 | 53 | 55 |

R. A. McCANCE and
ELSIE M. WIDDOWSON

more potassium than sodium (Bernstein, 1954). The dog, fox, and other carnivores are also different from other species in that they have higher concentrations of zinc in their eye tissues; the zinc is located in the tapetum lucidum cellulosum (Weitzel et al., 1955).

There are sexual differences in the amount of iron stored by the rat in its liver and kidneys, the female having much higher concentrations than the male (Widdowson and McCance, 1948). This has not yet been demonstrated in other species. Female rats also have more highly calcified bones than males of the same age (Zucker and Zucker, 1946).

Age has been shown to affect the concentration of copper in the liver in man, pig, sheep, and cat. In human liver at birth it is over 10 times that in the liver of the adult (Widdowson et al., 1972b). One must be careful to distinguish between the concentration and the total amount of a substance present in an organ like the liver which is likely to increase materially in size during growth to maturity. However, in man, the difference between the concentration of copper in the liver of the newborn and adult is so great that the whole liver at birth has more copper in it than at any subsequent age.

Coupled with the high concentration of copper in the liver at birth, the concentration of copper in the serum of the fetus is very low, and with it that of caeruloplasmin (Rosen, 1976; Smith, 1976; Walshe, 1976). The mean of the normal range for an adult is given by Markowitz et al. (1955) as 34 mg/100 ml of plasma. It is known to be higher in pregnancy, and Scheinberg et al. (1954) reported about 60 mg for maternal, and a mean of 6.5 mg/100 ml for cord serum. This relationship between the serum and the liver at birth is similar to that found in Wilson's disease.

Much that we know about the mineral composition of the body has been obtained by analysis of bodies and tissues after death. Minerals may seem dry as dust and unable to live, but they once formed parts of living matter and were as essential to it as all the complex organic molecules that often seem so much more exciting.

## 4. Organs and Tissues

This is only the beginning of the story on growth and development in the generally accepted sense of the terms, and much might have been written about the rest of it. To keep within the scope of their commitments the authors decided that they could only write about representative aspects which came within their field of experience or in which they had the advantage of expert advice and help. We have therefore selected the development only of some organs, biochemical systems, and individual functions and have concluded with a few words about the diversity among species in the way they develop their immunity.

### 4.1. The Skin

The skin is a good example of an organ that has a generalized structure with exceptional developments which are characteristic of a particular species or genus. All mammals have skin and man is no exception, and all skin consists of water, extracellular minerals and collagen, and cells, all of which are affected by age in a similar way (Widdowson and Dickerson, 1964). Thus the composition of the organ in most species is basically the same even if it undergoes special development as in the ox, hippopotamus, and rhinoceros, but it is usual to have the skin covered with hair and provided with glands. As hair grows it may become specially hardened for protection (as in the hedgehog) or localized parts of it highly sensitive to movement (as in some of the shrews and other rodents). It may also have peculiarities of color, texture, and distribution which change with age.

Some of these modifications of the hair and glands are instigated by the sexual and other hormones and only develop their special characteristics at puberty or at estrus. As we all know, the skin is the one part of the body which the world, and particularly members of the opposite sex, can see and feel and smell. Thus the pubic and axillary hair is developed in both the sexes in man under the influence of androgens from the adrenal gland and the facial and other bodily hair of the male by testosterone. Peptide hormones from the pars intermedia of the pituitary, controlled and modified by the paraventricular nucleus of the hypothalamus (Thody, 1974), are responsible for the production of sebum from the follicular glands of all species at all ages. These hormones modify the secretions at puberty and estrus to produce their own characteristic smell (Thody and Shuster, 1975). These melanocyte-stimulating hormones from the pars intermedia are also responsible for some of the physiological functions (Burton *et al.,* 1975; Goolamali and Shuster, 1975) and pathological pigmentation after sexual maturity in man (Gilkes *et al.,* 1975).

The skin is inevitably involved to a major extent in the provision of protection and thermal regulation in all animals, the latter most conspicuously perhaps in those with exocrine glands such as the horse and man, but one must remember that without any sebum to keep the hair oiled few animals would be able to live the lives they do in their present environments. Since it is so visible and accessible, more is probably known about the skin and its appendages in man than in any other animal, except perhaps in a limited way in skins of commercial value for leather, wool, or adornment.

## 4.2. Proteins of the Circulation

### 4.2.1. Hemoglobin

In the course of evolution hemoglobins have appeared in various phyla, including some of the lowest, although they are not as widespread a tool of nature as cytochrome (Barcroft, 1928). Given some heme and two pairs of unlike chains of amino acids with the necessary properties the thing is done so to speak—and all mammals have done it. In the course of development every human being produces at least three hemoglobins: the embryonic forms, of which there may be two, each consisting of two $\alpha$ chains and two $\epsilon$ chains; and the fetal form, which predominates in the circulation before birth and consists of $\alpha$ chains and $\gamma$ chains. Even adults still retain 1% of their total hemoglobin in the fetal form. There are two adult hemoglobins, 96% with the chain formula $\alpha_2 \beta_2$ and about 3% with the formula $\alpha_2 \delta_2$. Apart from the $\epsilon$ chains, the structure of these hemoglobins is known in great detail. Furthermore, many instances of amino acid substitution by gene mutations at various points in the chains are also known, along with the resulting changes of function (Huehns and Beaven, 1971; Wood, 1976).

Fetal hemoglobin has a higher affinity for oxygen than adult hemoglobin, and this shift of the dissociation curve to the left is of benefit to the fetus. Monkeys, goats, sheep, and cattle also have embryonic and fetal hemoglobins, but cats, pigs, and horses have no hemoglobin of the fetal type (Bailey and Zobrisky, 1968). In spite of this distribution, it is possible to establish some functional generalization in the way mammalian fetuses are provided with the suitable oxygen affinities. Both the pH of the red cell and the concentration of organic phosphates and carbon dioxide in the erythrocytes affect the oxygen affinity of the hemoglobin in them. It is due to these effects that the hemoglobin dissociation curve of an adult's red cells shifts to the left at high altitudes. In mammals these organic phosphates are compounds of the 2,3-diphosphoglycerate type which bind to the hemoglobin at particular sites on the non-$\alpha$ chains, and by so doing shift the dissociation curve to the right. Their place is taken by inositol hexaphosphate in reptiles and birds. The strength of the attachment varies with the amino acid sequence, and hence the affinity of that particular hemoglobin for oxygen. As already stated the horse has no fetal hemoglobin, but its erythrocytes contain no enzymes synthesizing 2,3-diphosphoglycerate and the hemoglobin in the cells behaves like fetal hemoglobin when the oxygen tension demands it.

Glucose-6-phosphate dehydrogenase is a red cell enzyme with somewhat variable characteristics transmitted by a gene carried on the X chromosome, and its absence, or deficiency, alters the metabolism of the glutathione in the red cells in such a way that with certain hemoglobins it alters their resistance to diseases, particularly malaria, but the interest in this is largely confined to human medicine. Indeed, the amount of hemoglobin in the blood, its importance in disease, and its ready accessibility has made man the spearhead of all hemoglobin investigations (Lehmann and Huntsman, 1974). Comparative studies in other species are of profound interest scientifically, but they have little practical value to human or veterinary medicine at present.

### 4.2.2. Erythropoietin

Once quantitative methods for determining the concentration of hemoglobin in circulating blood became available, it was recognized that this changed little in a healthy person. There were diseases in which it was too low and the sufferers were breathless, and one, polycythemia rubra vera, in which the numbers of red cells per $mm^3$, and the concentration of hemoglobin, were far too high. It was later noted that a rise in the number of red cells and in the hemoglobin was a normal event on moving from sea level to a high altitude and that high concentrations were the rule in residents of those areas. How was this all brought about?

It was thought at first that the lack of oxygen acted as a direct stimulus to the marrow, but Carnot and Deflandre (1906) suggested that the marrow might not be the sensing organ (for other references see Krantz and Jacobson, 1970; McCance, 1972a). Not till 1957 was it recognized that this organ was the kidney (Jacobsen *et al.*, 1957; Naets, 1958; Thorling, 1969; Krantz, 1970; Krantz and Jacobson, 1970). The kidney responds to hypoxia by secreting a glycoprotein termed erythropoietin or an enzyme precursor of it. Its molecular weight has been stated to be about ±65,000 (Goldwasser and Kung, 1968), but it now seems to differ according to the source from which the preparation is made (Giovannini and Bocchini, 1975). In adult dogs and rodents the kidney is preeminently its site of production (Kuratowska *et al.*, 1961; Krantz, 1970). It is also the most important site in man, but in all species there must be another site, or a similar agent, produced elsewhere, possibly in the liver or spleen, and this may become more active in the rodents after nephrectomy (Gordon *et al.*, 1967; Shaldon *et al.*, 1971). Some erythropoietin is produced even in man without kidneys or spleen, but its concentration in the tissues may not be regulated in the normal way (Erslev *et al.*, 1968). Others, however, have found that erythropoietin in nephrectomized rats and in man was liberated in response to the usual anoxia stimuli (Naets and Wittek, 1968; Mirand *et al.*, 1969; Fried *et al.*, 1969, 1971).

The juxtaglomerular apparatus or the glomerulus has been suggested as the part of the kidney responsible for the production of erythropoietin (Hartroft *et al.*, 1969; Busuttil *et al.*, 1971). The falling gradient of oxygen from cortex to medulla, however, makes it worth considering the latter as the part of the kidney most exposed to fluctuations of oxygen and, therefore, a possible site for the production of erythropoietin or its precursor. The prostaglandins may be involved, but this is not yet proven (Mujovic and Fisher, 1974). The active principle may be released as a precursor, as already stated, or with a lipid inhibitor (Erslev *et al.*, 1971). At all events erythropoietin, however and wherever produced, is intensely active and is excreted in the urine (Gordon *et al.*, 1967). In the light of this it is not surprising that substances have been described in plasma which can progressively modify its determinable activity (Lindemann, 1970; Editorial, 1972*a,b*). Erythropoietin acts on a stem cell that precedes the recognizable erythroblast or granulocytoblast, and the rest of the process of maturation then goes on without it. Its production rises in pregnancy and in response to thyroid hormones, adrenalin, and androgens, but estrogens do not actively stimulate its production.

Erythropoietin activity in the fetus and newborn is more complicated and there are undoubtedly species differences of one sort and another. In the mouse, rat, and sheep there is no evidence that erythropoietin plays any part in regulating red cell production in the yolk sac (Krantz and Jacobson, 1970). The production of red cells then shifts to the liver and spleen. In rats and mice these persist as the main sites until after birth; adult, not fetal, hemoglobin is produced, and the process is accelerated by erythropoietin which is probably not produced in the kidneys (Lucarelli *et al.*, 1968). The most recent suggestion is that the liver produces a substance which separates with the light mitochondrial fraction of the cells. A similar and possibly identical compound can be separated from the kidney. Both act on a substance already circulating in the plasma (as renin does), and erythropoietin is the result (Zucali and Mirand, 1975). If substantiated, this is a constructive thought. The production of red cells only becomes active in the marrow of the mouse and rat some time between birth and weaning, and removal of the kidneys 5 or 10 days after birth reduces the production of erythropoietin but makes little if any difference to the appearance of the marrow (Carmena *et al.*, 1968). One must always remember that in the first 10 days after birth in the rat the hypothalamic nuclei and tracts are being laid down and coordinated with the pituitary, sex glands, and all the other functional centers that take charge of somatic stability (Widdowson and McCance, 1975). In guinea pigs, which are one of the most mature animals at birth, the situation in the adult has already been reached by the time the fetus leaves the uterus.

In man erythropoietin has been found in cord blood as early as the 30th week of gestation (Halvorsen and Finne, 1968). This corresponds to about the 10th day after birth in the rat. There are traces in amniotic fluid, and its concentration there and in fetal serum rises when the fetus is poorly oxygenated (Finne, 1968; Krantz, 1970). It passes the placenta to a very slight extent in either direction in women (Krantz and Jacobson, 1970), and its production by the human fetus appears to be regulated autonomously regardless of the $O_2$ tension of the mother's blood; this was demonstrated experimentally in the fetal lamb by Zanjani *et al.* (1969).

Obscure though the details are, these observations on erythropoietin and the way it works are a contribution to the physiology of red cell production and hemoglobin formation. However, there are still many gaps in our knowledge (Krantz, 1970). Are the embryonic red cells produced without its aid in all species? Whence comes the erythropoietin that stimulates erythrocyte production in the liver and spleen of the mouse and rat, or indeed in the marrow of guinea pigs, sheep, and man until they are born—and later if the kidneys are removed. The evidence about adults has already been quoted and that about the newborn is equally convincing, for infants with bilateral renal agenesis are not born short of red cells or hemoglobin (Bain and Scott, 1960) and no evidence that they were came to light in a search through postmortem records at Cambridge (McCance, unpublished). According to Potter and Craig (1976), if such infants are born alive, death usually takes place within 48 hr, preceded by dyspnea and cyanosis.

Even if all these difficulties about species differences and sites of production had been ironed out, an outstanding developmental unknown remains to be discovered. What is it that brings about the replacement of one type of hemoglobin by another during development (Jonxis and Nijhof, 1969; Huens and Beaven, 1971; Paul, 1976; Wood, 1976)? Thurmon *et al.* (1970) showed that erythropoietin played a part in altering the type of hemoglobin formed in sheep red cells in anoxic states or at high altitudes. The production of the two hemoglobins in these sheep is under gene control, and directing the way the stem cells develop would be a new role for erythropoietin. If, however, as McClearn (1976) states, "we can accept as a working principle that developmental processes are under genetic influence," then the switchover from one hemoglobin to the next must also be under the same control. So many purebred strains of mice are now known that it might be worth finding if they all make the switch at the same developmental age, or if a strain can be bred which has a different timing.

### 4.2.3. Plasma Proteins

The proteins in the fetal plasma, like those in the red cells, are mostly synthesized by the fetus. The major exceptions are probably the IgG proteins, of which several classes exist. Their molecular weight is relatively small and is only of the order of 160,000 (Rosen, 1976). These cross the placenta by some highly specific transport mechanism. They also appear in smaller amounts in the amniotic fluid (Adinolfi, 1971).

Many of the proteins in fetal plasma appear to be synthesized by the liver. Ceruloplasmin and the albumin certainly are, as well as the proteins known as fetuins, which are only produced and used during fetal life. Calf fetuin was probably the first to be discovered (Pedersen, 1944), and the concentration in the plasma is high at birth. Goats, sheep, and pigs also produce a fetal protein in their plasmas, and so does man, the so-called $\alpha$-fetoprotein which has recently leaped into clinical prominence owing to its diagnostic significance both in fetal plasma and amniotic fluid. These compounds are all glycoproteins.

There may be similar proteins which bind to polysaccharides in the plasma of all developing mammals, but they will each almost certainly be species specific. From what we know about hemoglobin and myoglobin, moreover, variants are likely to turn up with slightly different properties. Indeed, they already have, for the transferrins of the fetus are not all the same as those of the adult, and those of individuals consist of a series of related globulins which are genetically determined and can be used as anthropological markers. Functionally the known transferrins are all similar (Adinolfi, 1971; Lehmann and Huntsman, 1974).

Since we have little, if any, really concrete evidence about what switches on and off the production of the various hemoglobins produced during fetal life (Wood, 1976), it would be rather optimistic at present to expect to have much evidence on this matter about the plasma proteins, which have been so much less explored.

## 4.3. The Development of the Kidney and Its Functions

The weight of the kidney as a percentage of that of the body tends to fall with development except during early embryonic life and in a few species in which there is a temporary rise immediately after birth. It was suggested by Mott (1973) that the rise took place only in those species that were born in an immature state, such as cats, dogs, and rabbits. In these animals the glomeruli and cortex are still undergoing development at birth when the kidney is faced with the task of maintaining the stability of the internal environment of an independent organism. There is a rise, however, in the pig (Widdowson and Crabb, 1976), which is roughly at the same stage of development as the baby at birth; there is no rise in the lamb (Mott, 1973).

The function of the kidney before birth has been a difficult problem in comparative development, and it concerns not only the function of the kidney but that of fetal fluids as well. The lamb has been studied in some detail by Alexander *et al.* (1958*a,b*) and by Alexander and Nixon (1961). The fetus in this species has a mesonephros which is functional for some time before the 90th day of gestational age and discharges a variably hypotonic urine into the allantoic sac, the contents of which are also hypotonic, although urea and creatinine tend to accumulate in it. Later in gestation the urine is passed into the amniotic sac in which nitrogenous end-products are also retained. The kidneys and fluids of the rabbit, calf, and pig develop in a similar way (Dickerson and McCance, 1957; Zweymuller *et al.,* 1959), but not so those of the rat, guinea pig, or man. At no time before birth, moreover, is the fetal kidney in sole control of the volume and composition of the fetal fluids, for this is also the responsibility of the fetal membranes and the placenta. Stanier (1971), for instance, has shown that sodium is reabsorbed from the renal pelvis of fetal pigs, and France *et al.* (1972) found that the same thing happens in the bladders of fetal lambs. In the last 20 years intricate internal systems have come to light. Both in rabbits and in man, for example, the lungs add a certain amount of fluid to that already in the sac (Jost and Policard, 1948; Carmel *et al.,* 1965). In man the volume of amniotic fluid depends not only upon the amounts passed into it by the kidney but also those removed from it by being swallowed by the fetus. Hence esophageal atresia leads to polyhydramnios and renal agenesis to oligohydramnios. The human fetus survives both genetic abnormalities, but in the latter is deformed by the pressure of the uterus. Fetal monkeys, lambs, and pigs survive bilateral removal of the kidneys; not so rabbits in which the blood pressure falls and the amniotic fluid disappears so that in this species the fetus is crushed by the contracting uterus (Berton, 1970) or dies from a profound fall of blood pressure (Mott, 1973). How the forces are balanced to maintain the desirable amount of fluid is not known. In the rat another complication is introduced by the vascularized trilaminar omphalopleure. Ligature of the vascular supply to this in the fetal rat or guinea pig leads to the whole structure becoming fibrous and white and the fetus grossly edematous (Le Goascogne, 1964, 1968; Everett, 1933; Noer and Mossman, 1947; Paul *et al.,* 1956; Petter, 1969).

With the possibility of finding two independent vascular supplies to the fetus and its membranes, and a kidney which may empty consecutively into two sacs or only into one, there are many opportunities for species differences and still more possibilities. Some are well known. Fructose, for instance, rather than glucose, is the main sugar circulating in the fetal blood and fluids of cattle, sheep, pigs, and hippopotami (Bernard, 1855). Other more recent discoveries have still to be fitted into the framework of knowledge.

### 4.3.1. Renin

Renin has been found in a number of fetal sites apart from the glomerular tufts in the kidney and the human chorion *leave,* among others (Brown *et al.,* 1964; Skinner *et al.,* 1968; Symonds *et al.,* 1968). Less appreciable amounts were found by these authors in human amniotic fluid and amnion. The work of North and Segal (1976) makes it seem unlikely that similar results will be obtained in guinea pigs, but renin has been located in the uterus of the rabbit (Ferris *et al.,* 1967; Lumbers, 1973), and there is much more in the plasma of newborn puppies than in that of an adult of the same species (Grainger *et al.,* 1971; Mott, 1975). The same applies to the fetal lamb (Broughton Pipkin *et al.,* 1974).

Hitherto interest in the function of the renin angiotensin system has tended to center on its role in blood pressure regulation and to a lesser extent on sodium transport and water metabolism (Vander, 1967; McCance, 1972*a,b*). These interests are still being pursued (Stella *et al.,* 1976). Since 1975, however, the experiments of Frederiksen *et al.* (1975) and Brumbach *et al.* (1976), although not very physiological (Blendstrup *et al.,* 1975), have indicated that the cells containing renin are sensitive osmometers, and the retention of renin in them depends upon active metabolism and a regulated cell volume which is interfered with by a fall of temperature, ethacrynic acid, and ouabain. Bertoncello *et al.* (1976) consider that its release depends upon renal innervation and vascular tone being normal but not on receptors transmitting through the vagus. With the known hypotonicity of the allantoic and amniotic fluids, the presence of renin in the human chorion, and the fact that ouabain prevents the normal operation of the rat's trilaminar omphalopleur (Petter, 1969), there seems to be some chance now of bringing order out of chaos in spite of the bewildering species differences which have prevented any generalizations about the discoveries already made. It may even be possible to bring the renin found in parts of the brain by Ganten *et al.* (1971) under the same umbrella.

Compared with all these complexities before birth, the development of the classical functions of the kidney after birth are more or less straightforward. The general principles were summarized by McCance (1972*a,b*) and again from a more historical point of view (McCance, 1977). Again, there are many species differences, but their place in nature has often been well defined (Schmidt Nielsen, 1958; 1975).

## 5. Placentation and Production of Immunity

If one goes further back in life to a study of placentation, one can find a range from the simple yolk sac placenta of birds and some marsupials to the hemochorial one of man. There is, moreover, a range of variations and species differences among them (Amoroso, 1961; Steven, 1975).

Throughout gestation the fetus has to be provided with all its supplies and disposal services and, sometime before birth or soon after, with enough passive immunity to carry it over till it can provide itself with its own armamentarium against infections and other invasions of its "person." Some account of how this is done was given by Brambell (1958) and again by Hemmings and Brambell (1961). The rabbit and guinea pig are provided with all the immune bodies the mother can pass to them before birth, and the transfer is made through the cells of the yolk sac placenta. This is embryologically an outgrowth of the gut. In these two animals, as in others, the absorption is selective and the selectivity varies with the species but never depends only on molecular size. In the ox, goat, sheep, pig, and horse the immune bodies are absorbed from the mother's colostrum in the first 36 hr or so of postnatal life. In the dog, mouse, and rat the antibodies are absorbed before birth and for a variable time after birth, 20 days in the case of the rat, i.e., practically throughout lactation. In man the antibodies that get to the fetus from the mother do so through the placenta, and these are mainly, if not entirely, those of the IgG class. Human colostrum contains other antibodies, but these are not absorbed from the gut, although the IgA group may have a function there in preventing ingested bacteria causing diarrhea. Any IgM found in cord blood was probably formed slowly in the fetus in response to fetal infection. The matter is complicated, for IgD molecules scarcely pass the placenta at all, but the levels in the maternal circulation rise considerably during pregnancy. The newborn and still more so the fetus react very slowly and badly to antigens, but cell-mediated responses can be detected as soon as small lymphocytes appear in the circulation (Miller, 1966; Rosen, 1976), so from that time onwards the mechanisms are there.

The great problem, however, in physiological immunity is this: How can an immunologically qualified mother, after she has conceived, come to tolerate a fetal allograft with a constitution partially paternal, and to nurture it throughout gestation instead of casting it off and so rejecting it? Amoroso and Perry (1975) may have come up with a unifying concept which would cover the whole range of vivipary. They suggest that the human trophoblast, and the analogous structures in other species, produce glycoproteins, the chemical structure of which is in no case yet known, but to which the names of human chorionic gonadotropin and human chorionic somatomammotropin (or human placental lactogen) have been assigned. To these should perhaps be added another, recently identified and termed pregnancy-specific $\beta_1$ glycoprotein (Towler *et al.,* 1976). These substances circulate in the maternal blood, cover the maternal surface of the trophoblast, and block the action of the maternal lymphocyctes, which would otherwise have treated the trophoblast as they would an allograft and rejected it out of hand. The steroids produced by the placenta are suspected of having some adjuvant action in all this, but the link up is not yet clear.

## 6. Conclusions

The search for an animal model to further our knowledge of man has become fashionable, and the phrase itself has become a modern cliché. It is safe to say that the search will make little difference to the work being done on animals, which is frequently undertaken for its own sake. It is interesting, however, to speculate how much would be known about human growth requirements and the physiology of growth had animals not been studied. It is indeed the value of a comparative approach that we have tried to show, for it is the one way which enables one to realize the importance of species differences and hence of picking the appropriate animal with which to solve the problem at hand or perhaps initiate some widespread biological principle. If those interested only in human nutrition want an animal model, they are crying for the moon unless they have the patience and the ability to find it for the particular purpose they have in mind.

## 7. References

Adinolfi, M., 1971, Genetic polymorphism and *in vitro* cultures of the ontogenesis of human fetal proteins, in: *The Biochemistry of Development* (P. Benson, ed.), pp. 224–247, Heinemann, London.

Adolph, E. F., 1972, Development of physiological functions, in: *Nutrition and Development* (M. Winick, ed.), pp. 1–25, Wiley, New York.

Alexander, D. P., and Nixon, D. A., 1961, The foetal kidney, *Br. Med. Bull.* **17:**112.

Alexander, D. P., Nixon, D. A., Widdas, W. F., and Wohlzogen, F. X., 1958a, Gestational variations in the composition of the foetal fluids and foetal urine in the sheep, *J. Physiol., London* **140:**1.

Alexander, D. P., Nixon, D. A., Widdas, W. F., and Wohlzogen, F. X., 1958b, Renal function in the foetal sheep, *J. Physiol., London* **140:**14.

Alpers, A., 1963, *Dolphins,* John Murray, London.

Altman, P. L., and Dittmer, D. S., (eds.), 1972, *Biological Data Book,* 2nd ed., Vol. 1, pp. 138–139, Federation of American Societies for Experimental Biology, Washington, D.C.

Amoroso, E. C., 1961, Histology of the placenta, *Br. Med. Bull.* **17:**81.

Amoroso, E. C., and Perry, J. S., 1975, The existence during gestation of an immunological buffer zone at the interface between maternal and foetal tissues, *Phil. Trans. R. Soc. Ser. B* **271:**343.

Anderson, A. C., 1970, Reproduction, in: *The Beagle as an Experimental Dog* (A. C. Anderson, ed.), pp. 31–39, Iowa State University Press, Ames, Iowa.

Asdell, S. A., 1965, *Patterns of Mammalian Reproduction,* 2nd ed, Constable, London.

Bailey, M. E., and Zobrisky, S. E., 1968, Changes in proteins during growth and development of animals, in: *Body Composition in Animals and Man,* pp. 87–125, Publication 1598, National Academy of Sciences, Washington, D.C.

Bain, A. D., and Scott, J. S., 1960, Renal agenesis and severe urinary tract dysplasia: A review of 50 cases with particular reference to the associated anomalies, *Br. Med. J.* **1:**841.

Barcroft, J., 1928, *The Respiratory Function of the Blood,* University Press, Cambridge.

Bernard, C., 1855, *Leçons de physiologie expérimentale appliquée à la médicine faites au Collège de France. Vingt et unième leçon,* Bailliere, Paris.

Bernstein, R. E., 1954, Potassium and sodium balance in mammalian red cells, *Science* **120:**459.

Berton, J. P., 1970, Effets de la néphrectomie bilatérale chez le foetus de lapin (survie et métabolisme hydrique), *C.R. Acad. Sci. Ser. D* **271:**219.

Bertoncello, I., Naughton, R. J., and Skinner, S. L., 1976, Sensitivity of renin secretion to volume depletion in the anaesthetized dog: Comparison between urinary drainage and slow haemorrhage, *J. Physiol., London* **259:**309.

Blendstrup, K., Leyssac, P. P., Poulsen, K., and Skinner, S. L., 1975, Characteristics of renin release from isolated superfused glomeruli *in vitro, J. Physiol., London* **246:**653.

Brambell, F. W. R., 1958, The passive immunity of the young mammal, *Biol. Rev.* **33:**488.

Brody, S., 1945, *Bioenergetics and Growth,* Reinhold, New York.

Brody, S., and Ragsdale, A. C., 1922, The equivalence of age in animals, *J. Gen. Physiol.* **5:**205.

Brody, S., Hogan, A. G., Kempster, H. L., Ragsdale, A. C., and Trowbridge, E. A., 1926, Growth and development with special reference to domestic animals [(1) Ragsdale, A. C., Elting, E. C., and Brody, S., Quantitative data. Weight growth and linear growth, pp. 1–40; (2) Brody, S., and Ragsdale, A. C., Age and other time changes in milk secretion, pp. 46–182], *Univ. M. Agric. Exp. Station Res. Bull.* **96:**182.

Broughton Pipkin, F., Lumbers, E. R., and Mott, J. C., 1974, Factors influencing plasma renin and angiotensin II in the conscious pregnant ewe and its foetus, *J. Physiol., London* **243:**619.

Brown, J. J., Doak, P. B., Davies, D. L., Lever, A. F., Robertson, J. I. S., and Tree, M., 1964, The presence of renin in human amniotic fluid, *Lancet* **2:**64.

Brumbach, L., Leyssac, P. P., and Skinner, S. L., 1976, Studies on renin release from isolated superfused glomeruli; effects of temperature, urea, ouabain and ethacrynic acid, *J. Physiol., London* **258**:243.

Burton, J. L., Shuster, S., and Cartlidge, M., 1975, The sebotrophic effect of pregnancy, *Acta Derm. Venereol.* **55**:11.

Busuttil, R. W., Roh, B. L., and Fisher, J. W., 1971, The cytological localisation of erythropoietin in the human kidney using the fluorescent antibody technique, *Proc. Soc. Exp. Biol. Med.* **137**:327.

Carmel, J. A., Friedman, F., and Adams, F., 1965, Fetal tracheal ligation and lung development, *Am. J. Dis. Child.* **109**:452.

Carmena, A. O., Howard, D., and Stohlman, F., 1968, Regulation of erythropoiesis: 12 Erythropoietin production in the newborn animal, *Blood* **32**:376.

Carnot, P., and Deflandre, C., 1906, Sur l'activité hémopoïétique des différents organes au cours de la régénération du sang, *C. R. Acad. Sci.* **143**:432.

Cupps, F. T., Anderson, L. L., and Cole, H. H., 1969, The estrus cycle, in: *Reproduction in Domestic Animals* (H. H. Cole and P. T. Cupps, eds.), pp. 218–219, Academic Press, New York.

Dickerson, J. W. T., and McCance, R. A., 1957, The composition and origin of the allantoic fluid of the rabbit, *J. Embryol. Exp. Morphol.* **5**:40.

Donaldson, H. H., 1895, *The Growth of the Brain,* Walter Scott, London.

Donaldson, H. H., 1908, A comparison of the albino rat with man in respect to the growth of the brain and of the spinal cord, *J. Comp. Neurol.* **18**:345.

Editorial, 1972*a*, Erythropoietin, *Br. Med. J.* **1**:263.

Editorial, 1972*b*, Haemopoiesis, *Lancet* **1**:1056.

Erslev, A. J., McKenna, P. J., Capelli, J. P., Hamburger, R. J., Cohn, H. E., and Clark, J. E., 1968, Rate of red cell production in two nephrectomised patients, *Arch. Intern. Med.* **122**:230.

Erslev, A. J., Kazal, L. A., and Miller, O. P., 1971, A renal lipid inhibitor of erythropoietin, *Trans. Assoc. Am. Physicians* **84**:212.

Everett, J. W., 1933, Structure and function of the yolk-sac placenta in *Mus norwegicus albinus, Proc. Soc. Exp. Biol. Med.* **31**:77.

Ferris, T. F., Gorden, P., and Mulrow, P. J., 1967, Rabbit uterus as a source of renin, *Am. J. Physiol.* **212**:698.

Finne, P. H., 1968, Erythropoietin production in fetal hypoxia and in anemic uremic patients, *Ann. N.Y. Acad. Sci.* **149**:497.

Foote, R. H., 1969, Physiological aspects of artificial insemination, in: *Reproduction in Domestic Animals* (H. H. Cole and P. T. Cupps, eds.), pp. 316–317, Academic Press, New York.

France, V. M., Saunders, N. R., and Stanier, M. W., 1972, Sodium transport in foetal sheep urinary bladder, *J. Physiol., London* **224**:23P.

Frederiksen, O., Leyssac, P. P., and Skinner, S. K., 1975, Sensitive osmometer function of juxtaglomerular cells *in vitro, J. Physiol., London* **252**:669.

Fried, W., Kilbridge, T., Krantz, S., McDonald, T. P., and Lange, R. D., 1969, Studies on extrarenal erythropoietin, *J. Lab. Clin. Med.* **73**:244.

Fried, W., Knospe, W. H., and Trobaugh, F. E., 1971, Effect of sex differences on extrarenal erythropoietin production, *Proc. Soc. Exp. Biol. Med.* **137**:255.

Ganten, D., Boucher, R., and Genest, J., 1971, Renin activity in brain tissue of puppies and adult dogs, *Brain Res.* **33**:557.

Gavan, J. A., 1953, Growth and development of the chimpanzee: A longtitudinal and comparative study, *Human Biol.* **25**:93.

Gilkes, J. J. H., Eady, R. A. J., Rees, L. H., Munro, D. D., and Moorhead, J. F., 1975, Plasma immuno reactive melantrophic hormones in patients on maintenance haemodialysis, *Br. Med. J.* **1**:656.

Giovannini, E., and Bocchini, V., 1975, Different molecular weight of the plasmatic erythropoietin compared with the renal and splenic factors in the rabbit, *Acta Haematol.* **53**:75.

Goldwasser, E., and Kung, K. H., 1968, Progress in the purification of erythropoietin, *Ann. N.Y. Acad. Sci.* **149**:49.

Goolamali, S. K., and Shuster, S., 1975, A sebotrophic stimulus in benign and malignant breast disease, *Lancet* **1**:428.

Gordon, A. S., Cooper, G. W., and Zanjani, E. D., 1967, The kidney and erythropoiesis, *Semin. Hematol.* **4**:337.

Grainger, P., Rojo-Ortega, J. M., Pérez, S. C., Boucher, R., and Genest, J., 1971, The renin angiotensin system in newborn dogs, *Can. J. Physiol. Pharmacol.* **49**:134.

Gregory, P. W., Guilbert, H. R., Shelby, C. E., and Clark, R. T., 1963, Growth of Hereford cows selected and rejected for breeding, *Growth* **27**:205.

*R. A. McCANCE and*
*ELSIE M. WIDDOWSON*

Grether, W. F., and Yerkes, R. M., 1940, Weight norms and relations for chimpanzee, *Am. J. Phys. Anthropol.* **27**:181.

Hafez, E. S. E., 1962, *The Behaviour of Domestic Animals,* Baillière, Tindall and Cox, London.

Halvorsen, S., and Finne, P. H., 1968, Erythropoietin production in the human fetus and newborn, *Ann. N.Y. Acad. Sci.* **149**:576.

Hanks, J., 1969, Growth in weight of the female African elephant in Zambia, *East Afr. Wildl. J.* **7**:7.

Hartroft, P. M., Bischoff, M. B., and Bucci, T. J., 1969, Effects of chronic exposure to high altitude on the juxtaglomerular complex and adrenal cortex of dogs, rabbits and rats. *Fed. Proc.* **28:** 1234.

Hayden, P., and Gambino, J. J., 1966, Growth and development of the little pocket mouse *Perognathus longimembris, Growth* **30**:187.

Hemmings, W. A., and Brambell, F. W. R., 1961, Protein transfer across the foetal membranes, *Br. Med. Bull.* **17**:96.

Huehns, E. R., and Beaven, G. H., 1971, Developmental changes in human haemoglobins, in: *The Biochemistry of Development* (P. Benson, ed.), pp. 175–203, Spastics International Medical Publications, Heinemann Medical Books, London.

Ireland, A., 1977, The dimensions of boxers, personal communication.

Jacobson, L. O., Goldwasser, E., Fried, W., and Plzak, L., 1957, Role of the kidney in erythropoiesis, *Nature* **179**:633.

Jonxis, J. H. P., and Nijhof, W., 1969, Factors influencing the switch over from fetal to adult haemoglobin in the first weeks of life, *Ann. N.Y. Acad. Sci.* **165**:205.

Jost, A., and Policard, A., 1948, Contribution expérimentale à l'étude du développement prénatal du poumon chez le lapin, *Arch. Anat. Microscc.* **37**:323.

Joubert, D. M., 1956, A study of pre-natal growth and development in the sheep, *J. Agric. Sci., Cambridge* **47**:382.

King, J., 1964, *Seals of the World,* British Museum of Natural History, London.

Krantz, S. B., 1970, Current status of erythropoietin, *Med. Clin. North. Am.* **54**:173.

Krantz, S. B., and Jacobson, L. O., 1970, *Erythropoietin and the Regulation of Erythropoiesis,* University of Chicago Press, Chicago.

Kuratowska, Z., Lewartowski, B., and Miehalak, E., 1961, Studies on the production of erythropoietin by isolated perfused organs, *Blood* **18**:527.

Laws, R. M., 1953, The elephant seal (*Mirounga leonina* Linn.). 1. Growth and age, *Falkland Islands Dependencies Survey Scientific Reports 8,* Her Majesty's Stationery Office, London.

Laws, R. M., 1956a, The elephant seal (*Mirounga leonina* Linn.). 2. General, social and reproductive behaviour, *Falkland Islands Dependencies Survey Scientific Reports 13,* Her Majesty's Stationery Office, London.

Laws, R. M., 1956b, The elephant seal (*Mirounga leonina* Linn.). 3. The physiology of reproduction, *Falkland Islands Dependencies Survey Scientific Reports 15,* Her Majesty's Stationery Office, London.

Laws, R. M., 1966, Age criteria for the African elephant, *East Afr. Wildl. J.* **4**:1.

Laws, R. M., Parker, I. S. C., and Johnstone, R. C. B., 1975, *Elephants and their Habitats,* Clarendon Press, Oxford.

Le Goascogne, C., 1964, Vascularisation de la vésicule ombilicale chez les rongeurs (rat et cobaye), *Bull. Assoc. Anat.* **127**:1065.

Le Goascogne, C., 1968, Influence de la ligature des vaisseaux vitellins sur le développment foetal du rat, *J. Physiol., Paris* **60**(Suppl. 2):483.

Lehmann, H., and Huntsman, R. G., 1974, *Man's Haemoglobins,* North Holland, Amsterdam.

Lindemann, R., 1970, Erythropoiesis inhibiting factor in urine, *Lancet* **1**:781.

Lubchenco, L. O., Hansman, C., Dressler, M., and Boyd, E., 1963, Intrauterine growth as estimated from liveborn birth-weight data at 24 to 42 weeks of gestation, *Pediatrics* **33**:793.

Lucarelli, G., Porcellini, A., Carnevali, C., Carmena, A., and Stohlman, F., 1968, Fetal and neonatal erythropoiesis, *Ann. N.Y. Acad. Sci.* **149**:544.

Lumbers, E. R., 1973, Renin and angiotensin II of extrarenal origin in the plasma of female rabbits, *J. Physiol.* **234**:94P.

Markowitz, H., Gubler, C. J., Mahoney, J. P., Cartwright, G. E., and Wintrobe, M. M., 1955, Studies on copper metabolism. 14. Copper caeruloplasmin and oxidase activity in sera of normal human subjects, pregnant women and patients with hepatolenticular degeneration and the nephrotic syndrome, *J. Clin. Invest.* **34**:1498.

McCance, R. A., 1972a, The role of the developing kidney in the maintenance of internal stability, *J. R. Coll. Physicians, London* **6**:235.

McCance, R. A., 1972*b,* The composition of the body; its maintenance and regulation, *Nutr. Abstr. Rev.* **42:**1269.

McCance, R. A., 1977, Perinatal physiology, in: *The Pursuit of Nature,* pp. 133–168, University Press, Cambridge.

McClearn, G. E., 1976, Experimental behavioural genetics, in: *Aspects of Genetics in Paediatrics* (D. Barltrop, ed.), pp. 31–39, Fellowship of Postgraduate Medicine, London.

Meyer, H., and Ahlswede, L., 1976, Über das intrauterine Wachstum und die Körperzusammensetzung von Fohlen sowie den Nährstoffbedarf tragender Stuten, *Übers. Tierernährg.* **4:**263.

Miller, J. F. A. P., 1966, Immunity in the foetus and newborn. *Br. Med. Bull.* **22:**21.

Minot, C. S., 1891, Senescence and rejuvenation, *J. Physiol.* **12:**7.

Mirand, E. A., Murphy, G. P., Steaves, R. A., Groenewald, J. M., and Deklerk, J. N., 1969, Erythropoietin activity in anephric, allotransplated, unilaterally nephrectomized, and intact man, *J. Lab. Clin. Med.* **73:**121.

Mott, J. C., 1973, The renin-angiotensin system in foetal and newborn mammals, in: *Foetal and Neonatal Physiology* (R. S. Comline, K. W. Cross, G. S. Dawes, and P. W. Nathanielsz, eds.), pp. 166–180, Cambridge University Press, Cambridge.

Mott, J. C., 1975, Place of the renin-angiotensin system before and after birth, *Br. Med. Bull.* **31:**44.

Moulton, C. R., 1923, Age and chemical development in mammals, *J. Biol. Chem.* **57:**79.

Mujovic, V. M., and Fisher, J. W., 1974, The effects of indomethacin on erythropoietin production in dogs following renal artery constriction. 1. The possible role of prostaglandins in the generation of erythropoietin by the kidney. *J. Pharmacol. Exp. Ther.* **191:**575.

Naets, J. P., 1958, The kidney and erythropoiesis, *Nature, London* **182:**1516.

Naets, J. P., and Wittek, M., 1968, Presence of erythropoietin in the plasma of one anephric patient, *Blood* **31:**249.

Napier, J. R., and Napier, P. H., 1967, *A Handbook of Living Primates,* Academic Press, London.

Noer, H. R., and Mossman, H. W., 1947, Surgical investigation of the function of the inverted yolk sac placenta in the rat, *Anat. Rec.* **98:**31.

North, P. M., and Segal, M. B., 1976, A study of the transport and permeability properties of the guinea pig amniotic membrane, *J. Physiol., London* **256:**245.

Paul, J., 1976, Haemoglobin synthesis and cell differentiation, *Br. Med. Bull.* **32:**277.

Paul, W. M., Enns, T., Reynolds, S. R. M., and Chinard, F. P., 1956, Sites of water exchange between the maternal system and the amniotic fluid of rabbits, *J. Clin. Invest.* **35:**634.

Pedersen, K. O., 1944, Fetuin, a new globulin isolated from serum. *Nature, London* **154:**575.

Perry, J. S., 1954, Growth and tusk weight in male and female African elephants. *Proc. Zool. Soc., London* **124:**97.

Petter, C., 1969, Production expérimentale d'hydramnios par administration locale d'ouabaine au niveau des membranes foetales de rat, *C. R. Soc. Biol.* **163:**1023.

Potter, E. L., and Craig, J. M., 1976, *Pathology of the Fetus and the Infant,* 3rd ed., Year Book Publishers, Chicago.

Quirang, D. P., 1939, Notes on an African elephant *(Elephas loxodonta Africana), Growth* **3:**9.

Rosen, F. S., 1976, Neonatal immunity, in: *The Physiology of the Newborn Infant,* 4th ed., (C. A. Smith and N. M. Nelson, eds.), pp. 736–752, Charles C. Thomas, Springfield, Illinois.

Scheffer, V. B., and Wilke, F., 1953, Relative growth in the northern fur seal, *Growth* **17:**129.

Scheinberg, H., Cook, C. D., and Murphy, J. A. 1954, Concentration of copper and ceruloplasmin in maternal and infant plasma, *J. Clin. Invest.* **33:**963.

Schmidt-Nielsen, B., 1958, Urea excretion in mammals, *Physiol. Rev.* **38:**139.

Schmidt-Nielsen, K., 1972, *How Animals Work,* Cambridge University Press, Cambridge.

Schmidt-Nielsen, K., 1975, *Animal Physiology: Adaptation and Environment,* Chapter 10, pp. 442–495, University Press, Cambridge.

Schultz, A. H., 1969, *The Life of Primates,* Weidenfeld and Nicolson, London.

Shaldon, S., Koch, K. M., Oppermann, F., and Patyna, W. D., 1971, Testosterone therapy for anaemia in maintenance dialysis, *Br. Med. J.* **3:**212.

Skinner, S. L., Lumbers, E. R., and Symonds, E. M., 1968, Renin concentration in human fetal and maternal tissues, *Am. J. Obstet. Gynecol.* **101:**529.

Slijper, E. J., 1962, *Whales,* Hutchinson and Co., London.

Small, G. L., 1971, *The Blue Whale,* Columbia University Press, New York.

Smith, C. A., 1976, Fetal and neonatal nutrition, in: *The Physiology of the Newborn Infant,* 4th ed. (C. A. Smith and N. M. Nelson, eds.), pp. 480–553, Charles C. Thomas, Springfield, Illinois.

Stanier, M. W., 1971, Osmolarity of urine from renal pelvis and bladder of foetal and post-natal pigs, *J. Physiol., London* **218:**30P.

Stella, A., Calaresu, F., and Zanchetti, A., 1976, Neural factors contributing to renin release during reduction in renal perfusion pressure and blood flow in cats, *Clin. Sci. Mol. Med.* **51**:453.

Steven, D. H., 1975, *Comparative Placentation: Essays in Structure and Function,* Academic Press, London.

Symonds, E. M., Stanley, M. A., and Skinner, S. L., 1968, Production of renin by *in vitro* cultures of human chorion and uterine muscle, *Nature, London* **217**:1152.

Tanner, J. M., 1962, *Growth at Adolescence,* 2nd ed., Blackwell, Oxford.

Thody, A. J., 1974, Plasma and pituitary MSH levels in the rat after lesions of the hypothalamus, *Neuroendocrinology* **16**:323.

Thody, A. J., and Shuster, S., 1975, Control of sebaceous gland function in the rat by $\alpha$-melanocyte stimulating hormone, *J. Endocrinol.* **64**:503.

Thompson, D. W., 1942, *On Growth and Form,* University Press, Cambridge.

Thorling, E. B., 1969, The history of the early theories of humoral regulation of erythropoiesis, *Dan. Med. Bull.* **16**:159.

Thurmon, T. F., Boyer, S. H., Crosby, E. F., Shepard, M. K., Noyes, A. N., and Stohlman, W., 1970, Haemoglobin switching in nonanaemic sheep. 3. Evidence for presumptive identity between the A → C factor and erythropoietin, *Blood* **36**:598.

Towler, C. M., Horne, C. H. W., Jandial, V., Campbell, D. M., and MacGillivray, I., 1976, Plasma levels of pregnancy-specific $\beta_1$-glycoprotein in normal pregnancy, *Br. J. Obstet. Gynaecol.* **83**:775.

Vander, A. J., 1967, Control of renin release, *Physiol. Rev.* **47**:359.

Walshe, J., 1976, personal communication.

Weitzel, G., Buddecke, E., Fretzdorff, A. M., Strecker, F. J., and Roester, V., 1955, Struktur der im *tapetum lucidum* von Hund and Fuchs enthaltenen Zinkverbindung, *Hoppe-Seyler's Z. Physiol. Chem.* **299**:193.

Widdowson, E. M., 1950, Chemical composition of newly born mammals, *Nature, London* **166**:626.

Widdowson, E. M., and Crabb, D. E., 1976, Changes in the organs of pigs in response to feeding for the first 24 hours after birth. 1. The internal organs and muscles, *Biol. Neonat.* **28**:261.

Widdowson, E. M., and Dickerson, J. W. T., 1964, The chemical composition of the body, in: *Mineral Metabolism* (C. L. Comar and F. Bronner, eds.) Vol. 2A, pp. 1–247, Academic Press, New York.

Widdowson, E. M., and McCance, R. A., 1948, Sexual differences in the storage and metabolism of iron, *Biochem. J.* **42**:577.

Widdowson, E. M., and McCance, R. A., 1975, New thoughts on growth, *Pediatr. Res.* **9**:154.

Widdowson, E. M., Crabb, D., and Milner, R. D. G., 1972*a*, Cellular development of some human organs before birth, *Arch. Dis. Child.* **47**:652.

Widdowson, E. M., Chan, H., Harrison, G. E., and Milner, R. D. G., 1972*b*, Accumulation of Cu, Zn, Mn, Cr and Co in the human liver before birth, *Biol. Neonat.* **20**:360.

Winick, M., and Noble, A., 1965, Quantitative changes in DNA, RNA and protein during prenatal and postnatal growth in the rat, *Dev. Biol.* **12**:451.

Wood, W. G., 1976, Haemoglobin synthesis during foetal development, *Br. Med. Bull.* **32**:282.

Zanjani, E. D., Horger, E. O., Gordon, A. S., Cantor, L. D., and Hutchinson, D. L., 1969, Erythropoietin production in the fetal lamb, *J. Lab. Clin. Med.* **74**:782.

Zucali, J. R., and Mirand, E. A., 1975, Extra-renal erythropoietin and erythrogenin production in the anephric rat, *Am. J. Physiol.* **229**:1094.

Zucker, T. F., and Zucker, L. M., 1946, Bone growth in the rat as related to age and body weight, *Am. J. Physiol.* **146**:585.

Zweymüller, E., Widdowson, E. M., and McCance, R. A., 1959, The passage of urea and creatinine across the placenta of the pig. *J. Embryol. Exp. Morphol.* **7**:202.

# II

## *Biometrical Methods in Human Growth*

# 5

# *Statistics of Growth Standards*

## *M. J. R. HEALY*

## *1. Assessment of Normality*

Perhaps the most basic question of human growth in practice is simply "is this child *normal?*" with respect to whatever aspect of its growth is being studied. This question introduces quite far-reaching difficulties, both logical and practical. To begin with, it is well recognized that the term "normal" carries two distinct meanings. The first aims to relate the child to some kind of perfect or ideal standard; the "norm" in this sense is regarded as a target, quite possibly an inaccessible one which no child actually reaches, and the opposite of "normal" is "subnormal." In its second sense, "normal" is roughly equivalent to "commonly occurring" or "ordinary." The "norm" is now simply the situation which most commonly occurs, and the opposite of "normal" is "abnormal." The distinction is very clear in relation to body weight. In ill-nourished communities, most "ordinary" weights may be well below any reasonable ideal standard, while in developed countries weights well below "ordinary" levels may be ideal in that they are associated with longevity and freedom from disease. It is important to insist that "normality" in human growth studies (as throughout most of medicine) has in practice the second of these two senses, so that the normality of a particular child (or more precisely of some measured aspect of that child) is assessed by the frequency of occurrence of his measurement in a *standardizing group,* a group of children who are (as it were) *normal by definition.* The normative overtone of closeness to an ideal is seldom absent, but it has to be supplied by the tacit assumption that the members of the standardizing group themselves approximate to the desired state.

This way of looking at normality automatically introduces the statistical notions of frequency and probability. This chapter gives a brief survey of the statistical problems which arise in assessing normality, especially those of constructing standards with which a given child can be compared. No attempt will be made to summarize the substantial literature on the subject, but fuller discussions of the problems will be found in Tanner (1952), Goldstein (1972), and Healy (1974).

---

*M. J. R. HEALY* • London School of Hygiene and Tropical Medicine, London, England.

Consider first the simple case in which we wish to assess the attained value of a measurement such as stature in a child of a given age, and suppose that we can define and get access to a suitable standardizing group of children, all of the same age and with known values of stature. We need a method of describing this collection of values in such a way that we can readily compare the given child's stature with it.

## 2.1. Frequency Curves and Centiles

Suppose that we plot against values of stature the proportions of children in the standardizing group whose statures fall short of those values. With a large group this will produce an S-shaped curve rising smoothly from 0% to 100% as shown in Figure 1. From this plot, we can see, for example, that 70% of the group have statures less than 140 cm, while 3% have statures less than 125 cm. Such a plot is called a *cumulative frequency curve,* and the statures read off from it as corresponding to given percentages are called its *centiles* (or percentiles, or fractiles)—in particular, 125 cm is the 3rd centile. Now, if the stature of our given child falls short of 125 cm, we know that he is shorter than at least 97% of the "normal" children of his age, and we can assert that his stature is to this extent abnormally small. The statistical problem of assessing normality becomes that of determining the centiles of a standardizing group.

This problem falls into two parts: that of obtaining the necessary data and that of their subsequent analysis. The first of these is largely nonstatistical and consists in the difficult practical issues of contacting the required subjects and taking the appropriate measurements, with due attention to techniques and quality control. An

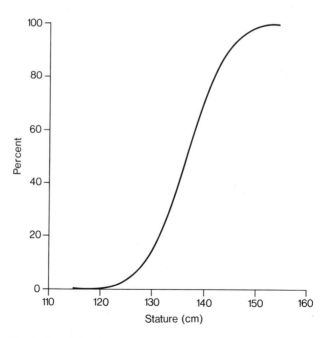

Fig. 1. Cumulative frequency curve for stature in 10-year-old boys.

important statistical question which does arise at this stage is that of sampling. The desired standardizing group (for example, all British children of a specified age) will usually be very large—often indefinitely so if no time constraints are imposed—and it will be quite unrealistic to measure all the children in it. Instead, a sample of children will be measured, and it is then important to ensure that the sample is adequately representative of the group as a whole. Methods for doing this are discussed in Chapter 6. The desirable size of the sample is also of obvious importance, and this will be discussed further below.

## 2.2. The Gaussian Distribution

Given the measurements on the sample, we can derive the centiles of the standardizing group (or population) in more than one way. First of all, we can represent the smooth curve in Figure 1 by a suitable algebraic formula. Such a formula will contain adjustable parameters, and we may estimate these from the sample data and so read off the centiles from the formula. This procedure will be somewhat more familiar if we replace the S-shaped curve of Figure 1 by the mathematically equivalent bell-shaped curve of Figure 2. Here the abscissa is still stature, but the ordinates now give the proportions of children in the population whose statures are equal to (strictly speaking, nearly equal to) the corresponding abscissa. A proportion at a given stature in Figure 1 is the same as the area under the curve to the left of the same stature in Figure 2. The particular curve given here has the common property that most of the measurements lie fairly close to a central value, with larger deviations being less common (this is almost inherent in our concept of normality). It also has the more specific property of symmetry about the central value, so that a given deviation below the center is just as common as the same deviation above. This specification describes quite well a number of anthropometric measurements, notably lengths such as stature and limb lengths. These measurements in fact fit a particular curve given by the rather fancy equation

$$y = \frac{1}{\sqrt{(2\pi)}\sigma} \exp\left[-\tfrac{1}{2}(x - \mu)^2\right]$$

with two parameters, $\mu$ and $\sigma$. As shown in Figure 2, the mean, $\mu$, locates the central value, and the standard deviation, $\sigma$, measures the scatter or spread of the other values about the center. The corresponding curve describes what is usually known as the Normal distribution, but we shall (for obvious reasons) avoid the usage and refer to it as the Gaussian distribution. It has been very thoroughly tabulated, and it is easy to read off from tables the tail area (or equivalently the

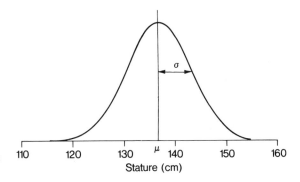

Fig. 2. Frequency curve for stature in 10-year-old boys.

Stature (cm)

ordinate of Figure 1) corresponding to any given multiple of $\sigma$, or *vice versa;* the 3rd centile, for example, is at 1.88 standard deviations below the mean. A convenient graphical version of these tables may be obtained from the equivalent of Figure 1 by stretching the $y$ scale away from the 50% point until the curve becomes a straight line (Figure 3). This suggests a quick way of checking whether a set of sample values may plausibly have been taken from a Gaussian population; the procedure involves three steps:

1. Arrange the $n$ sample values in ascending order, and call the $i$th value from the bottom $x_i$.
2. For $i = 1, 2, \ldots, n$, calculate the proportions $p_i = (i - \tfrac{1}{2})/n$ and the corresponding deviations $d_i$ read from Figure 3.
3. Plot the $x_i$ against the corresponding $d_i$.

The process is illustrated in miniature in Table I and Figure 4. It will be apparent that a sample from a Gaussian population will give points that lie fairly close to a straight line, while consistent curvature, for example, can be taken as evidence of lack of symmetry.

### 2.3. Estimating Gaussian Centiles

If then we satisfy ourselves that our standardizing population is reasonably close to a Gaussian form, it remains only to estimate the values of the parameters $\mu$ and $\sigma$. Writing $x_i$ for the sample values (now not necessarily in ascending order), we can do this by using the simple formulas

$$m = \Sigma x_i/n \qquad s = \sqrt{\Sigma(x_i - m)^2/(n - 1)}$$

where the sum in the formula for $s$ is most easily calculated as

$$\Sigma(x_i - m)^2 = \Sigma x_i^2 - (\Sigma x_i)^2/n$$

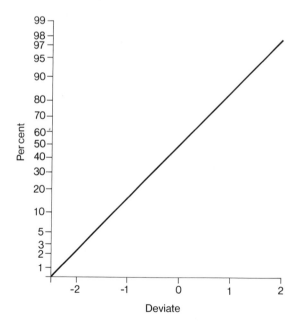

Fig. 3. Relation between proportions and deviates in a Gaussian distribution.

Table 1. Statures (cm) in a Sample of 10-Year-Old Boys[a]

Original data: 135.0 139.2 134.6 138.9 140.0 141.4
136.0 143.3 136.8 137.0 140.8 129.7

| $i$ | $x_i$ | $p_i$ | $d_i$ | $i$ | $x_i$ | $p_i$ | $d_i$ |
|---|---|---|---|---|---|---|---|
| 1 | 129.7 | 0.042 | −1.73 | 7 | 138.9 | 0.542 | 0.11 |
| 2 | 134.6 | 0.125 | −1.15 | 8 | 139.2 | 0.625 | 0.32 |
| 3 | 135.0 | 0.208 | −0.81 | 9 | 140.0 | 0.708 | 0.55 |
| 4 | 136.0 | 0.292 | −0.55 | 10 | 140.8 | 0.792 | 0.81 |
| 5 | 136.8 | 0.375 | −0.32 | 11 | 141.4 | 0.875 | 1.15 |
| 6 | 137.0 | 0.458 | −0.11 | 12 | 143.3 | 0.958 | 1.73 |

[a]$x_i$, the $i$th largest reading; $p_i = (i - \frac{1}{2})/12$; $d_i$ = deviation read from Fig. 3.

In our miniature example, we find

$$n = 12 \qquad m = 137.72 \qquad s = 3.68$$

and so, for example, the estimated position of the 3rd centile is at $m - 1.88s = 130.80$ cm.

It is worth noting at once that our assumption of a Gaussian distribution has allowed us to make a substantial extrapolation; we purport to say what 97 out of 100 statures would be like after measuring only 12! Clearly, were we to measure another 12 statures, we would be most unlikely to come up with the same answer. From our sample, we can only obtain an estimate of each centile, and these estimates will differ from the true values because of sampling error. We can, however, make some assessment of how big this sampling error is likely to be. Envisage the process of

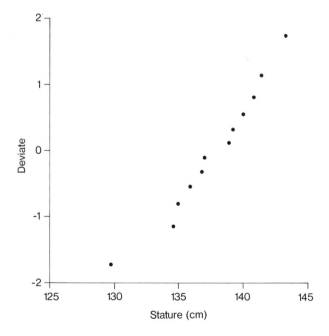

Fig. 4. Graphical check on Gaussianity; data from Table I.

drawing a sample of 12 over and over again from the same standardizing group, and suppose that from each sample we estimate the mean and standard deviation and hence a particular centile as $m + ks$ where $k$ is the appropriate multiplier (for the 3rd centile, $k = -1.88$). We can thus envisage building up a distribution of centile estimates—a so-called sampling distribution. It can be shown that, if the sample size $n$ is not too small, this distribution is roughly Gaussian with a mean roughly equal to the true centile and a standard deviation given by

$$\sigma \sqrt{(1 + \tfrac{1}{2}k^2)/n}$$

It can be seen from Figure 3 that in a Gaussian distribution 19 out of 20 values lie within two standard deviations of the mean. If we substitute $s$ for $\sigma$, we find that the estimated 3rd centile has a standard deviation of about 1.8, so that our small sample only entitles us to say that the true value is probably within the range 133.6 to 140.8 cm, an uncomfortably large range. Noticing that we must multiply the sample size by four in order to halve this range, we see at once that precise centile estimates require large samples.

### 2.4. Nonparametric and Graphical Methods

The methodology we have described relies heavily on the assumed gaussian shape of the standardizing population, and we must consider what to do when this assumption cannot be made. This is the case for weight, many circumferential or girth measurements, skinfolds, and perhaps a majority of biochemical and endocrinological measurements. One possibility is to rethink the scale of measurement—we can often find a simple arithmetic transformation of the original values (such as their logarithms) which produces a symmetric distribution close to the Gaussian form. However, a much more direct method immediately suggests itself. If we wish to estimate the 3rd centile from a sample of size 100, for example, we can simply arrange the sample values in ascending order and pick out the third from the bottom. With a smaller sample, some kind of interpolation will be needed; for a sample of 50, we might take a value halfway between the smallest and the next smallest. This nonparametric method makes no assumptions about the shape of the population, and it may well be asked why we do not always use it. To throw light on this, we must consider what will be the sampling error of centile estimates obtained in this way. As before, it can be shown that the sampling distribution is roughly gaussian and centered at the required population value, but the standard deviation for the $100p$th centile is now given by

$$\sqrt{p(1 - p)/(nf^2)}$$

where $f$ is the ordinate of the (noncumulative) distribution curve. If this curve is actually gaussian, then $f$ for the 3rd centile is equal to $0.0681/\sigma$ and the sampling standard deviation for this centile estimate comes to $\sqrt{6.27\sigma^2/n}$, as against $\sqrt{2.77\sigma^2/n}$ for the same centile estimated parametrically. Failing to use the gaussian property thus has the same effect on precision as throwing away over half the data, and it is worth going to some trouble to find, when necessary, a transformation which permits its use.

In searching for a suitable transformation, the plotting technique described above should play a major role. Assuming (as is usually the case) that the asymmetry is such that large deviations above the mean are more common than equally large deviations below, the points plotted as described will lie near a curve which is concave downwards. It is then always worth replacing each measurement by its logarithm and replotting; if the result is still a curve which is concave downwards (upwards), the situation can be improved by subtracting from (adding to) each measurement a suitable constant amount which can be found by trial and error. The objective should be to obtain a single transformation formula which brings close to gaussianity a wide range of data, relating perhaps to different sexes, age groups, racial groups, etc. Such an investigation will be extremely onerous, and for many workers a graphical approach will provide a satisfactory, if subjective, compromise between the Gaussian and nonparametric methods. It is simply necessary to plot the data against the Gaussian deviates as described and then to draw a smooth curve through the points by eye from which the centiles can be read off. A quite realistic idea of the variability of this procedure can be had by persuading one or more colleagues to draw their own independent curves. It is still worth transforming violently nongaussian data, since it is easiest to draw free-hand curves that do not depart too far from a straight line.

## 2.5. Standards for a Range of Ages

The discussion so far has assumed that (1) centiles were required for children of a single age, and (2) a standardizing group was available of children of exactly the required age. Neither assumption is realistic. Growth standards are usually needed over a range of ages, and the (necessarily large) standardizing group usually contains children whose ages are more or less evenly distributed across the range.

The first of these facts can be made use of, since the true age progression of a particular centile will presumably be along some kind of smooth curve. It is important that the data be sufficiently numerous to define the shapes of these curves satisfactorily; this is particularly so when the curve contains more or less sharp peaks or troughs, as with some of the biochemical aspects of fetal development. Rather as before, the successive estimated values for a particular centile can be smoothed by fitting a mathematical curve statistically or by a free-hand graphical technique. In the present context, the flexibility of the latter has much to recommend it.

When we come to assess the normality of a single child, we may usually assume that we know the child's age precisely; the implication is that we need centiles for a standardizing group of precisely the same age—exactly 10.00 years old, say—whereas the available data will usually relate to a group whose ages range evenly from 9.5 to 10.5 years. Since the children are growing during this period, it is not hard to see that ignoring the age span leads to estimated centiles that are too far apart (the bottom 3%, for instance, will contain predominantly children who have not actually reached their 10th birthday). In the Gaussian situation an allowance for this can quite easily be made; it consists in reducing the square of the estimated standard deviation (the variance, $s^2$) by the quantity $b^2/12$ where $b$ denotes the mean amount of growth over the age span concerned. Since the adjustment should be quite small, a rough estimate of $b$ is all that is required. When the centiles are estimated graphically, the correct action is far from clear. A possible strategy is as follows:

1. Find the unadjusted estimate $s$ of the standard deviation $x$ and the adjusted estimate $s'$ from the usual formulas.
2. "Shrink" the estimated centiles towards the 50% point by reducing each deviation by a factor $s'/s$.

M. J. R. HEALY

## 3. Velocity Standards

The discussion so far has described methods for assessing a single measurement taken at a single point in time. In a growing child, such a measurement integrates the whole of the growth process up to the time that it is made; accordingly it may give a good indication of the lifetime experiences of the child, but perhaps rather a poor one of the child's current growth status. For this second purpose we would like to know the rate at which the child is growing (in the dimension, be it stature, weight, or whatever, that concerns us) and to possess standards with which to compare this rate.

### 3.1. Cross-Sectional and Longitudinal Data

To measure a rate of growth, or growth velocity as it is more usually called, we must measure our child at two points of time and calculate the difference between the two measurements. It will be apparent at once that care in the measuring process is even more important than in the single measurement case; not only are there two errors of measurement involved, but their combined effect applies to a total quantity (the amount of growth) that will often be quite small.

To obtain standards for growth velocities, it is necessary for the children in the standardizing group to be measured at two points of time. Since this is sometimes misunderstood, an explanation in a little detail is in order. Suppose that the standardizing children are measured once only (such data are called cross-sectional), and for simplicity suppose that a substantial number are measured on their 10th birthdays and an equal number on their 11th birthdays. This provides just what we need for assessing the status of a 10-year-old or an 11-year-old child. It also provides a good assessment of the average amount of growth between the 10th and 11th birthdays; this amount is simply the difference between the average at 11 and 10 years. It completely fails to provide any information about the variability of the amount of growth, so that although we may be able to say that our child has an amount of growth that is below average, we are quite unable to say just how abnormal he is in this respect. To do this requires longitudinal data on the standardizing group, obtained by measuring children both on their 10th and 11th birthdays. Given such data, each child in the standardizing group has a known amount of growth over the year in question, and these figures can be used to provide a standard by the methods described above.

Obtaining longitudinal data is a major undertaking, even if they are only to cover a fairly short period, and it is not to be expected in normal circumstances that every child studied will be measured on every occasion. The groups of children available at two consecutive ages will thus usually overlap but will not coincide, giving rise to mixed longitudinal data. Although the cross-sectional element in such data cannot help with the assessment of variability in growth rate, it can contribute to the estimation of mean velocity, as pointed out above. The statistical methods involved (Patterson, 1950) are fairly complicated, but they have been described in the context of growth studies by Tanner (1951).

### 3.2. Interpretation of Velocity Standards

There are several practical difficulties in the use of velocity standards which may be mentioned here, although they are really not of a statistical nature. Longitudinal standardizing data obtained at yearly intervals actually provide annual growth increments rather than actual velocities, and can be used for assessing annual increments. An increment measured over 6 months can be doubled to render it comparable, but it is by no means certain that such figures will have the same variability as actual annual increments, especially since any contribution for measuring errors will be twice as large. A further problem with increments measured over less than a year is the noticeable seasonal differences in growth that have been found to occur (see for example Marshall and Swan, 1971; Marshall, 1975).

Velocity standards require special care when an acceleration in growth occurs at different ages in different children—the most obvious example is the effect on anthropometric measurements of the pubertal growth spurt. Towards the beginning of the appropriate period, some children will have started to accelerate and will develop high velocities while others will have the low velocities that precede acceleration. As a result, the variability of velocity at such an age is high and the associated standards quite insensitive. It is also true that the curve correctly describing average velocity as a function of age is not one which is actually followed by an actual child; this is illustrated in a famous diagram taken from Shuttleworth (1937) (Figure 5). This diagram shows how the peak of the average curve is flatter than those of the individuals, and also how the average curve becomes more realistic and the variability drastically less if the curves are replotted in such a way that their peaks coincide. This use of peak velocity age is one among many ways of transforming the age scale; clearly it can only be used retrospectively and when enough measurements have been taken to locate the velocity peak fairly precisely. Other possibilities are the use of menarcheal age (in girls) or bone age or some other age scale related to physical maturity (see Volume 2, Chapter 12).

### 3.3. Standards for Growth Curves

The essential statistical point about velocity standards is that a pair of measurements is summarized by a single quantity (their difference) which is readily interpretable and for which a standard can be constructed. This principle can be extended to take in series containing more than two measurements. Growth curves can be fitted to such measurements (see Chapter 7) and standards constructed for the parameters of these curves or for suitable combinations of them which may be easier to interpret. As an example, the logistic curve

$$y = a + b/(1 + e^{-(c+dx)})$$

provides a good fit to growth in stature during puberty (Marubini *et al.*, 1972; Tanner *et al.*, 1976*b*). The maximum velocity from this curve is $bd/4$, and this quantity is of interest in retrospective analyses of longitudinal studies. In principle it is a better estimate of the peak velocity for a given child than one obtained from the maximum increment over a year or a shorter period, since the latter averages the velocity over the period concerned and so will always tend to be an underestimate (compare Figure 5).

### 3.4. Regression Standards

It is sometimes possible to increase the usefulness of a standard for a particular measurement by taking into account the value of a second measurement. Suppose, for example, we aim to assess the birth weight of a child (Tanner *et al.* 1972). We

M. J. R. HEALY

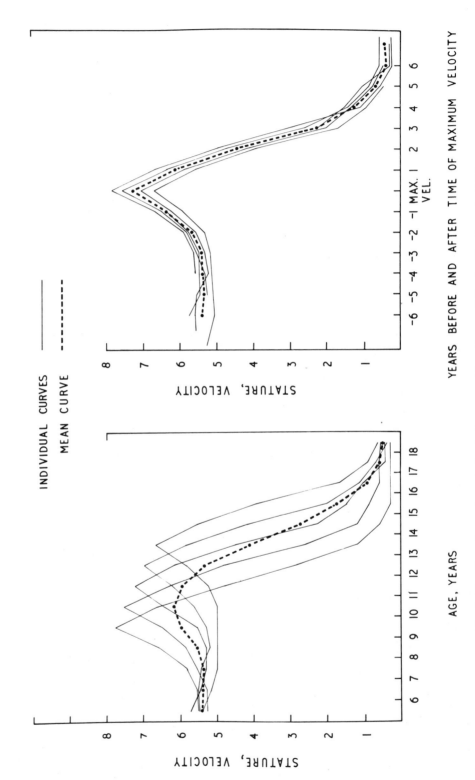

Fig. 5. The effect of averaging growth velocity curves. (From Tanner, 1962, after Shuttleworth, 1937.)

can derive a standard from a large sample of birth weights in the usual way; however, it is known that the birth weights of sibs are associated, in that the birth of a child who is above (or below) average weight makes it more likely that a subsequent sib will be born at a weight above (or below) average. Put another way, the centiles of the birth weights of children whose sibs all have a particular fixed birth weight are closer together than those of the overall birth weight distribution. A standard of this kind could be constructed by obtaining large samples of sib pairs and estimating the centiles of the second child's birth weight for a range of values of the first child's birth weight, just as previously we estimated centiles of stature for a range of values of age. A sufficiently large amount of data is, however, hard to come by, but we can make some progress if we adopt some plausible assumptions:

1. For a fixed value of the first birth weight $x$, the second birth weight $y$ is Gaussianly distributed with mean $\mu(y|x)$ and standard deviation $\sigma$ which does not depend on $x$.
2. $\mu(y|x)$ is a linear function of $x$,

$$\mu(y|x) = \alpha + \beta x$$

We can then estimate the parameters $\alpha$, $\beta$, and $\sigma$ from a sample of $n$ data pairs, using the formulas

$$\overline{x} = \Sigma x/n \qquad\qquad \overline{y} = \Sigma y/n$$

$$S_{xx} = \Sigma(x - \overline{x})^2 \qquad S_{xy} = \Sigma(x - \overline{x})(y - \overline{y})$$

$$b = S_{xy}/S_{xx} \qquad\qquad a = \overline{y} - b\overline{x}$$

$$s = \sqrt{S_{yy} - (S_{xy}^2/S_{xx})/(n - 2)}$$

and hence construct the standard in the usual way, using $m = a + bx$ for the mean for any given value of $x$ and $s$ for the standard deviation.

In the birth weight example, both $x$ and $y$ are birth weights, and it would be perfectly possible to construct a standard for the difference $(y - x)$; there is a quite close analogy with a velocity standard. The drawback of this approach has been pointed out by Billewicz (1972). Briefly, the difference $y - x$ may be unduly large (and negative) either because the second birth weight $y$ is small or because the first birth weight $x$ is large. As a result, the normal second children who fall below the 3rd centile on such a standard will predominantly be those from families with a tendency to high birth weight. The regression approach avoids this effect. On the other hand, the approach does assess the $y$ value taking the $x$ value as given; there is thus a tacit assumption that the $x$ value itself is normal, and this must be borne in mind when assessing $y$.

## 4. Standards for Events

We have discussed the construction of standards for continuous measurements, typified by anthropometric measurements such as height and weight, or biochemical measurements such as hormone levels. Some aspects of growth are more naturally assessed in terms of events (often referred to as *"milestones"*). To be useful, an event should occur in all normal children but at varying ages—a child in whom the event occurs early may then be said to be more mature, and the event is best regarded as a "maturity indicator." Typical events used for this purpose are the occurrence of menarche (in girls), the passage through defined stages of the secondary sex characteristics, the eruption of the temporary and permanent teeth, etc.

In assessing the height of a child against a height-for-age standard, we are led to statements like "this child has a height which is small for his age." With an event, such as menarche, we may be led to say "this girl has an age which is high for this event." Moreover, the event itself (and the exact age at which it occurs) will only be observable if the child is followed longitudinally and observed at frequent intervals. More usually, we can only know that the event has or has not occurred; in the first case, the age at which it occurred can sometimes be got by recollection.

The use of recall data has occasionally been studied. With a fairly important event like menarche, recall may be quite accurate over a few years, but with other events, notably those not usually observed by the subject, the method is not available. Instead we must see what can be done with standardizing data which records for a group of children covering a range of ages simply whether in each one the event has or has not occurred.

### 4.1. Standards for Age at Occurrence

With data of this kind, we can evaluate at any given age the proportion of the group in which the event has occurred. If we plot this against age, we will get a curve rising from 0% to 100%, and in practice, with menarche and several other events which have been studied, the actual curve is very close to the shape shown in Figure 1. This is just what would be produced if age at menarche was a continuous character like height and was distributed in Gaussian form (Figure 2) in the population. We can check this visually by converting the proportions to deviates using Figure 3 and plotting these against age, when an approximately straight line should result. In fact, we can use this line to estimate the mean and standard deviation of the underlying distribution, for the deviates plotted (unlike those arrived at earlier in this chapter) are statistically independent. It is important to realize that the plotted points will not all be equally precise. For a start, they may not each be based on the same total number of cases; apart from this, the distortion of the scale that leads from Figure 1 to Figure 3 has the effect of greatly decreasing the precision of points representing proportions near 0 or 1. The fitting process can thus tolerate larger discrepancies from the straight line in these regions than near the 50% point.

If these considerations are carefully borne in mind, a straight line fitted by eye to the transformed points may be satisfactory. The estimated mean is then the age corresponding to 50% (equivalent deviation 0), and the estimated standard deviation is most easily obtained as the age difference between deviations 0 and 1. A numerical technique for fitting is available and removes any element of subjectivity; it is straightforward but tedious, and rather long to describe here. Details, together with the necessary tables, can be found in Fisher and Yates (1957) or Finney (1952), and there are several standard computer programs available (for example, program BMDO3S in Dixon, 1973).

As indicated above, the use of an event-type standard requires a little care. If it is Gaussian with mean $m$ and standard deviation $s$, the "normal range" for the age of occurrence is approximately from $m - 2s$ to $m + 2s$. For menarche in present-day British girls, these ages are about $10\frac{1}{2}$ to $15\frac{1}{2}$ years. If a girl aged 16 has not yet menstruated, we can say that this event in her is occurring abnormally late; if a girl aged 14 has menstruated, on the other hand, she may or may not be abnormally advanced—without recall data we cannot say. Nor can we say that a premenarcheal girl of 15 is abnormally late—we can only hope to do this at age 15½ at the earliest.

Naturally, the fact that we observe only the occurrence of the event and not the time at which it occurs implies a loss of information, and standardizing groups for event-type standards need to be large. For example, if we study menarche in girls over a 5-year age span, the standard deviation of the estimated 97th centile is about $\sqrt{0.85/n}$ years, where $n$ is the total number of girls in the sample.

## 4.2. Sequences of Events, Maturity Scales

Clearly, information on a single event is of limited value. Much more can be had from sequences of events, such as successive stages of the pubic hair, or better still from parallel sequences, such as successive stages of the erupting teeth or of the wrist and hand bones as seen on X-rays. The problem of dealing with data of this sort is usually solved by constructing an "age" scale; thus we may assess the tooth age or bone age of a child by finding the standardizing age group whose mean the child most resembles. This solution brings two further problems with it. First, the measurement of resemblance is not straightforward, since it depends upon the importance ascribed to discrepancies in the different series of events used. Secondly, the standardizing group will be peculiar to its own geographical and social origin and to its date of measurement; secular changes in the rate of maturing leads to confusing statements such as "the average child of 12 now has a bone age of 11½." Both these can be overcome if the event data are combined to give a single maturity score for which standards can be produced in the usual way. A recent detailed discussion, with a scoring system for the bones of the wrist and hand, can be found in Tanner *et al.* (1976*a*) (see also Volume 2, Chapter 12).

## 5. References

Billewicz, W. Z., 1972, Within-family birth-weight standards, *Lancet* **2**:820.

Dixon, W. J. (ed.), 1973, *Biomedical Computer Programs,* University of California Press, Berkeley.

Finney, D. J., 1952, *Probit Analysis* (2nd ed.), University Press, Cambridge.

Fisher, R. A., and Yates, F., 1957, *Statistical Tables* (5th ed.), Oliver & Boyd, Edinburgh.

Goldstein, H., 1972, The construction of standards for measurement subject to growth, *Hum. Biol.* **44**:255–261.

Healy, M. J. R., 1974, Notes on the statistics of growth standards, *Ann. Hum. Biol.* **1**:41–46.

Marshall, W. A., 1975, The relationship of variations in children's growth rates to seasonal climatic variations, *Ann. Hum. Biol.* **2**:243–250.

Marshall, W. A., and Swan, A. V., 1971, Seasonal variation in growth rates of normal and blind children, *Hum. Biol.* **43**:502–516.

Marubini, E., Resele, L. F., Tanner, J. M., and Whitehouse, R. H., 1972, The fit of Gompertz and logistic curves to longitudinal data during adolescence, *Hum. Biol.* **44**:511–523.

Patterson, H. D., 1950, Sampling on successive occasions with partial replacement of units, *J. R. Stat. Soc. B* **12**:241–255.

Shuttleworth, F. K., 1937, Sexual maturation and the skeletal growth of girls age six to nineteen, *Monogr. Soc. Res. Child Dev.* **2**(5), 253 pp.

Tanner, J. M., 1951, Some notes on the reporting of growth data, *Hum. Biol.* **23**:93–159.

Tanner, J. M., 1952, The assessment of growth and development in children, *Arch. Dis. Child.* **27**:10–33.

Tanner, J. M., 1962, *Growth at Adolescence* (2nd ed.), Blackwell, Oxford.

Tanner, J. M., Lejarraga, H., and Healy, M. J. R., 1972, Within-family standards for birth-weight, *Lancet* **2**:1314–1315.

Tanner, J. M., Whitehouse, R. H., Marshall, W. A., Healy, M. J. R., and Goldstein, H., 1976*a*, *Assessment of Skeletal Maturity and Prediction of Adult Height,* Academic Press, London.

Tanner, J. M., Whitehouse, R. H., Marubini, E., and Resele, L. F., 1976*b*, The adolescent growth spurt of boys and girls of the Harpenden Growth Study, *Ann. Hum. Biol.* **3**:109–126.

# 6

# *Sampling for Growth Studies*

## *HARVEY GOLDSTEIN*

## *1. Introduction*

The feature of a growth study which introduces novel sampling problems is its duration over time. Thus, as well as dealing with the usual problems of human population sampling, we need to consider how to choose sample units to represent different occasions or ages. This first section will serve as a general introduction to the problems which can be encountered, and following sections will deal with methods for solving them. Although set in the context of surveys of human populations and in particular of children, many of the techniques described will also be applicable to experimental studies, for example, of rats or crop yields.

### *1.1. Definition of a Population in Time*

It is well known that the average growth characteristics of a geographically defined population change over time. For example, the mean height of British 7-year-olds increased by about 1.5 cm/decade in the first half of the twentieth century and this ''secular trend'' also affected the ages of occurrence of particular events such as menarche (Tanner, 1955). This historical process is therefore superimposed on the changes which occur with age, and although both age and historical time are measured in the same units, they are nevertheless logically distinct concepts. Hence, when defining a population, its historical origin needs to be specified, as well as its geographical location and its other defining characteristics.

### *1.2. Age Sampling*

It is convenient to consider individual subjects as being measured at one or more of a particular set of prespecified ages, for example, at exact whole years of age. Although in practice the measurements will often not be made precisely at the

---

*HARVEY GOLDSTEIN* • University of London, Institute of Education, London, England.

specified ages, we can still designate a set of ''target'' or intended ages. Where we wish to sample continuously over the age range, such a scheme can be approximated by considering a sufficiently large number of discrete ages and their relative spacing. This arises when we wish to make estimates for a complex growth curve, often spanning a wide age range, and for which there is no simple mathematical description available. We shall discuss this in the context of estimating cross-sectional growth standards.

When sampling over a set of occasions we need to consider how much of the sample is ''longitudinal,'' that is, which individuals are measured on more than one occasion, and what those occasions are. As far as possible we shall approach this by assuming that the optimum sampling scheme will have a partly longitudinal element, and then, in particular applications, determine the most efficient mixture of longitudinal and cross-sectional elements. Any study with a partly longitudinal element will be referred to as a ''mixed longitudinal study'' or more simply a ''longitudinal study.'' Where every individual is measured at each occasion we shall use the term ''pure longitudinal study.'' Likewise, where there is no longitudinal element, we shall use the term ''cross-sectional study.''

### 1.3. Losses from the Sample

In addition to the usual problems of nonresponse, when subjects due to enter a study cannot in fact be measured, a longitudinal study will normally suffer unscheduled losses as new occasions are reached. Thus, for example, individuals scheduled for three occasions may only attend the first or just the first and last ones. If the nonattenders have similar growth characteristics to the remainder, this will modify the sample design but not lead to serious biases. Where this is not the case, then biases can arise and may create serious problems which limit the usefulness of the results of the study. For this reason, considerable attention needs to be given to methods of tracking and locating subjects and securing their continuing cooperation with the study.

### 1.4. Complex Sampling

Most analytical statistical techniques assume that the sample is a simple random one. In many cases for reasons of cost, convenience, or efficiency, alternative strategies may be used which involve stratification, multistage sampling, etc. If we are concerned with estimating mean values for single occasions, the relative efficiencies of these different designs can fairly readily be calculated. The effects of complex sample designs on techniques such as regression analysis, however, are less well understood, although methods do exist for studying such effects.

### 1.5. Sampling for Growth Standards

Standards or norms for growth are representations of the frequency distributions of a measurement (or combination of measurements) for a given population at a series of ages. They are typically presented in terms of selected percentiles, for example, the 3rd, 10th, 50th, 90th, and 97th, plotted over a continuous age range. In most cases interest centers on the extreme percentiles, and the accurate estimation of these requires greater sample sizes than for the accurate estimation of, say, the mean or the 50th percentile.

Since any study of a human population takes place at a particular historical time, inferences about the population are strictly appropriate only to the time when the study was carried out. Hence we must assume some stability in the population being studied if we wish to make statements about the population at other times, past or future. For some characteristics of some populations such assumptions may be reasonable, particularly if supported by a series of studies at different times. With longitudinal studies, however, the period of the study itself, sometimes 20 years or more, may be long in relation to changes in the population. During this time the characteristics of the population being studied may have changed considerably, both in terms of average values, such as height at different ages, and also in terms of the relationships between variables. For example, a study of children from birth to the age of nine years may establish a relationship between standards of obstetrical care and physical growth. Subsequent changes, say in the availability of obstetrical care or in the demographic structure of the population, may considerably alter the nature of the relationship. For example, elective caesarian section in some populations is known to be largely confined to privileged groups. If for this reason it is also associated with faster growth, then changes in practice or policy which resulted in elective caesarian section being available to all sections of the population would tend to diminish this association.

To attempt to deal with such a situation, two strategies are available. First, we can plan a series of similar studies starting, say, at 3-year intervals, to determine how the average values and relationships change over time. Inevitably, however, so far as the relationships over time are concerned the most recent study will be out of date, by as much as nine years in the above example. Hence, we are left with having to extrapolate the trends from our previous studies. It should be noted that this problem is especially important in a longitudinal study and is of far less importance in a cross-sectional study where the only time delay is between collecting information and analyzing it. The second strategy is to collect additional information which may suggest changing patterns. For example, by studying the rate of caesarian section, its distribution in the population, and its relation to an immediate outcome of pregnancy such as birth weight, we might obtain indications of a changing relationship with subsequent growth.

Ultimately, however, our ability to utilize the results of an observational study will depend on our degree of understanding of the underlying causal relationships, and one function of observational studies is to advance our understanding of these. We discuss now a general framework which allows us to set a study within its historical context.

### 2.1. The Age–Time Plane

Figure 1 has two axes representing historical time and age, so that an individual who has a particular age at a particular time can be uniquely specified by a point. More generally we can specify whole populations in terms of their positions. For example, the area marked A, between the parallel lines at 45° to the axes, represents a "cohort" born around 1950 and studied at the ages indicated by the shaded areas labeled $a_1$–$a_4$ until 1965 when they were 15 years old. Likewise, the area marked B represents another cohort which was 9 years old around 1950 and was studied then and at 15 years of age.

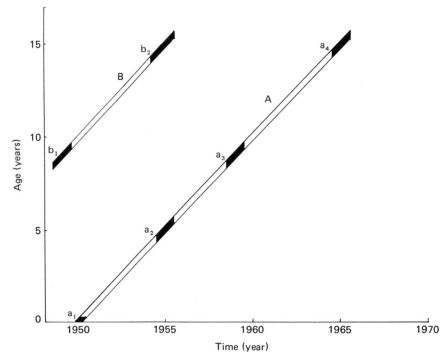

Fig. 1. Populations defined by age and time.

If we add a third perpendicular axis we can plot along this the mean value of a measurement such as weight, to give an age–time surface whose distance above the plane is the population mean weight. We can, in principle, study complex "multivariate" surfaces for many variables simultaneously, and it is the exploration of the properties of such surfaces which is the aim of statistical analysis.

### 2.1.1. Secular Trends

Various ways of specializing or simplifying the description of the age–time surface are available. We might, for example, assume that for a given age or ages no change in the average value of a variable occurred over time (these changes have already been referred to as secular trends). If this were the case, we could reasonably extrapolate findings in one time period to another time period. To study whether secular trends are occurring, we would compare measurements made at the same ages at different points in time. For example, in Figure 1 we might compare those measurements made at the occasions denoted by $b_1$ with those denoted by $a_3$ to see whether 9-year-olds in 1949 had the same characteristics as 9-year-olds in 1959. We might also wish to study whether the change in, say, mean weight between the ages of 9 and 15 years was the same for cohort B as for cohort A.

It is worth emphasizing that the difference between the means of subgroups may be just as responsive to a secular trend as the mean values themselves. This is illustrated by the following example. Table I gives estimates of mean stature of 11-year-old British boys and girls from two national surveys, the National Survey of Health and Development (Douglas, 1964) and the National Child Development Study (Davie *et al.*, 1972).

Table I.   Mean Height in cm at 11.0 Years [a] (Standard Errors in Parentheses)

| | Study and date of measurement | | | |
|---|---|---|---|---|
| | National Survey of Health and Development (1957) | | National Child Development Study (1969) | |
| Boys | 140.9 | (0.154) | 142.8 | (0.097) |
| Girls | 140.8 | (0.174) | 143.3 | (0.108) |
| Boys-Girls | 0.1 | (0.235) | −0.5 | (0.145) |

[a]The values have been corrected to a mean age of 11.0 years using the height velocity standards of Tanner *et al.* (1966). The unpublished data from the National Survey of Health and Development are quoted by kind permission of Dr. J. W. B. Douglas.

In the 12 years between the two surveys the mean value for boys has increased by 1.9 cm whereas that for girls has increased by 2.5 cm, the difference possibly being a result of earlier puberty which resulted in many of the girls, but not boys, already having entered their adolescent growth period by the age of 11 years. Thus we have an example of a secular trend operating to change both overall mean values and also the difference between subgroups of the population.

### 2.1.2. Cross-Sectional and Longitudinal Designs

The cross-sectional design samples different ages at the same point of time. For example, measurement occasions denoted by $a_2$ and $b_2$ in Figure 1 would constitute a cross-sectional study consisting of two age groups of 5- and 15-year-olds in 1955. A pure longitudinal study would sample the same individuals belonging, say, to cohort A at each occasion.

We have already used the term cohort to describe a study in which all the individuals born within certain time limits are followed up. The two studies referred to above are both cohort studies consisting of all the births in Britain in a single week. We have the choice, when studying a cohort, of either sampling different individuals from the same cohort at different ages or following the same individuals. The importance of this will become clear when we discuss how to obtain efficient estimates and how to deal with the problem of sample losses. For the moment we should note that if interest lies in growth changes, then whether the same or different individuals are used on different occasions, we are basically interested in age changes within well-defined cohorts. As we have already indicated, the presence of secular trends can distort the comparison of ages from different cohorts.

### 2.1.3. Superpopulations

So far we have used the word population rather loosely to describe a real population of individuals. Statisticians use the term to refer either to a finite or to a conceptually infinite set of individual units about which we wish to make inferences. Such inferences, for example, might be statements about subpopulation differences or mean values or measurements. In one sense a cohort study such as the British National Child Development Study is a finite population, defined by birth date. In practice, however, we would like to make statements not simply about a single week's births but about all children born at around that time. Because of possible secular trends the vagueness of the phrase "around that time" is inevitable, but this lack of precise definition does not preclude us from hypothesizing a "superpopulation" from which our actual sample has been drawn. It is, however, a superpopulation with changing characteristics, and it is these changes to which the term "secular trends" really refers. In the study of growth it is normal to conceive of the population of interest as infinite, and we shall assume in the remainder of this chapter that samples are selected from infinite populations.

### 2.1.4. Formal Designs

Finally, a brief word is necessary about various attempts which have been made to provide a formal model for different types of study design on the basis of the situation represented by Figure 1. Schaie (1965) and some subsequent authors (e.g., Buss, 1973) have suggested that three separate "effects" can be distinguished, namely an age, time, and cohort effect. As we have seen, a cohort may be defined by specifying appropriate times and ages. Likewise the scale of age may be defined by specifying appropriate times and cohorts, and time may similarly be defined in terms of ages and cohorts. It is clear that there are really only two underlying dimensions here (most conveniently considered as age and time) and to attempt to classify research designs simultaneously in terms of all three only leads to circular arguments. It also tends to complicate what is a relatively straightforward concept by attempting to link the different designs to hypotheses concerning genetic or cultural changes. The basic fallacy in the argument was pointed out by Baltes (1968), and it appears that much of the confusion on this issue centers around the failure to distinguish the "confounding" of three measurement scales which arise from the two-dimensional nature of the underlying classification, from the more usual "confounding" which can arise between logically independent variables within a particular experimental design.

### 2.2. Geographical Limits

To define a population in space at a particular point of time is in principle a straightforward matter, for example, by using national or local boundaries. Thus we may define the population of all 5-year-old children in a particular administrative region and devise suitable methods to sample and measure them. If we wish to follow up the population, either by remeasuring the same individuals or new individuals at a subsequent time, we immediately face a new problem. Because of death and mobility the predefined population will change, acquiring new members and losing others. Hence the definition of the population must extend over time.

With respect to time we may therefore define a population in four different ways as follows:

1. All those individuals who were living in the region at the start of the study and who were still living in the region at each subsequent measuring occasion.
2. The above group together with all those at each subsequent measuring occasion who had come to live in the region after the first occasion.
3. All those who were living in the region at the first occasion, wherever they happened to live at subsequent occasions.
4. All those individuals who were living in the region at the first occasion together with those who came to live in the region after the first occasion.

With each definition we are able to answer different questions about growth and are faced with different practical problems. If we restrict ourselves to the first definition then the practical problems of follow-up will be easiest, but we have to face the possibility that the children who leave the region have different growth characteristics from those who stay, which results in biases. We note however, that it is the population which suffers bias here and not, as is normally the case, just the sample. Indeed, we may be mainly interested in this biased population if the purpose of the study is to predict, say, those children who may require special services in the region at a subsequent age, so that it would not be appropriate to include those who move away. If, however, we wished to estimate changes in numbers of children requiring special services between the first and second occasion it would be appropriate to adopt the second definition.

Where we wish to answer questions about the growth of individuals, then we require longitudinal information and it would seem appropriate here to use the third definition. If we wish to prepare longitudinal growth standards starting at a particular age, then we should want to include all the children measured at this age. This would also allow us to compare the growth of those individuals who moved with those who stayed. The fourth definition encompasses the others and allows the most general comparisons to be made. It is also usually the most costly. The costs involved with definitions (3) and (4) clearly depend on local mobility factors which will vary greatly from country to country and from time to time. For example, Douglas and Bloomfield (1958) found that in Britain in the late 1940s during the first four years of their lives, 53% of children's families had not moved at all, 31% had stayed within their local authority area, and 14% had crossed the local authority boundary. Of those who crossed the local authority boundary, a third had moved completely out of the region. They also showed that relatively more of the nonlocal moves were among families where the child's father had a nonmanual occupation. Hence, in the light of known social class differences, definition (1) would produce different average growth estimates from definition (3).

## 3. Sampling Designs

Most of the theory of sample survey designs refers to a single sampling of a population. We begin, therefore, by outlining the main ideas behind the various designs and in subsequent sections see how these can be applied to growth studies. More comprehensive accounts of sample survey theory can be found in Moser and Kalton (1971) or Kish (1965).

### 3.1. Sample Survey Designs for Cross-Sectional Studies

At the basis of sampling theory is the notion of a simple random sample. This is a selection of individuals from a defined population or subpopulation where each individual member of the population has the same chance of selection. The choice of individuals is made using a set of random numbers, for example, by numbering sequentially every member of the population and using a table of random numbers or a computer-generated set of random numbers to select the sample members from this list. The basic sampling unit need not be an individual person but might be a household, school, etc. Simple random samples of whole populations are rare in surveys of human beings. For reasons we shall outline below, the population is often first divided into subpopulations, for example, geographically, within which random samples are then drawn.

The importance of randomness lies in its protection against biased selection.

One of the best known instances of a nonrandom selection which, with disastrous consequences, resulted in a biased selection, is the Literary Digest telephone poll of U.S. electors enquiring their voting intentions before the 1936 presidential election. With a sample of ten million individuals the wrong result was predicted. This happened because the sample was taken from the telephone directory which is clearly a nonrandom list of the total electorate. It is not just the mean value of a variable which may be biased because of nonrandom sampling, however, but also the dispersion or spread of values. For example, interviewers simply told to select random samples using subjective measures will often unconsciously choose "typical" individuals and completely avoid those with extreme characteristics.

If we wish to select a simple random sample from a large population we must first list all the individual units and then select a random sample of these. Such a list is called a sampling frame, and typical examples are electoral registers and birth registers. Unfortunately, for many real populations of interest in the study of growth, there is little or no possibility of obtaining an accurate list of this kind. Where overall population lists do not exist, we can adopt what are known as "multistage" sampling methods which involve compiling lists only for a few small subpopulations. Often, we have some preexisting knowledge about the population in terms of what we are measuring, for example that the average height varies between geographical regions. In such a case we can use this information to improve the precision of the estimates of population means or proportions by "stratifying" the populations.

### 3.1.1. Stratification

Suppose we wish to estimate the average height of 7-year-olds in a population, measuring a sample of 1000. It is well known that a child's height varies inversely with the number of children in his household, and we may consider using household size as a stratifying factor. We might choose say, five strata; households with one child, those with two, three, four, and five or more. We can do this by obtaining household size information for all 7-year-olds using, for example, census information or school records. Table II shows a typical distribution of children in the resulting strata.

A common method of sampling within strata is to choose a random sample within each stratum whose size is proportional to the known number of all individuals within that stratum. The third column of Table II shows how many of the 1000

*Table II. Distribution of the Number of Children in the Household[a]*

| Number of children in household | Children in each family size stratum (%) | Proportional allocation of sample numbers |
|---|---|---|
| 1 | 8.9 | 89 |
| 2 | 34.8 | 348 |
| 3 | 26.0 | 260 |
| 4 | 15.1 | 151 |
| 5+ | 15.2 | 152 |
| Total: | 100 | 1000 |

[a]Data obtained from Davie *et al.* (1972). Variance within each stratum = 36 cm². Variance between all individuals = 42 cm².

children would thus fall into each household group. If school records were available with household size information for each child, then we should be able to select random samples within each stratum from these records. This method of stratified random sampling is known as proportionate sampling, and with it we ensure that each population unit has the same probability of being selected. It also has the desirable feature that the best estimate of the mean height (or other variable) of the population turns out to be the same as if it were a simple random sample, namely the arithmetic mean of the individual measurements:

$$\overline{x} = \frac{1}{n} \sum_{i=1}^{k} n_i x_i = \frac{1}{n} \sum_{i=1}^{k} \sum_{j=1}^{n_i} x_{ij}$$

where $x_{ij}$ is the height of the $j$th child in the $i$th stratum, $n_i$ is the size of the $i$th stratum, $x_i$ is the mean of the $i$th stratum, and $n$ is the total sample size.

The variance of $\bar{x}$, however, is not the same as for a simple random sample, which is $\sigma^2/n$, where $\sigma^2$ is the between-individual variance. We have

$$\text{Variance } (\overline{x}) = \frac{1}{n^2} \sum_{i=1}^{k} n_i \sigma_i^2$$

where $\sigma_i^2$ is the variance between individuals within the $i$th stratum.

The ratio variance $(\overline{x})/(\sigma^2/n)$ is known as the "design effect" (deff). If less than or equal to 1, as is always the case for proportionate sampling, then for a given sample size we obtain an estimate of the mean at least as precise as that for simple random sampling. This gain in precision comes from the utilization of the information available about strata differences. The design effect will be equal to 1 only if there is no difference between strata. Thus we are assured that when we come to make estimates for other variables, some of which may not be related to the stratification factor, we will not obtain a precision worse than that for a simple random sample.

In the example in Table II we estimate the gain in precision as follows. Using the estimates for the within strata variance and the variance between all individuals in the population we have,

$$\text{Variance } (\overline{x}) = 0.036 \text{ cm}^2$$
$$\sigma^2/n = 0.042 \text{ cm}^2$$
$$\text{deff} = 0.857$$

Thus by using the stratified sample with only 857 individuals, we can achieve the same precision as a simple random sample of 1000 individuals. We shall return to this question of precision in a later section.

In the above example, if it were not feasible to obtain the household size information on each child prior to sampling, we could select a simple random sample and then apply what is known as "poststratification." Having selected the sample, we obtain the information about household size at the time of measuring and estimate the mean as

$$\overline{x} = \sum_{i=1}^{k} p_i \overline{x}_i$$

where $p_i$ is the known population proportion of individuals in the $i$th stratum. This procedure uses the prior information on the population in a slightly different way and gives an estimate which has about the same precision as the ordinary stratification procedure.

Although proportionate sampling has many advantages, there are circumstances where different strata sampling fractions, that is the proportions of the total population in each stratum, are desirable. This is relevant when stratum measures are important in themselves. For example, the British National Survey of Health and Development (1946 cohort) followed up only 25% of the original cohort of children whose fathers had manual occupations. This was done in order to retain, for a fixed total sample size, sufficient numbers of children from nonmanual backgrounds to provide reasonably precise estimates within that stratum. In other situations the cost of selecting one individual may differ between strata, so that in order to obtain the most precision for a given total cost, disproportionate sampling is required. The optimum sampling fractions are proportional to $\sigma_i/\sqrt{c_i}$, where $c_i$ is the average sampling cost per individual in the $i$th stratum. Such a situation may arise when measurers have to travel from a center so that with a geographical stratification the more distant regions will tend to be more expensive.

### 3.1.2. Multistage Sampling

In stratified sampling we need to have an adequate sampling frame of the population, and if this is unavailable the cost of constructing one may be very large. To overcome this difficulty we can select in stages, for example, by dividing a country into a large number of areas, selecting a sample of these and then forming a list of just the individuals in these areas. Such a procedure can also reduce the costs considerably by reducing the size of the sampling frame and concentrating field work on a few areas. We can add further stages if we wish, for example, by sampling schools within each area and classes within schools. We would then only need to list the children within each selected class and make a selection from this list. In many circumstances it is convenient to select all the individuals within this final stage unit, and this is known as "cluster sampling." The term clustering is also used more generally when only a sample of the individuals in the final stage units is selected.

The penalty we must usually pay for cutting our costs is a drop in precision, the design effect typically being greater than 1. This results from the fact that individuals within clusters are almost always more nearly alike than are all individuals in the population. This is a general attribute of geographically or institutionally defined clusters such as schools or classes.

The most important type of multistage sampling is where each population unit has the same probability of being selected so that, as with proportionate stratified sampling, we obtain a "self-weighting" sample. One of the most widely used methods of achieving this is known as sampling with probability proportional to size (pps). A sample of the first stage or primary sampling units (psu) is randomly selected, each one having a probability of selection which is proportional to the number of individuals within it. This is continued until the same number of individuals are selected from each final stage unit. This procedure is often combined with a preliminary stratification, and it is often possible to arrange the relative amounts of stratification and multistage sampling to produce an overall design effect close to 1.

In some situations we do not know the exact sizes of the psu's but we do have an estimate of their size. If we select with probability proportional to these estimates, then the above procedure will produce a "self-weighted" sample.

### 3.1.3. Complex Statistics from Complex Samples

So far we have discussed only the estimation of means or proportions using complex sampling procedures. We conclude this section with a few remarks on the use of more sophisticated statistical procedures, such as multiple regression, where complex sampling has occurred.

Nearly all techniques commonly used in statistical analysis assume that a simple random sample of independent elements is available. If we apply such techniques to data obtained from complex samples, treating them as if they were really simple random samples, significance tests will be distorted and we will obtain estimates of parameters which may be inconsistent. (An inconsistent estimate of a population value is one which does not approach the population value, however large the sample on which it is based. A biased estimate need not necessarily be an inconsistent one.) The problem arises most severely with multistage samples where the clustering of individuals within the final stage units, as mentioned above, leads to a small positive correlation between the individuals. With stratification, on the other hand, it may often be possible to incorporate strata differences into the statistical model, for example, as an analysis of variance factor.

Kish and Frankel (1975) have studied the effects of complex sampling on multiple regression and suggest that the usual estimates of the multiple regression coefficients (equally weighted if the sample is self-weighting) are consistent, and the design effect for the variance of each coefficient will usually be greater than 1 but less than the design effect for the mean value of the corresponding independent variable. This therefore allows us to correct the estimated standard errors for the coefficients using the design effect for the mean value, so giving a "conservative" confidence interval or significance test. The method these authors use to derive their results is called balanced repeated replications, and it involves carrying out separate analyses on different portions of the total sample. It can be applied to any kind of complex sampling, and in particular it may be applied to techniques for analyzing longitudinal data which arise from complex samples.

We shall deal with sampling for growth standards in a later section, but it should be noted that the estimation of population percentiles using a complex sample can be carried out as if the sample were a simple random one. (Naturally, where the sample is not self-weighting then the individuals are given their appropriate weights.) Actually the estimates so obtained are slightly biased, although consistent. For example, if we use the sample standard deviation to estimate the percentiles for a normally distributed variable, the average or expected value of the usual sample estimate of variance is

$$\left(\frac{n - d}{n - 1}\right) \sigma^2$$

where $\sigma^2$ is the true population variance, and $d$ is the design effect for the mean. Since $d$ is rarely greater than about 3, the bias is negligible when $n$ is 100 or more, and as we shall see in a later section, the recommended minimum sample size for percentile estimation with a simple random sample is several hundred. For descriptions of growth studies which use a complex sample, see Malina et al. (1974) and Jordan et al. (1975).

When we come to estimate the standard errors of the percentile estimates, the design effects may become more important. For details the reader should consult Kish (1965).

## 4. Sample Size Requirements

We have referred to the precision of estimates of mean values and used the variance of these estimates as a criterion for comparing different sampling schemes. For an unbiased estimate (and for a consistent estimate in large samples), the smaller its variance the nearer it is likely to be to the "true" population value. More precisely, we can define limits for the population value which should have a high probability of actually including the true value between them. The most common such limits are known as confidence limits, and are described as follows.

If $\bar{x}$ is the sample mean (or any other statistic), then 95% confidence limits, $a, b$ are two values such that the probability of the true population mean lying between them is 0.95. If the distribution of the measurement is normal, then for a simple random sample $a$ and $b$ are given by

$$a = \bar{x} - 1.96 S/\sqrt{n}$$
$$b = \bar{x} + 1.96 S/\sqrt{n}$$

Here $S$ is the estimated standard deviation of the measurement, $n$ is the sample size, and hence $S/\sqrt{n}$ is the standard deviation or standard error of the sample mean. The value 1.96 is chosen because 95% of the normal distribution lies in the symmetrical interval $-1.96$ to $+1.96$ or approximately $-2$ to $+2$. (If $n$ is less than about 30 we should use corresponding values from the $t$ distribution rather than the normal.) We may express this formally as

$$\text{Probability } (a \leqslant \mu \leqslant b) = 0.95$$

where $\mu$ is the true population mean.

Thus with these limits we can be sure that there is a 95% chance that our confidence interval includes the mean. Of course, we can be more stringent by asking for, say, a 99% confidence interval, but it will be wider than before, the corresponding normal distribution values being $-2.33$ and $+2.33$. Where the distribution is not normal or we are estimating a more complex statistic such as a ratio, then the method of calculating confidence limits will be different, although the principle is the same. The standard error of the sample estimate is not always used directly in the estimation of the confidence interval, but nevertheless remains a useful indicator of relative precision. Moreover, a useful feature of many estimates, as the sample size becomes large, is that we may regard them as having approximately normal distributions, so that the calculated standard error may be used to set confidence limits as above.

### 4.1. Significance Testing

Before going on to develop the above method for obtaining a specified precision we need to mention an alternative approach which may be applicable in certain circumstances.

Suppose we wish to study the average difference between two subgroups of the population, for example, the heights of children who have participated in one nutritional program and the heights of a comparable group who have participated in another. We may wish to ensure that we can determine which of the two programs resulted in a greater mean height. If we write $\overline{Z}$ for the difference between the two means, then we may carry out a conventional significance test of the null hypothesis that the expected value of $\overline{Z}$ is zero. Where we wish to detect a departure from a zero value of a predetermined size, this allows us to calculate the sample size necessary to do so with a given chance of success. In order words we can readily calculate the sample size necessary to obtain a "significant" result, say, 95% of the time, when the expected value of $\overline{Z}$ is, in fact, not equal to zero.

Much controversy has surrounded the theory of significance tests. Their main practical function is usually to help us make preliminary decisions, prior to studying the values of the estimates themselves. For example, in the above case the conventional "two-sided" significance test may be viewed as a test for the direction of the difference. In some circumstances it may well be reasonable to give first priority to determining the sample size with a view to giving a reliable decision about the direction of the difference. One of the difficulties with calculating sample sizes for significance tests, however, is the necessity for specifying the size of difference we want to detect. In some circumstances a prescribed difference may be determined from theoretical or practical policy considerations. Sometimes previous evidence may be available, or we may have to carry out a small pilot to obtain the required information. This does not arise when we only want to provide a confidence interval of a given length, although in both cases we need to have an estimate of the standard deviation. Assuming that significance tests may be regarded as preliminary steps in any analysis, we shall confine the following discussion to the precision of estimates of population values.

## 4.2. Specifying the Precision Required

If we leave aside sample design considerations, the obvious way of obtaining more precision or a narrower confidence interval is by increasing the sample size. If we know what precision we require, and have a good estimate of the standard deviation, then we can find the value of $n$, the sample size, which will give that precision. The precision we are seeking may be for a simple mean value, for the difference between two means, for the prediction of a child's or adult's height, for a percentile estimate, etc. Methods of carrying out the calculations for the first three estimates mentioned can be found in many elementary textbooks on statistics, and we shall only make general reference to them. We shall, however, go into more detail for percentile estimates.

To help us determine what size of confidence interval we should aim for, it is often useful to study the relative precision, obtained by dividing the confidence interval by an estimate of the mean value of the measurement. (The standard error divided by the mean value is often called the coefficient of variation.) For example, if the average height of a population of children of a given age is known to be about 120 cm and we have an estimate of the standard deviation in the population of 6.0 cm, Table III gives the 95% confidence interval width and relative width for different sizes of a simple random sample.

With a sample size of 500 we obtain a relative precision of about 0.4% whereas with a sample of only 10 the relative precision is about 2.5%. Such calculations can therefore give us a rough idea of the size of sample we might need, so long as we are prepared to commit ourselves to a level of relative precision. Similar calculations

*Table III.  Sample Sizes for Specified 95% Confidence Interval Widths*[a]

| | Sample size | | | | |
|---|---|---|---|---|---|
| | 10 | 50 | 100 | 5000 | 10,000 |
| Confidence interval width | 3.04 | 1.36 | 0.96 | 0.43 | 0.30 |
| Width relative to mean (%) | 2.53 | 1.13 | 0.80 | 0.36 | 0.25 |

[a]Population standard deviation = 6.0 cm. Mean = 120.0 cm.

can be carried out for the difference of two means representing, for example, the change between two ages, the divisor here being the mean value of the difference. When predicting the value of, say, adult height, we may divide the confidence intervals for predicted values by the mean predicted values.

With growth measurements an alternative method of arriving at an estimate of precision is to compare the confidence interval with the average growth over a convenient time interval, say a year. Thus in the above example, if the average growth in height over a year is 5.0 cm, then with 500 in the sample the confidence interval is about 1.0 months of growth whereas with a sample of only 10 it is about 7.3 months of growth.

### 4.2.1. Precision of Percentile Estimates

When constructing percentile standards for growth measurements, two aspects of sampling must be considered. The first, which we deal with in a later section, concerns the manner in which the sample is spread across the whole age range. The second is concerned with the sample size needed to obtain a stated accuracy at any given age, and we now deal with this.

Where the distribution of a variable is known to be normal, then we can use the estimated sample mean and standard deviation at the age concerned to give estimates of the percentiles. For example, a consistent estimate of the 97th percentile is

$$\bar{x} + 1.88S$$

and the standard error of this estimate is approximately $1.7S/\sqrt{n}$. For large samples this enables us to construct confidence intervals in the usual way for different values of $n$.

In practice, however, even for a variable so close to having a normal distribution as height, the assumption of a normal distribution ceases to provide good estimates of the extreme percentiles (see Goldstein, 1972). The simplest and, for large samples, the safest alternative procedure is an ordering method whereby all the sample values are arranged in ascending order of magnitude and we choose the value (or interpolated value) below which the required percentage of the measurements lie. If the sample does in fact come from a normal population, then the standard deviation of this estimate is approximately $2.5S/\sqrt{n}$, so that it gives confidence limits about 50% wider than those above. Table IV sets out the range of population percentiles covered by a 95% confidence interval for the 97th percentile for each method.

Whereas the sample mean and standard deviation give a considerable improvement for small sample sizes, for samples of 1000 or over either method is acceptable. The possibility of a large bias, which would occur if we used the mean and

Table IV.  *Percentile Range Covered on Average by a 95% Confidence Interval about the
Estimated 97th Percentile Where the Distribution Is Normal*

| | Sample size | | | | |
|---|---|---|---|---|---|
| | 100 | 200 | 500 | 1000 | 2000 |
| Simple ranking method | 91.6–99.1 | 93.5–98.7 | 95.2–98.2 | 95.8–97.9 | 96.2–97.7 |
| Using sample mean and standard deviation | 93.8–98.7 | 95.0–98.3 | 95.9–97.9 | 96.2–97.6 | 96.4–97.5 |

standard deviation for a population which does not have a normal distribution, might well outweigh the higher apparent accuracy.

We can improve the accuracy of the ordering method as follows. We plot the cumulative distribution of the points on a suitable scale (for example, normal probability paper) and then "smooth" the curve in the region of the percentile to be estimated. This can often be done adequately by eye. A detailed description of this method can be found in Goldstein (1972), and if we adopt it, then Table IV suggests that a sample size of about 500 ought to give acceptable accuracy.

## 5. Sampling with Respect to Age

When designing a growth study covering a specified age range, we need to consider the manner in which measurements are spread over time. Many ways of spreading the sample over the age range are possible, and we begin the discussion by assuming that measurements are made at a small number of discrete occasions. Later on we shall extend this to consider continuous sampling over the age range.

### 5.1. A Fixed Number of Occasions

A popular design for many growth studies consists of measuring each individual child once a year on his or her birthday. This provides a fixed set of occasions a year apart which is convenient and ensures that seasonal variations are eliminated. Extra occasions are often inserted, for example, every three months, at those times during growth, such as at adolescence, when rapid changes are taking place. We shall pursue this point and the topic of seasonal variation below.

To simplify matters, consider just two occasions labeled 1 and 2. Suppose also that we have decided to measure the same number of individuals ($n$) at each occasion and that we are interested in making the most precise estimates possible of the mean at each occasion and the difference between the means. In general, the measurements at the two occasions will be related, and we assume that the value of the measurement on the second occasion can be predicted from that on the first by the regression equation

$$x_2 = a + bx_1$$

where $x_2$ is the second occasion value and $x_1$ the first.

Suppose that $m$ individuals are measured on both occasions with $n - m$ measured at one occasion only. We may use the $m$ common measurements to estimate the values of $a$ and $b$ in the above equation in the usual way. Using these values, we can then apply the equation to those measurements taken on the first occasion only to predict their mean value on the second. We thus have three separate estimates for the mean on the second occasion. One is obtained from those measurements made at the second occasion only, another is obtained using the prediction from those measured only at the first occasion, and the other is obtained from the second occasion measurements of those measured on both occasions. We then find a weighted combination of these to give the most precise estimate of the overall mean at the second occasion. The reverse procedure can be adopted to obtain the most precise estimate of the mean at the first occasion. The most precise estimate of the difference between the means is then simply the difference of these two estimates. Looked at slightly differently, we are trying to find, for each occasion, a linear combination of the separate estimates of the mean which gives an overall estimate which has the greatest precision. There are four quantities which will enter into this combination: the two means at the first occasion—for those individuals who are and are not common, and likewise for the second occasion. The procedure can readily be extended to several occasions with different numbers sampled at each occasion. This method, with certain simplifying assumptions, was described in detail by Patterson (1950), and Gurney and Daly (1965) describe a more general procedure. Both these methods assume that the correlations between occasions and the variances within occasions are known. In practice they will often be calculated from the sample itself, and where the sample is large this will have little effect on the estimates. For small samples, however, and especially where the number of common individuals is small, it may be necessary to take this into account (see e.g., Jessen, 1942).

When all individuals are measured at both occasions, i.e., a pure longitudinal study, then the best estimates for each occasion are simply those which use the measurements made at that occasion. The same is true when there are no common units, since there is now insufficient information for relating the measurements on the two occasions.

Reverting to the two-occasion case, we can illustrate the effect of different sampling strategies. The variance of the best estimate of the mean on each occasion is

$$V = \frac{s^2}{n} \frac{(1 - qr^2)}{(1 - q^2 r^2)}$$

where $q$ is the proportion not common to each occasion, $s^2$ is the variance within occasions, and $r$ is the correlation between occasions. When $q = 1$ or 0 we have $V = s^2/n$, which is simply the variance of the mean of $n$ sample values. When $0 < q < 1$, then $(1 - qr^2)/(1 - q^2 r^2) < 1$, so that having some common elements give us a more accurate estimate of the mean. In fact the variance is smallest when

$$q = \frac{1 - \sqrt{1 - r^2}}{r^2}$$

The variance of the best estimate of change is $2s^2 (1 - r)/n(1 - qr)$. If $r$ is positive, as is often the case in growth studies, this variance is smallest when $q = 0$, that is all individuals are measured at both occasions. If $r$ is not positive, then we obtain the smallest variance when there are no common elements. Table V shows the optimum values of $q$ and corresponding precisions for estimating the occasions means and the difference of the means, for different values of $r$.

It is evident from this table that as the correlation between occasions increases, the gain from optimum allocation increases. The gain is much more striking for the

*Table V.  Relative Precision[a] of Optimum Values of q for Different Values of r*

| $r$ | Optimum value of $q$ | Relative precision (%) | Difference between occasion means, relative precision (%) | |
| --- | --- | --- | --- | --- |
| | | | $q$ = optimum value for occasion means | $q = 0$ |
| 0.1 | 0.50 | 100 | 106 | 111 |
| 0.3 | 0.51 | 102 | 121 | 143 |
| 0.5 | 0.54 | 107 | 166 | 200 |
| 0.7 | 0.58 | 117 | 193 | 333 |
| 0.9 | 0.70 | 139 | 370 | 1000 |

[a]Relative precision is the variance when $q = 1$ divided by the variance for the optimum value. It is therefore equal to the relative increase in sample size needed to maintain the same precision if separate individuals are used at each occasion as opposed to the optimum mixture of separate and common individuals.

estimate of the difference than for the means at each occasion. In particular, the last two columns show that when $r$ exceeds about 0.5, the optimum mixture of common and different individuals for estimation of the occasion *means* becomes very inefficient for the estimation of the *difference* between occasion means. When $r = 0.9$, the optimum mixture for the occasion means is to have 30% in common, but this is only about 37% as efficient as the optimum for the difference when all are common. On the other hand, when all are common, the resulting estimates for the occasion means are about 72% as efficient as the optimum. Thus, if we wish to use a sample both to estimate occasion means and occasion mean differences, we will need to achieve a compromise in terms of efficiencies. The compromise will depend on the relative importance we attach to each estimate. When $r$ is less than about 0.5, it would seem reasonable to choose all elements to be common. For growth measurements, however, the value of $r$ for yearly measurements will often be as high as 0.7–0.9, and we will therefore need to choose our priorities carefully. When the simple two-occasion situation is extended to several occasions with possibly unequal numbers, variances, etc., the optimum allocation procedures become more complicated, but the same general features still hold.

We have assumed so far that the relative cost of remeasuring an individual is the same as measuring a new individual. This is not necessarily true and ought to be taken into account. There may also be problems of differential bias, nonresponse, etc., when remeasuring individuals, which will also need consideration. We shall discuss these problems in a later section.

We have also assumed that the sample we are dealing with is a simple random sample from the population. Often in practice, however, the sample will be complex and involve several stages. In a two-stage design, for example, we have the choice of retaining or changing individuals, as well as retaining or changing primary sampling units. There may be considerable cost savings involved in retaining the same set of primary units on a second occasion and changing some of the individuals within the primary sampling units. A detailed discussion of such situations is beyond the scope of the present chapter, and the interested reader is referred to the papers by Singh (1968) and Chakrabarty and Rana (1974).

## 5.2. Sampling for Growth Standards

We discussed earlier the sample sizes needed to achieve a given precision for estimates of the extreme percentiles of a distribution at a given occasion. Typically, however, percentile growth standards are constructed for a wide age range, estimates usually being made at a large number of occasions within the range and these then joined up by smooth curves. The problem, therefore, is how such occasions are to be chosen, how many individuals should be sampled, and how the sample should be spread over the age range.

We start by assuming that percentiles are to be estimated with the same relative accuracy at all occasions or ages in the range. It follows that we should choose equal numbers at each occasion, leaving us with the problem of how to space out the occasions. For any given percentile, say the 50th, we can estimate the mean value of the measurement at each occasion but no estimates of the values between occasions can be made. Thus between any two adjacent occasions, the expected value of the 50th percentile will have to be interpolated from the values at the occasions themselves. We might do this simply by joining the means at the occasions by a straight line or, if other occasions are available, by a smooth curve through a set of three or more occasions. Assuming that the true 50th percentile curve is smoothly increasing or decreasing between the two adjacent occasions, it is clear that the maximum difference between the true and interpolated values is simply the difference between the values at the two occasions.

As the criterion for spacing occasions, we specify that this maximum difference is the same at each part of the age range, and this leads us to space occasions so that the change in value of the 50th percentile is the same between all adjacent occasions. This is equivalent to sampling occasions proportional to the rate of change of the 50th percentile value with age. Strictly speaking, this might lead to different sampling procedures if we choose different percentiles, but this is unlikely to be important in practice. Similar considerations apply to "velocity" standards where we will be led to sample proportional to the rate of change of the 50th velocity percentile.

A problem occurs at the approach to adulthood where growth slows down and eventually ceases. It will then be necessary to include an adult sample of sufficient size to give the required precision. A further problem occurs when several measurements with different patterns of change are made using the same sample, and some compromise will have to be used to determine the sampling fractions.

Table VI shows the relative sample sizes for yearly age groups based on the 50th percentile estimates of stature using British standards (Tanner *et al.*, 1966). The extra "adult" sample has not been included.

Rather than concentrate a sample at specific ages, it is more desirable to spread the sample along the whole age range. This is usually a more practical procedure and also enables us to estimate seasonal variations and provide better smoothing. We then proceed by estimating the percentiles at the center of narrow age intervals using a method which allows for the growth which takes place during the interval (Goldstein, 1972). The intervals are chosen as follows.

Ignoring the approach to adulthood, we choose a year when growth is slowest and select a large enough sample for that year to give precise estimates of the extreme percentiles. From Table VI we might therefore select the age of 7+. If we act in accordance with the suggestion made in an earlier section, we might take 500 children of each sex in this age group. To allow for seasonal effects estimation should be carried out over whole years so that a one-year age group is the smallest interval. Also, at this age, growth over one year is approximately linear, thus allowing a simple estimation procedure. The other age groups are then sampled accordingly. Where growth is fastest, namely under one year, the sample becomes relatively large and, since growth is changing very rapidly, a one-year interval will

Table VI.  Approximate Sampling Fraction
for Cross-Sectional Standards at Each Year
of Age from 0 to 19 Years

| Age | Boys (%) | Girls (%) |
|-----|----------|-----------|
| 0+ | 17 | 21 |
| 1+ | 8 | 9 |
| 2+ | 6 | 7 |
| 3+ | 5 | 7 |
| 4+ | 5 | 6 |
| 5+ | 5 | 5 |
| 6+ | 5 | 5 |
| 7+ | 5 | 5 |
| 8+ | 5 | 5 |
| 9+ | 5 | 5 |
| 10+ | 5 | 6 |
| 11+ | 5 | 6 |
| 12+ | 5 | 6 |
| 13+ | 6 | 4 |
| 14+ | 6 | 2 |
| 15+ | 4 | 1 |
| 16+ | 2 | - |
| 17+ | 1 | - |
| Total: | 100 | 100 |

be too large and much narrower intervals should be used, for example, one month. The seasonal effects here are likely to be very small relative to the change in growth itself and can probably be ignored.

In order to provide a smooth growth curve, we can form a moving average set of estimates by overlapping the age intervals. For example, we can choose the intervals 7.0–8.0, 7.25–8.25, 7.50–8.50, etc., estimating the percentiles at the centers of these intervals and subsequently joining up the resulting estimates 0.25 years apart.

In most longitudinal studies, where attempts are made to measure individuals at predetermined ages, there will usually be some variation about the intended or "target" age. Where this variation is small and random, it will increase the variability of the estimates but not invalidate the procedures described above. If the variation about the target occasion becomes large, however, serious problems may arise. First, even where the distribution of ages about the target age is symmetrical, the variability of the estimates may be considerably increased if growth is particularly rapid. More often than not, however, the distribution will be skewed, owing to most individuals being measured late, and there may also be an association between time of measurement and growth attained. Secondly, it is known that growth is more rapid at certain times of year (Marshall and Swan, 1971), so that differing time periods between occasions will lead to different estimates of growth rate.

Based on the data of Marshall and Swan (1971), Figure 2 shows two curves of average growth in height for two cohorts aged exactly 7.0 and 7.5 years at the beginning of the year. Over the one-year period it is assumed that they both increase by 6.0 cm so that there has been no secular trend operating over the half year separating the cohorts. We also see that the fastest rate of growth is in the spring. Thus an estimate of average growth rate based on a period of less than a whole year will give a biased estimate. For example, if we estimate growth rate using either cohort and based on the 6-month period from January to July, we obtain a value of

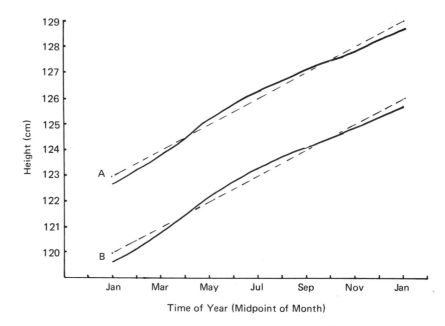

Fig. 2. Mean growth in height of two cohorts over one year. Cohort A is age 7.5 years and cohort B is age 7.0 years, in January.

7.2 cm per year, which is 20% too large. Likewise, if we based an estimate on the difference between the older cohort in April and the younger in July, we obtain a value of 5.4 cm per year, which is 11% too small.

We have similar problems when we wish to estimate growth attained. For example, the average height of children who are 7.5 years in January (from the older cohort) is 122.7 cm, whereas the average height of children who are 7.5 years in July (from the younger cohort) is 123.3 cm. This clearly has certain implications for the construction of standards for growth attained, and in particular for the spacing of measurements over time. A similar phenomenon has been observed with tests of school attainment where some average attainment scores actually appear to decrease between the spring and the summer (Goldstein and Fogelman, 1974).

If we wish to estimate such seasonal effects, it is clear that we need to measure individuals who are the same age at different times of the year from different cohorts, or alternatively individuals from the same cohort at different times of year. If we wish to avoid including any secular trends, the latter method is to be preferred. For periods of less than one year, we will be mainly concerned with estimating growth attained rather than rate of growth. We need not limit the sampling to specific occasions, but rather spread sampling uniformly over the whole year in order to obtain more detailed estimates of the growth pattern. If, in addition, we are also interested in change over a one-year period, then we can arrange for some or all of the individuals to be remeasured exactly one year later.

We have now arrived at a sampling scheme which is similar to one with yearly occasions, except that the occasions are now no longer at a specified age. Rather they are defined as any age within a year, subject to a uniform sample being spread over the year and measurements repeated at intervals of a year. In this way we largely solve both the problems outlined at the beginning of this section. Regarding the period between consecutive birthdays as an occasion, we can readily apply the theory outlined previously to give optimum sampling schemes. An alternative would be to measure all individuals at the same time of year, selecting from different cohorts so that the age distribution covered exactly one year, for example, 7.0 to 8.0 years. Finally, a combination of the two schemes could be used, sampling the same set of cohorts during the year. This approach seems to have been little tried in longitudinal growth studies, although it seems to possess advantages over the usual approach.

Before leaving this topic we mention briefly the problems which can arise from a failure to take account of seasonal effects in birth cohort studies. As an example, consider the National Child Development Study, where children were born during the second week of March 1958. The children born at this time had a wide range of gestation lengths and hence a wide time span during which conception occurred. For example, a baby of short gestation of 34 completed weeks will have been conceived two months after a late gestation of 43 weeks, that is at the end June 1957 as opposed to the end of April 1957. One consequence of this is that a change in conception rate between April and June 1957 will result in a biased estimate of the distribution of gestation length based on March 1958 births. If, as the evidence suggests, the conception rate was rising during this period, then the short gestations will tend to be slightly overrepresented in the sample. Another consequence arises from the fact that specific periods of pregnancy are associated with the season of the year. For example, the fourth month of pregnancy occurs in August 1957 for the long gestations, but in October 1957 for the 34-week gestations, and even later for the very short gestations. This early part of pregnancy is thought to be more vulnerable in terms of infections, etc., which may cause even minimal damage to the fetus, and since it is the short gestations which pass through the early part of pregnancy during the autumn when infections tend to be more prevalent, we might expect a resulting higher incidence of related difficulties and abnormalities in these births. Hence, an abnormality independent of gestation length *per se* may, in fact, be associated with it through the association of conception time with season of the year. Without very detailed and accurate information about pregnancy infections (and other relevant data) we shall be unable to unravel these factors. These considerations do not, however, appear to have caused serious problems in the National Child Development Study (Goldstein and Wedge, 1975).

## 6. Nonsampling Errors

Many stages intervene between the design of a study and the analysis of the data, and at each one errors may occur. Those which occur at the time of sampling and measuring often tend to be the most important and with the greatest potential for invalidating the results of analyses.

A nonrespondent is a selected individual for whom data are unavailable. We can distinguish three main situations which can give rise to this. First, it may not be possible to trace the individual, for example, because he has moved. Second, the individual may refuse to cooperate, and third the individual may become excluded from the study, for example, because of death.

It is clearly essential in a longitudinal study to be able to locate individual subjects when they are required for measurement. Methods for doing this efficiently will vary from study to study and from place to place. In some areas or countries, for example, centralized registers of individuals may be kept for health and welfare purposes, and these may be used to locate given subjects. In other areas population mobility may be very low, so that the effort allocated to tracing may be small. The following remarks will therefore need to be read bearing in mind the circumstances of any particular study.

Several authors (see, for example, Eckland, 1968) have suggested that, given sufficient resources, nearly all the subjects originally sampled can be traced, even after several years have elapsed. In practice, however, resources will not be unlimited and the problem is to know how many resources are sufficient to achieve a satisfactory rate of tracing. The question as to what constitutes a satisfactory tracing rate is not an easy one to answer, depending in particular on how atypical the untraced individuals are likely to be. In many studies failure to trace is associated with high geographical mobility so that the untraced may well possess relatively unusual characteristics. Many would regard a tracing rate of below 90% as poor, and a reasonable target would be to trace at least 95% of the original sample at each subsequent occasion.

The most efficient use of resources in tracing, as in other aspects of sample maintenance, is to begin with the method which uses least resources per individual. This may involve correspondence or a telephone call to the last known address, a search through a population register, or an approach to institutions such as schools. Very often this will locate the majority of subjects, especially in studies where the occasions are close together. The next stage usually involves pursuing individuals using whatever methods are available. One useful method is to visit the last known address and ask the occupants, neighbors, etc., for information they may possess. If information about a subject's relatives has been obtained at the first occasion, then it may be possible to contact one or more of these to establish the location of the subject. The importance of obtaining such data as well as detailed identification and family information on a subject at the first occasion cannot be overemphasized. Another approach which has sometimes been used successfully is to persuade the subjects themselves to notify the study whenever they move.

Once an individual who still qualifies as a member of the sample has been traced, we will seek to obtain measurements on him. At this stage we may possibly be met with a refusal or a failure to make contact, problems which occur in all types of surveys. Where there is an initial failure to make contact, we can call back several times until contact is made. Usually up to three times will be necessary to obtain a satisfactory response, and if this is expensive then we may choose to do it on a random subsample only. Using a subsample at least enables us to estimate the extent of any biases which may be caused by the nonresponse. Other methods have been suggested, and the interested reader should refer to Kish (1965).

Refusals to cooperate have to be dealt with at a much earlier stage of a longitudinal study. Subjects or their families who have an interest in the study are less likely to refuse to cooperate, and various methods of maintaining interest have been used. Financial incentives are often successful and are justified by the time a subject devotes to the study. With some studies there may be a "spin off" for a subject in terms of, for example, regular medical examinations. The feedback of the results of a study to all concerned is another valuable way of maintaining interest and contact, and a clear description of the structure of the study is important. All this needs to be coupled with an honest explanation of the reasons for the study to the subjects or their parents. Resources allocated to these aspects will usually be a worthwhile expenditure.

If the individuals who fail to provide information were a random subset of the sample, no bias would result although the design would become modified (see for example, Lehner and Koch, 1974). Unfortunately, in nearly all surveys the nonresponders have different characteristics from the remainder. If we knew how the nonresponders differed we could use the information from the remainder of the sample to make appropriate corrections to our estimates. For example, if we knew that the average age of the nonresponders was higher than the remainder, and that this was the sole factor responsible for the nonresponse, then we could weight up the remainder of the sample to correct for the bias in those variables related to age.

Most procedures for dealing with nonresponse bias are attempts to discover the characteristics in which the nonresponders are different from the remainder, to assess the likely effect of these differences and, where appropriate, to make suitable corrections. For example, if repeated callbacks are used to obtain measurements, we will often find that the more difficult the subject is to measure the more different are his characteristics. By extrapolating the relationships between difficulty to measure and the values of various characteristics, it may be possible to obtain a good estimate of the values of the characteristics of those who do not respond at all (see, e.g., Houseman, 1953).

With a longitudinal study we are, in some respects, in a worse position regarding nonresponse than with a single cross-sectional study, since nonresponse bias can occur at every occasion. As the study progresses the cumulative effect of nonresponse may lead to cumulative bias. This will affect both the estimates of cross-sectional parameters and also the estimates of change and relationships across time. Williams (1970), for example, quotes evidence from a longitudinal study showing that over time there is a tendency for large families to drop out of the study. Williams and Mallows (1970) suggest that for measurements which are highly associated with response, even an overall response rate as high as 90%, which is often regarded as satisfactory, can lead to very biased estimates of change. Unfortunately, it is often just those groups of the population who show the most change over time who are also the most mobile, better educated, etc., and more likely to be untraced, unavailable, or refuse to cooperate.

It is important to distinguish between these three groups of nonresponders. There is evidence (see, e.g., Labouvie *et al.,* 1974) that children whose parents refuse to cooperate on the second or subsequent occasions have school attainments at least as high as those with information, whereas those who are untraced and have no available data do worse. This is supported by evidence from the National Child Development Study (Goldstein, 1976) which shows that those children who refused to cooperate in the study at the age of 16 had reading and mathematics scores at 11 years, 2–5 months in advance of those with available data. Those who had been traced but had supplied no available data at 16 years were 7–8 months behind, and those who were untraced at 16 years were nearly a year behind at 11 years. In terms of height measured at 11 years, there were also some detectable differences between these groups. These were, however, relatively small, and although it would be rash to generalize from this study alone, it appears as if physical growth measurements might be less highly correlated with nonresponse than mental measurements. There was also an indication in this study that the refusers had a slightly faster rate of mental development before 11 years and those with no data a slightly slower rate of development, suggesting that similar differences might hold for the period from 11 to 16 years.

We see here how internal evidence, using the longitudinal elements of a study, may be used to study the effect of different response patterns. By itself it can say little about initial bias, nor can it give a complete picture of change bias. By sampling sufficient new individuals at each occasion, however, we can obtain further information about cumulative occasion bias and also investigate the bias in estimates of change. In any longitudinal study where nonresponse is likely to occur, it will be worthwhile to sample extra new individuals for this purpose. In some circumstances there may also be data from other sources which can be used to check for bias.

A type of longitudinal study which is of some practical interest in this connection is the "intervention" or "quasi-experimental" study, which in its simplest form compares two groups before and after an "intervention" of some kind has taken place. One group is given a "treatment," for example a nutritional supplementation program or a special educational program, and the other group is used as a control. The aim is to compare the status of the groups in terms of a growth or attainment measurement at a specified time after treatment, if necessary allowing for initial differences. Because the two groups are undergoing different experiences, this can often give rise to differential response rates and great care has to be taken to minimize this by careful tracing, call-backs, etc. Using the initial pretreatment measurements to study response bias will generally be less satisfactory here since the point of the program itself is to bring about large changes.

### 6.2. Repeated Response Bias

The repeated measurement of a sample of individuals may alter the natural history of their growth and development. This may occur in a physical growth study for example, because a set of detailed measurements may uncover growth irregularities or morbidity patterns which will receive treatment they would not otherwise have had. The increased awareness which usually results from taking part in a study may also affect the measurements being studied. Bailer (1975) shows that the length of exposure to a study affects both the cross-sectional and change estimates for certain measurements of employment. It is therefore worth considering setting aside portions of the sample measured on the first occasion, to be remeasured once only at later occasions in order to study the effect of repeated measurements on the estimates.

### 7. Secondary Sampling

Whereas most of this chapter has been concerned with overall sampling strategies, it will very often be necessary in a longitudinal study to defer final decisions regarding sampling on future occasions until, say, some results accumulate or the response rate becomes known. Often not very much can be said about such possibilities except that there should be an awareness of them from the outset. With large studies of the cohort type where large amounts of data are collected at relatively infrequent intervals, there is often a provision for special groups to be subsampled and studied more intensively. The decision to do this, however, can usually not be made at the outset. It may have to wait, for example, until there is relevant information to define subgroups, or there may arise new issues of theoretical or practical interest which can help to determine the course of the study. Conversely, we may often be justified in including measurements at the start of the study whose sole function is to provide basic data for possible later use. We might, for example, collect basic information concerning a study child's siblings to allow for a later study which would follow them up. Also we might collect data which had no immediate relevance to the study but which might become relevant later. Especially with long-term studies, it is often the case that the interests of researchers and policy makers will change during the course of the study. Some allowance for this seems desirable, provided of course that it does not result in a massive and uncritical collection of information.

## 8. Some Practicalities

Although not strictly part of a chapter on sampling, it will be useful to end with a few remarks on certain practical issues which confront those carrying out long-term growth studies. A more detailed discussion of some of these can be found in Wall and Williams (1970) and Crider *et al.* (1973).

As well as losing subjects, a long-term study is also prone to lose its research workers. It requires a considerable commitment to remain with one project for 10 to 20 years, although there are several examples where this has occurred. It becomes important, therefore, to formalize the organization of a study to a greater extent than is usual with single cross-sectional studies. An institutional commitment to a study is usually desirable so that accumulated experience can be utilized and consistent procedures can be maintained. It is also very necessary to prepare detailed documentation of decisions taken, coding instructions, and subject files. A full description of each measurement is necessary to ensure consistency over time and regular quality control is highly desirable (see for example, Jordan *et al.,* 1975). The rationale for particular decisions also needs to be made explicit for those who may come later, and there should be a proper record of analyses carried out. The data processing system needs to be flexible, able to be updated periodically, and carefully documented.

The discipline and organization which should be effected in any properly conducted study becomes vital for the full success of a longitudinal study, and careful attention to these matters will bring long-term benefits.

Acknowledgments

My thanks are due to Professor Michael Healy for his useful comments on an early draft. The writing of this chapter was partly supported by a grant from the Department of Health and Social Security to the National Children's Bureau.

## 9. References

Bailer, B. A., 1975, The effects of rotation group bias in estimates from panel surveys, *J. Am. Stat. Assoc.* **70:**23–30.
Baltes, P. B., 1968, Longitudinal and cross-sectional sequences in the study of age and generation effects, *Hum. Dev.* **11:**145–171.
Buss, A. R., 1973, An extension of developmental models that separate ontogenetic changes and cohort differences, *Psychol. Bull.* **80:**466–479.

Chakrabarty, R. P., and Rana, D. S., 1974, Multi-stage sampling with partial replacement of the sample on successive occasions, *Proc. Soc. Stat. Sect. Am. Stat. Assoc.* **69:**262–268.

Crider, D. M., Willits, F. K., and Bealer, R. C., 1973, Panel studies: Some practical problems, *Sociol. Methods Res.* **2:**3–19.

Davie, R., Butler, N. R., and Goldstein, H., 1972, *From Birth to Seven,* Longman, London.

Douglas, J. W. B., 1964, *The Home and the School,* McGibbon and Kee, London.

Douglas, J. W. B., and Blomfield, J. M., 1958, *Children under Five,* Allen and Unwin, London.

Eckland, B. K., 1968, Retrieving mobile cases in longitudinal surveys, *Public Opinion Q.* **32:**51–64.

Goldstein, H., 1972, The construction of standards for measurements subject to growth, *Hum. Biol.* **44:**255–261.

Goldstein, H., 1976, A study of the response rates of sixteen-year-olds in the National Child Development Study, in: *Britain's Sixteen-Year-Olds* (K. Fogelman, ed.), National Children's Bureau, London.

Goldstein, H., and Fogelman, K., 1974, Age standardisation and seasonal effects in mental testing, *Br. J. Educ. Psychol.* **44:**109–115.

Goldstein, H., and Wedge, P. J., 1975, The British National Child Development Study, *World Health Stat. Rep.* **28:**202–212.

Gurney, M., and Daly, J. F., 1965, A multivariate approach to estimation in periodic sample surveys, *Proc. Soc. Stat. Sect. Am. Stat. Assoc.* 202–257.

Houseman, E. E., 1953, Statistical treatment of the non-response problem, *Agric. Econ. Res.* **5:**12–18.

Jessen, R. J., 1942, Statistical investigation of a sample survey for obtaining farm facts, *Iowa Agric. Exp. Sta. Res. Bull. 304.*

Jordan, J., Ruben, J., Hernandez, J., Bebelagua, A., Tanner, J. M., and Goldstein, H., 1975, The 1972 Cuban National Child Growth Study as an example of population health monitoring: Design and methods, *Ann. Hum. Biol.* **2:**153–171.

Kish, L., 1965, *Survey Sampling,* Wiley, New York.

Kish, L., and Frankel, M. R., 1974, Inference from complex samples, *J. Roy. Stat. Soc. B.* **36:**1–37.

Labouvie, E. W., Bartsh, T. W., Nesselroode, J. R., and Baltes, P. B., 1974, On the internal and external validity of simple longitudinal designs, *Child Dev.* **45:**282–290.

Lehner, R. G., and Koch, G. G., 1974, Analyzing panel data with uncontrolled attrition, *Public Opinion Q.* **38:**40–56.

Malina, R. M., Hamill, P. V. V., and Lemenshaw, S., 1974, Body dimensions and proportions, White and Negro children 6–11 years, *Vital and Health Statistics Series 11,* p. 143, U.S. Dept. of Health, Education and Welfare.

Marshall, W. A., and Swan, A. V., 1971, Seasonal variation in growth rates of normal and blind children, *Hum. Biol.* **43:**502–516.

Moser, C. A., and Kalton, G., 1971, *Survey Methods in Social Investigation,* Heinemann, London.

Patterson, H. D., 1950, Sampling on successive occasions with partial replacement of units, *J. R. Stat. Soc. B* **12–13:**241–255.

Schaie, K. W., 1965, A general model for the study of developmental problems, *Psychol. Bull.* **64:**92–107.

Schlesselman, J. J., 1973, Planning a longitudinal study: Sample size determination, frequency of measurement and study duration, *J. Chron. Dis.* **26:**553–570.

Singh, D., 1968, Estimates in successive sampling using a multi-stage design, *J. Am. Stat. Assoc.* **63:**99–112.

Tanner, J. M., 1955, *Growth at Adolescence,* Blackwell, Oxford.

Tanner, J. M., Whitehouse, R. H., and Takaishi, M., 1966, Standards from birth to maturity for height, weight, height-velocity and weight velocity: British children, 1965, *Arch. Dis. Child.* **41:**545–471; 613–635.

Wall, W. D., and Williams, H. L., 1970, *Longitudinal Studies and the Social Sciences,* Heinemann, London.

Williams, W. H., 1970, The systematic bias effects of incomplete responses in rotation samples, *Public Opinion Q.* **32:** 593–602.

Williams, W. H., and Mallows, C. L., 1970, Systematic biases in panel surveys due to differential non-response, *J. Am. Stat. Assoc.* **65:**1338–1349.

# 7

# *Mathematical Handling of Long-Term Longitudinal Data*

## *ETTORE MARUBINI*

### *1. Preliminary Considerations*

In studying growth and development in man, information pertinent to several variables is gathered; the most commonly used are: height, weight, sitting height, leg length, biacromial and biiliac diameters, and head and chest circumferences.

Among the sampling designs generally followed in this branch of research—cross-sectional, mixed-longitudinal, and longitudinal (see Chapter 6)—the last is essential for studying the growth of individuals, not only to know values attained at specific ages, but also to appreciate the pattern of change over time. Longitudinal studies also are necessary for making accurate assessment of the means and standard deviations of growth increments for the construction of ''velocity'' standards which have important applications in clinical work (Anderson *et al.*, 1965; Tanner *et al.*, 1966).

Growth data are affected by daily rhythm and also by seasonal variations (Marshall, 1971; Whitehouse *et al.*, 1974). Thus the values collected at systematic, relatively long-spaced intervals from a regular oscillating course of growth can vary according to the phase of the cycle which happens to be hit. This point, in fact, plays an essential role in planning the time interval between two successive examinations. In the Harpenden Growth Study, for example, the children were measured every six months up to the beginning of adolescence, every three months throughout the adolescent growth spurt, and afterwards at longer intervals until they reached adulthood (Marubini *et al.*, 1972). According to Deming and Washburn (1963), it is advisable to have monthly measurements for the first six months after birth, with at least three measurements from 6 to 12 months. After the first year it is helpful to get them every three months, but satisfactory curves can be fitted to data collected at six-month intervals.

***ETTORE MARUBINI*** • Department of Biometry and Medical Statistics, University of Milan, Italy.

Most studies initially planned as longitudinal in effect become mixed studies when some children do not attend through illness, change of address, and so forth. There may be a further unplanned departure from the longitudinal design when children do not come for measurements at exactly the age specified, for instance, the birthday, but days or weeks early or late. This fact can bias the estimates of the means and standard deviations of the measured variables at the specific target age ($T$), and also the mean rates of change. It is a rather common practice to fit the value attained at $T$ by means of linear interpolation made from the absolute measurements before deriving the increments (Eichorn and Bayley, 1962; Anderson *et al.*, 1965). More recently Goldstein and Carter (1970) have suggested and evaluated a method of individual adjustment which can be used when measurements at three time points in an interval around $T$ are available. The essence of this method is to fit, by means of the Newtonian divided differences, a second-degree polynomial to pass through these points and then to read the value at the target age. This method does not assume a known form of growth curve for each subject (see Chapter 6).

## 2. Individual Longitudinal Records

As an example of longitudinal growth, in the left part of Figure 1 the distance curve for supine length from birth to two years and for height from two years upwards is shown. This curve informs the student about the value attained at a particular age. In the right part the corresponding velocity curve is drawn.

From the curves it is clear that length increases very rapidly during the first year after birth, while its velocity correspondingly decreases rapidly then and more slowly or even not at all during late childhood. After this point the velocity increases sharply during the adolescence growth spurt and then again decreases quickly as adulthood is approached.

This pattern of growth is not peculiar to height, but it is followed by several other anthropometric variables commonly measured (Simmons, 1944). The efficacy of these simple plots in the description of individual growth has been demonstrated by many students (see Israelsohn, 1960; Tanner, 1962; Hindley, 1972). However,

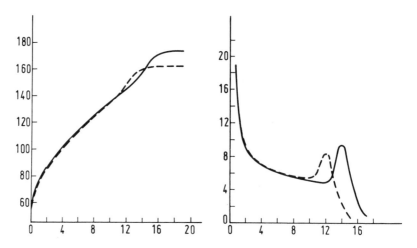

Fig. 1. (Left) Growth in supine length and height (cm) of boys (continuous line) and girls (intermittent). (Right) Rate of growth (cm/year) in supine length and height. Abscissa: age in years. (Reprinted from Tanner *et al.*, 1966.)

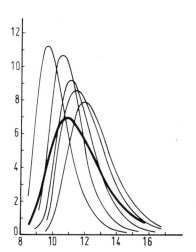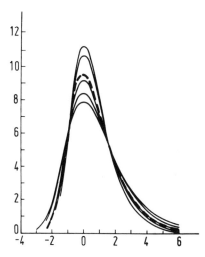

Fig. 2. (Left) Relation between five individual velocity curves and the velocity curve fitted to average values (thick line). (Right) The same five velocity curves plotted according to their time of maximum velocity and the mean-constant curve (intermittent line) (see text). Abscissa: age in years.

some words of caution were written by Van der Linden (1970) and Van der Linden *et al.* (1970), who suggested alternative plotting devices which may be useful in certain cases.

The sudden change in velocity at the onset of puberty occurs at different ages in different individuals, and therefore a growth curve drawn by averaging all the subjects' measurements collected at a given point of the time scale shows a smooth transition from one growth rate to the other and gives a misleading pattern of the individual growth process. Detailed discussions of unwarranted inferences which can be drawn from the mean curves are given by Sidman (1952) and by Estes (1956). As shown by Merrell (1931), and pointed out by Israelsohn (1960), the so-called *mean-constant* curve is the one normally required to characterize a population or its subset. This curve is obtained by fitting a given function to each subject's measurements, estimating the constants for each subject and then averaging the calculated constants. An example of the misleading information introduced by the cross-sectional treatment of longitudinal measurements can be seen in Figure 2 (Marubini, 1971) where, on the left, the fitted velocity curves (light lines) of five different girls are plotted against chronological age. The heavy line shows the curve, fitted by the same function, to the average values calculated at each age. It is evident that this line characterizes the average velocity curve very badly; in fact, the peaks have been smoothed and the adolescent spurt has been spread out along the time axis. On the right of the same figure the five curves have been plotted according to their time of maximum velocity and the heavy interrupted line shows the mean-constant curve; it appears to be a good representation of individual growth.

In this chapter, from this point up to the end of the section Computation Problems, the fitting of individual distance data arising from longitudinal studies will be dealt with, the velocity curves being obtained by taking first derivatives of the fitted distance curves. In addition, attention will be restricted to growth in a single variable without taking into account relative growth or allometry. In the terminology of Kowalski and Guire (1974), the univariate time series and the univariate one-sample data matrix will be considered.

ETTORE MARUBINI

The idea of fitting a curve to recorded data has always appealed to students of physical growth, the purposes being:

1. *Interpretative:* From biological considerations mathematical equations are derived and tested in order to know whether they are confirmed by the data.

2. *Descriptive:* The observed growth is "graduated" (i.e., followed) in a mathematically convenient way which should allow the student to estimate some important events during growth as well as to make objective comparisons between sets of data.

Models suitable for interpreting growth are rare and sometimes the constants of the curve cannot be identified with meaningful biological entities (Kohn, 1948). Weiss and Kavanau (1957) suggested an important model which is based on a concept of growth regulation by negative feedback and which is supported by much experimental evidence. The equation they derived has 14 constants which are meaningful in describing the growth of chickens, but there seems to have been little attempt to follow their approach for human growth data. "As is true for many other dynamic processes of research interest, no basic biological or physical principles as yet elucidated concerning human growth obviously give rise to any particular functional form" (Welch, 1970). The lack of specific principles clearly leading to functions peculiar to human growth has led most students to follow the humbler exercise of describing the growth process. Owing to the complexity of the human process as a whole, it has seemed fruitful to divide the curve into cycles and to fit each of them separately. According to Courtis (1937) a cycle of growth can be defined as a period of specific maturation during which the elements and forces acting upon the growth process are approximately constant.

After studying curves of many sets of data from different parts of the body (stature, shoulder and pelvic breadth, cephalic length and breadth, and weight), Count (1943) concluded that "the curve breaks up into three parts, *and no fewer,* as far as indicated by any formulation I have been able to attempt." For fitting the first part from birth to about first-molar-eruption time, a common kind of logarithmic function is suggested; from first-molar eruption until about second-molar-eruption time, there is a simple step-up of velocity implied in a linear function. In the third period a logistic function is fitted. Laird (1967), in studying the weight curve of man, also identified three growth phases which were fitted by means of the Gompertz and Gompertz–linear models connected by a straight line process with a distinctly faster rate than that of the first phase.

It is not clear how to determine the end points of the phases, and, moreover, it should be noted that these authors did not work with longitudinal data. Other workers (Deming, 1957; Israelsohn, 1960) preferred to divide growth, broadly speaking, into two periods—the preadolescent and the adolescent. This approach also carries the risk of wrong division.

Regardless of the number of cycles, the fitting of growth curves has followed two paths. In the first some students have used different mathematical functions with the following common properties: (1) simple form, (2) relatively reduced number of parameters, the meanings of which are clear and have a definite biological relevance, and (3) good agreement between the observed values and those predicted by the function. In the second, polynomials of different degree have been utilized.

A fairly general mathematical model may be specified as

$$Y = f(t) + e \qquad (1)$$

where the measured values $Y$ are thought of as being composed of a systematic part, $f(t)$, and a random component, the sum of seasonal and diurnal variations and of measurement errors, which follows a defined probability law.

### 3.1. Models Whose Parameters Have Biological Relevance

Table I reports some mathematical models which have been widely used for fitting human growth curves. A detailed discussion of the properties and the biological meaning of the constants of some of these functions can be found in the papers by Jenss and Bayley (1937), Deming (1957), Deming and Washburn (1963), and Marubini *et al.* (1971).

As an example of a constant whose biological meaning is easily understood, consider, in the Jenss and Bayley equation (1937), the quantity $\exp(-b)$; this is the ratio of the acceleration at any point of time to the acceleration one time unit previously. This is a pure number, nondimensional, that largely determines the shape of an individual growth curve. This parameter is therefore suitable for comparing the pattern of growth of different variables in the same child and the same variable in children belonging to different groups.

As far as the adolescent period is concerned, the suggested functions satisfy some basic requirements which can be realized by looking at the left part of Figure 1, i.e., a lower and an upper asymptote, a single point of inflection, and a different time origin for each child.

At present, in fitting adolescence curves, a difficult task facing the student is the estimation of the "take-off" age, i.e., the age of the beginning of the adolescent spurt, regardless of the approach used (Frisch and Revelle, 1971; Resele and Marubini, 1972).

The extent to which the logistic as well as the Gompertz curve are altered in their fits by placing the take-off point systematically earlier or later than a "real" estimate has been investigated by Marubini *et al.* (1972). A number of empirical points were excluded or included, as one would making an error in assessing the take-off point at the beginning of a curve-fitting exercise.

Table II shows the effect on peak velocity (PV) of height, sitting height, leg length, and biacromial diameter. Table III shows the analogous effect on age at PV. The tables show that errors of less than one year in choosing the "take-off" point do not have an important effect on the estimates of either the maximum velocity or the age at which this maximum arises. Furthermore, in their discussion, the authors affirmed that the logistic, on the whole, fitted the data better than the Gompertz. It was slightly more stable under errors of lower asymptote, and it produced fewer biased fits as judged by the runs test. Therefore, in spite of giving a slightly lower peak velocity than the Gompertz, these authors concluded that, on balance, the logistic approach is preferable. Subsequently by fitting the logistic function to the individual longitudinal records of the Harpenden Growth Study, Tanner *et al.* (1976) have studied the difference in pattern of growth of height, sitting height, leg length, and biacromial and biiliac diameter during adolescence in both boys and girls. In addition, the relationship between the estimates of the parameters and the development of pubertal characters was investigated.

Since, as was pointed out by Laird (1967) and Marubini *et al.* (1971), it is not possible, on the basis of fit alone, to favor one over any of the other functions that have been suggested, it appears that the choice of the kind of model to use may be regarded as, to some extent, a matter of taste. Day (1966) suggested an alternative

Table I. *Mathematical Functions Widely Used for Fitting Growth Data* [a]

| Function | Anthropometric variable | Growth period | Authors |
|---|---|---|---|
| $f(t) = (c + dt) - \exp(a - bt)$ | Length; weight | Birth–6 years | Jenss and Bayley (1937) |
| | Supine length; weight | Birth–8 years | Deming and Washburn (1963) |
| | Supine length; skull circumference; weight | Birth–48 weeks | Manwani and Agarwal (1973) |
| $f(t) = (a + bt) + c \log t$ | Skull dimensions | 6 months–5 yrs | Count (1942) |
| | Height | Birth–6 years | Count (1943) |
| | Supine length; cubit; foot length, biacromial bitrochanteric: weight; head length and breadth | Birth–5 years | Tanner *et al.* (1956) |
| $f(t) = at^b$ | Height | 2–10 years | Israelsohn (1960) |
| | Skeletal nose height | 5–12 years | Meredith (1958) |
| $f(t) = P + \dfrac{K}{1 + \exp[-b(t-c)]}$ (logistic function) | Height | 12 years–adulthood | Count (1943) |
| | Height | Adolescence | Marubini *et al.* (1971) |
| | Height; sitting–height; leg length; biacromial diameter | Adolescence | Marubini *et al.* (1972) |
| $f(t) = P + K \exp[e^{-b(t-c)}]$ (Gompertz function) | Length | Adolescence | Deming (1957) |
| | Weight | Adolescence | Laird (1967) |

[a] $t$ = age in decimals; $P$ = lower asymptote, i.e., value reached before the adolescent spurt; $a, b, c, d, K$ = constants to be estimated.

Table II. Mean Differences between PV Values (cm/yr) Obtained by Using Estimated Take-Off Age and Those Given by Altering Take-Off to Earlier or Later Ages[a]

| Starting points before and after estimated take-off age | Males | | | | Females | | | |
|---|---|---|---|---|---|---|---|---|
| | Height | Sitting height | Leg length | Biacromial diameter | Height | Sitting height | Leg length | Biacromial diameter |
| **Gompertz** | | | | | | | | |
| 12 months before | -0.48[c] (16)[d] | -0.33[c] (14) | -0.30[c] (16) | -0.08 (14) | -0.20 (12) | -0.05 (14) | -0.03 (12) | -0.04[b] (12) |
| 6 months before | -0.23[b] (19) | -0.12[c] (16) | -0.17[c] (19) | -0.04 (18) | -0.05 (12) | 0.02 (13) | 0.01 (12) | -0.02 (13) |
| 6 months after | 0.40[b] (21) | 0.04 (19) | 0.09[b] (21) | 0.02 (20) | -0.02 (14) | 0.08 (12) | -0.08 (14) | 0.03 (13) |
| 12 months after | 0.58[c] (21) | 0.18 (19) | 0.11 (20) | 0.11[b] (17) | -0.09 (14) | -0.21[b] (12) | -0.003 (12) | 0.03 (13) |
| **Logistic** | | | | | | | | |
| 12 months before | -0.30[c] (16) | -0.21[c] (14) | -0.21[c] (16) | -0.03 (14) | -0.09[b] (12) | -0.03 (14) | 0.01 (12) | -0.04 (12) |
| 6 months before | -0.16[c] (19) | -0.08[c] (16) | -0.11[c] (19) | -0.02[b] (18) | -0.02 (12) | 0.01 (13) | 0.01 (12) | -0.01 (13) |
| 6 months after | 0.15[c] (21) | 0.02 (19) | 0.15 (21) | 0.02 (19) | 0.08 (14) | 0.19 (12) | -0.01 (14) | 0.07 (14) |
| 12 months after | 0.74[c] (21) | 0.32[b] (19) | 0.08 (20) | 0.05 (14) | 0.27 (14) | -0.12 (9) | -0.09 (11) | 0.09 (13) |

[a] From Marubini et al., 1972.
[b] 0.01 < P < 0.05.
[c] P < 0.01.
[d] Numbers in parentheses indicate pertinent degrees of freedom.

ETTORE MARUBINI

Table III. Mean Differences between Age of PV Values (years) Obtained by Using Estimated Take-Off Age and Those Given by Altering Take-Off to Earlier or Later Age[a]

| Starting points before and after estimated take-off age | Males | | | | Females | | | |
|---|---|---|---|---|---|---|---|---|
| | Height | Sitting height | Leg length | Biacromial diameter | Height | Sitting height | Leg length | Biacromial diameter |
| **Gompertz** | | | | | | | | |
| 12 months before | $-0.36^c$ (16)[d] | $-0.12^b$ (14) | $-0.38^c$ (16) | $-0.11^c$ (14) | $-0.33^c$ (12) | $-0.19^b$ (14) | $-0.42^c$ (12) | $-0.36^c$ (12) |
| 6 months before | $-0.19^b$ (19) | $-0.04$ (16) | $-0.20^c$ (19) | $-0.02$ (18) | $-0.15^c$ (12) | $-0.15$ (13) | $-0.21^c$ (12) | $-0.22^c$ (13) |
| 6 months after | $0.00$ (21) | $0.03$ (19) | $0.13^c$ (21) | $-0.01$ (20) | $0.14^c$ (14) | $-0.06$ (12) | $0.24^c$ (14) | $-0.04$ (13) |
| 12 months after | $-0.01$ (21) | $-0.01$ (19) | $0.40^c$ (20) | $-0.18$ (17) | $0.26^c$ (14) | $0.26^c$ (9) | $0.61^c$ (12) | $0.10$ (13) |
| **Logistic** | | | | | | | | |
| 12 months before | $-0.16^c$ (16) | $0.01$ (14) | $-0.27^c$ (16) | $0.07$ (14) | $-0.19^c$ (12) | $-0.01$ (14) | $-0.31^c$ (12) | $-0.23^c$ (12) |
| 6 months before | $-0.06^c$ (19) | $0.06$ (16) | $-0.15^c$ (19) | $0.09$ (18) | $-0.07^c$ (12) | $-0.04$ (13) | $-0.16^c$ (12) | $-0.14^c$ (13) |
| 6 months after | $-0.04$ (21) | $-0.07$ (19) | $0.02$ (21) | $-0.16^c$ (19) | $0.01$ (14) | $-0.22^b$ (12) | $0.16$ (14) | $-0.20^c$ (14) |
| 12 months after | $-0.21^b$ (21) | $-0.30^b$ (19) | $0.27^c$ (20) | $-0.31^c$ (14) | $0.00$ (14) | $0.04$ (9) | $0.63^c$ (11) | $-0.14$ (13) |

[a]From Marubini et al., 1972.
[b]$0.01 < P < 0.05$.
[c]$P < 0.01$.
[d]Numbers in parentheses indicate pertinent degrees of freedom.

approach, sound from a methodological point of view, in order to choose one particular function among different ones suitable for fitting S-shaped curves. He showed that the choice in many cases is equivalent to obtaining an overall estimate (from all subjects) of an extra parameter.

Suppose that $f(t)$ be the generical function belonging to one of the asymptotic families mentioned in Table I. Day (1966) states that $f(t)$ is a solution of the second-order differential equation

$$\frac{d^2 f(t)}{dt^2} f(t) - (a + 1) \left( \frac{df(t)}{dt} \right)^2 + bf(t) \frac{df(t)}{dt} = 0 \qquad (2)$$

where $a$, the above cited extra parameter, is a characteristic of the growth process and hence is constant for the set of individuals studied, while $b$ is a scale factor specific to each subject.

How equation (2) might arise naturally in growth processes can be understood by remembering the left-hand part of Figure 1. This curve suggests that alongside the growth process $f(t)$ there is an aging process tending to inhibit growth.

When $b \neq 0$ it can be shown that:

1. If $a \neq 0$ from equation (2), the following equation can be derived:

$$\frac{df(t)}{dt} = \frac{b}{a} \ f(t) + C \ [f(t)]^{a+1} \qquad (3)$$

This is a Bernoulli's equation, whose solution can be written as follows:

$$f(t) = K \left\{ 1 + a \exp\left[ -b \ (t - c) \right] \right\}^{-1/a} \qquad (4)$$

This is the so-called "generalized logistic" function.

It should be noted that equation (3) is equivalent to the model developed by Bertalanffy (1957) for animal growth and subsequently extended by Richards (1959) to plant growth after discarding the limitations imposed by its theoretical background. From an empirical and physiological point of view, Bertalanffy (1957) argued that growth is the result of two processes, anabolism and catabolism, the effect of the former being directly proportional to weight. Certain properties of the growth model proposed by Bertalanffy have been elucidated by Fabens (1965).

From equation (4), for $a = -1$ and $a = 1$, respectively, the "monomolecular" (see Richards, 1959) and the logistic function can be derived.

2. If $a = 0$, equation (2) becomes

$$\frac{d^2 f(t)}{dt^2} f(t) - \left( \frac{df(t)}{dt} \right)^2 + bf(t) \frac{df(t)}{dt} = 0 \qquad (5)$$

The solution of equation (5) is the Gompertz function. Equation (5) corresponds to the growth model suggested by Laird et al. (1965) for the growth of the normal mammalian organism from the early embryonic period to young adulthood. Subsequently, as previously outlined, Laird (1967) resorted to model (5) for fitting human weight-growth data. This model is based on the empirical observation that the specific growth rate $(dW/W)$ tends to undergo an exponential decay with time.

When $b = 0$, the solutions of equation (2) are:
3. If $a \neq 0$,

$$f(t) = (K' + c't)^{-1/a} \tag{6}$$

which is a parabolic function;
  4. If $a = 0$,

$$f(t) = K'' \exp(c''t) \tag{7}$$

which is a simple exponential function.

In Laird's model the growth changes that can be explained ultimately in terms of the dynamics of cell proliferation are regarded as primary, with the accompanying metabolic changes relegated to a secondary role. In Bertalanffy's model, on the contrary, the metabolic changes that accompany the marked changes in size of the growing organism bear a mathematically defined relationship to the growth changes.

The formal relationship existing among Bertalanffy's, Laird's, and Day's models has been extensively discussed by Marubini and Cerina (1974). As far as this writer knows, Day's approach has never been used for growth studies in man.

Recently, Bock *et al.* (1973), to fit data for recumbent length of boys and girls selected from the Fels Study in the range of one year to maturity, used the following two-component logistic model:

$$f(t) = \frac{k_1}{1 + \exp[-b_1(t - c_1)]} + \frac{K - k_1}{1 + \exp[-b_2(t - c_2)]} \tag{8}$$

where $k_1$ is the upper limit of the preadolescent component; $b_1$ determines the initial slope of the preadolescent component; $c_1$ determines the location in time of the preadolescent component; $K$ is the mature recumbent length and the fitted curve is forced through it; $K - k_1$ is the contribution of the adolescent component to mature recumbent length; $b_2$ determines the slope of the adolescent component; and $c_2$ is the age at the maximum velocity of the adolescent component.

By using this model the risk of wrongly dividing the curve into different phases is avoided, and the complete pattern of growth is reduced to a simple functional form with five constants to be estimated, since $K$ is assumed known. Moreover, since the linear process discussed by Laird (1967) is accounted for by the simultaneous decrease of the first component and increase of the second, no further assumption is needed for introducing Laird's third component.

It is interesting to refer here to the most relevant points of the comparison made by Tanner *et al.* (1976) between the results they attained by fitting the logistic function during adolescence and those attained by Bock *et al.* (1973) in the same age interval by fitting the two-component logistic model. First of all this latter produced very low peak height velocities: 6.9 cm/yr (boys) and 6.4 cm/yr (girls), compared with 8.8 cm/yr and 8.1, respectively.

The Fels subjects' prepubertal curve asymptotes (supine length) averaged 149.7 cm in boys and 138.0 cm in girls. The Harpenden lower asymptotes (standing height) were 142.1 cm and 132.8 cm and the take-off points 146.1 cm and 137.9 cm. As expected, the agreement between prepubertal upper asymptote and graphic take-off values of Tanner *et al.* (1976) is quite close. The reader has to allow for supine length averaging about 1 cm greater than standing height, and additionally for Fels mature heights averaging some 6 cm for boys and 3.5 for girls more than those of Harpenden. The take-off values of these latter average 84% of adult height in both sexes and the Fels prepubertal asymptotes average 83% in boys and 82% in girls. The estimate of adolescent gain is thus higher in the two-logistic fit than in that

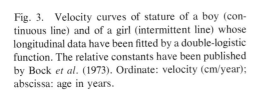

Fig. 3. Velocity curves of stature of a boy (continuous line) and of a girl (intermittent line) whose longitudinal data have been fitted by a double-logistic function. The relative constants have been published by Bock *et al.* (1973). Ordinate: velocity (cm/year); abscissa: age in years.

with one logistic (31.2 cm and 29.8 cm against 27.6 cm and 25.3 cm), but the two studies agree closely in finding only a small difference between male and female gains, the ratio being 1.05 in the Fels and 1.09 in the Harpenden data. Contrary to Bock *et al.* (1973), but in agreement with Deming (1957), the Harpenden data show the boys' spurt as lasting longer than the girls', as well as being greater at peak velocity.

The average residual variance given by Bock *et al.* (1973) is larger than that which can be found in the literature (Marubini *et al.*, 1971, 1972) as far as the fitting of the logistic to the adolescent growth curve is concerned. Moreover, for most of the individuals fitted, Bock found a definite time trend in the residuals from the fitted curves; thus the two-logistic model fares badly on the runs test.

In Figure 3, the derivative (i.e., velocity) curves of model (8) have been drawn by resorting to the constants given by Bock *et al.* (1973) for a boy (continuous line) and a girl (broken line). Also, if one takes into account the range of one year to maturity, as proposed by the authors, the pattern does not compare favorably with the one reported in the right-hand side of Figure 1. These considerations show that model (8) is questionable and suggest, as the authors pointed out, that some change in its composition might be introduced to yield a substantially better description of the growth pattern.

## 3.2. Polynomial Models

The lack of specific principles clearly leading to functions peculiar for describing human growth has led several students (for example, Vandenberg and Falkner, 1965; Welch, 1970; Goldstein, 1971) to follow the pragmatic approach of fitting polynomials in age:

$$f(t) = b_0 + b_1 t + b_2 t^2 + \cdots + b_n t^n \tag{9}$$

In this way the computational problems arising from the use of models which are nonlinear (see Draper and Smith, 1966) in the constants (all the functions reported in Table I with exception of the second) are precluded and, moreover, the average curve is equivalent to the mean constant curve. This may be a critical point in practice since, as has been previously shown, indiscriminate averaging travesties the pattern of growth by oversmoothing the peaks and masking the inherent interindividual variability, whose estimate is often one of the goals of the research. It should also be noted that several considerations are involved in choosing the appropriate degree of polynomial to fit data, including, for example, selection of values of the independent variable (Welch, 1970).

In order to investigate the reliability of this approach from the point of view of the goodness of fit, a third-degree polynomial has been fitted to height-growth data during adolescence. These are a subset of the data published by Tanner *et al.* (1976), and the results obtained are reported in Table IV.

Table IV. *Results of Fitting the Same Set of Longitudinal Data by Means of Third-Degree Polynomial and Logistic Model*

| Function fitted | | Third-degree polynomial | | | Logistic | |
|---|---|---|---|---|---|---|
| Sex | Number of subjects | Residual standard deviation (cm) | | % of significant runs tests ($\alpha = 0.025$) | Residual standard deviation (cm) | % of significant runs tests ($\alpha = 0.025$) |
| M | 40 | Min. | 0.30 | | Min. 0.19 | |
| | | Pooled | 1.08 | 37.5 | Pooled 0.39 | 0.00 |
| | | Max. | 1.65 | | Max. 0.81 | |
| F | 32 | Min. | 0.23 | | Min. 0.18 | |
| | | Pooled | 0.82 | 56.3 | Pooled 0.38 | 6.25 |
| | | Max. | 1.31 | | Max. 0.64 | |

As far as the logistic function is concerned, the overall residual standard deviations are very satisfactory considering they represent the effect of seasonal and diurnal variation as well as measurement error. They average 0.39 cm in boys and 0.38 cm in girls, compared with a SD of measurement error of 0.18 cm for the same measurer (Whitehouse *et al.*, 1974). When the third-degree polynomial is fitted, the SD increases about three times for boys and twice for girls. Even worse are the results of the runs test. It is also worth noting that the relationship between the size at PV (estimated by the logistic function) and final adult size is 0.91 for boys and 0.90 for girls and agrees very closely with the actual value of about 0.9 reported by Frisch and Revelle (1969). It follows that the use of polynomials to describe human growth should be avoided.

The previous discussion of fitting different mathematical models to each subject's growth data concerns stature (Deming, 1957; Anderson *et al.*, 1965; Marubini *et al.*, 1971; Bock *et al.*, 1973; Tanner *et al.*, 1976). Body weight is essentially a cubic power of linear dimensions, and its growth curve may reflect some differences which appear to be relevant from a biological point of view (Frisch and Revelle, 1971). It seems therefore questionable to fit the above-mentioned functions to longitudinal weight data (Bock *et al.*, 1973). An attempt to fit weight-growth curves in the preadolescent as well as in the adolescent period has been made by Tanner *et al.* (1966) by means of a graphical iterative approach.

Since there are now available longitudinal data, from birth to maturity, on weight, head circumference, biochemical and hematological variables, whose descriptive statistical summaries (McCammon, 1970) have already been published, further research is needed in order to find models most suitable for the description of relative growth processes.

## 4. Computation Problems

The use of polynomials guarantees unique least-square solutions for the estimates of the constants. If one follows the orthogonal polynomial approach (Forsythe, 1957), one gets very accurate coefficients which are considerably more robust, under deletions and additions of data, than those derived from standard polynomial curve-fitting programs.

In contrast, severe computational problems arise in estimating the parameters of the nonlinear models shown in Table I. Two different paths have been followed.

In the first, by taking advantages of certain characteristics peculiar to each function, some computationally convenient methods have been devised without evaluating their performances and the properties of the estimates yielded. Such are the methods used by the above-mentioned workers for fitting the Jenss and Bayley function, by Deming (1957) for fitting the Gompertz, and by Count (1943) for fitting the logistic. A detailed review of the computationally oriented methods for estimating the constants of the logistic function has been carried out by Barbensi (1965), who showed that, even when the data conform closely to the shape of the logistic, the various methods can yield markedly different estimates of the constants.

The other path allows the student to estimate the constants by means of the least-squares criterion, and this fitting is equivalent to maximum-likelihood estimation under the assumption of uncorrelated and normally distributed residuals. Since the functions are intrinsically nonlinear, the derived normal simultaneous equations cannot be solved directly, but only by resorting to iterative methods. Among these the Newton–Raphson method involves second derivatives of the fitted function, whereas the Gauss–Newton (Hartley, 1961) and scoring (Fisher, 1925) methods involve only first derivatives. These two latter have been recently shown to be identical when applied to nonlinear regression problems (Ratkowsky and Dolby, 1975).

Computer programs pertinent to most of the models previously mentioned are available in the literature. Causton (1969) wrote a program suitable for fitting the generalized logistic; Marubini and Resele (1971) for fitting the logistic as well as the Gompertz functions, and Wainer and Petersen (1972) for fitting the two-component logistic. All these models can be fitted by some general programs, if the user provides the function and its first and second derivatives (Morabito and Marubini, 1976). In programs with such a goal there are two crucial points: one concerns the convergence and the rate of convergence of the iterative process, and the other the computation of good starting values which allow the process to converge to a solution faster than would otherwise be possible.

An alternative method for minimization, the Nelder–Mead (Nelder and Mead, 1965) simplex procedure, must be cited. This is a robust direct-search procedure which, owing to the simplicity of application, lack of special requirements, and accuracy, has been found particularly suitable for the usual problems encountered in applied statistics, involving no more than (say) half a dozen constants (Olsson and Nelson, 1975). Unfortunately in its presently available form (Olsson, 1974), the Nelder–Mead procedure does not supply estimates of standard errors of the estimated constants. This fact, when contrasted with the accuracy of the estimates and the low cost of computing, appears to be a drawback of negligible relevance, particularly in research on human growth, the purpose of which is often to compare the results attained in different variables within the same subject or to compare the results attained in a given variable from different subjects.

Nelder (1961) gave a complete account of the least-squares procedures for fitting the generalized logistic function. In comparing the estimates and their variances and covariances with those obtained when the value of the $a$ parameter was assumed known, he found that the variances of $K$, $b$, and $c$ were considerably increased by the inclusion of $a$ and also that $a$ itself was poorly determined. Therefore it seems that the approach suggested by Day (1966) of computing the parameter $a$ on the entire set might be fruitful in order first to choose the most appropriate descriptive function.

## 5. Prediction of Adult Height

*ETTORE MARUBINI*

Data gathered in longitudinal research studies are of paramount concern in fulfilling the previously mentioned aims, as well as to perform and validate procedures suitable for predicting the adult height of boys and girls. The relevance of such a goal from a clinical point of view is easily understood, and this has been widely pointed out in literature since 1946, when Bayley published the first tables for predicting mature height from present height and bone age. Recently Roche *et al.* (1975) and Tanner *et al.* (1975) have published tables of regression weightings, estimated for age, which allow the prediction of adult height by means of several variables. These tables are the result of processing the serial data for the children in the Fels and Harpenden longitudinal studies, respectively. The method discussed in the first paper (referred to henceforth as the RWT method) utilizes the following variables: (1) present recumbent length (or adjusted stature); (2) present weight; (3) midparent adult stature; and (4) median Greulich–Pyle skeletal age of hand–wrist. These factors were selected from 78 initial variables by means of a rather sophisticated two-step selection procedure. The first step resorted to a principal-component analysis (see Morrison, 1967) and to a cluster analysis (Gruvaeus and Wainer, 1972) and the second one to the $C_p$ search procedure of Daniel and Wood (1971).

The variables retained in the regression equation by Tanner *et al.* (1975) (TWMC method) are: (1) present height; (2) bone age computed from the radius, ulna, metacarpals, and phalanges (RUS); and (3) chronological age. For girls aged 11.0–14.5 years, it is possible to allow for the age of menarche and, moreover, for both sexes, to allow for parents' heights by adding one-third of the difference between the midparent height and the average midparent height (168 cm in these data). The accuracy of prediction, evidenced by the correlation coefficients between the predicted and the actual adult statures, is slightly better with TWMC than with the RWT method. Considered as a coefficient of variation, the error of prediction reported by Tanner *et al.* (1975) is about ±2% of the mean for boys up to age 12, decreasing to ±1.7% at ages 13 and 14; and ±1.8% for girls up to age 11, decreasing to ±1.4% at ages 12 and 13 for premenarcheal and 1.1% for postmenarcheal girls.

The validation of the TWMC as well as the RWT methods has been carried out by means of the same sets of data that have been used to calculate the regression weightings; this tends to overestimate the accuracy of prediction.

Although the authors followed completely different approaches for selecting variables, the two methods have the three most important predictors in common. Moreover, the following common features are noteworthy:

1. For each age the partial regression coefficients are calculated by analyzing the data as if they had been collected in cross-sectional studies.

2. The set of regression weightings is inconsistent across time. Therefore the curve obtained by plotting the regression coefficients against time is smoothed in order to get weightings which are a continuous function of the child's age. To reach this aim RWT follows a very elegant approach such that the errors of smoothing, at a given age, are not correlated across predictors.

Although it might be desirable to improve the accuracy of the prediction, it is clear that the two above-mentioned longitudinal studies have shown the reliability of relevant predictors and have now made it easier to plan and carry out research appropriate for the collection of fruitful data in order to construct tables for predicting adult height in different countries.

# 6. Summary

Serial data on measurements of individuals, for example, of height, may be fitted by various curves. The generalized logistic and logistic Gompertz functions are described in single and two-component models. Computer programs suitable for fitting are referred to, and some results by different authors compared. The fits are generally good. Polynomial equations are shown not to fit the data so well. It is suggested that preliminary use of the generalized logistic to a whole set of data may be advisable to determine the particular form of curve to fit subsequently to each individual's data sets.

# 7. References

Anderson, M., Hwang, S., Green, W. T., 1965, Growth of the normal trunk in boys and girls during the second decade of life, *J. Bone Joint Surg.* **8**:1554.

Barbensi, G., 1965, L'adattamento della curva logistica, *Statistica* **25**:27; 225.

Bayley, N., 1946, Tables for predicting adult height from skeletal age and present height, *J. Pediatr.* **28**:49.

Bertalanffy, L. von, 1957, Quantitative laws in metabolism and growth, *Q. Rev. Biol.* **32**:217.

Bock, R. D., Wainer, H., Petersen, A., Thissen, D., Murray, J., and Roche A., 1973, A parametrization for individual human growth curves, *Hum. Biol.* **45**:63.

Causton, D. R., 1969, A computer program for fitting the Richards function, *Biometrics* **25**:401.

Count, E. W., 1942, A quantitative analysis of growth in certain human skull dimensions, *Hum. Biol* **14**:143.

Count, E. W., 1943, Growth patterns of the human physique: An approach to kinetic anthropometry. Part I, *Hum. Biol.* **15**:1.

Courtis, S. A., 1937, What is a growth cycle?, *Growth* **1**:155.

Daniel, C., and Wood, F., 1971, *Fitting Equations to Data, Computer Analysis of Multifactor Data for Scientists and Engineers,* Wiley, New York.

Draper, N. R., and Smith, H., 1966, *Applied Regression Analysis,* Wiley, New York.

Day, N. E., 1966, Fitting curves to longitudinal data, *Biometrics* **22**:276.

Deming, J., 1957, Application of the Gompertz curve to the observed pattern of growth in length of 48 individual boys and girls during the adolescent cycle of growth, *Hum. Biol.* **29**:83.

Deming, J., and Washburn, A. H., 1963, Application of the Jenss curve to the observed pattern of growth during the first eight years of life in forty boys and forty girls, *Hum. Biol.* **35**:484.

Eichorn, D. H., and Bayley, N., 1962, Growth in head circumference from birth through young adulthood, *Child Dev.* **33**:257.

Estes, W. K., 1956, The problem of inference from curves based on group data, *Psychol. Bull.* **53**:143.

Fabens, A. J., 1965, Properties and fitting of the von Bertalanffy growth curve, *Growth* **29**:265.

Fisher, R. A., 1925, Theory of statistical estimation, *Proc. Camb. Phil. Soc.* **22**:722.

Forsythe, G. E., 1957, Generation and use of orthogonal polynomials for data fitting by computer, *J. Soc. Ind. Appl. Math.* **5**:74.

Frisch, R. E., and Revelle, R., 1969, The height and weight of adolescent boys and girls at the time of peak velocity of growth in height and weight: Longitudinal data, *Hum. Biol.* **41**:536.

Frisch, R. E., and Revelle, R., 1971, The height and weight of girls and boys at the time of initiation of adolescent growth spurt in height and weight and the relationship to menarche, *Hum. Biol.* **43**:140.

Goldstein, H., 1971, The mathematical background to the analysis of growth curves. Proceedings of XIII International Congress of Pediatrics, Vienna, August 29–September 4, p. 39.

Goldstein, H., and Carter, B., 1970, Adjusted measurements in longitudinal studies. Compte Rendu de la X Réunion des Équipes chargées des Études sur la Croissance et le Développement de l'Enfant Normal, Centre International de l'Enfance, Davos.

Gruvaeus, G., and Wainer, H., 1972, Two additions to hierarchical cluster analysis, *Br. J. Math. Stat. Psychol.* **25**:200.

Hartley, H. O., 1961, The modified Gauss–Newton method for the fitting of nonlinear regression functions by least squares, *Technometrics* **3**:269.

Hindley, C. B., 1962, The place of longitudinal methods in the study of development, in: *Determinants of Behavioral Development* (F. J. Mönks, W. W. Hartup, and J. de Wit, eds.), pp. 23–50, Academic Press, New York.

Israelsohn, W. J., 1960, Description and modes of analysis of human growth, in: *Human Growth* (J. M. Tanner, ed.), pp. 21–42, Pergamon, Oxford.

Jenns, R. M., and Bayley, N., 1937, A mathematical method for studying the growth of a child, *Hum. Biol.* **9:**556.

Kohn, P., 1948, Increase in weight and growth of children in the first year of life, *Growth* **12:**150.

Kowalski, C. J., and Guire, K. E., 1974, Longitudinal data analysis, *Growth* **38:**131.

Laird, A. K., 1967, Evolution of human growth curve, *Growth* **31:**345.

Laird, A. K., Tyler, S. A., and Barton, A. D., 1965, Dynamics of normal growth, *Growth* **29:**233.

Manwani, A. H., and Agarwal, K. N., 1973, The growth pattern of Indian infants during the first year of life, *Hum. Biol.* **45:**341.

McCammon, R. W., 1970, *Human Growth and Development,* Charles C. Thomas, Springfield, Illinois.

Marshall, W. A., 1971, Evaluation of growth rate in height over periods of less than one year, *Arch. Dis. Chid.* **46:**414.

Marubini, E., 1971, I rilievi auxometrici; loro utilizzazione e possibilità di errore, *Prospet. Pediatr.* **1:**35.

Marubini, E., and Cerina, R., 1974, Considerazione sui modelli di uso comune nella interpolazione delle curve di crescita dell'uomo, *Statistica* **34:**557.

Marubini, E., and Resele, L. F., 1971, Computer program for fitting the logistic and the Gompertz function to growth data, *Comput. Programs Biomed.* **2:**16.

Marubini, E., Resele, L. F., and Barghini, G., 1971, A comparative fitting of the Gompertz and logistic function to longitudinal height data during adolescence in girls, *Hum. Biol.* **43:**237.

Marubini, E., Resele, L. F., Tanner, J. M., and Whitehouse, R. H., 1972, The fit of Gompertz and logistic curves to longitudinal data during adolescence on height, sitting height and biacromial diameter in boys and girls of the Harpenden Growth Study, *Hum. Biol.* **44:**511.

Meredith, H. V., 1958, A time series analysis of growth in nose height during childhood, *Child. Dev.* **29:**19.

Merrel, M., 1931, The relationship of individual growth to average growth, *Hum. Biol.* **3:**37.

Morabito, A., and Marubini, E., 1976, A computer program suitable for fitting linear models when the dependent variable is dichotomous, polichotomous or censored survival and non-linear models when the dependent variable is quantitative, *Comput. Programs Biomed.* **5:**283.

Morrison, D. F., 1967, *Multivariate Statistical Methods,* McGraw-Hill, New York.

Nelder, J. A., 1961, The fitting of a generalization of the logistic function, *Biometrics* **17:**89.

Nelder, J. A. M., and Mead, R., 1965, A simplex method for function minimization, *Comput. J.* **7:**308.

Olsson, D. M., 1974, A sequential simplex program for solving minimization problems, *J. Qual. Technol.* **6:**53.

Olsson, D. M., and Nelson, L. S., 1975, The Nelder–Mead simplex procedure for function minimization, *Technometrics* **17:**45.

Ratkowsky, D. A., and Dolby, G. R., 1975, Taylor series linearization and scoring for parameters in nonlinear regression, *Appl. Stat.* **24:**109.

Resele, L. F., and Marubini, E., 1972, Problemi connessi con l'uso dell'elaboratore elettronico in auxologia, *Appl. Biomed. del Calcolo Elettronico* **7:**187.

Richards, F. J., 1959, A flexible growth function for empirical use, *J. Exp. Bot.* **10:**290.

Roche, A. F., Wainer, H., and Thissen, D., 1975, *Predicting Adult Stature for Individuals,* S. Karger, Basel.

Sidman, M., 1952, A note on functional relations obtained from grouped data, *Psychol. Bull.* **49:**263.

Simmons, K., 1944, The Brush Foundation Study of child growth and development. II. Physical growth and development., *Monogr. Soc. Res. Child Dev.* **9**(37):1.

Tanner, J. M., 1962, *Growth at Adolescence,* Blackwell, Oxford.

Tanner, J. N., Healy, M. J. R., Lockhart, R. D., MacKenzie, J. D., and Whitehouse, R. H., 1956, Aberdeen growth study. I. The prediction of adult body measurements from measurements taken each year from birth to 5 years, *Arch. Dis. Child.* **31:**372.

Tanner, J. M., Whitehouse, R. H., and Takaishi, M., 1966, Standards from birth to maturity for height, weight, height velocity, and weight velocity: British Children, 1965, *Arch. Dis. Child.* **41:**454.

Tanner, J. M., Whitehouse, R. H., Marshall, W. A., and Carter, B. S., 1975, Prediction of adult height from height, bone age, and occurence of menarche, at ages 4 to 16 with allowance for midparent height, *Arch. Dis. Child.* **50:**14.

Tanner, J. M., Whitehouse, R. H., Marubini, E., and Resele, L. F., 1976, The adolescent growth spurt of the boys and girls of the Harpenden Growth Study, *Ann. Hum. Biol.* **3:**109.

Van der Linden, F. P. G. M., 1970, The interpretation of incremental data and velocity growth curves, *Growth* **84**:221.

Van der Linden: F. P. G. M., Hirschfeld, W. J., and Miller R. L., 1970, On the analysis and presentation of longitudinally collected growth data, *Growth* **34**:385.

Vanderberg, S. G., and Falkner, F., 1965, Hereditary factors in growth: Twin concordance in growth, *Hum. Biol.* **37**:357.

Wainer, H., and Peterson, A., 1972, GROFIT: a FORTRAN program for the estimation of parameters of a human growth curve. Research Memorandum no. 17, Department of Education Statistical Laboratory, University of Chicago, Illinois.

Weiss, P., and Kavanau, J. L., 1957, A model of growth and growth control in mathematical terms, *J. Gen. Physiol.* **41**:1.

Welch, Q. B., 1970, Fitting growth and research data, *Growth* **34**:293.

Whitehouse, R. H., Tanner, J. M., and Healy, M. J. R., 1974, Diurnal variation in stature and sitting-height in 12- to 14-year-old boys, *Ann. Hum. Biol.* **1**:103.

# III

## Genetics

# 8

# Introduction to Genetic Analysis

## JEAN FRÉZAL and
## CATHERINE BONAÏTI-PELLIÉ

### 1. Qualitative and Quantitative Variations

Hereditary characters may be classified as qualitative or quantitative. Qualitative characters occur in populations in forms which can be objectively distinguished from each other. For instance, people belong to blood group A, B, etc.; the distribution of such characters is discontinuous. On the other hand, quantitative characters, i.e., measurable ones, are distributed on a continuous scale. Where there is no clear-cut distinction between classes of individuals, the definition of classes remains more or less arbitrary and it is always possible to find measures or individuals situated between two classes.

Qualitative characters are determined by single gene effects, and their alternative forms, the phenotypes, may be linked to one or another genotype. Hereditary determination of quantitative characters is more elaborate. Part of the quantitative variation may be associated, like qualitative ones, with identifiable gene differences. For instance, Penrose, in his paper on the measurement of pleiotropic effects in phenylketonuria (1952), showed that although phenylalanine levels in the blood of phenylketonurics and normal individuals are very clearly separated, the distribution of head size or hair color in these two groups shows a considerable overlap. As pointed out by Cavalli-Sforza and Bodmer (1971), the difference in the relation between a gene's individual effect and the total amount of variation with respect to the character is the essential feature which distinguishes qualitative from quantitative variation.

This example also suggests that genes involved in quantitative variation behave in the same way as the genes responsible for mendelian inheritance, a point which may be further exemplified by the data quoted by Harris (1975) on the level of activity of red cell acid phosphatase. If this activity is determined in a series of randomly selected individuals in the general population, the overall distribution appears to be continuous and unimodal, as is the distribution of many other enzyme

*JEAN FRÉZAL and CATHERINE BONAÏTI-PELLIÉ* • Clinique et Unité de Recherches de Génétique Médicale, Hôpital des Enfants-Malades, Paris.

activities. However, in this particular case, the overall distribution is simply a summation of a series of separate distributions corresponding to each of six discrete phenotypes which are determined by three allelic genes.

The general conclusions to be drawn from these examples are that continuous variation may not be of an essentially different pattern of genetic determination than qualitative discontinuous ones and does not necessarily imply the interplay of many genes at many loci. Furthermore it illustrates the difficulty of genetic analysis when discrete phenotypes cannot be discriminated.

When methods for discriminating phenotypes are not available, the genetic analysis of quantitative variation is based on the assumption that it results from a combination of environmental variation and the effects of many genes. In contrast to the study of qualitative characters which depends on the analysis of segregations (differences), the genetic analysis of quantitative traits rests on the study of resemblance between relatives and on the testing of phenotypic correlation against genotypic correlation.

Growth is a process which can be characterized and analyzed by measurable parameters, and it appears intuitively reasonable to assume that each parameter is influenced by a number of loci, none of which has a major effect, as well as by environmental factors. The involvement of many loci and more generally of a balanced genetic information in the growth process is also suggested, *a contrario*, by disturbances observed in many chromosomal aberrations.

Therefore, from the geneticist's point of view, growth may be considered as a quantitative character and treated by the statistical methods in use for their study. However, it is possible to identify effects of specific loci upon growth mainly through the disabling consequences of mutations, although it is hardly possible to infer from these observations and the effects of mutations the role of normal alleles at these loci in the overall pattern of growth.

## 2. Monofactorial Inheritance

Unifactorial characters are due to single genes, and the pattern of inheritance of them is accounted for by the location of the gene's locus on the X chromosome or on an autosome and by the relationship between the alleles which occupy the corresponding loci of the two homologous chromosomes. A character which is manifest when the determinant allele is in a single copy, in other words in a heterozygous state, is said to be dominant. Contrariwise, a character which is only expressed when the determinant allele is in duplicate, i.e., in a homozygous state, is said to be recessive. Sometimes the three genotypes which can be constituted with two allelic genes may be distinguished from each other. In this circumstance, the trait is said to be intermediary.

It is noteworthy that these classical notions of dominance, recessivity, and intermediacy refer to characters, not to gene action. According to the concept of molecular biology, each gene, that is an informative segment of DNA coding for one polypeptide chain, is transcribed on its own account without interference with the allele carried by the homologous chromosome.

### 2.1. Autosomal Dominant Inheritance

#### 2.1.1. Rules

Autosomal dominant characters are easy to ascertain from the pattern of a pedigree. As a rule, carriers are born from one parent similarly affected, the other parent being normal. Among the sibs of the carrier, one of two (on the average) is also a carrier. Children of the carriers are in equal proportion either carriers or normals. Finally children of normal sibs of carriers are unaffected. All these features are simply understood if one considers that carriers for autosomal dominant characters are heterozygotes for one gene, say $A$ ($Aa$) and were born from the marriage between a heterozygote $Aa$ (carrier) and a homozygote $(aa)$ normal (Figure 1):

$$Aa \times aa \rightarrow \tfrac{1}{2}Aa + \tfrac{1}{2}aa$$

### 2.1.2. Exceptions

There are some exceptions to these rules. Some of them are the results of statistical bias in the ascertainment of the cases. Others result from biological phenomena among which are mutation, penetrance, expressivity, epistasis, or lethality.

Mutations may be evoked when a child carrying an otherwise regular dominant is born from normal parents. For instance, in eight of ten cases, achondroplasia appears *de novo* in a family as a result of a mutation. In such cases, paternal age is, on the average, relatively greater than usual.

Penetrance and expressivity are classical but ambiguous concepts because they refer, as does dominance, to characters, not to gene action. Statistically, penetrance is the ratio of the heterozygous carriers of the trait to the total heterozygotes (carriers and unaffected). It is estimated from the proportion of carriers to the expected ratio of affected children.

Expressivity means that an allele may manifest itself in different ways from patient to patient. However, differences of expressivity and penetrance are most probably due to events, genetically or environmentally determined, which are quite independent of the gene determining the trait.

Epistasis defines interaction between nonallelic genes. On theoretical grounds, one can distinguish interaction at the genetic level via regulator genes or at the metabolic level.

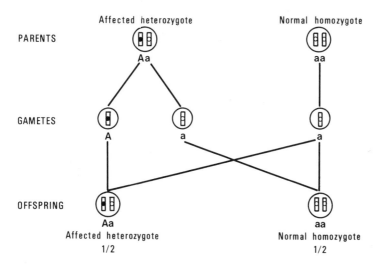

Fig. 1. Autosomal dominant inheritance: offspring of affected individuals.

Lethality, when dominant, would obviously be sporadic in families. Partial lethality as well as low penetrance leads to departure from expected ratios.

### 2.1.3. Examples

Many examples of dominant characters have been described. In McKusick's catalog (1975), 415 items are recorded as definitely dominant, 528 as likely. According to one WHO expert group, the overall frequency of dominant characters amounts to 1–10 per 1000 liveborn.

Dominant characters are mostly morphologic anomalies affecting several tissues or organs, like achondroplasia, osteogenesis imperfecta, or the Marfan syndrome. There are very few examples of biochemical characters which behave as dominant. They concern unstable hemoglobins and the several hemoglobins M, which are responsible for the so-called dominant methemoglobinemia, or errors of metabolism of unknown origin, such as the several types of periodic paralysis or the acute intermittent and variegata types of porphyria in which amino levulinic synthetase activity is raised without this abnormality being considered as the primary defect. As pointed out by Harris (1975), dominant inheritance of a disease due to an enzyme deficiency is most likely to occur where the enzyme in question happens to be rate-limiting in the metabolic pathway in which it takes part.

Taking all things into consideration, it appears that dominant characters are poorly understood from the molecular biology point of view and in all likelihood they represent remote effects of gene action.

This is one of the reasons why so-called dominant mutations are more likely to be collections of still indistinguishable disorders, each being the consequence of a quite different mutation, rather than distinct entities. For this reason it is still impossible to uncover these entities, as is so perfectly done with hemoglobinopathies.

## 2.2. Autosomal Recessive Inheritance

### 2.2.1. Rules

Carriers of recessive characters are usually born from apparently normal parents who are more often consanguineous than randomly selected parents. The proportion of affected sibs among the carriers is one out of four. As a rule, children of carriers are unaffected.

These features are well explained if one considers that carriers are homozygous for a gene, say, $a$ ($aa$), and were born from heterozygous parents (Figure 2):

$$Aa \times Aa \to \tfrac{1}{4}\, aa + \tfrac{3}{4}\, [A]$$
$$[A]\ \text{dominant phenotype}\ A = \tfrac{2}{3}\, Aa + \tfrac{1}{3}\, AA$$

The consanguinity increases the probability of occurrence of recessive characters; the consanguineous parents have a greater expectation of having inherited a gene identical by mendelian descent from their common ancestor(s) than are two subjects taken at random in the general population to carry isoactive alleles.

If the gene is not too uncommon or in small isolated populations, marriages may take place between a homozygous carrier and a heterozygote, leading to pedigrees mimicking dominant inheritance.

Finally people carrying the same trait, for instance, deafness, may marry. If they are homozygous for the same gene, either identical or isoactive, all their

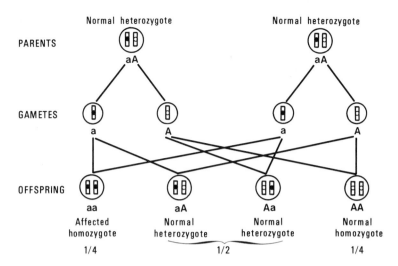

Fig. 2. Autosomal recessive inheritance: affected offspring born from normal heterozygous parents.

offspring are affected. Actually deduction about the relationship of the parents'
genes is only possible after the birth of a child. This is exemplified by hereditary
deafness where many affected pairs of parents give birth to hearing offspring
because each of them is homozygous for a different gene. This observation under-
lines again the heterogeneity of hereditary characters in man.

### 2.2.2. Ascertainment of Families

In recessive conditions, the families are ascertained through the affected
children, heterozygosity of the parents being deduced from this finding. Heterozygous parents (and families) having produced by chance only unaffected offspring are
then omitted. Therefore the sample is said to be incomplete. If the segregation ratio
is estimated from the sibships so ascertained, it generally overestimates the true
proportion. Several methods are available for eliminating this bias, the choice of
which depends on the mode of ascertainment. With incomplete selection, there are
three possibilities:

*Truncate Selection.* Every affected individual is independently ascertained,
and the probability of ascertainment of a family is independent of the number of
affected children above one. Therefore, the probability of selection of cases is unity
(1), as if the population was exhaustively examined, and the distribution of affected
children among selected families is a truncated binomial one (with the omission of
the class with zero affected).

*Single Selection.* Single selection is the opposite situation, where the probability of ascertainment approaches zero. In that case, a family has no chance to be
ascertained more than once, and the probability of ascertainment is proportional to
the number of affected in a family.

*Multiple Selection.* In actual situations even if the probability of ascertainment
is low, families may be independently ascertained more than once.

### 2.2.3. Methods

One can distinguish two kinds of methods:

*A Priori Method.* The first is the *a priori* method, which can only be used in
the case of truncate selection and consists of comparing the calculated and observed

number of affected under the *a priori* assumption that the segregation ratio is actually 0.25.

*A Posteriori Methods.* The other methods are said to be *a posteriori* methods and consist in finding out the true segregation ratio without any assumption, thereafter comparing it to the ratio expected on the hypothesis of recessive inheritance, i.e., 0.25.

The simplest *a posteriori* method is called the sib method, which in the case of single selection is as efficient as the maximum likelihood one. It simply consists of estimating the segregation ratio among the sibs of the index case or proband (that is to say, the patient through whom the family has been ascertained):

$$p = \frac{R - N}{T - N}$$

where $R$ is the total number of affected, $T$ is the total number of children, and $N$ is the number of index cases (of families).

The method may be simply adapted to multiple selection, if the number of independent ascertainments per family is correctly recorded. In such cases, the families are counted as many times as there are probands (proband method). The proband method is the method of choice for a first analysis of the data. However, if the probability of ascertainment does not approach zero, it underestimates the true segregation ratio.

Fisher's maximum likelihood method was first used by Haldane in the case of truncate selection. It can be extended to other situations, if the probability of ascertainment is known.

These methods have been criticized, especially by Morton (1958), on the ground that, although they provide estimates of $p$ based on certain assumptions, they do not permit testing the validity of these assumptions. Furthermore they provide no basis for separation of sporadic and familial cases.

Sporadic cases are different from isolated ones in the sense that they are not the result of segregation for the gene known to cause the condition. They can have a variety of causes such as mutations or phenocopies resulting from environmental causes or manifestation of polygenic complexes. The simplest way of separating familial from sporadic cases is to separate families with two and more affected (familial or multiplex cases) from families with one affected (simplex) which are a mixture of sporadic and isolated cases.

On this basis elaborate methods have been devised by Morton (1959) to derive the segregation ratio using maximum likelihood estimation, internal consistency, number of discrete entities among familial cases, proportion of sporadic cases, and so on. Practical description of the methods may be found in Cavalli-Sforza and Bodmer (1971, 1965), or Li (1955, 1961).

### 2.2.4. Examples

From McKusick's catalogs it appears that the number of identified recessive characters is lower than the number of dominants; this difference is perhaps not genuine but due to the greater difficulty in ascertaining recessive characters. According to WHO, their incidence would be 1–2 per 1000 live born.

Many characters involving growth behave as recessives, for instance, osteochondrodysplasias such as achondrogenesis, diastrophic and metatrophic dwarfisms, progeria, Cockayne syndrome, bird-headed dwarfism, etc.; metabolic errors such as mucopolysaccharidosis and mucolipidosis; disorders of calcium/phosphorus metabolism such as pseudodeficient rickets, hypophosphatasia; and endocrine disorders such as adrenogenital syndrome (Maroteaux, 1974).

In many of these errors of metabolism, the enzyme deficiency has been discovered. Generally it consists in a total or near-total loss of activity which is generally attributed to a defect of synthesis (a quantitative defect). However, more and more examples are known of qualitative defects which perhaps represent the general case. By reference to structural defects of the hemoglobin molecule, one can speculate that substitution of one amino acid, as a result of a base-pair change on the DNA, could lead to a complete loss of function if it affects a critical point, such as catalytic, allosteric, or prosthetic sites; if it impairs the capacity to form polymers; or if it decreases the stability of the molecule.

### 2.2.5. Heterozygote Detection

Inborn errors are recessive characters, and the heterozygotes are healthy. However, they usually have a partial deficiency which is demonstrable in numerous instances. As a general rule (although there are a few exceptions), the average level of activity in the heterozygotes is intermediate between the very low levels seen in the homozygotes and the levels found in randomly selected controls. It should also be noted that there is always considerable variation about the mean with overlap of the distributions. This variation is due both to nongenetic and genetic factors. It may be due to the greater or lesser efficiency of the normal allele, as exemplified previously by red cell acid phosphatase. It may also be due to variation in genes at other loci.

## 2.3. Sex-Linked Recessive Inheritance

### 2.3.1. Rules

Sex-linked recessive characters which are controlled by rare genes are observed in males born from normal parents, the mother being a healthy carrier. Among the sibs, one boy out of two is affected; sisters are normal, but one out of two is herself a carrier. The children of an affected patient are healthy, but all the daughters are carriers. Affected relatives are restricted to the maternal side.

The nontransmission of the trait from father to son is an essential feature of sex-linked inheritance. In the case of lethal characters, where the patients do not give birth to offspring, it is impossible to distinguish by pedigree analysis sex-linked inheritance from autosomal dominant sex-limited transmission. This distinction could be made indirectly either by linkage studies, if the locus of the gene does not segregate independently from a sex-linked marker, or by comparative pathology owing to the homology of X chromosomes in mammals (Ohno, 1974). For instance the testicular feminization locus may be assigned to the X chromosome because it is sex-linked in the mouse.

The carriers of a sex-linked character are said to be hemizygous; they are born from the marriage between a heterozygous woman and a normal man (Figure 3):

$$xX \times xY \rightarrow \tfrac{1}{4}\, xX + \tfrac{1}{4}\, XX + \tfrac{1}{4}\, xY + \tfrac{1}{4}\, XY$$

If an affected male marries a normal female, he gives birth to two kinds of offspring:

$$xY \times XX \rightarrow \tfrac{1}{2}\, XY + \tfrac{1}{2}\, xX$$

JEAN FRÉZAL and
CATHERINE
BONAÏTI-PELLIÉ

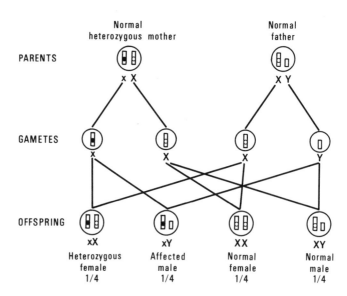

Fig. 3. Sex-linked recessive inheritance: offspring of a heterozygous woman.

To give birth to affected girls, an affected male has to marry a heterozygous female, a very unlikely possibility if the gene is uncommon. Therefore occasional manifestation of rare sex-linked traits in females may be accounted for by alternative explanations: genetic heterogeneity with an autosomal gene leading to similar disturbances, as exemplified by limb girdle myopathy or Hurler (autosomal) vs. Hunter (sex-linked) mucopolysaccharidosis or Willebrand disease vs. hemophilia; chromosomal aberration, "female" of XO constitution; and unusual manifestation of the gene as the result of inactivation of the X chromosome bearing the normal allele in critical areas (Lyon, 1961).

### 2.3.2. Examples

Sex-linked recessive inheritance is easily identifiable. Therefore many such characters have been described—93 definite and 78 likely ones are quoted in McKusick's catalog, and the incidence of these traits would amount to one in 2000 live born.

Besides hemophilias A and B, one can mention the Duchenne de Boulogne and Becker types of muscular dystrophy, and spondyloepiphyseal dysplasia tarda. Incontinentia pigmenti is probably a lethal sex-linked recessive character. Several mutations affecting enzymes controlled by sex-linked loci have been described, for instance, Fabry's disease ($\alpha$-galactosidase), Hunter's disease (iduronosulfatase), Lesch-Nyhan disease (HGPRT), and G6PD deficiency. In these instances it is possible to detect the heterozygous females by enzymatic determination.

### 2.4. Sex-Linked Dominant Inheritance

Sex-linked dominant inheritance is indistinguishable from autosomal dominant inheritance among the offspring of female patients. The difference appears from the study of affected males. There must be no example of father-to-son transmission. All the girls are carriers, but they may be unaffected if the penetrance is not complete. Finally if the trait is uncommon, the incidence is twice as great in females as in males. Hypophosphatemic-resistant rickets is a good example of a dominant sex-linked character; a similar mutation has been described in the mouse.

## 3. Quantitative Genetics

The criteria characterizing growth, whether they are birth weight, adult size, or growth rate, are quantitative characters and thus they are measurable. In this case, distinct phenotypic classes cannot be isolated, hence a study of this type needs a technique of analysis which is essentially statistical.

We will see step-by-step how the concept of quantitative genetics was created (the whole theory requiring a mathematical approach) and also the methods of analysis in man with some examples corresponding to the different stages of growth.

### 3.1. Concept of Quantitative Genetics

Mendel first introduced the concept of quantitative genetics with his famous experiment on red-purple and white flowering beans: the $F_1$ generation was of an intermediate color, which suggested codominance, but the $F_2$ generation, derived from crossing two plants of the $F_1$ generation, would give not only red-purple and white flowers, similar to the initial ones, but at the same time a whole series of intermediate colors, instead of the three usual phenotypes in the case of two codominant alleles. Mendel thought that these results came from the segregation of many independent elements.

Mendel's idea was taken up by Fisher (1918) in his outstanding study, in which he built a mathematical model for inheritance of quantitative characters. He showed that the only difference that distinguishes it from qualitative characters was the number and not the nature of the genes underlying the characters studied.

Since Fisher, the concept of quantitative genetics has not changed; only the techniques of analysis have become more refined.

### 3.1.1. Distribution of Quantitative Characters

If we suppose that numerous factors determine a character and that the effect of each of them is weak, then the variable measuring this character has to follow a law of normal distribution, provided that an appropriate scale of measurement is chosen: this is a fundamental property of this law. The characters allowing the measurement of growth stages are distributed in this way (Figure 4).

These factors can be genetic, each represented by several loci and having several alleles. In this case, the character is called polygenic. On the other hand, environmental factors can contribute to the realization of the character, and it is then called multifactorial.

### 3.1.2. Theoretical Analysis of a Quantitative Character

If $P$ is the phenotype value, or the measure of a quantitative character (for instance, size in centimeters), this value is the sum of the genotypic value $G$, due only to the genes in an individual, and the deviation due to the environment, $E$, that

JEAN FRÉZAL and
CATHERINE
BONAÏTI-PELLIÉ

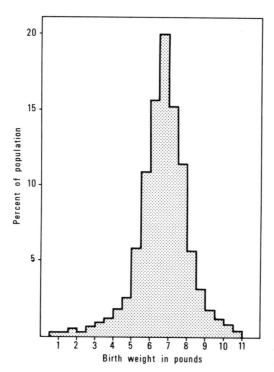

Fig. 4. The distribution of birth weight among 13,730 children. (Based on data from Karn and Penrose, 1951.)

corresponds to the different influences of the environment on the phenotypic value (for instance, feeding conditions),

$$P = G + E \qquad\qquad (1)$$

provided that the interactive effect between the genotype and the environment is small compared with the actual effects of each of these two factors. $E$ is defined such that its average equals 0; the average genotypic value is equal to the average phenotypic value.

*The Case of One Locus.* We will divide $G$ value in taking, first of all, the easiest case of a locus occupied by two alleles $A_1$ and $A_2$ with their respective frequencies $p$ and $q$ in a given population. We will choose the genotypic values $i, j,$ and $k$ for the $A_1A_1$, $A_1A_2$, and $A_2A_2$ genotypes, such that the average value of the character in the population is zero. This is expressed by

$$p^2i + 2pqj + q^2k = 0$$

(The population is supposed to be in panmictic equilibrium).

To understand things better, we can imagine that $i, j,$ and $k$ are the contributions in centimeters (for the locus in question) to the size of the individual.

In general, the effect of a genotype is not the sum of the effects of all the alleles of which it is made up; there is a certain interaction between them. We will divide the genotypic value into two effects, an additive effect of the genes, $s$ for $A_1$, $t$ for $A_2$, and a dominance effect $d_{11}$, $d_{12}$, and $d_{22}$ for $A_1A_1$, $A_1A_2$, and $A_2A_2$, respectively, such that

$$i = 2s + d_{11}$$
$$j = s + t + d_{12}$$
$$k = 2t + d_{22}$$

The value of these effects is defined in such a way that the average of the squares of the dominance effects in the population is a minimum. The resolution of this system therefore gives the following values:

$$s = pi + qj$$
$$t = pj + qk$$

If we put $s - t = \alpha$, we have

$$s = q\alpha$$
$$t = -p\alpha$$

$\alpha$ being the average effect of gene substitution.

Dominance effects are therefore:

$$d_{11} = q^2(i - 2j + k)$$
$$d_{12} = pq(2j - i - k)$$
$$d_{22} = p^2(i - 2j + k)$$

If we put $\delta = j - (i + k)/2$, which is the difference between the heterozygote value and the arithmetic average of the homozygote value, we have

$$d_{11} = -2q^2\delta$$
$$d_{12} = 2pq\delta$$
$$d_{22} = -2p^2\delta$$

It should be noted that the covariance between the additive effects and the dominance effects in the population is zero:

$$\begin{aligned}
\text{cov}(AD) &= (p^2 \times 2s \times d_{11}) + 2pq(s + t)d_{12} + (q^2 \times 2t \times d_{22}) \\
&= p^2(2\alpha q)(-2q^2\delta) + 2pq(q - p)\alpha(2pq\delta) + q^2(-2\alpha p)(-2p^2\delta) \\
&= 0
\end{aligned}$$

The term $\delta$ gives the degree of dominance between the $A_1$ and $A_2$ alleles: if $\delta = 0$, there is no dominance; if $\delta = (i - k)/2$, the dominance is complete; if $\delta$ is between these values, then the dominance is partial; and if $\delta$ is outside of this range, then there is an overdominance.

We therefore define the breeding value of a genotype as the sum of the additive effects of the genes of which it is made up. This is, in effect, the only part of the genotypic value which can be transmitted by an individual, since an individual cannot transmit his genotype to his offspring, but only one gene. Thus the dominance effect, coming from a special combination of two alleles, could not be found in his offspring.

Table I shows the partitioning of the genotypic value into its different components.

Table I. Values of Genotypes in a Two-Allele System[a]

|  | Genotype | | |
| --- | --- | --- | --- |
|  | $A_1A_1$ | $A_1A_2$ | $A_2A_2$ |
| Frequency | $p^2$ | $2pq$ | $q^2$ |
| Genotypic value | $i$ | $j$ | $k$ |
| Breeding values | $2s$ | $s + t$ | $2t$ |
|  | $2(pi + qj)$ | $pi + j + qk$ | $2(pj + qk)$ |
|  | $2q\alpha$ | $(q - p)\alpha$ | $-2p\alpha$ |
| Dominance deviation | $-2q^2\delta$ | $2pq\delta$ | $-2p^2\delta$ |

[a]From Jacquard, 1970.

JEAN FRÉZAL and
CATHERINE
BONAÏTI-PELLIÉ

*The Case of Several Loci.* We can generalize these results to cover several loci: if we call $A$ the sum of breeding values and $D$ the sum of the deviations due to the dominance of the different loci, the genotypic value $G$ of the character is equal to the sum of the values $A$ and $D$ and to another effect caused by the interaction between different loci (also called epistasis represented by $I$). Then we have the relation

$$G = A + D + I \qquad (2)$$

Thus the total phenotypic value can be divided into four elements through the combination of the relations (1) and (2):

$$P = A + D + I + E \qquad (3)$$

Up to now, we have supposed that the interaction between the genotype and the environment was negligible, but this is not always the case. Haldane (1946) studied this problem thoroughly. Suppose that we have two genotypes $A$ and $B$, giving, for example, different weights, located in two possible environments $X$ and $Y$ corresponding to good and bad feeding, respectively. If the weight is higher overall in $A$, whatever the feeding is, and if good feeding gives a higher weight than bad feeding for the two genotypes, then it is nevertheless possible that the superiority of $A$ to $B$ could be more important in the environment $X$. In other words, the genotype $A$ takes better advantage of good feeding than genotype $B$, and reciprocally the weight superiority in the environment $X$ is greater with genotype $A$ than genotype $B$. In this case, we have to add to relation (3) a term corresponding to the interaction between genotype and environment $I_{GE}$, so that now we have

$$P = A + D + I + E + I_{GE} \qquad (4)$$

*Partitioning of Variance.* In the case of a quantitative character, the study of its variation is obviously our first concern. In effect, if it is generally impossible in practice to determine the value of the different effects that we have defined, there are nevertheless methods to estimate the respective components of the variation attributable to these effects.

The phenotypic variance $V_P$ can be divided into

$$V_P = V_G + V_E + V_{GE} \qquad (5)$$

$V_G$ being the genotypic variance, $V_E$ the environmental variance, and $V_{GE}$ the variance caused by the interaction between genotype and environment; this last component is often supposed to be equal to zero, as we have seen.

The genotypic variance can itself be divided into

$$V_G = V_A + V_D + V_I + V_{AD} + V_{DI} + V_{IA} \qquad (6)$$

$V_A$ being the additive variance, $V_D$ the dominance variance, $V_I$ the epistasis variance; and $V_{AD}$, $V_{DI}$, and $V_{IA}$ the variances caused by the interactions between these different effects. The epistasis variance is generally assumed equal to zero to simplify the model; as seen before, the covariance between the additive effect and the dominance effect is equal to zero.

Thus we obtain the simplified relation

$$V_G = V_A + V_D \qquad (7)$$

*Heritability.* We have seen that the breeding value due to the additive effects of the genes was the only part of the genotypic value which could be transmitted. So the ratio of the additive genetic variance to the phenotypic variance, $V_A/V_P$, represents the fraction of the total variation which can effectively be transmitted and which contributes to the resemblance between parents and offspring.

This ratio is called heritability in the narrow sense as opposed to heritability in the broad sense (ratio $V_G/V_P$, i.e., the totality of the genotypic contribution to the phenotypic variation). It is the first concept which is the most used and is symbolized by $h^2$.

We will see later how we can estimate the heritability of a character. But it is worthwhile noting from now on that, contrary to Mendel's segregation of a monofactorial character which is the same in either circumstance, the heritability of a quantitative character is essentially a function of the population studied, since it depends on gene frequency and on the effect of the environment.

We have known for a long time that in animal species very different values are obtained according to the environment in which the animal lives. We can observe that the more homogeneous the environment is, the higher are the values obtained. On the other hand, only small differences are observed between animal populations.

### 3.2. Methods of Analysis in Human Genetics

Generally, methods tend to establish the part played by heredity and environment in determining a quantitative character and to analyze each of these two components as precisely as possible. They are essentially the resemblance between relatives and the twin method.

### 3.2.1. Resemblance between Relatives

*General Case.* The degree of resemblance between relatives is essentially due to the additive genotypic variance. We will see which parts of the different components of the genotypic variance enter into this resemblance, according to the family relationship between the individuals.

The measure of this resemblance is given by the regression coefficient $b$ or the interclass correlation coefficient between parents and offspring $r$ and by the intraclass correlation coefficient $r_i$ between sibs. Below are some examples of correlations calculated in man for metric characters (after Pearson and Lee, 1903):

|  | Parent–offspring | Sibs |
|---|---|---|
| Stature | 0.51 | 0.53 |
| Span | 0.45 | 0.54 |
| Length of forearm | 0.42 | 0.48 |

The covariance (numerator of regression and correlation) of parents and offspring is composed only of the additive variance.

Let $P_p$ be the phenotypic value of one parent, then

$$P_p = A_p + D_p + E_p$$

and $P_o$ the phenotypic value of the offspring, then

$$P_o = A_o + D_o + E_o$$

The covariance of $P_p$ and $P_o$ is

$$\text{cov}\,(P_o, P_p) = \text{cov}\,(A_o, A_p)$$

the other terms being equal to zero.

*JEAN FRÉZAL and*
*CATHERINE*
*BONAÏTI-PELLIÉ*

Since $A_o = \frac{1}{2}A_p$ (the child receives half the genes from one of his parents)

$$\text{cov}\,(P_o, P_p) = \frac{1}{2}\,\text{cov}\,(A_p, A_p) = \frac{1}{2}V_A$$

Let us consider now the covariance of midparent (arithmetical mean of the two parents) and offspring; if $P_p$ and $P_m$ are the phenotypic values of the parents, the value for the midparents $P_{pm}$ is

$$P_{pm} = \frac{1}{2}\,(P_p + P_m)$$

So

$$\text{cov}\,(P_o, P_{pm}) = \frac{1}{2}\,[\text{cov}\,(P_o, P_p) + \text{cov}\,(P_o, P_m)]$$
$$= \frac{1}{2}\,[\frac{1}{2}V_A + \frac{1}{2}V_A] = \frac{1}{2}V_A$$

Therefore, the covariance of offspring and midparent is the same as of offspring and one parent.

As for the covariance of sibs, a part is due (as above) to the additive genotypic variance, $\frac{1}{2}V_A$, but another part is due to the dominance variance, since sibs have both parents in common, and a pair of sibs have a quarter chance of having the same genotype for any locus:

$$\text{cov}\,(P_G, P_G) = \frac{1}{2}V_A + \frac{1}{4}V_D$$

From these results, it is easy to estimate the value of the heritability $h^2$ that is equal to the ratio $V_A/V_P$.

In the case of the parent–offspring regression, we have

$$b = \frac{\text{cov}\,(P_o, P_p)}{V_{P_p}} = \frac{1}{2}\,\frac{V_A}{V_P} = \frac{1}{2}h^2$$

The heritability is equal to twice the regression coefficient of offspring on parent.

The result is a little different in the case of the midparent–offspring regression, for the denominator is $\frac{1}{2}V_P$ instead of $V_P$, so that the heritability is exactly equal to the regression coefficient.

In a general way, the regression coefficient $b_r$ of offspring of $r$th degree relative is equal to

$$b_r = (\frac{1}{2})^r\,h^2$$

These results appear in Table II.

We have dealt with the easiest case, where the population was panmictic and where the resemblance between relatives was due only to the genes that they had in

Table II.  *Phenotypic Resemblance between Relatives and Estimation of Heritability*[a]

|  | Covariance | Regression ($b$) or correlation ($r_i$) |
|---|---|---|
| Offspring and one parent | $\frac{1}{2}V_A$ | $b_{p,o} = \frac{1}{2}\dfrac{V_A}{V_P} = \frac{1}{2}h^2$ |
| Offspring and midparent | $\frac{1}{2}V_A$ | $b_{pm,o} = \dfrac{V_A}{V_P} = h^2$ |
| Sibs | $\frac{1}{2}V_A + \frac{1}{4}V_D$ | $r_i = \dfrac{\frac{1}{2}V_A + \frac{1}{4}V_D}{V_P} = \frac{1}{2}h^2 + \frac{1}{4}\dfrac{V_D}{V_P}$ |
| $r$th degree relative | $(\frac{1}{2})^r V_A$ | $b_r = (\frac{1}{2})^r\dfrac{V_A}{V_P} = (\frac{1}{2})^r h^2$ |

[a]From Falconer, 1960.

Table III.   Partitioning of Variance of Birth Weight[a]

| Cause of variation | Percentage of total variance |
|---|---|
| Genetic | |
|    Additive | 15 |
|    Nonadditive | 1 |
|    Sex | 2 |
|       Total genotypic | 18 |
| Environmental | |
|    Maternal genotype | 20 |
|    Maternal environment | 24 |
|    Age of mother | 1 |
|    Parity | 7 |
|    Intangible | 30 |
|       Total environmental | 82 |

[a]From Penrose, 1954.

common. Now, we can see how the problem can be dealt with when these hypotheses are not verified.

*Common Environment.* Genetic causes are not the only reasons for resemblance between relatives: the members of the same family who live together share a common environment, in particular with respect to feeding which has, as we know, a great influence on growth. This means that some environmental factors can provoke differences between unrelated individuals, but not between the individuals belonging to the same family. In other words, there is a component of environmental variance that contributes to the variance between means of families but not to the variance within the families and which therefore contributes to the covariance of relatives. This component of the variance due to the common environment is symbolized by $V_{EC}$. The other part of the environmental variance is denoted by $V_{EW}$ and corresponds to the differences between individuals which are independent of their relationship.

There can be various effects resulting from this common environment. In the case of growth, an obvious factor is the common feeding of the members of the same family. For fetal growth, we easily understand that sibs have a resemblance only because they were in the same maternal environment during this period. This maternal effect can be found again in the analysis of birth weight, as many authors have shown, in particular Penrose (1954) (see Table III).

*Assortative Mating.* Up to now we have supposed that the population was panmictic, which means that the spouse's choice is not influenced by the character studied. In fact, we know very well that for some characters, such as size, there is a certain resemblance between married people, due to the homogamy for this character. The consequences of such a phenomenon have been analyzed by Fisher (1918), Reeve (1955), and Crow and Felsenstein (1968); they are extremely complex, so we will only give their conclusions. As the variance of the phenotypic value of the midparent is increased, so the covariance of sibs is increased. However, the regression coefficient of offspring on midparent is not very much modified and can be considered as a good measurement of $h^2$.

### 3.2.2. The Twin Method

There are two varieties of twins. Identical, or monozygotic, twins are produced from the division of an ovum and possess exactly the same hereditary makeup; on the other hand, nonidentical, or dizygotic, twins are produced from two different ova and are comparable to two sibs from a genotypic point of view.

Thus the differences between monozygotic twins only come from the environment, while the ones observed between dizygotic twins come from both genotype and environment.

We also have to take into account that twins develop in a common environment from conception to birth, and if we admit that this common environment is the same one for the two types of twins, we can divide the variance between twins into two components, variance between pairs and variance within pairs, as indicated in Table IV.

If we consider now the intraclass correlation coefficient between twins, equal to the ratio of the variance between pairs to the total variance, it is equal to

$$\frac{V_A + V_D + V_{EC}}{V_P}$$

for monozygotic twins, and to

$$\frac{\frac{1}{2}V_A + \frac{1}{4}V_D + V_{EC}}{V_P}$$

for dizygotic twins, the difference between the two being

$$\frac{\frac{1}{2}V_A + \frac{3}{4}V_D}{V_P}$$

So this difference represents an upper limit of half of the heritability value and is equal to it if the genotypic variance is only additive.

Table V gives some examples of the correlation coefficients between twins for weight, size, and birth weight.

The main limitation of this method is that there is a strong probability that the environment of monozygotic twins is more homogeneous and similar than the environment in which dizygotic twins develop.

### 3.2.3. Growth, A Dynamic Phenomenon

Growth is a continuous phenomenon from conception to maturity, and the circumstances acting on growth can change in the different stages of development, so that the relative importance of genotype and environment can vary during the development of an individual. It is essential to take this into account in the methods of analysis and to consider, in particular, the resemblance between relatives as a variable function of time, as Tanner (1962) did, and more recently Rao et al. (1975)

Table IV. Composition of Components of Variance between and within Pairs of Twins [a]

| | Between pairs | Within pairs |
|---|---|---|
| Identical twins | $V_A + V_D + V_{EC}$ | $V_{EW}$ |
| Nonidentical twins | $\frac{1}{2}V_A + \frac{1}{4}V_D + V_{EC}$ | $\frac{1}{2}V_A + \frac{3}{4}V_D + V_{EW}$ |
| Difference | $\frac{1}{2}V_A + \frac{3}{4}V_D$ | $\frac{1}{2}V_A + \frac{3}{4}V_D$ |

[a] From Falconer, 1960.

| Table V. | Resemblance between Twins | | |
|---|---|---|---|
| | Correlation coefficients | | |
| | Identical twins | Nonidentical twins | Difference |
| Weight[a] | 0.92 | 0.63 | 0.29 |
| Height[a] | 0.93 | 0.64 | 0.29 |
| Birth weight[b] | 0.67 | 0.58 | 0.09 |

[a]From Newman *et al.*, 1937.
[b]From Robson, 1955.

(Figures 5 and 6). The parent–offspring correlation increases with the age of the offspring, faster for father–child than for mother–child, and the correlation between sibs decreases with the age difference between them, both for size and weight. These variations point to a predominant maternal effect at birth, decreasing with growth, and then allowing the child genotype to express itself.

In domestic animals, similar observations are obtained: heritability increases with the animal's age. It is particularly small during the suckling period and becomes higher after weaning (Dickerson, 1954). In some cases the dam–offspring correlation is higher than the sire–offspring correlation; this is due to the relationship existing between the dam's milk production and its weight.

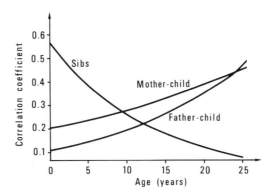

Fig. 5. Temporal trends in familial correlations for normalized height. (From Rao *et al.*, 1975.)

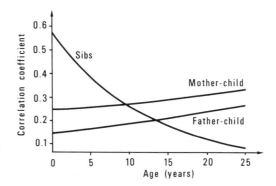

Fig. 6. Temporal trends in familial correlations for normalized weight. (From Rao *et al.*, 1975.)

JEAN FRÉZAL and
CATHERINE
BONAÏTI-PELLIÉ

## 4. Final Remarks

We have reviewed in this paper the two classical approaches of genetic analysis, the single-factor one in which there is a well-understood mechanism—or at least an identifiable genetic determinant—and the multifactorial one in which a number of unidentifiable determinants, either genetic or environmental, with weak interactions, are assumed.

Considering complex characters like height or patterns like growth, it is intuitively clear that they imply the interplay of many factors. Therefore they cannot be analyzed except by the multifactorial approach. It is no less clear that these procedures definitely preclude any identification or analysis of the numerous pathways and genes involved.

Elucidation of metabolic and genetic pathways rests on the study of discrete determinants: identification of their alternative forms, the alleles, mapping of their loci, determination of linkage relationships with other loci involved in the same pathway, and characterization of transcription and translation products. For a large part, recent advances in human genetics must be credited to these procedures of segregation analysis and molecular biology.

In the future, in order to bypass the gap between the two approaches, one can emphasize the study of quantitative characters when they are handled by one or a small number of enzymatic pathways. In this case, the variation is conceivably determined by the enzymes produced by a few loci, especially when one of the enzymes is rate limiting. Uric acid level and gout are probably good candidates for such analysis as well as, although less surely, cholesterolemia and diabetes. This aim is clearly out of reach for characters the determination of which involves many interacting pathways. That is to say, it is unwise to hope to get a comprehensive view on the genetic aspects of growth.

Single and multifactorial characters do not rest on different basic mechanisms. In both cases, the genetic determinants are segments of DNA coding a specific polypeptide, entering or composing a structural protein, an enzyme or an informative (regulatory) protein. The unique differences between them rest on the number of determinants and the capacity to identify them.

It is sometimes said that each of the many factors involved in height determination are minor effects and additive, and add, say, half a centimeter to height. This formulation is clearly a simplification required for the algebraic treatment. From the consideration of the unity of the basic mechanisms, one can deduce that phenotypic variation is the consequence of functional differences between the products of alleles as exemplified by the recent studies of allelic diversity, i.e., enzymatic polymorphisms and rare variants (Harris, 1975).

In our opinion and in the present state of affairs, it would be most rewarding to focus research on the elucidation of discrete pathways, rather than computing further correlations and heritability estimates.

It must be again underlined that heritability is a confusing or misleading concept which is specific to a set of data. Heritability does not refer to the part of nature and nurture in the determination of a character in a single individual, but to the variation of the character in a given population. Therefore heritability could be high or low for the same character in different populations, depending entirely on the environment. The point is that high heritability does not preclude any action on the part of the milieu, as Pearson (see Edwards, 1970) thought. Although heritability of height is high, it is perfectly clear that it is particularly prone to nutritional and environmental influences. It is worthwhile to keep this point in mind when studying growth.

## 5. References

Cavalli-Sforza, L. L., and Bodmer, W. F., 1971, *The Genetics of Human Populations,* W. H. Freeman, San Francisco.

Crow, J. F., 1965, Problems of ascertainment in the analysis of family data, in: *Genetics and the Epidemiology of Chronic Diseases* (J. V. Neel, M. W. Shaw, and W. J. Schull, eds.), pp. 23–44, Public Health Service, Washington, D.C.

Crow, J. F., and Felsenstein, J., 1968, The effect of assortative mating on the genetic composition of a population, *Eugen. Q.* **15:**85.

Dickerson, G. E., 1954, Hereditary mechanisms in animal growth, in: *Dynamics of Growth Processes* (E. J. Boell, ed.), pp. 242–276, Princeton University Press, Princeton.

Edwards, J. H., 1970, Analysis of pedigree data, in: *Advances in Human Genetics* Vol. 1 (H. Harris and K. Hirschhorn, eds.), pp. 1–34, Plenum Press, New York.

Falconer, D. S., 1960, *Introduction to Quantitative Genetics,* Oliver and Boyd, Edinburgh.

Fisher, R. A., 1918, The correlation between relatives on the supposition of Mendelian inheritance, *Trans. R. Soc. Edinburgh* **52:**399.

Haldane, J. B. S., 1932, A method for investigating recessive characters in man, *J. Genet.* **28:**251.

Haldane, J. B. S., 1946, The interaction of nature and nurture, *Ann. Eugen.* **13:**197.

Harris, H., 1975, *The Principles of Human Biochemical Genetics,* 2nd ed., North-Holland, Amsterdam.

Jacquard, A., 1970, *Structure Génétique des Populations,* Masson, Paris.

Karn, M. N., and Penrose, L. S., 1951, Birth weight and gestation time in relation to maternal age, parity and infant survival, *Ann. Eugen.* **16:**147.

Li, C. C., 1955, *Population Genetics,* University of Chicago Press, Chicago.

Li, C. C., 1961, *Human Genetics,* McGraw-Hill, New York.

Lyon, M. F., 1961, Gene action in the X chromosomes of the mouse, *Nature* **190:**372.

Maroteaux, P., 1974, *Les Maladies Osseuses de l'Enfant,* Flammarion, Paris.

McKusick, V. A., 1975, *Mendelian Inheritance in Man,* 4th ed., Johns Hopkins Press, Baltimore.

Morton, N. E., 1958, Segregation analysis in human genetics, *Science* **127:**79.

Morton, N. E., 1959, Genetic tests under incomplete ascertainment, *Amer. J. Hum. Genet.* **11:**1.

Newman, H. H., Freeman, F. N., and Holzinger, K. J., 1937, *Twins: A Study of Heredity and Environment,* University of Chicago Press, Chicago.

Ohno, S., 1974, Regulatory genetics of sex differentiation, in: *Birth Defects* (A. G. Motulsky and W. Lenz, eds.), p. 148, Excerpta Medica, Amsterdam.

Pearson, K., and Lee, A., 1903, On the laws of inheritance in man. I. Inheritance of physical characters, *Biometrika* **2:**357.

Penrose, L. S., 1952, Measurement of pleiotropic effects in phenylketonuria, *Ann. Eugen.* **16:**134.

Penrose, L. S., 1954, Some recent trends in human genetics, *Proc. 9th Int. Congr. Genet. Pt. I,* p. 520.

Rao, D. C., MacLean, C. J., Morton, N. E., and Yee, S., 1975, Analysis of family resemblance. V. Height and weight in northeastern Brazil, *Am. J. Hum. Genet.* **27:**509.

Reeve, E. C. R., 1955 (Contribution to discussion), *Cold. Spring Harbor Symp. Quant. Biol.* **20:**76.

Robson, E. B., 1955, Birth weight in cousins, *Ann. Hum. Genet. (London)* **19:**262.

Tanner, J. M., 1962, *Growth at Adolescence,* 2nd ed., Blackwell, Oxford.

# 9

# *The Genetics of Human Fetal Growth*

## *D. F. ROBERTS*

## *1. Introduction*

Study of the genetics of human fetal growth and development is at once fascinating
and frustrating. It is fascinating because, in this secret and critical period of prenatal
life, the genetic endowment of the individual comes to be realized in terms of the
differentiation of anatomical structures, biochemical activities, physiological stabilization, and growth to the point of readiness for separate postnatal life. At that time
are established the rudiments of subsequent development of most of the characters
that become manifest in the uniqueness of the later individual. It is fascinating too
because of the delicacy of the processes by which, from a single cell and the
information contained in it, a multiplicity of different cell types, tissues, and organs
all eventually come to appear, at the proper times, in the proper places, and with the
proper function. It is frustrating because of the difficulties attendant upon its study.
Only during the last few years has there been marked progress in the study of
human fetal development generally, thanks largely to new techniques of noninterventive examination, and in Britain and other countries to the liberalization of the
abortion laws which meant that unprecedented amounts of fetal material became
available. Thus the body of knowledge relating to the norms of human development
before birth is steadily increasing. By the nature of the material, however, the
classical methods of genetic analysis cannot be applied; there are few data relating
to more than one member of a sibship, none by which characters in pairs of relatives
in different degrees can be compared (with the notable exception of birth weight,
the final manifestation of fetal growth, which is reviewed in Chapter 10.) However,
with the recent technical developments in this field, whereby the fetus can be
examined by ultrasonic scan, serial examinations should allow the collection, for
familial correlation, of data on growth in size of the body and its parts, its skeletal
development, and the onset of specific functions; in more and more pregnancies

***D. F. ROBERTS*** • Department of Human Genetics, University of Newcastle upon Tyne, England.

amniocentesis is being carried out, so that data on biochemical variants present in the amniotic fluid will accumulate.

For the present, however, knowledge of the genetics of human fetal growth is obtained indirectly. Some fundamental biological mechanisms established on other organisms can be extrapolated to man. Models derived from techniques, facts, and theories accumulated in the study of human polymorphisms are of direct relevance. In clinical genetics, the genetic component in the etiology of a wide range of disorders is established and frequently the biochemical defects responsible are known. Hence this knowledge of genes causing maldevelopment provides a template against which one can envisage the corresponding alleles responsible for normal development; it provides much of the material in the following synthesis.

The genetic information of DNA that is transmitted on the parental chromosomes to the new organism at the moment of conception, and which is so accurately reproduced in subsequent cell divisions, controls the structures of all the proteins making up and made by the new organism and regulates their synthesis and their interaction with other substances, so that the biochemical makeup of each individual ante- and postnatally is essentially a reflection of the genetic constitution. Hence not only inherited differences in normal physical, physiological, or mental characteristics, or in the presence of particular abnormalities, are likely to be a consequence of differences in enzyme or protein synthesis (Harris, 1970), but so too is the sequence of developmental differentiation and individual variations in it. However, cell differentiation and development does not occur in a vacuum; it is a complex of gene-controlled synthetic processes proceeding in the environment provided by the cell cytoplasm, where it is subject to other influences, for example from extrachromosomal DNA in other organelles of the cell. Indeed some of the organelles, especially the mitochondria and membranes, into which the macromolecules are organized act as templates for their further assembly. Further instructions come via the highly organized, cyclical chemical reaction systems, in which the product of one reaction forms the substrate for the next. Extrachromosomal instructions appear to be especially important in the earliest stages of development immediately after fertilization, which are particularly heavily influenced by the maternal genotype through the genetic information carried in the cytoplasm. Upon fertilization the sperm consists of little more than the chromosomes with negligible other material, whereas the maternal contribution embodies the full cytoplasmic material as well as the chromosomes.

The single cell that is the fertilized egg gives rise to different kinds of cells, different tissues, different organs. This development is brought about through a sequence of changing populations of cells organized in increasingly complex patterns. Cytogenetics shows that from the behavior of the chromosomes at each mitotic cell division, daughter cells contain virtually identical sets of genes. Yet as division succeeds division, divergence of cells occurs until they have differentiated into types as different as those of muscle, pigment, or nerve. Visibly quite different in appearance and function, the cells appear to behave as though different sets of genes had been given to them at earlier cell divisions. This is clearly not so. It is not the genes or the chromosomes that are distributed unequally. Rather it is the cell that manifests only a small fraction of its genes, in the differing chemical environments that have developed. The potentialities, certainly of the initial cells, are far greater than the actualities that they express.

# 2. A Model for Genetic Control

## 2.1. Normal Hemoglobin Sequence

A model of how genetic control of fetal development may be exercised is provided by the hemoglobins. The globin of the normal adult hemoglobin molecule consists of two pairs of different polypeptide chains, that is, two $\alpha$ chains and two $\beta$ chains. A different gene specifies each chain, in which the sequence of constituents, respectively 141 and 146 amino acid residues, is controlled by the DNA base-pair sequence. To each chain is attached a heme group so that the molecular weight of the total molecule is approximately 66,000. There is also a minor hemoglobin fraction (Hb $A_2$) consisting of two $\alpha$ chains and two $\zeta$ chains, the latter again differing in structure from the $\alpha$ and $\beta$. But these hemoglobins do not persist throughout the whole life span. There is instead a steady developmental sequence. During the first 10 weeks of embryonic life, there exist unique hemoglobins which contain quite different polypeptide chains ($\epsilon$ and $\zeta$), the latter appearing to be an early form of $\alpha$ chain. The principal ones are Gower 1 and 2 consisting of $\epsilon_4$, ($\zeta_2 \epsilon_2$). Then during early embryonic development $\alpha$ and $\gamma$ chains begin to be synthesized, and there occur certainly in some individuals hemoglobin Portland 1 ($\gamma_2 \zeta_2$), and then the principal fetal hemoglobin, hemoglobin F ($\alpha_2 \gamma_2$), which the adult form ($\alpha_2 \beta_2$) gradually replaces during later fetal and early postnatal life. These chains are synthesized in different tissues. This means that since all cells contain identical genetic information, specific genes must be switched on and switched off at the appropriate point in development. That this is so is also indicated by the fact that fetal hemoglobin production, taking place primarily in the liver and spleen, can be artificially reactivated in adult bone marrow (Hall and Motulsky, 1968).

There are two genes coding for $\alpha$ globin, one for $\beta$, one for $\delta$, and two for $\gamma$ chains. How the genes governing the various chains are interrelated is not yet firmly established. Zuckerkandl (1964), employing the Jacob-Monod model, proposed that the closely linked $\beta$ and $\delta$ chain genes are included in the one operon and depend on the one operator gene; that the $\gamma$ chain gene is under the control of a separate operator gene or genes and its activity is also influenced by physiological conditions. More recent views envisage more complex control involving the presence, for example, of sensor, producer, and receptor genes (Britten and Davidson, 1969), or promotor and initiator genes and address loci with destabilizer elements (Paul, 1972). But whatever the precise control system, the model of sequential switching on and off of specific genes remains valid.

## 2.2. Variant Forms

Besides temporal variation, other variants occur at a given stage in the life cycle. In all human populations there are polymorphisms, in that there are variant forms of substances, due to minor alterations in the base-pair sequence and present in an appreciable number of individuals. Hemoglobin provides an example. Variant forms occur in adult hemoglobin (for example, hemoglobin S or hemoglobin C reach very high frequencies in some African populations).

In view of the widespread occurrence of polymorphisms in adult populations, a similar occurrence of polymorphic variants in substances active in fetal life appears very likely and indeed they occur in fetal hemoglobin. Many variants in sequence within the $\alpha$, $\beta$, and $\gamma$ chains giving rise to structural variants of hemoglobin are well known. Most have a relatively minor biological effect, but some are more severe; few homozygous Hb S infants survive. More severe, too, are conditions in which the switch mechanism is involved. Suppression or deficiency of production of a specific globin chain gives pure multimers ($\alpha_4$, $\beta_4$, or $\gamma_4$), deleterious if present in more than trace frequencies on account of their malfunction (e.g., in oxygen transport or ability to combine with haptoglobin). One mutation causes failure of activation of the $\beta$ and $\delta$ chains, so only fetal hemoglobin is synthesized even postnatally. Some chromosomal defects interfere with chain synthesis: in trisomy 13 disappearance of embryonic and fetal hemoglobin is delayed.

Glucose-6-phosphate dehydrogenase (G6PD), an enzyme involved in the shunt pathway of carbohydrate metabolism, requires a single structural gene for its synthesis, this time located on the X chromosome. The monomeric chain has a molecular weight about three times that of a hemoglobin chain, consists of about 350 amino acids, and over 60 variant forms are known, some reaching high frequencies in Mediterranean and African populations. One group of variants spontaneously causes chronic hemolytic disease, another group causes hemolytic reactions only in specific environmental conditions.

Of such fetal variants, those that are severe in effect may prevent fetal survival; others milder in effect may influence, for example, physiological efficiency or growth rates, and so may contribute to the variations among normal individuals in fetal growth and development.

### 2.3. Implications

It therefore appears that embryonic and fetal development occur as a result of switch mechanisms, switching on and off the activity of particular genes, at specific times in development, and producing specific substances. It follows that variations in fetal development as well as intrinsic malformations may arise by (1) variation in efficiency or failure of the switching mechanism, (2) variation in its timing, and (3) variations or errors in the substances synthesized. Search for variant substances, or deficiency of substances, in association with fetal developmental stages in different tissues appears well worthwhile. So too is their exploration in specific malformations. Of the latter there are few initiated antenatally in which the precise biochemical defect is known, even among the number that are already well established as being under simple genetic control. Partly this is due to the inadequate amount, despite the recent change of U.K. law legalizing abortion, of early fetal material available for study although, with the passage of time, more fetuses in which defects are identifiable will become available. But most fetuses from voluntary pregnancy terminations will be normal. Their investigation gives valuable information first on the normal sequence of appearance of substances and their activities at different stages of fetal growth so that one may establish the pattern of normal development, against which findings from defective fetuses can be compared, and secondly on the occurrence of variant forms of substances, against which minor variations in fetal development can be compared.

Some insight can be gained in this work from the much greater knowledge of biochemical changes and defects in postnatal life and their effects, since fetal and postnatal phases of growth are temporally continuous in many respects and many processes initiated in fetal life continue in the postnatal period of development. But there are a number of other difficulties. There are the problems of whether variant substances are true biological variants or whether they are due to artifacts of preparation of the material, storage of specimens, or indeed the result of the particular mode of pregnancy termination employed. One needs to be sure that the substances are in fact present at the time of differentiation and development of the relevant fetal tissues. Finally, it is necessary to know that the variant forms are genetic, and here it is essential to draw to the attention of obstetricians the importance of the investigation of successive fetuses that may become available from the same mother.

## 3. Enzyme Heterogeneity

The existence of multimolecular forms of enzymes, suggested over 30 years ago by Desreux and Heriott (1939) with the enzyme pepsin, was confirmed by technological advances, and today it is established that many enzymes, if not the majority, possess multiple forms. These differ in varying degrees in their physiological and biochemical properties and therefore may be relevant in the field of embryonic and fetal development.

Markert and Møller (1959) coined the term isoenzyme to describe those different molecular species with closely related enzymatic and physical properties. The subsequent disagreement over definition and classification of the term was partly resolved by the International Commission (I.U.B., 1972) which recommended that

1. The term "multiple forms of the enzyme" should be used as a broad term covering all proteins possessing the same enzyme activity and occurring naturally in a single species.
2. The term "isoenzyme" or "isozyme" should apply only to those multiple forms of enzyme arising from genetically determined differences in primary structure and not to those derived by modification of the same primary sequence.

From the Commission's classification, based on mode of production (Table I), it appears that groups 1, 2 and 3 are "true" isoenzymes, while those in groups 4–7 should be simply called "multiple forms" since these represent essentially mechanisms of intracellular modification of the same polypeptide chain; their biological significance is not thereby diminished, of course, since they are just as much part of the intracellular milieu.

*Table I.   Enzyme Heterogeneity*

| Group | Reasons for multiplicity |
|-------|--------------------------|
| 1 | Genetically independent proteins |
| 2 | Heteropolymers (hybrids) of 2 or more polypeptide chains, noncovalently bound |
| 3 | Genetic variants (allelic): Common polymorphic Rare mutants |
| 4 | Proteins derived from one polypeptide chain |
| 5 | Proteins conjugated with other groups |
| 6 | Polymers of a single subunit |
| 7 | Conformationally different forms |
| 8 | Artifacts of preparation |

Of the genetic groups (1–3), group 1 may be illustrated by the malate dehydrogenase enzymes of the mitochondrial and cytosol fractions of liver cells described by Kaplan (1961). The two types of protein have the same enzymic activity, but are structurally quite different and each is under separate genetic control. Group 2 may be illustrated by lactate dehydrogenase in which (like adult hemoglobin) two different types (A and B) of polypeptide chains occur, each under the control of a different gene. The whole normal enzyme molecule is composed of four of these polypeptide chains; its electrophoretic mobility is determined by the number of chains of each type, so that there are five different isoenzymes of lactic dehydrogenase separable on electrophoresis (one with four A chains, one with three A and one B, etc.). In group 3 occur the common polymorphic variants such as erythrocyte acid phosphatase or phosphoglucomutase, in which each allele is responsible for the production of a particular band, or group of bands, on the zymogram produced by electrophoresis, most of which can be shown to have identifiable variation in physiological or biochemical properties. In this group there also occur the rare mutants, such as the variant forms of A or B chains of lactate dehydrogenase.

For fetal development all structurally different proteins are of potential relevance, irrespective of how they are brought about. The isoenzymes of group 3 have been particularly useful in the demonstration of qualitative differences and of the quantitative variation in activity that so frequently occurs. Epigenetic control may be equally important. For example, epigenetic modification, besides bringing about the folding of the protein chains into a variety of three-dimensional conformations each with its own characteristic properties, may produce variations in polymer size by the aggregation of a variety of small molecules (steroids, nucleotides) as in glutamate dehydrogenase; by deletion or addition of part of the peptide chain, as in the proteolytic removal of an inhibiting peptide fragment in the activation of trypsinogen, or in the oxidation or reduction in lactate dehydrogenase producing interconvertible conformational isozymic forms (group 7). There may be attachments of a small molecular fraction to the enzyme itself, as in the sialic acid residues attached to alkaline phosphatase which create a variety of distinct forms each with a potentially different role in cell metabolism.

## 4. Tissue Specificity of Isoenzymes

Involvement of isoenzymes in fetal differentiation implies their tissue specificity. This is well established and can be illustrated by lactate dehydrogenase where different tissues exhibit a preponderance of either the isoenzymes with predominantly A subunits or those with predominantly B subunits, and this, Harris (1970) suggests, is attributable to variations in activity of the two genes responsible in different cell types. LDH 5 with four A subunits and LDH 1 with four B subunits differ in their kinetic properties, particularly in the inhibition of the latter by the accumulation of the substrate pyruvate. LDH 1 is predominant in tissues such as heart and brain, in which aerobic metabolism is continuous, while LDH 5 is predominant in voluntary muscle, where anaerobic metabolism can occur. In fetal tissues, LDH 5 tends to predominate, and it has been suggested that the relatively low tissue oxygen tension in the fetus may have a controlling influence, stimulating production of the A subunits and leading to the predominance of LDH 5.

Such tissue-specific patterns, well illustrated in some inborn errors (Table II), influence the clinical features of the disorder (and the methods for diagnosis).

Table II.   Spatial Variation of Isoenzymes as Revealed by Inborn Errors of Metabolism

| Disorder | Enzyme | Enzyme normal in | Enzyme deficient in | References |
|---|---|---|---|---|
| Hexokinase deficiency | Hexokinase | WBC | RBC | Valentine et al., 1967; Keitt, 1969; Necheles et al., 1970 |
| Pyruvate kinase deficiency | Pyruvate kinase | WBC | RBC | Bigley and Koler, 1968; Bigley and Stenzel, 1968; Harris, 1970 |
| Phosphofructokinase deficiency | PFK | RBC (1 isoenzyme) | Muscle | Waterbury and Frenkel, 1969, 1972 |
| Chronic granulomatous disease | Glutathione peroxidase | RBC | WBC | Necheles et al., 1969 |
| Fructose intolerance | Aldolase | Muscle | Liver | Penhoet and Rajkumar, 1966; Rutter et al., 1968; Hers and Joassin, 1961; Froesch et al., 1963 |
| McArdle's disease | Phosphorylase | Liver | Muscle | Schmid et al., 1959; Robbins, 1960; Dawson et al., 1968 |
| Hers' disease | Phosphorylase | Muscle | Liver, WBC | Hers, 1959; Williams and Field, 1961; Hers and Van Hoof, 1968 |
| Phosphorylase kinase deficiency | Phosphorylase kinase | Muscle | Liver, WBC | Hug et al., 1969, 1970; Huijing and Fernandes, 1969 |
| Myeloperoxidase deficiency | Myeloperoxidase | Eosinophils | Neutrophils, monocytes | Lehrer and Cline, 1969; Salmon et al., 1970 |
| Oculocutaneous albinism | Tyrosinase | Adrenals | Melanocytes | Fitzpatrick and Quevedo, 1966 |
| Hypophosphatasia | Alkaline phosphatase | Intestine | Bone, liver, kidney | Rathburn, 1948; McCance et al., 1956; Danovitch et al., 1968 |
| Primary hyperoxaluria (type 1) | Alphaketoglutarate | Mitochondria | Cytoplasm | Crawhall and Watts, 1962; Koch and Skokstad, 1967 |

Where the error affects fetal metabolism, it is already possible in a number of instances to detect the disorder antenatally. But these cases are of more fundamental importance, in addition to their practical value, for they show the role of the enzyme in normal development. For example, the enzyme phosphorylase is responsible for the separation of sugar groups from the storage molecule glycogen, in liver and skeletal muscle particularly. Two clinically distinct conditions are attributed to an inherited defect of this enzyme. McArdle's disease (Cori type 5) is characterized by the development of severe muscle cramps on exercise, and enzyme assay shows almost complete absence of muscle phosphorylase. In Hers' disease (Cori type 6) there is excessive deposition of glycogen in the liver associated with hypoglycemia and acidosis, and here liver phosphorylase has been shown to be inactive. Hence separate loci are responsible for normal phosphorylase production in muscle and liver; mutation at these different loci is responsible for the specific symptoms of the two types of glycogenesis.

The enzyme aldolase has three structurally distinct forms (aldolase A, B, and C) which are found particularly in muscle, liver, and brain, respectively. These isoenzymic proteins are tetramers and evidently differ in the structure of their characteristic polypeptide subunits, each of which is presumably determined by a separate gene locus. Each isoenzyme is responsible for the division of fructose sugars into two 3-carbon components in the pentose shunt pathway (Rutter *et al.*, 1968). In the inherited defect hereditary fructose intolerance this division of fructose does not occur, leading to profound hypoglucosemia when it is ingested. When the tissues are examined, this metabolic defect is seen to be present only in the liver where aldolase B is found, and the other isoenzymes act as normal.

The enzymes pyruvate kinase and hexokinase both show two isoenzymic forms produced by separate loci in red and white blood cells. A chronic hemolytic anemia has been attributed to a genetic defect in each of these enzymes, and in both cases this has been found to be due specifically to loss of the red cell isoenzyme; the white cell isoenzyme is spared.

That different tissues possess distinct isoenzymes, one of which is specifically affected by a mutation, does not necessarily mean that these multiple forms must be coded for by separate gene loci. Tissue-specific patterns may arise by any of the mechanisms outlined in Table I, as well as by the switch procedures (p. 252) activating different loci in different cell types. For example, the defects in some patients in the past with clinical signs closely resembling true Hers' disease, and identified as such on the basis of high liver glycogen and low liver phosphorylase levels, were subsequently shown to be due to a deficiency, not in phosphorylase production, but in phosphorylase activation by the enzyme phosphorylase kinase. In these patients, now referred to as Cori type 8, further work thus led to refinement of the location of the disorder, a mutant kinase allele whose product also shows tissue specificity; it also demonstrated a further complexity in the series of genes controlling the normal development of this metabolic sequence.

### 4.1. Subcellular Variation

It is now well established that there exist considerable differences in the metabolism of the cytoplasm and the numerous organelles within it, and a number of the ''inborn errors'' are now known to be specifically localized. Thus, among the glycolysis defects, the enzyme deficiency in Von Gierke's disease is microsomal, that in Gaucher's disease is lysosomal, that in propionic acidemia mitochondrial. If different isoenzymes were found in the various subcellular fractions, then different tissue fractionation would be necessary before any conclusions could be reached about the possible role of an enzyme in the pathogenesis of a disorder and hence in normal development. In hyperoxaluria type 1, Crawhall and Watts (1962) and Koch and Skokstad (1967) found that the deficient enzyme ($\alpha$-ketoglutarate: glyoxylate carboligase) showed normal activity in the mitochondrial fraction of the liver, kidney, and spleen, i.e., the inherited defect responsible for the raised oxalate production affected only the cytoplasmic isoenzyme. This situation does not appear to be common, but this may be due to the fact that most enzyme deficiencies have been investigated in whole tissue homogenates.

## 5. Temporal Variation during Development

Involvement of isoenzymes in fetal differentiation or malformation also implies availability or variation in activity at critical stages in development. Here again, information is not extensive. The presence and activity of a particular enzyme variant may be demonstrated by its interaction with an environmental insult; in the first example this was treatment by a drug which, like many others, freely crosses the placenta to the fetus and produces a fetal defect. The enzyme glucose-6-phosphate dehydrogenase which is active antenatally (Messina, 1972) possesses many allelic variants. Perkins (1971) reported on a stillborn black male, carrying the gene for the A⁻ variant of G6PD on the X chromosome, which was delivered at 36 weeks with severe erythroblastosis and anemia, 2½ weeks after the mother had taken a course of sulfisoxazole for a urinary tract infection. This drug, the only etiological factor of note in the case, produces hemolysis in G6PD-deficient adults so that the fetus appears to have suffered a drug-induced hemolytic crisis due to the presence at the critical time of an allelic variant of G6PD. Others, of course, among the rare variants of G6PD are so severe as to cause spontaneous congenital hemolytic disease, and cannot be used in this way. But this approach has indicated how short are the critical periods for the teratogenic effects of some drugs. The investigations of the thalidomide tragedy showed that no malformations occur when the drug was taken less than 34 or more than 50 days after the last menstrual period, quite a restricted vulnerable period; exposure at days 39–44 brings hypoplasia or absence of the upper limbs, at days 42–48 of the lower limbs, and at days 41–43 anomalies of the heart, so that in general the teratogenic exposure to thalidomide precedes by about two weeks the completion of the developmental event. Hence the substances (whatever they may be) involved in the metabolism of such drugs may only be available to the fetus, or the relevant tissues, for limited periods.

Many isoenzymes show quantitative and qualitative changes with development. In many, this involves an increase in the number of multiple forms due to increased intracellular complexity and development of mature function. These changes apply particularly to those multiple forms which arise as a result of posttranscription modifications within the cell. A few, however, possess isoenzymes which disappear later in development. The isoenzymes of alkaline phosphatase which are specific to the placenta provide an obvious example, but a similar disappearance is also seen with intrinsic fetal material.

Phosphoglycerate mutase (PGAM), found in muscle, brain, and other tissues, shows tissue-specific isoenzymes. There is a striking transition in the PGAM electrophoretic pattern of human skeletal muscle during fetal development

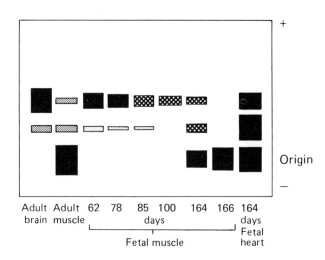

+

Origin

−

Adult Adult  62  78  85  100   164  166  164
brain muscle           days              days
         |_____|  Fetal
                  Fetal muscle           heart

Fig. 1. Developmental transition of phosphoglycerate mutase in fetal muscle, explicable by switching off the gene for type B and switching on that for type M.

(Figure 1). Young fetal muscle (62 days) contains almost exclusively a single-banded type B; then there appears a band of intermediate mobility (80–100 days) signifying production of type M subunits, and finally a 3-banded pattern becomes recognizable (115–164 days). Thereafter to term the M-type pattern of the adult predominates. Creatine phosphokinase shows a similar replacement of early B-type dominance by adult M-type patterns. Examination of the same fetal muscle specimens shows that the developmental transitions are not synchronous with, but are in advance of, those of PGAM. The developmental transition of phosphoglycerate mutase in human fetal skeletal muscle represents a two-step process, turning off the gene for type B (brain type) PGAM (whose disappearance then depends on its half-life) and turning on the gene for type M (muscle type) PGAM. That the transition represents the maturation of the muscle cells, and not a shift in cell population from predominantly fibroblastic connective tissue to myoblastic, is shown by the asynchrony of the developmental transitions of PGAM and CPK (Omenn and Hermodson, 1975).

For a quantitative example, the activity of G6PD rises from a low level in early fetal life to a peak at 4–5 months, falling to the adult level by birth, and the enzyme molecule responsible for the early peak is distinct from the adult type, i.e., is an isoenzyme (Messina, 1972). In other systems the change is less dramatic. Thus, with lactate dehydrogenase in the brain, a relative excess of the M-type subunit during fetal life and persisting three months after birth results in a changed electrophoretic pattern with the band LDH 3 being dominant (Bonavita and Guarneri, 1962).

Some, at least, of these developmental enzyme variations must be causally related to the profound biochemical and structural changes which occur during differentiation. In some cases, where an early-acting variant produces a defect, its presence may manifest as a positive family history of lethal congenital defect but no obvious postnatal biochemical abnormality, so that a proportion of unexplained fetal deaths may well be due to defects and variations in enzyme systems operating exclusively in early and very critical periods of life. Moreover there is no reason to regard isoenzymic modifications in fetal life as being unidirectional, for it is possible that some substances may be synthesized and available to the fetal tissues for a limited period, just as some atavistic structures are formed only to become modified into others; indeed some variant substances that appear in fetal life may not be causally involved at all in differentiation. The following example illustrates some of the possible complexities.

## 6. Nonspecific Esterases and Development—An Example

The esterase enzymes which hydrolyze various esters are closely related to the enzyme lipase but differ in that their substrates have much shorter acyl chains, i.e., their side chains contain fewer carbon atoms. The complex esterase system includes several discrete enzymes with distinct physiological substrates, particularly the cholinesterases, and also the "nonspecific esterases," for which no natural substrate can be defined and whose role remains obscure. Masters and Holmes (1975), in a study of phylogenetic variation in esterases, suggested that their heterogeneity derives from gene duplication during evolution and is enhanced by epigenetic processes such as polymerization of the gene product. For investigation of the nonspecific esterases, the enzymes are made to act on artificial esters of naphthyl (Nachlas and Seligman, 1949), liberating free naphthyl which reacts with a diazonium salt to produce a water-insoluble pigment. In electrophoresis the deposition of this pigment marks the position of the esterase bands. Inhibitors such as E600, DFP, PCMB, and eserine are used to subdivide the numerous electrophoretic zones present in most tissues. At present the two major groupings are the E600-resistant arylesterases and the carboxylesterases which are sensitive to this inhibitor. In our study of their role in fetal development, three artificial esters were employed: α-naphthyl acetate (NA), α-naphthyl propionate (NP), and α-naphthyl butyrate (NB) with acyl chains containing 2, 3, and 4 carbons, respectively.

### 6.1. The Brain

Figure 2 column A shows the typical electrophoretic pattern of nonspecific esterases in the grey matter of the adult human brain. The 16 major bands marked agree with those described elsewhere (Bernsohn, 1964; Bernsohn et al., 1964; Barron, 1961; Barron et al., 1962, 1966, 1972; Barron and Bernsohn, 1963, 1968). These NA bands can be divided into two groups: the cathodal and lower anodal bands with band 8 are carboxylesterases, and bands 5, 6, 9, and 10 are arylesterases. The remaining bands are probably due to substrate overlap with cholinesterase and lipases and need not be discussed further. In addition to the bands shown, three or four additional bands lie between 8 and 9 which are visualized with the 3-carbon substrate NP. Like band 8, these resemble the carboxylesterase group.

The early fetal pattern in Figure 2 column B shows clear differences from the adult brain pattern. The fetal brain shows first an apparent absence of carboxylesterases and secondly the presence of a dense zone of activity which will be referred to as the "F" band. When a series of brains extending throughout prenatal life was examined, it was found that the fetal pattern described was universally present in fetuses from 8 up to about 20 weeks. Thereafter, the carboxylesterases showed a progressive rise in activity, reaching adult levels by birth, while the fetal band continued at about the same activity until 25 weeks in the cortex and fell gradually thereafter, reaching a low level by birth and disappearing by about 3 months postnatal age. Relating these temporal changes to the histological development of the brain, the rise in activity of the cathodal and lower anodal bands is associated with the period of neuronal maturation, axon and dendrite growth, and glial cell

D. F. ROBERTS

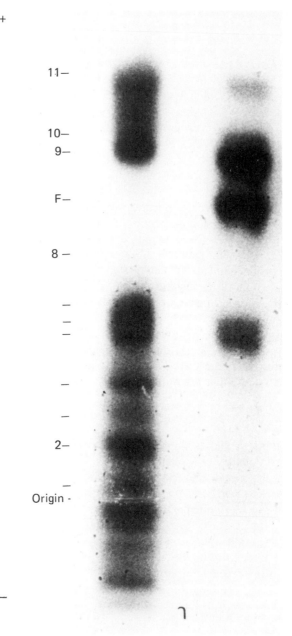

Fig. 2. Nonspecific esterases in human brain. A: normal adult, B: early fetus.

proliferation. Since these enzyme bands show a widespread distribution in different tissues and species, however, and since this period of brain development is associated with widespread biochemical changes, it seems likely that there is no direct causal relationship between these events and that this case provides a good example of developmental isoenzyme changes associated with tissue and cellular maturation.

This explanation cannot be applied to band 8, however, which together with the NP-specific bands began to show increased activity in the last 10 weeks of pregnancy, rising to a high level by 6 months postnatal age. These NP-specific bands as a whole are specific to the white matter of the central nervous system and are found in higher primates only. They have been extensively studied (Barron and Bernsohn, 1968; Barron *et al.*, 1972) and have been found to react best with the 3- and 4-carbon substrates; only the most proximal band reacts with naphthyl acetate—band 8. Highly sensitive to organophosphorus inhibitors such as DFP and able to hydrolyze 5-carbon esters, they fall midway between the other esterases and the lipases.

There are clues to the developmental significance of this isoenzyme group. These polypeptides seem to be closely associated with CNS myelin in their distribution. Their appearance in the cortex immediately precedes myelination according to the timing described by Yakovlev and LeCours (1967), and Barron demonstrated that the activity of this group declined in the second postnatal year, after the majority of the cortex has completed its myelogenetic cycle. Wender and Kozik (1971) found, on histological examination of developing brain, a diffuse esterase activity in maturing neuroglia with accentuated activity in the myelination clusters of oligodendroglia. A similar raised activity was seen in the nerve fibers, and this persisted into adult life while activity in the neuroglia subsided. It seems reasonable to suggest that this group of isoenzymes is closely involved in the production and maintenance of CNS myelin in man. Support for this suggestion comes from pathology involving demyelination; the NP-specific bands were absent in the plaque tissue of multiple sclerosis cases (Barron and Bernsohn, 1965; Barron *et al.*, 1963), and perhaps of more interest in the context of fetal growth, in a case of diffuse cerebral sclerosis, a condition which involves massive demyelination of the CNS in early childhood.

### 6.1.1. The F Band

Further study of the F band provides information on its importance in fetal development. Its biochemical properties show it to be similar to the arylesterases, being able to hydrolyze NA, NP, and NB, i.e., up to 4-carbon esters, and being resistant to the inhibitor E600; in addition the band shows a relative resistance to the inhibitor PCMB. Developmentally, this band persists at a constant level in the cortex until 25 weeks and thereafter fades. This timing can be correlated with the cessation of neuroblast division in the cortex (Dobbing and Sands, 1970). In adult brain, by using detergent to extract bound esterases, a band with the same mobility is faintly visible in the zymogram from gray matter, where the cortical neurons are located, but not in white matter. Furthermore, the electrophoretic data of Kokko (1965) on spinal ganglia, which contain a high concentration of neuronal tissue, also show a band with the same mobility and inhibitor characteristics as the F band. It appears that this isoenzyme may be involved in the division of the immature neurons.

In some cases very close to the leading edge of the F band there appeared, after storage of the specimen, an additional band which could not hydrolyze NP, i.e., was specific to the 2-carbon substrate NA (Figure 3, columns A and B). This phenomenon may be explained on the basis of the subunit concept suggested by Choudhury (1972) to account for the various properties of the nonspecific esterases. On this hypothesis the F band would comprise two subunits, labeled C2 and C4 on the basis of the maximum length of side chain they can accommodate. On storage these

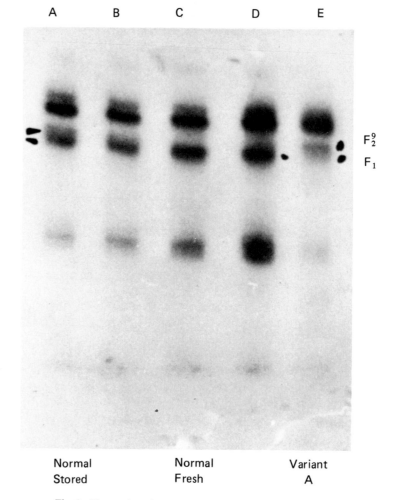

Fig. 3. Electrophoretic variations in fetal brain esterase pattern.

dissociate, and the free C2 subunits migrate further on electrophoresis, so producing the extra band. This suggestion was supported by the findings on isoelectric focusing. It was found that three bands corresponded to the fetal esterase (Figure 4), which were labeled V, VII, and VIII. It was shown that with a fresh sample band V was prominent, while an older sample showed reduced activity at position V and raised activity at VII and VIII. Since the latter reacted only with NA and band VII showed a prominent reaction with NP, it appeared that band V represented the complete enzyme and the other bands the two subunits.

During this study of the developing fetal brain, three specimens were found in which the electrophoretic properties of the F band were different. The first variant observed, variant A, is shown in column E of Figure 3; the normally dominant $F_1$ band is virtually absent, while the $F_2$ band is relatively prominent and, as in the other cases, was found to react only with the 2-carbon substrate NA. The electrophoretic distinction was supported by isoelectric focusing, with bands V and VII absent and continued activity at position VIII (column A, Figure 4). These changes were attributed to some defect in the C4 component of the F band. Short-run electrophoresis (Figure 5) demonstrated the presence of a defective unstable protein; this was taken to represent an abnormal C4 subunit which, because of its altered properties, was unable to combine with the C2 subunits. All this evidence strongly suggests genetic determination of this variant, in which case this 12-week fetus is either heterozygous or homozygous for a mutant allele at the C4 locus. The two other specimens found (variants B and C) also showed greater $F_2$ and less $F_1$ activity than normal and the presence of the mutant C4 band, but their activity

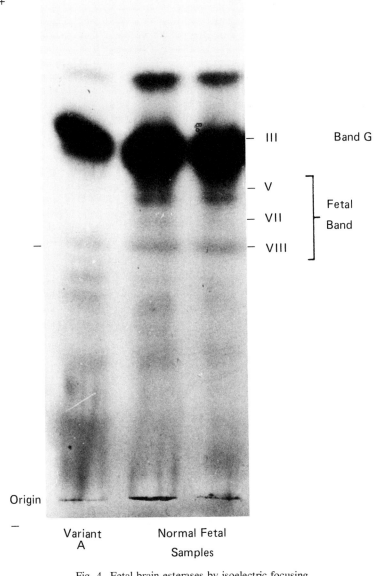

Fig. 4. Fetal brain esterases by isoelectric focusing.

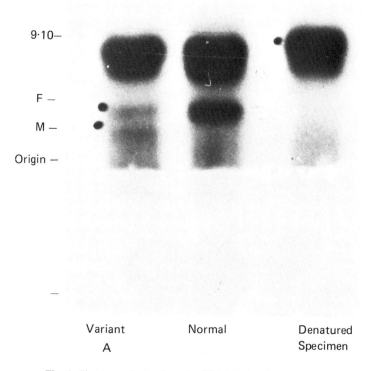

Fig. 5. Short-run electrophoresis of fetal brain esterases.

differences were less extreme. They may represent, on this genetic hypothesis, either heterozygotes for alleles of quantitatively different effect from that carried by variant A, or heterozygotes for the same allele as that for which variant A is homozygous. Figure 6 summarizes the electrophoretic findings relating to the fetal esterase band.

In view of the association between the F band and dividing neuroblasts, an F band defect would be likely to interfere with normal neuron division, and since the fetal brain in the first half of pregnancy is almost entirely made up of neuroblasts, any interference would probably affect total brain weight. The fetus with variant A had a brain weight of 2.3 g, which was more than 8 standard deviations below the age mean of 5.5 g (i.e., rather less than 50% of normal). Brain weights in variants B and C were approximately 75% of the mean, falling on the lower 95% confidence limit. In all three cases the external size of the head remained within normal limits, the extra space being filled by CSF. These findings appeared to support the explanation of a genetically determined defect in a fetal isoenzyme specifically involved in neuron division. Final support for the theory was provided by retrospective data relating to the social history of variant A. The mother of this fetus was found to have 14 sibs of whom 6 were mentally subnormal, 4 being housed

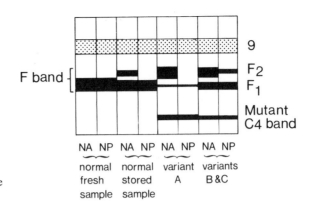

Fig. 6. Diagrammatic summary of the F band variant in fetal esterases.

permanently in institutions. An unspecified number of the remaining sibs were described as being "borderline IQ," and the mother showed evidence of being in the lower range of normal IQ and being somewhat emotionally immature having had two abortions in two years to different consorts.

On clinical grounds, the affliction in this family would be described as "simple mental retardation" owing to the absence of any obvious biochemical defect or associated structural deformity, this being the classification into which almost 50% of all retardates are placed (Walton, 1971). Dodge and Adams (1971) observed that such retardates are often born of dull or retarded parents, so that the term "familial" was frequently justified; no physical abnormality, biochemical defect, or gross brain growth defect is usually evident in these people. However, the biochemical association with neuron division suggested in the present study means that the etiological factor is specific to nerve cells. These represent less than 10% of the total weight at birth, and variation of the order here described is not detectable in the postnatal period. As Davison and Dobbing (1968) and Dobbing and Sands (1970) have shown, if neuron division is not completed in the given time, a permanent deficit results, and this would affect the higher centers most severely. Further, the lower cerebral structures such as the brain stem are not affected in these individuals.

### 6.2. The Kidney

Comparison of nonspecific esterases in other tissues with those in the brain is also rewarding. Kidney tissue was obtained from 26 fetuses, including at least 2 from each week of gestational age from the 10th to the 22nd, and these were compared with neonatal, postnatal, and adult specimens acquired within 24 h of death from postmortem cases with no history or evidence of renal disease (Papiha *et al.*, 1976). The NA bands in the cortex and medulla of adult human kidney show seven anodal and four cathodal molecular species, by comparison with eleven and five, respectively, in the adult human brain. Whereas the fetal brain pattern showed the differences from the adult brain reviewed above, the zymogram of the fetal kidney at 12 weeks was more like that of the adult kidney. The cathodal bands C2, C3, and C4 were faintly visible, and while anodal bands A1 to A6 could not be resolved so sharply, there was quite evident activity in these regions. Particularly interesting was band A7 which, instead of the single heavily staining mass seen in the adult kidney, in the fetus showed a clear triplet formation, designated as bands

A7a, b, and c (Figure 7). The close similarity of the fetal kidney to the adult pattern was also shown when a Tris maleate buffer was used instead of a borate, and when naphthyl propionate and naphthyl butyrate substrates were used.

Developmental progression appears to be different in the two fetal tissues. All fetal brain samples from 10 to 20 weeks of gestational age show a consistent esterase zymogram (apart from the variant specimens noted above); then a transition period starts after 25 weeks, and by 28 weeks the pattern is clearly changing. By 34 weeks the zymogram shows qualitative similarity to the adult pattern, and by 38 weeks there is very little difference from the adult zymogram, either qualitatively or quantitatively. On the other hand, the kidney shows a different type of pattern. Even in the earliest kidney studied, in the 10th week of gestational age, the zymogram was essentially similar to the adult kidney pattern apart from some qualitative changes. Between 11 and 21 weeks gestational age, the qualitative differences disappear, and after 21 weeks the intensity of the bands increases until in the neonate there is hardly any difference from the adult.

The role of nonspecific esterases in the growth and development of the human fetus is clearly very important. The molecules with esterolytic activity start to appear in the human embryo even when it is only 10–11 mm in crown–rump length (Yamamura *et al.*, 1965). The present results show a definite progression in the pattern of esterase isozymes during the early phase of fetal development, but there is a difference in the nature of the mechanism controlling the biosynthesis of esterases in the two morphologically distinct cell populations during ontogenesis.

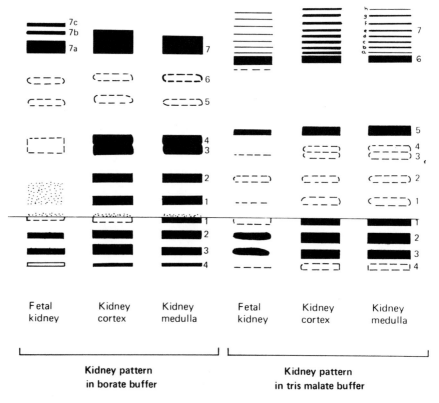

Fig. 7. Esterase patterns in early fetal kidney and adult kidney cortex and medulla.

The brain cell populations show the presence of a tissue-specific enzyme and other differences from the adult in the early phases of development. In contrast, the kidney cells show considerable similarity between fetal and adult pattern. The changes in the esterase pattern reflect differences occurring simultaneously at the cellular level during their differentiation in the tissue, and it has been argued that the brain F band is exclusively associated with neuroblast division. Similar phases of development occur in the kidneys also (Potter, 1972). In period 1 there is active ampullary division with continual nephron induction. In period 2 the ampullary division ceases, and the kidney metanephrons start to function and produce urine at a very low rate by 12–13 weeks of gestation (Vernier and Smith, 1968). To meet the requirements of developing kidney function, the biochemical functional components also have to develop faster. It is therefore not surprising that the esterase isoenzyme picture is complete by 11 weeks of gestational age.

Although the overall pattern is quite similar in adult and fetal kidney, some fetal esterase bands, for example band A6 in Tris maleate buffer, show higher activity than in the adult. There is some similarity of these particular kidney bands with the F band of the fetal brain, and this similarity of the kidney A6 band may also be associated with the maintenance of early cellular division and proliferation of metanephrons but, unlike the brain, the gene controlling this band continues to function during cellular maturation and in the fundamental cellular metabolism.

The association between the appearance of particular esterase heteromorphs and particular stages of cellular differentiation, both in brain and kidney, demonstrates the close relationship between biochemical and morphological ontogeny. Evidence is accumulating, for example, from cell hybridization studies (Shows, 1972) which indicate the separate genetic control of different esterase heteromorphs. Functionally, these may respond differently to changes of cell environment, as is indicated by the fact that certain tissues develop characteristic esterase patterns during a particular stage of cellular differentiation. Thus the vast heterogeneity in the isoenzyme system that exists in different tissues provides an ideal mechanism by which to adapt to the changing molecular environment during ontogeny.

### 6.3. General

While further work is necessary to corroborate these findings on nonspecific esterase, they have been presented in some detail to provide illustrations of some of the many ways in which isoenzymes may be involved in both normal and abnormal development. Thus an enzyme or one of its multiple forms may owe its appearance during development to tissue maturity and therefore not be involved in the process of differentiation. On the other hand, some multiple forms appear to be specifically involved in development and either disappear completely or show much-diminished activity once a particular stage is passed. Just as morphological and functional stages in different tissues do not necessarily coincide, but may be earlier or later according to physiological requirements, so too does isoenzyme development vary in time from one tissue or organ to another. It seems very likely that a closer search for variants in such "fetal" isoenzymes will provide a biochemical answer to a number of hitherto unexplained inherited defects. Certainly the study of variations in appearance, timing, and activity of fetal isoenzymes in different tissues will relate development and differentiation to biochemical processes and thus bring the understanding of their genetic control one step nearer.

*D. F. ROBERTS*

The relevance in fetal existence of the human blood groups was first noted on account of hemolytic disease of the newborn, and the associated disorders to which they give rise, as a direct result of interaction of genetic factors in the mother and fetus. Here the life span of the infant's red cells is reduced by the action of specific antibodies derived from the mother across the placenta, and hence indicates that the fetus can make the antigen to which the mother is reacting. The blood groups are not direct gene products, their chemical constitution is in most cases unknown, and they have no known positive function. But by analogy with the previous section, their early appearance in fetal life suggests an important biological role. The A and B receptors can be detected (although not fully developed) on the red cells of 5- to 6-week-old embryos, and A, B, and H substances derived from the fetus are found in the amniotic fluid from the 9th week onwards. IgG globulins which can cross the placenta are already present in the fetus before the 12th week of gestation, and the rhesus antigens are established in the fetus by about the 6th week. Although other blood group antibodies very rarely cause hemolytic disease, there is now an impressive body of evidence to show how early the other blood group antigens are established in the fetus, for example K (10 weeks), k (6–7 weeks), $Kp^a$ (16 weeks), S and s (12 weeks), M and N (9 weeks), $Fy^a$ and $Fy^b$ (6–7 weeks), $Jk^a$ (11 weeks), $Jk^b$ (7 weeks), and many others.

Of other polymorphic systems usually used in population studies, transferrin bands are well established by the 17th week (Parker and Bearn, 1962) and have been detected as early as eight weeks (Gitlin and Biasucci, 1969). Transferrin activity rises steadily through fetal life, and it appears that specific fetal components exist (Parker *et al.*, 1963) which disappear shortly after birth.

Haptoglobin (Hp) types in early fetal life have been difficult to detect, although Hirschfeld and Lunell (1962) found evidence for haptoglobin synthesis in fetuses from 17 to 30 weeks old. It required the development of immunological techniques to detect haptoglobin present in small amounts in the sera of 4 out of 50 fetuses between 8 and 16 weeks old (Gitlin and Biasucci, 1969). Activity is very low. The group-specific components (Gc) in fetuses were first observed by Hirschfeld and Lunell (1962) in 25-week fetuses, but again the development of radial immunodiffusion techniques was necessary before the small amounts present in early fetuses could be detected (Toivanen *et al.*, 1969); the concentration increases during prenatal development.

Of the components of complement, C4 and C3 appear to have zero activity in early fetuses, but they have been detected from the age of 15 weeks onward and thereafter the concentration of the two proteins increases steadily. They and the commoner C'3 variants do not cross the placenta (Propp and Alper, 1968).

The studies of such polymorphisms in human fetuses have provided useful information on the quantitative development of some and have demonstrated the importance of immunoelectrophoretic procedures for the accurate estimations of the small amounts that may exist. They have shown how maternofetal interaction can affect the development of and indeed damage the fetus. They have provided useful information on the stages of gestation at which new proteins are released into fetal circulation and on the origins and sites of synthesis of the proteins present in fetal blood (Adinolfi, 1971). The polymorphic variants have provided useful information on which proteins cross the placenta, especially where that of the fetus differs from that of the mother. While the role, if any, of such genetic characters in fetal growth and development is still unexplained, on the principle of "early emergence, early function" that is suggested in the foregoing it is likely to be important.

## 8. Mechanisms

The discussion so far has been concerned with how gene products, regardless of whether they are primary or derived, are involved in human fetal growth. This restriction has been deliberate, for it is in this area in man that most is known. But there are other areas of developmental genetics in which human studies shed virtually no light, in particular that relating to the mechanisms by which selective expression of genes is brought about. This is obviously important, for whatever regulates gene function is ultimately responsible for cell differentiation and thus the morphological changes that occur in embryonic and fetal growth. In this area, information comes mainly from organisms other than man and thus may or may not be applicable, but it is so important that it is worth briefly summarizing the findings.

There are a number of points, in the normal process by which a cell acts on genetic instructions, at which variation in activity can be controlled:

1. There may be differential replication of some genes; certainly in the oocytes of some organisms there are many multiple copies of the genes for ribosomal RNA, each producing a given number of rRNA molecules per unit time, i.e., there is gene amplification.

2. There may be altered rates of synthesis of RNA, i.e., there are different rates of transcription, RNA molecules being produced at different rates not only at different developmental stages, but also at different rates in different tissues. This appears to be the mechanism by which the change in proportion of the human globin $\gamma$ chain after birth is brought about.

3. There may be regulation of the passage of the RNA from the nucleus to the cytoplasm.

4. The mRNA once synthesized may function a number of times before being degraded and thus enhance the synthetic activity of the cell.

5. There may be regulation of the activity of the RNA in the cytoplasm to change the rates of protein synthesis and kinds of protein produced at different stages.

6. There is evidence from isotopic labeling that the rate of degradation, as well as the rate of synthesis (and hence the half life and the quantity per unit tissue), of some enzymes varies and is characteristic of the tissue from which the cells derive.

7. The mRNA may be stored in an inactive form, accumulating for periods when bursts of activity are required; this would demand stabilizing mechanisms to prevent its degradation. Such stabilization has been demonstrated in the continuing process of red blood cell formation. The stem cells contain no hemoglobin, and in their initial multiplication through the proerythroblast stages synthesis of RNA occurs. At the next stage, the reticulocytes, no RNA is synthesized, and yet it is at this stage that protein synthesis, particularly of hemoglobin, becomes very active. Therefore the messenger RNA responsible for this must have been produced before its actual utilization and protected thereafter. There seems little doubt that the hemoglobin template RNA is synthesized during or before the basophilic erythroblast stage and therefore survives for at least two days before it is required. It is stabilized by becoming associated with ribosomes to form polysomes, in which

form it remains stable until it is utilized for protein synthesis. There is similar evidence for the stabilization of messenger RNA in the cells of the lens of the eye. The fibers that make up the lens of the eye are formed during terminal differentiation of an epithelial cell, and the templates necessary for protein synthesis are synthesized in epithelial cells and stabilized thereafter.

8. Hormones may have an important role in gene control during fetal development, for they are potential gene regulators, and those of fetal origin become differentially available during antenatal life. They can stimulate mRNA synthesis by acting as derepressors, hence stimulating transcription, and may also control translation. They are organ-specific, rather than simply gene-specific, in effect, and influence the synthesis of groups of (seemingly unrelated) enzymes. They may be a factor in the variation in isoenzyme activity that occurs between organs in fetal life.

9. Epigenetic modification of structure after polypeptide chain production commonly occurs (p. 254), which permits an increase in the number of isozymic forms of the enzyme, each with characteristic kinetic properties and probably with a distinctive role to play in cell metabolism.

Mechanisms controlling the implementation of genetic instructions thus may operate at all points from the replication of the genes, their transcription, translation, and final epigenetic modifications of their primary products. In light of the very fine regulation of steady-state level that can be brought about by varying rates of synthesis and rates of degradation, the many additional mechanisms must mean ample potential for control of the delicate processes of differentiation and development. Man certainly has a sufficient number of genes for these mechanisms themselves to be genetically controlled. He has enough DNA in his cells to code for at least 1000 times as many proteins as have been identified. The function of this excess is as yet unknown. The excess is partly due to a linear multiplicity of genes—some are represented singly and others a very large number of times per cell. Partly it must represent nonstructural regulator or controller genes demonstrated in lower organisms but not yet convincingly in man, although in view of the greater complexity of man and other mammals their existence appears highly likely. In addition to the DNA information, there is the more complex chromosome structure, with a variety of basic (histone) and acidic proteins interdigitating with the DNA to form the chromatin; their ratio varies in successive stages in cell differentiation and in different cell types, and again a relation to gene function is suggested.

### 8.1. Genes and Differentiation

Many of the characteristics of the developing embryo are encoded in the genes. The sequence of nucleotides making up the DNA and other essential information may be stored in metabolic systems and in macromolecular aggregates. However, for the organization of cells into tissues and organs there are a number of embryological mechanisms, upon which harmonious integration of diverse developmental pathways depends. One such mechanism is embryonic induction, and another is morphogenetic movement, both largely epigenetic. The former brings about specific types of differentiation, the latter relocates cells with respect to each other. Genes therefore may influence shape and order in the organism by involvement with morphogenetic movements or with the release of inductive stimuli, and there are many congenital malformations in man which may result from errors in the instructions either from major genes or from others in polygenic systems, e.g., spina bifida, cleft palate, anencephaly, and agenesis of the corpus callosum.

As an example of genetic control of morphogenetic movement closer to the structural gene–protein relationship and explicable by switch mechanisms, the embryonic cell migrations that occur in the undifferentiated matrix are preceded by hyaluronate production concentrated along the migration path (Toole and Trelstad, 1971). This appears to provide a suitable substratum for the migrating cells, and perhaps inhibits potential inductive interaction between them and their neighbors. When they reach their final location hyaluronidase is produced, removing the hyaluronate and allowing condensation and differentiation. Interference with synthesis of collagen, possibly a matrix component and serving as a site for the attachment of migrating cells, inhibits migrating heart cells (Fitzharris and Markwald, 1974).

As an example of genetic involvement in embryonic induction, it is by induction that the lens of the eye develops from the fetal surface ectoderm. The primary optic vesicles, which are outgrowths of the brain, come into contact with the ectoderm, which becomes thickened at that point. The anterior part of the optic vesicle invaginates and forms an optic cup with double walls; the thickened lens plate also invaginates to enclose a cavity called the lens vesicle which separates from the surface ectoderm. The primary lens fibers which fill the cavity of this lens vesicle arrive through elongation of the cells of its posterior wall. This, the most translucent central part of the adult lens, is known as the embryonic nucleus and is formed during the first three months. The secondary lens fibers, originating from cells in the equatorial region, grow around the embryonic nucleus and meet at the sutures of the so-called fetal nucleus, formed between the third month and the last month before birth. The process continues after birth, with the apposition of the secondary lens fibers, leading to the development of the cortex, and in the third decade of life the adult nucleus is formed in this cortex. There are various forms of cataract, which relate to particular sections of the lens; e.g., nuclear cataract is found at the center of the embryonic nucleus, anterior polar cataract is traceable to a defect of the capsular epithelium with impairment of contact between this and the lens fibers. The genetic specificity of the different cataract types shown by family studies indicates the presence of a multiplicity of "normal" genes controlling the different stages of development of the normal lens.

An important detail of genetic control of differentiation relates to cell death. Some cells at specific locations in the embryo are instructed to die at specific times as a part of the normal development of the organism. Cell death is an essential part of morphogenesis, for instance, in the separation of the fingers and toes from one another, of the lip from the gums, of the eyelids, and in the formation of ducts in many organs. Failure of cell death to occur at the proper time produces congenital abnormalities, for example, syndactyly, development of embryonic tumors, and the fact that many such abnormalities are genetic indicates the importance of corresponding "normal" genes programming for cell death in normal morphogenesis.

## 9. Chromosomes and Fetal Development

Investigation of the role of chromosomes in human fetal growth and development has been mainly devoted to chromosomal abnormality, the pathology it produces and its effects on survival, while normal chromosome variation has been relatively neglected and its effects remain unknown. But by the same argument as in the foregoing, knowledge of the pathology induced by specific chromosome defects can be used as an indicator of the location of chromosomal material affecting normal development.

Comparison of the occurrence of different types of chromosomal aberrations at birth and in spontaneous abortuses shows how chromosomal imbalance of different types affects survival and development. In abortuses of the first trimester, triploidy forms the commonest single abnormality, accounting for nearly 8% of total abortuses; it is usually incompatible with fetal development, for most triploids abort as an empty sac early in pregnancy, although a few do survive into later prenatal life. The next most frequent specific aberration is the loss of a sex chromosome, giving the XO karyotype, accounting for about 7% of early trimester abortuses, and the estimated prenatal loss of all XO conceptuses is something over 99%. The heterogeneous group of autosomal trisomies as a whole accounts for the major proportion of chromosomal abnormalities, 50% of all the anomalies that occur in the early spontaneous abortions. Many of these are found in abortuses that are very rare indeed in live births. For example, trisomies of the A, B, and C chromosome groups show a virtually 100% prenatal loss, and this is perhaps not surprising, for they are so large that the presence of an extra chromosome may seriously interfere with cell division, or they carry so much genetic information (in the three chromosomes of group A, 9%, 8%, and 7%, respectively, of the total) that any imbalance is incompatible with survival. But, surprisingly, trisomy of a chromosome in the small F group also carries a virtually 100% prenatal loss, and here the argument of chromosome size and the amount of genetic information cannot apply, for the F group chromosomes, 19 and 20, carry little more information than 21 and 22, the G group, in which 14% of trisomies survive to live birth. It appears that balance in the F group must be critical for cell division or on them must be situated particular loci that are critical in early development. A similar inference is to be drawn with respect to chromosome 16, in which trisomy is again lethal. From the mechanisms of cell division error by which autosomal trisomy is brought about, one would expect autosomal monosomy to be at least as common, but surprisingly it occurs only in approximately 0.25% of early abortuses. This means that too little chromosome material, as well as too much, is incompatible with survival, again either through its interference with cell division or the lack of genetic material, and this applies over the whole chromosomal range, except for the XO monosomy. Chromosome balance is obviously important for early development, and it also seems likely that loci critical in embryonic development are distributed throughout the autosomal array.

In view of the differential antenatal loss, fetuses with some chromosomal abnormality who survive to term may represent a somewhat biased sample, but their examination is of interest. For example, Polani (1964) compared the weights at term of infants with trisomy D, E, and G, and showed that growth retardation was most severe in E trisomy and least in G. In infants with partial translocations, and hence some excess material, of the D group, growth retardation is less than in those with the full trisomy (i.e., with a greater excess), and the same holds for the E group. From this type of comparison, the genetic control of intrauterine growth is seen to involve loci widely distributed through the autosomes. But this does not apply to abnormalities involving excess material of the X and Y chromosomes, where there are instances of 3, 4, and more X chromosomes, and 2 Y chromosomes, without any alteration in intrauterine growth. Only the XO aberration is consistenly associated with reduced growth rate. Its involvement is confirmed by examination of a series of fetuses of different ages; crown–rump length in XO fetuses is severely retarded compared with chromosomally normal fetuses of the same gestational age (Wright, 1976).

The presence of a chromosomal abnormality not only severely retards the overall growth of the fetus but also interferes with differentiation of the developing embryo. Of spontaneously aborted fetuses with those autosomal trisomies not found at birth (i.e., trisomy A, B, C, E16, F), no specific anatomical pattern of abnormalities emerges, except that development is markedly arrested at a very early stage or is too abnormal for subsequent viability. The spontaneously aborted fetuses with the trisomies that are known at birth show the characteristic clinical features. The same does not hold for the XO abortuses, in which the abnormalities range from empty intact or ruptured sacs, through a variety of apparently unrelated abnormalites, to outwardly normal fetuses. Usually XO live births appear virtually normal outwardly, and it is the internal sex organs that are primarily affected. In early XO abortuses, the ovaries appear grossly normal and contain germ cells, and only after 12 weeks gestation do the germ cells begin to degenerate (Singh and Carr, 1966). This, of course, is related to the normal development of sex; fetuses with a Y chromosome develop into males, those without into females, with a few rare exceptions. Although chromosomal sex is established at conception, it is only at about the 8th week of gestation that the gonadal sex begins to be distinguishable microscopically. By this time the Y chromosome, or rather the masculinizing locus on its short arm, has begun to exert a clear effect on the differentiating cells and their metabolism, for Leydig cells, which secrete testosterone, can be discerned between the medullary cords, so that the male gonad is recognizable as a testis while the female gonad still shows no differentiation. Serum and gonad concentrations of testosterone are already clearly raised in male fetuses by 11–17 weeks (Reyes *et al.,* 1973, 1974). Besides the hormonal stimulation of development of the male organs derived from the Wolffian duct system, the testis secretes another hormone that suppresses development of female organs from the Mullerian duct system. The fetal testis is a highly active endocrine organ, particularly as regards steroidogenesis, in contrast to the fetal ovary (Challis *et al.,* 1976), and there are clear sex differences in other related hormones. Male differentiation is complete by the fourth month.

## 9.1. Location of Genetic Information

The characteristics of trisomic patients, due to the chromosomal imbalance brought about by the extra chromosome material, suggests that genetic material on G group chromosomes is involved in the control of the following:

Muscle tone
Cartilage and ligament elasticity and joint flexibility
Intelligence and mental development
Growth in height and trunk–limb proportions
Skull proportions
Cranial thickness
Sinus development
Shape and rate of growth of the ear
Development of ear lobe
Development of iris peripherally and of the lens
Development of the inner epicanthic folds, the slant of the palpebral fissures
Dental spacing and size

Growth of phalanges and metacarpals
Palmar skin creases
Placement of axial triradius
Development of dermal patterns on the digits and on the soles
Development of the heart
Quality and amount of the hair
Genital development
Growth of the pelvis

Genes on E group chromosomes (17 and 18) are involved in the control of

Placental size
Development of skeletal muscle and muscle tone
Intelligence
Response to sound
Development of occiput and lateral growth of anterior skull
Position and shape of auricles
Development of palate and jaw
Dermatoglyphic patterns on digits, and nail development
Number of ossification centers in sternum
Development of pelvis
Heart development
Genital development

Genes on D group chromosomes are involved in the control of

Development of the forebrain and the olfactory and optic nerves
Intelligence and mental development
Development of organs of Corti (hearing)
Size of cranium, slope of forehead, size of sagittal suture and fontenelles
Size of eyes, development of iris and retina
Closure of lip and palate
Ear shape and position
Amount of skin and presence of capillary hemangiomata
Position of axial triradii, palmar skin creases, shape of fingernails, number of
    digits
Development of ribs and pelvis
Development of heart
Persistence of fetal-type hemoglobin
Genital development

A key localization here appears to be control of the early development (3 weeks) of the prechordal mesoderm. This is necessary for morphogenesis of the midface and exerts an inductive role on the forepart of the brain; hence a single defect here will produce the various defects of midface, eye, and forebrain as seen in this trisomy (D).

These generalizations are based on the combination of defects that manifest in the majority of patients with trisomies designated 21, 18, and 13, syndromes which have been known for a long time and for which details of many patients are available. In the earlier days of chromosome studies, the unequivocal identification of the chromosome involved was not always possible, so that this summary of locations is in terms of the chromosome group rather than the specific chromosome by number. Also, even the closest morphological examination could not identify where an apparent deletion had occurred, so early hopes of drawing chromosome

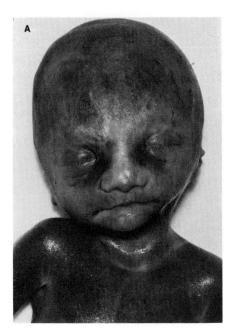
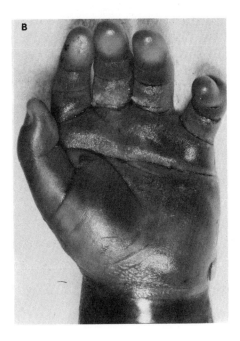

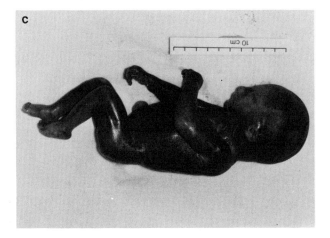

Fig. 8. Fetus 1(a,b,c): partial trisomy 9 with monosomy 22.

Figs. 8–10. Distinctive patterns of malformation in syndromes due to chromosomal anomalies are established early in fetal life. They suggest that genetic information controlling development of specific features is localized on the chromosomes involved.

maps from the phenotype data in trisomy and deletion syndromes were overoptimistic. Moreover, many of the clinical features are generalized, and must be controlled by a number of genes and not a single locus; for example, developmental retardation is an accompaniment to many different chromosomal aberrations, and in fact is a most usual result of chromosome imbalance. Loss or gain of a chromosome segment means loss or gain of many loci, at some of which may be sited structural genes, at others genes regulating the activity of other genes on other chromosomes. Today, however, with the development of modern techniques of banding and other methods of examining chromosome fine structure, identification of particular chro-

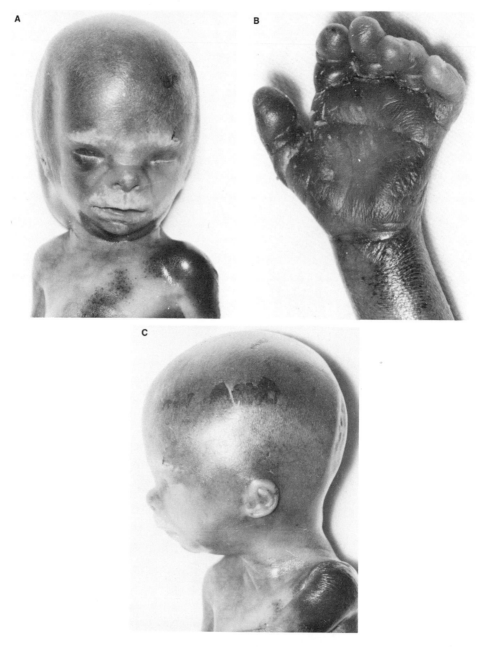

Fig. 9. Fetus 2(a,b,c): trisomy 21 (mongol).

mosomes, and in many cases of the sites of location of breakage, inversion, or translocation, is a routine procedure in a good laboratory. It can be confidently expected that the next few years will produce surveys of adequate series of patients with a specific chromosome defect, and comparison of their phenotypes will allow the location of chromosomal material affecting normal development more comprehensively than is yet possible.

In the last few years, however, the assignation of a number of genes to a chromosome has been made by a combination of classical methods of pedigree and linkage analysis and the recent methods of somatic cell hybridization. Table III

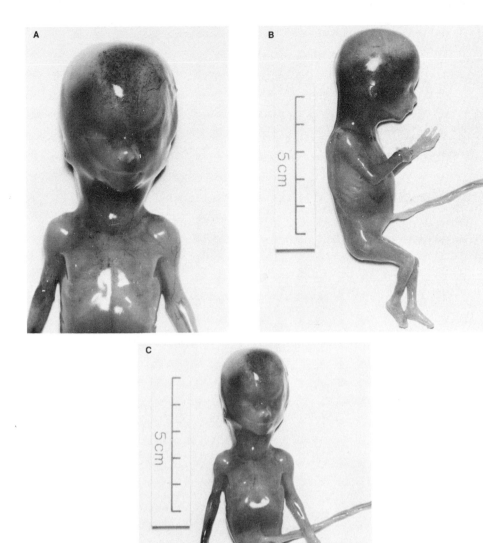

Fig. 10. Fetus 3(a,b,c): normal fetus of gestational age 12 weeks.

shows the chromosome location of some 40 enzymes, blood groups, serum proteins, and miscellaneous entities. Some of these have been mentioned in this
discussion as being involved in fetal growth, e.g., the lactic dehydrogenase loci on
chromosomes 11 and 12, and the glucose-6-phosphate dehydrogenase locus on the
X chromosome. As in the earlier discussion, the localization of a gene controlling a
mendelian disorder means that at that locus occurs the allele responsible for the
normal condition, e.g., the location of hemophilia and Christmas disease on the X
chromosome means that the genes involved in the normal production of factors VIII
and IX, respectively, are located there.

*Table III.   Chromosome Location of Genes*

| Chromosome | Character |
|---|---|
| 1 | Auriculo-osteodysplasia |
| | Duffy blood groups |
| | Elliptocytosis$_1$ |
| | Fumarate hydratase |
| | UDP-glucose pyrophosphorylase |
| | Guanylate kinase |
| | Pancreatic amylase |
| | Peptidase C |
| | Phosphoglucomutase$_1$ |
| | 6-Phosphogluconate dehydrogenase |
| | Phosphopyruvate hydratase |
| | Rhesus blood groups |
| | 5-S ribonucleic acid region |
| | Salivary amylase |
| | Uncoiler 1 |
| | Zonular pulverulent cataract |
| 2 | Galactose-1-phosphate uridyltransferase |
| | Interferon$_1$ |
| | Isocitrate dehydrogenase$_1$ |
| | Malate dehydrogenase$_1$ (NAD-dependent) |
| | Red cell acid phosphatase$_1$ |
| 5 | Interferon$_2$ |
| 6 | C$_3$ proactivator |
| | C$_2$ component of complement |
| | C$_4$ component of complement |
| | Chido blood group |
| | Histocompatibility locus A = HLA locus LA |
| | B = HLA locus FOUR |
| | C = Locus AJ |
| | D = mixed lymphocyte response |
| | LD-2 = minor mixed lymphocyte response locus |
| | Malic enzyme$_1$ [soluble malate dehydrogenase (NADP-dependent)] |
| | Phosphoglucomutase$_3$ |
| | Tetrameric form of indophenol oxidase-B (mitochondrial superoxide dismutase) |
| 7 | Mannose phosphate isomerase |
| | White cell pyruvate kinase$_3$ |
| 10 | Hexokinase |
| | Soluble glutamic oxaloacetic transaminase$_1$ |
| 11 | Esterase A$_4$ |
| | Lactic dehydrogenase A |
| | Lysosomal acid phosphatase |
| | Puck's cell surface antigen$_1$ |
| 12 | Citrate synthetase |
| | Glycine auxotroph (seryl hydroxmethylase) |
| | Lactic dehydrogenase B |
| | Mitochondrial malate dehydrogenase |
| | Peptidase B |
| | Triose phosphate isomerase |
| 13 | Ribosomal RNA region |

(*continued*)

Table III (continued)

| Chromosome | Character |
| --- | --- |
| 14 | Nucleoside phosphorylase |
| | Ribosomal RNA region |
| 15 | Ribosomal RNA region |
| 16 | Adenosine phosphoribosyltransferase |
| | α-Haptoglobin |
| 17 | Galactokinase |
| | Thymidine kinase |
| 18 | Peptidase A |
| 19 | Glucose phosphate isomerase |
| | Polio receptor site |
| 20 | Adenosine deaminase |
| 21 | Antiviral protein |
| | Dimeric indophenol oxidase |
| | Ribosomal RNA region |
| 22 | Ribosomal RNA region |
| X | Becker's muscular dystrophy |
| | Clotting factor 8 |
| | Clotting factor 9 |
| | Deuteranopia |
| | Duchenne muscular dystrophy |
| | α-Galactosidase |
| | Glucose-6-phosphate dehydrogenase |
| | Hypoxanthineguanine phosphoribosyltransferase |
| | Ichthyosis |
| | Ocular albinism |
| | Phosphoglycerate kinase |
| | Phosphorylase kinase |
| | Protanopia |
| | Retinoschisis |
| | Xg blood groups |
| Y (short arm) | Masculinizing locus |

## 10. Conclusion

The genetic control of human fetal growth and development cannot be studied directly by the classical methods of genetic analysis. However, from the study of human polymorphisms, from knowledge of fundamental biological mechanisms extrapolated from other organisms, and particularly from the extensive knowledge of clinical genetics in which the genes known to cause maldevelopment provide information on the corresponding alleles responsible for normal development, much indirect information has accumulated.

Embryonic and fetal development is brought about through a sequence of changing populations of cells organized in increasingly complex patterns. The sequence of hemoglobins in human development provides a model of how genetic control may be exercised. This demonstrates that embryonic and fetal development occurs as the result of switch mechanisms switching on and off the activity of particular genes at specific times in development and producing specific substances. Normal individual variations in fetal development, as well as intrinsic malformations, may thus arise by (1) variation in efficiency or failure of the switching mechanism, (2) variation in its timing, and (3) variations or errors in the substances synthesized.

Therefore the various multiple forms of enzymes and other proteins are of great relevance for fetal development, irrespective of whether they are under direct genetic control or modified by epigenetic processes. Essential for isoenzyme involvement in fetal differentiation is tissue specificity, and this is demonstrable both at the tissue level and at the subcellular level. Similarly essential is temporal variation since for developmental relevance an isoenzyme needs to be available at the appropriate stage in development, and this too is demonstrable with reference to switch mechanisms. A detailed example, the role of nonspecific esterases in fetal brain and kidney development, shows how some isoenzymes are associated with tissue and cellular maturation rather than directly involved in differentiation, while others appear more specifically involved, e.g., the NP-specific bands with the myelination process, and the F band in the division of immature neurons. The developmental progression in fetal brain and kidney is different, and this difference is related to the differing functional requirements of the developing tissues. The example, however, besides illustrating the complexity of the interrelationships of genetic isoenzymes and developmental processes, also demonstrates the close relationship between biochemical and morphological ontogeny and shows how the vast heterogeneity that exists in the isoenzyme system provides an ideal mechanism by which the organism can adapt to the changing molecular environment during ontogeny.

The emergence in fetal life of polymorphisms well established at the population level provides useful information on the quantitative development of some substances and on the stages of gestation at which new proteins are released into fetal circulation. Although their role in fetal development is still unexplained, from the earlier argument it is likely to be important.

Information on the mechanisms by which selective gene expression is brought about, so that the different gene products are available to be involved in human fetal growth, is derived mainly from other organisms. It appears that mechanisms controlling the implementation of genetic instructions may operate at all points from the replication of the genes, their transcription, translation, and final epigenetic modifications of their primary products. Thus, there is ample potential for the genetic control of the delicate processes of differentiation and development, including genetic control of morphogenetic movement and of embryonic induction.

Knowledge of the role of chromosomal variation in fetal development relates primarily to pathology, and little is known of the developmental effects of chromosome polymorphisms. Some chromosomal aberrations and imbalances are incompatible with embryonic development: some lead to fetal death, others produce characteristic malformations and retardation of growth. Genetic control of intrauterine growth is seen to involve loci widely distributed through the autosomes. From the details of pathology associated with specific chromosomal aberrations, the location of some of the genetic material involved in the control of normal development is suggested, while more specific information on gene location comes from recent developments in somatic cell hybridization.

Altogether, the present state of knowledge of the genetics of human fetal growth and development is exciting, with a number of fundamental principles established, a host of half glimpses of delicate processes and their control to stimulate further enquiry, and the challenge of the necessity to use indirect methods of analysis and argument. The outlook is bright for considerable increase in that knowledge as the result of the development of new noninterventive methods of fetal examination and new laboratory procedures.

ACKNOWLEDGMENTS

I am greatly indebted to my colleagues Dr. S. S. Papiha for many hours of companionable laboratory work and stimulating discussion, Messrs. J. Burn, C. K. Creen, and T. Walls for their technical help, and Miss E. V. Davison for the fetal photographs.

## 11. References

Adinolfi, M. A., 1971, Genetic polymorphism and *in vitro* cultures in studies of the ontogenesis of human fetal proteins, in: *The Biochemistry of Development* (P. Benson, ed.), pp. 224–247, Heineman, London.

Barron, K. D., 1961, Starch gel electrophoresis of brain esterases, *J. Histochem. Cytochem.* **9:**139.

Barron, K. D., and Bernsohn, J., 1963, Separation and properties of human brain esterases, *J. Histochem. Cytochem.* **11:**139.

Barron, K. D., and Bernsohn, J., 1965, Brain esterases and phosphatases in multiple sclerosis, *Ann. N.Y. Acad. Sci.* **122:**369.

Barron, K. D., and Bernsohn, J., 1968, Esterases of developing human brain, *J. Neurochem.* **15**(4):276.

Barron, K. D., Bernsohn, J., and Doolin, P. F., 1966, Subcellular localization of multiple forms of brain esterases, *J. Histochem. Cytochem.* **14:**817.

Barron, K. D., Bernsohn, J., and Hess, A. R., 1962, Electrophoretic analysis of human brain esterases, *Acta Neurol. Scand. Suppl.* **1:**61.

Barron, K. D., Bernsohn, J., and Hess, A. R., 1963, Abnormalities in brain esterases in multiple sclerosis, *Proc. Soc. Exp. Biol. N.Y.* **113:**521.

Barron, K. D., Bernsohn, J., and Mitzen, E., 1972, Nonspecific esterases of human peripheral nerve and centrum ovale, *J. Neuropathol. Exp. Neurol.* **31:**562.

Bernsohn, J., 1964, Multiple molecular forms of brain hydrolase, *Int. Rev. Neurobiol.* **7:**297.

Bernsohn, J., Barron, K. D., and Norgello, H., 1964, 'Bound' and soluble esterases in human brain, *J. Biochem.* **91:**240.

Bigley, R. H., and Koler, R. D., 1968, Liver pyruvate kinase (P.K.) isoenzymes in a P.K.-deficient patient, *Ann. Hum. Genet.* **31:**383.

Bigley, R. H., and Stenzel, L. P., 1968, Tissue distribution of human pyruvate kinase isoenzymes, *Enzymol. Biol. Clin.* **9:**10.

Bonavita, V., and Guarneri, R., 1962, Lactic dehydrogenase isoenzymes in the nervous tissue: (1) The reaction of isozymes with DPN analogues and their inhibition by sodium meta-bisulfite, *Biochem. Biophys. Acta* **59:**634.

Britten, R. J., and Davidson, E. H., 1969, Gene regulation for higher cells: A theory, *Science* **165:**349.

Challis, J., Robinson, J., Rurak, D. W., and Thorburn, G. D., 1976, The development of endocrine function in the human fetus, in: *The Biology of Human Fetal Growth* (D. F. Roberts and A. M. Thomson, eds.), pp. 149–194, Taylor & Francis, London.

Choudhury, S. R., 1972, The nature of non-specific esterases: A subunit concept, *J. Histochem. Cytochem.* **20:**507.

Crawhall, J. C., and Watts, R. W. E., 1962, The metabolism of [1-¹⁴C]-glyoxylate by the liver mitochondria of patients with primary hyperoxaluria and non-hyperoxaluric subjects (1962), *Clin. Sci.* **23:**163.

Danovitch, S. H., Baer, P. N., and Laster, L., 1968, Intestinal alkaline phosphatase activity in familial hypophosphatasia, *N. Engl. J. Med.* **278:**1253.

Davison, A. N., and Dobbing, J., 1968, *Applied Neurochemistry,* Chapter 6, The developing brain (rat), p. 612, Contemporary Neurology Series 4/5, Blackwell, Oxford.

Dawson, D. M., Spong, F. W., and Harrington, J. F., 1968, McArdle's disease, lack of muscle phosphorylase, *Ann. Intern. Med.* **69:**229.

Desreux, V., and Heriott, R. M., 1939, Existence of several active components in crude pepsin preparation, *Nature (London)* **144:**287.

Dobbing, J., and Sands, J., 1970, The timing of neuroblast multiplication in developing human brain, *Nature (London)* **226**:639.

Dodge, P. R., and Adams, R. D., 1971, Developmental abnormalities, in: *Harrison's Principles of Internal Medicine* (M. M. Wintrobe, ed.), p. 365,

Fitzharris, T. P., and Markwald, R. R., 1974, Structural components of cardiac jelly, *J. Cell Biol.* **80**:235.

Fitzpatrick, T. B., and Quevedo, W. C., 1966, in: *The Metabolic Basis of Inherited Disease*, 1st ed. (J. B. Stanbury, J. B. Wyngaarden and D. S. Fredrickson, eds.), 324, McGraw-Hill, New York.

Froesch, E. R., Wolf, H. P., Baitsch, H., Prader, A., and Labhart, A., 1963, Hereditary fructose intolerance: An inborn defect of hepatic fructose-1-phosphate splitting aldolase, *Am. J. Med.* **34**:151.

Gitlin, D., and Biasucci, A., 1969, Development of $\gamma$G, $\gamma$A, $\gamma$M, $\beta$1C/$\beta$1A, C'1 esterase inhibitor, ceruloplasmin, transferrin, hemopexin, haptoglobin, fibrinogen, plasminogen, $\alpha_1$-antitrypsin, orosomucoid, $\beta$-lipoprotein, $\alpha_2$-macroglobulin and prealbumin in the human conceptus, *J. Clin. Invest.* **48**:1433.

Hall, J. G., and Motulsky, A. G., 1968, Production of fetal haemoglobin in marrow cultures of human adults, *Nature (London)* **217**:569.

Harris, H., 1970, *The Principles of Human Biochemical Genetics*, North-Holland, Amsterdam.

Hers, H. G., 1959, Etudes enzymatiques sur fragments hépatiques: Application à la classification des glycogenoses, *Rev. Int. Hepatol.* **9**:35.

Hers, H. G., and Hoof, F. van, 1968, *Carbohydrate Metabolism and its Disorders* (F. Dickens, P. J. Randle, and W. T. Whelan, eds.), Vol. 2, p. 151, Academic Press, New York.

Hers, H. G., and Joassin, G., 1961, Anomalie de l' aldolase hépatique dans l'intolerance au fructose, *Enzymol. Biol. Clin.* **1**:4.

Hirschfeld, J., and Lunell, N. O., 1962, Serum protein synthesis in fetus: Haptoglobins and group-specific components, *Nature (London)* **196**:1220.

Hug, G., Schubert, W. K., and Chuck, G., 1969, Deficient activity of phosphorylase kinase and accumulation of glycogen in the liver, *J. Clin. Invest.* **48**:704.

Hug, G., Schubert, W. K., and Chuck, G., 1970, Liver glycogenosis and phosphorylase kinase deficiency, *Am. J. Hum. Genet.* **22**:484.

Huijing, F., and Fernandes, J., 1969, X-Chromosomal inheritance of liver glycogenosis with phosphorylase kinase deficiency, *Am. J. Hum. Genet.* **21**:275.

I.U.B. Commission on Biochemical Nomenclature, 1972, The nomenclature of multiple forms of enzymes, *Biochem. J.* **126**:769.

Kaplan, N. O., 1961, in: *Mechanism of Action of Steroid Hormones* (C. A. Villee and A. A. Engel, eds.), Pergamon, Oxford.

Keitt, A. S., 1969, Haemolytic anaemia with impaired hexokinase activity, *J. Clin. Invest.* **48**:1997.

Koch, J., and Skokstad, E. L. R., 1967, Deficiency of 2 oxoglutarate: Glyoxylate carboligase activity in primary hyperoxaluria, *Proc. Natl. Acad. Sci. U.S.A.* **57**:1123.

Kokko A., 1965, Histochemical and cytophotometric observations on esterases in the spinal ganglion of the rat, *Acta Physiol. Scand.* **65**(Suppl):261.

Lehrer, R. I., and Cline, M. J., 1969, Leukocyte myeloperoxidase deficiency and disseminated candidiasis: The role of myeloperoxidase in resistance to *Candida* infection, *J. Clin. Invest.* **48**:1478.

Markert, C. L., and Møller, F., 1959, Multiple forms of enzymes: Tissue ontogenetic and species specific patterns, *Proc. Natl. Acad. Sci. U.S.A.* **45**:753.

Masters, C. J., and Holmes, R. S., 1975, *Haemoglobin, Isoenzymes and Tissue Differentiation*, North-Holland/American Elsevier, Amsterdam.

McCance, R. A., Fairweather, D. V., Barrett, A. M., and Morrison, A. B., 1956, Genetic, clinical biochemical, and pathological features of hypophosphatasia, *Q. J. Med.* **25**:523.

Messina, A. M., 1972, G6PD: Developing human liver, *Proc. Soc. Exp. Biol. N.Y.* **139**:778.

Nachlas, N. M., and Seligman, A. W., 1949, The histochemical demonstration of esterase, *J. Natl. Cancer Inst.* **9**:415.

Necheles, T. F., Maldonado, N., Barquet-Chediak, A., and Allen, D. M., 1969, Homozygous erythrocyte glutathione-peroxidase deficiency: Clinical and biochemical studies, *Blood* **33**:164.

Necheles, T. F., Rai, U. S., and Cameron, D., 1970, Congenital nonspherocytic haemolytic anaemia associated with an unusual hexokinase abnormality, *J. Lab. Clin. Med.* **76**:593.

Omenn, G. S., and Hermodson, M. A., 1975, Human phosphoglycerate mutase: Isozyme marker for muscle differentiation and for neoplasia, in: *Isozymes III—Developmental Biology* (C. L. Markert, ed.), p. 1005, Academic Press, New York.

Papiha, S. S., Walls, T. J., and Burn, J., 1976, Nonspecific esterases in human fetal tissue differentiation: An essay in developmental genetics, in: *The Biology of Human Fetal Growth* (D. F. Roberts and A. M. Thomson, eds.), p. 253, Taylor & Francis, London.

Parker, W. C., and Bearn, A. G., 1962, Studies on the transferrins of adult serum, cord serum and cerebrospinal fluid: The effect of neuraminidase, *J. Exp. Med.* **115**:85.

Parker, W. C., Hagstrom, J. W., and Bearn, A. G., 1963, Additional studies on the transferrins of cord serum and cerebrospinal fluid: Variation in carbohydrate and prosthetic groups, *J. Exp. Med.* **118**:975.

Paul, J., 1972, General theory of chromosome structure and gene activation in eukaryotes, *Nature (London)* **238**:444.

Penhoet, E., and Rajkumar, T., 1966, Multiple forms of fructose diphosphate aldolase in mammalian tissue, *Proc. Natl. Acad. Sci. U.S.A.* **56**:1275.

Perkins, R. P., 1971, Hydrops fetalis and stillbirth in a male G6PD deficient foetus possibly due to maternal ingestion of sulfisoxazole, *Am. J. Obstet. Gynecol.* **111**:379.

Polani, P. E., 1964, Chromosome anomalies, *Annu. Rev. Med.* **15**:93.

Potter, E. L., 1972, *Normal and Abnormal Development of the Kidney,* Yearbook Medical Publishers, Chicago.

Propp, R. P., and Alper, C. A., 1968, C′3 synthesis in the human fetus and lack of transplacental passage, *Science* **162**:672.

Rathburn, J. C., 1948, Hypophosphatasia—a new developmental anomaly, *Am. J. Dis. Child.* **75**:822.

Reyes, F. I., Winter, J. S. D., and Faiman, C., 1973, Studies on human sexual development, I Fetal gonadal and adrenal sex steroids, *J. Clin. Endocrinol. Metab.* **37**:74.

Reyes, F. I., Boroditsky, R. S., Winter, J. S. D., and Faiman, C., 1974, Studies on human sexual development, II Fetal and maternal serum gonadotropin and sex steroid concentrations, *J. Clin. Endocrinol. Metab.* **38**:612.

Robbins, P. W., 1960, Immunological study of human muscle lacking phosphorylase, *Fed. Proc. Fed. Am. Soc. Exp. Biol.* **19**:193.

Rutter, W. J., Rajkumar, T., Penhoet, E., Kochman, M., and Valentine, R., 1968, Aldolase variants: Structure and physiological significance, *Ann. N.Y. Acad. Sci.* **151**:102.

Salmon, S. E., Cline, M. J., Schultz, J., and Lehrer, R. I., 1970, Myeloperoxidase deficiency, *N. Engl. J. Med.* **282**:250.

Schmid, R., Robbins, P. W., and Taut, R. R., 1959, Glycogen synthesis in muscle lacking phosphorylase, *Proc. Natl. Acad. Sci. U.S.A.* **45**:1236.

Shows, T. B., 1972, Genetics of human–mouse somatic cell hybrids. Linkage of human genes for LDH-A and esterase $A_4$, *Proc. Natl. Acad. Sci. U.S.A.* **69**:348.

Singh, R. P., and Carr, D. H., 1966, The anatomy and histology of XO embryos and fetuses, *Anat. Record* **155**:369–383.

Toivanen, P., Rossi, T., and Hirvonen, T., 1969, The concentration of Gc globulin and transferrin in human fetal and infant sera, *Scand. J. Haematol.* **6**:113.

Toole, B. P., and Trelstad, R. L., 1971, Hyaluronate production and removal during corneal development in the chick, *Dev. Biol.* **26**:28.

Valentine, W. N., Oski, F. A., Paglia, D. E., Baughan, M. A., Schneider, A. S., and Naiman, J. L., 1967, Hereditary haemolytic anaemia with hexokinase deficiency. Role of hexokinase in erythrocyte aging, *N. Engl. J. Med.* **276**:1.

Vernier, R. L., and Smith, F. G., Jr., 1968, Fetal and neonate kidney, in: *Biology of Gestation, The Fetus and Neonate* (N. S. Assali, ed.), Vol. II, Chapter V, Academic Press, New York.

Walton, J. N., 1971, *Essentials of Neurology,* 3rd ed., Pitman, London.

Waterbury, L., and Frenkel, E. P., 1969, Phosphofructokinase deficiency in congenital non-spherocytic anaemia, *Clin. Res.* **17**:347.

Waterbury, L., and Frenkel, E. P., 1972, Hereditary nonspherocytic hemolysis with erythrocyte phosphofructokinase deficiency, *Blood* **39**:415.

Wender, M., and Kozik, M., 1971, Histoenzymology of cerebral white matter in the developing rat brain, *J. Hirnforsch.* **13**:223.

Williams, H. E., and Field, J. B., 1961, Low leukocyte phosphorylase in hepatic phosphorylase deficient glycogen storage disease, *J. Clin. Invest.* **40**:1841.

Wright, E. V., 1976, Chromosomes and human fetal development, in: *The Biology of Human Fetal Growth* (D. F. Roberts and A. M. Thomson, eds.), pp. 237–252, Taylor & Francis, London.

Yakovlev, P. I., and LeCours, A. R., 1967, The myelogenetic cycle of regional maturation of the brain, in: *Regional Development of the Brain in Early Life* (A. Minkowski, ed.), Blackwell Scientific Publishers, Oxford.

Yamamura, H. F., Takano, T. U., and Okada, T., 1965, Antigenic molecules with esterase activity contained in human embryo, *Embryologia* **8**:319.

Zuckerkandl, E., 1964, Controller-gene diseases, *J. Mol. Biol.* **8**:128.

# 10

# *The Genetics of Birth Weight*

## *ELIZABETH B. ROBSON*

### *1. Effects of Single Gene Loci on Birth Weight*

Genes causing metabolic disturbances leading to frank disease are the most obvious candidates for identifiable single loci involved in birth-weight determination. The rarity of most of the disorders means that data are few and statistically inadequate in all but a few conditions. In cystic fibrosis, an autosomal recessive condition, Boyer (1955) found a reduction of about 260 g in birth weight in a series of 173 cases, and Hsia (1959) a rather smaller reduction of just over 100 g. Hsia's cases were selected differently, being infants surviving 28 days or more, and so the more severely affected infants were excluded. Furthermore, he used English hospital cases as controls for his Chicago cases. Since mean birth weights had increased in England by about 60 g for surviving infants (Jayant, 1966) in the period between the collection of the control data and the 1950s, it is more likely that the reduction in birth weight in cystic fibrosis is better estimated by Boyer's value of over 200 g. In galactosemia, another autosomal recessive condition, due to a deficiency of the enzyme galactose-1-phosphate uridyltransferase, a study of 27 families by Hsia and Walker (1961) showed that affected babies weighed 300 g less than their unaffected siblings. In phenylketonuria, a condition due to deficiency of phenylalanine hydroxylase activity, Saugstad (1972) has shown a reduction in birth weight of about 100 g.

It is perhaps of little general interest that metabolically abnormal babies have abnormal birth weights, but what *is* remarkable is the observation of Saugstad that in phenylketonuria the unaffected siblings have birth weights that are higher than average. She has shown in a large Norwegian family material that the siblings of affected cases are over 300 g heavier than controls, despite being of slightly lower mean gestational age. Hsia (1959) made the same type of observation in cystic fibrosis, the unaffected siblings being about 140 g heavier than the controls. The effect was not statistically significant, however, and he attributed it in part to the inadequacy of the control population, suggesting that Chicago babies are normally heavier than those in London. However, Boyer's data show the same effect using

---

***ELIZABETH B. ROBSON*** • Galton Laboratory, University College, London.

local controls, the unaffected siblings being about 150 g overweight. A similar interpretation could be placed upon Hsia and Walker's (1961) findings in galactosemia where the affected siblings had a mean weight of 3200 g and the unaffected 3500 g, although control values are not given. Whether the increased birth weight is the effect of maternal or fetal genotype is not clear, for all the mothers of affected cases are necessarily heterozygous at the locus involved, and the unaffected siblings are heterozygous, or normal, in the ratio of 2 to 1. For none of the three conditions discussed are there birth weight data which distinguish the genotype of the unaffected siblings, and so it is impossible to tell whether the heterozygous effect is fetal or maternal. Whatever the cause, however, the increase in birth weight brings these infants near to the survival optimum (Karn and Penrose, 1951) and so improves the chances of the abnormal allele remaining in the population despite the selection against the affected homozygotes.

Another autosomal recessive disease with an even more dramatic effect on birth weight is Bloom's syndrome. This disease is characterized by telangiectatic erythema, sun sensitivity, and stunted growth. Affected children have a mean birth weight of only 1988 g compared with a control of 3450 g (German, 1969), but have no signs of developmental prematurity. The biochemical defect in this disease is unknown, but at the cellular level a high incidence of chromosomal rearrangements of a characteristic type can be demonstrated in tissue cultures of blood and skin cells. Although such abnormal cells can survive *in vitro,* they would be expected to suffer a high mortality *in vivo* and such a high level of cell wastage could account for the low weight achieved at term.

Apart from genes causing disease, the most widely studied loci from a biochemical point of view are those determining structural variants of proteins, many of them enzymes. About a third of such loci are polymorphic; that is, they show commonly occurring variation. Theoretical considerations in population genetics have raised questions about the maintenance of this level of variation, and one theory requires heterozygotes to be at a selective advantage compared with either type of homozygote. This advantage may depend on one or many factors acting from conception to the end of the reproductive period, and in most cases there are no clues about where to look first. Since birth weight is highly correlated with survival during the neonatal period, it is certainly a reasonable character to consider in relation to heterozygous advantage.

Enzymes specific to fetal life might be thought the most likely to affect birth weight, and of these placental alkaline phosphatase fulfills the criterion of being polymorphic. Analyses of quite extensive surveys both in Europe and India have failed to show any significant effect of the various fetal genotypes on birth weight or placental weight (Beckman *et al.,* 1969; Beck and Ananthakrishnan, 1974; Beckman, 1974; Chakraborty *et al.,* 1975). No other studies of birth weight in relation to enzyme polymorphisms have been reported, but there are more data available on ·the blood group polymorphisms. In a critical paper on a sample of 6381 mothers and babies in California, Reed (1967) has examined the relationship between birth weight and seven different blood group systems, considering not only the fetal genotype, but also the interactions with parental genotypes, compatibility or incompatibility. Reed concluded that the number of apparently significant effects detected by multiple regression analysis was no greater than the expected number of type I errors (errors where the null hypothesis is rejected when it is true). Nevertheless, it is still possible that real effects exist, and Reed discusses what magnitude they could have without being recognizable. In a sample of 5000, the size of the Caucasian population in his investigation, effects of 4% for infrequent and 2% for frequent phenotypes might not be detected. It is probably safe to say, therefore, that large effects of blood group phenotypes, of the order of 10% of the mean or more, do not occur. The amount of technical time and cost involved in such large investigations, which even so can detect only relatively crude effects, means that we are unlikely to get the answer to the question of whether many polymorphic loci affect birth weight in a biologically meaningful way.

Although the ABO blood groups of mother or fetus have no significant effect on birth weight, there are contradictory reports on their effect on placental weight, a correlated variable. Jones (1968) found that women of group O had lighter placentae than other women in a sample of 3688 Englishwomen. Group O women are more often incompatible with their fetuses than other women, and Jones suggested that the reduction in placental weight was due to immunological factors. On the other hand, Toivanen and Hirvonen (1970) found an increase in placental weight, but only in males in first pregnancies, associated with incompatibility. No effect at all was found by Seppälä and Tolonen (1970) and Hohler et al. (1972). These investigations were stimulated by the development of hypotheses in the mid-1960s, mainly stemming from experiments on mice, that immunological factors played a large part in stimulating fetal growth (Billington, 1964). It has not proved possible to confirm some of the earlier results (for review see McLaren, 1975), and McLaren believes that the observations which have been confirmed can be equally well interpreted in terms of heterosis (hybrid vigor). The general thoughts on immunological factors were timely since they seemed to provide an acceptable explanation for the high level of genetic diversity being currently demonstrated for proteins in general, and for histocompatibility systems in particular. There are no data on birth weight and HLA in humans, but the contradictory nature of the observations on ABO do not suggest that supporting evidence for the hypothesis will be found in man.

## 2. Chromosomal Abnormalities and Birth Weight

Marked reductions in birth weight have been observed in infants with abnormal numbers of chromosomes. Babies with Down's syndrome, trisomy 21, have birth weights below normal when compared with controls of similar gestational age (Polani, 1974). Infants having mosaicism with a mixture of normal and trisomic cells are underweight, but are significantly heavier than nonmosaic infants with Down's syndrome. Data on Patau's and Edwards' syndromes, trisomies 13 and 18, respectively, are less abundant, but the findings of Chen et al. (1972) and Polani (1974) clearly demonstrate an even greater diminution of birth weight. The mean birth weights in the data of Chen et al. are 3324 g for control newborns, 3007 g for trisomy 21, 2680 g for trisomy 13, and 2213 g for trisomy 18. The most dramatic reduction in birth weight is, however, found in babies with a deletion of the short arm of chromosome 4. In a sample of 23 cases, with a mean gestational age of 40 weeks, the mean birth weight was only 1929 g. The absolute values for the newborn are particularly difficult to interpret when it is remembered that they probably represent only 3% of all conceptions carrying autosomal chromosome abnormalities. The reasons why these should proceed to term while the vast majority are spontaneously aborted are unknown but may not be unconnected with size at certain critical points in development.

Abnormalities of the sex chromosomes have a similar effect on birth weight to that described for the autosomes (Chen et al., 1971; Barlow, 1973; Polani, 1974). XO babies, phenotypic females who lack a second X chromosome, are lighter than normal, as are babies with additional X chromosomes. The reduction in birth weight appears to be roughly proportional to the number of extra Xs. Additional Y chromosomes do not appear to affect birth weight significantly. Cases with sex chromosome abnormalities, like autosomal cases, show a high rate of prenatal lethality.

Although there is ample evidence that abnormal amounts of DNA cause a lowering of mean birth weight, there is as yet no explanation of the processes involved. At the level of gene action the effect could be locus-specific and due to the presence of reduced or increased amounts of particular gene products. In monosomy this would be similar to the situation obtaining in heterozygotes for alleles with very low activity. Any visible deletion, however, leading to monosomy would involve hundreds of such gene loci, although of course not all of them would necessarily affect birth weight. In trisomy there might be excessive amounts of particular gene products if all three alleles at a particular locus are expressed. There is evidence as regards structural loci that all three alleles are expressed (Marimo and Gianelli, 1975), but the situation with regard to regulatory loci is obscure. Neither is the functional significance of excess of a product understood. Since abnormalities of five different chromosomes have been shown to affect birth weight, there are, on this hypothesis, more than five specific loci involved and almost certainly very many more. Since their functions are unlikely to be completely independent, the imbalance between gene products arising from alleles present in normal as opposed to abnormal numbers would be a further complicating factor.

Viewed at the physiological level there is evidence for a lowering of the rate of cell multiplication in humans with trisomy which might account for their reduced birth weight. Mittwoch (1967) has shown that the process of DNA synthesis may be slowed down in cultured cells from patients with Down's syndrome, and Naeye (1967) has reported that there are subnormal numbers of cells in several different organs, not only in newborn trisomy 21 but also in 13 and 18 trisomy. Since it is likely that specific chromosomal regions differ in their activity at different developmental stages, a reduction in the rate of cell division, produced for whatever reason by the presence of extra chromosomal material, could produce different and specific developmental abnormalities according to the stage at which the affected chromosome is particularly active.

Humans with triploidy surviving to term are very rare, but they have been described and are gestationally immature and underweight, with a reduced number of relatively large cells with large nuclei. Again the rate of cell division has been shown to be slower than normal (Mittwoch and Delhanty, 1972). Cases with triploidy do not suffer from imbalance of the autosomes, having three copies of each, but can have imbalance of the sex chromosomes, being XXY or XYY. Such cases ought, therefore, to provide critical data on the relative importance of genic imbalance as opposed to the total amount of DNA. Since there are only 14 well-documented cases in the literature which have survived 28 or more weeks of gestation (Polani, 1974), any deductions which are drawn can hardly be more than speculation. Since the XXX cases are very underweight, by about 30%, it seems clear that the reduction in birth weight is not caused by genic imbalance. The same conclusion may be drawn from the apparent lack of difference between XXX and XXY triploidy. On balance it seems likely that the reduction in birth weight found in babies with abnormal numbers of chromosomes is due to the increased *amount* of DNA in their cells, and this may have its effect by lengthening the cell cycle by its sheer physical presence.

Even if the rate of cell division is reduced, why is the final birth weight reduced when mammals in general have considerable powers to regulate their size? Mouse embryos formed by fusing two blastocysts are soon of normal weight (Bowman and McLaren, 1970), and even those which start off at a disadvantage, having been experimentally derived from only 1 of 2 blastomeres in the dividing ovum, have achieved normal weight by the 12th day of gestation (Tarkowski, 1959). It must be that regulation is possible only by normal diploid cells and that chromosomally abnormal cells cannot compensate in a similar manner.

The effect on the speed of the cell cycle may be simply a reflection of the amount of DNA to be replicated, but it could possibly be due to the composition of the particular chromosome involved. Chromosomes 13, 18, and 21 have rather more constitutive heterochromatin than the other autosomes, and the X chromosomes consist entirely of facultative heterochromatin. Heterochromatin, as opposed to euchromatin, is generally supposed to be genetically inert and to act nonspecifically by its very quantity. There is evidence from plants, which can tolerate extra chromosomes better than mammals, that the supernumerary B chromosomes in rye, which are partly heterochromatic, affect a wide range of morphological and physiological characters, including the rate of cell division (Adonoadu and Rees, 1968).

If heterochromatin reduces the rate of cell division, then one might expect that cell division in the human XO would be faster than in a normal female, where one of the Xs is heterochromatic. Angell (1969) showed that this is indeed so, as she demonstrated that 45,X cells show a proliferative advantage over normal cells when fibroblasts are cultured from sex chromosome mosaics. Barlow (1972) also has shown that 45,X cells have a shortened life cycle. However, the 3% of 45,X conceptions which survive to term are not heavier than normal, but about 16% lighter. It may be that the increased rate of cell division causes a lethal upset in the normal process of development, and Polani (1974) suggests that the few who survive may be those with the very lowest normal growth potential.

## 3. Biometric Approach to the Genetics of Normal Birth Weight

Most of the situations considered so far have involved abnormal infants, and although such an approach can be very informative in elucidating some aspects of the determination of birth weight, it provides only a partial picture. The study of the birth weight of normal infants has been approached using the classical methods devised by Fisher (1918) for the analysis of quantitative characters, which assume that many gene loci are involved. The findings are summarized as correlation coefficients between relatives; these are listed in Table I for sibs, half-sibs, and cousin pairs and in Table II for twins. These are the only relationships for which adequate data are available. The difficulty of collecting reliable information on more than one generation means that there are no estimates of the parent–child correlations, and consequently some standard procedures, such as the estimation of dominance deviations based on the differences between parent–child and sib-pair correlations, cannot be carried out.

Donald (1939) was the first to point out that the likeness of sib pairs ($r = +0.54$) could be explained on the hypothesis that birth weight is genetically determined, probably by both fetus and mother. He argued that the normal distribution of birth weight is what one would expect as the result of a purely random assortment of

Table I. Correlations for Birth Weight between Full Sib, Half Sib, and Cousin Pairs

| Description of sample | Number of sib pairs | r | Reference |
|---|---|---|---|
| Full sibs | | | |
| Edinburgh, live births, no adjustment for sex | | | |
| Parity 0 and 1 | 454 | 0.53 | Donald, 1939 |
| Parity 0 and 2 | 135 | 0.44 | |
| Parity 1 and 2 | 191 | 0.62 | |
| London, survivors,[a] no adjustment for sex | | | |
| Parity 0 and 1 | 891 | 0.409 | Karn et al., 1951 |
| Parity 0 and 2 | 228 | 0.439 | |
| Parity 1 and 2 | 314 | 0.472 | |
| London, survivors, sibships, not all unrelated, adjusted for sex and parity | 386 | 0.502 | Robson, 1955 |
| Japan, survivors, | | | |
| Adjacent sibs | 367 | 0.523 | Morton, 1955 |
| 1 sib apart | 654 | 0.425 | |
| 2 sibs apart | 153 | 0.363 | |
| Adjacent parity with 1st cousin parents | 442 | 0.481 | |
| Birmingham, all live births | 5042 | 0.50 | Record et al., 1969 |
| Aberdeen, survivors | | | Billewicz (1972a) |
| Adjusted for parity, sex and gestation | | 0.534 | |
| Adjusted but excluding para 0 | | 0.563 | |
| Adjusted for parity and sex | 4229 | 0.494 | |
| Adjusted but excluding para 0 | | 0.525 | |
| Unadjusted | | 0.467 | |
| Unadjusted but excluding para 0 | | 0.512 | |
| Australia, survivors, adjusted for sex, parity, gestation, and maternal height, 198 siblings in 41 families | | 0.56 | Tanner et al., 1972 |
| London, survivors, adjusted for sex, parity, gestation, and maternal height, 49 siblings in 15 families | | 0.53 | Tanner et al., 1972 |
| Half sibs | | | |
| Japan, maternal, adjacent sibs | 30 | 0.581 | Morton, 1955 |
| Survivors, paternal | 168 | 0.102 | |
| First cousins | | | |
| Great Britain, survivors, adjusted for sex and parity | | | |
| Whose mothers are sisters | 554 | 0.135 | Robson, 1955 |
| Whose fathers are brothers | 288 | 0.015 | |
| Others | 675 | 0.013 | |

[a]Survivors are defined as those infants reaching 3–4 weeks of extrauterine life.

genetic and environmental factors with additive interaction. Arguing solely from a formal genetic point of view, the observed correlation of approximately 0.50 is in agreement with the theory that birth weight is determined solely by the fetal genotype. This seems to be intuitively unlikely, and of course sibs share a great deal of common prenatal environment as well as genetic similarity. Several estimates of the sib-pair correlation have been obtained since the original report, and the values vary between 0.409 and 0.62. The absolute values differ according to the mode of selection of the data, as detailed in Table I, and increase, as expected, as various sources of variation are allowed for. The highest values are obtained for sib pairs

*Table II.   Correlations for Birth Weight between Twins of Like or Unlike Sex*

|  | Description of sample | Like sex | | Unlike sex | | |
|---|---|---|---|---|---|---|
|  |  | Number | r | Number | r | Reference |
| (i) | London, all births | 201 | 0.721 | 103 | 0.673 | Karn, 1952 |
| (ii) | London, all births | 163 | 0.710 | 92 | 0.636 | Karn 1953 |
| (iii) | Rome, all births | 1232 | 0.803 | 355 | 0.760 | Karn, 1954 |
| (iv) | London, survivors[a] only from (i) and (ii) |  |  |  |  |  |
|  | M | 146 | 0.578 | 157 | 0.585 | Penrose, 1954 |
|  | F | 133 | 0.664 |  |  |  |
| (v) | Japan,[b] survivors only | 220 | 0.557 | 40 | 0.655 | Morton, 1955 |
| (vi) | Italy (Pavia), all births | 159 | 0.765 | 78 | 0.792 | Fraccaro, 1957 |
| (vii) | India (Trivandrum), survivors only |  |  |  |  |  |
|  | M | 85 | 0.70 | 58 | 0.66 | Namboodiri and |
|  | F | 78 | 0.78 |  |  | Balakrishnan, 1959 |
| (viii) | Oxford, survivors only |  |  |  |  |  |
|  | M | 152 | 0.688 | 202 | 0.634 | Corney and Robson, |
|  | F | 134 | 0.655 |  |  | unpublished data |

[a]Survivors are defined as those infants reaching 3–4 weeks of extrauterine life.
[b]Note the low rate of dizygotic twinning in Japan.

where the firstborn are omitted. This is in agreement with the generally held view that there are sources of variation affecting the first child which do not affect later infants. The most recent and very large samples, those of Record *et al.* (1969) and Billewicz (1972*b*), confirm that 0.5 is a reasonable estimate of the sib-pair correlation coefficient.

The interpretation of the sib-pair correlation coefficient is impossible in the absence of further data which distinguish between genetic and environmental sources of variation. Morton's estimates of the correlation between half-sibs (Table I), derived from the large Japanese material collected by the Atomic Bomb Casualty Commission, show that maternal half-sibs are just as alike as full sibs, whereas paternal half-sibs are much less alike (Morton, 1955). He interpreted this finding as meaning that the resemblance in birth weight of sibs is largely attributable to the maternal constitution or environment, not to genetic similarity.

Correlation coefficients are available for only two other relationships, twins (Table II) and first cousins (Table I). These pairs share different degrees of genetic and environmental relationship, and Penrose devised a scheme for partitioning the total variance in birth weight using these different correlations (Table III). The "balance sheet" can be read along the bottom line of the table. Maternal heredity contributes 20% of the total variance, the general maternal environment 18%, and the immediate maternal environment 6%. The fetal genotype itself contributes only about 16%. The other known sources of variation are relatively unimportant, while unknown intrauterine environment still accounts for more variation than any single known factor. Despite the ingenuity of the approach, the values attributed to the different factors are no more reliable than the estimates of the correlations on which they are based. There are, in addition, various theoretical objections to the method of analysis, especially if the total variances in the different groups are not the same. In practice they never are, and in the particular values used by Penrose the variance is reduced by excluding nonsurvivors. Nevertheless, the approach is an interesting one, and its various components may now be more critically examined in the light of newer estimates of the various operative factors.

The estimates of importance of maternal and fetal genotype in determining birth weight derived from cousin-pair correlations have not improved since the original report in 1955. There is a significant positive correlation ($r = +0.135$)

Table III. Penrose's Model for Partitioning Variance in Birth Weight, Based on Correlations between Relatives[a]

| | Maternal heredity | Maternal environment | | | | Fetal genotype | | | Unknown intrauterine | Total | |
|---|---|---|---|---|---|---|---|---|---|---|---|
| | | General | Immediate | Age | Parity | Without dominance | Dominance | Sex | | Covariance | Variance |
| **Theoretical covariance:** | | | | | | | | | | | |
| Monozygotic twins | 1 | 1 | 1 | 1 | 1 | 1 | 1 | 1 | 0 | | |
| Dizygotic twins | 1 | 1 | 1 | 1 | 1 | ½ | 0 | 0 | 0 | | |
| Sibs | 1 | 1 | 0 | 0 | 0 | ½ | 0 | 0 | 0 | | |
| Children of sisters | ½ | 0 | 0 | 0 | 0 | ⅛ | 0 | 0 | 0 | | |
| All other first cousins | 0 | 0 | 0 | 0 | 0 | ⅛ | 0 | 0 | 0 | | |
| Theoretical variance | 1 | 1 | 1 | 1 | 1 | 1 | 1 | 1 | 1 | | |
| **Observed covariance:** | | | | | | | | | | | |
| Monozygotic twins | 0.22 | 0.20 | 0.07 | 0.01 | — | 0.16 | 0.01 | — | 0 | 0.67 | 1.00 |
| Dizygotic twins | 0.22 | 0.20 | 0.07 | 0.01 | — | 0.08 | 0 | — | 0 | 0.58 | 1.00 |
| Sibs | 0.22 | 0.20 | 0 | 0 | — | 0.08 | 0 | — | 0 | 0.50 | 1.00 |
| Children of sisters | 0.11 | 0 | 0 | 0 | — | 0.02 | 0 | — | 0 | 0.13 | 1.00 |
| All other first cousins | 0 | 0 | 0 | 0 | — | 0.02 | 0 | — | 0 | 0.02 | 1.00 |
| Observed variance (excluding sex and parity) | 0.22 | 0.20 | 0.07 | 0.1 | | 0.16 | 0.01 | | 0.33 | 1.00 | 1.00 |
| | | | | | (0.07) | | | (0.02) | | (0.09) | (0.09) |
| Proportional variance[b] | 0.20 | 0.18 | 0.06 | 0.01 | 0.06 | 0.15 | 0.01 | 0.02 | 0.30 | 1.00 | 1.00 |

[a] Table from Robson, 1955.
[b] Variation due to parity and sex was excluded by adjustment from the observed total correlations; an estimate of the proportional influences of these factors is included in the final estimate.

between the cousins where the mothers are sibs, but not between other first cousins (Table I). The correlation between the other first cousins is +0.014, and although not significant, it may be used as a first approximation to estimate the part played by fetal genotype, since these pairs are otherwise independent. If birth weight were entirely determined by fetal genotype, then the expected correlation would be 0.125. The observed $r$ of 0.014 thus suggests that fetal genotype plays only a small part in determining birth weight, something of the order of 11%. First cousins whose mothers are sisters, however, are more highly correlated, suggesting that there the similarity of maternal genotype is an important factor. The difference between the correlations for maternal and other types of first cousins is +0.12, and since the genetic correlation between sisters is 0.5, this suggests a value of 24% for the contribution of the maternal genotype.

The value of 0.5 used for the sib-pair correlation may be a slight underestimate when the values given in Table I are considered. A value omitting the firstborn may be preferable for use in some comparisons, but when comparison is being made with the cousin pairs, where the firstborn are included in the data, the values of Record *et al.* (1969) and Billewicz (1972*b*) are in good agreement with 0.5. The increase of the sib correlation over the maternal first cousin correlation, +0.37, can be used to measure the effect of the maternal environment constant between pregnancies. The covariance is partly used up by the doubling in the maternal genetic resemblance, and the fourfold increase in fetal genetic resemblance, as shown in the breakdown of the theoretical covariance at the top of Table III, but this still leaves roughly 20% as the share of the general maternal environment not gentically determined.

The values of the twin correlations were initially the most unreliable estimates, and unfortunately also critical ones to the argument. The genetic relationship between sibs and dizygotic twins is the same, so that any difference between them is due to factors specific to twinning or the particular pregnancy. Monozygotic and dizygotic twins differ in their genotype but share all other factors, except any related to the type of twinning, and so any difference between them can be used to obtain a second estimate of the importance of the fetal genotype. There have been a number of studies of twins which consider birth weight, and these are listed in Table II. Unfortunately most of these give weights classified by sex only, since the data are derived from routine hospital records, whereas what is needed is classification by zygosity. Most of the reports show a slightly higher correlation for like-sexed pairs, but not consistently so. In the data of Penrose, when survival is taken into account, the survivors have lower correlations than total births due to the high mortality of very light twins which contribute heavily to the correlation coefficient. Penrose used these types of data to estimate the correlation between monozygotic and dizygotic twins. As the unlike-sexed pairs consist entirely of dizygotic twins while the like-sexed pairs consist of both monozygotic (MZ) and dizygotic (DZ) pairs, the difference between the MZ and DZ correlations is probably greater than the observed difference between the two values of $r$. The difference may be estimated by using Weinberg's rule to give the probable zygosity composition of the like-sexed group; then, having an estimate of the number of DZ twins in that group and assuming that they have the same correlation as the unlike-sexed DZ pairs, a value for the correlation between monozygotic twins can be found. In the case of the London data in Table II, line (iv), these values are 0.667 for MZ twins and 0.585 for DZ twins.

The difference of 0.08 between DZ twins and sibs is attributable to immediate factors in the maternal environment. However, if a different value from 0.5 is used for the sib correlation, say 0.56 which allows for length of gestation and excludes parity 0, thus making the sibs much more directly comparable with DZ twins, the immediate maternal environment, apart from these factors, seems almost totally unimportant.

The difference between MZ and DZ twins, 0.09, provides a second estimate of the effect of the fetal genotype of 18%, which is in tolerably good agreement with the value of 11% (+2% for sex) obtained from the first cousin estimates. The values of *r* are critical for this interpretation and are in disagreement with the findings of Morton (1955). In his data the like-sexed twins, who are predominantly MZ in Japan, are not significantly different from the unlike-sexed twins nor from adjacent sibs, and so he concludes that the fetal genotype plays no part in determining birth weight. The main criticism of Penrose's estimates, however, lies in the assumption underlying the estimation of MZ and DZ correlations from data on like- and unlike-sexed pairs, that is, that like-sexed DZ twins have the same correlation as unlike-sexed DZ. Furthermore, it is known that MZ twins are heterogeneous with respect to placentation and that marked differences in birth weight may occur in those sharing the same placenta. Indirect estimates of *r*, such as those of Penrose, are therefore very dubious indeed. Direct estimates of *r* for MZ and DZ twins have recently been made from the data of the Oxford Twin Survey (Corney *et al.*, 1972) where zygosity was determined on a newborn twin sample using blood groups and biochemical markers (Table IV). It is immediately clear that unlike- and like-sexed DZ pairs are not equally correlated, so that Penrose's adjustment overestimates the difference between MZ and DZ pairs. Indeed when comparing like-sexed pairs, MZ have the lower correlation, thus apparently excluding the fetal genotype from playing a major part in the determination of within-pair differences in birth weight. Similar results have been obtained by Wilson (1976) in the Louisville Twin Study, where zygosity was determined by the use of blood groups. The correlations for DZ twins are 0.70 for like-sexed pairs, and 0.68 for all pairs, in a sample of 195 DZ pairs. Again the lowest correlation is for the MZ pairs, 0.61 for a sample of 159 pairs. Unfortunately placentation is not known for the Louisville Study, and so the only further data on this problem are those of Corney and Robson in Table IV. When the MZ group is considered in terms of placentation, it can be seen that the dramatically low correlations are shown by the group with monochorionic placentation, a known physiological hazard. When dichorionic MZ pairs are compared with DZ pairs they *are* more highly correlated, but the difference is only about +0.04, about half that

*Table IV. Correlation between Birth Weights of Twin Pairs of Known Zygosity in the Oxford Survey (Corney and Robson, unpublished), Survivors Only*

| | Monozygotic | | | | Dizygotic | |
| | Monochorionic | | Dichorionic | | Dichorionic | |
| Sex of twin pair | Number | *r* | Number | *r* | Number | *r* |
|---|---|---|---|---|---|---|
| Male–male | 54 | 0.537 | 18 | 0.882 | 80 | 0.700 |
| Female–female | 53 | 0.574 | 9 | 0.623 | 72 | 0.738 |
| All like sexed | 107 | 0.541 | 27 | 0.758 | 152 | 0.719 |
| Unlike sexed | | | | | 202 | 0.634 |
| All pairs | | | | | 354 | 0.668 |

estimated by Penrose. This value is only the very roughest approximation, as the proportion of twins in the MZ dichorionic class is so small, but more data will soon be available from other prospective twin surveys in Great Britain. If confirmed, it would have amused Penrose to know that he was right for the wrong reasons. The variances in Table III were made to add up, using the MZ–DZ estimate for the fetal genotype, by rounding up the estimate obtained from the cousin pairs. Now, using the smaller difference between MZ and DZ twins, a value of about 10% fits tolerably well with both methods. All the twin correlations of Corney and Robson are rather higher than those of Penrose, thus again allowing for some effect of immediate maternal environment, even using a sib correlation of +0.56. It also has the effect of reducing the remaining variance to be attributed to unknown intrauterine factors. Factors associated with monochorionic twinning itself are obviously a major cause of dissimilarity in MZ twins.

## 4. Summary

The part played by the fetal genotype in determining birth weight is small, probably of the order of 10%, while the maternal genotype plays a rather more important role at around 24% of the total variance. However, maternal environment and unknown intrauterine factors are both more important than maternal or fetal genotype. Morton's observations on inbreeding (1958) are in agreement with this conclusion. Data collected in Japan show that the mean birth weight of the children of unrelated parents is 3074 g and that of the children of first cousins 3046 g. This is certainly a small difference, but it is significant. It is presumably due to the effect of increased homozygosity in the fetus and could in turn be due either to the effects of single genes, manifesting as deleterious recessives, or to the effect of increased homozygosity over many gene loci of small effect. The size of the effects of single genes described earlier in this section would, if characteristic of many loci, leave little room for the polygenes often invoked in quantitative theory. It has become apparent in the last few years that there is a great deal of redundancy in human DNA, and that as a consequence the number of gene loci may not be as enormous as was once assumed. This also brings into question the assumption that quantitative characters showing a normal distribution are due to the segregation of a really large number of genes. It has been shown for the human enzyme red cell acid phosphatase that 60% of the normal variation can be attributed to three alleles at one locus polymorphic for electrophoretic variants (Spencer *et al.*, 1964; Eze *et al.*, 1974), that the classic quantitative human character total finger ridge count may be determined by a fairly small number of loci (Holt, 1968), and in *Drosophila* that only five loci account for more than 85% of the difference between high and control lines for sternopleural bristle number (Spickett and Thoday, 1966). So it may be that the number of genes involved in determining birth weight is not impossibly large and that their individual recognition will not be the impossible task it has generally been assumed to be.

Although our understanding of the genetics of birth weight is still relatively imprecise, it has been considered to be sufficiently important in practice to warrant the preparation of within-family standards for birth weight (Tanner *et al.*, 1972). The possibility of error in using such a system, due to the problem of small family size particularly, is discussed by Billewicz (1972*a*), who feels that such tables should be used, like most things, with caution.

ELIZABETH B. ROBSON

Adonoadu, U. W., and Rees, H., 1968, The regulation of mitosis by B chromosomes in rye, *Exp. Cell Res.* **52**:284.

Angell, R. R., 1969, Cytogenetic and genetic studies in Turner's syndrome and allied conditions in man, Ph.D. thesis, University of London.

Barlow, P. W., 1972, Differential cell division in human X chromosome mosaics, *Humangenetik* **14**:122.

Barlow, P., 1973, The influence of inactive chromosomes on human development. Anomalous sex chromosome complements and the phenotype, *Humangenetik* **17**:105.

Beck, W., and Ananthakrishnan, R., 1974, Placental alkaline phosphatase types in Germany, *Humangenetik* **25**:127.

Beckman, G., 1974, Relationship between placental alkaline phosphatase phenotype and placental weight, *Hum. Hered.* **24**:291.

Beckman, L, Beckman, G., and Mi, M. P., 1969, The relation between human placental alkaline phosphatase types and some perinatal factors, *Hum. Hered.* **19**:258.

Billewicz, W. Z., 1972a, Within-family birth weight standards, *Lancet* **2**:820.

Billewicz, W. Z., 1972b, A note on birth weight correlation in full sibs, *J. Biosoc. Sci.* **4**:455.

Billington, W. D., 1964, Influence of immunological disparity of mother and foetus on size of placenta in mice, *Nature* **202**:317.

Bowman, P., and McLaren, A., 1970, Viability and growth of mouse embryos after *in vitro* culture and fusion, *J. Embryol. Exp. Morphol.* **23**:693.

Boyer, P. H., 1955, Low birth weight in fibrocystic disease of the pancreas, *Pediatrics* **16**:778.

Chakraborty, R., Das, S. R., Roy, M., Mukherjee, B. N., and Das, S. K., 1975, The effect of parity on placental weight and birth weight: Interaction with placental alkaline phosphatase polymorphism, *Ann. Hum. Biol.* **2**:227.

Chen, A. T. L., Chan, Y.-K., and Falek, A., 1971, The effects of chromosome abnormalities on birth weight in man. I. Sex chromosome disorders, *Hum. Hered.* **21**:543.

Chen, A. T. L., Chan, Y.-K., and Falek, A., 1972, The effects of chromosome abnormalities on birth weight in man. II. Autosomal defects, *Hum. Hered.* **22**:209.

Corney, G., Robson, E. B., and Strong, S. J., 1972, The effect of zygosity on the birth weight of twins, *Ann. Hum. Genet., London* **36**:45.

Donald, H. P., 1939, Sources of variance in human birth weight, *Proc. R. Soc. Edinburgh* **59**:91.

Eze, L. C., Tweedie, M. C. K., Bullen, M. F., Wren, P. J. J., and Evans, D. A. P., 1974, Quantitative genetics of human red cell acid phosphatase, *Ann. Hum. Genet., London* **37**:333.

Fisher, R. A., 1918, The correlation between relatives on the supposition of Mendelian inheritance, *Trans. R. Soc. Edinburgh* **52**:399.

Fraccaro, M., 1957, A contribution to the study of birth weight based on an Italian sample twin data, *Ann. Hum. Genet., London* **21**:224.

German, J., 1969, Bloom's syndrome. I. Genetical and clinical observations in the first twenty-seven patients, *Am. J. Hum. Genet.* **21**:196.

Hohler, C. W., Bardawil, W. A., and Mitchell, G. W., 1972, Placental weight and water content relative to blood types of human mothers and their offspring, *Gynecol. Obstet.* **40**:799.

Holt, S. B., 1968, *The Genetics of Dermal Ridges,* Charles C. Thomas, Springfield, Illinois.

Hsia, D. Y.-Y., 1959, Birth weight in cystic fibrosis of the pancreas, *Ann. Hum. Genet., London* **23**:289.

Hsia, D. Y.-Y., and Walker, F. A., 1961, Variability in the clinical manifestations of galactosemia, *J. Pediatr.* **59**:872.

Jayant, K., 1966, Birth weight and survival: A hospital survey repeated after 15 years, *Ann. Hum. Genet., London* **29**:367.

Jones, W. R., 1968, Immunological factors in human placentation, *Nature* **218**:480.

Karn, M. N., 1952, Birth weight and length of gestation of twins, together with maternal age, parity and survival rate, *Ann. Eugen., London* **16**:365.

Karn, M. N., 1953, Twin data: A further study of birth weight, gestation time, maternal age, order of birth, and survival, *Ann. Eugen., London* **17**:233.

Karn, M. N., 1954, Data of twins born in Italy 1936–1951, *Acta Genet. Med. Gemell.* **3**:42.

Karn, M. N., and Penrose, L. S., 1951, Birth weight and gestation time in relation to maternal age, parity and infant survival, *Ann. Eugen., London* **16**:147.

Karn, M. N., Lang-Brown, H., MacKenzie, H., and Penrose, L. S., 1951, Birth weight, gestation time and survival in sibs, *Ann. Eugen., London* **15**:306.

Marimo, B., and Gianelli, F., 1975, Gene dosage effect in human trisomy 16, *Nature* **256**:204.

McLaren, A., 1975, Antigenic disparity: Does it affect placental size, implantation or population genetics? in: *Immunobiology of Trophoblast* (R. G. Edwards, C. W. S. Howe, and H. Johnson, eds.), pp. 255–276, Cambridge University Press, Cambridge.

Mittwoch, U., 1967, DNA synthesis in cells grown in tissue culture from patients with mongolism, in: *Mongolism* (G. E. W. Wolstenholme and R. Porter, eds.), pp. 51–61, Churchill, London.

Mittwoch, U., and Delhanty, J. D. A., 1972, Inhibition of mitosis in human triploid cells, *Nature (London), New Biol.* **238:**11.

Morton, N. E., 1955, The inheritance of human birth weight, *Ann. Hum. Genet., London* **20:**125.

Morton, N. E., 1958, Empirical risks in consanguineous marriages. Birth weight, gestation time, and measurements of infants, *Am. J. Hum. Genet.* **10:**344.

Naeye, R. L., 1967, Prenatal organ and cellular growth with various chromosomal disorders, *Biol. Neonat.* **11:**248.

Namboodiri, N. K., and Balakrishnan, V., 1959, A contribution to the study of birth weight and survival of twins based on an Indian sample, *Ann. Hum. Genet., London* **23:**334.

Penrose, L. S., 1954, Some recent trends in human genetics, *Caryologia* **6**(Suppl):521.

Polani, P. E., 1974, Chromosomal and other genetic influences on birth weight variation, in: *Size at Birth,* CIBA Foundation Symposium 27 (New Series), pp. 127–164, North-Holland, Amsterdam.

Record, R. G., McKeown, T., and Edwards, J. H., 1969, The relation of measured intelligence to birth weight and duration of gestation, *Ann. Hum. Genet., London* **33:**71.

Reed, T. E., 1967, Research on blood groups and selection from the child health and development studies, Oakland, California. I. Infant birth measurements, *Am. J. Hum. Genet.* **19:**732.

Robson, E. B., 1955, Birth weight in cousins, *Ann. Hum. Genet., London* **19:**262.

Saugstad, L. F., 1972, Birth weights in children with phenylketonuria and in their siblings, *Lancet* **1:**809.

Seppälä, M., and Tolonen, M., 1970, Histocompatibility and human placentation, *Nature* **225:**950.

Spencer, N., Hopkinson, D. A., and Harris, H., 1964, Quantitative differences and gene dosage in the human red cell acid phosphatase polymorphism, *Nature* **201:**299.

Spickett, S. G., and Thoday, J. M., 1966, Regular responses to selection 3. Interaction between located polygenes, *Genet. Res.* **7:**96.

Tanner, J. M., Lejarraga, H., and Turner, G., 1972, Within family standards for birth-weight, *Lancet* **2:**193.

Tarkowski, A. K., 1959, Experiments on the development of isolated blastomeres of mouse eggs, *Nature* **184:**1286.

Toivanen, P., and Hirvonen, T., 1970, Placental weight in human foeto-maternal incompatibility, *Clin. Exp. Immunol.* **7:**533.

Wilson, R. S., 1976, Concordance in physical growth for monozygotic and dizygotic twins, *Ann. Hum. Biol.* **3:**1.

# 11

# The Genetics of Adult Stature

## C. O. CARTER and W. A. MARSHALL

## 1. Introduction

Adult stature is continuously and near Normally distributed. This indicates that the determining factors are likely to be multiple—multiple genetic, multiple environmental, or a mixture of both. There are indications that the major sources of variation within populations in developed countries are genetic. The most compelling evidence is the well-known resemblance in stature of genetically identical monozygotic (MZ) twins, even when reared apart. A difference of more than 2 cm between such twins is unusual, except in association with disease in one of the pair. In contrast, dizygotic (DZ) like-sex twins reared together may show marked differences in height, consistent with their having, on average, only half their genes in common.

Studies of familial resemblance for stature are compatible with the hypothesis that the genetic source of variation in the population depends on several, perhaps many, gene loci. Further, there are indications that different genetic factors influence growth in height at different ages. For example, longitudinal studies show that the correlation of a child's height with his own full adult height increases rapidly up to age 3, remains fairly constant up to age 8 in girls and 11 in boys, shows a temporary decrease during the adolescent growth spurt, and then increases again towards unity as full adult stature is reached (Tanner *et al.*, 1956; Tuttenham and Snyder, 1954; Tanner and Marshall, unpublished). In contrast, the correlation of monozygotic twins for height is high by early childhood, 0.85 by 1 year, 0.94 by 4 years (Wilson, 1976), and remains consistently above 0.9 throughout the school years and into adulthood (Fischbein, 1977). There are no indications of dominance effects at the gene loci controlling stature since the limited data which are available suggest that the child on midparent regression for adult stature is linear (see below). Also, there is no evidence that genes on the X or Y chromosomes have any major effect since the correlation coefficients for son and father, son and mother, daughter and father, and daughter and mother do not differ significantly (see below).

---

*C. O. CARTER* • Institute of Child Health, London. *W. A. MARSHALL* • University of Technology, Loughborough, Leicestershire, England.

The secular trend of increasing height in many countries during the past century indicates that environmental factors play some part in the normal variation of height. While the increase in stature at school age, or even at the age of 18 years, is largely due to earlier maturity, there has also been an undoubted increase in full adult stature. However, this increase is possibly coming to an end in the developed countries where nutrition is near optimal for growth (see Volume 3, Chapter 16).

While roughly a score of monogenically determined conditions are known to cause short stature (Carter and Fairbank, 1974), these are all rare disorders and insufficient to cause any apparent bimodality of the distribution of stature, although they give classical mendelian patterns within families.

## 2. Simple Additive Polygenic Model

The inheritance of much continuous variation in plants and animals may be described by the simple additive polygenic model which is based on the assumption that, on an appropriate measuring scale, the effects of different genes are additive. This implies that the effect of being heterozygous for two allelic genes is approximately half way between the effects of being homozygous for either allele (i.e., there is no dominance). Similarly having gene A at one locus and gene B at another locus gives an A + B effect and does not tend to give the effect of A or B alone (i.e., there is no epistasis). The appropriate scale will often be that of the unaltered direct measurement, especially when the distribution in the population is Normal. The plausibility of a polygenic model has been increased in recent years as antigenic, and chemical techniques have demonstrated that at many, perhaps the majority, of gene loci there is polymorphism, i.e., two or more alleles at the locus have a fairly high frequency in the general population. Polymorphism was first demonstrated for the ABO and then for other blood antigens and has now been shown to be present at many of the numerous gene loci responsible for the production of specific enzymes.

Polymorphism at even a few gene loci with additive effect will tend to give a Normal distribution. This is the result of mendelian segregation. Figure 1 demonstrates this. Figure 1A shows the distribution given by a single gene locus at which there are two additive alleles, each with a gene frequency of 0.5. Figure 1B shows the distribution given by similar genes with similar gene frequencies at each of two gene loci on different chromosomes, and Figure 1C shows that given by three similar pairs of allelic genes on different chromosomes with similar gene frequencies. The distribution in Figure 1C already approaches the Normal and would

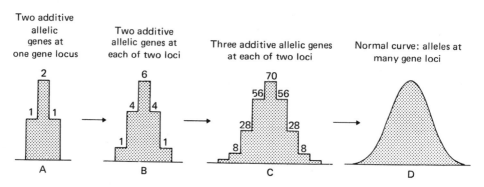

Fig. 1. Additive polygenic inheritance tends to give a Normal distribution.

approach it still closer if the remaining discontinuity of the distribution were to be blurred by small environmental effects. The shortest individuals are those homozygous for the alleles reducing height, and the tallest are those homozygous for the alleles increasing height above the population average.

The family resemblances to be expected from the model may be readily deduced from the number of genes that each degree of relative has in common with the index individual or "propositus." A parent, child, or sib of a propositus is a first-degree relative with, on average, half his or her genes in common with the propositus. A parent transmits only half his or her genes to the propositus. Thus a father transmits only half his chromosomes to any individual child, and so on average only half the genes which make the father deviate from the population mean will be transmitted to the child. In the absence of assortative mating, the genes the child gets from the mother will be representative of those of the general population and tend to give the mean height. Therefore assuming no change in the population variance for height between the parental and filial generations, the expected regression coefficient of son on father would be 0.5. Similarly, when a son deviates from the population mean, on average only half the genes responsible will have been derived from the father, and so the regression of father on son and the correlation coefficient of father on son (the geometrical mean of the two regression coefficients) will also be 0.5. Similarly after correction for any phenotypic expression of the genes in males and females, the cross sex regression and correlation coefficients will also be 0.5. Sibs will also, on average, have half their genes in common with each other, and therefore the expected regression coefficient of sib on sib and the correlation coefficient of sibs is 0.5.

For second-degree relatives, for example, uncles and nephews, grandfathers and grandsons, and half-sibs with a quarter of their genes in common, the correlation of regression coefficients expected from the model is 0.25. For third-degree relatives, for example, first cousins, great-uncles, and great-nephews, with one-eighth of their genes in common the expected coefficients are 0.125. For monozygotic co-twins the expected coefficients are 1.0 and for dizygotic twins the same as for sibs, 0.5. The estimated child on midparent regression is 1.0 since all the child's genes come from his parents, but the regression of midparent on child is 0.5 since an individual child has only half the genes present in his two parents. Hence the midparent–child correlation is $\sqrt{0.5 \times 1}$, or 0.71.

Dominance deviation is said to occur when subjects who are homozygous for one allele bear a closer resemblance to heterozygous subjects than to those who are homozygous for the alternative allele. Dominance deviation does not alter the MZ correlations, but reduces the regression coefficients of child on parent and of child on midparent, since it is not transmitted from parent to child. The effect is rather less on sib–sib regression and correlation coefficients since sibs share a quarter of the dominance deviation. Dominance also tends to make the child on midparent regression nonlinear, with the regression coefficients lowered at the lower and upper ends of midparental distribution (Penrose, quoted by Roberts, 1961).

Genes on the X chromosomes cause the correlation between fathers and daughters to be the same as that between mothers and sons since father and daughter share an X chromosome, as do mother and son. The correlation between mothers and daughters is lower because they share only one of their two X chromosomes and that between father and son is even lower because they have no X chromosome in common although they share a Y chromosome. A major gene on the Y chromosome will raise the father–son correlation above those for the other parent–child combinations.

C. O. CARTER AND
W. A. MARSHALL

The effect of assortative mating is to increase the parent–child and sib–sib correlations by an amount which is close to half the genotypic correlation of the two parents.

## 3. Familial Correlations Observed and Their Compatibility with the Polygenic Model

Let us now consider whether or not the limited data which are available are in keeping with the simple additive polygenic model and what they tell us about the proportion of variance for stature in a particular population which is due to genetic variation. It is this proportion which is called heritability of stature. "Broad" heritability is a measure of the contribution of all kinds of genetic variation and "narrow" heritability of the contribution of purely additive genetic variation to the variance. Only the latter is transmissible from parent to child and hence narrow heritability is the measure of the expected phenotypic response to selection.

Estimates of heritability are made in several ways (Falconer, 1964):

1. If there is no significant effect of "common family environment," the correlation of MZ twins is a measure of broad heritability. If there is a significant effect of common family environment, then the MZ twin correlation will give an inflated estimate of heritability, although the inflation will be least in samples of monozygotic twins reared apart from birth. If there is dominance or epistatic deviation from additive inheritance, this will be shared by MZ twins and so this correlation could give an inflated value of narrow heritability but not of broad heritability.

Two most informative studies of adolescent and adult twins reared apart have been reported. One is the old American study of Newman, Freeman, and Holzinger (1937); the other is the more recent British study of Shields (1962). The American series included 50 MZ and 50 DZ pairs reared together and 19 MZ pairs reared apart. The correlations for stature with age effects eliminated were: MZ together, 0.93; DZ together, 0.65; MZ apart, 0.97. While the MZ twins reared apart did not in general have strikingly different childhood environments, it is noteworthy that they were somewhat more alike than the MZ twins reared together.

In the British series of 41 pairs of adult MZ twins who had been reared together and 39 pairs who had been reared apart the correlations were: MZ together, 0.96; MZ apart, 0.82. Here was the expected somewhat lower correlation for MZ twins reared apart.

Both series indicate a high heritability for adult stature. In the case of the DZ twins (but not the MZ) the correlation will be raised by assortative mating in the parents for stature, which in white American and European populations is of the order 0.2–0.3 (Spuhler, 1968).

2. If there is no common family environmental effect, the child on midparent regression provides a measure of narrow heritability which is independent of dominance and epistasis and which is also not affected by assortative mating. This estimate will, however, be inflated by any environmental effects common to both parents and child.

There is unfortunately little data for parents and their full-grown offspring on this most informative regression coefficient. Studies have been made on prepubertal children (Mueller 1976), but these will not reflect the resemblance of parent and child for genes which are not turned on till adolescence.

In a small unpublished series of 48 fully grown children and their parents, all professionally measured, from the Institute of Child Health, London (Tanner and Marshall, unpublished data), the child on midparent regression coefficient was 1.05. However, because of the small numbers, this coefficient has the relatively large standard error of 0.13. Susanne (1976) found a child on midparent regression of 0.81.

3. Twice the regression of child on one parent or the corresponding correlation is also a measure of narrow heritability, but it will be inflated by common family environment and by assortative mating. In the literature concerning larger studies, only correlations have been reported.

Pearson and Lee (1903) reported a parent–child correlation from their large English series in which the data were provided by students who had measured their relatives. The correlations were: father–son, 0.51; mother–daughter, 0.51; father–daughter, 0.51; mother–son, 0.49; parent–child, 0.51.

Again, Durnin and Weir (1952) reported a series based on Scottish students who measured their parents. Tanner and Israelsohn (1963) report the following correlations calculated from these measurements: father–son, 0.55; mother–daughter, 0.52; father–daughter, 0.42; mother–son, 0.37. In Susanne's Belgian series (1975), all professionally measured, the correlations were similar: father–son, 0.54; mother–daughter, 0.47; father–daughter, 0.52; mother–son, 0.53; parent–child, 0.51; midparent–child, 0.63.

These correlations for adult offspring and their parents may be compared with the expectedly somewhat lower correlations for 9-year-old children and their parents in a large series from four European cities (Tanner *et al.*, 1970): father–son, 0.42; mother–daughter, 0.44; father–daughter, 0.44; mother–son, 0.43.

Correlations of this order were already present by age 3 years and changed little between 3 and 9 years. Mueller (1976) summarizes other studies of these correlations.

Twice the sib–sib correlation is also a useful measure of heritability but will be inflated by common family environment which is likely to be greater for sibs than for parents and child and will include a quarter of any dominance deviation. The estimate will also be raised by assortative mating.

Galton in 1889 summarized earlier reports of studies on the resemblance of sibs. His pioneering work was of much importance in the history of quantitative genetics. However, he did not personally measure his subjects for the family studies, but interested correspondents supplied the measurements of their own relatives. It is noteworthy that Galton's estimate of the factor 1.08 to convert the height of adult women to those of the equivalent male is precisely that appropriate to adults today.

Galton's successor, Pearson, together with Lee (1903), reported a very much larger series based on several hundred families. Again the subjects were not professionally measured, but students were instructed how to measure members of their own families. The correlations found were: sib–sib, 0.53; brother–brother, 0.51; sister–sister, 0.54; brother–sister, 0.55.

Howells (1948) found a brother–brother correlation of 0.47 based on 94 pairs and quotes a figure of 0.57 from Bowles (1932) based on 79 pairs.

Susanne (1975) has recently reported a Belgian series of 125 families all professionally measured. The correlations found were: sib–sib, 0.59; brother–brother, 0.52; sister–sister, 0.57; brother–sister, 0.61.

The effect of assortative mating is to raise the regression coefficient of a child on one parent and also the sib–sib correlation by about one half the genotype correlation between parents; the proportion of the phenotypic correlation between parents that is genetic is not readily estimated.

4. Four times the correlation of second-degree and eight times the correlation of third-degree relatives are also measures of heritability which are likely to be less affected by common family environment and the dominance deviation than those estimated from sib correlations. The correlation of maternal and paternal half-sibs will provide information about the importance of purely external, e.g., intrauterine, influences. There are no satisfactory data on this correlation. A greater resemblance of mother and child than father and child would suggest a maternal influence, but there is no indication of such a greater resemblance in published data for adults.

5. Twice the difference between MZ and DZ correlation gives an estimate of heritability which is not inflated by common family environment, provided that this is similar in degree for MZ and DZ twins. It will, however, be markedly affected by assortative mating, since this will raise the concordance of DZ twins but cannot raise that of MZ twins, thus leading to an underestimate of heritability. On the other hand, this estimate of heritability will also include about three quarters of the dominance deviation.

## 4. Conclusions

The available data on adult stature from American and Western European populations are best explained by the hypothesis that the major part of variation is largely due to additive polygenic inheritance, but that environmental factors make some contribution to the variation. The classic example of additive polygenic inheritance is the total fingerprint ridge count. Table I shows a comparison of a ''best estimate'' of the true correlations for stature calculated from the observed phenotypic data described above with the theoretical correlations resulting from simple additive polygenic inheritance and with the correlations for total fingerprint ridge count (Holt, 1961). The ridge count correlations are close to the theoretical values, with some suggestion of dominance effects from the child on midparent regression which is rather less than the MZ twin correlation.

Only rough estimates of heritability of stature may be made from the data available. The best estimate of broad heritability given by the correlation of MZ twins reared apart from the series of Newman *et al.* (1937) and Shields (1962) is probably about 0.85, a little below that for MZ twins reared together. The best estimate of narrow heritability is given by the child on midparent regression where

*Table I.   Comparison of Correlation Coefficients for Adult Stature (Unbroken Families) with Those for Total Fingerprint Ridge Count and the Theoretical Value*

|  | Theoretical | Fingerprint ridge count | Adult stature |
|---|---|---|---|
| Husband–wife | 0.00 | 0.05 | 0.2–0.3 |
| Monozygotic twins | 1.00 | 0.95 | 0.94 |
| Dizygotic twins | 0.50 | 0.49 | 0.65 |
| Sib–sib | 0.50 | 0.50 | 0.57 |
| Parent–child | 0.50 | 0.48 | 0.51 |
| Midparent–child | 0.71 | 0.66 | 0.63 |

the child has not been reared by his own parents. On the two small series available, where a child has been reared by its own parents, the child on midparent correlation appears to be of the order 0.8–0.9. Assuming a true heritability of 0.8 and a simple additive inheritance, then the sib–sib and parent–child correlations would be expected to be 0.4. They are, in fact, higher than this for unbroken families, but this is readily attributable to the small, but significant, degree of assortative mating and the common family environment. Any such estimate of heritability applies to the populations studied, that is, American and northwest European. The heritability of adult height would almost certainly be less in a community with greater environmental, particularly nutritional, differences within the population.

# 5. References

Bowles, G. T., 1932, *New Types of Old Americans at Harvard,* Harvard Univ. Press, Cambridge.

Carter, C. O., and Fairbank, J., 1974, *The Genetics of Locomotor Disorders,* Oxford University Press, London.

Durnin J. V. G. A., and Weir, J. B. de V., 1952, Statures of a group of university students and their parents, *Br. Med. J.* **1:**1006.

Falconer, R. S., 1964, *Introduction to Quantitative Genetics,* Oliver and Boyd, Edinburgh.

Fischbein, S., 1977, Growth of twins during puberty, *Ann. Hum. Biol.* **4:**417.

Galton, F., 1889, *Natural Inheritance,* Macmillan, London.

Holt, S., 1961, Quantitative genetics of finger print patterns, *Br. Med. Bull.* **17:**247.

Howells, W. W., 1948, Birth order and body size, *Am. J. Phys. Anthropol. NS* **6:**449.

Mueller, W. H., 1976, Parent–child correlations for stature and weight among school aged children. A review of 24 studies, *Hum. Biol.* **48:**379.

Newman, H. N., Freeman, F. N., and Holzinger, K. J., 1937, *Twins: A Study of Heredity and Environment,* University of Chicago Press, Chicago.

Pearson, K., and Lee, A., 1903, On the laws of inheritance in man. *Biometrika* **2:**359.

Roberts, J. A. F., 1961, Multifactorial inheritance in relation to human traits, *Br. Med. Bull.* **17:**241.

Shields, J., 1962, *Monozygotic Twins,* Oxford University Press, London.

Spuhler, J. N., 1968, Assortative mating with respect to physical characteristics, *Eugen. Q.* **15:**128.

Susanne, C., 1975, Genetic and environmental influences on morphological characteristics, *Ann. Hum. Biol.* **2:**279.

Susanne, C., 1976, Personal communication.

Tanner, J. M., and Israelsohn, W., 1963, Parent–child correlations for body measurements of children between the ages of one month and 7 years, *Ann. Hum. Genet.* **26:**245.

Tanner, J. M., Healy, M. J. R., Lockhart, R. D., MacKenzie, J. D., and Whitehouse, R. H., 1956, Aberdeen growth study. I. The prediction of adult body measurements from measurements taken each year from birth to 5 years, *Arch. Dis. Child.* **31:**372.

Tanner, J. M., Goldstein, H., and Whitehouse, R. H., 1970, Standards for children's height at 2–9 years allowing for height of parents, *Arch. Dis Child.* **45:**755.

Tuttenham, R. D., and Snyder, M. A., 1954, Physical growth of Californian boys and girls from birth to 18 years, *Univ. Calif. Publ. Child Dev.* **1:**183.

Wilson, R. S., 1976, Concordance in physical growth for monozygotic and dizygotic twins, *Ann. Hum. Biol* **3:**1.

# 12

# *Genetics of Maturational Processes*

## STANLEY M. GARN and
## STEPHEN M. BAILEY

## 1. Introduction

The genetics of human maturational events, like the conventional genetics of dimensional development, has long been inferred from two different, quasi-experimental procedures. The first of these is the population comparison. The second of these involves family-line analysis. Together, under optimal conditions, they provide a partial indication of the extent to which specific maturational events differ in timing, between natural populations, on a primarily genetic basis.

There is no lack of maturational events to consider, either in the broadest sense or strictly limited to those discrete events linked to gametogenesis. Birth is a maturational event, to begin the listing. The attainment of each Gesell item marks a separate maturational event. The age at appearance of some 79 postnatal ossification centers, 20 deciduous teeth, 28–32 permanent teeth, and the age at epiphysial union of some 60-odd secondary centers—each is a maturational event. So are the discrete stages of areolar and breast development, of axillary and pubic hair growth, of parafrontal balding, sebaceous gland secretion, beard and moustache hair, or appearance of magenta-colored adolescent striae on the buttocks, abdomen, and breasts. So is the age at menarche.

As depicted in the standard atlases of the hand, foot, knee, elbow, and hip, each bony center goes through a series of discrete steps as bone mineral makes performed bone radiographically visible. If these steps be called "maturation," then there are hundreds of additional maturational stages to be considered (both in population and family-line comparisons). If the formation stages of the 52 teeth (from the appearance of the follicle to apical closure of each root) be considered as maturation, then there are an additional 240–300 discrete maturational items, for each maturing human being.

**STANLEY M. GARN and STEPHEN M. BAILEY** • Center for Human Growth and Development, University of Michigan, Ann Arbor, Michigan.

For a full description of maturational timing, therefore, there are perhaps 1000 discrete items—of bones (and bony centers), of teeth (including both formation and movement), of the skin and its appendages (and including the pigmentary changes in so-called sexual skin). This number includes, more or less, evidences of maturation of the apocrine and eccrine glands, production of sebum and smegma, changes in the endometrium of the uterus, and in Leydig-cell and Sertoli-cell activity.

Despite this wealth of potential information on the genetics of normal human maturation, reflecting innumerable maturational events through which each human being inevitably goes, what we have for sure is meager, and what we have at the textbook level is often questionable. The best-documented two-generational maturational information is on mother–daughter similarities in the age at menarche, which stands up well to critical examination, as we shall see. The best or perhaps securest sibling data we have (on any series of maturational events) is on formation timing of 28–32 permanent teeth, simply because the teeth are rather less affected—in the timing of formation—either by nutritional alteration or other environmental effects. But for many other maturational events, population comparisons are eminently unsatisfactory (lacking nutritional control), and sibling comparisons and even twin comparisons are less than perfect indications of the probable role of the genes.

## 2. Investigative Limitations

To a very large extent our present lack of knowledge of the genetics of human maturational timing is due to the paucity of longitudinal or semilongitudinal maturational information with attention to each discrete maturational event. There is also the lack of family-line and sibling information sufficient to provide brother–brother, sister–sister, and sister–brother comparisons in adequate numbers for the correlation coefficients to carry quantitative meaning or to allow testing for X and Y linkage, or at least the influences of sex. Of the major United States longitudinal studies—Boston, Berkeley, Denver, Yellow Springs (Fels), and Cleveland (Brush Foundation), only the radiographic data at the Fels Institute are sufficient in sample size and number of sibling pairs to provide beginning indications.

To be sure, these long-term longitudinal studies were not established to provide sibling comparisons; they were conceived as models for the study of "normal" boys and girls, with norms as the intended purpose. In the 1930s, the time of their inception, the need for hundreds of exact-age brother–brother, sister–sister, and brother–sister comparisons was not perceived. Later European longitudinal studies mostly followed the original American models. That is why we do not have the amount of maturational information on siblings that we would like, after all these years.

More recent studies and surveys in Canada and elsewhere contain, in basic form, maturational data of value, but they are often expressed as less-than-primary information. A "bone age" or a "dental age" contains within it the basic information, but unfortunately not bone by bone or tooth by tooth. Nutritional survey data, although not longitudinal, also contain sibling information and maturational information of relevance, provided that they are given appropriate attention and considered analytical treatment. But it should now be evident why, in the last quarter of the 20th century, we do not have much of the information we would like on those thousand-odd maturational events that every child goes through.

Population comparisons have long been used to provide some information on human developmental genetics. But, as we have increasingly become aware, population comparisons can be faulty indeed. Even if the chronological ages are correct and verified from birth records and the anthropometric dimensions carefully taken and meticulously checked, conventional population differences must be regarded with a grain of salt. This is even more true when the measures are maturational in nature and intended to explore population differences in genetically determined maturational timing.

Different populations, studied half a world apart, are simply not comparable in nutritional status from birth through the end of growth. Breast-fed Zulus in South Africa, weaned onto "mealies," are not comparable to Ukrainians in Cleveland (bottle fed from the start, and ingesting 60 g of quality protein per day from infancy on). The Bundi diet and the Boston diet simply do not allow meaningful population comparisons of maturational timing. Hong Kong Chinese (with rather low intakes of animal protein), Mayan-speaking Guatemalan Indians subsisting on maize, Japanese in Hawaii, and Pakistanis in London cannot be compared directly to provide hard information on the genetics of maturational timing.

Within a demographic population, there are still large socioeconomic differences in both nutritional knowledge and dietary practices. Inevitably, the children of the rich are larger and earlier to mature than the sons and daughters of the poor. The children of the rich are longer and fatter, with a higher weight-to-length ratio, longer legs relative to stature, and advanced in osseous development and in the age at menarche. Under these circumstances we can compare age-at-appearance of ossification centers and discrete (radiographic) tooth formation steps in different populations. But the differences do not necessarily reflect differences in the genes.

Even within a single geographical area, such as a modern megacity, children of different geographical origins do not enjoy comparable access to food, fuel, and fiber. The disparity may be economic, limiting food and promoting crowding. It may be educational—the problem of nutritional knowledge. But such differences, pertaining both to immigrant groups and rural migrants to the urban scene, place severe limitations on the meaning of population comparisons even within the same demographic or political unit.

As we shall see, careful attention to socioeconomic variables provides a partial resolution of the problem. We shall make comparisons variously holding income, education, and occupation constant within a geographical area. It is then possible to say somewhat more about population differences in maturational timing, other variables held under reasonable control. Still, we must realize that different groups prefer infants and children of different levels of fatness, and in so doing affect maturational timing more by virtue of esthetics than by the units of heredity.

## 4. The Limits of Family-Line Comparisons

In man, as in other life forms, similarities between related individuals provide useful indications as to magnitude and mechanisms of genetic control. The key values of ordinary interest are fourfold: (1) the extent to which correlations between related individuals are significantly different from zero, (2) the extent to which like-sexed and unlike-sexed parent–child and sibling correlations are similar or different, (3) the extent to which monozygotic twin correlations exceed dizygotic twin correlations, and (4) the extent to which sibling $r$s exceed those of parent vs. child. From the numbers so gained we may ordinarily infer (1) inheritance, (2) autosomal vs. sex-chromosomal inheritance, and (3) the estimated extent of "heritability" (Table I and references therein).

Against these theoretical expectations, however mathematically based, there is the very real problem that human beings assort themselves (and are assorted) in remarkably nonrandom ways. Parent and growing child share a cultural milieu, a socioeconomic status (SES), and common attitudes toward food and eating. So an 0.35 parent–child maturational correlation may reflect far more than genetic expectation. Brothers and sisters share more than genes in common and $r$s as high as 0.6, far higher than expectation. Dizygotic twins tend to similarities well in excess of the theoretical maximum of 0.50, and with $r_{DZ}$ often suspiciously high compared with $r_{MZ}$. (Parents of twins will immediately spot the reasons.)

The comparison of parents and their adopted "children" and of genetically unrelated "siblings" immediately points the problem. For adoptive parents and their adopted children come to resemble each other to a rather embarrassing degree, assuming the genetic hypothesis. Indeed, adopted (unrelated) "siblings" also come to resemble each other, far more than sheer chance would allow. From the comparison of adoptive parents and their adoptive children and of adopted (unrelated) "siblings" we begin to realize the limitations of the conventional measures just listed (Garn *et al.*, 1976; 1977). A statistically significant parent–child or sibling correlation does not necessarily reflect the work of the genes. The simple comparison of $r_{MZ}$ and $r_{DZ}$ or $\sigma_{MZ}$ and $\sigma_{DZ}$ does not necessarily provide a true measure of "heritability."

There are other limitations, of course, among them sample size, which can both diminish and exaggerate parent–child and sibling correlations and may lead to erroneous inferences as to X-linked and Y-linked inheritance and possible nutritional effects (cf. Figure 1). There is the tendency to read genetic meaning into age-increasing parent–child and sibling similarities (cf. Rao *et al.*, 1975), yet these may

Table I.  *Estimates of Heritability and Mode of Inheritance as Ascertained from Correlations between Relatives*[a]

| Measure of comparison | Examples or formulas | Genetic comments |
|---|---|---|
| **Correlations between relatives** | | |
| Parent-child correlations | $r_{FaDa}$ $r_{FaSo}$ $r_{MaDa}$ $r_{MaSo}$ | For nondominant inheritance $r \to 0.50$. For X-linked inheritance $r_{FaDa} \to 0.50$, $r_{FaSo} \approx 0.0$ Y-linked $r_{FaSo} \approx 1.0$. |
| Sibling correlations | $r_{SS}$ $r_{BB}$ $r_{SB}$ | For nondominant inheritance $r_{SS}$, $r_{BB}$, $r_{SB} > 0.25 \to 0.50$. For X-linked inheritance $r_{SS} \to 0.75$, $r_{BB} \to 0.50$. For Y-linked inheritance $r_{BB} \to 1.0$, $r_{SS} \approx 0.00$ |
| Twin correlations | $r_{MZ}$ $r_{DZ}$ | Assuming complete genetic control $r_{MZ} \approx 1.0$, $r_{DZ} \approx r_{BB}$, $r_{SS}$, etc. |
| **Estimates of "heritability"** | | |
| Full-sib heritability | $h^2 = \dfrac{2(\sigma_F^2 + \sigma_M^2)}{\sigma_t^2}$ | $h^2$ will vary according to dominance and mode of inheritance. |
| Parent–child heritability | $h^2 = \dfrac{b}{R}$ | $R$, the inbreeding coefficient of offspring, will be ½ for single parent–child $h^2$ and 1 for midparent–child $h^2$ estimates. (From Falconer, 1960). |
| Twin heritability | $h^2 = \dfrac{\sigma_{DZ}^2 - \sigma_{MZ}^2}{\sigma_{DZ}^2}$ | Traditional estimate of $h^2$ using monozygotic and dizygotic twins. |
| | $h^2 = \dfrac{r_{MZ} - r_{DZ}}{1 - r_{DZ}}$ | Holzinger's original formula using correlation coefficients ($r$). |

[a]cf. Falconer, 1960; Li, 1961, pp. 195–204; Cavalli-Sforza and Bodmer, 1971; and Mather and Jinks, 1971.

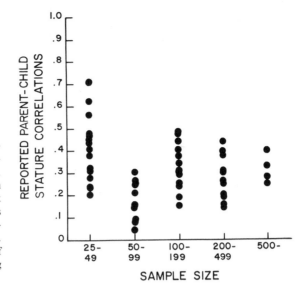

Fig. 1. Effect of sample size on apparent magnitudes of parent–child correlations in stature. In this example, plotting values directly from Table I in Mueller (1976), it is apparent that small-sample correlations yield values from <0.1 to 0.7, whereas large-sample $N$s yield $r$s much closer to 0.3. The danger of misinterpretation of individual parent–child and sibling correlations is thus evident.

be duplicated in adoptive (rather than biological) parent–child and sibling comparisons.

Without belaboring the point, it is scarcely surprising that people who live together may come to resemble each other, for reasons beyond the genes. Yet it is also obvious that biologically related individuals share genes in common leading to dimensional, developmental, and maturational resemblances. Just as population comparisons carry the need to minimize socioeconomic differences (among them income, education, and occupation), family-line comparisons indicate the need to separate the effects of living together from shared genetic control. When it comes to the genetics of maturational timing, the need for new, novel, and quasi-experimental designs is more than apparent.

## 5. Maturity Status at Birth

The problem of genetically determined differences in human maturational timing is evident at birth in sex differences and population differences for birth size, gestation length, size for length of gestation, and the various skeletal and neurological indications of maturity status. Since in all groups considered skeletal maturity is advanced in girls, despite generally smaller dimensions, we may take this feature as a characteristic of the XX as compared with the XY. However, population differences in gestation length, in the proportion of neonates of short gestation length, and in size for gestation length present problems of explanation. Although they are considered extensively in another chapter in this volume, there are cogent reasons to examine them separately here.

If we plot the distribution of gestation lengths for American white and American black neonates, as in Figure 2, we find the two distribution curves displaced, with the curve for the blacks to the left of the curve for the whites. Alternatively, if we calculate the prevalence of "prematurity" as defined by gestation lengths of 37 weeks or less, we find a great excess of gestation-defined prematurity among the blacks. Depending upon the sample, and on various exclusions, the prevalence of

**STANLEY M. GARN
and STEPHEN M.
BAILEY**

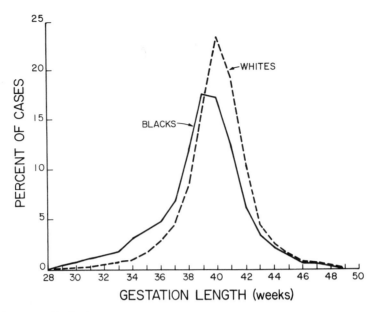

Fig. 2. Frequency distribution of gestation lengths of 11,749 black live-born male singletons (solid line) and 11,544 white live-born male singletons (dashed line). As further corrected for parity, maternal age, income level, etc., short gestation lengths are more common in black newborns, suggesting greater developmental maturity at constant length of gestation (National Institute of Neurological and Communicative Disease Collaborative Survey data, unpublished).

such prematurity may be twice as great in black neonates as in whites (Table II). Either there is a very real black–white difference in the course of prenatal development, such that black infants are ready for birth in a shorter period of time, or the social, economic, and nutritional backgrounds of black mothers combine to precipitate labor earlier. (The black–white comparison is alone made here because the mothers are of comparable body size, in contrast with various groups of Asiatic ancestry.)

Given data on income and maternal occupation and education, it is possible to "match" the mothers, by income groupings, occupational groupings, and educational groupings, to see whether such matching minimizes or even eliminates the apparent black–white differences in the prevalence of prematurity. And, as we find, there are very real socioeconomic determinants of "prematurity," as defined by weight or length of gestation. Gestational prematurity decreases somewhat with

*Table II.   Percent of Short-Gestation-Length Singletons by Various Maternal Categories[a]*

| NINCDS sample | Black neonates | | White neonates | |
|---|---|---|---|---|
| | $N$ | Percent ≤ 37 weeks | $N$ | Percent ≤ 37 weeks |
| Total sample | 23,515 | 25.5 | 22,361 | 11.5 |
| Parity 1 (only) | 6,125 | 25.3 | 6,755 | 10.9 |
| Maternal age 20–35 | 15,624 | 25.9 | 16,842 | 11.6 |
| High school education | 6,352 | 22.7 | 6,546 | 10.7 |

[a]Calculated from NINCDS data.

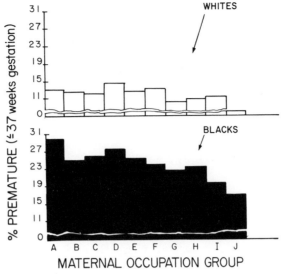

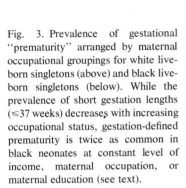

Fig. 3. Prevalence of gestational "prematurity" arranged by maternal occupational groupings for white live-born singletons (above) and black live-born singletons (below). While the prevalence of short gestation lengths (≤37 weeks) decreases with increasing occupational status, gestation-defined prematurity is twice as common in black neonates at constant level of income, maternal occupation, or maternal education (see text).

increasing income. It decreases as we progress from the never-employed or the unemployed through to the managerial and professional levels (Figure 3). It decreases, most dramatically, as maternal educational level increases from less than high school through to college and beyond. Gestational prematurity (gestation lengths less than 37 weeks) is least common in the progeny of the best-educated women whether black or white (Table III). Yet at every income level, every occupational level, and every educational level, prematurity (<37 weeks) is approximately twice as prevalent in black neonates than in white.

Smoking (during pregnancy) has been shown to be a major determinant of birth size of the offspring. And smoking, once an affectation of upper-class women, is now more of a lower-class habit in the United States. So we have explored the effect of maternal smoking on the prevalence of gestation-defined prematurity in blacks and whites, with virtually no effect on the results described above. Corrected for smoking, the large black–white difference in the prevalence of prematurity is still there. Thus, despite all the corrections and matching we have been able to make, the prevalence of prematurity (gestation-defined) remains far larger in blacks (Figure 4).

There is, moreover, compelling evidence that black neonates and white neonates differ in various ways at comparable gestation lengths. Black neonates of both sexes tend to be more advanced or more "mature" at the same gestation lengths, with black boys then about as mature as white girls (cf. Kelley and Reynolds, 1947). Moreover, there are impressive black–white differences in body size for constant

Table III. Percent of Short-Gestation-Length Neonates by Maternal Educational Level[a]

| | Through 6th grade | 7th through 9th grade | 10th through 11th grade | High school graduate | College and/or professional |
|---|---|---|---|---|---|
| Black neonates | 27.7 | 26.7 | 27.0 | 22.7 | 20.5 |
| White neonates | 14.5 | 13.5 | 13.0 | 10.7 | 8.5 |

[a] 21,656 black singletons and 21,112 white singletons.

STANLEY M. GARN
and STEPHEN M.
BAILEY

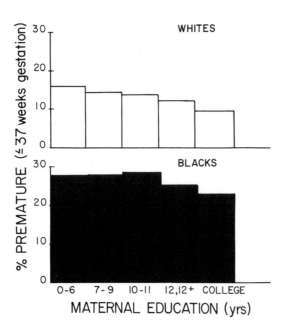

Fig. 4. Gestational ''prematurity'' (≤37 weeks) by maternal educational level in white live-born singletons (above) and black live-born singletons (below) after correction for maternal smoking habits. Although maternal smoking affects both placental weight and birth-weight, elimination of the smoking variable does not alter the characteristically large black–white difference in the prevalence of short gestation lengths.

gestation length, in the range of 36–42 weeks gestation length. Generally speaking, white infants are far heavier than black infants of the same sex for comparable weeks of gestation, as shown in Figure 5. Week-for-week black male neonates are about as large as white female neonates, at which time they are approximately comparable in skeletal (ossification) status.

With the three different socioeconomic variables thus taken into consideration (and maternal size reasonably comparable), several statements may now be made about these black–white differences in maturity status at birth. First, we must

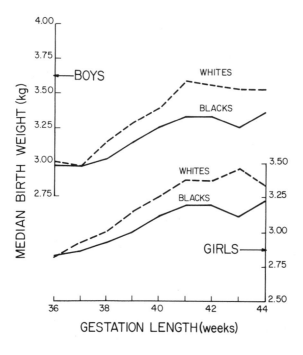

Fig. 5. Weight for length of gestation in the 36- to 44-week range in black live-born singletons (solid line) and white live-born singletons (dashed line) and in boys (above) and girls (below). For constant gestation length newborn girls weigh less than boys and—to an equal extent—black neonates weigh less than white neonates (NINCDS data, unpublished).

acknowledge the socioeconomic factors in total-population data, with maternal education especially important, noting that the prevalence of "prematurity" is somewhat more comparable in blacks and whites if maternal education is taken into account. At the same time, the prevalence of short gestation lengths (<37 weeks) is still far higher in black neonates, after correcting for maternal age, education, parity, and even smoking during pregnancy.

Moreover, when we investigate the prevalence of gestation-defined prematurity in the progeny of cross-racial marriages as contrasted with neonates resulting from intraracial matings, we find evidence that the offspring of black mothers and white fathers tend to be of lower birthweight and higher prevalence of gestation-defined prematurity than is true for reciprocal crossings, i.e., white mothers with black fathers. These preliminary results, newly shown in Table IV, are in accordance with Penrose's hypothesis of a "maternal" effect, with the long-observed black–white differences in length of gestation apparently inherent in the reproductive organs of the mother as contrasted with the developmental programming of the offspring.

The evidence we have here, from some 40,000 live singleton births, suggests that black fetuses become "mature" at a somewhat earlier gestation length than do white fetuses of the same sex if birth is taken as a functional measure of maturity. This statement is made after careful matching for various measures of socioeconomic status (SES) and for smoking during pregnancy. It is in accordance with embryological and fetal evidence and with radiographic (osseous) measures of maturity; it is also in accordance with evidence for advanced dental and skeletal maturity of black infants, children, and adolescents. In order of magnitude, the black–white difference in developmental maturity at birth closely resembles the sex difference in developmental maturity at birth, and it is notable that for comparable gestation lengths black male neonates approximate white female neonates in weight for gestation length.

## 6. Genetic Evidence for Tooth Formation and Eruption Timing

In 1942, Steggerda and Hill (1942) published a most important paper describing population differences in tooth-emergence timing. Moreover, they showed that virtually all groups studied (Navajos, Mayan-speaking Mexicans, American blacks, etc.) were advanced over American whites in timing in the age at eruption (or more correctly *emergence*) of 28 secondary or permanent teeth. Over the 30 years that has ensued, their paper has remained contentious, and a great many opinions were expressed to the effect that such differences were not so. After all, the differences

Table IV. Gestational "Prematurity" (≤ 37 Weeks) in Intraracial and Interracial Matings[a]

| Parental combination | | Total series | | Parity > 1 | |
| --- | --- | --- | --- | --- | --- |
| Mother | Father | N | Percent ≤ 37 weeks | N | Percent ≤ 37 weeks |
| Black | Black | 9,901 | 26.2 | 7,100 | 26.4 |
| Black | White | 34 | 20.6 | 26 | 19.2 |
| White | Black | 98 | 17.3 | 65 | 16.9 |
| White | White | 9,433 | 11.7 | 6,107 | 12.0 |

[a]All live-born singletons. For details see Garn *et al*. (1977*b*).

**STANLEY M. GARN**
**and STEPHEN M.**
**BAILEY**

were in the opposite direction from what one might at first expect. Dahlberg and Menegaz-Bock (1958) wrote a carefully considered critique of population comparisons in tooth-emergence timing, pointing out the many problems arising from inconsistent age identification, differing class intervals, different measures of central tendency, etc.

Analysis of tooth-emergence timing is indeed complicated by these considerations and by dental caries and extractions, which may both speed and delay eruption of selected permanent teeth. The most severe limitation, of course, is sample size. In order to obtain reliable median ages at emergence, for 28 permanent teeth in cross-sectional samples, literally thousands of boys and girls need to be examined. Otherwise the very real sampling problems encountered by Hierneaux (1968) will be repeated.

It is in the context of recent nutrition surveys in the United States that we now have definitive emergence information, and with sufficient socioeconomic information to know the extent to which the income variable is a complicating factor. We can, therefore, summarize pertinent knowledge of tooth emergence derived from the Ten-State Nutrition Survey and the Pre-School Nutrition Survey as follows:

1. Socioeconomic status does have a perceptible bearing on tooth-emergence timing, but its influence is relatively small (Garn *et al.*, 1973*a*).

2. Individuals of largely African ancestry are systematically advanced over individuals of European derivation in the emergence timing of 28 permanent teeth (Garn *et al.*, 1972*c*, 1973*a;* Infante, 1975).

3. The later-erupting deciduous teeth also demonstate earlier emergence in black infants (Infante, 1975).

4. A correction for the degree of white admixture (approximately 20% by most recent estimates) would suggest a still larger difference in emergence timing for the Gold Coast, Ivory Coast, and Slave Coast ancestors of American blacks.

Such evidence, as given in Table V, does support the existence of such population differences in emergence timing, as originally reported by Steggerda and Hill, and creates renewed interest in earlier studies on Formosans (Hurme, 1946), Aleuts (Garn and Moorrees, 1951), and others. At the same time they reiterate the warning of Dahlberg and Menegaz-Bock against the uncritical comparisons of mean

*Table V.  Evidence Bearing on Sex Differences and Black–White Differences in Dental "Maturity"*

| Variables affecting dental "maturity" | Comment |
| --- | --- |
| 1. Nutritional and socioeconomic effects on tooth formation and emergence (Garn *et al.*, 1965, 1973*b;* Infante and Russell, 1974). | Both nutritional status as measured by fatness and per-capita income affect tooth formation and emergence timing, but far less than is true for ossification timing and skeletal ("bone") age. |
| 2. The sex difference in tooth formation and emergence (Garn *et al.*, 1958, 1960, 1967; Infante, 1974). | Girls are advanced over boys in postnatal tooth formation and emergence (except for M3) and in later-forming and emerging deciduous teeth. |
| 3. Black–white differences in tooth emergence (Steggerda and Hill, 1942; Garn *et al.*, 1972*c*, 1973*a*, 1975; Infante, 1975). | Blacks are advanced over whites in later-emerging deciduous teeth and all permanent teeth, even without correction for socioeconomic status. Despite this, caries rates (DMFT scores) are lower in blacks. |

and median values culled from the literature. Small samples, different age group-ings, lack of exact ages, and different methods of computation scarcely allow us the liberty of such bibliographic compilations as are too often made.

In addition to such population comparisons of eruption times, we also have sibling comparisons and twin comparisons of tooth-formation timing (using serial oblique jaw radiographs). Such sibling comparisons reveal the cross-sex and like-sex sibling correlations of the expected order of magnitude, excellent monozygotic twin correspondences, etc. (Garn *et al.*, 1960). Since the nutritional influence is small, even in extremes of fatness (Garn *et al.*, 1965), and since the socioeconomic component appears to be small, these sibling and twin correspondences are proba-bly not much inflated (cf. Figure 6).

The greater proportion of interpersonal variance in tooth-formation timing and also tooth-emergence timing would, therefore, appear to be genetic in nature. Moreover, there appear to be valid population differences in emergence timing (for all the permanent teeth) that transcend socioeconomic differences. However, adop-tion studies, including cross-racial adoptions, are clearly needed before we can calculate tooth emergence $h^2$ values with complete confidence.

If the details of permanent tooth formation (as observed oblique jaw radio-graphs) are taken as maturity measures, then these details—ranging from appear-ance of the follicle to apical closure—are largely under genetic control. It is true that

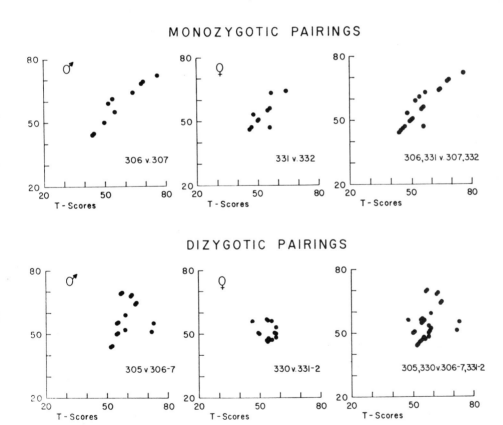

Fig. 6. Scattergrams showing higher monozygotic pair resemblances than dizygotic pair resemblances in permanent tooth formation as ascertained in oblique-jaw radiographs. As shown, the monoxygotic pairs are in far closer agreement at specific stages of tooth formation than is true for the dizygotic pairs compared.

STANLEY M. GARN
and STEPHEN M.
BAILEY

tooth-formation timing is advanced in the obese and retarded in the lean, but far less so than postnatal ossification timing. (Even in such endocrinopathies as hypothyroidism and hyperthyroidism, tooth-formation timing is far less affected than are various measures of skeletal "maturity.") So conventional sibling correlations and monozygotic twin correlations provide a reasonable estimate of the extent of genetic control of tooth-formation timing, with perhaps 90% of interpersonal variance so explained. Averaging methods expressing "dental age" as the average for available teeth may be expected to yield similar estimates.

If the details of tooth emergence are similarly analyzed, tooth by tooth, and with samples of adequate size, then at least three generalizations may be made concerning dental "maturation." First, there is a socioeconomic effect, although a small one, with the more affluent expectedly advanced in tooth emergence and the poor (expectedly) slightly delayed. Second, there is a very clear genetic effect, as shown by twin and sibling correlations, with perhaps 85–90% of emergence variability so explained in samples where dental caries and premature loss of deciduous teeth and some permanent teeth is not a complicating factor. Finally, there are real population differences in tooth-emergence timing, with at least the later-emerging deciduous teeth and all permanent teeth advanced in American blacks over their age-matched white peers. Indeed, it would appear that tooth emergence and dental "maturity" is advanced in most groups of Asiatic, Amerindian, and African origin or ancestry as compared with Europeans or Americans of European descent.

## 7. Population Differences in Postnatal Ossification Timing

Population differences in postnatal skeletal maturation (and more specifically postnatal ossification timing) have long provided tantalizing hints that some populations differed, in the timing of skeletal development, on a genetic basis. But such hints were confused by the fact that the evidence appeared contradictory and that different measures were often employed and on groups differing in socioeconomic status. The Brush Foundation radiographs, employed in the familiar Gruelich–Pyle (1969) norms, were taken on a uniquely affluent and dimensionally large group of children, quite different from the various English samples incorporated into the later Tanner-Whitehouse norms (Tanner et al., 1966).

Apart from these problems in reconciling "standards" because of sampling differences, and differences arising from assumptions incorporated into the scoring procedures, there was one set of differences that was clearly puzzling. American studies of pediatric patients and clinically healthy boys and girls placed American black boys and girls ahead of their white peers, not only at birth and in the first year, but for some years thereafter, and even into adolescence. African studies, from South Africa, East Africa, and the Congo, tended to show African blacks as advanced over whites in early skeletal development, but falling behind (i.e., delayed) at later ages. In various African studies the delay was pinpointed as taking place by (1) the end of the first year, (2) the end of the second year, or (3) the end of the third year. Clearly, black boys and girls cannot be both advanced and retarded in skeletal maturity on a purely genetic basis, nor genetically ahead in the U.S.A. and genetically retarded in most of Africa (cf. Table VI).

We now have enough data from tens of thousands of children in the U.S.A. to reconcile these apparent discrepancies. And although the skeletal measures differ, the results are much the same. Using the median age at ossification for 28–30

Table VI.   *Apparent Black–White Differences in Postnatal Ossification Timing*

| Authors | Findings |
|---|---|
| **African studies** | |
| Jones and Dean, 1956 | Though skeletally advanced over whites at birth, |
| Massé and Hunt, 1963 | black boys and girls tend to fall behind white boys |
| Tobias, 1958 | and girls by 1.0, 2.0, or 3.0 years, depending upon |
| Falkner *et al.,* 1958 | sample comparisons. |
| Beresowski and Lundie, 1952 | |
| See also Garn *et al.,* 1972*b* | |
| **North American studies** | |
| Hess and Weinstock, 1925 | Even without income correction and especially |
| Dunham *et al.,* 1939 | after income correction American (black) boys |
| Christie, 1949 | and girls are advanced over American white |
| Malina, 1969 | children in postnatal ossification timing and |
| Garn *et al.,* 1972*b* | skeletal "maturity" ratings from birth onward. |
| Roche *et al.,* 1974 | |
| Roche *et al.,* 1975 | |
| Roche *et al.,* 1976 | |

postnatal ossification centers of the hand and wrist, American black boys and girls tend to be advanced over their white age peers. This generalization is drawn from the data of the Ten-State Nutrition Survey of 1968–1970, the Pre-School Nutrition Survey, and Head-Start radiographic data. Even before income matching, and especially after income matching, the black boys and girls (with some 80% of African genes) tend to be advanced in the ossification centers through to those latest to become radiographically visible (Garn *et al.,* 1972*b*, 1973*b*).

Data from the National Health Examination are not only differently collected (on a national probability basis) but quite differently analyzed (using a unisex variant of the Gruelich–Pyle standards). Nevertheless, the blacks so analyzed prove to be skeletally advanced over the whites. In actuality, although the National Health Examination data are national probability in scope (including each economic level in proportion), the blacks are still of lower (lesser) socioeconomic status than the whites. Unless American blacks, with some 20% of white genes and derived from the Ivory, Gold, and Slave Coasts, represent some unique sampling of Africans, two conclusions are inescapable. First, black Africans and their genetic relatives in the U.S.A. are genetically advanced over whites in postnatal ossification timing, or in "bone age," as variously ascertained. Second, the blacks incorporated into the various African studies must be subject to increasing developmental delay as the result of differential access to calories and nutrients and restricted access to medical care.

## 8. Population Differences in Postnatal Ossification Sequence

Separate from the question of ossification timing and its genetic determination is the question of ossification *sequence* or order. After all, the ossification sequence could be fixed and invariable, but with the age at radiographic appearance uniformly earlier or later, or there could be variations in both ossification sequence and timing. It could be that the sequence (or order) is the same in both sexes, with timing the principal distinction, or both sequence *and* timing could differ between boys and girls.

STANLEY M. GARN
and STEPHEN M.
BAILEY

The mean age at ossification (or better, the median age at ossification), although differing between the sexes, suggests a similar sequence, a point emphasized by Gruelich and Pyle (1969). Yet the order or sequence of the mean ages at ossification is not the same as the order or sequence of ossification centers as seen in individuals. By way of example, there are children in whom the hamate precedes the capitate, reversing the 1–2 order of the means, and there are boys and girls in whom the triquetral precedes most of the centers of the hand, or in whom the trapezium and trapezoid are unusually late. This kind of observation was made early in the history of radiography (more than 60 years ago) by Pryor, who observed sibling similarities in ossification order as well (Pryor, 1907). It was emphasized by Christ (1961) in his studies of South African children, and it was the subject of several of our earlier studies, using serial radiographs from the Fels Research Institute Longitudinal studies (Garn and Rohmann, 1960; Garn et al., 1961, 1967). Dreizen and his associates have raised the question of whether atypical ossification sequences might be the result of chronic or acute malnutrition or specific nutrient deficiency (Dreizen et al., 1964).

It is, of course, possible to show that unusual ossification sequences are often found in siblings, as we did with brother–brother, sister–sister, and brother–sister pairs. Given two-generational longitudinal radiographic data, it is also possible to demonstrate one-parent/child similarities in the extent of unusual or atypical ossification sequences or orders (Garn and Rohmann, 1962), suggesting rather simple modes of inheritance (cf. Garn and Shamir, 1958). And there is, as expected, higher monozygotic than dizygotic twin concordance with respect to ossification sequence or order—circumstantial evidence as to probable genetic control and akin to twin similarities in the *order* of formation, calcification, and movement of the teeth.

Apart from the laborious but necessary ascertainment of individual ossification sequences from serial longitudinal radiographs (possible only with complete series of films rather closely spaced), there is another approach to the problem. This is practical with cross-sectional, single-film data, provided that the usable $N$ is large, $N \approx 1000$ or more. It involves the computer identification and tabulation of every possible hand–wrist ossification sequence for $N'$ ossification centers, for $N$ individuals in whom at least one such center is radiographically visible, but not more than 27. For 28 postnatal ossification centers there are 756 theoretically possible dichotomous (present–absent/absent–present) ossification sequences for the hand alone. Amazingly, more than 600 of the 756 theoretically possible dichotomous hand–wrist ossification sequences can be demonstrated in 3000–5000 apparently normal boys and girls.

While it is possible to provide frequencies for each ossification sequence (capitate–hamate/hamate–capitate, etc.), it is simpler here to refer to sequences observed or seen and sequences not observed and not seen. It is possible to picture sequences observed or seen in a simple $28 \times 28$ matrix, denoting those observed in black and those not observed in white (Figure 7). One can "read" the matrix, to ascertain what sequence we have observed and what sequence we have not, but the black areas and the white areas on the matrix tell the elementary story. Of the 624 dichotomous (present–absent/absent–present) hand–wrist sequences or orders actually observed or seen, there are some that are peculiar to girls and some that are restricted to boys. The sexes are not quite identical in respect to individual dichotomous ossification sequences or orders, and the matrix of sequences or orders differs somewhat between whites and blacks and between boys and girls (Garn et al., 1972a, 1975).

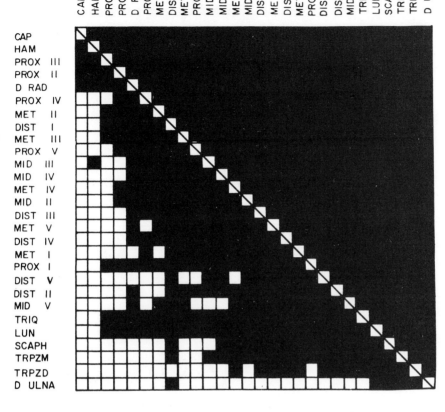

■ 626 Observed (2972 White Ss M+F)

Fig. 7. Dichotomous (present–absent/absent–present) hand–wrist ossification sequences observed in 2972 apparently normal white boys and girls. Black squares indicate 626 sequences actually observed; white squares denote sequences not observed, and the entire matrix indicates the problem of skeletal-maturity assessment by conventional averaging methods. Similar problems apply to tooth-emergence sequence variability, where genetically unique sequences must also be taken into consideration.

If, then, we turn to grossly abnormal individuals, with congenital malformation syndromes, skeletal abnormalities, karyotypic abnormalities, and growth defects, we discover that many sequences not previously observed or seen are now present. With such patients much of the blank or white area of the matrix becomes black. Indeed, it is then simple to show that some 97 dichotomous ossification sequences not observed or seen in "normal" white boys and girls are observed in the congenital malformation syndromes. Abnormal children, we can certainly say, do evidence ossification sequences peculiar to them and not seen or observed in the normal group (Figure 8).

A particular example is *diastrophic dwarfism,* originally described by Lamay and Maroteaux (1960). Perhaps better than any other congenital malformation syndrome or developmental abnormality, diastrophic dwarfism evidences a surprising number of sequences not seen in normal individuals and unique to itself. Indeed, some 33 diastrophic dwarfs (but only 12 in the appropriate age range for ossification

STANLEY M. GARN
and STEPHEN M.
BAILEY

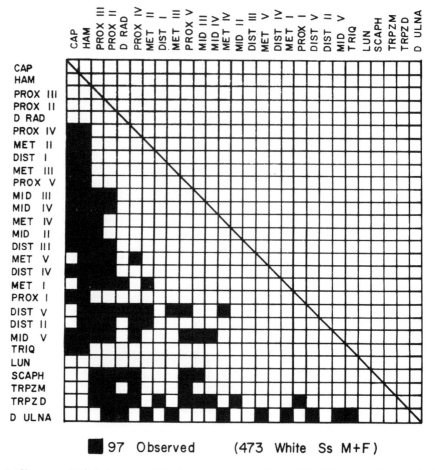

■ 97 Observed    (473 White Ss M+F)

Fig. 8. Shown are 93 dichotomous ossification sequences not observed in 2972 apparently normal white children but observed in 473 children with congenital malformation syndromes. Differences in sequence as well as in timing characterize these malformation syndromes, emphasizing the problem of defining skeletal "maturity," and showing novel problems in the investigation of maturity *per se* (cf. Pryor, 1907; Christ, 1961).

sequence determination) evidenced 23 sequences unique to this condition (Figure 9). Other congenital malformation syndromes also make their contributions. For some (such as homocystinuria) a single center in a unique sequence is virtually diagnostic. And, then, there are additional karyotypic abnormalities that are characterized not only by abnormal sequences but by centers not seen in normal individuals. In these situations we have genetically determined and chromosomally determined sequences not only unique to the disorder but involving centers unique to the disorder as well!

From studies such as these, the existence and extent of genetically determined ossification sequence variability and genetic control of ossification sequence (or order) become abundantly clear. Comparisons of dichotomous ossification sequences in monozygotic and dizygotic twins further prove the point (Hertzog *et al.*, 1969).

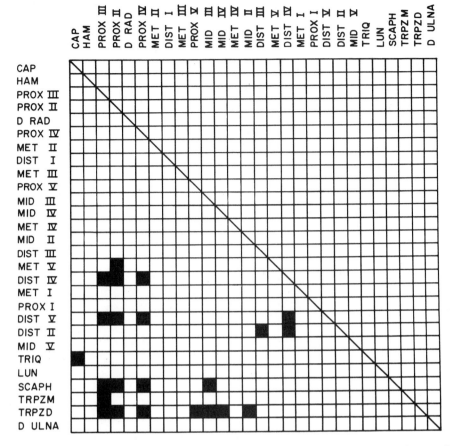

Fig. 9. Matrix of dichotomous (present–absent/absent–present) ossification sequences unique to diastrophic dwarfism. As shown here, and in 12 cases of suitable age, there are 33 ossification sequences not observed in clinically normal boys and girls and actually unique to this type of dwarfism (Garn, Poznanski and Shaw, unpublished data, 1977).

## 9. Genetics of Sexual Maturation

Information on population differences and family-line similarities in sexual maturation is ordinarily of limited value, the timing of sexual maturation being much influenced by nutrition. Indeed, differences in maturational timing between obese and lean girls may be as much as 4 years (cf. Garn and Haskell, 1960), and for this reason the simple comparison of age at menarche in different populations provides little real indication of the extent to which this developmental event is truly genetically determined. Even the familiar mother–daughter correlations in the age at menarche are somewhat suspect (since mothers and daughters also tend to similarity in both socioeconomic status and level of fatness). Moreover, the "secular" or generational trend in menarcheal timing may be a complicating factor if this trend is not taken into proper account.

These complications may be avoided if there is information on the level of fatness of both mother and daughter and with data sufficient in size and completeness to afford socioeconomic correction. Moreover, if the study goes beyond the simple reporting of age at menarche, giving attention to other developmental measures and developmental variables in the sons as well as the daughters of early-maturing and late-maturing mothers, then far more useful information will inevitably result.

Taking mother–daughter menarcheal correlations for some 550 mother–daughter pairs from the Tecumseh (Michigan) Project of the University of Michigan School of Public Health into account, the mother–daughter $r$s approximate 0.25—a value much like that reported by other investigators (Popenoe, 1928; Kantero and Widholm, 1971), and this value is automatically corrected for secular trend by using normalized $Z$ scores for menarcheal age of the mothers and their daughters alike (Table VII).

Actually, the degree of fatness in common does not much affect the mother–daughter correlation. Despite the 0.21 mother–daughter correlation in fatness, as shown in Table VII, the partial (fatness-corrected) mother–daughter correlation in age at menarche is virtually the same as the raw-order correlation. Thus, the value of 0.25 represents a fair estimate of the *true* mother–daughter similarity in menarcheal timing, given a sample of sufficient size ($N \geqslant 500$) and with the fatness factor filtered out. Naturally, there is the very real question as to whether age at menarche of the mother is more than a maturational indicator and whether it is a major developmental and maturational variable that may be reflected in the dimensional, developmental, and maturational progress of their children. Moreover, a second question arises as to whether early-maturing and late-maturing women differ systematically well beyond the age at menarche.

Looking at the sons and the daughters of early-maturing women (those earlier than the 15th percentile for age at menarche) and the late-maturing women (those at and beyond the 85th percentile, decade-corrected), very real and major developmental and dimensional differences can be seen. Both the sons and the daughters of the early-maturing mothers tend to be systematically heavier, not only in childhood and adolescence but also in adulthood, as is clearly summarized in Figure 10. And both the sons and the daughters of early-maturing mothers tend to be advanced in age of attainment of the adductor sesamoid while the children of late-maturing mothers tend to be correspondingly delayed in this measure of osseous maturity (Figure 11). Without question, therefore, as shown in these radiographic data from the Ten-State Nutrition Survey, the factors affecting menarcheal timing in women are reflected in the growth and development and maturational timing of their children of both sexes.

Table VII.   Two-Generational Menarcheal Correlations Corrected for Fatness[a]

| Correlations for menarche | Correlation designation | $N$ | $r$ |
|---|---|---|---|
| Mother–daughter menarche | $r_{12}$ | 550 | 0.25 |
| Mother–daugher fatness | $r_{34}$ | 546 | 0.21 |
| Mothers' fatness vs. menarche | $r_{13}$ | 548 | −0.11 |
| Daughters' fatness vs. menarche | $r_{24}$ | 548 | −0.14 |
| Mother–daughter menarche corrected for fatness | $r_{12\cdot34}$ | 546 | 0.23 |

[a]Data from examination round 1 of the Tecumseh Project of the University of Michigan School of Public Health.

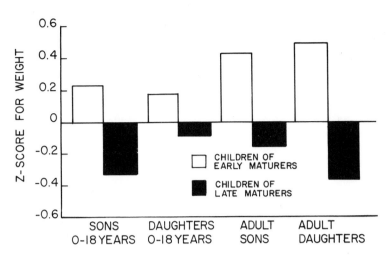

Fig. 10. Body weight in the children of early-maturing women (menarcheal age in the lowest 15th percentile) and in the children of late-maturing women (menarcheal age in the 85th percentile and beyond). Here shown for 170 sons, 191 daughters, 61 adult sons, and 51 adult daughters, children of early-maturing mothers tend to weigh more and children of late-maturing mothers tend to weigh less (Unpublished data analysis from the Tecumseh Project of the University of Michigan using high-school education as a restriction so as to eliminate educational and socioeconomic extremes.

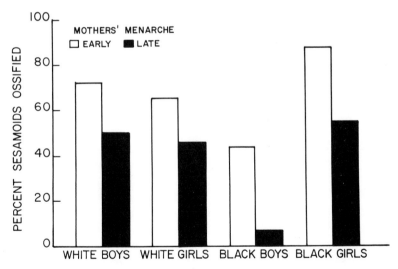

Fig. 11. Relationship between maternal menarcheal age and adductor sesamoid ossification in sons and daughters of early-maturing mothers (white bars) and late-maturing mothers (solid bars). As shown, the percent of children with adductor sesamoid ossification rdiographically present at a given age is greater in the children of the early-maturing mothers and less in the sons and daughters of late-maturing mothers, showing that factors that control age at menarche in women are reflected in the osseous maturity of their children of both sexes (radiographic data from the Ten-State Nutrition Survey; the key ages are 13, 11, 12, and 11, respectively).

It is of further interest, therefore, that early-maturing women (through the 15th percentile for age at menarche) and later-maturing women (beyond the 85th percentile for their cohort group) differ systematically long after the age at menarche, even through the 7th and 8th decades. Taking the data on their children of both sexes, from infancy onward, it is possible to make three generalizations from these studies of many thousands of women, and including both blacks and whites, separately tabulated. First, the age at menarche (although nutritionally mediated in part) has a reasonable genetic component, as indicated by the fatness-corrected mother–daughter correlations further corrected for secular (generational) trend. Second, maternal age at menarche is reflected in the dimensional and maturational progress of their offspring of both sexes, so that the maturational effect is not sex-limited, although it may be partially X-linked. Third, early-maturing and late-maturing women differ at all later ages in both stature and in fatness so that the simple age at menarche is part of a constellation of developmental differences.

Thus, the age at menarche emerges as a key maturational variable, relevant to dimensional as well as developmental differences, not only in early-maturing and late-maturing women themselves and through the later decades, but to the dimensional and developmental progress of their offspring of both sexes, from infancy through adulthood. Although the age at menarche can still differ markedly between sisters, anthropometric and radiographic comparisons of their offspring offer the possibilities of ''pedigree'' studies, if not of formal genetic analysis.

## 10. Strategies for Research

From what we have shown so far, it follows that both conventional population comparisons and conventional family-line and twin analyses may obscure the extent to which maturational timing is under genetic control. Even with the best of income matching, two sympatric populations may still differ in attitudes toward food and eating and so differ in maturational timing on a purely nongenetic basis. Even the correction of monozygotic (MZ) twin similarities by dizygotic (DZ) twin similarities may be insufficient if monozygotic twin and dizygotic twins are treated differently during development. And, of course, many observed sibling similarities are of a magnitude far greater than genetic theory would allow.

Population comparisons can perhaps be improved by borrowing the ''desert island'' model, which for practical purposes comes down to the biracial orphanage situation, cross-racial adoptions, and the progeny of biracial marriages. In the biracial orphanage, children of two differing population groups are reared together under common conditions. In cross-racial adoptions, matched for income level, we also have the human equivalent of cross-rearing, which may be improved if there are at least two children in each adoptive family, one from each of two population groups. In cross-racial or biracial marriages, given two population groups, A and B, we can further explore the two parental possibilities $F_A M_B$ and $F_B M_A$ insofar as maturational timing of the ''hybrid'' offspring is concerned (cf. Table VIII).

Within families, adoptions offer the possiblity of ascertaining the nongenetic extent of maturational control. Given two adopted children in each family, excluding adopted relatives, the extent of nongenetic similarity can be measured. This model can also be applied to a pair of children, one adopted and one not, as long as they are genetically unrelated. It may be applied to two biological children of different parents living together as the result of divorce and remarriage.

With larger families, involving at least a pair of biological full sibs and an

Table VIII.   *Strategies in Human Developmental Genetics*

| Conventional strategy and purpose | New and necessary strategy |
|---|---|
| 1. Husband–wife correlations (to measure assortative mating or "homogamy"). | Husband–wife correlations at the time of marriage or after divorce or separation. |
| 2. Parent–child correlations (to measure one-parent and midparent vs. child resemblances). | Parent–child correlations in adult and separated children, in adopted children, and in biological vs. adopted pairings. |
| 3. Sibling correlations (to measure resemblances and, with 2, to estimate "heritability"). | Comparison of separated siblings, adult siblings, genetically unrelated (adopted) "siblings" and in biological vs. adopted sibling pairs. |
| 4. Twin correlations, MZ and DZ, to estimate $h^2$. | Comparison of adult twins, separated MZ and DZ twins, twins, and biologically unrelated adopted siblings and half-siblings, children of twins. |
| 5. Population comparisons (to ascertain population differences). | Cross-racial adoptions, biracial or cross-racial marriages, institutional comparisons, cross biracial matings. |

unrelated adopted "sibling," the ideal analysis is possible. This can take place in the course of simple adoption (to "complete" a family) or following divorce and remarriage (but avoiding half-sibs).

Lest the maturational comparison of genetically unrelated "siblings" living together or of genetically related siblings living apart seem impractical, let us not forget that one out of every ten families currently affords these very investigative possibilities. Recently, in fact, we have explored comparisons of parents and their adoptive children and even families containing two or more children, one biological and one adoptive (Garn *et al.,* 1976, 1977*a*). And, lest the maturational comparisons of parents and their biological children at identical ages in childhood and adolescence seem impractical, just this possibility exists with available data from existing longitudinal growth studies, as we demonstrated for hand–wrist ossification 15 years ago (cf. Garn and Rohmann, 1962, and Table IX). Accordingly, strategies for research on the genetics of maturational timing are both numerous and rewarding.

Table IX.   *Magnitude of One Parent–Child Similarities in Number of Hand–Wrist Ossification Centers* [a]

| Exact age | No pairs | Correlation coefficient ($r$) |
|---|---|---|
| 1.0 | 27 | 0.50 |
| 1.5 | 42 | 0.34 |
| 2.0 | 53 | 0.37 |
| 2.5 | 52 | 0.30 |
| 3.0 | 54 | 0.36 |
| 3.5 | 45 | 0.25 |
| 4.0 | 35 | 0.19 |
| 4.5 | 35 | 0.33 |
| 5.0 | 22 | 0.32 |

[a]From Garn and Rohmann, 1962.

**STANLEY M. GARN**
**and STEPHEN M.**
**BAILEY**

## 11. Conclusions and Summary

When it comes to maturational events, there are at least 1000 discrete phenomena that can be or may be investigated in brother pairs, sister pairs, and brother–sister pairings. These include various dichotomous (present–absent) stages of osseous, dental, psychomotor, and sexual maturation. A smaller number of events may be compared in parents and their children, until truly longitudinal two-generational data become available. What we know now, however, is remarkably limited, in large part because longitudinal growth studies have been relatively few, with too few sibling pairs to make such comparisons truly practicable. Moreover, conventional parent–child and sibling comparisons fail to take into account the developmental and maturational consequences of living together, as apart from genes held in common. And conventional population comparisons, too, rarely include adequate corrections for socioeconomic variables (among them income, education, and occupation).

Present evidence strongly suggests genetic control of dental maturation (tooth formation, tooth emergence, and apical closure). For the teeth the nutritional effect and the economic effect—although present—are seemingly minimal. Present evidence also favors partial genetic control of postnatal ossification timing, "remodeling," and epiphyseal union. Although there are both nutritional and socioeconomic effects, often to a considerable degree, skeletal maturation, like dental maturation appears to be advanced in blacks (of largely African origin) in the U.S.A. even before socioeconomic matching. Analyses also show that the ossification sequence (or order) is also under genetic control, quite apart from postnatal ossification timing *per se*.

Menarcheal timing is affected by secular trend, besides being income-related, nutritionally mediated, and earlier in the obese and later in the lean. But the mother–daughter and sister–sister similarities in the age at menarche emerge as being separately and independently under genetic control and of major relevance to growth and development. Maturity at birth is similarly complicated, with such variables as smoking and maternal education involved. Yet, when all such variables are held constant, maturity status at birth appears to be greater in blacks than in whites, even taking birth weight relative to gestation length into account.

What we know about the genetics of maturational timing is limited in amount and further limited by conventional research designs. Because individuals from different populations are rarely comparable in social and economic status and because members of a "family" share a common milieu (even when they are genetically unrelated), new designs and new strategies are necessary to unravel the descriptive genetics of human maturational events and to prepare for the day when formal genetic analysis is truly possible.

## 12. References

Beresowski, A., and Lundie, J. K., 1952, Sequence in the time of ossification of the carpal bones in 705 African children from birth to six years of age, *S. Afr. J. Med. Sci.* **17**:25.
Cavalli-Sforza, L. L., and Bodmer, W. F., 1971, *The Genetics of Human Populations,* W. H. Freeman, San Francisco.
Christ, H. H., 1961, A discussion of causes of error in the determination of chronological age in children by means of x-ray studies of carpal-bone development, *S. Afr. Med. J.* **35**:854–857.
Christie, A. V., 1949, Prevalence and distribution of ossification centers in the newborn infant, *Am. J. Dis. Child.* **77**:355.

Dunham, F. C., Jenss, R. M., and Christie, A. V., 1939, A consideration of race and sex in relation to the growth and development of infants, *J. Pediatr.* **14**:156.

Dreizen, S., Spirakis, C. N., and Stone, R. E., 1964, Chronic undernutrition and postnatal ossification, *Am. J. Dis. Child.* **93**:122–172.

Falconer, D. S., 1960, *Introduction to Quantitative Genetics,* Ronald Press, New York.

Falkner, F., Pernot-Roy, M. P., Habich, H., Sénécal, J., and Massé, G., 1958, Some international comparisons of physical growth in the first two years of life, *Courrier* **8**:1.

Garn, S. M., 1964, *The Genetics of Normal Human Growth,* Dell'Istituto Gregorio Mendel, Rome.

Garn, S. M., and Haskell, J. A., 1960, Fat thickness and developmental status in childhood and adolescence, *Am. J. Dis. Child.* **99**:746–751.

Garn, S. M., and Moorrees, C. F. A., 1951, Stature, body build and tooth emergence in the Aleutian aleut, *Child. Dev.* **22**:262–270.

Garn, S. M., and Rohmann, C. G., 1960, Variability in the order of ossification of the bony centers of the hand and wrist, *Am. J. Phys. Anthropol.* **18**:219–230.

Garn, S. M., and Rohmann, C. G., 1962, Parent–child similarities in hand–wrist ossification, *Am. J. Dis. Child.* **103**:603–607.

Garn, S. M., and Shamir, Z., 1958, *Methods for Research in Human Growth,* Charles C. Thomas, Springfield, Illinois.

Garn, S. M., Lewis, A. B., Koski, K., and Polacheck, D. L., 1958, The sex difference in tooth calcification, *J. Dent. Res.* **30**:561–567.

Garn, S. M., Lewis, A. B., and Polacheck, D. L., 1960, Sibling similarities in dental development, *J. Dent. Res.* **39**:170–175.

Garn, S. M., Rohmann, C. G., and Robinow, M., 1961, Increments in hand–wrist ossification, *Am. J. Phys. Anthropol.* **19**:45–53.

Garn, S. M., Lewis, A. B., and Kerewsky, R. S., 1965, Genetic, nutritional and maturational correlates of dental development, *J. Dent. Res.* **44**:228–242.

Garn, S. M., Rohmann, C. G., Blumenthal, T., and Silverman, F. N., 1967, Ossification communalities of the hand and other body parts, *Am. J. Phys. Anthropol.* **27**:75–82.

Garn, S. M., Sandusky, S. T., Miller, R. L., and Nagy, J. M., 1972a, Developmental implications of dichotomous ossification sequences in the wrist region, *Am. J. Phys. Anthropol.* **37**:111–115.

Garn, S. M., Sandusky, S. T., Nagy, J. M., and McCann, M. B., 1972b, Advanced skeletal development in low-income negro children, *J. Pediatr.* **80**:965–969.

Garn, S. M., Wertheimer, F., Sandusky, S. T., and McCann, M. B., 1972c, Advanced tooth emergence in negro individuals, *J. Dent. Res.* **51**:1509.

Garn, S. M., Sandusky, S. T., Nagy, J. M., and Trowbridge, F. C., 1973a, Negro–caucasoid differences in permanent tooth emergence at a constant income level, *Arch. Oral Biol.* **18**:609–615.

Garn, S. M., Sandusky, S. T., Rosen, N. N., and Trowbridge, F., 1973b, Economic impact on postnatal ossification, *Am. J. Phys. Anthropol.* **37**:1–4.

Garn, S. M., Poznanski, A. K., and Larson, K. E., 1975, Magnitude of sex differences in dichotomous ossification sequences of the hand and wrist, *Am. J. Phys. Anthropol.* **42**:85–89.

Garn, S. M., Bailey, S. M., and Cole, P. E., 1976, Similarities between parents and their adopted children, *Am. J. Phys. Anthropol.* **45**:539–543.

Garn, S. M., Cole, P. E., and Bailey, S. M., 1977a, Effect of parental fatness levels on the fatness of biological and adoptive children, *Ecol. Food Nutr.* **7**:91–93.

Garn, S. M., Shaw, H. A., and McCabe, K. D., 1977b, Effects of socioeconomic status (SES) and race on weight-defined and gestational prematurity in the U.S.A., in: *The Epidemiology of Prematurity* (D. Reed and F. Stanley, eds.) Urban and Schwartenberg, Baltimore.

Gruelich, W. W., and Pyle, S. I., 1969, *Radiographic Atlas of Skeletal Development of the Hand and Wrist,* 2nd ed., Stanford University Press, Calif.

Hertzog, K. P., Falkner, F., and Garn, S. M., 1969, The genetic determination of ossification sequence polymorphism, *Am. J. Phys. Anthropol.* **30**:141–144.

Hess, A. F., and Weinstock, M., 1925, A comparison of the evolution of carpal centers in white and negro newborn infants, *Am. J. Dis. Child.* **29**:347.

Hiernaux, J., 1968, Ethnic differences in growth and development, *Eugen. Q.* **15**:12–21.

Hurme, V. O., 1946, Decay of the deciduous teeth of Formosa Chinese, *J. Dent. Res.* **25**:127–136.

Infante, P. F., 1974, Sex differences in the chronology of deciduous tooth emergence in white and black children, *J. Dent. Res.* **53**:418–421.

Infante, P. F., 1975, An epidemiological study of deciduous tooth emergence and growth in white and black children of southeastern Michigan, *Ecol. Food Nutr.* **4**:117–124.

Infante, P. F., and Russell, A. L., 1974, An epidemiologic study of dental caries in preschool children in the United States by race and socioeconomic level, *J. Dent. Res.* **53**:393–396.

Jones, P. R. M., and Dean, R. F. A., 1956, The effects of kwashiorkor on the development of the bones of the hand, *J. Trop. Pediatr.* 2:51.

Kantero, R.-L., and Widholm, O., 1971, Correlations of menstrual traits between adolescent girls and their mothers, *Acta Obstet. Gynecol. Scand.* **50:**30–36.

Kelley, H. J., and Reynolds, L., 1947, Appearance and growth of ossification centers and increases in the body dimensions of white and negro infants, *Am. J. Roentgenol.* **57:**477.

Lamay, M., and Maroteaux, P., 1960, Le nanisme diastrophique, *Presse Med.* **68:**1977.

Li, C. C., 1961, *Human Genetics,* McGraw-Hill, New York.

Malina, R. M., 1969, Skeletal maturation rate in North American negro and white children, *Nature* **223:**1075.

Massé, G., and Hunt, E. E., Jr., 1963, Skeletal maturation in the hand and wrist in West African children, *Hum. Biol.* **35:**3–25.

Mather, K., and Jinks, J. L., 1971, *Biometrical Genetics,* Chapman and Hall, London.

Mueller, W. H., 1976, Parent–child correlations for stature and weight among school aged children: A review of 24 studies, *Hum. Biol.* **48:**379–397.

Popenoe, P., 1928, Inheritance of age of onset of menstruation, *Eugen. News* **13:**101.

Pryor, J. W., 1907, The hereditary nature of variation in the ossification of bones, *Anat. Rec.* **1:**84–88.

Rao, D. C., MacLean, C. F., Morton, N. E., and Yee, S., 1975, Analysis of family resemblance. V: Height and weight in northeastern Brazil, *Am. J. Hum. Genet.* **27:**509–520.

Roche, A. F., Roberts, J., and Hamill, P. V. V., 1974, Skeletal maturity of children 6–11 years (United States), DHEW Publication No. (HRA) 75-1622 (Vital and Health Statistics, Series 11, Number 140).

Roche, A. F., Roberts, J., and Hamill, P. V. V., 1975, Skeletal maturity of children 6–11 years: Racial, geographic area and socioeconomic differentials, DHEW Publication No. (HRA) 76-1631 (Vital and Health Statistics, Series 11, Number 149).

Roche, A. F., Roberts, J., and Hamill, P. V. V., 1976, Skeletal maturity of children 12–17 years (United States), DHEW Publication No. (HRA) 76-0000 (Vital and Health Statistics, Series 11, Number 160).

Rowe, N. H., Garn, S. M., Clark, D., and Guire, K. E., 1976, The effect of age, sex, race and economic status on dental caries experience of the permanent dentition, *Pediatrics* **57:**457–461.

Steggerda, M., and Hill, T. J., 1942, Eruption time of teeth among Whites, Negroes and Indians, *Am. J. Orthodont. Oral Surg.* **28:**361–370.

Tobias, P. V., 1958, Some aspects of the biology of the Bantu-speaking African, *Leech* **28:**3.

Tanner, J. M., Whitehouse, R. H., and Takaishi, M., 1966, Standards from birth to maturity on height, weight, height velocity and weight velocity: British children, 1965, *Arch. Dis. Child.* **41:**454–471; 613–635.

# IV
## *Prenatal Growth*

# 13

# *Anatomy of the Placenta*

## *DOUGLAS R. SHANKLIN*

## *1. The Normal Discoid Singleton Placenta*

### *1.1. Gross Form*

#### *1.1.1. Postimplantation Period*

The future shape of the discoid placenta is not visible in the early implantational stages. The earliest stage seen by this author was about ten days postconceptional, that is some three to five days postimplantation (Figure 1). The trophoblast was a loose aggregation of cells resembling tissue culture explants. The trophoblast was seemingly random, with many open spaces and an irregular, hemorrhagic zone at the interface with the endometrial stroma. Delamination of prototrophoblastic cells from these anchors serve as the placental mesenchyme. This gives rise to angioblasts, fibroblasts, and placental reticuloendothelial cells. Just after this same period, generally, but in some areas earlier, the distinction between cytotrophoblast and syncytial trophoblast becomes apparent (Figure 2). The open spaces or lacunae eventually coalesce into the intervillous space, with communication to the maternal vascular system at the sites of hemorrhagic leaks. This previllous stage is seen generally about the inner cell mass, the future embryo, which also shows axial and orientational differentiation during this period.

#### *1.1.2. Definite Discoid Shape*

The first step towards discoid shape comes with the differentiation of the amnion from the chorion. This occurs about the twelfth day according to Hertig (1968). Although it will be several weeks before the discoid form is developed, this act sets the stage, as the mesodermal components of the chorion and amnion are the framework for the discoid placenta. Future growth is also conditioned by the secretion of amniotic fluid and the compression of the decidua reflexa on the side of

***DOUGLAS R. SHANKLIN*** • Chicago Lying-in Hospital, and University of Chicago, Illinois.

DOUGLAS R.
SHANKLIN

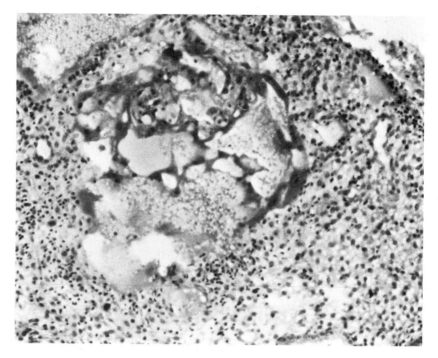

Fig. 1. Early implantation. Implanting trabecular trophoblast, blood lakes, and early embryonic cell mass. The surrounding endometrium shows an early decidual metaplasia. H&E, 100×.

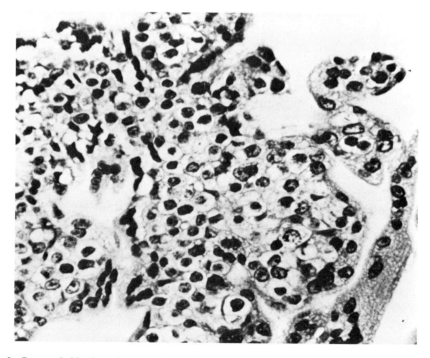

Fig. 2. Cytotrophoblastic nodule with marginal zone of syncytial trophoblast. The degree of pleomorphism is normal for early trophoblast. H&E, 200×.

the blastocyst away from the embryonic (later umbilical) stalk. A fully discoid form has been seen as early as six weeks postconceptional age. The rarity of intact specimens before this time has limited observation; it is possible that recognizable discoid form may occur earlier in some cases. Future growth and development is dependent upon interaction among the fetus, the uterus, and the placental tissues themselves. The relative contribution of each, and that of fluid balance in the amniotic space, is not known.

### 1.1.3. At Maturity

The placenta grows progressively in surface dimensions and weight up to about 35 weeks gestation, when growth stops, for all practical purposes. Small differences cannot be detected by present means, and if there are increments in the size and weight of the placenta after 35 weeks they are lost in the much greater absolute and relative growth of the fetus. The typical weight of a term placenta, after trimming the cord and membranes away, is between 425 and 550 g (Figure 3). The actual weight is subject to several influences, the most important of which is the amount of

Fig. 3. Typical term placenta, fetal surface. Note the mildly eccentric cord insertion. The white material beneath the amniochorion is degenerate thrombus in the subchorial space.

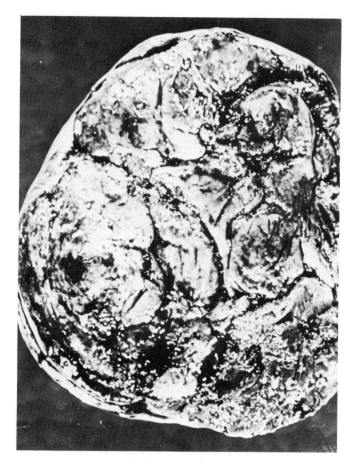

Fig. 4. Typical term placenta, maternal view. Same specimen as in Figure 3. These cotyledons are well defined. The crevices between the cotyledons are coated by decidua. This extends only a short distance into the substance.

trapped maternal blood. A postpartum specimen that has been allowed to stand for 6–8 hr is close to basal weight. The surface measurements of the chorioamnion vary from 12 × 12 to 20 × 17 cm. When laid flat, the term placenta is about 2.0–2.5 cm thick. In the uterus it is up to twice this, largely because of maternal blood content, especially during uterine systole. The outer margin is generally ovoid, smooth, and with only minor irregularities. The surface is slightly gray-red to deep purple-red, again depending on the maternal blood content. On the maternal surface the cotyledons may or may not be well marked (Figure 4). If they are, about 15–30 macrocotyledons will be present. Each cotyledon is a collection of the villi, stems, and stalks deriving from a set of major surface umbilical arteries and veins. The arborization of the vascular placenta begins at the cord insertion. In almost all specimens the two umbilical arteries have an anastomosis at the level of the chorionic plate. From this point the vein and the arteries branch out and pair off. Each pair or set then gives rise to penetrating vessels which enter the macrocotyledons and eventually the microcotyledons which are the functional units of the placenta. Three to eight microcotyledons may be found per macrocotyledon. A total of 100 is commonplace.

## 1.2. Histogenesis

### 1.2.1. Trophoblast

Very soon after fertilization, the primitive blastocyst differentiates into an inner cell mass and the outer trophoblast. The zona pellucida disappears and implantation occurs. During implantation there is further differentiation into cyto-trophoblast and syncytial trophoblast. The cytotrophoblast is that which immediately surrounds the mesenchymal core of the villi. It is composed of fairly large, pale staining cells, with prominent, vesicular nuclei. Mitosis is scant to absent. Histochemical stains have demonstrated that a variety of materials are made or stored in the cytotrophoblast, including chorionic gonadotropin. A PAS-positive basement membrane is laid down between these cells and the mesenchyme. Outside the cytotrophoblast is a syncytial layer of variable nucleation. No cell boundaries are seen here but the boundary with the cytotrophoblast is distinct. The syncytium is commonly vacuolated. Special techniques have suggested that the steroid hormones of the placenta are made here. As time passes, fewer cytotrophoblastic cells can be recognized. A roughly reciprocal relation exists between the decline of the cytotrophoblast and the growth and expansion of the syncytial layer. This fact alone is evidence for a conversion of one into the other. This is unproven by observation, however. During the third trimester the syncytial nuclei tend to cluster and to assume a polypoid profile called "knots" (Figure 5). In some term placentas this is striking, and the extent of anucleate, bare cytoplasm is often astonishing.

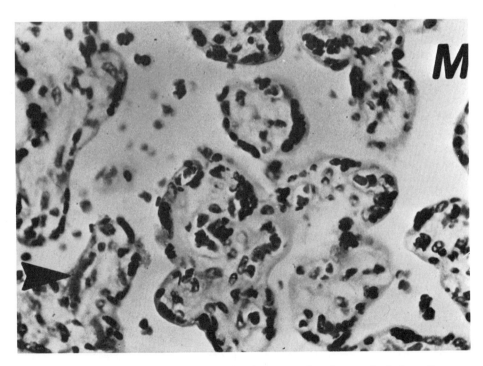

Fig. 5. Late third-trimester tertiary villi. There are a few maternal erythrocytes in the intervillous space (M). Fetal erythrocytes are in the villous capillaries (arrow and other sites). Syncytial nuclei tend to cluster. Cytotrophoblasts are hard to identify, but may be found at term. H&E, 400×.

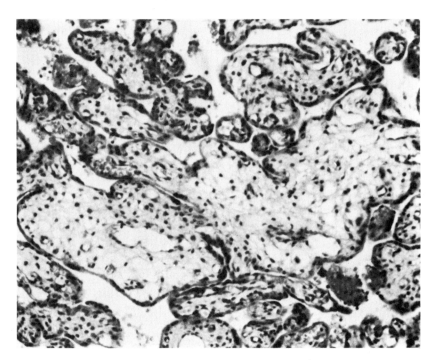

Fig. 6. Early third-trimester, secondary and tertiary villi. Note several "fibrinoid" excrescences. The central zone is a compound secondary villous and stem. The lacunae of "edema" demark the stromal cells better. Higher magnification will aid in identification of stromal cells as fibroblasts or macrophages. H&E, 100×.

### 1.2.2. Stroma

An early derivative of the trophoblast, the stroma, forms the connective tissue cores of the villi, stems, and stalks. In so doing it also gives rise to the villous vascular system and probably also to the reticuloendothelial system, known in the villi as Hofbauer cells or macrophages. The basic stromal cell has a small darker nucleus with ill-defined cytoplasmic boundaries, sometimes fusiform, sometimes stellate (Figure 6). In the mature villus a distinction cannot always be made between the resting macrophages and the stromal cells. While not a proof of their interconversion, this common observation at least begs the question of their relationship.

### 1.2.3. Vascularization

In the protomesenchyme some cells acquire a denser cytoplasm and appear to elongate, with two or three stellate projections. Several of these commonly appear in a given focus under the microscope and interconnect, surrounding a space or lumen. As these lacunae enlarge other cellular elements appear in them, possibly arising from and differentiating away from the endothelial precursors. These are, in turn, the early vessels (blood lakes or blood islands) and the erythroblastic islands in the stream. As early as 7 weeks nucleated erythrocytes of surprisingly mature form are present in these lakes. Coalescence of the lakes leads to a true vascular system, in conjunction with the developing fetal circulation. In the absence of this circulation angiogenesis will arrest at an early stage and erythropoiesis will not go beyond a few rare stem cells. Ultimately there are two vascular systems in the villi. There is the arterial–capillary–venous circuit and the paracapillary system of Bøe (1953). This latter is a delicate network of small capillaries near the trophoblast of small villi and stems.

### 1.2.4. The Reticuloendothelial System

The origin of the macrophagic cells of Hofbauer in the placenta is uncertain. Their orgin *de novo* in the villous mesenchyme is suggested by the early stage at which they are seen, often before there is a fully competent circulation to all parts of the placenta. On the other hand they are more commonly seen in association with lacunae of hydroptic villi, but hydroptic villi with a circulation as contrasted to the avascular hydrops seen in such conditions as chorionic molar pregnancy. As such the metastatic method of formation of lymph nodes in the fetal body, described by Kyriazis and Esterly (1970), may play some role in the populating of villi by specialized RE cells. This is one of many areas in which much more needs to be known for a fully functional understanding of the placenta.

### 1.2.5. The X Cell System

In the mature decidua, in both plate and septae, are polygonal, sharply defined cells with a single nucleus. The chromatin is vesicular. These are often isolated or clustered and commonly mingle with decidual cells from which they can easily be distinguished. The decidua is made up of larger, paler cells, often less well defined (Figure 7). The X cells often are found emmeshed in maternal fibrin within the decidual plate. They are usually without mitoses. Their origin is a mystery; their

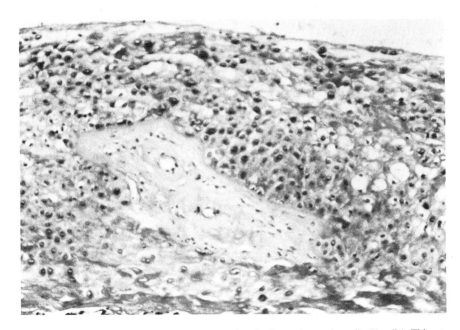

Fig. 7. Uteroplacental artery in decidua. Note occasional polygonal granular cells (X-cells). This artery is essentially normal. In certain vascular diseases lesions may be found in these vessels. These are usually left behind in the retained decidua after delivery. H&E, 100×.

function is unknown. Benirschke and Driscoll (1967) have summarized the matter by pointing out: (1) transitions from trophoblast are not seen; (2) they are not seen in trophoblastic tumors; and (3) their location is fairly constant at the interface. They consider these points as indicating a maternal origin. The corona radiata of the granulosa has been suggested. There is no current evidence to suggest an endocrine role, although their granularity is suggestive of a potential for rich mitochondrial or microsomal activity.

### 1.2.6. The Formed Membranes

The fusion of amnion to chorion on the surface of the early placenta serves, in part, to limit the growth to a discoid form. The reflected membranes, which take off from the lateral edge of the discoid placenta, complete the formation of the fetal sac, within which is the amniotic fluid. At term these membranes range up to 50 g or so. During placental separation a variable amount of decidua comes away with them. From the fluid surface outward, the membranes contain the following layers: (1) amniotic epithelium; (2) amniotic connective tissue; (3) chorionic connective tissue; (4) chorion, often scanty, but sometimes with regressed, hyalinized, or seemingly intact but avascular villi; and (5) maternal decidua, with or without a lymphocytic infiltration. In the formed membranes will be the rupture site through which the fetus will pass during labor.

### 1.2.7. The Umbilical Cord

The umbilical cord is the mature result of elongation of the embryonic stalk (body stalk), which incorporates the allantois. The allantoic vessels contribute to the formation of the umbilical vessels, possibly in whole rather than part. The cord is coated by a layer of attentuated amnion but contains specialized mesenchyme known as Wharton's jelly. Basically this is a lacunar collection of interstitial matrix apparently laid down by mast cells and other elements of the cord. A few macrophages can be found also. Two arteries with well developed elastica (Figure 8) and one vein with a thick internal elastic membrane (Figure 9) are present. The cord varies enormously in diameter at term, the principal reason being the amount of Wharton's jelly. It ranges from 0.5 to 2.0 g/cm length. The cord averages 50–60 cm long but extreme forms at term from 14 to 144 cm have been observed. The average length of cords in premature births parallels the uterine growth.

### 1.2.8. Tissue Interaction

The best examples of this, during histogenesis, lie within the reticuloendothelial system. A competent fetal circulation is an aid to the vascularization of villi and to the formation of the blood islands. A further possible example of this might be the role of the maternal lymphocytic system in limitation of the placental site. The lymphocytes found in the decidua late in pregnancy can be seen to migrate toward the placental site in the first 6–10 weeks across the interstices of the myometrium. Finally, some form of reactivity of the Hofbauer cells seems to be the normal mode during the first trimester. This takes the form of a highly granular, well-defined cytoplasm which distinguishes the macrophages from the other connective tissue cells. A possible function for this is a role in the immune recognition/suppression action of early pregnancy. Curiously these cells are not often seen in the second

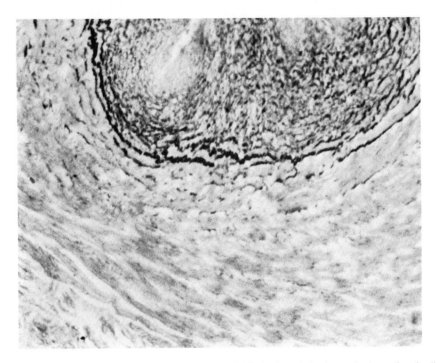

Fig. 8. Umbilical artery, cross section. Note the essential limitation of elastica to the inner (longitudinal) muscle with accentuation at the junction. Verhoeff–Van Gieson, 100×.

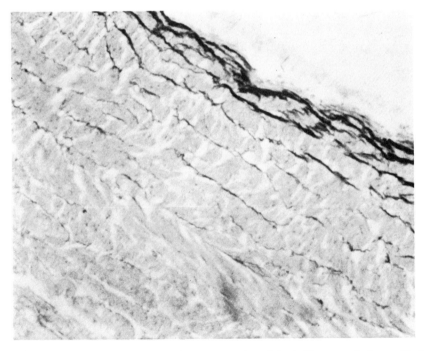

Fig. 9. Umbilical vein, cross section. There is a concentration of elastic fibers near the lumen. Verhoeff–Van Gieson, 100×.

trimester and they appear again in the third trimester under the influence of several kinds of pathologic injury. Examples of the latter include erythroblastosis fetalis, luetic placentitis, and, segmentally, fetal heart failure. General form and appearance all suggest that these are derived from the same cell. Trophoblastic and mesenchymal interaction occurs in the sense of leading to coordinated growth, but specific influences have not been investigated.

## 2. "Normal" Variations in Form

### 2.1. Multiple Pregnancies

#### 2.1.1. Minor Effects

Twins and other multiple pregnancies are characterized by small to large reductions in average size and early onset of labor. This size reduction is accompanied by a correspondingly smaller placenta. The crude index of placental–fetal weight ratio (P : F) indicates a minor deviation from expected ranges and averages in the case of twins. Other mild effects are in the placement of the cord insertion. It is generally held that the cord insertion more or less represents the original implantation site. The fetus grows away from it with elongation of the cord, and consistent with opportunity, the placental disk also grows away from this site. At minimum it represents a frame of reference. Twin placentas show variations in the cord insertion site with some consistency of pattern according to the kind of twin placentation.

There is a dispute in the literature over the relation between zygosity (number of fertilized ova) and choriality (number of chorionic disks) as applied to twins or other multiple pregnancies in the human species. The descriptive anatomist may despair at correlating choriality with such entities as monospermic, monozygous twinning; dispermic monovular pairs; trizygotic, monovular, quintuplet sets (e.g., the Dionnes), and so on. The further distinctions, suggested in the previous sentence, between zygosity and ovularity must be kept in mind. Using twin sacs as the prototypical series, the actual process is greatly simplified by the ready observation that human twin placental sacs take only three basic forms. These are the unfused dichorionic, diamniotic placentas (totally separate) (Figure 10), the fused dichorionic, diamniotic placentas (with fusion of the disks) (Figure 11), and the monochorionic placenta, either diamniotic or monoamniotic (Figure 12). Each will be described in turn.

The unfused dichorionic, diamniotic twin placentas are easily recognized. Each is essentially like a singleton placenta, slightly reduced in scale commensurate with the weight of the fetuses. A fusion of membranes is commonplace, and this can occur on the surface of one or both of the discoid masses. The cord insertions are essentially random as in singleton placentas.

The fused, dichorionic, diamniotic placenta is characterized by the presence of a thick line of chorionic fusion somewhere near and across the middle zone of a placenta larger than a singleton. If the assignable areas are used to estimate the weight share for each fetus, the P : F ratios fall into the range of twins with unfused placentas. The cords often insert close to the fusion line. The fusion line is a four-layered membrane consisting of two amnions and two chorions (Figure 13).

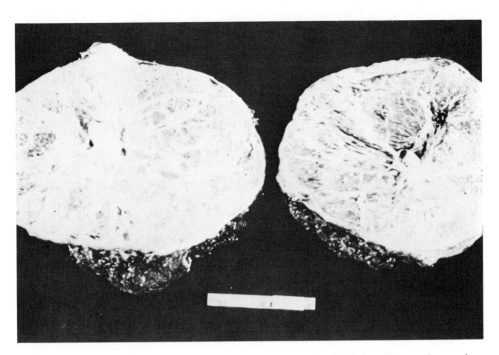

Fig. 10. Nonfused dichorionic twin placentas. Occasionally there may be fusion of the membranes along a line across one or both discoid masses.

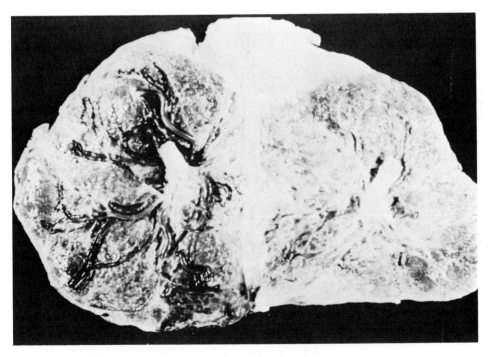

Fig. 11. Fused dichorionic twin placentas. There is a rather straight line of fusion. As in the specimen in Figure 10, careful dissection will yield four layers (see also Figure 13).

DOUGLAS R.
SHANKLIN

Fig. 12. Monochorionic diamniotic twin placenta. The common double amnion has been removed. A white paper rectangle is adjacent to the top edge indicating one cord insertion. In the mid-left field is another similar indicator for the second insertion. The white arrows indicate an artery-to-artery anastomosis. The rectangle in the lower right is near a field of typical dispersion of surface vessels.

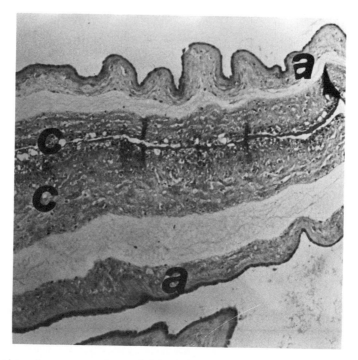

Fig. 13. Four-layered common membrane of dichorionic, diamniotic fused membrane. The trophoblastic edges of each chorion (c) are fused together in the middle. Each amnion (a) also has a connective tissue layer and an epithelial surface.

The monochorial placenta has a single, usually smooth surface on which sit two amniotic sacs (Figure 14). Rarely a single amniotic sac is present. These dual amnions do not produce a thick line of fusion and actually can often be readily moved about on the chorionic surface. There is a special intimacy of the circulation in monochorionic twinning, which, if extreme, has profound pathological consequences. The cords often insert at far ends of the specimen but can be most anywhere.

### 2.1.2. Monochorial Parabiosis

It has been claimed by Benirschke that all monochorial twin placentas show vascular anastomoses. Certainly when such anastomoses are large they have important physiological significance. In our study of 26 monochorial twin placentas and three sets of triplets of mixed zygosity which included a monochorial twin pair, we could not confirm their universal presence (Shanklin, 1970). Six of the placentas did not have an adequate test of possible anastomosis largely because they were fixed before examination. Another one was too badly torn. Of the remaining 22 there were 16 with clearly demonstrated shunts between fetal circulations. There were two others in which a shunt was found but was trivial in the sense that it was hard to demonstrate. There was one in which no shunt could be demonstrated despite a technically suitable injection at multiple sites and from both sides. Although this negative evidence is not conclusive, it suggests that, despite the prospects on a theoretical basis, monochorial twin placentas may occur in which no anastomosis is present. In the twin sets within the triplets we found two with easily identified shunts and one in which no shunt could be demonstrated despite excellent study conditions. Among the six that could not be subjected to injection study, upon

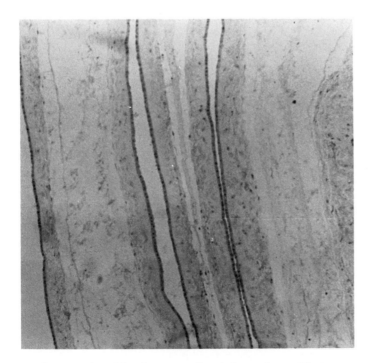

Fig. 14. "Jelly-roll" preparation of fused common diamniotic membrane. The two connective tissue layers readily separate, more so than the fused chorions shown in Figure 13.

dissection one showed a probable shunt and three showed no vascular configuration of this type. The other two were not dissected. This greater lack of anastomosis in those previously fixed is probably a consequence of this action and represents an artifact. Of the 21 which had definite shunts, seven or 33.3% clinically manifested the parabiotic syndrome.

Two others showed some disparity of birth weights or blood counts, but these were of minor degree. Naeye (1965) and others have adequately reported on the significance of this syndrome in the human, and this information will not be repeated here. We would like to emphasize that, in general, actual low birth weight relative to gestational age is fairly uncommon in parabionts. Despite this, significant shunting is a major contributor to perinatal death. There were seven cases with complete information. Only two infants survived out of 14. Both of these came from the same set. Furthermore, they were gestationally mature at 41 weeks. They weighed, respectively, 2960 and 2100 g.

## 2.2. Placenta Extrachorialis

A common and interesting form of placenta is that termed by Scott (1960) placenta extrachorialis (PE). In the mature form this is characterized by a marginal rim of variable width averaging 1–2 cm inside the actual physical perimeter (Figure 15). The chorioamnion reflects from this rim, and the superior (fetal) surface of this zone is bereft of chorion, either cellular or membranous. The surface in this area is a modified decidua. The clinical importance is that trimester hemorrhage occurs from this rim area in more severe examples of placenta extrachorialis. The bleeding is maternal but this may influence uterine blood flow to the intervillous space. Fibrin is commonly laid down here, often from these hemorrhages, and may be bulky. An older term for severe PE is circumvallate placenta, and an older term for mild forms

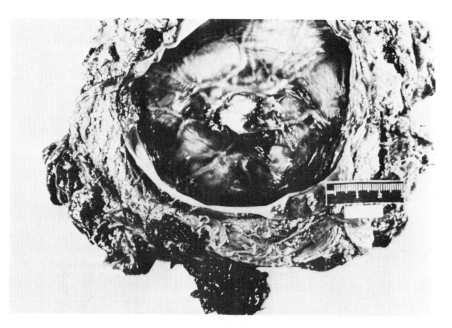

Fig. 15. Placenta extrachorialis. There is abundant fibrin deposit along the rim, giving rise to the impression the discoid surface is even more constricted. Fresh blood clot is adherent at the lower edge.

is circummarginata. The frequency and incidence varies with the report. Figures as high as 25% have been reported, and some series, taking all degrees, have suggested an incidence of 50% of all term placentas. We have previously mentioned that surface arteries and veins sort themselves out into dyads which penetrate the chorion at intervals. Those that supply the peripheral cotyledons generally penetrate at an average of 1.8 cm from the outer edge. In placenta extrachorialis there are fewer such outer dyads, and they average a greater distance from the edge. This is evidence for Goodall's (1934) secondary growth hypothesis for the formation of PE. This must occur quite early in pregnancy because the frequency of PE is relatively constant throughout the interval of 13–40 weeks.

The P : F ratio of conceptuses with PE is normal and the gestational profile is normal, indicating no especial effect on fetal growth or development, once compensatory growth has occured. That certain zygotes have a predilection for PE is shown by analysis of twin data. In a large series in which the frequency in singleton births was 20%, a group of over 120 twin sets had only 8.3% PE. Re-analysis according to the type of choriality showed that nonfused dichorionic, diamniotic twin placentas had the same frequency of PE as singleton births. Fused dichorionic and monochorionic had none. This surprising segregation of effects is evidence that shared circuits, a form of hemometakinesis, offers an alternative prospect for adjustment of the growth in mass of placenta and fetus, possibly preventing the formation of the extrachorial rim by second-order growth. This is of course a sharing of the maternal circuit since fused dichorial twin placentas do not have a common fetal flow.

## 2.3. Variations with Little or No Effect on the Fetus or Newborn

### 2.3.1. Accessory Lobe

A moderately common (up to 5% in some series) minor anomaly, accessory lobe, refers to the presence of a cotyledon outside of the profile of a discoid shape (Figure 16) (Shanklin, 1958). The amount of membrane between the accessory and the main mass varies. If closer than 2.0 cm it is usually termed simply accessory; if more than 2.0 cm it is called a succenturiate lobe (Figure 17). These cotyledons are often less well developed than those on the main mass of the placenta. Their importance is limited to such unusual conditions as ruptured vasa (when the connecting vein is torn at membrane rupture) or as the means of presentation as placenta previa. Both events are rare. Ordinarily, however, these lobes do not cover the internal cervical os, and these vessels, being rather thick walled, do not rupture.

### 2.3.2. Bilobate or Reniform Placenta

The most common abnormal outline of a single discoid form is bilobate–reniform placenta. In bilobates there are notches or grooves on opposite sides of the placenta which tend to divide it into two lobes (Figure 18). The reniform placenta is the same situation with a single notch creating a shape similar to the profile of the kidney or, in the popular sense, heart-shaped. In a series of 6500 we found 135 examples or 2.1%. There were 70 bilobate and 65 reniform placentas. The gestation was known with precision for 128 of these and the preterm delivery rate was 26.5% up to 38 weeks of gestation, a figure identical to the control population from the same hospital. There were some interesting associated conditions. There were 11

DOUGLAS R.
SHANKLIN

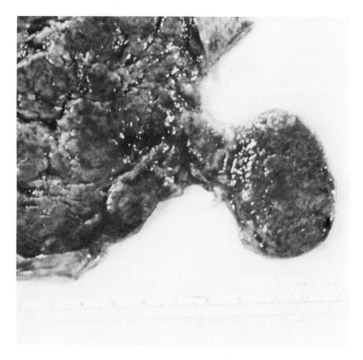

Fig. 16. Accessory lobe, maternal view. Note that villous cotyleden fills the bridge to the extra lobe.

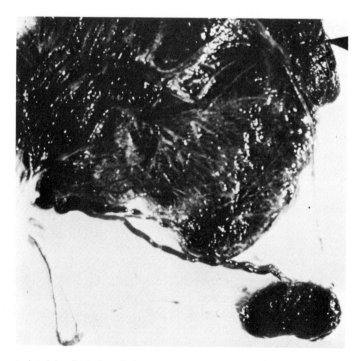

Fig. 17. Succenturiate lobe, fetal view. Only a narrow vascular pedicle attaches the accessory lobe to the main mass.

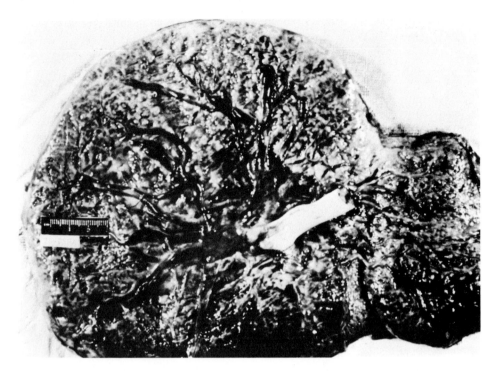

Fig. 18. Bilobate placenta, fetal view. The cord insertion site is not always at the isthmus. In this case the lesser lobe is about 25% of the total mass.

abdominal deliveries, an expected coincidence of abnormal cord insertion, and an expected incidence of placenta extrachorialis but a rate of infarction twice that of the control series. This increase was not statistically significant. The P : F ratios were in the mid-normal range. The principal clinical importance of these rather large placentas is the presence of a larger than average placental site for postpartum hemorrhage. Torpin and Hart (1941) have shown convincingly that bilobate reniform placentas are the consequence of implantation in the superior or lateral sulcus of the uterine cavity. Not all sulcus implantations are bilobate, but the evidence suggests that all bilobates are sulcus implants. The bilobate form is developed because of the tendency for the blastocyst to implant on both sides of the angle creating the lateral notches and the bilobate form. As the uterus expands with the progression of pregnancy, the sulcus is obliterated. The notches persist, yielding the bilobate pattern.

### 2.3.3. Minor Changes

Other trivia that have no effect on the fetus include such common things as small mucinous cysts. These are often found in the cytotrophoblastic cell columns which anchor the placenta to the decidual plate. The commonplace varices of the umbilical vein are also innocuous as a rule. Rarely they rupture. When they do, it is fetal bleeding which occurs and this has profound consequences. However, the most massive example seen by this author, some 125 ml of blood under the surface amnion, was associated with neonatal survival following prompt transfusion. Other variations in outward shape are legion; quadrilaterals, triangles, trapezoids, and umbilicated outlines have all been observed. These have no intrinsic meaning as far as can be discovered.

### 2.4. Variations with Significant Effect on the Fetus or Newborn

#### 2.4.1. Anomalous Cord Insertions

In a series of 6500 placentas, four different types of anomalies of insertion of umbilical cord were recognized (Shanklin, 1970). Two of these were discovered during the course of this experience, and incidence figures are available for less than the total number of cases in these instances. The four anomalies may be defined as follows:

*Velamentous.* This is insertion of the cord vessels onto the reflected membranes beyond the placental disk (Figure 19).

*Battledore.* This is when the cord vessels insert onto the edge of the placental disk.

*Pedicle.* The cord splits up into individually definable vessels at some distance above the surface. It is theoretically possible that pedicle insertions may appear at battledore and velamentous locations. These combinations are rare.

*Chordee.* This is a fold of chorioamnion, usually somewhat triangular, which rises from the surface of the placental disk and attaches along the terminal segment of the cord. The appearance is such as to suggest a limitation on mobility of the umbilical cord about the insertion site.

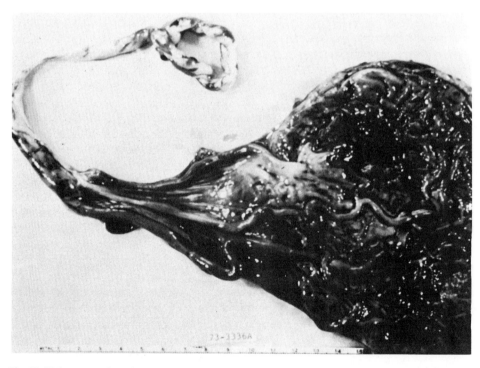

Fig. 19. Velamentous insertion of cord. Here the insertion site is in the outer reflected membranes 9 cm from the lateral edge of the placenta. Note that the vessels diverge widely through the membranes before reaching the surface zone.

Two hundred sixteen abnormal insertions were found. Including twin placentas, there were some combinations of one sort or another in five cases. Two hundred six had a single anomalous condition, or 3.36%. The birth weight was accurately recorded for 203 of these latter infants, and for each gestation period the average weight was at the low end of the principal weight range as reported by Gruenwald (1963) and others. There is an increased frequency of these anomalous insertions in multiple pregnancies, accounting for approximately 6.9% as compared to 1.51% in singleton placentas. Furthermore, the coincidence of abnormal insertions as in the cases of twins was that of a chance occurrence. Low birth weight is commonplace and may actually result from the abnormal vascular dynamics of extreme forms of battledore and velamentous insertions in particular.

## 2.4.2. Chorangioma

These uncommon lesions suggest a mechanism for the development of low-birth-weight syndrome and represent a growth anomaly which can be considered an anatomic variant only because of the representation of the essential features of normal histology. Chorangiomas are also known as hemangiomas and are derived from fetal vascular cells at the villus and stem levels. They take a wide variety of sizes and shapes and may be found at almost any location in the placental system. For example, they have been described on a pedicle off the umbilical cord, within the chorionic plate on the surface, and within the substance of the villous cotyledons. In this latter location they may be on the margin or centrally placed. Their importance lies in the fact that they represent arteriovenous shunts, depending upon the relative proportions of large sinuses which communicate between the fetal artery and fetal vein and on size. Even very large or extensive and complex chorangiomas do not of themselves displace a sufficient amount of placental tissue to cause a functional inadequacy. Their impact is rather upon the competency of the vascular circulation. Frequencies in large series have ranged up to 1 : 833. X cells have not been found in them, further support for the maternal origin of X cells.

## 2.4.3. Placenta Membranacea

A number of gross malformations of the placenta have been suggested as causative of the premature onset of labor. Among these, one can be accepted with partial qualification. This is placenta membranacea. In these the placenta totally or almost totally surrounds the fetus and by its physical presence may inhibit further uterine growth. The relatively small membrane area available is placed under unusual stress during further distention and growth, and it conceivably could rupture prematurely. We have seen three good examples in a series of over 5000 placentas. One was delivered at 18 weeks, one at 30 weeks, and the other by elective section at 39 weeks.

## 2.4.4. Infrequent Variations

Any such list must include the extreme form of monoamniotic, monochorial twinning with a parasitic twin, the acardiacus syndrome. This seems to be the consequence of monochorial twinning, with parabiosis and umbilical artery aplasia. Umbilical artery aplasia itself is often significant as a marker for various anomalies, especially of the genitourinary tract.

*DOUGLAS R.*
*SHANKLIN*

## 3. Common Findings Properly Treated as Lesions

These are variations in form which are usually related to one or more types of vascular incident. In those local forms of interruption in maternal flow, we find thrombi of the central lake, infarctions (often paradecidual), exaggerated fibrin deposit about villi, abruptio placenta, subchorial thrombohematoma, and sickling in the genetically appropriate situation.

Some of these may become generalized or will occur on a large scale. Of the greatest importance to the fetus is abruption, next subchorial thrombohematoma (Shanklin and Scott, 1975), and then severe sickle impaction. These are but clear reminders that in the final analysis the placenta is a vascular organ.

## 4. Summary

The human placenta is a remarkable organ showing interesting variations, but in most instances it is well designed and fully adequate to the proper nurture of the fetus. There is a progressive change in the mass of the placenta towards term and a progressive decline in the relative mass of the placenta against that of the fetus. The increasing pattern of arborization and the closer proximity of fetal capillaries to the syncytial surface are all interpreted as serving the function of increased metabolic exchange during a period of marked protein synthesis and growth in later fetal life. This progression suggests the possibility that an inadequacy could develop through failure of proper growth and development of the placenta itself. In fact, such events usually seem to occur with concomitant and compensated decrease in fetal growth. This has been well documented by Gruenwald (1963) as far as the third trimester is concerned, but Hardwick (1969) has also shown convincingly that the same consequences occur in the second trimester. The present evidence suggests that both placental and fetal diminution of growth occur in response to deprivation of either maternal vascular supply or through other limiting factors in maternal environment. Few placental lesions specifically cause difficulty in fetal life. Perhaps the principal of these is the hemangioma or chorangioma which is described above.

The study of placental development is not complete. A great deal remains to be determined about its ultrastructural capacities, its functional digressions, and its own metabolic requirements, particularly with respect to maternal vascular supply. Since it is not readily accessible to the experimental approach, a great deal of what is known is based upon interpretation of clinical events and the assembly of large series of cases demonstrating variations of normal as well as pathological processes. Experimental data from animal placental systems is not readily transferred to the human condition because of the profound differences in vascular dynamics and the anatomy of the barrier. Finally, the most critical periods of early fetal development come at a time when the discoid placenta is itself just forming. A possible benefit not anticipated by the proponents of abortion on demand is the increased frequency with which trained observers now have available histologic material from the early second and late first trimester, including examples of *in situ* placentas and well-formed fetuses delivered by hysterotomy. Hopefully, these observations will lend themselves to a better understanding of the interrelation between placental development and fetal growth.

Benirschke, K., and Driscoll, S. G., 1967, *The Pathology of the Human Placenta,* 512 pp, Springer-Verlag, New York.

Bøe, F., 1953, Studies on the vascularization of the human placenta, *Acta Obstet. Gynecol. Scand.* **23**(Suppl. 5): 1–92.

Goodall, J. R., 1934, Circumcrescent and circumvallate placentas, *Am. J. Obstet. Gynecol.* **28:**707–722.

Gruenwald, P., 1963, Chronic fetal distress and placental insufficiency, *Biol. Neonat.* **5:**215–265.

Hardwick, D. F., 1969, Weight and length of fetuses in relation to age, *J. Reprod. Med.* **2:**98–100.

Hertig. A. T., 1968, *Human Trophoblast,* p. 99, Charles C. Thomas, Springfield, Illinois.

Kyriazis, A. A., and Esterly, J. R., 1970, Development of lymphoid tissues in the human embryo and early fetus, *Arch. Pathol.* **90:**348–353.

Naeye, R. L., 1965, Organ abnormalities in a human parabiotic syndrome, *Am. J. Pathol.* **46:**829–842.

Scott, J. S., 1960, Placenta extrachorialis (placenta marginata and placenta circumvallata), *J. Obstet. Gynaecol. Brit. Emp.* **67:**904–918.

Shanklin, D. R., 1958, The human placenta: A clinicopathological study. *Obstet. Gynecol.* **11:**129–136.

Shanklin, D. R., 1970, The influence of placental lestions on the newborn infant, *Pediat. Clin. North Am.* **17:**25–42.

Shanklin, D. R., and Scott, J. S., 1975, Massive subchorial thrombohaematoma (Breus' mole), *Br. J. Obstet. Gynaecol.* **82:**476–487.

Torpin, R., and Hart, B. F., 1941, Placenta bilobata, *Amer. J. Obst. Gynecol.* **42:**38–49.

# 14

# *Physiology of the Placenta*

## JOSEPH DANCIS and HENNING SCHNEIDER

### 1. Introduction

Retention of the fetus within the mother offers many obvious survival advantages. A wealth of experimentation towards achieving this objective is already found among fishes, and early forms of the placenta can be identified (Amoroso, 1952). Placentation achieves its most complex development in the mammal. Two of the fetal membranes which are easily recognizable in egg-laying animals, the chorion and allantois, are extensively modified to provide the fetal contribution to the placenta. The maternal component of the placenta varies in the different species. The uterus, which is developed for retention of the embryo, also contributes the maternal vasculature. Within the placenta, maternal and fetal vessels lie in intimate juxtaposition, facilitating the exchange of nutrients and waste products.

Mammalian placentas, despite their common origin from the chorion and allantois, reveal anatomical and functional diversity. It is essential that this feature be kept firmly in mind by those whose interest is primarily in the human placenta. Much of recorded information of placental function is derived from the study of animals. Applicability of the results to the human must be considered tentative. Two examples drawn from species that are favorites among experimental placentologists will serve as illustrations. Brambell, in a brilliant series of studies, demonstrated that in the guinea pig transport of antibodies is through the yolk sac which is retained throughout pregnancy and modified to form a specialized membrane, the splanchnopleure (Brambell *et al.,* 1951). In the human, the yolk sac disappears early in embryonic life and the chorioallantoic placenta is the route of transfer for maternal antibodies (Dancis *et al.,* 1961). In the sheep, the placenta efficiently converts maternal glucose into fructose, which is then made available to the fetus. The human placenta retains only vestigial remnants of this functional capacity, the circulating carbohydrate in the fetus being almost exclusively glucose (Hagerman and Villee, 1952).

***JOSEPH DANCIS and HENNING SCHNEIDER*** • New York University School of Medicine, New York, New York.

JOSEPH DANCIS and
HENNING SCHNEIDER

In this chapter particular consideration will be given to the study of the human placenta. Nevertheless, frequent reference to observations in animals will be necessary when they represent the only available information. The physiology of the placenta will be discussed under three major headings: transport, endocrinology, and immunology.

## 2. Placental Transport

### 2.1. General Considerations

Membranes define the boundaries of biological structures, from organelles to organisms, and selectively control the passage of materials. In this respect, the placenta may be considered a membrane placed at the periphery of fetal tissues, facing the "exterior," the maternal blood. As noted above, the placenta does, in fact, derive embryologically from fetal membranes, the chorion and allantois. A relatively constant concentration of nutrients is presented to the placenta by the maternal circulation as a result of the homeostatic mechanisms that maintain the composition of maternal blood within narrow limits. The highly developed defense mechanisms of the mother are linked in this way to the protection and maintenance of her fetus.

The efficiency of transfer of materials between maternal and fetal circulations is affected by the vascular arrangements within the placenta (Meschia *et al.,* 1967). These vary among mammalian species. In the sheep, the direction of flow in the maternal and fetal circulation is in parallel or concurrent. At equal flow rates, no more than half of a rapidly diffusible substance can be transferred from one circulation to another, contrasting with the more efficient countercurrent arrangement found in the rabbit where the blood in fetal and maternal vessels flows in opposite directions. The design in the human, with a pulsatile flow from maternal spiral arterioles into open sinusoids, does not permit such simple characterization. The maternal bloodstream passes many villi containing fetal capillaries, and the flow direction in the villi varies relative to the maternal flow. This arrangement has been called "multivillous" (Bartels *et al.,* 1962), and its efficiency is believed to lie between the classical arrangements of concurrent and countercurrent. An unanswered question is whether all of the maternal and fetal blood directed towards the placenta actually reaches the area of exchange or is "shunted" into the venous circulation, reducing the efficiency of transfer.

An early anatomist, Grosser, classified placentas according to the number of cell layers interposed between maternal and fetal circulations (Table I). The classification is an oversimplification in that "thick" placentas commonly have "thin" areas, presumably specialized to permit rapid exchange. However, the classification persists in the literature and appears to have some counterpart in function, correlating with the rate of exchange of sodium. The transfer of sodium across sheep placenta, which retains a full complement of cellular layers between the maternal and fetal circulations, is relatively slow as compared to the human, where the maternal structures within the placenta have been eroded and the fetal villi are directly exposed to maternal blood (Flexner and Gellhorn, 1942).

The "efficiency" of a placenta may be defined by the exchange per unit of blood between maternal and fetal circulations of diffusible materials. The efficiency of exchange of "flow-limited" materials, such as antipyrine or tritiated water, is

Table I.  Classification of Placentation According to Number of Tissues Interposed between Maternal and Fetal Circulations (There is a Progressive Erosion of Maternal Tissue)[a]

| | Placental barrier | | | | | | |
| | Maternal tissues | | | Fetal tissues | | | |
| Classification | Endothelium | Connective tissue | Epithelium | Epithelium | Connective tissue | Endothelium | Species |
|---|---|---|---|---|---|---|---|
| Epitheliochorial | + | + | + | + | + | + | Pig, horse |
| Syndesmochorial | + | + | − | + | + | + | Ruminants |
| Endotheliochorial | + | − | − | + | + | + | Carnivora |
| Hemochorial | − | − | − | + | + | + | Rodents, apes, man |

[a]After E. D. Amorosa. *Gestation, Transactions of the First Conference,* Corlies, Macy & Co., Inc., New York 1954, as taken from: A. Grosser, *Fruhentwicklung, Eihantbildung, und Placentation des Menschen und der Saugetiere,* Vol. 5, J. F. Bergman, Munich, 1927.

determined by the flow rate and flow pattern. More slowly diffusible materials, such as sodium, are classified as "membrane-limited" because exchange rates are influenced by characteristics of the membrane as well as blood flow. The most efficient placenta for diffusible materials would be expected to have a thin membrane and a countercurrent arrangement of maternal and fetal circulations, disregarding for the moment other characteristics of membrane structure that affect diffusion rates. Using these criteria, the human placenta is neither the most nor the least efficient. The tacit assumption that the most efficient placenta is the best has led to considerable controversy in the past. Obviously, all variations of the placenta that are now existent are effective in that they foster intrauterine growth and development of the fetus. Efficiency, as defined here, may not rank high among the priorities developed under evolutionary pressures.

Protein binding of materials in the blood also affects rates of placental transfer, as illustrated in simplified form in Figure 1. The protein-bound fraction is in kinetic equilibrium with the physically dissolved fraction. By reducing the concentration of

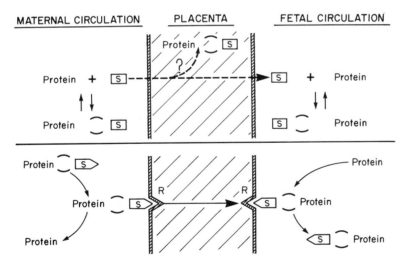

Fig. 1. Effect of protein binding on placental transfer. The scheme in the upper panel appears to be more generally applicable. Protein binding in maternal perfusate and placenta reduces concentration of inbound substrate (S), lowering the diffusion pressure and transfer rate. Availability of protein binding in fetal circulation enhances transfer. (Dancis *et al.,* 1976). In the lower panel, protein binding makes the substrate more accessible to a receptor site on the placenta. This mechanism has been described for the transferrin–iron complex (Awai *et al.,* 1975*b*).

JOSEPH DANCIS and
HENNING SCHNEIDER

the latter, protein binding lowers the diffusion pressure across the placenta (see below) and transfer rate falls. With substances with limited solubility in water solutions, such as oxygen and the free fatty acids, protein binding to hemoglobin and to albumin, respectively, vastly increases the amount transported to the placenta. The net result is an increase in the amount transferred across the placenta. The effect of protein binding on placental transfer may be expected to vary with the substrates and the characteristics of the binding. In many instances, its effect has not been clearly defined.

## 2.2. Placental Transport Mechanisms

The general discussion so far has emphasized factors influencing the presentation of materials to the placental membrane for transfer, what has been called the "maternal supply line" (Gruenwald, 1975). There are actually a series of membranes that must be traversed in the transit from maternal to fetal circulations, beginning with the cellular membrane of the syncytiotrophoblast and proceeding to the endothelial lining of the fetal blood vessels. It is usual, but not accurate, to consider them as a unit.

The mechanisms by which substances are transferred across membranes are generally classified as simple diffusion, facilitated diffusion, and active transport. Simple diffusion is viewed as an exchange of molecules across a membrane as the result of random molecular movements, with a net transfer from the higher to the lower concentration. The factors that influence the diffusion rate have been described mathematically by Fick:

$$\text{Diffusion rate} = \frac{DA(C_1 - C_2)}{\text{thickness of membrane}}$$

where $D$ = diffusion coefficient; $A$ = area of exchange; and $C_1 - C_2$ = concentration gradient across the membrane.

The diffusion coefficient is a factor that describes the ease with which a particular substance crosses a particular membrane and is related to the physicochemical characteristics of both. In general, there is an inverse relation between molecular size, polarity, and electrical charge of a substance and its diffusion coefficient.

Facilitated diffusion resembles simple diffusion in that there is no energy requirement and transfer does not take place against an electrochemical gradient. Transfer is more rapid than expected by simple diffusion, and it is assumed that there is a transport system within the membrane accelerating the rate of transfer. Active transport is capable of transfer against an electrochemical gradient and requires energy. The transport mechanisms in both are stereospecific.

A summary of the factors affecting placental transfer is presented in Table II. It will be noted the same factors that influence delivery to the placenta on the maternal side are operative in removing materials into the fetal circulation.

## 2.3. Oxygen Transfer

A review of the subject of oxygen transfer will serve to illustrate many of the points discussed in the previous section. It is generally believed that oxygen is

Table II. Major Factors Affecting the Placental Transfer of a Diffusible Substance

| Maternal | Placental | Fetal |
|---|---|---|
| 1. Amount delivered to the placenta<br>  a. Blood concentration: exogenous<br>  and endogenous supplies:<br>  homeostatic mechanisms<br>  b. Flow-rate in intervillous space:<br>  hemodynamic factors in mother:<br>  local circulatory factors<br>2. "Shunting" and arteriovenous mixing<br>  in intervillous space | 1. Area of diffusing membrane<br>2. Diffusion pressure: difference in concentration<br>  in intervillous space and in villous capillary<br>3. Diffusion resistance<br>  a. Characteristics of transferred material: size,<br>  charge, polarity, etc.<br>  b. Characteristics of membrane:<br>  physiochemical composition: thickness | 1. Blood concentration<br>2. Hemodynamic factors:<br>  systemic and local<br>3. Flow characteristics (pool,<br>  cross- current, etc.) |

"From G. B. Avery, ed., *Neonatology. Pathophysiology and Management of Newborn*, J. B. Lippincott Company, Philadelphia, 1975.

transferred across the placenta by simple diffusion[1] and that the placental membrane offers little diffusion resistance. The oxygen gradient at the site of oxygen exchange is probably much less than the reported gradient of 19–24 mm Hg derived from measurements in larger tributary vessels (Longo *et al.*, 1967). Oxygen transfer under normal conditions may be considered "flow limited." The factors controlling the transfer are the amount delivered to the placenta in the maternal circulation (concentration × flow rate), the circulatory arrangement within the placenta, and the efficiency of the fetal circulation in removing the oxygen (Table II). There are two important modifying factors: protein binding and placental consumption of oxygen.

Hemoglobin increases the carrying capacity of blood from 0.3 ml/dl to over 15 ml/dl. The hemoglobin-bound oxygen is in kinetic equilibrium with the small amount of physically dissolved oxygen. It is the latter, however, that actually determines the diffusion pressure across the placenta and is the immediately transferrable oxygen. The hemoglobin-bound oxygen serves as a reservoir which increases the amount available for transfer from the maternal circulation. The conditions for protein binding in the fetal blood also increase the efficiency of transfer to the fetus. The hemoglobin concentration in fetal blood is generally about 2 g/dl higher than in the maternal, and fetal hemoglobin demostrates a more "avid" binding for oxygen. By the latter is meant that the oxygen binding in fetal blood at a given oxygen tension is higher than in maternal blood. This effect is produced by the interaction of hemoglobin and 2,3-diphosphoglycerate, a metabolite within the erythrocyte. The affinity for oxygen of adult hemoglobin is reduced, whereas fetal hemoglobin is unaffected (Bauer *et al.*, 1968). Transfer of $O_2$ into the fetal circulation is further enhanced by the reverse transfer of carbon dioxide, the binding of $O_2$ to hemoglobin being increased by an increase in pH (Bohr effect).

As a large, metabolically active organ situated more proximally to the maternal circulation than the fetus, the placenta exerts a prior claim on maternal oxygen, and these requirements are not insignificant. Oxygen consumption by the human fetus at term has been estimated at 7.4 ml/min/kg (Bartels, 1970). Placental oxygen consumption as measured *in vivo* in the rabbit and sheep is approximately 10.7 and 10.4 ml/min/kg (Campbell *et al.*, 1966; Faber and Hart, 1966), respectively. Estimates of $O_2$ consumption for human placenta derived from *in vitro* studies of placental slices (Friedman and Sachtleben, 1960; Mackay, 1958; Villee, 1953*b*) and from perfusion of the umbilical circulation (Nyberg and Westin, 1957) were 3–4 ml/min/kg, a figure

[1]It has been reported that cytochrome $P_{450}$ functions as a carrier in the transfer of oxygen across the placenta (Gurtner and Burns, 1975). If confirmed some of the following statements may have to be modified, although probably not radically changed.

that appears to be far too low. Recent investigations in which the maternal intervillous space was perfused with blood yielded estimates of 10 ml/min/kg (Challier *et al.*, 1976). If this latest figure is accepted, approximately one fifth of the maternal oxygen supplied to the conceptus at term is preempted by the placenta. Crude as these estimates are, they serve to emphasize that maintaining placental function does not come cheaply in terms of consumption of $O_2$ and other nutrients.

## 2.4. Carbon Dioxide Transfer

Carbon dioxide is transported in blood in three forms, which are in equilibrium with each other; physically dissolved (8%), as bicarbonate (62%), and bound to hemoglobin (30%) (Roughton, 1954). Release from fetal hemoglobin is facilitated by oxygen uptake. The diffusion constant is 20 times higher than for oxygen. Transfer rates are largely determined by circulatory factors.

## 2.5. Water

The exchange of water across the placenta is very rapid. Net transfer occurs in response to osmotic gradients. The gradual accumulation of body water by the fetus as it grows is probably effected by the creation of osmotic forces within the fetus as the result of tissue synthesis. It has not been possible to demonstrate a consistent physiological gradient across the placenta, an observation that is not surprising in view of the rapid response of water to differences in osmotic pressure (Bruns *et al.*, 1963, 1964; Dancis *et al.*, 1957*b*).

Hemodynamic gradients may also cause a net transfer of water. Increases in intracapillary pressure, particularly within the fetal placental vessels, produce a fetomaternal flow. If the pressure is sufficiently high, protein and cellular elements leak into the maternal circulation (Dancis *et al.*, 1962). One might visualize such events as occurring under natural conditions with pressure on the umbilical cord and with uterine contractions, although direct observations are lacking.

## 2.6. Electrolytes

### 2.6.1. Univalent Ions

The transfer of sodium across the placenta appears to be by simple diffusion. Early work with radioactive sodium demonstrated a correlation among mammalian species between the rate of exchange and the thickness of the barrier between maternal and fetal circulations (Flexner and Gellhorn, 1942). A similar phenomenon has been described in the course of human pregnancy with increasing exchange rates per gram of placenta as the placenta progressively vascularizes. A fall off occurs in the last weeks before term, attributed to hyalinization of villi and the deposition of fibrinoid (Flexner *et al.*, 1948).

Fetal potassium levels which reliably reflect the physiological concentration have been difficult to obtain because of the rapidity with which they are altered by experimental intervention. The initial reports of high levels were undoubtedly artifactual. More careful sampling techniques have yielded results that approximate the maternal (Earle *et al.*, 1951). In the rat, the fetal levels are actually lower than maternal, conforming to predictions of passive distribution according to transplacental electrochemical gradients (Fantel, 1975).

That different mechanisms are involved in the placental transfer of sodium and potassium is suggested by nutritional studies in pregnant rats. The administration of sodium-poor diets during pregnancy caused a parallel reduction in sodium concentration in maternal and fetal plasma. Potassium deprivation reduced the level in maternal plasma and maternal muscle, whereas the fetal blood levels remained relatively constant (Dancis and Springer, 1970).

Chloride exchange rate, as measured in the subhuman primate, is similar to that of sodium (Battaglia *et al.,* 1968). Iodide is actively transported to the fetus against a gradient (Logothetopoulos and Scott, 1956).

### 2.6.2. Divalent Ions

Transfer of divalent ions to the fetus is frequently against a gradient, indicating an active transport mechanism. Calcium concentration in the human is higher in fetal blood than in maternal in both the protein-bound and ultrafiltrable fractions (Delivoria-Papadopoulos *et al.,* 1967). Exchange across the placenta, in the subhuman primate is bidirectional. Fetal magnesium in the rat is also higher than maternal, but there is no consistent gradient in the human (Dancis *et al.,* 1971; Lipsitz, 1971).

Iron transfer has been studied in the rabbit and rat. Transfer is undirectional from mother to fetus, against a gradient. At term over 90% of iron turnover in maternal plasma is directed towards the fetus (Bothwell *et al.,* 1955). Following fectectomy, the placenta concentrates within itself an amount of iron equivalent to that normally found in both placenta and fetus (Mansour *et al.,* 1972). The placental mechanism controlling transfer has been recently clarified by experimental work in the rat. Iron is delivered to the placenta bound to transferrin. Transferrin has two binding sites for iron per molecule, labeled A and B. It has been shown that specific receptors present in hematopoietic tissues bind with iron from site A, while storage cells contain receptors which react with iron bound to site B (Awai *et al.,* 1975*a*). When the binding sites of transferrin were differentially labeled with $^{59}Fe$ and $^{55}Fe$, uptake by placenta and transfer to the fetus was selectively from site A (Awai *et al.,* 1975*b*), suggesting that the placenta contains the same receptors as immature red blood cells (see Figure 1).

The transfer of zinc in the rabbit is also against a gradient with the rate of uptake under placental control. The transfer rate is slower than for iron and is bidirectional (Terry *et al.,* 1960).

### 2.7. Carbohydrates

Maternal glucose is the major metabolic fuel for the fetus and yet, being a highly polar compound, it may be expected that diffusion will be relatively slow. This is, in fact, true of some of the hexoses such as fructose (Davies, 1955; Holmberg *et al.,* 1956). The transfer of glucose is facilitated by a stereospecific (Folkart *et al.,* 1960) placental transport mechanism. Transfer is "downhill," maternal levels at term being approximately 95 mg/100 ml and the concentration in cord blood, 75 mg/100 ml (Dawes, 1968). It is assumed that the lower fetal levels are maintained by placental and fetal consumption of glucose. It is possible that "shunting" by preventing complete exposure of maternal blood to the placental membrane contributes to the differences.

The concept that glucose is the predominant fuel for fetal aerobic metabolism

*JOSEPH DANCIS and*
*HENNING SCHNEIDER*

has been challenged recently. Experimental data obtained in sheep using indwelling catheters in maternal and fetal vessels indicate that fetal glucose uptake would provide substrate for only 50% of total oxygen consumption and that amino acids as well as lactate may be important additional substrates for aerobic metabolism in the sheep (Boyd *et al.*, 1973; Burd *et al.*, 1975; Gresham *et al.*, 1971). Whether this is true for the human is unknown. As mentioned previously, the sheep placenta metabolizes glucose differently.

The concentration of glycogen is high in human placenta in the first trimester of pregnancy but not at term (Villee, 1953*b*). Insulin increases the activity of placental glycogen synthesis, and possibly of glucose transport (Demers *et al.*, 1972). The enzymes associated with glucose metabolism display differences in activity in the course of gestation (Sakurai *et al.*, 1969). An early suggestion made by Claude-Bernard that placental glycogen may provide supplementary supplies of glucose for the fetus received support from *in vitro* studies demonstrating release of glucose to the medium by incubated placental slices (Villee, 1953*a*). The concept has been challenged by investigators unable to demonstrate changes in gluconeogenesis in the course of gestation, or a glucose-6-phosphatase capable of releasing glucose (DiPietro *et al.*, 1967).

## 2.8. Transfer of Amino Acids

The fetus synthesizes most of its proteins from amino acids received from the mother. The only maternal proteins that are known to be delivered to the fetus in amounts significant to the fetal economy are the IgG fraction of gamma globulin, and possibly serum albumin (see Section 2.14). Placental proteins that are synthesized for export, for example, the peptide hormones, are secreted preferentially into the maternal circulation.

Studies with placental slices have revealed a pattern of uptake and intracellular concentration that generally conforms with the several amino acid transport systems described in other organs. Energy is required, but uptake continues under relatively hypoxic conditions (Longo *et al.*, 1973). The placenta has been compared somewhat vaguely to the kidney because it provides a route of excretion for the fetus. The role of the kidney towards amino acids, however, is recovery and conservation from urine. The placental function consists in uptake of amino acids from maternal blood and transfer to the fetus. This does not require the high-affinity transport mechanisms which become active postnatally in the kidney tubules (Baerlocher *et al.*, 1970; Webber and Cairns, 1968). The placental uptake mechanisms compare more closely with those described in embryonic kidneys which are of low affinity but high capacity (Schneider and Dancis, 1974).

Amino acid uptake provides substrate for placental metabolism and for transfer to the fetus. The latter function has been studied *in vivo* and with perfusion techniques. The concentration of each amino acid in the fetal circulation is maintained, with few exceptions, at a level higher than in the maternal (Ghadimi and Pecora, 1964; Glendening *et al.*, 1961; Van Slyke and Meyer, 1913; Young and Prenton, 1969). The net transfer to the fetus is therefore against a concentration gradient and requires an active transport process. That such a mechanism exists in placenta has been demonstrated with perfusion of isolated placenta. Leucine, in its physiological L form, was transferred at a 30% faster rate than the D form and established a gradient in the fetal circuit (Schneider *et al.*, 1969). In the guinea pig, the concentration in fetal blood never exceeds that within the placenta, indicating that the transplacental gradient is established by the uptake process rather than an efflux mechanism (Hill and Young, 1973).

There are some interesting exceptions to these generalizations. L-Cystine is relatively poorly transferred across the placenta and does not establish a gradient. Gaull has attempted to relate this feature and the absence of cystathionase in the fetus to the particular metabolic requirements of the fetus during gestation (Gaull *et al.*, 1973). Glutamic and aspartic acids are transferred slowly and are subject to extensive metabolic alteration in the placenta, particularly to their respective amides. This, too, may interrelate with fetal metabolism (Dierks-Ventling *et al.*, 1971; Steglink *et al.*, 1975).

Improved transport has been reported following amino acid deprivation *in vitro* with human placental slices and *in vivo* in the guinea pig, possibly providing a defense mechanism in the event of maternal protein depletion (Young and Widdowson, 1975). It has also been reported that maternal starvation in rat reduces amino acid transfer (Rosso, 1975).

## 2.9. Lipids

Lipids are readily soluble in organic solvents and only sparingly soluble in water. These characteristics have an important influence on placental transport. Increasing solubility in organic solvents correlates with ease of diffusion across lipoprotein membranes suggesting that lipids should be rapidly transferred across the placenta without the aid of active transport mechanisms. The low water solubility restricts the amount that can be transported in plasma in physical solution for presentation to the placenta. The capacity of the plasma to transport lipids is often greatly increased by binding the lipids to plasma proteins. The characteristics of the binding and the kinetic relationships to physically dissolved substrate and possibly the placenta membrane may exert a major influence on transfer rates (Figure 1). In most instances, these remain theoretical considerations that have not been experimentally quantified.

Some indication of the contribution of maternal lipids to fetal adipose tissue has been obtained from analyses of maternal and newborn subcutaneous fat (Hirsch *et al.*, 1960). The similarity between the two in the premature infant suggests a significant maternal contribution early in gestation. The differences in composition when the infant is at term suggest that the demand for rapid deposition of fat in the last trimester is not satisfied by placental transfer of lipids and requires *de novo* synthesis.

### 2.9.1. Free Fatty Acids

These circulate in plasma bound to albumin. The concentrations are low, but the turnover rates are fast so that large amounts of free fatty acids are potentially available to the fetus. The contribution to the fetal economy appears to vary among the mammalian species (Hershfield and Nemeth, 1968; Kayden *et al.*, 1969; Van Duyne *et al.*, 1962). Its role in the human is not certain. The results of perfusion studies in placentas at term are consistent with a low transfer rate, inadequate to meet the requirements for fat deposition (Dancis *et al.*, 1973). The direct effect of protein binding is to reduce the rate of transfer (Dancis *et al.*, 1976). However, by increasing the concentration of free fatty acids circulating in the maternal plasma, the net effect of protein binding is an increase in the amount transferred to the fetus.

Phospholipids, cholesterol, and the neutral triglycerides circulate in the lipo-protein fraction of plasma in relatively large complexes. The mechanism by which these materials are made available at the cell membrane for transfer and metabolism have not been completely elucidated. One would suspect from the nature of the transport complex that placental transfer would be slow. The transfer rate of cholesterol is slow, and yet significant amounts of fetal cholesterol are derived from the mother because the turnover rate in the fetus is also slow (Goldwater and Stetten, 1947; Kayden *et al.,* 1969; Pitkin *et al.,* 1972). Phospholipids do not reach the fetus intact. They are hydrolyzed by the placenta and resynthesized by the fetus and possibly by the placenta (Biezenski, 1969; Popjak, 1954).

## 2.10. Steroids

The synthesis, transport, and metabolism of estrogens provide an interesting illustration of the interplay among mother, placenta, and fetus (Figures 2 and 3). The complexities of estrogen synthesis are described in some detail in Section 3. The newly synthesized, unconjugated estrogens are transferred rapidly across

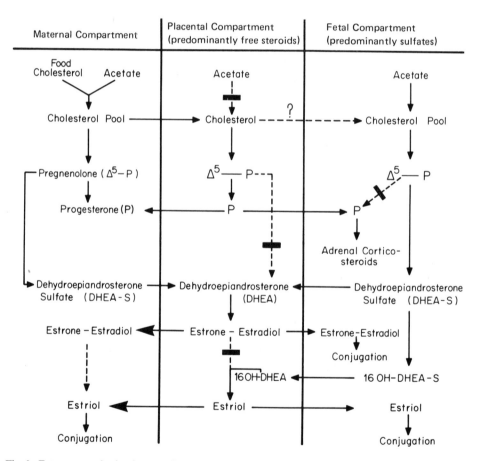

Fig. 2. Estrogen synthesis: the coordinated effort utilized in the synthesis of estrogens is diagrammed. The horizontal bars indicate metabolic steps that cannot be accomplished in either placenta or fetus (from Buster and Abraham, 1975, *Obstet. Gynecol.* **46(4):**492, with permission).

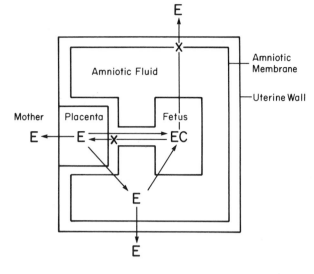

Fig. 3. Routes for disposal of placental estrogens. Placental estrogens (E) may reach the fetus through the umbilical circulation or via the amniotic fluid. Effective conjugation, primarily to sulfates (EC), at the "portals of entry" (liver, skin, lung, gastrointestinal tract) protects the fetus against high levels of unconjugated estrogens (Diczfalusy *et al.*, 1963). For excretion, the sulfate conjugates are cleared by sulfatases (X) at the exit routes (placenta and membranes) permitting the readily transferred unconjugated estrogens to reach the maternal circulation for conjugation and excretion.

membranes, including the placenta, easily entering maternal and fetal circulations (Levitz *et al.*, 1967) Although the bulk of placental estrogens are secreted into the maternal system, significant amounts reach the fetus. The "portals of entry" into the fetus (the liver, skin, lung, gastrointestinal tract) are well endowed with sulfokinase (Diczfalusy *et al.*, 1963) and, as recently shown, fetal tissues also contain glucuronyltransferase (Mikhail *et al.*, 1963). Sulfate and glucuronide conjugates are less physiologically active, and therefore the mechanisms of fetal conjugation may act as a protection against excess of estrogen. The conjugates are highly polar and are transferred across membranes very slowly. Accumulation of sulfates in the fetal compartment is avoided because the "portal of exit" (the placenta and amniotic membranes) are rich in sulfatase, capable of cleaving the conjugates (Katz *et al.*, 1965). The absence of significant activity of glucuronidase in the membranes is reflected in a relative accumulation of glucuronide-containing conjugates in amniotic fluid (Young *et al.*, 1974).

Progesterone and testosterone are also transferred rapidly across the placenta (Haskins and Soiva, 1960; M. Levitz, W. L. Money, and J. Dancis, unpublished observations). Paradoxically, there appears to be limited transfer of cortisol, only 25% of fetal cortisol being derived from the mother (Beitins *et al.*, 1973). This is probably the result of rapid conversion to cortisone by the placenta rather than a restriction in the diffusion of cortisol.

### 2.11. Bilirubin

Experimental studies in the guinea pig and monkey (Bashore *et al.*, 1969; Schenker *et al.*, 1964) have demonstrated that, similar to the estrogens, unconjugated bilirubin diffuses readily across the placenta, whereas the polar conjugates do not. Clinical observations suggest that the same is true in the human (Dancis, 1959). Significant hyperbilirubinemia is seldom evident in the infant before birth, but may quickly become manifest after delivery, particularly in erythroblastosis, strongly suggesting that the placenta provides an effective route of excretion. From this perspective the inability of the fetus to synthesize effectively the poorly transferrable glucuronide conjugates before birth provides an advantage. After birth, when the liver becomes the physiological route of excretion, glucuronyltransferase activity is rapidly developed.

### 2.12. Vitamins

There is little information concerning the rates of placental transfer of vitamins in the human. It is believed that the fat-soluble vitamins are transferred by simple diffusion at relatively slow rates. Inadequate attention has been given to the effect of protein binding. Vitamin A exists in plasma as retinol and as the provitamin, carotene. Maternal and fetal levels of retinol vary independently, whereas there is good correlation in carotene concentrations, with the fetal level much lower than the maternal. This has led to the suggestion that vitamin A is transferred to the fetus as carotene which is then converted to retinol (Barnes, 1951; Lund and Kimble, 1943). There is also a good correlation between maternal and cord levels of 25-hydroxy-vitamin $D_3$ (25-hydroxycholecalciferol) (Hillman and Haddad, 1974). There is no similar information concerning 1,25-dihydroxy-vitamin $D_3$ or the parent compound. Cord blood levels of vitamin E in the human fetus are lower than in the mother (Straumfjord and Quaife, 1946). Placental transfer of vitamins E (Sternberg and Pascoe-Dawson, 1959) and K (Taylor *et al.*, 1957) has been demonstrated in rats, but the design of the experiments permits no conclusion concerning the rates or mechanisms of transfer.

The fetal blood levels of water-soluble vitamins are higher than the maternal, suggesting active transport processes. It has been assumed without supporting evidence that transport rates are rapid. Vitamin C is found in blood as reduced L-ascorbic acid and as oxidized dehydroascorbic acid. In the human, the levels of the latter are approximately equivalent in maternal and fetal blood, whereas the ascorbic acid concentration is higher in cord blood (Raiha, 1959). Studies in the guinea pig indicate rapid transfer of dehydroascorbic acid with fetal conversion to ascorbic acid. It has been reported that the isolated perfused human placenta is capable of independently establishing a gradient with L-ascorbic acid (Hensleigh and Krantz, 1966). Vitamin $B_1$ (thiamine) is almost twice as high in cord blood than in the mother (Slobody *et al.*, 1949). Vitamin $B_2$, measured as free riboflavin, is approximately 4 times higher in cord blood than maternal blood, whereas flavin adenine dinucleotide (FAD) is higher in maternal blood (Lust *et al.*, 1953). It has been suggested that the placenta is relatively impermeable to both forms of the vitamin and that FAD is taken up by the placenta from maternal blood, converted to riboflavin, and released to the fetus. Vitamin $B_6$ (pyridoxol) blood levels are lower in the pregnant woman than in the nonpregnant, and fetal levels are higher than in the maternal. The partition of available vitamin appears to be in favor of the fetus (Contractor and Shane, 1970). Similar observations have been made for vitamin $B_{12}$ (Boger *et al.*, 1957; Hellegers *et al.*, 1957). Folic and folinic acid, as measured by microbiological assays, are higher in cord blood than in maternal blood (Grossowicz *et al.*, 1960).

### 2.13. Transfer of Hormones

The question is commonly raised as to whether circulating maternal hormones will cross the placenta in amounts that significantly affect fetal function and growth. The answer for the protein hormones appears to be negative. Although transfer can be demonstrated if assay methods are sufficiently sensitive, as has been shown with insulin (Gitlin *et al.*, 1965), the amounts transferred are inadequate to meet fetal requirements. The same appears to be true of thyroid hormones, permitting congenital hypothyroidism to occur in spite of normal maternal thyroid function.

In contrast, the transfer of steroids is rapid (Levitz *et al.*, 1967). Maternally administered steroids may affect the fetus, as documented by the unfortunate accidents of masculinization of the fetus in offspring of mothers treated with progestins (Williams, 1971). More recently, adrenocortical steroids are being administered late in pregnancy to effect maturation of the fetal pulmonary surfactant system (Liggins and Howie, 1972).

Insulin receptors have been identified on trophoblastic membrane (Haour and Bertrand, 1974). Their role has not yet been clarified. In slice experiments an increase in uptake of amino acids could be shown in response to insulin (Dancis *et al.*, 1968*b*).

## 2.14. Transfer of Proteins and Antibodies

The retention of the mammalian fetus *in utero* protects it from external pathogens. During gestation, its immunological apparatus develops but is relatively unutilized. At birth, the newborn is at risk from sudden exposure to an inclement environment to which it may not be able to react sufficiently rapidly. During this period, maternal antibody formed in response to the same environmental pathogens that threaten the fetus are transferred to the fetus providing a temporary umbrella of protection.

The route through which such transfer occurs differs among the mammalian species. In the guinea pig the transfer is before birth but by a complex path involving secretion into the uterine cavity and absorption through an everted yolk sac, the splanchnopleure (Brambell *et al.*, 1951). In the mouse, some antibody transfer takes places before birth and more is transferred through mother's milk for some weeks after birth. In the ungulate, antibody is transferred through maternal colostrum shortly after birth. In the human, transfer is primarily before birth through the chorioallantoic placenta (Dancis *et al.*, 1961). Transfer is selective, limited almost entirely to the IgG fraction. After birth, IgA antibody is transferred to the newborn through mother's milk. This antibody is not absorbed into the infant's circulation and is apparently intended for local protection against enteric pathogens.

Transfer is believed accomplished by pinocytosis, an engulfing of antibody molecules by infolding of the membrane. Specific receptors are located on the trophoblast cell membrane which bind to the IgG fraction of gamma globulin, explaining its selective transmission (Gitlin, 1974). Albumin is also transferred to the fetus in easily detectable amounts but at rates lower than IgG.

The significance of transferred protein to the fetal economy is determined by the relation of placental transfer rate to fetal utilization rate. Both IgG and albumin have half-lives measured in weeks. The entire fetal complement of IgG is derived from the mother. Fetal liver is capable of synthesizing albumin early in gestation so that the fetus is not dependent on transplacental sources (Dancis *et al.*, 1957*a*). The half-lives of other plasma proteins are considerably shorter, with maternal plasma proteins probably not contributing significantly to fetal levels. Proteases within the placenta may act as an additional barrier.

Hemopoietic stem cells have been demonstrated in mouse placenta. Under suitable experimental conditions, the cells mature into the erythrocytic, leukocytic, and megakaryocytic series (Dancis *et al.*, 1968*a*). It is possible that these cells may serve as a first line of immunological defense for the fetus or may provide specialized cells with Fc receptors believed to be engaged in antibody transport. Receptors capable of binding to the Fc fragment of gamma globulin are found on macrophages and lymphocytes. A select population of mouse trophoblast also possess such receptors (Elson *et al.*, 1975). Binding to these receptors may be the initial step in pinocytosis.

### 2.15. Fetal Excretion

Most placental transport mechanisms are bidirectional. Fetal utilization of nutrients maintains the net transfer of these materials toward the fetus, while the accumulation of the end-products of fetal metabolism (e.g., carbon dioxide, urea, uric acid, creatinine) is prevented by transfer in the reverse direction. Only bilirubin (see above) and urea have been adequately studied. The placenta, like other membranes, is permeable to urea. The clearance in monkeys is 15 ml/kg fetal weight (Battaglia *et al.*, 1968).

## 3. Endocrinology

The placenta synthesizes four hormones: chorionic gonadotropin (HCG), choriosommatotropin (HCS)—also called placental lactogen (HPL), progesterone, and estrogens. Suggestions have been made that it also synthesizes other hormones such as ACTH, thyroid-stimulating hormone, and thyroid-releasing hormone (Genazzani, 1974, 1975; Gibbons *et al.*, 1975; Hennen *et al.*, 1969; Hershman and Starnes, 1971). These reports await further confirmation. In its capacity to synthesize this wide variety of hormones, the placenta outperforms all other endocrine glands. The site of synthesis is the syncytiotrophoblast. The two peptide hormones, HCG and HCS, are secreted into the maternal circulation, establishing much higher levels in the mother than in the fetus. The steroids are released into both maternal and fetal circulations. Estrogens and progesterone are identical in structure with those synthesized by the gonads, and the peptides are close analogs to the pituitary glycoprotein hormones.

### 3.1. Steroids

The synthesis of estrogens and progesterone increases rapidly during pregnancy, reaching very high levels by term. Early in gestation, the corpus luteum is the site of synthesis but by seven weeks placental synthesis of steroids is adequate to maintain pregnancy even after removal of the corpus luteum (Csapo, 1973). Neither the placenta nor the fetus can independently perform all the essential synthetic steps. The interdependence of the productive effort is indicated in Figure 2.

Progesterone is produced by the placenta from maternal serum cholesterol (Hellig *et al.*, 1970). The fetal adrenal is incapable of synthesizing progesterone because of a relative deficiency in $3\beta$-OH-dehydrogenase. Placental progesterone serves as the precursor for fetal synthesis of corticosteroids, androstenedione, and testosterone (Solomon and Fuchs, 1971; Taylor *et al.*, 1974). Progesterone secreted into the mother has systemic and local effects contributing to retention of the fetus during gestation. It is excreted by the mother as pregnanediol.

Synthesis of the three estrogens, estrone, estradiol, and estriol, increases progressively during pregnancy. Estriol is produced in greatest quantity, a unique characteristic of the human. The placenta contributes to the synthesis of all the estrogens but is unable to accomplish this function *de novo*. Estrogen and estradiol are synthesized from dehydroepiandrosterone (DHEA), derived from mother and fetus (Figure 2) (Siiteri and MacDonald, 1966). The importance of the fetal supply is documented by pregnancies complicated by anencephalic monsters with atrophic adrenals, in which estradiol production is reduced to approximately half (MacDonald and Siiteri, 1965). Conversion of maternal DHEA to estradiol has been confirmed and quantified using isotopic techniques (Siiteri and MacDonald, 1963). A significant decrease in the clearance of DHEA injected into women who later in pregnancy developed signs of toxemia has been related to a reduction of blood flow through the intervillous space (Gant *et al.*, 1971).

The placenta requires an hydroxylated precursor for the synthesis of estriol, which is supplied almost exclusively by the fetus in the form of 16-OH-dehydroepiandrosterone sulfate (16-OH-DEAS). The sulfate radical is removed by a placental sulfatase (Pulkinnen, 1961) which facilitates the uptake of 16-OH-DHEA, which the placenta then converts into estriol. The estriol is released into the maternal system to be disposed of in the urine after conjugation in the liver.

Estrogens have multiple effects on maternal metabolism and probably contribute directly to fetal welfare by developing and maintaining uterine circulation (Rosenfeld *et al.*, 1976). Clinicians, recognizing that estriol synthesis depends on fetal and placental metabolism, have monitored pregnancies by measuring plasma and urinary estriol levels (Beischer and Brown, 1972). An interesting deviation without important clinical repercussions results from the absence of placental sulfatase (Cedard *et al.*, 1971; Oakey *et al.*, 1974). The placenta is unable to metabolize the fetal DHEA sulfate and maternal estriol levels are very low. The babies have been normal.

### 3.2. Protein Hormones

Human chorionic gonadotropin (HCG) is composed of two noncovalently bound subunits. The $\alpha$ subunit is believed to be nonspecific and very similar to that in human luteinizing hormone, follicle stimulating hormone, and thyroid-stimulating hormone (Vaitukaitis and Ross, 1973). Biological specificity appears to reside in the $\beta$ subunit. The molecular weight of HCG is roughly 40,000; the $\alpha$ unit, 15,000; and the $\beta$ unit, 23,000 (Bellisaro *et al.*, 1973; Carlsen *et al.*, 1973). The metabolic half-life is calculated as 11 hr (Yen *et al.*, 1968), much longer than that of LH and FSH, a feature attributed to its high sialic acid content (Van Hall *et al.*, 1971).

HCG is synthesized by the syncytiotrophoblast (Dreskin *et al.*, 1970; Midgley and Pierce, 1962) and secreted into the maternal circulation. A $\beta$-subunit-specific radioimmunoassay permits differentiation between HCG and LH, and the diagnosis of pregnancy can be made as soon as 8–10 days following ovulation, about the time of implantation (Braunstein *et al.*, 1973). The early development of HCG synthesis is believed to be responsible for the maintenance of the corpus luteum of pregnancy. HCG production peaks at 60–70 days and then falls off rapidly to a low plateau which remains for the rest of the pregnancy. The highest levels are attained after corpus luteum function has begun to wane (Tulchinsky and Hobel, 1973). The significance of this pattern of production which is characteristic for HCG synthesis is not understood. HCG suppresses the response of lymphocytes to mitogens, suggesting that it may **contribute** to the prevention of immunological rejection by the mother (see Section 4).

Clinically, HCG serves as a biological indicator for the presence of trophoblastic tissue, used in the diagnosis of pregnancy and in the diagnosis and treatment of trophoblastic diseases.

Human chorionic somatomammotropin (HCS) is a single-chain polypeptide with a molecular weight of 20,000 and is remarkably similar in amino acid sequence and structure to human growth hormone (Andrews, 1969; Chan *et al.*, 1971; Sherwood, 1971). Like HCG, HCS is synthesized in the syncytiotrophoblast (Delkonikoff and Cedard, 1973) and secreted into the maternal circulation establishing a large maternal–fetal gradient (Kaplan *et al.*, 1968). The metabolic half-life is only 20 min. The amount synthesized increases throughout pregnancy, correlating with placental size (Sdenkow *et al.*, 1969).

HCS has a prolactin-like activity (Josimovich and MacLaren, 1962), a weak growth-promoting effect, and potentiates human growth hormone (Kaplan and Grumbach, 1964). Its biological effect in pregnancy appears to be the reverse of insulin, reducing uptake and consumption of glucose in the mother and mobilizing her fatty acids. The net effect of these changes in maternal metabolism would be to make glucose more readily available to the fetus. Fluctuating secretion of insulin by the mother may serve to modify the more constant effects induced by HCS (Josimovich *et al.*, 1973).

Clinically, plasma levels of HCS have been used to monitor high-risk pregnancy, reflecting placental function.

## 4. Immunology of Pregnancy

Mammalian reproduction has provided a provocative puzzle for immunologists. The fetus is genetically distinct from the mother because of the paternal chromosomal complement, and yet the fetus is not rejected by maternal immunological mechanisms. Many plausible explanations have been offered. It is now believed that several factors contribute to the protection of the fetus (Table III). The relative importance of each factor, and whether all are necessary for the continuation of pregnancy, is not known. There is evidence that the mother does respond immunologically to the fetus, but that cellular-mediated immunity is controlled. The placenta plays a pivotal role in these considerations (Beer and Billingham, 1974).

The placenta, by virtue of its location and structure, serves as a barrier to the

Table III.   Theories Propounded to Account for the Invulnerability of the Fetoplacental
Unit to Rejection[a]

| Altered immunological responsiveness of mother | Uterus | Trophoblast | Fetus |
|---|---|---|---|
| Nonspecific<br>  1. Placental protein or steroid hormones<br>  2. Plasma factors<br>Specific<br>  1. Antibody-mediated suppression<br>  2. Adult tolerance | 1. Complete separation of fetal and maternal circulations<br>2. Immunologically privileged site (decidual tissue) | 1. Nonantigenic<br>2. Physiologic barrier<br>3. Local immunosuppression by hormones (HCG and HCS) | 1. Antigenic immaturity<br>2. Serum factors (fetuin) |

[a] From Beer and Billingham, 1974.

migration of cells between the maternal and fetal circulations, thereby reducing the opportunity for sensitization and for the entry of foreign immunocompetent cells. The barrier is imperfect; leukocytes transverse the placenta in both directions (Schroder, 1974a,b), but cellular traffic is limited, particularly from mother to fetus.

The trophoblast, of fetal origin and directly exposed to the maternal circulation, has been considered to be immunologically inert. Large numbers of trophoblast are shed into the maternal circulation and lodge in the maternal lung where they may persist without eliciting a visible host response (Thomas *et al.*, 1959). Transplantation of the ectoplacental cone of the mouse to a homologous recipient has also escaped rejection, whereas the fetal primordia at the same stage of gestation did not (Simmons and Russel, 1963). Prior treatment of the ectoplacental cone with neuraminidase produced an immunological reaction on transplantation suggesting that a protective nonantigenic coat of sialomucin had been removed (Currie *et al.*, 1968). A similar buffer role had been suggested for the "fibrinoid" layer described in human placentas, although it would appear to be too discontinuous to be very effective (Wynn, 1973).

A variety of studies point to suppression of the immune response. Human chorionic gonadotropin (Adcock *et al.*, 1973; Muchmore and Blaese, 1974; Schiff *et al.*, 1974) and somatomammotropin (Contractor and Davies, 1973) suppress the response of lymphocytes to phytohemagglutinin and might cause local inhibition of the immune reaction. Tolerance for an exchange of skin grafts between mother and fetus has been demonstrated in the human (Peer *et al.*, 1958) and in experimental animals (Rogers *et al.*, 1960). The trophoblast shed into the maternal circulation could, theoretically, play a role in modifying the immune response. A mechanism for the specific suppression of cellular immunity in the mother is offered by the phenomenon of "enhancement" described in experimental oncology. The development of specific humoral antibody to tumor may protect it from cellular rejection, presumably by masking antigenic sites. An observation consistent with this hypothesis in pregnancy has been described using *in vitro* colonies formed from mouse embryonic cells and lymphocytes obtained from the mother. Inhibition of colonies by the lymphocytes was nullified by maternal serum (Hellstrom *et al.*, 1969). The nature of the protective material in the serum was not identified. More recently, maternal IgG has been eluted from the basement membrane of human trophoblast where it was immunologically bound, thus blocking antigenic sites (Faulk, 1974).

That sensitization of the mother to the fetus poses a real as well as a theoretical threat has been demonstrated with a variety of experimental animals (Beer and Billingham, 1973). Abortion and runt disease have been caused by adoptive or active immunization of the mother against her fetuses. Clinical examples in the human have been suggested (Kadowaki *et al.*, 1965) but are difficult to prove.

The interesting suggestion has been advanced that fetomaternal histoincompatability and the ensuing immunological interaction may be advantageous to the fetus. Enlarged placental and fetal size have been demonstrated in crosses of inbred strains of mice with a small but significant selective advantage in the progeny (Beer *et al.*, 1975).

## 5. Conclusion

An organ as versatile and complex as the placenta requires a summary statement to synthesize the information presented in the preceding pages.

JOSEPH DANCIS and
HENNING SCHNEIDER

The placenta was designed solely to make intrauterine pregnancy, and thus fetal growth, possible and is discarded by the fetus and mother when that role is completed. Appearing on the evolutionary scene long after the development of two key biological functions, sexual reproduction and immunological rejection of foreign invaders, an essential role of the placenta was to modify the maternal immune response so that retention of the fetus is permitted. This is accomplished through a series of complementary mechanisms. The placenta also contributes to the maintenance of pregnancy by developing into a large endocrine organ, capable of synthesizing an impressive variety of hormones in unprecedented amounts. The hollow, contractile uterus is initially more suited for expulsion than retention. Extensive modification under the influence of placental hormones is required to convert it into an hospitable environment for the fetus, as well as to maintain it in its quiescent state for the duration of pregnancy. Maternal metabolism is altered in many significant ways by placental hormones secreted directly into the maternal circulation, presumably to favor the fetus in obtaining its nutrients.

Feeding the fetus is a primary function of the placenta. This is accomplished within limited space by forming a very large epithelial absorptive surface of villi and microvilli directly exposed to the nutrient source, the maternal blood. Nutrients are selectively transferred into the fetal blood by a variety of mechanisms which often resemble those described in other organs but which are modified for this specific function. Modifications must be expected when one considers that the placental membrane is unique in that transfer is from and to protein-rich plasma circulating past both surfaces of the membrane.

The placenta also serves as the main route of excretion of fetal metabolites, supplemented to some extent by discharge of urine and pulmonary secretions into amniotic fluid. Although clearly an essential function, transfer from fetal to maternal circulations has been less intensively studied, partly because of technical difficulties.

As a very active metabolic organ, the placenta also requires considerable support in the form of oxygen and nutrients, exacting its toll from the materials extracted from maternal blood. A wealth of enzymes within the placenta utilize these materials to maintain the organ and perform its many functions. The recent rediscovery of similarities between metabolism in the placenta and in tumors, at the enzyme level, has led to a surge of interest by biochemists and oncologists.

However, an understanding of the physiology of the placenta requires a broader rather than a finer focus. Diczfalusy *et al.* (1963) are responsible for coining the term "fetoplacental unit," stressing that steroid synthesis requires the coordinated action of both placenta and fetus. To interpret many placental functions, the concept must be expanded to include the mother. These three partners in pregnancy, the mother, the placenta, and the fetus, interact as a single organism, an important feature that the research scientist, the human biologist, and the physician must keep in mind.

## 6. References

Adcock, E. W., III, Teasdale, F., August, C. S., Cox, S., Meschia, G., Battaglia, F. C., and Naughton, M. A., 1973, Human chorionic gonadotropin: Its possible role in maternal lymphocyte suppression, *Science* **181**:845.

Amoroso, E. C., 1952, Placentation, in: *Marshall's Pysiology of Reproduction* (A. S. Parkes, ed.), pp. 127–311, Spottiswoode, Ballantyne, London.

Andrews, P., 1969, Molecular weight of human placental lactogen investigated by gel filtration, *Biochem. J.* **111**:799.

Awai, M., Chipman, B., and Brownes, E. B., 1975*a*, *In vivo* evidence for the functional heterogeneity of transferrin bound iron. I. Studies in normal rats, *J. Lab. Clin. Med.* **85**:769.

Awai, M., Chipman, B., and Brownes, E. B., 1975*b*, *In vivo* evidence for the functional heterogeneity of transferrin bound iron. II. Studies in pregnant rats, *J. Lab. Clin. Med.* **85**;785.

Baerlocher, K. E., Scriver, C. R., and Mohyuddin, F., 1970, Ontogeny of iminoglycine transport in mammalian kidney, *Proc. Natl. Acad. Sci. U.S.A.* **65**:1009.

Barnes, A. C., 1951, The placental metabolism of vitamin A, *Am. J. Obstet. Gynecol.* **61**:368.

Bartels, H., 1970, Prenatal respiration, in: *Frontiers of Biology,* Vol. 17 (A. Neuberger and E. L. Tatum, eds.), pp. 1–187, North-Holland, Amsterdam.

Bartels, H., Moll, W., and Metcalfe, J., 1962, Physiology of gas exchange in the human placenta, *Am. J. Obstet. Gynecol.* **84**:1714.

Bashore, R. A., Smith, F., and Schenker, S., 1969, Placental transfer and disposition of bilirubin in the pregnant monkey, *Am. J. Obstet. Gynecol.* **103**:950.

Battaglia, F. C., Behrman, R. E., Meschia, G., Seeds, A. E., and Bruns, P. D., 1968, Clearance of inert molecules, Na and Cl ions across the primate placenta, *Am. J. Obstet. Gynecol.* **102**:1135.

Bauer, C., Ludwig, I., and Ludwig, M., 1968, Different effects of 2,3-diphosphoglycerate and adenosine triphosphate on the oxygen affinity of adult and foetal human haemoglobin, *Life Sci.* **17**:1339.

Beer, A. E., and Billingham, R. E., 1973, Maternally acquired runt disease, *Science* **179**:240.

Beer, A. E., and Billingham, R. E., 1974, Immunologic coexistence in the maternal-fetal relationship, in: *Modern Perinatal Medicine* (L. Gluck, ed), p. 83, Year-Book Medical Publishers, Chicago.

Beer, A. E., Billingham, R. E., and Scott, J. R., 1975, Immunogenetic aspects of implantation, placentation and feto-placental growth rates, *Biol. Reprod.* **12**:176.

Beischer, N. A., and Brown, J. B., 1972, Current status of estrogen assays in obstetrics and gynecology, *Obstet. Gynecol. Surv.* **27**:303.

Beitins, I. Z., Bayard, F., Ances, I. G., Kowarski, A., and Migeon, C. J., 1973, The metabolic clearance rate, blood production, interconversion and transplacental passage of cortisol and cortisone in pregnancy near term. *Pediatr. Res.* **7**:509.

Bellisario, R., Carlson, R. B., and Bahl, O. P., 1973, Human chorionic gonadotropin. Linear amino acid sequence of the $\alpha$ subunit, *J. Biol. Chem.* **248**:6796.

Biezenski, J. J., 1969, Role of placenta in fetal lipid metabolism. I. Injection of phospholipids double labeled with $C^{14}$ glycerol and $P^{32}$ into pregnant rabbits, *Am. J. Obstet. Gynecol.* **104**:1177.

Boger, W. P., Bayne, G. M., Wright, L. D., and Beck, G. D., 1957, Differential serum vitamin $B_{12}$ concentrations in mothers and infants, *N Engl. J. Med.* **256**:1085.

Bothwell, T. H., Pribella, W. F., Mebust, W., and Finch, C. A., 1958, Iron metabolism in the pregnant rabbit: Iron transport across the placenta, *Am. J. Physiol.* **193**:615.

Boyd, R. D. H., Morris, F. H., Meschia, G., Makowski, E. L., and Battaglia, F. C., 1973, Growth of glucose and oxygen uptake by fetuses of fed and starved ewes, *Am. J. Physiol.* **225**:897.

Brambell, F. W. R., Hemming, W. A., and Henderson, M., 1951, *Antibodies and Embryos,* Athlone Press, London.

Braunstein, G. D., Grodin, J. M., Vaitukaitis, J. L., and Ross, G. T., 1973, Secretory rates of human chorionic gonadotropin by normal trophoblast, *Am. J. Obstet. Gynecol.* **115**:447.

Bruns, P. D., Linder, R. O., Drose, V. E., and Battaglia, F., 1963, The placental transfer of water from fetus to mother following the intravenous infusion of hypertonic mannitol to the maternal rabbit, *Am. J. Obstet. Gynecol.* **86**:160.

Bruns, P. D., Hellegers, A. E., Seeds, A. E., Jr., Behrman, R. E., and Battaglia, F. C., 1964, Effects of osmotic gradients across the primate placenta upon fetal and placental water contents, *Pediatrics* **34**:407.

Burd, L. J., Jones, M. D., Jr., Simmons, M. A., Makowski, E. L., Meschia, G., and Battaglia, F. C., 1975, Placental production and fetal utilization of lactate, *Nature* **254**:710.

Campbell, A. G. M., Dawes, G. S., Fishman, A. P., Human, A. J., and James, G. B., 1966, The oxygen consumption of the placenta and fetal membranes in the sheep, *J. Physiol. (London)* **182**:439.

Carlsen, R. B., Bahl, O. P., and Swaminathan, N., 1973, Human chorionic gonadotropin. Linear aminoacid sequence of the $\beta$ subunit, *J. Biol. Chem.* **248**:6810.

Cedard, L., Tehobronsky, C., Guglielmina, R., and Mailhac, M., 1971, Insuffisance oestrogénique paradoxale au cours d'une grossesse normal par défaut de sulfataseplacentaire, *Bull. Fed. Soc. Gynecol. Obstet. Lang. Fr.* **23**:16.

Challier, J. C., Schneider, H., and Dancis, J., 1976, *In vitro* perfusion of human placenta. V. Oxygen consumption, *Am. J. Obstet. Gynecol.* **126**:261–265.

Chan, H. L., Dixon, J. S., and Chung, D., 1971, Primary structure of the human chorionic somatomammotropin (HCS) molecule, *Science* **173**:56.

Contractor, S. F., and Davies, H., 1973, Effect of human chorionic somatomammotropin and human chorionic gonadotropin on phytohaemagglutinin-induced lymphocyte transformation, *Nature (London), New Biol.* **243**:284.

Contractor, S. F., and Shane, B., 1970, Blood and urine levels of vitamin $B_6$ in the mother and fetus before and after loading of the mother with vitamin $B_6$, *Am. J. Obstet. Gynecol.* **107**:635.

Csapo, A. J., 1973, The effect of $E_2$ replacement therapy on early pregnant lutectomized patients, *Am. J. Obstet. Gynecol.* **117**:987.

Currie, G. A., Van Dominick, W., and Bagshawe, K. D., 1968, Effects of neuraminidase on the immunogenecity of early mouse trophoblast, *Nature* **219**:191.

Dancis, J., 1959, Aspects of bilirubin metabolism before and after birth, *Pediatrics* **24**:980.

Dancis, J., and Springer, D., 1970, Fetal homeostasis in maternal malnutrition: Potassium and sodium deficiency, *Pediatr. Res.* **4**:345.

Dancis, J., Braverman, N., and Lind, J., 1957*a*, Plasma protein synthesis in the human fetus and placenta, *J. Clin. Invest.* **36**:398.

Dancis, J., Worth, M., Jr., and Schneidau, P. B., 1957*b*, Effect of electrolyte disturbance in the pregnant rabbit on the fetus, *Am. J. Physiol.* **188**:535.

Dancis, J., Lind, J., Oratz, M., Smolens, J., and Vara, P., 1961, Placental transfer of proteins in human gestation, *Am. J. Obstet. Gynecol.* **82**:167.

Dancis, J., Brenner, M., and Money, W. L., 1962, Some factors affecting the permeability of guinea pig placenta, *Am. J. Obstet. Gynecol.* **84**:570.

Dancis, J., Jansen, V., Gorstein, F., and Douglas, G., 1968*a*, Hematopoietic cells in mouse placenta, *Am. J. Obstet. Gynecol.* **100**:1110.

Dancis, J., Money, W. L., Springer, D., and Levitz, M., 1968*b*, Transport of amino acids by placenta, *Am. J. Obstet. Gynecol.* **101**:820.

Dancis, J., Springer, D., and Cohlan, S. Q., 1971, Fetal homeostasis in maternal malnutrition. II. Magnesium deprivation, *Pediatr. Res.* **5**:131.

Dancis, J., Jansen, V., Kayden, H. J., Schneider, H., and Levitz, M., 1973, Transfer across perfused human placenta. II. Free fatty acids *Pediatr. Res.* **7**:192.

Dancis, J., Jansen, V., and Levitz, M., 1976, Transfer across perfused human placenta. IV. Effect of protein binding on free fatty acids, *Pediatr. Res.* **10**:5.

Davies, J., 1955, Permeability of the rabbit placenta to glucose and fructose, *Am. J. Physiol.* **181**:532.

Dawes, G. S., and Shelley, H. J., 1968, Carbohydrate metabolism in the fetus and newborn, in: *Carbohydrate Metabolism* (F. Dickens, ed.), pp. 87–121, Academic Press, New York.

Delivoria-Papadopoulos, M., Battaglia, F. C., Bruns, P. D., and Meschia, G., 1967, Total protein-bound and ultrafiltrable calcium in maternal and fetal plasmas, *Am. J. Physiol.* **213**:363.

Demers, L. M., Gabbe, S. G., Villee, C. A., and Greep, R. D., 1972, The effects of insulin on human placental glycogenesis, *Endocrinology* **91**:270.

Delkonikoff, L. K., and Cedard, L., 1973, Localization of human chorionic gonadotropic and somatomammotropic hormones by the peroxidase immunohistoenzymologic method in villi and amniotic epithelium of human placentas (from six weeks to term), *Am. J. Obstet. Gynecol.* **116**:1124.

Diczfalusy, E., Tillinger, K. F., Wiqvist, N., Levitz, M., Condon, G. P., and Dancis, J., 1963, Disposition of intra-amniotically administered estriol-16-$^{14}$C and estrone-16-$^{14}$C sulfate by women, *J. Clin. Endocrinol. Metab.* **23**:503.

Dierks-Ventling, C., Cone, A. L., and Wapnir, R. A., 1971, Placental transfer of amino acids in the rat. I. L-Glutamic acid and L-glutamine, *Biol. Neonat.* **17**:361.

DiPietro, D. L., Gutierrez-Correa, J., and Thaidigman, J. H., 1967, Glucose metabolism by human placental villi, *Biochem. J.* **103**:246.

Dreskin, R. B., Spicer, S. S., and Greene, W. B., 1970, Ultrastructural localization of chorionic gonadotropin in human term placenta, *J. Histochem. Cytochem.* **18**:862.

Earle, D. P., Bakwin, H., and Hirsch, D., 1951, Plasma potassium level in newborn, *Proc. Soc. Exp. Biol. Med.* **76**:756.

Elson, J., Jenkinson, E. J., and Billington, W. D., 1975, Fc receptors on mouse placenta and yolk sac cells, *Nature* **255**:412.

Faber, J. J., and Hart, F. M., 1966, The rabbit placenta as an organ of diffusional exchange. Comparison with other species by dimensional analysis, *Circ. Res.* **19**:816.

Fantel, A. G., 1975, Feto-maternal potassium relations in the fetal rat on the twentieth day of gestation, *Pediatr. Res.* **9**:527.

Faulk, W. P., 1974, Immunological studies of the human placenta. Characterization of immunolgobulins on trophoblastic basement membranes, *J. Clin. Invest.* **54**:1011.

Flexner, L. B., and Gellhorn, A., 1942, The comparative physiology of placental transfer, *Am. J. Obstet. Gynecol.* **43**:965.

Flexner, L. B., Cowie, D. B., Hellman, L. M., Wilde, W. S., and Vosburgh, G. J., 1948, The permeability of the human placenta to sodium in normal and abnormal pregnancies and the supply of sodium to the human fetus as determined with radioactive sodium, *Am. J. Obstet. Gynecol.* **55**:469.

Folkart, G. R., Dancis, J., and Money, W. L., 1960, Transfer of carbohydrates across guinea pig placenta, *Am. J. Obstet. Gynecol.* **80**:221.

Friedman, E. A., and Sachtleben, M. R., 1960, Placental oxygen consumption *in vitro*. I. Baseline studies, *Am. J. Obstet. Gynecol.* **79**:1058.

Gant, N. F., Hutchinson, H. T., Siiteri, P. K., and MacDonald, P. C., 1971, Study of the metabolic clearance rate of dehydroepiandrosterone sulfate in pregnancy, *Am. J. Obstet. Gynecol.* **111**:555.

Gaull, G. E., Raiha, N. C. R., Saarikoski, S., and Sturman, J. A., 1973, Transfer of cyst(e)ine and methionine across the human placenta, *Pediatr. Res.* **7**:908.

Genazzani, A. R., 1974, *In vitro* synthesis of an ACTH like hormone and HCS by placental and amniotic cells, *Experientia* **30**:430.

Genazzani, A. R., 1975, Immunoreactive ACTH and cortisol plasma levels during pregnancy, *Clin. Endocrinol.* **4**:1.

Ghadimi, H., and Pecora, P., 1964, Free amino acids of cord plasma as compared with maternal plasma during pregnancy, *Pediatrics* **33**:500.

Gibbons, J. M., Mitnick, M., and Chieffo, V., 1975, *In vitro* biosynthesis of TSH- and LH-releasing factors by the human placenta, *Am. J. Obstet. Gynecol.* **121**:127.

Gitlin, J. D., 1974, Protein-binding by specific receptors on human placenta, murine placenta and suckling murine intestine in relation to protein transport across these tissues, *J. Clin. Invest.* **54**:1155.

Gitlin, D., Kumate, J., and Morales, C., 1965, On the transport of insulin across the human placenta, *Pediatrics* **35**:65.

Glendening, M. B., Margolis, A. J., and Page, E. W., 1961, Amino acid concentrations in fetal and maternal plasma, *Am. J. Obstet. Gynecol.* **81**:591.

Goldwater, W. H., and Stetten, DeW., Jr., 1947, Studies in fetal metabolism, *J. Biol. Chem.* **169**:722.

Gresham, E. L., James, E. J., Raye, J. R., Battaglia, F. C., Makowski, E. L., and Meschia, G., 1971, Production and excretion of urea by the fetal lamb, *Pediatrics* **50**:372.

Grossowicz, N., Aronovitch, J., Rachmilewitz, M., Izak, G., Sadovsky, A., and Bercovici, B., 1960, Folic and folinic acid in maternal and foetal blood, *Br. J. Haematol.* **6**:296.

Gruenwald, P., 1975, Introduction, the supply line of the fetus: Definitions relating to fetal growth, in: *The Placenta and Its Maternal Supply Line* (P. Gruenwald, ed.), pp. 1–17, Medical and Technical Publishing, Lancaster, Great Britain.

Gurtner, G., and Burns, B., 1975, Physiological evidence consistent with the presence of specific $O_2$ carrier in the placenta, *J. Appl. Physiol.* **39**:728.

Hagerman, D. D., and Villee, C. A., 1952, Transport of fructose by human placenta, *J. Clin. Invest.* **31**:911.

Haour, F., and Bertrand, J., 1974, Insulin receptors in the plasma membranes of human placenta, *J. Clin. Endocrinol. Metab.* **38**:334.

Haskins, A. L., and Soiva, K. U., 1960, The placental transfer of progesterone-4-$^{14}$C in human term pregnancy, *Am. J. Obstet. Gynecol.* **79**:674.

Hellegers, A., Okuda, K., Nesbitt, R. E. L., Jr., Smith, D. W., and Chow, B. F., 1957, Vitamin $B_{12}$ absorption in pregnancy and in the newborn, *Am. J. Clin. Nutr.* **5**:327.

Hellig, H., Gattereau, D., Lefebvre, C., and Bolte, E., 1970, Steroid production from plasma cholesterol to placental progesterone in humans, *J. Clin. Endocrinol.* **30**:624.

Hellstrom, K. A., Hellstrom, J., and Brawn, J., 1969, Abrogation of cellular immunity to antigenetically foreign mouse embryonic cells by a serum factor, *Nature* **224**:914.

Hennen, G., Pierce, J. G., and Treychet, P., 1969, Human chorionic thyrotropin: Further characterization and study of its secretion during pregnancy, *J. Clin. Endocrinol. Metab.* **29**:581.

Hensleigh, P. A., and Krantz, K. E., 1966, Extracorporeal perfusion of the human placenta. I: Placental transfer of ascorbic acid, *Am. J. Obstet. Gynecol.* **96**:5.

Hershfield, M. S., and Nemeth, A. M., 1968, Placental transport of free palmitic and lineolic acids in the guinea pig, *J. Lipid Res.* **9:**460.

Hershman, J. M., and Starnes, W. R., 1971, Placental content and characterization of human chorionic thyrotropin, *J. Clin. Endocrinol. Metab.* **32:**52.

Hill, P. M. M., and Young, M., 1973, Net placental transfer of free amino acids against varying concentrations, *J. Physiol.* **235:**409.

Hillman, L. S., and Haddad, J., 1974, Human prenatal vitamin D metabolism I: 25 hydroxyvitamin D in maternal and cord blood, *J. Pediatr.* **84:**742.

Hirsch, J., Farquhar, J., Ahrens, E. H., Jr., Peterson, M. L., and Stoffel, W., 1960, Studies of adipose tissue in man. A microtechnic for sampling and analysis, *Am. J. Clin. Nutr.* **8:**499.

Holmberg, N. G., Kaplan, B., Karvonen, M. J., Lind, J., and Malm, M., 1956, Permeability of human placenta to glucose, fructose and xylose, *Acta Physiol. Scand.* **36:**291.

Josimovich, J. B., and MacLaren, J. A., 1962, Presence in the human placenta and term serum of a highly lactogenic substance immunologically related to pituitary growth hormone, *Endocrinology* **71:**209.

Josimovich, J. B., Levitt, M. J., Grumbach, M. M., Kaplan, S. L. and Vinik, A., 1973, Human chorionic somatomammotropin (HCS), in: *Methods in Investigative and Diagnostic Endocrinology* (S. A. Berson, ed), pp. 787–822, North Holland, Amsterdam.

Kadowaki, J., Thompson, R. J., Zuelzer, W. W., Wooley, P. V., Jr., Brough, A. J., and Gruber, D., 1965, XX-XY lymphoid chimaerism in congenital immunological deficiency syndrome with thymic alymphoplasia, *Lancet* **2:**1152.

Kaplan, S. L., and Grumbach, M. M., 1964, Studies of a human and placental hormone with growth hormone like and prolactin like activities, *J. Clin. Endocrinol. Metab.* **24:**80.

Kaplan, S. L., Gurpide, E., Sciarra, J. J., and Grumbach, M. M., 1968, Metabolic clearance rate and production rate of chorionic growth hormone-prolactin in late pregnancy, *J. Clin. Endocrinol. Metab.* **28:**1450.

Katz, S. R., Dancis, J., and Levitz, M., 1965, Relative transfer of estriol and its conjugates across the fetal membranes *in vitro, Endocrinology* **76:**722.

Kayden, H. J., Dancis, J., and Money, W. L., 1969, Transfer of lipids across the guinea pig placenta, *Am. J. Obstet. Gynecol.* **104:**564.

Levitz, M., Condon, G. P., Dancis, J., Goebelsmann, U., Eriksson, G., and Diczfalusy, E., 1967, Transfer of estriol and estriol conjugates across the human placenta perfused in situ at midpregnancy, *J. Clin. Endocrinol. Metab.* **27:**1723.

Liggins, G. C., and Howie, R. N., 1972, A controlled trial of antepartum glucocorticoid treatment for prevention of the respiratory distress syndrome in premature infants, *Pediatrics* **50:**515.

Lipsitz, P. J., 1971, The clinical and biochemical effects of excess magnesium in the newborn, *Pediatrics* **47:**501.

Logothetopoulos, J. H., and Scott, R. F., 1956, Active iodide transport across the placenta of the guinea pig, rabbit and rat, *J. Physiol.* **132:**365.

Longo, L. D., Power, G. G., and Foster, R. E., II, 1967, Respiratory function of the placenta as determined with carbon monoxide in sheep and dogs, *J. Clin. Invest.* **46:**812.

Longo, L. D., Yuen, P., and Gusseck, D. J., 1973, Anaerobic glycogen dependent transport of amino acids by the placenta, *Nature (London)* **243:**531.

Lund, C. J., and Kimble, M. S., 1943, Plasma vitamin A and carotene of the newborn infant; with consideration of fetal–maternal relationships, *Am. J. Obstet. Gynecol.* **46:**207.

Lust, J., Hagerman, D. D., and Villee, C. A., 1953, The transport of riboflavin by human placenta, *J. Clin. Invest.* **33:**38.

MacDonald, P. C., and Siiteri, P. K., 1965, Origin of estrogen in women pregnant with an anencephalic fetus, *J. Clin. Invest.* **44:**465.

Mackay, R. B., 1958, Studies of the oxygen consumption of fresh placental tissue from normal and abnormal pregnancies, *J. Obstet. Gynaecol. Br. Emp.* **65:**791.

Mansour, M. M., Schulert, A. R., and Glasser, S. R., 1972, Mechanism of placental iron transfer in the rat, *Am. J. Physiol.* **222:**1628.

Meschia, G., Battaglia, F. C., and Bruns, P. D., 1967, Theoretical and experimental study of transplacental diffusion, *J. Appl. Physiol.* **22:**1171.

Midgley, A. R., Jr., and Pierce, G. B., Jr., 1962, Immunohistochemical localization of human chorionic gonadotropin, *J. Exp. Med.* **115:**289.

Mikhail, G., Wiqvist, N., and Diczfalusy, E., 1963, Oestriol metabolism in the previable human foetus, *Acta Endocrinol.* **42:**519.

Muchmore, A. V., and Blaese, R. M., 1974, Immunoregulatory effects of human chorionic gonadotropin (HCG), *Fed. Proc.* **33:**750.

Nyberg, R., and Westin, B., 1957, The influence of oxygen tension and some drugs on human placental vessels, *Acta Physiol. Scand.* **39:**216.

Oakey, R. E., Cawood, M. L., and MacDonald, R. R., 1974, Biochemical and clinical observations in a pregnancy with placental sulfatase and other enzyme deficiencies, *Clin. Endocrinol.* **3:**131.

Peer, L. A., Bernhard, W., and Walker, J. C., Jr., 1958, Full-thickness skin exchanges between parents and their children, *Am. J. Surg.* **95:**239.

Pitkin, R. M., Connor, W. E., and Lin, D. S., 1972, Cholesterol metabolism and placental transfer in the pregnant rhesus monkey, *J. Clin. Invest.* **51:**2584.

Popjak, G., 1954, The origin of fetal lipids, *Cold Spring Harbor Symp. Quant. Biol.* **19:**200.

Pulkinnen, M. O. L., 1961, Arylsulfatase and the hydrolysis of some steroid sulfates in developing organisms and placenta, *Acta Physiol. Scand.* **52**(Suppl. 180):1–92.

Raiha, N., 1959, On the placental transfer of vitamin C. An experimental study on guinea pigs and human subjects, *Acta Physiol. Scand.* **45**(Suppl. 155):1.

Rogers, B. O., Raisbeck, A. P., Ballantyne, D. L., Jr., and Converse, J. M., 1960, The genetics of skin homografting in rats between brothers, sisters, parents and grandparents, *Trans. Int. Soc. Plastic Surg., Second Congr.,* pp. 421–436, Livingstone, London.

Rosenfeld, C. R., Morris, F. H., Battaglia, F. C., Makowski, E. L., and Meschia, G., 1976, Effect of estradiol-17$\beta$ on blood flow to reproductive and non reproductive tissues in pregnant ewes, *Am. J. Obstet. Gynecol.* **124:**618.

Rosso, P., 1975, Maternal malnutrition and placental transfer of $\alpha$-aminoisobutyric acid in the rat, *Science* **187:**648.

Roughton, F. J. W., 1954, Respiratory functions of blood, in: *Respiration Physiology in Aviation* (W. M. Boothby, ed), pp. 51–102, Air University USAF School of Aviation Medicine, Randolph Field, Texas.

Sakurai, T., Takagi, H., and Hisoya, N., 1969, Metabolic pathways of glucose in human placenta: Changes with gestation and with added 17$\beta$-estradiol, *Am. J. Obstet. Gynecol.* **105:**1044.

Schenker, S., Dawber, N. H., and Schmid, R., 1964, Bilirubin metabolism in the fetus, *J. Clin. Invest.* **43:**32.

Schiff, R. J., Mercier, D., and Buckley, R. H., 1974, Effects of gestational hormones on human lymphocyte responses *in vitro, Fed. Proc.* **33**(Pt. 1):749.

Schneider, H., and Dancis, J., 1974, Amino acid transport in human placental slices, *Am. J. Obstet. Gynecol.* **120:**1092.

Schneider, H., Panigel, M., and Dancis, J., 1969, Informal comments on perfusion studies, in: *Transactions of the Fifth Rochester Trophoblast Conference* (C. J. Lund and J. W. Choate, eds.), pp. 189–196, University of Rochester School of Medicine and Dentistry, Rochester, New York.

Schroder, J., 1974*a,* Fetal leucocytes in the maternal circulation after delivery. I. Cytological aspects, *Transplantation* **17:**346.

Schroder, J., 1974*b,* Passage of leucocytes from mother to fetus, *Scand. J. Immunol.* **3:**369.

Sdenkow, H. A., Saxena, B. N., Daria, C. L., and Emerson, K., Jr., 1969, Measurement and pathophysiologic significance of human placental lactogen, in: *Foeto-Placental Unit* (A. Peale and C. Finzi, eds.), p. 340, Exerpta Medica, Amsterdam.

Sherwood, L. M., 1971, A comparison of the structure and function of human placental lactogen and human growth hormone, Proceedings Second International Symposium on Growth Hormone, Milan.

Siiteri, P. K., and MacDonald, P. C., 1963, The utilization of circulating dehydroisoandrosteronesulfate for estrogen synthesis during human pregnancy, *Steriods* **7:**713.

Siiteri, P. K., and MacDonald, P. C., 1966, Placental estrogen biosynthesis during human pregnancy, *J. Clin. Endocrinol. Metab.* **26:**571.

Simmons, R. L., and Russel, P. S., 1963, Antigenicity of mouse trophoblast, *Ann. N.Y. Acad. Sci.* **99:**717.

Slobody, L. B., Willner, M. M., and Mestern, J., 1949, Comparison of vitamin B$_1$ levels in mothers and their newborn infants, *Am. J. Dis. Child* **77:**736.

Solomon, S., and Fuchs, F., 1972, Progesterone and related neutral steroids, in: *Endocrinology of Pregnancy* (F. Fuchs and A. Klopper, eds.), pp. 66–100, Harper & Row, New York.

Stegink, L. D., Pitkin, R., Reynolds, W. A., Filer, L. J., Boaz, D. P., and Brummel, M. C., 1975, Placental transfer of glutamate and its metabolites in the primate, *Am. J. Obstet. Gynecol.* **122:**70.

Sternberg, J., and Pascoe-Dawson, E., 1959, Metabolic studies in artherosclerosis. I: Metabolic pathway of $^{14}$C labelled alphatocopherol, *Can. Med. Assoc. J.* **80:**266.

Straumfjord, J. V., and Quaife, M. L., 1946, Vitamin E levels in maternal and fetal blood plasma, *Proc. Soc. Exp. Biol. Med.* **61:**369.

Taylor, J. D., Millar, G. J., and Wood, R. J., 1957, A comparison of the concentration of $^{14}$C in the tissues of pregnant and nonpregnant female rats following the intravenous administration of vitamin $K_1$-$^{14}$C and vitamin $K_3$-$^{14}$C, *Can. J. Biochem. Physiol.* **35:**691.

Taylor, T., Coutts, J. R. T., and Macnaughton, M. C., 1974, Human fetal synthesis of testosterone from perfused progesterone, *J. Endocrinol.* **60:**321.

Terry, C. W., Terry, B. E., and Davies, J., 1960, Transfer of zinc across the placenta and fetal membranes of the rabbit, *Am. J. Physiol.* **198:**303.

Thomas, L., Douglas, G. W., and Carr, M. C., 1959, The continual migration of syncytial trophoblasts into the maternal circulation, *Trans. Assoc. Am. Phys.* **72:**140.

Tulchinsky, D., and Hobel, C. J., 1973, Plasma human chorionic gonadotropin, estrone, estradiol, estriol, progesterone and 17α-hydroxyprogesterone in human pregnancy, *Am. J. Obstet. Gynecol.* **117:**884.

Vaitukaitis, J. L., and Ross, G. T., 1973, Recent advances in evaluation of gonadotropic hormones, *Annu. Rev. Med.* **24:**295.

Van Duyne, C. M., Havel, R. J., and Felts, J. M., 1962, Placental transfer of palmitic acid-1-$^{14}$C in rabbits, *Am. J. Obstet. Gynecol.* **84:**1069.

Van Hall, E. V., Vaitukaitis, J. L., Ross, G. T., Hickman, J. W., and Ashwell, G., 1971, Immunological and biological activity of HCG following progressive desialylation, *Endocrinology* **88:**456.

Van Slyke, D. D., and Meyer, G. M., 1913, The fate of protein digestion products in the body. III. The absorption of amino acids from the blood by the tissues, *J. Biol. Chem.* **16:**197.

Villee, C. A., 1953a, Regulation of blood glucose in the human fetus, *J. Appl. Physiol.* **5:**437.

Villee, C. A., 1953b, The metabolism of human placenta *in vitro, J. Biol. Chem.* **205:**113.

Webber, W. A., and Cairns, J. A., 1968, A comparison of the efflux rates of AIB from kidney cortex slices of mature and newborn rats, *Can. J. Physiol. Pharmacol.* **46:**765.

Williams, D. J., 1971, Abnormalities of the genito-urinary system, in: *Congenital Abnormalities in Infancy* (A. P. Norman, ed.), p. 199, Blackwell, Oxford.

Wynn, R. H., 1973, Fine structure of the placenta, in: *Handbook of Physiology,* Section 7: Endocrinology, Vol. II, Female Reproductive System, Part 2 (R. O. Greep, ed.), p. 261, American Physiological Society, Washington.

Yen, S. S. C., Llerena, O., Little, B., and Pearson, O. H., 1968, Disappearance rates of endogenous luteinizing hormone and chorionic gonadotropin in man, *J. Clin. Endocrinol. Metab.* **28:**1763.

Young, B. K., Jirku, H., Slyper, A. J., Levitz, M., Kelley, W. G., and Yaverbaum, S., 1974, Estriol conjugates in amniotic fluid of normal and Rh-isoimmunized patients, *Clin. Endocrinol. Metab.* **39:**842.

Young, M., and Prenton, M. A., 1969, Maternal and fetal plasma amino acid concentrations during gestation and in retarded fetal growth, *J. Obstet. Gynaecol. Brit. Commonw.* **76:**333.

Young, M., and Widdowson, E. M., 1975, The influence of diet deficient in energy or in protein, on conceptus weight, and the placental transfer of a non-metabolisable amino acid in the guinea pig, *Biol. Neonat.* **27:**184.

# 15

# *Fetal Measurements*

## *D. A. T. SOUTHGATE*

### *1. Introduction—The Scope of Fetal Measurements*

This chapter reviews the range of fetal measurements that have been employed in the characterization of human fetal growth. These measurements can be considered to fall into two categories. First, anthropometric measurements have classically formed the basis of the description of fetal growth (Scammon and Calkins, 1929; Streeter, 1951) as they have for growth after birth. Table I lists some of the more common fetal measurements. The second category includes both chemical and to a lesser extent biochemical measurements which have been used to describe growth in terms of chemical and biochemical parameters and to relate these to the development of physiological function (Widdowson, 1968). In the main these studies have as their aim a better understanding of the physiology of the neonate and the improvement of the clinical care of the neonate (Corner, 1960) or the prognosis for physical and mental development postnatally (Ounsted and Ounsted, 1973); in other studies the major interest has been a better understanding of the physiology of fetal growth.

Measurements made during the course of obstetrical procedures with few exceptions are only approximate estimates of fetal size (Usher and McLean, 1969), and it is difficult to determine the precision or reliability of these estimates. Studies of fetal measurements from radiographs cannot be considered ethical for routine measurements on the normal fetus, but ultrasonic measurements do not appear to suffer from this objection.

This chapter will therefore deal exclusively with measurements on the fetus outside the uterus, and this immediately raises the important issue of the sources of fetal material on which these measurements are based and the extent to which these measurements can be regarded as documenting "normal" growth. This must be discussed in association with the question of the determination of fetal age.

### *1.1. Sources of Fetal Material*

The early series of fetal measurements (Scammon and Calkins, 1929; Streeter, 1951) was based entirely on material obtained accidentally, either as a result of

---

***D. A. T. SOUTHGATE*** • Dunn Nutritional Laboratory, Medical Research Council, University of Cambridge, England.

*Table I.  Summary of Measurements on the Fetus*

| Anthropometric | Chemical and biochemical |
|---|---|
| Age | Whole body composition and tissue composition in terms of water, |
| Body weight | total nitrogen, protein, fat, inorganic constituents |
| Crown–rump length | Cell number |
| Crown–heel length | Amino acid composition |
| Foot length | Fatty acid composition |
| Head circumference | Calcification of long bones |
| Biparietal diameter | Trace elements |
| Occipitofrontal diameter | |
| Thorax circumference | |
| Bideltoid diameter | |

some obstetrical event which lead to premature delivery, or in some of the classical accounts of early human development from accidents involving the death of the mother. In some series the fetuses were preserved in formaldehyde (or some other histological fixative), and this resulted in some hardening and shrinkage. In general measurements on these fetuses are smaller than those made on fresh fetuses. These fetal series were supplemented with measurements on large numbers of preterm infants both live- and stillborn. These measurements, often collected as part of the normal neonatal medical care of the infants concerned, provided extensive data in the weeks immediately before "normal" term (Lubchenko *et al.*, 1963; Gruenwald, 1966; Thomson *et al.*, 1968; Usher and McLean, 1969) but the data were more sparse and scattered for the earlier fetal ages.

During the last ten years the change of legal and other attitudes towards the termination of pregnancy has meant that larger series of early fetal material has become available for study (Birkbeck *et al.*, 1975*a,b*). There still remains a period of gestation from around 20 to 28 weeks where fetal material and therefore measurements are limited in number.

The fetal measurements described below have therefore been applied to material derived either from preterm deliveries of live- or stillborn infants, usually delivered per vagina, or in the case of the young fetuses as a result of termination of pregnancy.

This immediately raises the question as to whether this fetal material can be considered normal. In the early (termination) fetal material, while some of this is probably derived from young mothers or mothers who for medical reasons may be unfit to carry a child to term, there is no real evidence to suggest that this material should not be considered normal.

Some doubts must inevitably surround fetuses delivered during midterm as these are usually associated with some obstetrical accident (placenta abruptio, for example) where it is uncertain whether or not the circumstances leading to the delivery could have resulted in abnormal fetal development during the time immediately before the delivery. The later fetal series, derived from measurements on preterm infants, may include infants that have suffered growth retardation *in utero* (light for gestational age or small for dates) (Thomson and Billewicz, 1976), and the possible presence of two populations of small infants in these series must be borne in mind when examining the early data. The mean fetal growth curves constructed from these data appear to fit the extensive data on normal term infants, and this is suggestive evidence that measurements on these fetal series are valid as an index of normal fetal growth *in utero*.

Any discussion of fetal measurements must consider the assignation of a developmental or chronological age to the fetal material. There is no general agreement on the anatomical or physiological development stage which marks the commencement of the fetal stage. Table II sets out in summary the chronological stages of human development. The fetal stages are not clearly separable, and in most cases the measurements on the human fetus apply to material from about 60 to 70 days gestation, (see below), that is they apply to the third fetal stage, a period that covers about half the total gestation and during which the fetus grows from a weight of 50 g to the normal term infant of around 3.5 kg.

True conceptional age, that is age from the time of fertilization of the ovum, is virtually impossible to define. Even in laboratory animals where the time of mating can be observed the assignation of conceptional age is associated with a significant degree of uncertainty (Jensch *et al.*, 1970). This appears to be true in man even where the pregnancy is the result of a single act of intercourse. The difficulty in precise definition of the time of ovulation and the period of viability of sperm in the female reproductive tract combine to make conceptional age impracticable as a method of recording fetal age. The convention of calculating gestational age from the time of the last menstrual period is practicable in most circumstances, although it is inevitably associated with error and uncertainty, particularly where the mother has an irregular menstrual cycle.

Gestational age as calculated in this way (American Academy of Pediatrics, 1967) is greater than conceptional age by 12–14 days and probably has an uncertainty of the order of ±5 days (Usher and McLean, 1969) although this is difficult to quantify.

All series include some fetuses where the probable error is of the order of 28 days due to a mistaken date of the last menstrual period. Gruenwald (1966) attempted to correct his data on the basis of the existence of two populations of fetuses born after 35 weeks. This correction is, however, not acceptable to all workers, and the potential distortion of data by errors of this kind remains a

*Table II.   Human Development Stages* [a]

| Period | Time after conception | Stages present | Conception age (days) | Approximate weight at end of stage (g) |
|--------|----------------------|----------------|----------------------|----------------------------------------|
| Ovum | First week | Cleavage | 1–3 | |
| | | Blastocyst | 4–5 | |
| | | Implantation | 7 | |
| Embryo | 7–56 days | Gastrula | 7–8 | |
| | | Neurula | 20 | |
| | | Tail-bud embryo | 29 | |
| | | Complete embryo | 35–37 | |
| | | Metamorphosing embryo | 38–56 | |
| Fetal | 56–term | First fetal | 56–70 | 5 |
| | | Second fetal | 70–140 | 45 |
| | | Third fetal | 140–268 (term) | 3400 |

[a] After Bodemyer, 1968; Hamilton *et al.*, 1956.

problem. This uncertainty must not be ignored in the use of weight-for-gestational-age charts and is of greatest significance during the periods of very active growth.

It is also possible to see fetal material assigned ages by reference to some anthropometric measurement, and in this context one must guard against the circular use of data unless other criteria of age are also available (American Academy of Pediatrics, 1967).

## 2. Anthropometric Measurements

### 2.1. Fetal Weight

The weight of the fetus has been measured as an index of fetal size from the earliest investigators to the present time. It remains one of the most fundamental measurements and one where the technique of measurement is less prone to observer error than possibly any other measurement.

The only major technical problem associated with this measurement is the evaporative loss of water from the surface of the fetus. This can be critical in the very early fetus, and precautions are necessary to minimize loss.

Some variability in fetal weights can be attributed to the proportion of other parts of the conceptus left with the fetus. The level of sectioning the umbilical cord and the amount of amniotic fluid that can be left on the surface of the fetus can produce significant variability, and standardization and description of technique, particularly with the younger fetus, is desirable.

The proportion of blood in the fetal–placental unit which should be included in the fetal weight can be problematical and is difficult to standardize in series that include fetal material from terminations and live and stillborn fetuses. This is possibly more important when one considers placental weights where undrained weights are often quoted.

### 2.1.1. Sources of Data

Many extensive series of fetal weights are available in the literature. Many series, of which that of Thomson *et al.* (1968) is a good example, are drawn mainly from fetal material from the later part of gestation. As more earlier material has become available, the fetal growth curve has been extended earlier in gestation with greater confidence (Birkbeck *et al.*, 1975a). The series of Lubchenko *et al.* (1963), Gruenwald (1966), and Usher and McLean (1969) are very extensive and provide the best available body weight/gestational age comparisons, although they apply to the later weeks of gestation (25 weeks to term).

### 2.1.2. Relationship between Body Weight and Gestational Age

Most series show an exponential relationship of weight with gestational age with a distinctive fall in the rate of growth in the period from approximately 36 weeks (262 days) to term. This deceleration in the rate of growth is possibly associated with deterioration in placental function (Gruenwald, 1966). Figure 1 shows a typical fetal growth curve; the example shown is that of Usher and McLean (1969), and other series show divergences. For example, in the series of Thomson *et al.* (1968) the weights are higher from 32 to 38 weeks, and in the Lubchenko *et al.* (1963) series the weights are lower after 35 weeks.

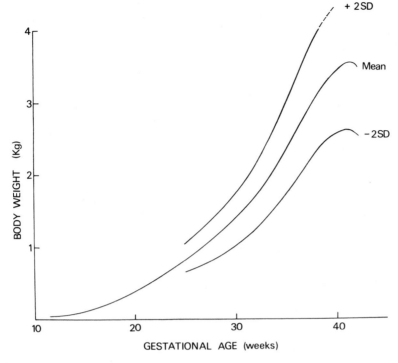

Fig. 1. The relation between fetal weight and gestational age. 25–42 weeks: smoothed mean ± two standard deviations; redrawn from Usher and McLeans' (1969) data. 10–25 weeks: regression line of series studied by Birkbeck *et al.* (1975a,b) and Hey, Southgate, and Widdowson (in preparation)

A number of equations have been derived to describe the normal growth in body weight. These all have an exponential component, and Roberts (1906) suggested that the relationship involved a third-power relationship. Balthazard and Dervieux (1921), Hugget and Widdas (1951), and others have also evolved relationships of this kind; some of these are summarized in Table III.

The curves of the early fetal weights suggest that fetal growth is unlikely to be described by a single function throughout the fetal period, and one should perhaps bear in mind the cautionary words of Snedecor (1962) on the matter of mathematical descriptions of biological events in suggesting that this is only justifiable if the biological laws controlling the data are sufficiently understood to suggest the appropriate equation.

### 2.1.3. Limitations of Body-Weight Data

Changes in body weight during gestation provide a useful index of growth: body weight is, however, a crude statistic, and it does not provide information of the developmental maturation of the fetus nor does it provide more than a measure of one aspect of growth.

*Table III. Equations Relating Body Weight to Gestational Age*

| | |
|---|---|
| Balthazard and Dervieux, 1921 | $Age = 19.4 \sqrt[3]{W}$ |
| Huggett and Widdas, 1951 | $W^{1/3} = 0.063\ (A - 33)$ |
| Birkbeck, 1976 | $Log_{10}W = 5.9679 \log_{10}A - 10.2576$ |
| Hey, Southgate, and Widdowson, in preparation | $W = (0.06346\ A - 2.3571)^3$ |

Furthermore it is affected by the age and parity of the mother (Thomson *et al.*, 1968), and in the larger series (Lubchenko *et al.*, 1963; Thomson *et al.*, 1968) there are clearly distinguishable differences between male and female fetuses. Comparison between series must allow for the stature and socioeconomic status of the populations being compared and for environmental factors such as height above sea level (Lubchenko *et al.*, 1963; Gruenwald, 1966).

Body weight is a useful indicator of the incidence of neonatal survival of the infant, but its value without reference to the stage of maturity or the gestational age of the child is limited. It is reasonably clear that infants born with a birth weight lower than the normal full-term infant fall into two categories: those born preterm but of normal weight for gestational age and those who are small for gestational age. The prognosis for survival of these two classes of infants depends to some degree on the category in which they fall, and their effective management in the neonatal period can depend very greatly on the classification to which they belong (Goldstein and Peckham, 1976).

Other fetal measurements are necessary to define the growth and developmental stage of the fetus in addition to body weight, and these are discussed in the later parts of this chapter.

## 2.2. Growth in Length

In this area two broad types of measurement have been used. The first comprises linear measurements of the head and body and the second linear measurements of the upper and lower limbs.

### 2.2.1. Crown–Rump Length

This measurement is the one most usually performed after measurement of fetal weight. Technically the posture of the fetus in the early stages presents problem in the definition of the measurement, and even with later fetuses some standardization of technique is very desirable. If the changes in this parameter in the early fetus are to be related directly to those of the later fetuses, the back of the fetus must be straightened and the flexure of the head reduced before measurement. This is most easily done if a measuring board is used.

Early fetuses are fragile and the tissues are readily torn; if other measurements are to be made, some care is necessary when handling the fetuses.

### 2.2.2. Crown–Heel Length

Crown–rump measurements in the fetus can be considered to be derived as measurements of the embryological type and as such are useful. However, the limbs are an important part of the fetus and a measurement which incorporates some index of limb growth is possibly of greater value.

Crown–heel measurements fall within this category and have been used in many of the anthropometric studies of fetuses.

The definition of the measurement and of a standardized technique specifically for the fetus has not been agreed upon, but Birkbeck (1976) suggests that the procedure recommended for infants in Weiner and Lourie (1969) should be used. This is probably the most reproducible technique; in the early fetus this will involve reducing the normal flexure and, as with all fetuses, the fetus must be handled with care if tissue tearing and dislocation of limbs is to be avoided. Even the mildest traction can damage the fetal hip articulation.

### 2.2.3. Other Measurements of Linear Growth

Measurements of the limbs in the intact fetus are difficult to standardize. Much of the useful evidence on fetal growth has been obtained on dissected material and will be discussed later in the section on body and tissue composition.

Radiographic estimation of limb bone lengths is a possibly useful technique in this area, but careful standardization of the technique as essential if reliable measurements are to be taken from the X-ray plate or film, and its greatest potential would seem to lie in the characterization of stages of ossification in the long bones (Birkbeck, 1976).

Foot length has been used by several authors as an index of linear growth (Usher and McLean, 1969; Dolhay et al., 1973). The size of the measurements itself implies a greater relative significance of measurement error. The relation between the growth of the foot and gestational age appears to be similar in the various series available, but the correlation with gestational age is lower and Birkbeck (1976) considers that this is a measurement which should be used when the fetus is damaged and other measurements are impracticable.

### 2.2.4. Relationship with Other Fetal Measurements

The two parameters, crown–rump (CR) and crown–heel (CH), have been studied in many fetal series. The series suggest that there is an essentially linear relationship between both CR and CH and gestational age. The growth lines fitted or calculated for the early fetuses are generally in reasonable agreement (Birkbeck, 1976), but there is some evidence that CH measurements have a biphasic relationship (Figure 2).

The figure illustrates one of the major anomalies which arises from the use of cross-sectional data to construct fetal "growth" curves. The fetuses delivered after normal term, i.e., 40 weeks, are frequently smaller than infants delivered at term. A loss of weight due to a failing placenta can be regarded as a possibility; shrinkage of the fetus is not a possibility. It is probable that the fetuses delivered after normal term include a high proportion of pathological fetuses or that the errors in assessment of fetal age are more evident.

All series must be expected to include fetuses where the probable error is of the order of one menstrual cycle (say 28 days) due to a mistaken date of the last menstrual cycle. It could be argued that this is probably the most common error in the assessment of fetal age, and in fetuses delivered after a normal 40 weeks overassessment of fetal age is a likely cause of this anomaly.

The two measurements are related to one another, and the ratio of CR and CH falls with gestational age as the limbs become proportionately larger.

### 2.3. Growth in Volume

Measurements of growth in volume of the fetus have been concerned with estimating the growth of the head, thorax and, to a much smaller extent, limbs.

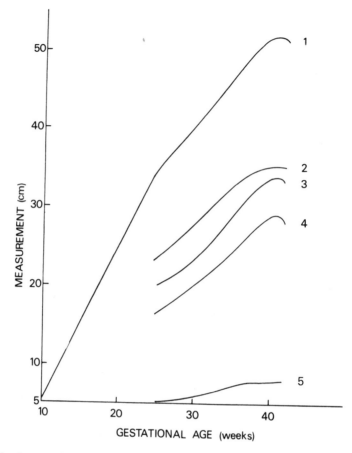

Fig. 2. Relation between five measurements in the same fetal series. Drawn from data of Usher and McLean (1969) and Birkbeck (1976). 1. Crown-heel measurements; 2. head circumference; 3. chest circumference; 4. abdominal circumference; 5. Foot length.

## 2.3.1. Measurements of the Head

The occipitofrontal diameter was used by some earlier authors (Scammon and Calkins, 1929; Kiseki, 1933), who described a linear relationship with crown–heel length. Biparietal diameter (Birkbeck, 1976) bears a similar relationship to crown–heel length.

Vertical head height has also been measured in several series (Scammon and Calkins, 1929; Wich, 1972; Birkbeck, 1976). Here again a linear relationship with crown–heel length is observed.

The measurements of the fetal head do not present any serious technical difficulties, but there is need for careful standardization. In general the use of calipers for the measurement is essential.

Other authors have preferred the use of head circumference, primarily because of the suspected relationship between this parameter and normal brain growth and development (Lubchenko *et al.*, 1963; Usher and McLean, 1969; Brandt, 1976). The technique is relatively simple and usually is taken as the maximum value found for the occipitofrontal circumference measured horizontally with a nonstretch measuring tape. The values show an exponential relationship with gestational age, and there is some evidence for a sex relationship, the male fetus tending to have slightly greater head circumferences than the female fetus, which is in accord with other measurements (Lubchenko *et al.,* 1963).

### 2.3.2. Measurements of the Thorax

Scammon and Calkins (1929) and Kiseki (1933) measured bideltoid diameter and obtained a linear relationship between their measurements and crown–heel length. In a recent study a similar relationship was observed (Birkbeck, 1976) with a slightly lower regression line, but Wich (1972) found a very divergent regression. This may possibly be related to technique as there does not seem to be any attempt at standardization of measurement.

Diameter of the thorax measured at the level of the nipples has been used in a few series, but few extensive series are available (Usher and McLean, 1969).

### 2.4. Value of Anthropometric Measurements in the Fetus

The need for careful standardization of anthropometric measurements is accepted by most workers in the field, if valid comparisons between the results obtained in different series are to be made. This need for standardization is applicable in fetal work, and careful standardized series can form the basis for the evaluation of environmental factors and maternal nutrition on fetal growth.

Central to this need for standardization is the need to codify the fetal age. As discussed earlier, true conception age is difficult, if not impossible, to obtain. Postmenstrual age (gestational age) has the merit of usually being available, even if associated with some uncertainty, and it would be ideal if all authors reported the ages of their fetuses. However, in view of the problems and uncertainties associated with the assignment of fetal age, a system of giving fetal growth measurements quoted against body weight, for example, has much to commend it.

Comparisons between different series show a similar type of growth response with age, but differences in the selection of material, the estimation of fetal age, and slight variations in measuring techniques all combine to make direct comparisons difficult.

Figure 2 shows the relation between five measurements in one fetal series, that of Usher and McLean (1969). Four curves have the same general form and show a feature of cross-sectional data which is difficult to visualize in terms of the growing fetus, that is, the reduction in the measurement after term.

## 3. Chemical and Biochemical Measurements

Although the numbers of chemical and biochemical measurements on fetus are limited in comparison with the large number of anthropometric measurements, in one particular area—measurements of whole body composition in the growing fetus—they are much more numerous than for any other period of human growth. Many studies have included measurements of the growth of individual organs, and these measurements have provided information on the nutritional requirements for fetal growth and on the chemical maturation of the body and organs (Widdowson and Dickerson, 1960). These measurements complement and increase the value of anthropometric measurements in the characterization of growth. In the fetus, where the maturation of function in organs has an important bearing on the prognosis for survival of the infant, these measurements have additional significance. For the purpose of this chapter, the presentation will be limited to a description of the range of observations that have been made. The interpretation of the clinical significance of these measurements is discussed elsewhere.

### 3.1. Whole Body Composition

Many studies have reported measurements of whole body composition; the series published before 1951 were collected by Kelly *et al.* (1951)—at this time body composition had been measured in 95 fetuses. The analytical measurements had, however, been limited to water, total nitrogen, and some inorganic constituents. Since that time a more extensive and sophisticated approach to the chemical characterization of growth has developed (Widdowson, 1968, 1970). In this approach the interest has moved toward the measurement of a wider range of constituents and to studies on the chemical maturation of individual organs. These studies have begun to provide evidence for the net nutritional requirements for fetal growth (Shaw, 1973).

These chemical and biochemical measurements are more time-consuming than anthropometric measurements, and this (together with the practical difficulties of whole-body analysis and the supply of fetal organs) has meant that the series for which information of this kind is available are much smaller. Furthermore, the material on which these series are based was, and to a great extent still is, limited to fetuses delivered as a result of an obstetrical or other incident (Widdowson and Spray, 1951; Apte and Iyengar, 1972; Southgate and Hey, 1976). More recently, series including fetuses from terminations have been described, and it is arguable that such material is more "normal" than fetal material where some biological factor has resulted in a spontaneous preterm delivery. In the early studies interest centered around relatively crude analytical parameters.

### 3.1.1. Water Content

This has been measured in most fetal series and shows that water content falls progressively during gestation and that earlier fetuses may contain up to 950 g/kg water (Kelly *et al.*, 1951). The early fetuses are very gelatinous, and it is usually difficult to decide, when removing the fetal membranes, when the amniotic fluid has been blotted from the fetus and when water is being drawn from the fetal tissues. Great care is necessary to prevent the fetus from drying out during measurements or dissection if accurate estimates of water content are to be obtained. The technique used to measure water is of some interest, and prolonged drying in an air oven at 100°C can produce degradation of fat and a dry residue which is unsuitable for any subsequent measurements; freeze-drying is a much less damaging method and, while not removing all the water, leaves a residue suitable for many other determinations.

Associated with this change in total body water, there is a change in the distribution of water between the intracellular and extracellular compartments, and measurements of this partitioning have shown that the extracellular compartment declines during gestation (Widdowson, 1968).

### 3.1.2. Total Nitrogen and Protein

The decrease in the fetal water content is accompanied by a rise in fetal solids, and a substantial and important component of this is protein. Kelly *et al.* (1951)

found that the accumulation of total nitrogen during gestation could be described by a simple regression line. Other workers since then (Widdowson and Spray, 1951; Southgate and Hey, 1976) have shown a similar type of relationship but that this was accompanied by a change in the nitrogen content of the tissues.

Total nitrogen has been used to calculate values for total protein by the use of the conventional factor (6.25). This, however, is a crude approximation and does not take into consideration the fact that the fetus contains nitrogenous substances other than protein. The early fetus is particularly rich in glycosaminoglycans (Breen *et al.*, 1970); furthermore proportions of the fetal proteins in the developing fetus are changing so that one factor applied to total nitrogen might be expected to result in anomalies.

Measurements of amino acid content of the human fetus have only been made on a few isolated fetuses, and measurements of this kind could provide information on the net requirements for the essential amino acids by the growing fetus.

### 3.1.3. Fat

In most of the earlier series of analytical measurements of the fetus, fat was not measured and the importance of fat measurements in the growing fetus was illustrated by the work of Widdowson and Spray (1951).

The human fetus at term contains a high proportion of fat (180 g/kg), and the major portion of this fat is deposited in the last few weeks of gestation (Figure 3). Comparisons of body composition measurements during growth are therefore limited unless they are made on a fat-free basis (Widdowson and Dickerson, 1964).

While total body fat deposition is exponential, deep body and subcutaneous fat are deposited at different rates (Southgate and Hey, 1976) and this factor must be considered when the characteristics of the light-for-gestational age infants are reviewed.

Measurements of fatty acid composition (Roux *et al.*, 1971) of fetal lipid suggest that during the fetal period white adipose tissue does not vary greatly in

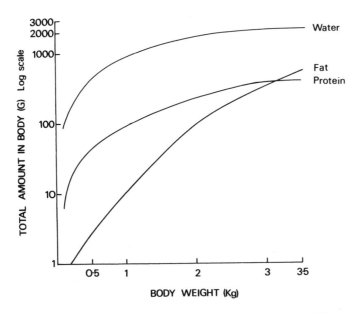

Fig. 3. Composition of whole body at different weights. Drawn from data of Widdowson (1968).

composition, except when the mothers have been consuming a diet rich in polyunsaturated fat (Widdowson *et al.*, 1975). Brown adipose tissue has a distinctive fatty acid composition.

### 3.1.4. Inorganic Constituents

The deposition of inorganic constituents, especially in the skeletal tissues, has attracted a considerable amount of interest. Widdowson and Dickerson (1964) reviewed the subject in great depth, and the general features of the range of measurements discussed by those authors has not been altered by work done since the review was completed as far as whole body measurements of the major inorganic constituents are concerned. Some measurements of trace elements were reviewed by Widdowson (1968), but since this review greater attention has been given to this, probably because of the improved analytical techniques especially in atomic absorption spectrophotometry (Widdowson *et al.*, 1972*a,b*).

### 3.2. Measurements of Individual Organs

The weights of individual organs in the developing fetus have been measured by several authors (e.g., Potter, 1961). The growth of the liver, heart, and kidneys has the same general form as body weight against age; it is an exponential relationship (Figure 4). In the series studied by Southgate and Hey (1976) the proportion of the body weight contributed by these three organs was substantially the same from

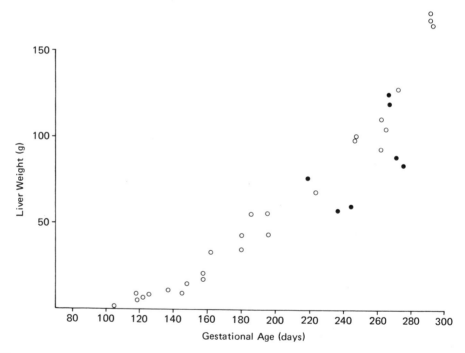

Fig. 4. Growth of the liver: The figure shows the values in the series studied by Southgate and Hey (1976). The open circles are values obtained from fetuses whose body weights lay close to the 50th centile of Thomson *et al.* (1968). The solid circles are for values from fetuses which were at or below the 10th centile, i.e., those, that were light for gestational age.

100 days gestation to term. Insufficient information is available, however, to define the normal limits of variation. Recently considerable attention has been given to the measurements of brain weights in the fetus (Dobbing, 1970). Brain weights plotted against age from the Southgate and Hey series are shown in Figure 5. The growth curve for brain has an apparently more complex form, but some exponential character is still evident. The light-for-gestational-age infants in this series (with one exception) did not have brains that were smaller than the normal-weight fetuses.

In the more recent of these series, the gross composition of the organs has been measured and were reviewed by Widdowson (1968). In addition to gross chemical composition, i.e., measurement of protein and fat, many studies have attempted to correlate biochemical parameters with growth of these organs. Studies of skeletal

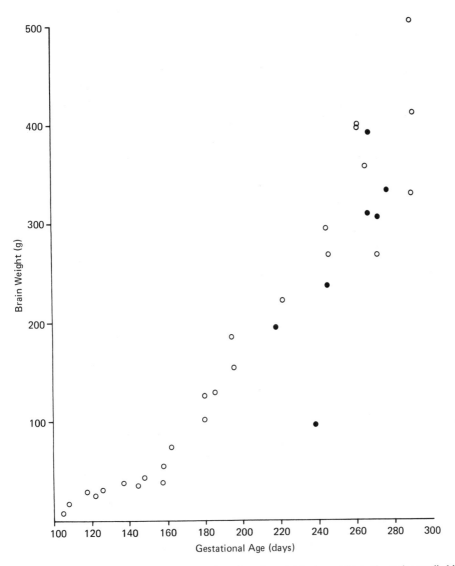

Fig. 5. Growth of the brain: The figure shows the values for total brain weight in the series studied by Southgate and Hey (1976). The open circles refer to fetuses of normal body weights and the solid circles to those that were light for gestational age.

muscle and long bones in a limited series of fetus were reported (Dickerson, 1962; Dickerson and Widdowson, 1960). Particular interest has centered around measurements of trace metals in the liver especially, and these have shown the importance of fetal stores of iron and copper (Widdowson *et al.*, 1972*b*; Iyengar and Apte, 1972).

The importance of considering measurements of tissue composition in relation to fat-free weight was emphasized by Widdowson (1970), but some work has omitted this variable for analytical or other reasons. Rapid methods for the measurement of fat in tissues are available (Atkinson *et al.*, 1972; Woodward *et al.*, 1976), and it is hoped that this parameter will be measured routinely in the future.

Many biochemical measurements on individual tissues have been made and some of these special areas are discussed briefly below.

### 3.2.1. Cell Number

In many studies of fetal tissues during growth the distinction between hyperplasia and hypertrophy of the tissue has a potential bearing on the functional development of the tissue, and measurements of cell number have become of great interest in studies of fetal tissues.

The premise that the DNA content of the diploid nucleus is constant at 6.2 pg/nucleus has wide acceptance, and the concept that cell number can be estimated from DNA content is sound except for liver where polyploidy is quite common (Harrison, 1951; Epstein, 1967). The measurement of DNA itself in some tissues can be difficult, and the direct application of a method such as that of Zamenhof *et al.* (1964), can occasionally produce anomalous values. Care is needed in the selection of the most appropriate extraction and hydrolysis conditions (Munro and Fleck, 1966).

The relationship between the increase in cell number, organ growth, and gestational age is complex, and Widdowson *et al.* (1972*a*) suggested that expression on a logarithim to base 2 basis was useful. More studies are needed to develop this comparison fully.

### 3.2.2. Calcification

The calcification and growth of the long bones in the fetus were studied in a small series by Dickerson (1962), who also examined the changed in protein in the bone with particular reference to the calcium–collagen ratio. Detailed studies in a larger series (Birkbeck and Roberts, 1971) has, in general, confirmed these findings. In the series studied by Southgate and Hey (1976) linear growth of the femur and to a lesser extent the humerus was less affected in fetuses which were very light for gestational age than many other parameters.

The interpretation of the results of the gross composition of fetal bones is complicated by the differential growth of the different anatomical regions and by the different compositions of these regions. Dickerson (1962) found that it was advisable to dissect the bone into these regions before analysis. At a purely technical level it may be useful to comment on the rapidity with which fetal bones dry out during dissection, making measurements of fresh weight somewhat difficult. The X-ray measurement of ossification centers has some usefulness in the fetus (Birkbeck, 1976) and is discussed in detail in Vol. 2, Chapter 12.

Several detailed biochemical studies on the complex lipid constituents of brain tissue have been made, following the discussion on the effects of malnutrition *in utero* on subsequent brain function (Dobbing, 1970). The relationship between these constituents and the functional maturation of brain tissue is difficult to assess and depends on the correlation of biochemical compositional data with some functional index of nervous tissue.

Studies of whole brain composition are of limited value, and in most recent studies some dissection into anatomical and functional regions has been preferred.

## 4. Conclusions

This chapter has briefly reviewed the scope of fetal measurements. Three clear principles can be drawn from this discussion:

First, in all fetal series there is a need to define the material carefully with respect to age and the criteria used in admitting a fetus to the series. Where fetal series include the products of spontaneous preterm delivery, some criteria of normality must be established if the data from fetal series is to be used to judge normal fetal growth.

Second, anthropometrical and other measurements must be made according to a standardized protocol, and measurements of the early fetus must take into account the posture of the fetus *in utero* and that in these fetuses the posture adopted for measurements in later life can only be achieved with damage to the fetus. Ideally measurement made on the fetus should include variables that can be measured *in utero* by ultrasonic techniques.

Third, chemical and biochemical measurements on the fetus are based on limited series, and the extent of normal biological variation is not known in many instances. More detailed work in this area could usefully be biased towards obtaining more information on the development of physiological function in fetal life.

## 5. References

American Academy of Pediatrics, 1967, Committee on fetus and newborn, *Pediatrics* **39**:935–939.

Apte, S. V., and Iyengar, L., 1972, Composition of the human foetus, *Br. J. Nutr.* **27**:305–312.

Atkinson, T., Fowler, V. R., Garton, G. A., and Lough, A. K., 1972, A rapid method for the accurate determination of lipid in animal tissues, *Analyst* **97**:562.

Balthazard, V., and Dervieux, O., 1921, *Ann. Med. Legale* **1**:37; quoted by Huggett and Widdas, 1951.

Birkbeck, J. A., 1976, Metrical growth and skeletal development of the human fetus, in: *The Biology of Human Fetal Growth* (D. F. Roberts and A. M. Thomson, eds.), pp. 39–68, Taylor & Francis, London.

Birkbeck, J. A., and Roberts, J. A., 1971, Skeletal composition in the early human fetus, *Biol. Neonate* **19**:465–471.

Birkbeck, J. A., Billewicz, W. Z., and Thomson, A. M., 1975*a,* Human fetal measurements between 50 and 150 days of gestation, *Ann. Hum. Biol.* **2**:173.

Birkbeck, J. A., Billewicz, W. Z., and Thomson, A. M., 1975*b,* Foetal growth from 50 to 150 days of gestation, *Ann. Hum. Biol.* **2**:319.

Bodemer, C. W., 1968, *Modern Embryology,* Holt, Rinehart and Winston, New York.

Brandt, I., 1976, Dynamics of head circumference growth before and after term, in: *The Biology of Human Fetal Growth* (D. F. Roberts and A. M. Thomson, eds.) pp. 109–136, Taylor & Francis, London.

Breen, M., Weinstein, H. G., Johnson, R. L., Veis, A., and Marshall, R. T., 1970, Acidic glycosamino-glycans in human skin during fetal development and adult life, *Biochem. Biophys. Acta* **201**:54–60.

Corner, B. 1960, *Prematurity,* Cassell, London.

Dickerson, J. W. T., 1962, Changes in the composition of the human femur during growth, *Biochem. J.* **82**:56–61.

Dickerson, J. W. T., and Widdowson, E. M., 1960, Chemical changes in skeletal muscle during development, *Biochem. J.* **74**:247–257.

Dobbing, J. 1970, Undernutrition and the developing brain: The relevance of animal models to the human problem, *Am. J. Dis. Child.* **120**:411–415.

Dolhay, B., Batar, I., and Papp, Z. 1973, Correlation of the distance between the heel and the big toe with weeks of gestation, biparietal diameter and body weight, *Am. J. Obstet. Gynecol.* **117**:142.

Epstein, C. J., 1967, Cell size, nuclear content and the development of polyploidy in mammalian liver, *Proc. Natl. Acad. Sci. U.S.A.* **52**:327.

Goldstein, H., and Peckham, C., 1976, Birth weight, gestation, neonatal mortality and child development, in: *The Biology of Human Fetal Growth* (D. F. Roberts and A. M. Thomson, eds.), pp. 80–101, Taylor & Francis, London.

Gruenwald, P., 1966, Growth of the human fetus 1. Normal growth and its variation, *Am. J. Obstet. Gynecol.* **94**:1112–1119.

Hamilton, W. J., Boyd, J. D., and Mossman, H. W., 1956, *Human Embryology, Prenatal Development of Form and Function,* 2nd ed., Heffers, Cambridge, England.

Harrison, M. F., 1951, Relation between polyploidy and the amounts of deoxynucleic acid per nucleus in the liver and kidney of adult rats, *Nature, London* **168**:248–250.

Hey, E. N., Southgate, D. A. T., and Widdowson, E. M., in preparation.

Huggett, A. St. G., and Widdas, W. F., 1951, The relationship between mammalian foetal weight and conception age, *J. Physiol.* **114**:306–317.

Hytten, F. E., and Leitch, I., 1971, *The Physiology of Human Pregnancy,* 2nd ed., Blackwell Scientific Publications, Oxford.

Iyengar, L., and Apte, S. V., 1972, Nutrient stores in human foetal livers, *Br. J. Nutr.* **27**:313–317.

Jensch, R. P., Brent, R. L., and Barr, M., Jr., 1970, The litter effect as a variable in tetratologic studies in the albino rat, *Am. J. Anat.* **122**:185–192.

Kelly, H. J., Sloan, R. E., Hoffman, W., and Saunders, C., 1951, Accumulation of nitrogen and six minerals in the human fetus during gestation, *Hum. Biol.* **23**:61–74.

Kiseki, T., 1933, Relationship of length of body of fetus to length of trunk and spine, width of shoulder and hip, and various diameters and circumferences of the head, *Mitt. Med. Ges. Tokio* **47**:100.

Lubchenko, L. O., Hansman, C., Dressler, M., and Boyd, E., 1963, Intrauterine growth as estimated from liveborn birth weight at 24 to 42 weeks of gestation, *Pediatrics* **32**:793–800.

Lubchenko, L. O., Hansman, C., and Boyd, E., 1966, Intrauterine growth in length and head circumference as estimated from live births at gestational ages from 26 to 42 weeks, *Pediatrics* **37**:403–416.

Munro, H. N., and Fleck, A., 1966, Recent developments in the measurement of nucleic acids in biological materials, *Analyst* **91**:78.

Ounsted, C., and Ounsted, M., 1973. *On Fetal Growth Rate (Its Variations and their Consequences),* Spastics International Medical Publications, Heinemann, London; J. B. Lippincott, Philadelphia.

Potter, E. L., 1961, *Pathology of the Fetus and Infant,* 2nd ed., Year Book Medical Publishers, Chicago, p. 14.

Roberts, R. C., 1906, *Lancet* **1**:170, 295 quoted by Huggett and Widdas, 1951.

Roux, J. F., Takeda, Y., and Grigorian, A., 1971, Lipid concentration and composition in human fetal tissue during development, *Pediatrics* **48**:540–546.

Scammon, R. E., and Calkins, L. A., 1929, *The Development and Growth of the External Dimensions of the Human Fetus,* University of Minnesota Press, Minneapolis.

Shaw, J. C. L., 1973, Special problems of feeding very low birth weight infants, in: *Nutritional Problems in a Changing World* (D. F. Hollingsworth and M. Russell, eds.), p. 115, Applied Science Publishers, London.

Snedecor, G. W., 1962, *Statistical Methods,* Iowa State University Press, Ames, p. 471.

Southgate, D. A. T., and Hey, E. N., 1976, Chemical and biochemical development of the human fetus, in: *The Biology of Human Fetal Growth* (D. F. Roberts and A. M. Thomson eds.), pp. 195–209, Taylor & Francis, London.

Streeter, G. L., 1951, *Developmental Horizons in Human Embryos,* Carnegie Institution of Washington, Washington, D.C.

Thomson, A. M., and Billewicz, W. Z., 1976, The concept of the light for dates infant in: *The Biology of Human Fetal Growth* (D. F. Roberts and A. M. Thomson eds.), pp. 69–79, Taylor and Francis, London.

Thomson, A. M., Billewicz, W. Z., and Hytten, F. E., 1968, The assessment of fetal growth, *J. Obstet. Gynaecol. Brit. Commonw.* **75:**903–916.

Usher, R. U., and McLean, F., 1969, Interuterine growth of live-born infants at sea level: Standards obtained from measurements in 7 dimension of infants born between 25–44 weeks of gestation, *J. Pediatr.* **74:**901–910.

Weiner, J. S., and Lourie, J. A., 1969, *Human Biology—A Guide to Field Methods,* IBP Handbook No. 9, Blackwell Scientific Publications, Oxford, p. 7.

Wich, J. 1972, z badan nad rozwogem plodowym czloirieka, *Mater. Pr. Antropol.* **83:**249; quoted by Birkbeck, 1976.

Widdowson, E. M. 1968, Growth and composition of the fetus and newborn in: *The Biology of Gestation;* Vol. 2, (N. S. Assali, ed.), pp. 1–49, Academic Press, New York.

Widdowson, E. M. 1970, Harmony of growth, *Lancet* **1:**901–905.

Widdowson, E. M., and Dickerson, J. W. T., 1960, The effect of growth and function on the chemical composition of soft tissues, *Biochem. J.* **72:**30–43.

Widdowson, E. M., and Dickerson, J. W. T., 1964, Chemical composition of the body in: *Mineral Metabolism,* Part 2 A (C. L. Comar and F. Bronner, eds.), pp. 1–247, Academic Press, New York.

Widdowson, E. M., and Spray, C. M. 1951, Chemical development *in utero, Arch. Dis. Child.* **26:**205–214.

Widdowson, E. M., Crabb, D. E., and Milner, R. D. G., 1972a, Cellular development of some human organs before birth, *Arch. Dis. Child.* **47:**652–655.

Widdowson, E. M., Chan, H., Harrison, G. E., and Milner, R. D. G., 1972b, Accumulation of Cu, Zn, Mn, Cr and Co in the human liver before birth, *Biol. Neonate* **20:**360–367.

Widdowson, E. M., Dauncey, M. J., Gairdner, D. M. T., Jonxis, J. H. P., and Pelikan-Filipova, M., 1975, Body fat of British and Dutch infants, *Br. Med. J.* **1:**653–655.

Woodward, C. J. H., Trayhurn, P., and James, W. P. T., 1976, Rapid estimation of carcase fat by Fosslet specific gravity technique, *Br. J. Nutr.* **36:**567–570.

Zamenhof, S., Bursztyn, H., Rich, K., and Zamenhof, P. J., 1964, The determination of deoxyribonucleic acid and of cell number in brain, *J. Neurochem.* **11:**505–509.

# 16

# Implications for Growth in Human Twins

## FRANK FALKNER

## 1. Introduction

The concept of growth as a continuum from conception is vital to its effective study. Realization of this has led, comparatively recently, to intense interest in, and study of, the prenatal period. Thus, the contribution that human twins can make to the study of human growth is considerably enhanced if attention is paid to the conditions of their prenatal growth in relating appropriate factors to their later patterns of growth and development.

Twins are useful for growth research because:

1. They are implanted in the same uterus at the same time. Thus each twin has the same gestational age; and importantly so at birth in spite of possible greatly differing birth weights.
2. The placentation is related to prenatal nutrition and blood supply deriving from a common maternal source.
3. Two fetuses are growing in the same environment and space usually occupied by one fetus.
4. Dizygous (DZ), nonidentical twins are growing as siblings, sometimes of different sex, in the same conditions outlined in 1, 2, and 3; as are
5. Monozygous (MZ), identical twins with identical (theoretically) genotypes.

These factors govern the evaluation of within-pair differences and similarities as twins grow, and play a part in determining the final outcome of their growth patterns.

## 2. Embryology and Placentation

Approximately two thirds of all twins are DZ and one third MZ. DZ twins occur after the independent release of two ova and fertilization of these ova by two

---

**FRANK FALKNER** • Fels Research Institute, Wright State University School of Medicine, Yellow Springs, Ohio, and University of Cincinnati School of Medicine.

separate sperm. MZ twins derive from a single fertilized ovum and division of the embryo at some stage in its development after fertilization.

The blastocysts of each DZ twin implant in the tissues of the uterine wall and two membranes, the chorion and the amnion, surround the fetus. The placentas of these DZ twins are separate if the blastocytes implant far apart on the uterine wall, but fused if the implantations are close together. If the placentas are fused, there are four membranes separating the twin fetuses: amnion, chorion, chorion, amnion.

In monozygous twins, the ovum divides, producing a pair derived from a single diploid set of chromosomes. If no abnormalities during replication occur, the twins will be genetically identical. If the division of the ovum occurs before the blastocyst implants, the two blastocysts may implant at different placental sites on the uterine wall. These MZ twins will then have separate amniotic and chorionic membranes, as do all DZ twins. And, as in the case of DZ twins, the placentas may be separate or fused.

It is more common, however, for the ovum to divide after blastocyst implantation. Such an MZ pair will have two amnions but only one chorion. Table I gives a selected list of references, where zygosity determination has been judged reliable, showing that an overall estimate of the number of MZ twins having dichorial (DC) placentation is about 30%. The range of percentages shown in the table is large because a reliable estimate is hard to obtain without reasonably large samples of *consecutive* twin births, and rarely has a study achieved this.

Another, and much rarer, type of placentation takes place in MZ twins when division occurs in the embryonic disk after amniotic differentiation. This results in the twins having a single chorion and single amnion. These twins are commonly conjoined and rarely survive. The frequency of monoamniotic twins in all MZ twins is approximately 4%.

## 3. Vascular Anastomosis

One important facet of monochorionic (MC) placentation is that vascular anastomoses occur between the parts of the placenta supplying each twin. Hence, nutritional and other communications exist via these vascular connections. Indeed, Benirschke and Driscoll (1967) found almost all MC placentas to possess vascular shunts of some kind, and the transfusion syndrome (where one twin transfuses blood to the other via a placental anastomosis) may occur when an artery-to-vein shunt exists of sufficient magnitude; perhaps one third of all MC placentas show some degree of the syndrome.

Table I.  Percentage of MZ Twins Having DC Placentation
According to Various Authors

| Author | Dichorial (%) |
|---|---|
| Benirschke (1961) | 30 |
| Potter (1963) | 40 |
| Edwards and Cameron (1967) | 28 |
| Corney *et al.* (1968) | 20 |
| Falkner (1966) and Wilson (1976) | 39 |
| The Collaborative Perinatal Study (Niswander and Gordon, 1972) | 45 |
| Robson (1976) | 20 |
| Mean | 31.7% |

Vascular anastomoses are not found across dichorionic (DC) placentas even when these are fused. It must be recorded, however, that while they are not found after birth, blood chimeras have been reported (Benirschke and Driscoll, 1967) in twins with DC placentas showing that, even if rarely, there must have been some vascular connection across the DC placentas at some time previously, even if evidence has disappeared by birth. An extremely rare case was also reported by these authors where a small artery-to-artery and vein-to-vein anastomosis was demonstrable in MZ twins with DC placentas at birth. The implication thus is still that MZ twins with MC placentas should be regarded differently from MZ twins with DC placentas—and obviously from DZ twins of like sex who always have DC placentas. Since placental "nutrition" presumably has an influence upon fetal and early postnatal growth, many substances and agents may be transferred from one twin's placental part to the other's in the case of MZ twins with MC placentas. It has been suggested that the placental tissue mass supplying each twin, denoting a capacity to supply nutrients, could be regarded subsequently as a monument for this factor. It was shown (Falkner, 1966) that the wet and dry weights of placentas, or placental parts, were significantly correlated with the corresponding twins' birth weights and that this was so for MZ and DZ twins with MC or DC placentas.

In a series of 92 twins, the within-pair difference in birth weight averaged 326.0 g in MZ-MC twins, and 227.8 g in MZ-DC twins (Falkner, 1966). This suggests that separated placental parts lead to a greater tendency for MZ twins to be *more* similar in birth weight.

## 4. The Multifactorial Nature of Prenatal Influences

At this stage it is profitable to consider the growth and development of one twin pair, exemplifying the multifactorial nature of prenatal influences upon growth.

The monozygous male twins shown in Figure 1 were born near term. The

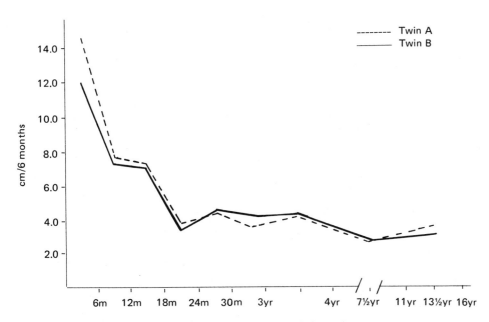

Fig. 1. Six-monthly increments in length (in centimeters).

placenta was monochorionic-diamniotic. Routine study of the placenta revealed that there was a marked difference in mass between the placenta part supplying one twin and that part supplying the other. The first-born twin (A) weighed 1460 g at birth and the second-born (B) 2806 g. The corresponding placenta part supplying twin A weighed 258.5 g (wet), 37.40 g (dry), and measured 20 × 7.5 × 2.0 cm; twin B's placenta part weighed 551.0 g (wet), 85.90 g (dry), and measured 30 × 15.5 × 2 cm. In addition to the above inequality, there was a small arteriovenous anastomatic shunt resulting in a mild transfusion syndrome in the direction of twin A "transfusing" twin B. The above measures are highly related with the birth weights of the twins.

Both were full-term infants (FTI), yet one is a small-for-date infant (SFDI)—"premature" (PTI) by an old definition. Since both are genetically identical it is interesting to follow their postnatal growth pattern.

Table II shows the anthropometric measures at various ages. The within-pair weight difference was reduced in the first 9 months and was lowest of all at 12 months. Although the within-pair subcutaneous tissue measurement difference remained more or less constant, at one year twin A was slightly above twin B for the only time. After one year the within-pair difference relative to total body size remained about the same; at 11 years there was a 5.1-kg difference, and at 16 years, 8.0 kg.

Length and stature growth patterns are illuminating. In the first 9 months twin A's rate of growth exhibited marked catch-up, and he grew significantly faster in length than twin B, as if to try and make up the birth length deficit of 7.0 cm.

Figure 1 shows this, and also illustrates the fact that since that time the twins continued to grow in length at approximately the same rate—at least until 16 years of age. The smaller twin did not experience sufficient catch-up growth to reach his brother's height curve. With only two measurements, at ages 11 and 16, since 4 years of age, the adolescent growth spurt and its onset is of course obscured. There are, however, acceptable observations that twin A is somewhat advanced over twin B in puberty status, and it is interesting that twin A's velocity curve for height rose above twin B's, albeit by a small amount in the 11-year to 16-year period. Thus, whatever their actual individual ages for the onset of puberty and their peak height velocities, their difference in size remains notable at 16 years.

There is no essential difference in their IQ scores, and their overall achievement in school has been the same for both while in separate classes.

Head circumference measurements exhibit the pattern of growth expected from an infant of low birthweight (ILB)—namely, very rapid growth compared to a full-term infant with normal size limits reached in the early months. Twin A was 4.0 cm smaller in this measure than twin B at birth, yet by 9 months he became (even if slightly) larger. From then on he has maintained a head circumference very similar to that of his twin. Other anthropometric measures follow, in general, the overall body-size indications already presented.

By 2½ years the twins should have been virtually free of perinatal influences and onto their genetic growth curves. Is there (at 16 years when their adolescent growth spurt is approaching its end) now an irreversible difference in their length and size due to the factors already described? Presumably twin A is following the growth pattern of a SFDI, and the illustration of this pair is contributory because we can hold the genetic growth factor constant between the two.

Cruise (1973) showed that both PTI and SFDI grew much faster than FTI in the first 9 months, and that given good neonatal care, PTI continued catch-up growth

Table II. Anthropometric Measures of a Monozygous Twin Pair at Various Ages

| Age / Twin | Birth A | x[a] | B | 9 months A | x | B | 12 months A | x | B | 18 months A | x | B | 24 months A | x | B | 30 months A | x | B | 3 years A | x | B | 4 years A | x | B | 11 years A | x | B | 16 years A | x | B |
|---|---|---|---|---|---|---|---|---|---|---|---|---|---|---|---|---|---|---|---|---|---|---|---|---|---|---|---|---|---|---|
| Weight (kg) | 1.46 | 1.35 | 2.81 | 6.94 | 1.13 | 8.07 | 8.21 | 0.32 | 8.53 | 8.6 | 1.5 | 10.1 | 9.7 | 1.25 | 10.95 | 10.36 | 1.46 | 11.82 | 11.25 | 1.71 | 12.96 | 13.28 | 2.25 | 15.53 | 25.57 | 5.11 | 30.68 | 50.56 | 7.96 | 58.52 |
| Length or stature (cm) | 43.0 | 7.0 | 50.0 | 64.9 | 3.1 | 68.0 | 68.7 | 2.9 | 71.6 | 75.9 | 2.7 | 78.6 | 79.8 | 2.2 | 82.0 | 84.2 | 2.4 | 86.6 | 87.8 | 3.0 | 90.8 | 96.2 | 3.0 | 99.2 | 130.81 | 5.08 | 135.89 | 161.92 | 5.38 | 167.30 |
| Head circumference (cm) | 30.0 | 4.0 | 34.0 | 44.6 | −0.4 | 44.2 | 45.6 | −0.9 | 44.7 | 47.1 | 0.3 | 47.4 | 48.2 | 0.1 | 48.3 | 48.6 | 0.2 | 48.8 | 49.2 | 0.0 | 49.2 | 49.6 | 0.4 | 50.0 | | | | 55.24 | 0.64 | 55.88 |
| Max. calf circumference (cm) | | | | − | 2.5 | − | − | 0.4 | − | − | 2.0 | − | − | 1.9 | − | − | 2.2 | − | − | 2.3 | − | − | 2.0 | − | | | | | | |
| Subcutaneous tissue (mm) | | | | − | 0.2 | − | − | −1.4 | − | − | 1.3 | − | − | 2.6 | − | − | 0.8 | − | − | 1.2 | − | − | 1.0 | − | | | | | | |

[a] x = within-pair difference of twin B over twin A.

and had reached the size of FTI by three years of age. SFDI, however, did not continue catch up and had not achieved FTI size by three years. Since intrauterine growth retardation, perhaps due to prenatal "fetal malnutrition," is thought to be associated with SFDI, the above example would tend to support this view. However, recently Brandt (see Volume 2, Chapter 19) has demonstrated that two subdivisions of SFDI can exist—those who do catch up and those who do not. The prenatal influences and their evaluation may assume great importance relative to the outcome of growth and development for SFDI.

In fact, it seems to be a comparatively uncommon situation for MZ twins to have grossly differing birth weights. Babson and Phillips (1973) found in such situations that, as in the case of the twin pair just described, the much smaller twin at birth does not exhibit sufficient catch-up to achieve the size of his larger twin. Buckler and Robinson (1974), however, followed a pair of MZ-MC twins, the smaller twin having a birth weight 45% less than that of her larger twin. Rapid catch-up of the smaller twin occurred and continued, and by 10 years both twins were similar in size and mental ability. It is interesting to note that the MC placenta part supplying the smaller twin was 87% of the weight of the larger twin's part. The corresponding percentage of the smaller twin's placental part in the twins described earlier in this text was 46%—the smaller twin's birth weight being 52% of the larger twin. This might indicate that placental function, perhaps indicated by placental mass, could be related to differing growth outcome for MZ twins; it is also possible that there is a "critical level" of size for a placenta or placenta part below which growth deficit occurs which may be irreversible. Buckler and Robinson's twin pair also supports the view mentioned earlier that SFDI may be divisible into two subgroups. One of this pair clearly exhibited sufficient and prolonged catch-up.

## 5. Prenatal and Postnatal Growth of Twins

Twin growth is studied, in part, in an effort to determine the relative importance of environmental and genetic influences upon growth and development.

Standards for twins growth have been given by Wilson (1974*a,b*). When compared to singletons, twins showed a deficit in size at birth that increased with gestational age. Twins born at 33-plus weeks were equal to the 36th centile for singletons, but at term the twins average size was below the singleton's 5th centile. Wilson also gave standards for weight, height, and head circumference from birth to four years. He compared high- and low-birth-weight pooled groups and found the mean weight gain was similar for the first two years. By age four years, over 20% of the low-birth-weight twins were larger than the average for high-birth-weight twins. After a marked size deficit at birth compared to singletons, there was notable catch-up for six months but only little thereafter. Wilson makes the interesting suggestion that prenatal adipose cell replication may be diminished by the nutritional sharing required in a twin pregnancy since twins nearly caught up with singletons in height by four years but remained lower in weight.

In an effort to examine in more detail some of the general questions raised up to this point, data were studied from a multiple birth cohort in the Collaborative Perinatal Study of the National Institute of Neurological Diseases and Stroke (NINDS) (e.g., Niswander and Gordon, 1972), and from the Louisville Twin Study (e.g., Wilson, 1974*b*).

The analyses of these data were focused around two major comparisons. First, are there any implications of the monochorionic vs. dichorionic condition among monozygotic twins? Since, as has been noted, there is always a vascular connection in the placenta of MZ-MC twins, and not in MZ-DC twins, the growth and development of these two kinds of twins could be different. Specifically, do these chorionic groups differ with respect to weight, length, and head circumference between birth and four years of life? (We did not analyze beyond this age since the sample size became small). The second consideration was a comparison between MZ and DZ twins, all of whom had the dichorionic condition. Specifically, do MZ and DZ twins having two chorions differ in within-pair similarity with respect to weight, length, and head circumference between birth and four years of age?

## 5.1.1. Longitudinal Sample

Information concerning within-pair similarities as a function of zygosity and chorionic condition can be obtained by restricting the sample to those twin pairs who have complete longitudinal data across several ages. The liability of this approach is that inevitably the sample size is reduced by demanding complete data. Given the data base available, the ages having the least missing data were birth, 4, 12, and 48 months of age. Consequently, the entire data base of the Collaborative Study was searched for twin pairs having complete data at these four ages for each of the three dependent measures (e.g., weight, length, and head circumference). Since there was a marked tendency for a pair to have all three dependent measures (or none) available at a given age, very few pairs were eliminated for not having all dependent variables at a specific age.

## 5.1.2. Nondevelopmental Comparisons

The research focus dictated two *a priori* (but nonorthogonal) comparisons be made within this general design. That is, MZ-DC vs. MZ-MC twin pairs constituted one comparison, while MZ-DC vs. DZ-DC twin pairs constituted the other. These contrasts were tested against error terms derived from the entire analysis.

The statistical procedure we have used has been suggested for data of this type by Bock and Vandenberg (1968), McCall *et al.* (1973), and McCall (1970, 1972). This approach uses as the basic data the orthogonal polynomial contrast scores for each pair of twins as determined on their algebraic within-pair differences at the four ages. That is, for any given twin pair there exists a within-pair difference at birth, 4, 12, and 48 months of age. Orthogonal polynomial trend comparisons (Winer, 1971) are calculated for each twin pair, taking into consideration the unequal spacing of the ages. This analysis yields a test of the difference between MZ and DZ twin pairs (for example) with respect to the average of their three comparisons which may be interpreted as a test of the comparability of developmental trends in within-pair similarity between MZs and DZs. Similarly, the difference in within-pair developmental trends for MZ and DZ pairs may depend on the sex, race, or the combination of sex and race of those pairs. In short, this procedure provides a test of whether a difference exists in the developmental pattern of within-pair similarity as a function of MZ vs. DZ conditions and their possible dependencies on sex and/or race. Similar analyses can be performed with respect to mono-chorionic vs. dichorionic twin pairs.

### 5.1.3. Cross-Sectional Samples

The above analyses were performed only on the subset of twin pairs that possessed complete longitudinal data on the three growth measures of interest. In the present case, whereas there were 90 pairs of twins having complete data which fell into one of the race, sex, and chorionic-zygosity combinations, there were 179 pairs of twins available if weight at birth only was analyzed. Therefore, restricting the sample to twin pairs having complete longitudinal data cut the number of available subjects in half at a given age. Consequently, comparisons were made between zygosity and between chorionic condition for the two sexes and races separately for each variable at each age. This supplemented the above information by providing more stable estimates of average within-pair similarity at each age and more sensitive tests of significance.

In addition to weight, length, and head circumference, the gestational age of the subjects was also available. This variable was analyzed in all the appropriate ways described above, and no differences for chorionic or zygosity comparisons were observed.

### 5.2. Chorionic Comparisons

The following analyses represent comparisons between MZ-MC vs. MZ-DC twins for the dependent variables of weight, length, and head circumference. For each variable, the developmental analyses based upon the longitudinal sample will be considered first and labeled "developmental comparisons." This will be followed by the general comparisons between groups over the entire 0 to 4-year period based upon the longitudinal sample and the within-age tests based upon the cross-sectional sample, both reported under "nondevelopmental comparisons."

### 5.2.1. Weight

*Developmental Comparisons.* If the chorion condition is associated with differences in growth of the twins prenatally, such differences would be expected to be most marked at birth. Table III presents the average within-pair absolute difference in birth weight for each chorion condition of MZ twins. The within-pair differences for MZ-MC twins are 29% larger than for MZ-DC pairs, and this difference is not qualified by the race or sex of the pair.

The chorion × race × sex analysis on the longitudinal sample revealed a race × chorion difference in age trends ($F_{mult}$ (3, 76) = 2.66, $P$ = 0.05) for average within-pair differences in weight. Table IV presents the average within-pair absolute

*Table III.   Chorionic Mean Within-Pair Differences in Birth Weight (Grams) for MZ Twins*

| Chorionic condition | N | Birth weight difference (g) | Difference (%) |
|---|---|---|---|
| DC | 25 | 282.32 | 29%; $F(1,78)$ = 3.36, |
| MC | 30 | 394.99 | $P$ = 0.07 |

Table IV.  *Chorionic Mean Within-Pair Differences in Weight in Grams (Percent) as a Function of Race and Age for MZ Twins*

| Race | Chorionic condition | N | Weight (g)[a] | | | |
|------|---------------------|---|---------------|---|---|---|
| | | | Birth | 4 Months | 12 Months | 48 Months |
| White | Di | 5 | 227.00 (9.5) | 577.60 (10.5) | 799.80 (9.0) | 1668.80 (10.7) |
| | Mono | 16 | 342.00 (14.4) | 355.13 (6.4) | 369.06 (4.1) | 455.38 (2.9) |
| Black | Di | 20 | 296.15 (12.4) | 395.80 (7.2) | 402.90 (4.5) | 683.00 (4.4) |
| | Mono | 14 | 455.57 (19.1) | 627.86 (11.4) | 567.21 (6.4) | 1022.21 (6.5) |

[a]Numbers in parentheses are percent difference within pairs.

weight differences as in Table III, subdivided according to race at each of the four ages. In addition, the table presents in parentheses the percent of average individual subject weight represented by this within-pair difference. Table IV shows that the difference between MC and DC twins is more emphatic among whites than blacks. White DC twins maintain approximately a 10% weight difference between birth and 4 years, while MC white twins differ 14.4% in birth weight but only approximately 3% by age four. Among blacks, there is a tendency for MC twins to show slightly larger differences at each age, but both groups seem to level off in mean difference by one year of age (in contrast to the white MC group).

*Nondevelopmental Comparisons.* The longitudinal sample revealed a general race × chorionic condition difference in within-pair similarity in weight when averaged over the four ages [$F(1,78) = 7.89$, $P = 0.006$], indicating that the difference between MC and DC twin-pair similarity in weight was greater for white than black pairs. However, the developmental analyses described above showed that such chorionic × race differences varied over age, and there was not an obvious difference between chorionic groups which was consistent at each age. Table IV shows that the differences between groups seemed to solidify by 4 years of age, and consequently one might expect within-age differences to be confined to the later years. In fact, such was the case. The only significant within-age group differences occurred at 4 years where there was a race × chorionic group difference for the longitudinal sample [$F(1,78) = 4.39$, $P = 0.039$] and a similar suggestion for the cross-sectional sample [$F(1,99) = 3.71$, $P = 0.057$]. In addition, in the cross-sectional sample MC twins were more similar within pairs than DC twins, but this was true only for males [$F(1,99) = 3.83$, $P = 0.053$], a tendency which was not as statistically emphatic among the longitudinal sample [$F(1,78) = 3.44$, $P = 0.067$].

These data suggest that both white and black MZ-MC twins differ notably more in birth weight than MZ-DC twins, but the MZ-MC twins become more similar in weight to one another as they develop between birth and 4 years—possibly as genetic factors eventually override prenatal and placental influences. The MZ-DC white twins appear to maintain a steady within-pair weight difference from birth to 4 years.

### 5.2.2. Length

It should be noted here that length measurement of the newborn and infant may involve considerable error when a large number of measurers are involved, as in the Collaborative Study.

*Developmental Comparisons.* There were no differences between chorionic groups in terms of their developmental trends in within-pair similarity of length.

*Nondevelopmental Comparisons.* For the longitudinal sample, MC twins were more similar in length than DC twins among the whites (1.47 vs. 2.08 cm), but just the opposite was true among blacks (1.78 vs. 1.24) when data were averaged across all four ages [$F(1,78) = 7.68$, $P = 0.007$]. However, such a difference did not always emerge at each age for either sample, nor were the means within an age always in this direction, and the differences were not always this great.

Although there may be differences in within-pair similarity relative to length as a function of chorionic condition and race, these differences here are neither large nor consistent, and thus these data are ambiguous in this regard.

### 5.2.3. Head Circumference

*Developmental Comparisons.* There were no differences between MZ-MC and MZ-DC twins with respect to developmental changes in within-pair similarity in head circumference between birth and 4 years of age.

*Nondevelopmental Comparisons.* In both the longitudinal and complete samples, MZ-DC twins were *more* similar in head circumference at birth than MZ-MC pairs. For the developmental sample, the difference was 1.30 vs. 0.68 cm [$F(1,78) = 6.74$, $P = 0.011$], while for the complete sample the differences were 1.16 vs. 0.73 cm [$F(1,161) = 4.74$, $P = 0.031$]. No other differences emerged at this or later ages.

### 5.2.4. Conclusions for Chorionic Comparisons

General physical growth can depend in part on prenatal chorionic condition. With respect to weight, MZ-MC pairs become progressively more similar in weight (for example, 14.4–2.9%) between birth and 4 years of age, having at birth notable dissimilarity, while MZ-DC pairs maintain in general similar within-pair differences across the birth-to-4-year period. However, this occurs only for twins of the white race. Differences at any single age do not emerge until 4 years, and then they are more emphatic for male than female pairs. There appear to be no consistent differences in length, but within-pair differences for head circumference are almost twice as great for MZ-MC than MZ-DC twins at birth; such differences are not significant thereafter. This suggests that an MZ-MC twin with a deficit at birth exhibits catch-up early in postnatal life.

## 5.3. Zygosity Comparisons

Comparisons between MZ-DC and DZ-DC twin pairs with respect to weight, length, and head circumference were made.

### 5.3.1. Weight

*Developmental Comparisons.* There was a general age trend difference in within-pair similarity in weight in which MZ-DC twins showed a progressive shift towards proportionally greater similarity while DZ-DC twins showed an irregular developmental pattern [$F_{mult}(3,76) = 3.08$, $P = 0.033$]. However, this general result was qualified by interactions separately with race and with sex. The means for these two interactions are presented in Table V.

Quite in contrast to the chorionic growth trend differences, zygosity differences appear to occur principally for blacks [$F_{mult}(3,76) = 3.80$, $P = 0.014$] and for females [$F_{mult}(3,76) = 2.69$, $P < 0.05$], rather than for whites and males. That is,

Table V.  *Zygosity Mean Within-Pair Differences in Weight in Grams (Percent) as a Function of Race, Sex, and Age for Dichorionic Placentation*

| | Zygosity | N | Weight (g)[a] | | | |
| | | | Birth | 4 Months | 12 Months | 48 Months |
|---|---|---|---|---|---|---|
| Race | | | | | | |
| White | MZ | 5 | 227.00 (9.5) | 577.60 (10.5) | 799.80 (9.0) | 1668.80 (10.7) |
| | DZ | 16 | 256.88 (10.8) | 474.07 (8.6) | 776.01 (8.7) | 1869.13 (12.0) |
| Black | MZ | 20 | 296.15 (12.4) | 395.30 (7.2) | 402.90 (4.5) | 683.00 (4.4) |
| | DZ | 19 | 355.00 (14.9) | 387.16 (7.0) | 877.26 (9.9) | 1858.68 (11.9) |
| Sex | | | | | | |
| Male | MZ | 7 | 461.71 (19.4) | 685.43 (12.4) | 778.71 (8.7) | 1833.86 (11.7) |
| | DZ | 18 | 340.17 (14.3) | 378.11 (6.8) | 860.39 (9.7) | 1767.11 (11.3) |
| Female | MZ | 18 | 212.56 (8.9) | 333.11 (6.0) | 368.11 (4.1) | 927.67 (5.9) |
| | DZ | 17 | 278.35 (11.7) | 478.53 (8.7) | 799.82 (9.0) | 1965.42 (12.6) |

[a]Numbers in parentheses are percent difference within pairs.

black and female MZ twins showed progressively greater similarity in weight with increasing age, whereas their DZ counterparts displayed something of a U-shaped developmental trend.

*Nondevelopmental Comparisons.* The direction and significance of within-age comparisons in zygosity were nearly identical for the longitudinal and cross-sectional samples, and once again, within-age comparisons were significant only at the older ages.

MZ-DC pair members were more similar for weight at 12 months [$F(1,78) = 5.49$, $P = 0.022$] and at 48 months [$F(1,78) = 8.25$, $P = 0.005$], but the latter effect was especially true for females [$F(1,78) = 4.44$, $P = 0.038$].

While chorionic differences in within-pair similarity in weight were largely confined to white and male pairs, zygosity differences in within-pair similarity in weight were largely confined to blacks and females. MZ-DC blacks and MZ-DC females tended to show progressively greater within-pair similarity between birth and 48 months relative to DZ pairs, such differences reaching within-age significance at the older ages.

### 5.3.2. Length

*Developmental Comparisons.* There were no significant differences in the developmental trends for within-pair similarity in length.

*Nondevelopmental Comparisons.* For the longitudinal sample MZ-DC twins were generally more similar in length between birth and 4 years than were DZ twins [0.91 vs. 0.183 cm; $F(1,78) = 7.69$, $P = 0.007$]. Within-age comparisons for the cross-sectional sample revealed that this zygosity difference was significant at 12 months [$F(1,125) = 6.76$, $P = 0.01$] and at 48 months [$F(1,99) = 8.43$, $P = 0.005$], but the latter effect appeared to be more emphatic for blacks [$F(1,99) = 5.89$, $P = 0.017$].

Thus MZ-DC twins tend to be more similar in length to one another than do DZ twins throughout this age period, although within-age tests were not significant until 12 and 48 months of age, the effect being slightly greater for blacks at 48 months.

### 5.3.3. Head Circumference

*Developmental Comparisons.* There were no differences in the growth trends for within-pair similarities in head circumference for MZ-DC and DZ twin pairs.

*Nondevelopmental Comparisons.* For the longitudinal sample, the within-pair difference of head circumference was nearly twice as great for DZ-DC compared to MZ-DC twins [0.911 vs. 0.545 cm; $F$ (1.78) = 7.87, $P$ = 0.006], and this difference was significant beyond the 0.035 level for each within-age comparison from birth through 48 months for the cross-sectional sample [$F(1,161) = 4.53$, $F(1,134) = 5.76$, $F(1,127) = 7.97$, $F(1,99) = 14.44$]. There were no other effects.

It appears MZ-DC twin pairs are more similar with respect to head circumference than are DZ pairs, and the difference is significant at each of the four ages between birth and 48 months.

It is generally concluded that while MZ twins tend to be more consistently similar in weight, length, and head circumference after birth and up to 48 months of age than are DZ twins, there are additional differences in these growth measures as a function of MC vs. DC condition. The chorionic differences are less consistent than the zygosity contrasts, and the data for chorionic differences with respect to length are ambiguous.

In addition, differences between zygosity and chorionic condition changed with development in the case of weight but not for length or head circumference. MZ twins progressively became proportionally more similar to one another, whereas other groups remained the same or showed irregular growth trends. For head circumference, DC pairs were more similar than MC, but only at birth.

### 5.4. The Louisville Twin Study

Birth weight, estimated gestational age, and placental weight were available on twins from the Louisville Twin Study. Consequently, it was possible to compare zygosity within the DC condition as well as chorion status within the MZ condition for white twin pairs from this sample.

The placental weights were wet weights. Separate placentas pose no problem. When fused they were divided at the fusion plane. Monochorionic placentas were divided into the respective parts for each twin by cleavage along the vascular equator plane.

The statistical comparisons consisted of a consideration of DC twin pairs in which MZ twins were compared with DZ pairs. A test was also performed to determine if the differences between these MZ and DZ twins differed as a function of the sex of the twin pair (only like-sexed pairs were considered). Similarly, among MZ twins, MC pairs were compared with DC pairs, and this difference was assessed for possible dependencies on sex. These comparisons were made for the average absolute difference in birth weight within pairs, the average gestational age, and the average absolute difference in within-pair placental weight. Table VI shows the sample sizes involved.

*Table VI.   The Louisville Sample Sizes*

| Zygosity | Sex | $N$ |
|---|---|---|
| MZ–MC | Male | 20 |
| MZ–DC | Male | 14 |
| MZ–MC | Female | 12 |
| MZ–DC | Female | 7 |
| DZ–DC | Male | 23 |
| DZ–DC | Female | 24 |

Table VII. *Chorionic Mean Within-Pair Differences in Birth Weight (Grams) for MZ Twins*

| Chorionic condition | N | Birth weight difference |
|---|---|---|
| DC | 21 | 359 |
| MC | 32 | 313 $F(1.94) = 0.044$ |
| | | $F < 1.0$ |

### 5.4.1. Gestational Age

There was no zygosity difference in gestational age, $[F(1,80) < 1.00]$ nor was there a difference between MZ and DZ twins in gestational age as a function of sex $[F(1,80) = 1.07, P = 0.30]$.

Similarly, there was no difference between chorionic conditions for gestational age $[F(1,80) < 1.00]$ or for the interaction with sex $[F(1,80) < 1.00]$.

Thus, as was the case in the Collaborative Study, there were no differences in gestational age as a function of zygosity or chorionic condition.

### 5.4.2. Placental Weight

There were no significant results for the average within-pair absolute difference in placental weight as a function of zygosity or chorionic condition and no interactions with sex (all $Fs < 1.00$).

### 5.4.3. Birth Weight

There were no significant differences as a function of zygosity $[F(1,94) < 1.00]$ or the interaction of zygosity and sex $[F(1,94) = 2.48, P = 0.12]$. Table VII shows that unlike the Collaborative Study, MZ-MC twins have a very slightly lower birth-weight difference than MZ-DC twins, although the difference is not significant.

### 5.4.4. Placental Weight and Birth Weight

Table VIII shows the general correlation between actual placenta and birth weight is approximately 0.36–0.61, and there is no consistent or sizable difference in the level of this relationship as a function of zygosity or chorion. There are no differences in the *r*s for MZ-MC vs. MZ-DC. Thus, regardless of chorionic condition or zygosity, there is a strong positive relationship between birth weight and placental weight.

In both the Collaborative and the Louisville studies lack of differences for zygosity is consistent across studies. In the Collaborative Study, zygosity differences in weight were not present at birth, although within-pair differences diverged for MZ vs. DZ with development. However, even these developmental contrasts were most pronounced for blacks (and females), whereas the Louisville data are derived from a white sample.

Earlier in the text it was suggested that MZ-MC twins were likely to be more different in size than MZ-DC twins at birth. In general, the studies bear this out, although for birth weight MZ-MC and MZ-DC twins were similar.

Robson (1976) provides some supportive data and also raises some important

Table VIII.  Correlations between Placental and Birth Weights

|  | r | df | P |
|---|---|---|---|
| Birth weight twin A/placenta twin A |  |  |  |
| MZ–MC | 0.41 | 24 | 0.025 |
| MZ–DC | 0.41 | 15 | 0.05 |
| DZ–DC | 0.49 | 42 | 0.005 |
| Birth weight twin B/placenta twin B |  |  |  |
| MZ–MC | 0.58 | 24 | 0.005 |
| MZ–DC | 0.36 | 15 | n.s. |
| DZ–DC | 0.61 | 42 | 0.005 |

questions. Her data show 27 MZ-DC twins had an average birth-weight difference of 261 g, while in the 107 MZ-MC twins, the difference was 323 g (20% difference). She points out the need for selection criteria in regard to survival. Amongst male pairs, one of whom died, the within-pair birth-weight differences were 2.2 times greater than in male pairs both of whom survived. Amongst female pairs, no such effect was seen. Together with the need to note the constitution of MZ samples with regard to condition, for example, whether there is a higher proportion of MZ-DC than expected, these considerations show the need for care in interpretation of results and that the multifactorial nature of the influences upon growth are more complex than has been considered. In the MZ-MC condition, for example, it may be necessary to add further factors such as the MZ-MC placental status with regard to the shunt–transfusion syndrome. A possible "critical size" factor of placenta, or placenta part, has already been mentioned.

## 6. General Commentary

### 6.1. Zygosity Determination

In deriving information from studies of human twins, it is usually necessary to be sure of the correct diagnosis of zygosity. In the past one could often be left in doubt as to the correctness of the classification. Very careful examination of the placenta(s) will diagnose monozygosity if the placenta is monochorionic-diamniotic (or indeed monoamniotic). And it will be helpful in the case of like-sexed twins having a dichorionic placenta. But to be sure of the MZ vs. DZ state—particularly in the latter case—for all practical purposes it is necessary to blood-type the twins. Wilson (1970), in a definitive study, described the use of 22 or more antigens in over 700 pairs of twins and showed that there was about a 3% chance that twins concordant for all the major blood groups may be DZ; this chance is even less if parental blood groups are also available.

### 6.2. Gestational Age

In addition to the last-menstrual-period method of estimation—if known—an interesting study by Keet et al. (1974) shows that estimation of gestational age for individual twins by the method of Dubowitz et al. (1970) gave no significant difference between the mean estimated gestational age of the lighter and heavier birth-weight twins even in a group where one twin was over 20% lighter than the other twin. Since the birth weight of the heavier twin is assumed appropriate for gestational age, this is important when studying the smaller twin (particularly if MZ) as a SFDI.

birth-weight twins even in a group where one twin was over 20% lighter than the other twin. Since the birth weight of the heavier twin is assumed appropriate for gestational age, this is important when studying the smaller twin (particularly if MZ) as a SFDI.

## 6.3. Concordance

Since the study of twins, particularly the within-pair difference in MZ twins, is often for the purpose of assessing the weightings of environmental and genetic factors, it is presumed that MZ twins are genetically completely identical. A word of caution is necessary here. Chromosome studies have shown, increasingly of late, that MZ twins are not always genetically identical, and this can explain the lack of concordance in some traits and inheritable disorders in MZ twins. Nielsen (1967) reviewed the literature and found that 23 MZ pairs were reported, one or both of whom had a chromosomal abnormality, including five pairs with one twin with Down's syndrome and one pair where one twin was female and the other male. Study of chromosomes in a large group of MZ twins is much needed.

Smith (1974) summarizes the view that concordance rates in twins can be very informative about the genetic role, but they are not likely to be critical in the discrimination of modes of different inheritance.

## 6.4. Longitudinal Study of Physical Growth

As part of the Louisville Twin Study, Vandenberg and Falkner (1966) used a fitted growth curve for each individual twin and a method of using parameters of this curve to summarize the information for that individual in a few numbers. The concordance rates in height growth for MZ and like-sexed DZ twins was very similar for the first few months only. This is not surprising since it has been shown (e.g., Tanner *et al.,* 1956) that the "genetic target" curve is not reached until 2–3 years when some prenatal and previous influences on growth have waned. It was found that—at least for height—hereditary influence appears to be of paramount importance only for the deceleration of the growth rate (taking effect later), that it is of moderate importance for growth rate, but not significantly important for the initial status.

Wilson (1976) supports these general contentions by showing that in his large sample of twins the early strong prenatal influences upon growth are gradually neutralized, and postnatal growth moves firmly in a direction determined genotypically. Wilson (1976) also showed that MZ twins were less concordant for birth weight (the chorionic placental factor was not studied) than DZ but that this was in some measure due to a few MZ pairs who had markedly great within-pair birth-weight differences (see earlier in this chapter). However, by one year of age, MZ twins had become more concordant for size and DZ twins moved further apart, and this trend continued to 4 years. He notes the complex perinatal influences upon birth size and the rapid conveyance of each twin onto his own genetic growth curve.

## 6.5. The Future

It must now be clear that interpretation of results from twin studies have to be made with caution, especially since twins are not representative of the general population. Nevertheless, more and more information is forthcoming concerning prenatal influences on growth (and their complex multifactorial nature) and concerning that often-neglected factor of prenatal life—the placenta. This will allow careful planning and research design for twin studies, for it is also clear that so much may be learned from them, especially in unraveling nature–nurture influences upon growth. MacMillan *et al.* (1973) provide a good example of the multidisciplinary longitudinal approach needed for future studies. They found that placental concentration of chorionic somatomammotropin (HCS) and total content of HCS were related to birth weight of twins—the maternal HCS levels being, of course, common to both twins. This suggests a valid index of placental function. They also found HCS placental concentration was predictive of length at birth and at 2 years, suggesting that HCS production may represent an influence independent of general placental function. Here is a current stimulus and example of the need to study the influence of placental nutrition, size, and function upon growth pre- and postnatally both in twins and in singletons and to continue study of the human twin as a means to explore other influences upon known growth.

ACKNOWLEDGMENTS

Part of this study was supported by NINDS Contract Number 3221045. I acknowledge with pleasure the encouragement given me by my colleagues on the Genetic Task Force of the NINDS Collaborative Study.

I am very grateful indeed to Dr. Ronald S. Wilson, Director of the Child Development Unit and Louisville Twin Study, University of Louisville School of Medicine, and his staff. He graciously allowed me access to data only some of which had been collected during my tenure there. Dr. Robert McCall, Senior Scientist, The Fels Research Institute, gave me invaluable statistical guidance and help. Dr. E. B. Robson, The Galton Laboratory, University College London, kindly allowed me to see a preliminary analysis of her Oxford data and gave me valuable suggestions. Pamela Hogarty, Robert Engstrom, and Cynthia Kennedy performed the computer analyses at the Fels Research Institute and are accorded grateful thanks.

## 7. References

Babson, S. G., and Phillips, D. S., 1973, Growth and development of twins dissimilar in size at birth, *N. Engl. J. Med.* **289:**937.

Benirschke, K., 1961, Twin placenta in perinatal mortality, *N.Y. State J. Med.* **61:**1499–1508.

Benirschke, K., and Driscoll, S. G., 1967, *The Pathology of the Human Placenta,* Springer, Berlin.

Bock, R. D., and Vandenberg, S. G., 1968, Components of heritable variation in mental test scores, in: *Progess in Human Behavior Genetics* (S. G. Vandenberg, ed.), pp. 233–260, Johns Hopkins Press, Baltimore.

Buckler, J. M. H., and Robinson, A., 1974, Matched development of a pair of monozygous twins of grossly different size at birth, *Arch. Dis. Child.* **49:**472–476.

Corney, G., Robson, E. B., and Strong, S. J., 1968, Twin zygosity and placentation, *Ann. Hum. Genet.* **32:**89–96.

Cruise, M. O., 1973, A longitudinal study of the growth of low birth weight infants, *Pediatrics* **51:**620–628.

Dubowitz, L. M. S., Dubowitz, V., and Goldberg, C., 1970, Clinical assessment of gestational age in the newborn infant, *J. Pediatr.* **77:**1.

Edwards, J. H., and Cameron, A. H., 1967, as quoted in: *The Placenta in Twin Pregnancy* (S. J. Strong and G. Corney, eds.), p. 24, Pergamon Press, New York.

Falkner, F., 1966, General considerations in human development, in: *Human Development* (F. Falkner, ed.), pp. 10–39, Saunders, Philadelphia.

Keet, M. P., Jaroszewicz, A. M., and Liebenberg, A. R., 1974, Assessment of gestational age in twins, *Arch. Dis. Child.* **49**(9):741–742.

MacMillan, D. R., Brown, A. M., Matheny, A. P., and Wilson, R. S., 1973, Relationships between placental concentrations of chorionic somatomammotropin (placental lactogen) and growth: A study using the twin method, *Pediatr. Res.* **7**:719–723.

McCall, R. B., 1970, IQ pattern over age: Comparisons among siblings and parent–child pairs, *Science* **170**:644–648.

McCall, R. B., 1972, Similarity in developmental profile among related pairs of human infants, *Science* **178**:1004–1005.

McCall, R. B., Appelbaum, M., and Hogarty, P. S., 1973, Developmental changes in mental performance, *Monogr. Soc. Res. Child. Dev.* **38**(150):1–84.

Nielsen, J., 1967, Inheritance in monozygotic twins, *Lancet* **2**:717–718.

Niswander, K. R., and Gordon, M., 1972, *The NINDS Collaborative Perinatal Study: The Women and Their Pregnancies,* Saunders, Philadelphia.

Potter, E. L., 1963, Twin zygosity and placental form in relation to the outcome of pregnancy, *Am. J. Obstet. Gynecol.* **87**:566–577.

Robson, E. B., 1976, personal communication on preliminary analysis of Oxford data.

Smith, C., 1974, Concordance in twins: Methods and interpretation, *Am. J. Hum. Genet.* **26**:454–466.

Strong, S. J., and Corney, G., 1967, *The Placenta in Twin Pregnancy,* Pergamon, Oxford.

Tanner, J. M., Healy, M. J. R., Lockhart, R. D., Mackenzie, J. D., and Whitehouse, R. H., 1956, The prediction of adult body measurements from measurements taken each year from birth to 5 years, *Arch. Dis. Child.* **31**:372–381.

Vandenberg, S. G., and Falkner, F., 1966, Hereditary factors in human growth, *Hum. Biol.* **37**:357–365.

Wilson, R. S., 1970, Bloodtyping and twin zygosity, *Hum. Hered.* **20**:30–56.

Wilson, R. S., 1974a, Twins: Measures of birth size at different gestational ages, *Ann. Hum. Biol.* **1**:57–64.

Wilson, R. S., 1974b, Growth standards for twins from birth to four years, *Ann. Hum. Biol.* **1**:175–188.

Wilson, R. S., 1976, Concordance in physical growth for monozygotic and dizygotic twins, *Ann. Hum. Biol.* **3**:1–10.

Winer, B. J., 1971, *Statistical Principles in Experimental Design,* 2nd ed., McGraw-Hill, New York.

# 17

# Association of Fetal Growth with Maternal Nutrition

## JACK METCOFF

### 1. General Considerations—Some Associations between Maternal Nutrition and Fetal Growth

A complete bibliography of all papers dealing with maternal nutrition and fetal growth in both animals and man is beyond the scope of this chapter. Recognizing that animal investigations may be relevant, I have attempted to concentrate on information derived from studies in human beings. References cited may be incomplete in some cases. Usually I have emphasized recent reports for the perspective they provide.

The most rapid period of fetal growth is in the 20 weeks between 12 and 36 weeks of gestation (i.e., postmenstrual age). Between 32 and 36 weeks, the rate of fetal weight gain reaches its peak at 200–225 g/week. The rate declines thereafter (Brenner *et al.*, 1976). Normal full-term gestation is usually considered to be 40 ± 2 weeks. If born before 38 weeks, the infant is considered premature; if born after 42 weeks, postmature (Lubchenko, 1976).[1] Mean birth weights for babies are smaller in most developing countries than in Europe, the U.S.A., or Canada. This may be due to genetic factors, a shorter period of high-velocity growth, or perhaps related to nutrition of the mother as a child, prenatally, and/or during pregnancy, or to infections and diseases complicating pregnancy.

Human fetal nutritional requirements for normal growth are inferred either from comparisons between maternal diet during pregnancy and the size or condition of the baby at birth or from animal experiments. It is generally assumed that a nutritionally excellent diet consumed by the mother during pregnancy is good for the fetus, while a poor diet will embarrass fetal nutrition and growth. Common

[1]WHO standard for gestation is still 39 ± 2 weeks, and infants born before 37 weeks are considered premature.

---

*JACK METCOFF* • Health Sciences Center, University of Oklahoma, Oklahoma City, Oklahoma.

sense, some anecdotal experiences, nutritional catastrophes of war, and the impact of poverty and long-standing dietary inadequacy leading to lower birth weights of term babies are used to support these assumptions. Nutritional deficits among pregnant women in developing societies, like some in Central and South America, have been related to the greater frequencies of low-birth-weight infants (Arroyave, 1975; Lechtig *et al.*, 1975). A generation of socioeconomic improvement in a Japanese population (Gruenwald, 1966) and protein and energy food supplementation for pregnant women (Habicht *et al.*, 1974; Jacobson, 1974; Lechtig *et al.*, 1975) show a positive effect of food supplementation on baby size and may reduce the frequency of low-birth-weight infants.

Nonetheless, neither the macro and micro requirements nor the interrelationships between them needed to support optimal growth of the human fetus are known. Furthermore the type and extent of placental processing or synthesis of nutrients being transported from mother to fetus are inadequately understood. How maternal, placental, and fetal hormones modulate metabolism of nutrients needed for cell replication, differentiation, growth, and metabolism is not clear. Moreover, placental structure, composition, and function undergo continuous change while fetal mass increases 100 times in the last 6 months of pregnancy. The enzyme-mediated pathways modulating the use of nutrients for growth emerge at different rates and in different sequences in each fetal organ. The success of mechanisms controlling and integrating these processes is evidenced by the fact that most pregnancies are successful.

Nutrition before and during pregnancy, and possibly the nutritional status of the mother's own mother in the past (Thomson and Billewicz, 1957; Ounsted and Ounsted, 1966), may influence fetal growth. Malnutrition prior to mating and through gestation of the $F_0$ generation of rats may lead to fetal malnutrition (FM) in the $F_2$ generation (Zamenhof *et al.*, 1971). In colonies of rats maintained for 12 generations on diets marginally deficient in protein, the proportion of fetally mal-nourished offspring was 10 times as high as found among the well-nourished control colony (Stewart *et al.*, 1975). The timing of nutritional insults and the ability of mother and fetus to adapt to suboptimal nutritional intakes are also important. The type of nutritional limitation, e.g., protein, energy, iron, zinc, folate, etc., may influence fetal growth. Efforts to relate maternal nutrition to fetal development are further confounded by the criteria used to assess fetal development. A few carefully done studies involving limited samples have contributed highly suggestive results favoring the relationship. The available information on this subject has been summarized periodically (National Research Council, 1970; Giroud, 1970; Pitkin *et al.*, 1972, Osofsky, 1975; Rush, 1975).

An influence of nutrition during pregnancy upon the condition of the infant at birth was reported more than 30 years ago by Burke *et al.* (1943). These workers found a statistically significant relationship between the diet of the mother during pregnancy and the condition of her infant at birth and within the first two weeks of life. The diet history included three-day food records, food expenditures, family food consumption patterns, and cross-check analyses to validate the diet histories. The studies were carried out from the fourth through the ninth months of pregnancy. Only 14% of the 216 women consumed excellent or good diets, 23% fair to poor diets, and 17% had diets which were evaluated as poor to very poor. Forty percent of the women in the study were considered malnourished, based on dietary intake standards, during the period when the fetus undergoes rapid growth and development. They concluded that women with poor to very poor diets would "undoubtedly have an infant whose physical condition will be poor." All stillborn infants, all infants who died within a few days of birth (with one exception), most infants who had marked congenital defects, all premature, and all "functionally immature" infants were born to mothers whose diets during pregnancy were very inadequate. At that time, babies were considered "premature" or "immature" if the birth weight was less than 5 pounds. The average birth weight of the babies born to mothers with very poor diets averaged 5 lb, 13 oz, while that of mothers with excellent or good diets was 8 lb, 8 oz.

Smith (1947) reported that the "famine winter" of 1944–1945 in Holland adversely affected birth weights of babies born during that period. In a detailed, historical cohort-type retrospective study involving one area which had suffered from famine and two areas unaffected during that period, Stein and Susser (1975) reviewed available records and found that prenatal exposure to famine reduced postpartum maternal weight (4.3%), birth weight (9%), birth length (2.5%), head circumference (2.7%), and placental weight (15%). Third-trimester exposure accounted for all the famine effects. Fetuses whose mothers were exposed to famine during the first or second trimesters were not significantly reduced in size. The effects on the size of the fetus were apparent only when the caloric intake was less than the official 1500-calorie ration. This unfortunate geopolitical catastrophe of war, like animal experiments, demonstrated that severe nutritional deprivation during pregnancy impairs fetal growth.

At Vanderbilt University, 2129 pregnancies were evaluated by Darby *et al.* (1953) and by McGanity *et al.* (1954) with respect to maternal nutrition and pregnancy outcome. The Vanderbilt experience showed that a wide range of laboratory findings was compatible with a normal outcome of pregnancy. No advantages were evident for dietary intakes in excess of recommended allowances or for vitamin supplementation. Fetal malnutrition had not been described at that time, and all babies whose birth weights were less than 5.5 pounds were considered prematurely born, irrespective of estimated gestational age. Thomson (1959) observed that birth weight of infants rose and the incidence of low-birth-weight babies decreased as the energy value of the mothers' diets increased. Later, when the data were statistically adjusted to control for the fact that women with greater energy intakes were taller, heavier, healthier, and from a better socioeconomic class, it appeared that social class and stature among the women studied by Thomson were more important influences on weight of the baby at birth than diet in pregnancy. Recent studies in Guatemala among village Indian women of uniformly low socioeconomic state indicate that supplemental protein and energy intakes during pregnancy are associated with increased birth weights and decreased morbidity and mortality among newborns (Habicht, 1974; Lechtig *et al.,* 1975). Higgins (1974) also found a positive association between infant birth weight, maternal weight gain, and protein and energy intakes during pregnancy. Studies of the effect of food supplementation during gestation on the outcome of pregnancy appear to indicate a positive relationship (Jacobson, 1974).

Poor maternal nutrition also has been associated with an increased incidence of toxemia (preeclampsia) of pregnancy (Brewer, 1966). Toxemia of pregnancy may be associated with intrauterine growth retardation or fetal malnutrition. While some of the pathophysiologic features of these diseases peculiar to pregnancy are known, the etiologies are not. Chronic vascular disease and hypertensive syndromes of pregnancy often are cited. These limit delivery of blood containing nutrients to placenta and fetus. Poor nutrition, particularly in teenagers and preeclamptics, as well as increased effects of aldosterone and angiotensin II, have been implicated. Poverty, chronic stress, and heredity also are thought to play some role.

Clear-cut relationships between maternal nutrition during pregnancy and fetal growth are difficult to show especially in industrialized societies where obvious malnutrition is uncommon and diet variability among pregnant women is great. Rush (1975) has emphasized that techniques for assessment of maternal nutritional status are highly unsatisfactory and that universally acceptable standards for nutrient intake and balance are lacking. Fetal growth may be the best method to assess the consequences of maternal nutrition during pregnancy. The nutritional effects on fetal growth must be separated from other influences such as socioeconomic status, smoking, alcohol and other drugs, infections and other complications of pregnancy, the influence of parental size and race, the sex and duration of gestation of the baby.

## 2. Nutrition and the Course of Pregnancy

### 2.1. Maternal Weight Gain and Baby Size

Several authors have reported a high correlation between maternal weight gain and birth weight (Beilly and Kurland, 1945; Love and Kinch, 1965; O'Sullivan *et al.*, 1965; Thomson *et al.*, 1968), while others have shown a positive association between both prepregnancy weight and weight gain, with birth weight (Eastman and Jackson, 1968; Niswander *et al.*, 1969; Ademowore *et al.*, 1972). An analysis of data derived from 24,395 white and 2133 black pregnancies showed a positive, almost linear, correlation between maternal weight gain during pregnancy and term birth weight, with prepregnancy weight up to 160 pounds having an additive positive influence on birth weight. The women of this study were army wives, and all delivered at the Brooke General Army hospital. They were mostly young, healthy, and of low parity. Weight gain was estimated from the difference between the prepregnancy weight recalled by the mother and the weight at the last prenatal visit, within 2 weeks of delivery. The incidence of low-birth-weight term infants decreased as either prepregnancy weight or weight gain increased (Simpson *et al.*, 1975).

Eighty percent of the white and 74% of black women gained between 11 and 30 lb. Ten thousand one hundred forty-four white and 742 black women had prepregnancy weights below 120 pounds, and delivered 389 (3.83%) and 37 (5.0%) babies with birth weights <2501 g, respectively. Two thousand nine hundred twenty-four white and 283 black women gained less than 11 lb during pregnancy. They had 4.7% and 6.4% of babies with birth weights <2501 g. Women with the lowest prepregnancy weights (<120 lb) and least weight gain (<11 lb) had the largest proportion of low-birth-weight infants. The neonatal mortality rate was 23.9/1000, among white low-birth-weight babies, 6 times higher than the neonatal death rate among the other white babies (3.86/1000). Low birth weight did not increase neonatal mortality of the black infants, which was low (2.34/1000).

Total weight gain during normal pregnancy among nonpoverty populations of industrialized societies ranges from 10 to 12 kg. The curve of weight gain is hyperbolic, increasing steeply from about the 12th to 14th week of gestation until term. Less than 1 kg is gained during the first trimester, about 3 kg in the second, and about 6 kg in the third. The growing fetal components (fetus, placenta, and amniotic fluid) account for about 50% of the weight gain in the second trimester and about 90% in the third. The components of maternal and fetal weight gain during pregnancy have been described by Hytten and Leitch (1971) and others. The approximated distribution of component weight gains among fetal and maternal tissues is shown in Table I. The accountable gains total 9–10 kg at term and are less than the 11–12 kg which may be gained by women with free access to desired diets. The additional 1–3 kg could be either lean tissue or fat. Measurements of skin-fold thickness (Taggart *et al.*, 1967) or of body specific gravity (McCartney *et al.*, 1959) suggest it is largely fat.

Extracellular, and possibly intracellular, fluids increase appreciably during pregnancy. For an average woman weighing 55–60 kg, with an extracellular fluid equivalent to 11–12 liters, an increment of 3 liters, characteristic of pregnancy, represents a 25% expansion. The expanded volume may account in part for the tendency to increased blood pressure and apparent reduction observed in some plasma components, e.g., albumin, zinc, amino acids, etc., during pregnancy. On the other hand, increasing concentrations of gamma globulin, cholesterol, and copper indicate absolute increases in the total quantities of these substances in the extracellular body fluids.

Women with large total body water and plasma volumes (usually associated with edema) may produce larger babies. There appears to be a positive correlation between the total intravascular protein content and birth weight (Duffus *et al.*, 1971).

## 2.2. Role of Specific Nutrients for Fetal Growth

Trends for some specific nutrients (e.g., diet protein and calorie intakes, and plasma levels of proteins, cholesterol, vitamin A, carotene, zinc, and copper) derived from a current prospective study of pregnant Oklahoma women are given in Table II.

### 2.2.1. Energy

Maternal energy intake, metabolism, and expenditure are difficult to measure, hence are difficult to relate to fetal growth, reflected in birth size. Energy needs for

*Table I.  Components of Weight Gain in Pregnancy*[a]

| | Weight gained (kg) | Nitrogen (g) | Protein (g) | Fat (g) |
|---|---|---|---|---|
| Fetoplacental(2,3) | | | | |
|   Fetus | 3.4 | 41(1) | 256(1) | 360(1) |
|   Placenta | 0.6 | 9(1) | 60(1) | 4(1) |
|   Amniotic fluid(2) | 1.0 | 0.5(2) | 3 | 0.5 |
|   Subtotal | 5.0 | 50.5 | 319 | 364.5 |
| Maternal(2,3) | | | | |
|   Uterus | 1.0 | 26.6[b] | 166 | 3.9 |
|   Breasts | 0.5 | 13.0[b] | 81 | 12.2 |
|   Blood volume + interstitial fluid (i.e. ECF) | 1.3 | 21.6[b] | 135 | 19.6 |
| | 1.7 | — | — | — |
|   Subtotal | 4.5 | 61.2 | 382 | 35.7 |
|   Total | 9.5 | 111.7 | 701 | 400.2 |

[a]Modified from (2) Hytten & Leitch (1971), (3) Pitkin *et al.* (1972), and (1) Widdowson (1968).
[b]Estimated from protein figures.

Table II. Nutrient Trends During the Last Half of Pregnancy[a]

| Item | Trend[b] | $n^c$ | $t^c$ |
|---|---|---|---|
| Diet protein (g/kg/day) | $Y = 1.20 + 0.010 \ (X - 31.74)$ | 34 | 2.37 |
| Zn ($\mu$g/dl) | $Y = 61.15 - 0.376 \ (X - 31.58)$ | 54 | 3.43 |
| Fe ($\mu$g/dl) | $Y = 85.70 - 2.31 \ (X - 31.62)$ | 52 | 6.14 |
| Cu ($\mu$g/dl) | $Y = 218.98 + 1.904 (X - 31.50)$ | 55 | 5.41 |
| Albumin (g/dl) | $Y = 3.15 - 0.031 \ (X - 31.60)$ | 53 | 8.17 |
| Carotene (mg/dl) | $Y = 121.04 - 0.647 (X - 31.95)$ | 30 | 2.29 |
| Vitamin C | $Y = 0.991 - 0.020 \ (X - 31.24)$ | 20 | 2.36 |
| $\delta_1$-Globulin (g/dl) | $Y = 0.355 + 0.005 \ (X - 31.60)$ | 53 | 3.95 |
| $\alpha_2$-Globulin (g/dl) | $Y = 0.803 + 0.008 \ (X - 31.60)$ | 53 | 2.49 |
| $\beta$-Globulin (g/dl) | $Y = 1.070 + 0.008 \ (X - 31.60)$ | 53 | 3.51 |
| $\gamma$-Globulin (g/dl) | $Y = 0.879 + 0.009 \ (X - 31.60)$ | 53 | 2.86 |

[a]Unpublished data (Metcoff et al., 1974).
[b]Trends for the last half of pregnancy were derived by pooled regression analysis. Each woman had 3 measurements at approximately 24 and 32 weeks of gestation and at 3 days postpartum. $Y$ = concentration of nutrient at 24 weeks; $X$ = concentration of nutrient at any time between 24 and 40 weeks.
[c]$n$ = number of mothers; $t$ = student's $t$.

the pregnant woman must be defined not only in relation to her basal needs and her level of activity, but also must provide for growth of the fetoplacental unit, apposition of her own tissue stores, and expansion of her body fluids with a resultant increased metabolic load. Adequacy of energy supply during pregnancy is usually inferred by estimating weight gain (change from known prepregnancy weight) and/or assessment of dietary intake, and sometimes by birth weight of the baby. Each of these indicators has limitations.

Estimates of weight gain often depend on stated (recalled), rather than measured, prepregnancy weight. More precise estimates of weight gain may be made from actual measurements during early pregnancy and at term. The "term" weight, as reported in the literature actually may be taken at any time within the last two weeks of pregnancy and may be associated with variable amounts of edema. Adjustments for the excess water are not usually made. Standardized measurements of skin-fold thickness at specific sites (e.g., triceps) may provide reasonable estimates of water and fat in the subcutaneous tissues and skin if 15- and 60-sec measurements are obtained, as suggested by Brans et al. (1974) for newborns. Constant caliper pressure for 60 sec apparently squeezes excess water from the tissue.

Diet recall histories may give unreliable estimates of energy and protein intake. It has been estimated that nine 24-hr diet recalls for total calories, and 27 for protein, would be needed to achieve a 95% probability that 90% of the individuals have calculated mean values within 20% of their true means (Balogh et al., 1971). More studies are needed to improve the reliability of diet recalls as a method of estimating energy and protein consumption. Currently they can only be considered approximations of actual dietary intakes.

Birth weight or size of the baby may not be a valid indicator for adequacy of maternal nutrition, even assuming that maternal nutrition regulates fetal growth, unless the measurements are adjusted for nonnutritional variables which can modify baby size, e.g., race (white babies are bigger than black babies), sex (males are bigger than females), parity (second and later babies are larger than first), gestational age (1–2 weeks difference in estimate of gestational age may be highly significant), maternal size (taller, heavier, older mothers have bigger babies than smaller, very young mothers), socioeconomic factors (low family income and cigarette smoking may be important adverse variables affecting baby size), etc. Classical statistical techniques can be used to adjust for these nonnutritional variables but have seldom been used. In a recent report, multiple regression analysis was used to isolate variables significantly and independently related to birth weight from 44 maternal and environmental factors (Philipps and Johnson, 1977). In their study of 47 subjects, birth weight of the infant was positively related to the number of weeks of gestation, overall dietary quality, delivery weight of the mother squared, her age squared, and the number of previous pregnancies squared; and negatively related to weeks of gestation squared, iron and protein intake, age of the mother, the number of cigarettes she smoked per day, and the number of people in the household. Nine factors accounted for up to 86% of the variance in birth weight. When the quality of the diet (nutrient adequacy ratio derived from diet intake of 12 nutrients as percentage of the RDA) was included, the prediction equations were improved by 6–8%.

Although supportive evidence is somewhat limited, most countries have recommended additional energy and protein allowances for pregnant women (Tables III and IV). Studies by Emerson et al. (1972) suggest that the increased energy cost of pregnancy is due entirely to the products of conception and to the adaptive changes in the mother. About 27,000 ± 3000 kcal (113 ± 13 MJ) additional energy, or about 100 kcal (0.42 MJ)/day, are required for the entire gestational period. The extra calories are not evenly distributed during the interval, but more or less follow the pattern of fetal growth and maternal weight gain. According to Emerson et al. energy balance can be maintained by 9 kcal (0.04 MJ)/day during the first trimester, 84 kcal (0.35 MJ) for the second, and 216 kcal (0.90 MJ) for the third, added to the basal pregravid intake of about 2000 kcal (8.4 MJ). The energy cost of physical activity during pregnancy was more or less constant at 550–600 kcal (2.30–2.51 MJ) in pregnant women living in a metabolic ward. Pregravid energy expenditure in the resting state averaged about 1470 kcal (6.15 MJ), with a mean oxygen consumption of 200 ml/min. While the total oxygen consumption increased progressively during pregnancy, the oxygen consumption per kg remained constant at about 4 ml/min. From Emerson's data, an addition of 100 (0.42), 200 (0.84), or 300 1.26) kcal (MJ)/day above the average 2000 kcal (8.4 MJ)/day during each successive trimester apparently should suffice to meet usual energy requirements, unless there are exceptional energy demands. The National Research Council (1974) recommends 2000 kcal (8.4 MJ)/day for women between 23 and 50 years of age, plus an additional 300 kcal (1.26 MJ)/day throughout pregnancy. This would provide more energy than appears to be indicated by Emerson's data.

Several "intervention" studies which gave food supplements during pregnancy have been carried out or are currently in progress in Guatemala, Taiwan, Montreal, Los Angeles, and New York. Preliminary data from these studies have been summarized in a National Academy of Science Research Council (U.S.) workshop report (Jacobson, 1974) and some recent reports (see Katz et al., 1975). In general, these studies indicate that mothers receiving supplemental energy and/or protein intakes during pregnancy will have heavier babies than unsupplemented mothers.

The prevalence of low-birth-weight babies (<2500 g), including both prematurely born and underweight term babies, ranges between 13 and 43% among lower socioeconomic classes in many countries. Since small babies are at high risk for increased neonatal and postnatal morbidity and mortality, and may have impaired physical and mental development, the high prevalence of low-birth-weight babies is a major public health problem. In many poor communities, maternal nutrition may be poor, but gross malnutrition is not obvious. To test the effect of improved maternal nutrition on birth weight in poor communities, the INCAP group supplemented the diet of pregnant women in four Guatemalan rural villages. Two types of food supplements were used. Two villages received a refreshing cool drink providing calories, vitamins, and iron (Fresca), while two received a gruel providing approximately 3 times as many calories, with similar amounts of vitamins and iron, but also containing about 6 g protein/100 ml (Atole). The average unsupplemented daily dietary intake during pregnancy in these villages approximated 1500 calories and 40 g of protein. The average mother weighed 49 kg at the end of the first trimester and gained a total of about 7 kg during pregnancy. The average birth weight of babies in such rural Guatemalan villages was about 3000 g and one third of the babies weighed <2500 g at term. The INCAP group found a consistent association between supplemental calories (including protein calories) and increased birth weight: 29 g for every additional 10,000 calories consumed throughout the pregnancy. For mothers consuming >20,000 supplemental calories, 9% of the babies weighed less than 2500 g at term; while for mothers consuming <20,000 supplemental calories, the incidence was 19%. The effect of supplemental calories was evident after statistically controlling for the maternal home diet, height, head circumference, parity, duration of disease during pregnancy, socioeconomic status, gestational age, and missing data. Lechtig *et al.* (1975) concluded that caloric supplementation during pregnancy *caused* an increase in birth weight and reduced the incidence of low-birth-weight term babies in their population. This important study indicates that nutritional supplementation during pregnancy, including carbohydrate and protein calories, may reduce the numbers of low-birth-weight babies and secondarily reduce the high neonatal and postnatal mortality in poor communities when the average daily intake otherwise might not exceed 1500 calories.

### 2.2.2. Protein

*Nitrogen Balance.* Based on balance studies, Calloway (1974) found a linear relation between energy intake and nitrogen balance. Nitrogen apparently is used at 25–30% efficiency in practical diets of pregnant women. Calloway reported that the average retention during pregnancy for 70 balances was 51 ± 40 SD mg/kg/day at mean intakes of 52 ± 9 kcal (0.28 ± 0.04 MJ)/kg and 271 ± 93 mg/kg nitrogen (1.7 g protein/kg/day). Summarizing 273 nitrogen-balance studies during pregnancy reported by different investigators, Calloway concluded that after 20 weeks of gestation the average nitrogen balance was 1.57 ± 1.72 (SD) g/day for women between 48 and 72 kg. Correcting for unmeasured losses, amounting to about 0.5 g/day, the corrected average nitrogen balance for the *last half* of pregnancy was 1.1 g/day. This is equivalent to a deposition of about 92 g of nitrogen or 572 g of protein during that interval. Since about 50 g of nitrogen can be accounted for by fetus and placenta (Table I), 42 g of nitrogen must have been incorporated in maternal tissues during the last 20–24 weeks of pregnancy. This would be equivalent to 1.26 kg of lean tissue. About ⅔ of the maternal tissue increment occurs during the last half of pregnancy. Extrapolating from Hytten and Leitch's data in Table I, 1.2 kg of lean tissue (61 g N) would be deposited, consistent with the N balance estimate derived from Calloway's data. About 7% of protein intake is used for energy. Thus at 2000 kcal (8.4 MJ) intakes with 14 g of nitrogen, dietary protein could contribute 18% of dietary energy. At higher energy intakes, relatively less diet protein would be used for energy.

*Protein Intake.* Protein allowances for pregnant women differ from country to country, although presumably the same data were used to derive the values. Reference body weights range from 45 kg (India) to 58 kg (Australia and U.S.). The protein allowance for nonpregnant women, assuming 60–70% utilization, is between 0.7–1.00 g/kg/day. For pregnancy, 0.09 are added (0.22 g/kg/day in U.K.) WHO is more generous, recommending an additional 0.75 g/kg. The total recommended daily protein intakes during pregnancy range from 48 g (Canada) to 75 g (Guatemala) (Table III). Recommended daily allowances for U.S. women are given in Table IV.

It is usually assumed that dietary protein is used with 60–70% efficiency. This is questionable. Based on Calloway's analysis of the published nitrogen balance data, the efficiency of nitrogen utilization may be only 25–30%, not 60–70%. There is, however, considerable variation in the efficiency of nitrogen utilization by different women. Nitrogen balance can be estimated from protein intake:

$$\text{N balance (g/day)} = -1.73 + 0.30 \left( \frac{\text{Protein intake, g/day}}{6.25} \right)$$

Where no protein is fed, urinary nitrogen losses are about 1.0 g/day.

In spite of the accuracy and care with which the reported nitrogen balances have been done, they only represent fragments of the gestational period and are rarely applicable to the individual pregnant woman not living in a metabolic ward. Calloway points out that only a few studies (12 subjects, 96 balances) recorded nitrogen balance continuously throughout the last 10 weeks of pregnancy. In those cases, birth weights did not correlate with nitrogen balance.

*Energy plus Protein Calories.* In a prospective study of the relation between maternal nutrition and fetal development, 24-hr recall and 3-day food records of dietary intake were obtained as one of the indicators of maternal nutrition. Food models were used to help estimate portion sizes. Among 262 mothers (to December 1976) with dietary histories obtained at about 24 weeks of pregnancy, 156 had delivered babies and had complete data. For these mothers the mid-pregnancy intakes averaged per kilogram per day $32 \pm 12$ (SD) kcal and $1.2 \pm 0.5$ g protein. There was a high degree of association between the protein and the calorie intakes, with a correlation coefficient of 0.786. For these mothers there appeared to be a clear association between the diet protein and calorie intake and the adjusted birth weight of the infant when a regression based on these relations was used to estimate

*Table III. Daily Protein and Energy Allowances for Women in Several Countries[a]*

| Country and year | Reference body weight (kg) | Protein (g/kg)[b] | | Total (g) | Calories (kcal/kg) | | Total kcal |
|---|---|---|---|---|---|---|---|
| | | Nonpregnant | Pregnant[c] | | Nonpregnant | Pregnant[c] | |
| Australia, 1965 | 58 | 1.00 | 1.14 | 66 | 36 | 39 | 2250 |
| Canada, 1964 | 56 | 0.70 | 0.85 | 48 | 43 | 52 | 2900 |
| Colombia, 1955 | 55 | 1.09 | 1.31 | 72 | 36 | 40 | 2200 |
| Guatemala, 1969 | 55 | 1.18 | 2.36 | 75 | 36 | 40 | 2200 |
| India, 1968 | 45 | 1.00 | 1.22 | 55 | 49 | 56 | 2500 |
| Philippines, 1970 | 49 | 1.12 | 1.32 | 65 | 39 | 47 | 2300 |
| United Kingdom, 1969 | 55 | 1.00 | 1.09 | 60 | 40 | 44 | 2400 |
| United States, 1974 | 58 | 0.79 | 1.31 | 76 | 36 | 41 | 2400 |
| FAO/WHO, 1972 | 55 | 0.75 | 1.01 | 55 | 40 | 46 | 2550 |
| Average: | 54 | 0.96 | 1.29 | 63 | 39 | 45 | 2411 |

[a]Modified from Calloway, 1974.
[b]Includes correction for assumed 60–70% utilization of protein.
[c]Calculated from standard allowance, irrespective of body weight.

Table IV.   Food and Nutrition Board, National Academy of Sciences–National

| | Age (years) | Weight | | Height | | Energy (kcal)[c] | Protein (g) | Fat-soluble vitamins | | | |
|---|---|---|---|---|---|---|---|---|---|---|---|
| | | | | | | | | Vitamin A activity | | Vita-min D (IU) | Vitamin E activity[f] (IU) |
| | | (kg) | (lb) | (cm) | (in) | | | (RE)[d] | (IU) | | |
| Infants | 0.0–0.5 | 6 | 14 | 60 | 24 | kg × 117 | kg × 2.2 | 420[e] | 1400 | 400 | 4 |
| | 0.5–1.0 | 9 | 20 | 71 | 28 | kg × 108 | kg × 2.9 | 400 | 2000 | 400 | 5 |
| Children | 1–3 | 13 | 28 | 86 | 34 | 1,300 | 23 | 400 | 2000 | 400 | 7 |
| | 4–6 | 20 | 44 | 110 | 44 | 1,800 | 30 | 500 | 2500 | 400 | 9 |
| | 7–10 | 30 | 66 | 135 | 54 | 2,400 | 36 | 700 | 3300 | 400 | 10 |
| Males | 11–14 | 44 | 97 | 158 | 63 | 2,800 | 44 | 1,000 | 5000 | 400 | 12 |
| | 15–18 | 61 | 134 | 172 | 69 | 3,000 | 54 | 1,000 | 5000 | 400 | 15 |
| | 19–22 | 67 | 147 | 172 | 69 | 3,000 | 54 | 1,000 | 5000 | 400 | 15 |
| | 23–50 | 70 | 154 | 172 | 69 | 2,700 | 56 | 1,000 | 5000 | | 15 |
| | 51+ | 70 | 154 | 172 | 69 | 2,400 | 56 | 1,000 | 5000 | | 15 |
| Females | 11–14 | 44 | 97 | 155 | 62 | 2,400 | 44 | 800 | 4000 | 400 | 12 |
| | 15–18 | 54 | 119 | 162 | 65 | 2,100 | 48 | 800 | 4000 | 400 | 12 |
| | 19–22 | 58 | 128 | 162 | 65 | 2,100 | 46 | 800 | 4000 | 400 | 12 |
| | 23–50 | 58 | 128 | 162 | 65 | 2,000 | 46 | 800 | 4000 | | 12 |
| | 51+ | 58 | 128 | 162 | 65 | 1,800 | 46 | 800 | 4000 | | 12 |
| Pregnant | | | | | | +300 | +30 | 1,000 | 5000 | 400 | 15 |
| Lactating | | | | | | +500 | +20 | 1,200 | 6000 | 400 | 15 |

[a]From *Recommended Dietary Allowances, 8th ed., National Academy of Sciences, Washington, D.C., 1974.*
[b]The allowances are intended to provide for individual variations among most normal persons as they live in the United States under usual environmental stresses. Diets should be based on a variety of common foods in order to provide other nutrients for which human requirements have been less well defined. See text for more detailed discussion of allowances and of nutrients not tabulated. See Table I (p. 6) for weights and heights by individual year of age.
[c]Kilojoules (kJ) = 4.2 × kcal.
[d]Retinol equivalents.
[e]Assumed to be all as retinol in milk during the first six months of life. All subsequent intakes are assumed to be half as retinol and half as β-carotene when calculated from international units. As retinol equivalents, three fourths are as retinol and one fourth as β-carotene.

birth weight. Of the 156 babies, 27 had adjusted birth weight residuals which were more than 1 standard deviation below the mean value and were considered smaller than they should have been considering maternal factors, baby sex, and gestational age. A similar number of babies were more than 1 SD above the expected mean and were considered large for gestational age. Seventeen of the small babies were born to 17 of 66 mothers whose energy and protein intakes were low; only five were born to 69 mothers with normal or high values for protein and energy intake; three of the babies had mothers with high calorie but low protein intakes. The proportion of small babies delivered to mothers having low protein-energy intakes was significant. To the extent that the 3-day diet records and recall histories actually indicated diet intake among these mothers, this study demonstrates an association between low protein and calorie intake and low birth weight, after the latter is adjusted for possible conditioning variables. This prospective study supports the association between maternal protein–energy intake during pregnancy and birth weight (Crosby *et al.,* 1977).

## 2.2.3. Amino Acids

Levels of most essential amino acids (AA) fall during pregnancy (Young, 1969). Metabolism of the glucogenic amino acids may be particularly relevant for fetal growth. Churchill *et al.* (1969) found a highly significant relationship between

| Water-soluble vitamins | | | | | | | Minerals | | | | | |
|---|---|---|---|---|---|---|---|---|---|---|---|---|
| Ascor-bic acid (mg) | Folacin[g] ($\mu$g) | Nia-cin[h] (mg) | Ribo-flavin (mg) | Thia-min (mg) | Vita-min $B_6$ (mg) | Vita-min $B_{12}$ (mg) | Cal-cium (mg) | Phos-pho-rus (mg) | Iodine ($\mu$g) | Iron (mg) | Mag-nesi-um (mg) | Zinc (mg) |
| 35 | 50  | 5   | 0.4  | 0.3  | 0.3 | 0.3 | 360  | 240  | 35  | 10  | 60  | 3  |
| 35 | 50  | 8   | 0.6  | 0.5  | 0.4 | 0.3 | 540  | 400  | 45  | 15  | 70  | 5  |
| 40 | 100 | 9   | 0.8  | 0.7  | 0.6 | 1.0 | 800  | 800  | 60  | 15  | 150 | 10 |
| 40 | 200 | 12  | 1.1  | 0.9  | 0.9 | 1.5 | 800  | 800  | 80  | 10  | 200 | 10 |
| 40 | 300 | 16  | 1.2  | 1.2  | 1.2 | 2.0 | 800  | 800  | 110 | 10  | 250 | 10 |
| 45 | 400 | 18  | 1.5  | 1.4  | 1.6 | 3.0 | 1200 | 1200 | 130 | 18  | 350 | 15 |
| 45 | 400 | 20  | 1.8  | 1.5  | 2.0 | 3.0 | 1200 | 1200 | 150 | 18  | 400 | 15 |
| 45 | 400 | 20  | 1.8  | 1.5  | 2.0 | 3.0 | 800  | 800  | 140 | 10  | 350 | 15 |
| 45 | 400 | 18  | 1.6  | 1.4  | 2.0 | 3.0 | 800  | 800  | 130 | 10  | 350 | 15 |
| 45 | 400 | 16  | 1.5  | 1.2  | 2.0 | 3.0 | 800  | 800  | 110 | 10  | 350 | 15 |
| 45 | 400 | 16  | 1.3  | 1.2  | 1.6 | 3.0 | 1200 | 1200 | 115 | 18  | 300 | 15 |
| 45 | 400 | 14  | 1.4  | 1.1  | 2.0 | 3.0 | 1200 | 1200 | 115 | 18  | 300 | 15 |
| 45 | 400 | 14  | 1.4  | 1.1  | 2.0 | 3.0 | 800  | 800  | 100 | 18  | 300 | 15 |
| 45 | 400 | 13  | 1.2  | 1.0  | 2.0 | 3.0 | 800  | 800  | 100 | 18  | 300 | 15 |
| 45 | 400 | 12  | 1.1  | 1.0  | 2.0 | 3.0 | 800  | 800  | 80  | 10  | 300 | 15 |
| 60 | 800 | +2  | +0.3 | +0.3 | 2.5 | 4.0 | 1200 | 1200 | 125 | 18+ | 450 | 20 |
| 80 | 600 | +4  | +0.5 | +0.3 | 2.5 | 4.0 | 1200 | 1200 | 150 | 18  | 450 | 25 |

[f]Total vitamin E activity, estimated to be 80 percent as α-tocopherol and 20 percent other tocopherols. See text for variation in allowances.

[g]The folacin allowances refer to dietary sources as determined by *Lactobacillus casei* assay. Pure forms of folacin may be effective in doses less than one fourth of the recommended dietary allowances.

[h]Although allowances are expressed as niacin, it is recognized that on the average 1 mg of niacin is derived from each 60 mg of dietary tryptophan.

[i]This increased requirement cannot be met by ordinary diets; therefore, the use of supplemental iron is recommended.

maternal blood total amino acids levels (alpha amino N) and fetal outcome as measured by birth weight, cranial volume, and mental and motor Bayley scores of the infants at 8 months of age. The fetus is dependent for energy principally upon glucose transported across the placenta. The glucogenic amino acids such as alanine provide a continuous and major source of glucose for the fetus. The free AA entering the gluconeogenic pathway via pyruvate (alanine, threonine, glycine, serine, and cystine) account for 30–40% of the total plasma free AA usually measured in nonpregnant women. At 24 weeks of gestation and at term, they account for 35–50% of the plasma free AA. The maternal plasma AA pattern (aminogram) at approximately 24 weeks of pregnancy (Table V) appears to be related to fetal growth (McClain *et al.*, 1976). Generally cigarette-smoking mothers have lower plasma AA levels (Crosby *et al.*, 1977). However, even when mid-pregnancy AA levels are statistically controlled for smoking, a combined set of several AA correlates with the size of the baby born some 16 weeks later (Metcoff *et al.*, 1976). Recently, McClain and Metcoff (1977) have shown a relationship between maternal plasma free AA levels at 25 weeks of gestation and cord blood AA levels at term. Both maternal 25-week AA and cord blood AA levels were correlated with the size of the baby at birth.

The metabolism of AA is closely related to gluconeogenesis. If glucose require-ments of mother, placenta, and fetus are to be met during pregnancy, maternal gluconeogenic processes must keep pace with demands. During a short 3- to 4-day fast, maternal blood glucose levels fall, while urea nitrogen excretion, reflecting hepatic gluconeogenesis, fails to increase sufficiently (Felig and Lynch, 1970; Felig

Table V.   Free Amino Acid Concentrations in Plasma of 79 Women at 22.8 ± 3.1 Weeks of Gestation[a]

|  | μmol/liter | |
|---|---|---|
|  | m | SD |
| Glutamic acid[b] | 377 | 80 |
| Alanine | 361 | 71 |
| Proline | 188 | 63 |
| Threonine[c] | 179 | 43 |
| Glycine | 178 | 46 |
| Valine[c] | 178 | 36 |
| Lysine[c] | 168 | 45 |
| Tryptophane[c] | 161 | 57 |
| Serine[c] | 132 | 51 |
| Arginine | 119 | 56 |
| Leucine[c] | 118 | 36 |
| Histidine[c] | 93 | 18 |
| Phenylalanine[c] | 68 | 20 |
| Isoleucine[c] | 66 | 18 |
| Tyrosine | 62 | 16 |
| Cystine[c] | 49 | 33 |
| Aspartic acid | 43 | 19 |
| Ornithine | 33 | 13 |
| Methionine[c] | 23 | 10 |

[a]From McClain and Metcoff (1977).
[b]Glutamic acid + glutamine.
[c]Essential amino acid.

*et al.*, 1970). The gestational hypoglycemia occurred in spite of hypoinsulinemia and hyperketonemia, which should have enhanced hepatic gluconeogenesis. Felig *et al.* (1972) found that pregnancy exaggerates the hypoalaninemic and hyperglycinemic effect of starvation. The hyperaminoacidemia of starvation may be a consequence of hypoinsulinemia. Presumably lack of key endogenous substrates, rather than altered intrahepatic processes, limit hepatic gluconeogenesis in pregnancy.

Alanine is the primary endogenous glucogenic substrate released by muscle and extracted by the liver during starvation. Levels of this amino acid fall during starvation in nonpregnant (Felig *et al.*, 1969) and pregnant (Felig *et al.*, 1972) subjects. Since, quantitatively, alanine concentrations are among the highest in plasma, a fall in plasma alanine may impair hepatic gluconeogenesis and lead to maternal hypoglycemia. As maternal plasma alanine levels fall, the level in the amniotic fluid also falls. When alanine was infused, maternal plasma alanine and glucose levels rose significantly.

During a prolonged (84- to 90-hr) fast (prior to abortion) in pregnant women at 16–22 weeks gestation, hypoglycemia, hyperketonemia, and an increase in glycerol and free fatty acid levels were noted. The concentration of glucose in the amniotic fluid also fell, while acetoacetate and hydroxybutyrate increased 10- to 30-fold, approximating maternal plasma levels (Kim and Felig, 1972). These data indicate that under extreme conditions of food deprivation, at least, glucose is less available to the fetus.

Felig found that glycine, threonine, and serine, like ketones, rose during prolonged starvation in pregnancy. Hyperglycinemia has been reported in protein-depleted mothers and newborns (Linblad *et al.*, 1969).

Amino acid deficiencies, *per se,* may be of less practical importance than amino acid imbalances. Relative excess of one or several amino acids may interfere with the utilization of others for growth or cell function (Harper *et al.*, 1970). For example, alanine, intimately associated with gluconeogenesis, attains one of the highest plasma concentrations of all free amino acids during pregnancy. Alanine, at high concentrations, is an allosteric inhibitor of the rate-limiting glycolytic enzyme, pyruvic kinase (Mameesh *et al.*, 1976), hence might limit the utilization of glucose by mother and fetus. Imbalance of other AA conceivably could adversely modify glucose utilization and protein synthesis by fetal tissues. Relevant data from humans are not available.

Amino acids are not necessarily exchanged freely and equally in both directions across the human placenta. The fetus tends to retain tryptophan, aspartic acid, crysteine, methionine, and phenylalanine out of the AA reaching it near term, while releasing relatively large proportions of glutamic acid and glutamine into the efferent umbilical blood (Velazquez *et al.*, 1976).

### 2.2.4. Bulk Minerals

*Sodium.* The minimal sodium requirement needed during pregnancy to support fetal and placental growth is unknown. The fetus, at term, contains about 280 mEq Na (Widdowson, 1974). Assuming that placental transport of water (containing Na at 140 mEq/liter) approximates 129.6 liters/24 hr (Dancis *et al.*, 1961) and the efficiency of Na transport is 25%, then 4536 mEq of Na reach the fetus daily near term. Since the fetal accretion of Na averages about 1 mEq/day, 99.98% of the transported Na is returned to the mother. Thus, unless hyponatremia is extreme, or some extraordinary limitation in placental blood flow occurs, and transport of Na across the placenta almost ceases, the maternal ECF concentration of Na provides a vast excess for fetal growth.

Use of diuretics during pregnancy may produce hyponatremia and alter placental blood flow, but it is difficult to see how the vast excess of Na available to the fetus could be significantly reduced. Nonetheless, hyponatremia has been observed in the newborn following thiazide-induced diuresis (Lindheimer and Katz, 1973).

The toxemic patient is edematous and retains abnormal amounts of sodium. However, the babies of toxemic mothers usually are not excessively edematous, but they may have fetal malnutrition which usually is characterized by decreased total body water and sodium content.

*Potassium.* According to Hytten and Leitch (1971) about 200 mEq of K are accumulated by the fetus (154 mEq), placenta (42 mEq), and amniotic fluid (3 mEq). An additional 120 mEq are deposited in the enlarging maternal uterus, breasts, and body fluids. The total of 320 mEq represents an average rate of accretion of 1.1 mEq K/day during pregnancy. The cytoplasmic K/N ratio would be 2.2 mEq/g. Calloway (1974) reported average potassium accretion by eight teenage pregnant girls as 3.4 ± 0.9 SD mEq/day, as determined by $^{40}$K whole-body counter determinations. The observed N retention was 2.42 ± 1.8 SD g/day. The estimated K/N ratio was 1.4 or, corrected for unmeasured N loss of 0.5 g/day, 1.8, a rather low figure. From Widdowson's data (1974*b*) the K/N ratio in the developing fetus during the last half of gestation ranges from 2.3 to 2.9 mEq/g; the fetus accretes 150 mEq K to 40 weeks. Assuming approximate urine excretion of 2–30 mEq and fecal loss of 10 mEq daily by the mother, about 50 mEq K would be required to satisfy the observed balance of 3–4 mEq/day of which 0.5 mEq/day would be deposited in the fetus, and 0.2 mEq/day in the placenta.

Assuming transport of 129.6 1 of $H_2O$ containing 5 mEq/liter of K across the placenta daily, with efficiency of K transport of 25%, then 162 mEq would be available to the fetus daily. Like Na, this is a vast excess relative to K accretion by the fetus.

*Calcium and Phosphorus.* The calcification of the cartilaginous fetal skeleton begins at about the 8th week. From Widdowson's data (1974*b*), by 26 weeks, the 1000-g fetus contains 6.0 g Ca. This increases to 30 g by 40 weeks, and about 98% of this is in the bones. Although the National Research Council recommends an additional 0.4 g calcium/day during pregnancy, the total recommended intake of 1.2 g calcium/day is contained in a quart (0.97 liter) of milk (Table IV).

Phosphates are the principal anionic component of cells. At term the fetus contains about 17 g of phosphorus, and 80% of this is in the bone. Widdowson's data show that the Ca/P ratio at term is between 1.7 and 1.8. While the optimal quantities of Ca and P for pregnant women have not been determined, it is likely that intakes of foods containing good quantities of Ca and Mg, without excessive phosphate, will suffice.

Calcium metabolism is altered extensively in pregnancy in response to changes in hormonal balance and fetoplacental exchanges (Pitkin, 1975*a*). Bone turnover is increased progressively, as a consequence of increasing production of human chorionic somatomammotropin. Bone resorption is inhibited by estrogen with a resultant increased parathyroid hormone secretion, which, in turn, increases intestinal absorption and urinary excretion of Ca, to sustain serum levels. Calcitonin secretion also increases and inhibits excessive bone resorption. The net effect appears to be progressively positive Ca balance, provided Ca and vitamin D intakes are adequate. Serum total Ca levels fall progressively during pregnancy, but this results from a progressive fall in serum albumin, since about one half the serum Ca is bound to albumin. As serum albumin falls, largely by dilution with increased plasma and extracellular fluid volumes, the bound Ca also decreases. The remaining Ca is unbound, ionized ($Ca^{2+}$), and does not change appreciably during pregnancy (Pitkin, 1975). PTH enhances the absorption of $Ca^{2+}$. The placenta transports $Ca^{2+}$ from the maternal to fetal side, but neither PTH nor calcitonin cross the placenta, although vitamin D and its metabolites do. Fetal levels of $Ca^{2+}$, $PO_4^{2-}$, and calcitonin are higher than maternal levels; vitamin D levels are similar; PTH levels are lower. The vitamin D levels in mother and fetus depend on maternal dietary intake of vitamin $D_3$ and its subsequent hydroxylation to the metabolically active form 25-$OHD_3$ in liver and then to $1,25(OH)_2D_3$ by the kidney (Tsang, 1976).

## 2.2.5. Trace Minerals

*Iron.* Iron in the term fetus amounts to 280–300 mg (Widdowson, 1968). As the maternal blood volume expands, increased hemopoiesis to sustain hemoglobin concentration requires about 500 mg Fe. Thus, about 1 g of ferrous iron is required to prevent anemia. Iron stores in American women average approximately 300 mg. The usual diet of pregnant American women provides only about 10–15 mg iron/day, and only about 10–20% of this is absorbed. A significant portion of the 1–3 mg absorbed is subsequently lost in excreta. Unless additional iron is provided, anemia is a common complication of pregnancy (Pitkin *et al.*, 1972). Anemia is preventable by administration of 30–60 mg/day of supplemental ferrous iron to all pregnant women.

*Zinc.* Zinc intake may be marginal in pregnancy. Zinc is essential for nucleic acid synthesis (Sanstead and Rinaldi, 1969; Sandstead *et al.*, 1972; Terhune and Sandstead, 1972; Fosmire *et al.*, 1975), protein synthesis (Hsu *et al.*, 1969), and probably for the utilization of dietary protein (Somers and Underwood, 1969). Zinc is a metal constituent of many enzymes having important metabolic regulatory functions. Women should accumulate approximately 750 $\mu$g of zinc daily during the latter two thirds of pregnancy so that the zinc requirements of the fetus can be met (Sandstead, 1973). Analysis of self-selected diets of college women and high school girls (White, 1969) indicates that many of the women were not consuming sufficient zinc to meet the estimated requirements for pregnancy. As yet, it is unknown how well the above estimates relate to the larger population of pregnant women in this country. Plasma zinc is bound to albumin. Since albumin does not cross the placenta, presumably some zinc is not covalently bound to protein, for 50–60 mg accumulate in fetal tissue by term. About half this amount is contained in the liver of the fetus, either bound to proteins or incorporated with metalloenzymes.

Plasma zinc levels decrease substantially during pregnancy, from a normal range of 80–100 $\mu$g/100 ml to a mean of 61 $\mu$g/100 ml (Table II, Metcoff *et al.*, 1976). The decrease occurs along with the decrease in serum albumin, the slopes of the regression lines being almost identical. Whether the total quantity of zinc in maternal plasma falls is unknown. It is possible that the total pool of zinc is not reduced during pregnancy, but extracellular water (and plasma) concentrations are diluted by the expanded fluid volume. The plasma zinc concentration of the newborn may be within the normal nonpregnant adult range, while zinc levels in mothers are depressed (Henkin *et al.*, 1971).

Jameson (1976) reported a linear decrement of plasma zinc during gestation and an association between plasma zinc levels, baby size, and incidence of congenital malformations. Favier *et al.* (1972) described a linear relation between the levels of zinc in amniotic fluid and birth weight of the baby in 53 mother–baby pairs. The level of zinc in the amniotic fluid was less influenced by the duration of gestation (when the sample was obtained as early as 27 weeks) than by the ultimate weight of the baby born after 38 weeks. Neither age nor parity of the mother affected the results. The presence of meconium in the amniotic fluid greatly increased the zinc level. There were no significant differences between boys and girls for regression of amniotic fluid zinc level on birth weight. The combined equation was

$$y = 0.194x - 325.57, \qquad \pm 2.01 \text{ SE}, \qquad r = 0.769$$

where $x$ = zinc concentration in $\mu$g/liter and $y$ = birth weight in grams. The decrease in plasma zinc concentration with pregnancy has been found to be greater in rural women from a low socioeconomic stratum in Iran than in women from the city (Sarram *et al.*, 1969). Presumably, this latter finding was related to the diet of the women, zinc content of rural diets being lower. The incidence of congenital malformation of the nervous system is higher in countries where the commonly consumed diet is low in zinc and similar to that of the Iranian and Egyptian villagers (Sever and Emanuel, 1972). These investigators have speculated that some of these anomalies may be related to maternal zinc deficiency.

In experimental animals, prenatal zinc deficiency is teratogenic. Congenital malformations in zinc-deficient embryos may result from impaired synthesis of nucleic acids (Hurley, 1976, 1977). Recently, Sandstead *et al.* (1977) reported that zinc deficiency during the last trimester of pregnancy in rhesus monkeys adversely affected mother–infant interactions during the suckling period. The infants were more timid, more dependent, less active, and less exploratory. In rats, the postnatal effects of prenatal zinc deprivation from days 14–19 of gestation led to abnormal behavior in adult life. Male rats showed impaired avoidance conditioning responses, while females showed increased aggression to shock (Halas *et al.*, 1975).

*Copper.* The fetus at term contains about 14 mg of copper, half being in the liver (Widdowson, 1974*b*). Apparently a copper–protein complex is formed, mainly within liver cell mitochondria. At term, the newborn liver contains one and a half times more copper than the adult liver, and the concentration (per 100 g tissue) is more than 10 times that of the adult.

Serum copper levels almost double during pregnancy, and this is thought to be caused by estrogen secretion. The Cu concentration in cord blood is about one fifth that in maternal blood at term (Schenker *et al.*, 1972). Copper content of the placenta was found to be constant, irrespective of placental weight (Nassi *et al.*, 1970). Blood copper levels gradually increase in the fetus from 4 months to term and are correlated with rising ceruloplasmin levels (Canzler *et al.*, 1972). Low serum copper levels may indicate inadequate hepatic synthesis of ceruloplasmin. The Oklahoma study shows significantly rising values of plasma Cu between 24 weeks and term (Table II), but the increase is not as great as that observed by Schenker.

In contrast to zinc, the levels of copper, lead, and magnesium in the amniotic fluid do not change appreciably during pregnancy (Favier, 1972). A genetic disease of children associated with kinky hair, slow growth, early mortality, and cerebral degeneration is known to involve defective copper metabolism (Menkes, 1972). Abnormalities in brain, skin, hair, and increased neonatal mortality have been observed in the offspring of copper-deficient rats (Hurley and Bell, 1975). Impaired myelination, blindness, and ataxia have been observed in lambs of copper-deficient ewes (Underwood, 1971).

*Chromium.* Recently interest has been focused on this trace metal because of its role in carbohydrate metabolism. It acts as a cofactor for insulin, apparently by increasing the affinity of cell membrane receptors for insulin. Chromium in foods may be complexed with nicotinic acid and three amino acids to form a "glucose-tolerance factor" (GTF). Neither the precise structure nor function of this factor is known to date (Mertz, 1976). While plasma and hair chromium levels do not change appreciably in a single pregnancy, repeated pregnancies lead to decreased chromium stores in hair and might indicate gradual depletion (Hambridge, 1971, 1974). Children with protein–calorie malnutrition (PCM) may have impaired glucose tolerance which can be normalized by increasing chromium intake (Gurson and Saner, 1971). GTF is contained in brewer's yeast, meats, whole grains, and cheese. The chromium requirement for man is unknown but might approximate 5–60 $\mu$g/day (Mertz, 1976). Gestational diabetes may be related to maternal chromium deficiency, as suggested by impaired chromium response to glucose load. (Davidson and Burt, 1973).

*Iodine.* Maternal iodine deficiency may lead to cretinism in goitrogenic regions, which can be prevented by maternal iodine supplementation (Clements, 1960). Since some women in goiter regions do not produce cretins, it is likely that there is some interaction between nutrient deficiency of iodine, susceptibility to goitrogenic factors, and genetic characteristics leading to cretinism in the neonate (Hurley, 1976).

*Manganese.* The role of manganese in human fetal development, and the requirement for the trace mineral during pregnancy, are unknown. Manganese is a component of metalloenzymes and functions as an enzyme activator for many types of metabolically important enzymes, e.g., kinases, thioesterases, peptidases, decarboxylases, adenyl cyclase, etc. A manganoprotein complex, pyruvate carboxylase, catalyzes the pyruvate-to-oxalacetate step, with ultimate formation of phosphoenolpyruvate, in gluconeogenesis; however, it is not known whether the Mn–pyruvate carboxylase complex is biologically significant (Utter, 1976). In several species of experimental animal (chick, guinea pig, pig, rat, and mouse), the most striking effect of prenatal manganese deficiency is postnatal ataxia, with incoordination, lack of equilibrium, and head retraction. Deficiency of manganese during days 14–18 of gestation in the rat will produce these changes and is characterized by abnormal and reduced otolith formation in the inner ear (Erway *et al.*, 1971). Genetic background may modify the effect of manganese deficiency, at least in mice (Hurley and Bell, 1974) and in mink (Erway and Mitchell, 1973). A major effect of manganese deficiency may be interference with the synthesis of mucopolysaccharides (Hurley *et al.*, 1968). The role of $Mn^{2+}$ in mucopolysaccharide biosynthesis is still uncertain. It may be important for synthesis of glycoproteins, by activating glycosyltransferases (Utter, 1976).

### 2.2.6. Vitamins

There is little evidence that vitamin supplementation is required for pregnant women consuming normal diets meeting the currently recommended daily allowances (RDAs) (Table IV). Vitamin $B_6$ may be an exception, especially in teenagers consuming erratic and poor diets (Kaminetsky *et al.*, 1973). Folate deficiency is not rare in pregnancy. To prevent this, supplements of 200–400 $\mu$g folic acid/day have been recommended.

*Fat-Soluble Vitamins.* Fat-soluble vitamins, A, D, K, and E, are stored principally in the liver, which serves as a reservoir during pregnancy. There is usually more danger from excessive intake than from deficiency during pregnancy.

Vitamin A (retinol) levels during pregnancy are variable, but levels of its precursor, carotene, fall progressively during pregnancy (Metcoff *et al.*, 1976). Both vitamin A and carotene cross the placenta, but fetal levels are lower than maternal levels (Clausen, 1938). Birth weight has been correlated with maternal levels of carotene at mid-pregnancy (Crosby *et al.*, 1977), but the significance of this association is unknown. Both deficiency and excess of vitamin A are teratogenic in experimental animals (Hurley, 1976). Congenital renal anomalies in a human infant have been ascribed to overenthusiastic intake of vitamin A (5–10 times RDA) by a pregnant mother (Bernhardt and Dorsey, 1974). Other congenital defects similar to those noted in experimental animals have been noted in human infants and ascribed to maternal vitamin A deficiency (McLaren, 1968) or excess (Bernhardt and Dorsey, 1974) during pregnancy.

Congenital vitamin D deficiency may impair dentition and increase the incidence of caries (Mellanby and Coumoulos, 1944). Excessive vitamin D during gestation in rats retards fetal and postnatal growth and osteogenesis (Portliege, 1972; Ornoy *et al.*, 1972). In rabbits, prenatal vitamin D intoxication produces supravalvular aortic stenosis, similar to the congenital defect of children (Friedman and Mills, 1969).

Vitamin E deficiency causes pregnant rats to abort, but no such relationship has been confirmed in human pregnancy. Alpha-tocopherol (vitamin E) is carried in the plasma by lipoproteins, and lipoprotein levels increase during pregnancy, perhaps accounting for the gradually increasing levels of alpha-tocopherol. Since the placental transport of the vitamin is limited, fetal serum levels of vitamin E are low relative to maternal levels. Prematurely born infants may have very low plasma levels of alpha-tocopherol. There is some suggestive evidence that a mild hemolytic anemia occurring in such infants is responsive to vitamin E therapy.

Vitamin K occurs naturally in plants ($K_1$) or is synthesized by intestinal bacterial flora ($K_2$). Deficiency is extremely rare in pregnancy. Oral administration of $K_1$ is not generally recommended, but it may be useful during the last week of pregnancy where hemorrhagic disease of the newborn is suspected.

*Water-Soluble Vitamins.* Thiamin ($B_1$), pyridoxine ($B_6$), riboflavin, pantothenic acid, $B_{12}$, folacin, and ascorbic acid (vitamin C) are stored only to a limited extent and are promptly excreted in the urine. Their blood concentrations decrease during pregnancy. The progressive decline may result from the gradual utilization of these vitamins in the fetus coupled with dilution by expansion of maternal body water and increased urinary excretion.

Erythrocyte transketolase activity is now used as a measure of thiamin sufficiency. Low levels of activity of this enzyme have been reported in 25–30% of pregnant women sampled (Heller *et al.*, 1974*a*). Low $B_1$ blood levels were found in 18–30% of teenage pregnancies studied by Kaminetsky *et al.* (1973) and were unaffected by vitamin supplements. No correlations have been observed between the maternal blood level of thiamine (or transketolase activity) and the outcome of pregnancy.

Pyridoxine deficiency is associated with low blood levels of $B_6$ and increased urinary excretion of xanthurenic acid following a test dose of tryptophan. By these criteria, pregnancy is commonly associated with biochemical evidence of $B_6$ deficiency. Blood levels of $B_6$ fall progressively during pregnancy to about ¼ of non-pregnancy levels (Coursin and Brown, 1961), while xanthenuric acid excretion increases 10–15 times (Wachstein and Graffeo, 1952). Supplementation with pyridoxine normalizes both parameters (Cleary *et al.*, 1975; Brown *et al.*, 1961). Pitkin (1975*b*) suggests that these biochemical changes in pregnant women reflect physiologic adjustments of some kind, rather than vitamin $B_6$ deficiency. Since fetal levels of the vitamin are derived from the mother, and the requirement for $B_6$ may be increased by high-protein diets, Pitkin advises a modest increase of $B_6$ intake during pregnancy.

Riboflavin excretion in the urine decreases (Brezezinski *et al.*, 1952) while erythrocyte glutathione reductase activity (an estimate of riboflavin utilization) increases progressively during pregnancy (Heller *et al.*, 1974*b*). These findings have been taken as biochemical evidence for gestational riboflavin deficiency. While fetal death or premature birth once were thought to be related to riboflavin deficiency, recent studies have not substantiated this.

Although maternal pantothenate deficiency is known to produce brain and eye malformations in the offspring of several animal species, similar observations have not been made in humans. Infants delivered to teenage mothers who had low urinary excretion and blood levels of pantothenate had lower cord blood pantothenate levels than did infants of pantothenate-normal mothers (Cohenour and Calloway, 1972).

Little is known concerning the requirements for $B_{12}$ during pregnancy, although serum levels, like those of other water-soluble vitamins, decline progressively during pregnancy. Although $B_{12}$-deficient pregnant rats may have offspring with low birth weight and hydrocephalus (Woodard and Newberne, 1967; Newberne and Young, 1973), similar associations have not been described in humans.

An association between maternal folate deficiency and congenital malformation (Fraser and Watt, 1964), including central nervous system malformation (Hibbard and Hibbard, 1966), has been reported, but more recent studies have not confirmed the early reports (Scott *et al.,* 1970; Hall, 1972). However, the duration and degree of folate deficiency may have differed. The increased erythropoiesis of pregnancy and the folate needed for DNA synthesis by rapidly replicating fetal and placental cells are thought to increase folate requirements during pregnancy (Kitay, 1969). However, current opinion indicates that folate deficiency during pregnancy is likely to produce maternal megaloblastic anemia but does not seriously impair fetal survival, growth, or development (Pitkin, 1976).

Ascorbic acid levels, like those of other water-soluble vitamins, decline progressively during pregnancy, with fetal levels higher than maternal levels (Hamil *et al.,* 1947; Mason and Rivers, 1971). An increased incidence of premature births was related to low intakes and low serum levels of ascorbic acid (Martin *et al.,* 1957). The possible association needs to be reexamined. The ready transfer of ascorbic acid across the placenta and the concentration of this relatively potent oxidizing agent by the fetus suggests that megadoses of this vitamin during pregnancy might have adverse effects. In pregnant experimental animals large doses of ascorbic acid adversely affect the fetus and may cause fetal death (Samborskaya and Ferdman, 1966).

Currently recommended daily allowances for nutrients during pregnancy are given in Tables III and IV. More detailed discussion of the possible teratogenic effects of nutrients is given by Hurley (1976).

# 3. Nutrition and Growth of the Fetoplacental Unit

## 3.1. Placental Nutrition

Development of the organism depends on cell replication, differentiation, and growth. In eukaryotic diploid cells, DNA is confined largely to the nucleus and is a measure of cell number. Enesco and Leblond (1962), Cheek (1968), Winick and Noble (1965), and many others have used the DNA content of a tissue not only to estimate cell number but also as an indicator of cellular replication or hyperplasia. The protein content of cells provides an estimate of the cellular mass, and RNA content indicates cytoplasmic mass. Cell growth involves an increase of cytoplasmic mass. Hence the growth phase of the cell may be identified by an increase in the protein/DNA ratio. The ratio of protein to DNA is an index of cell size and the RNA/DNA ratio of functional cytoplasmic mass per cell. Organ growth, like growth of the organism, is characterized by three more or less sequential cell development phases: hyperplasia, hypertrophy, and cessation of hyperplasia with continued hypertrophy. These sequential processes overlap, but can, in general, be followed by measurements of DNA, protein, and RNA in tissues.

Cell growth processes can be better understood by measurements of protein and RNA synthesis, an indication of the health of the cell. There are many different types of mammalian transfer, ribosomal, and messenger RNAs. Increased synthesis of ribosomal RNA suggests cell proliferation. On the other hand, increased synthesis of some types of messenger RNA suggests some final steps in cell differentiation or possibly in the transformation of cells by external (i.e., environmental) factors.

The process of cell differentiation is much more complex. The synthesis of specific proteins, such as enzymes and structural proteins, specifically determines the functional differentiation of the cell. Busch *et al.* (1964) have proposed that every cell of developed organisms contains various groups of operons. These operons or polyoperons, actively transcribe messenger RNAs which provide the information for specific aspects of cell metabolism, specialized functions of the cell, cell growth, and cell division. Enzymes regulating a single pathway in a specific type of cell may be encoded in one polyoperon, thereby comprising a genome. DNA formed within a cell nucleus does not appear to have a predetermined potential. DNA of a liver cell could, under proper circumstances, support the development of muscle or nerve cells. This has been demonstrated by transfer of nuclear material from one organ to a different organ of the developing embryo.

### 3.1.1. Biochemical Aspects of Fetoplacental Composition

Several aspects of the fetomaternal unit which are relevant to nutrition are discussed below. Details of placentology are described in Chapter 14.

The nutrient supply to the growing fetus depends on the quantity and composition of maternal blood reaching the placenta and upon the capability of the placenta to concentrate, synthesize, and/or transport essential nutrients from the maternal to the fetal side. Unused nutrients and metabolites also are transferred from the fetal to the maternal side. The nutrients are delivered to the fetus by the umbilical vessels. Circulatory alterations in the mother, such as those caused by hypertension or congenital heart disease, may either limit or increase the quantity of blood reaching the placenta per unit time. Diseases affecting the mother and the placenta, such as infections, may damage or limit transport capabilities. Hormonal effects, such as those occurring in diabetes, may alter placental functions. The DNA content of the placenta is related to the weight of the child at birth, being greater for children of larger size. Similarly the protein/DNA ratio and the RNA/DNA ratio increase with increasing birth weight. Placental DNA content does not appear to increase after 36 weeks of gestation (Winick, 1967). Adenine nucleotide levels, and activities of the energy-related enzymes adenylate kinase and pyruvate kinase, also increase in proportion to birth weight of the baby (Rosado *et al.,* 1972). Protein synthesis, estimated from measurements of [$^{14}$C]leucine incorporation, also increase in proportion to placental size, gestational age, and birth weight of the child (Laga *et al.,* 1972; Metcoff *et al.,* 1973). RNA polymerase activity of the placenta is inversely proportional to the birth weight of the baby and decreases with gestational age (Metcoff *et al.,* 1973). Chatterjee *et al.* (1976) have shown that synthesis of the placental hormones, human placental lactogen (HPL) and human chorionic gonadotropin (HCG), is achieved by free and membrane-bound polyribosomes, with the latter being several times more active in term placentas, but not in early (10–20 week) placentas. They concluded that the decreasing levels of HCG (from about 12–14 weeks to term) and increasing levels of HPL (from about 6–36 weeks) reflected the relative abundance of mRNAs being synthesized by the two types of placental polyribosomes during gestation. Cystine aminopeptidase (oxytocinase) also is synthesized by the placenta. Measurements in maternal serum of HPL (Hobbins, 1975), oxytocinase (Pathak *et al.,* 1974), and oxytocinase combined with urinary estrogen excretion (Petrucco *et al.,* 1973) have been correlated with poor fetal growth.

Although the hyperplastic phase of placental growth stops by 36 weeks, a higher fetal growth rate continues. How the two growth processes are interrelated and why placental hyperplasia stops, while fetal growth increases, is not known, nor are the control mechanisms responsible for these phenomena clear. At term, the ratio of fetal body weight to total placental weight ranges from 6 to 8. There is a positive correlation between placental mass, especially "functional" mass, and baby size at birth. The ratio of baby weight to placental weight is not constant, and the correlation, while significant, is not very high. The relationship may be confounded by crude measurements of an untrimmed placenta containing various amounts of adherent blood. Even when corrected for retained blood (blood-free placental weight), the ratio increases from 6 at 32 weeks to 9 at 42 weeks (Garrow and Hawes, 1971).

### 3.1.2. Transport and Synthesis of Nutrients

Carbohydrate is the principal metabolic fuel of the fetus and is provided in continuous supply by transfer of glucose from the mother through the placenta.

The placenta appears relatively impermeable to lipids, which circulate as constituents of large molecular size, e.g., lipoproteins, such as cholesterol, phospholipid, and triglycerides. Some fatty acids can cross the placenta, but apparently do so slowly and in small amounts. It is generally thought that the low levels of free fatty acids in the fetus and their slow, limited transfer across the placenta are consistent with the thesis that fat is not a main source of energy in the human fetus (Dancis *et al.*, 1973).

Protein molecules ranging in size from albumin to gamma globulins pass from the maternal to the fetal circulation via the placenta; however, relatively few intact maternal serum proteins are transferred to the human fetus during the first third or half of gestation. Such transfer is more efficient later in pregnancy (Dancis *et al.*, 1961). While protein synthesis is a prominent feature of placental metabolism, there is little or no evidence about whether proteins synthesized by the placental trophoblast are transported into the fetal circulation and utilized by the fetus. It is assumed that cell growth in the fetus results from fetal synthesis of amino acids into protein molecules.

The concentration of free amino acids generally is higher in fetal than in maternal blood. Cyst(e)ine appears to be unique. Maternal concentrations are higher than those of the fetus. The placenta in some way limits cyst(e)ine transport, even after intravenous loads (Gaull *et al.*, 1973). *In vitro* studies of placental amino acid uptake using the nonmetabolizable amino acid analog, alpha-aminoisobutyric acid, revealed that that process required protein synthesis and aerobic metabolism (Smith *et al.*, 1973). However, the uptake process appears to have two components: one to increase transport activity, which is demonstrated by preincubation of villous tissue in a salt–glucose solution; and the other to limit uptake, which occurs when amino acids, like alanine, are present in high intracellular concentrations. (Smith and Depper, 1974). Some recent evidence suggests that there is a differential bidirectional transport of some amino acids across the placenta. Aspartic acid, tryptophan, methionine, phenylalanine, serine, cysteine, lysine, glycine, and threonine appear to be significantly retained by fetal tissues as evidenced by venous–arterial differences in cord blood at term. Four amino acids, arginine, glutamic acid, proline, and glutamine, have higher concentration on the arterial than on the venous side of cord blood, indicating release of these amino acids from fetal tissues. Glutamic acid and glutamine were released in highest concentrations from fetal tissues. The highest relative retentions by the fetus were for tryptophan, aspartic acid, and phenylalanine (Velasquez *et al.*, 1976).

In the fetomaternal system, the levels of essential amino acids were significantly reduced in venous, as compared to arterial, blood of the mother, and they were higher in arterial cord blood. Amniotic fluid levels of essential amino acids were about equivalent to those in arterial blood of the mother and cord. Nonessential amino acids were higher in cord blood than in maternal arterial or venous blood. The highest concentrations of essential and nonessential amino acids were found in the placenta.

The ratio of essential to nonessential amino acids was significantly greater in the placenta than in other components of the fetomaternal system. This suggests either "trapping" of amino acids by the placenta (since the essential amino acids presumably were derived from maternal sources), a concentration of amino acids from maternal sources, or utilization of some nonessential amino acids by placenta and/or fetus. Levels of urea and creatinine rise in amniotic fluid during gestation, exceeding maternal blood levels which remain relatively constant (Lind et al., 1971). This suggests maturation of fetal renal function and a nonequilibration of amniotic fluid with maternal plasma late in pregnancy. Since levels of urea and creatinine are greater than normal in preeclamptic patients, it has been suggested that placental transport functions may be involved in the selective retention of these metabolites on the fetal side (Roopnarinesingh et al., 1971).

Water-soluble vitamins such as thiamine, folate, and $B_{12}$ readily cross the placenta. Vitamin C apparently crosses the human placenta readily in the form of dehydroascorbic acid, rather than ascorbic acid (Raiha, 1958). Vitamin A has been found in varying amounts in human fetal liver but is lower in fetal than in maternal blood. How vitamin A is transported across the placenta is not clear, nor is the mechanism of transport of vitamin D understood. It is not known whether the human fetal kidney can synthesize 1,25-dihydroxycholecalciferol or whether it is transported across the placenta.

With respect to inorganic ions, Widdowson (1974a) has pointed out that by term 500–1000 times as much sodium reaches the human fetus from the maternal circulation by placental transport as is required for growth. Of course most of this returns to the mother's circulation. Fetal swallowing of amniotic fluid and subsequent absorption and processing of the metabolites also contributes to the molecular disparity between amniotic fluid and maternal plasma. Subsequent excretion of a dilute urine has been considered a possible explanation for the lower osmolality and sodium concentration of amniotic fluid relative to maternal plasma at term (Lind et al., 1971). Five precent of the total calcium in the maternal plasma and 10% of plasma phosphorus are taken up by the fetus every hour in the last 3 months of gestation. This does provide a potential drain on maternal sources which usually can be easily continuously replenished by a quart of milk (or Ca and P equivalent food) daily, as mentioned previously.

The mechanism for transport of trace minerals is not clear since many of these are bound to specific plasma proteins. After the sixth month of gestation, the concentration of iron in fetal plasma rises, reaching levels nearly three times those of the mother at term and indicating effective transport to and retention by the fetus. Although the copper content of the placenta appears to be constant irrespective of placental weight, blood copper levels gradually increase in the fetus from 4 months to term and are correlated with rising ceruloplasmin levels (Canzler et al., 1972); Schenker et al., 1972). Serum copper levels in the mother almost double during pregnancy. This is thought to be caused by estrogen secretion. The copper concentration in cord blood is about one fifth that in maternal blood at term, suggesting that the placental transport mechanism may have saturation kinetics.

### 3.2.1. Normal Growth Patterns and Body Composition

*Developmental Biology.* There is little information concerning nucleic acid synthesis by the human fetus. DNA and RNA synthesis by human organs between 7 and 14 weeks of gestation were studied using incorporation of [³H]thymidine and [³H]uridine, respectively, by Mukherjee *et al.* (1973). Unique patterns of nucleic acid synthesis were observed in each of the seven organs studied (liver, kidney, heart, skin, spleen, lung, and brain). The patterns for DNA and RNA synthesis were mirror images of each other, indicating a pattern which could be interpreted as hyperplasia and hypertrophy of cells. In the brain, thymidine incorporation peaked at 7 weeks and then gradually declined, while uridine incorporation gradually increased, exceeding thymidine incorporation at 12 weeks.

The processing of nutrients along metabolic pathways is catalyzed by enzymes. The emergence, or ontogenesis, of the specific enzyme proteins is a critical feature for the utilization of nutrients by the growing fetus. Details of enzyme ontogenesis are considered elsewhere in this treatise (see Chapter 19). A few points relevant to fetal nutrition follow.

There are at least six factors which influence enzyme levels in developing tissues. These include: (1) synthesis and degradation, (2) hormonal regulation, (3) substrate induction and product repression, (4) some "tissue-specific" factors, (5) environmental factors, and (6) other factors, including vitamins, micronutrients, and drugs.

Certain key enzymes of glucose metabolism are present in the human fetus at 10–20 weeks of gestation. During this period of active fetal growth when glucose is the main source of energy, there seems to be an increase in the enzymes of glycolysis in the brain (Hahn and Skala, 1970). During that interval, phosphofructokinase undergoes a fivefold increase in cerebral cortex, but does not change in the placenta or the liver. Pyruvate kinase increases fourfold in the cerebral cortex, again without significant change in placenta or liver. Glucose-6-phosphate dehydrogenase increases in the liver, but does not change in cerebral cortex or placenta. Glyceraldehydephosphate dehydrogenase increases in the placenta. Tyrosine transaminase and phosphoenolpyruvate carboxykinase can be induced in human fetal liver before 16 weeks of gestation (Kirby and Hahn, 1973). Addition of cyclic AMP to the *in vitro* study system increases the activities of these enzymes by 35–50%. In the fetus, the liver is more of a receiver than a giver, according to Hahn and Skala (1970).

While the pathways of carbohydrate metabolism are developing as early as 10 weeks, the liver only starts to synthesize fatty acids at about 16 weeks of gestation in the human. Although glucose is normally the major fetal substrate, lipids and ketone bodies may play a role in fetal development, for example, in the synthesis and availability of fetal adipose tissue lipid and the formation of various phospholipids involved in myelin formation, lung surfactant, etc. The fetus also can use beta-hydroxybutyrate and other ketone bodies for energy, even in the presence of adequate glucose. The ability of the fetus to synthesize ketones late in pregnancy reflects both fetal maturity and tissue adequacy of the triglyceride reserves (Harding *et al.*, 1975).

Enzymes for the formation of nonessential amino acids are not fully developed in the fetus. The majority of the amino acids essential for rapid growth of the fetus are readily transported through the placenta from the maternal blood. The transulfuration pathway for cystine biosynthesis does not appear until after the 37th week of gestation. It has been hypothesized by Gaull *et al.* (1973) that the inability of the human fetus to synthesize cystine may facilitate DNA synthesis by preventing inhibition of the methionine-activating enzyme, which catalyzes the first step in the transulfuration pathway, leading to the formation of homocysteine. Homocysteine is utilized in the formation of a precursor of thymidylate, which is a precursor of DNA.

Tyrosine, a nonessential amino acid, can be synthesized in normal adult liver by hydroxylation of phenylalanine, an essential amino acid. Tyrosine metabolism in the fetal liver begins with transamination of alpha-ketoglutarate by transaminase. The collection of enzymes for metabolism of tyrosine, known as the "tyrosine-oxidizing system," has a very low activity in human fetal liver. It has been known for many years that premature infants metabolize tyrosine poorly, but this defect is transient and can be overcome by addition of ascorbic acid. Presumably, vitamin C activates the enzyme *p*-hydroxyphenylpyruvate oxidase. The urea-synthesizing enzymes, carbamylsynthetase, ornithine transcarbamylase, arginine synthetase, and arginase, are present in the human fetal liver at an early stage of development. The presence of the mesonephros in the human fetus also is associated with the ability to synthesize urea in the kidney. The early ontogenesis of urea cycle enzymes undoubtedly contributes indirectly to the gradually rising levels of urea and creatinine, hence osmolality, in the amniotic fluid.

In the human fetal forebrain, neuroblast formation is more than two thirds completed between 10 and 20 weeks of gestation, based on DNA-P measurements (Dobbing and Sands, 1973). This precedes the mid-gestational brain growth spurt and late increase in dendrite and synaptosome complexity. The period of rapid brain growth may be most vulnerable to adverse events, like malnutrition (Dobbing, 1968), viral infections, intensive radiation, etc. Between 10 and 18 weeks of gestation, neuroblast development might be adversely affected. Thereafter, from mid-pregnancy to the end of the first or second postnatal year, neurological development, particularly dendritic arborization and synaptosomal connections, could be impaired. Late in pregnancy, myelination may be affected, although Dobbing estimates that this process continues for several years postnatally. In recent observations, Ballabriga's group found a sharp increase in neutral glycosphingolipids, sulfatides, and sphingomyelin in brain of liveborn infants after the 32nd week of gestation, compared to patterns found between 22 and 29 weeks (Conde *et al.*, 1974). During this interval there was a parabolic increase in the $n - 3/n - 6$ ratio of ethanolamine phosphoglycerides in the cerebrum. However, they noted a linear increase in the ethanolamine phosphoglycerides with increasing gestational age from 22 to 44 weeks (Martinez *et al.*, 1974).

Endocrine factors relating to growth and development of the fetus and maintenance of pregnancy are discussed in Chapter 19.

*Normal Fetal Growth.* Studies of human intrauterine growth usually are based on anthropometric measurements of infants born at various gestational ages. Gestation is usually estimated in weeks from the first day of the last menstrual period before presumed conception. This conceptional age and the period of actual embryonal and fetal growth may differ by 1–3 weeks. Critical assessment of fetal growth patterns depends upon an accurate knowledge of gestational age. Estimates based on reported dates of the last menstrual period (LMP) are highly susceptible to error, particularly if the LMP is not ascertained within 12–16 weeks of conception. Using basal body temperature curves, Boyce *et al.*, (1976) defined the last day of the hypothermic phase as the day of ovulation, and hence could determine conceptional age. In a sample of 317 women whose pregnancies were dated by this method, they found that 1% of infants were undetected postmature infants (>280 days + 14 days), 17% were premature infants (<245 + 14 days), and 6% were at term, but were incorrectly classified as premature infants. This method for determining the conceptional age, however reasonable, does not seem practical at present. Numerous centile charts have been constructed which relate fetal measurements to gestational age (e.g., Lubchenko *et al.*, 1963; Gruenwald, 1966; Tanner and Thomson, 1970; Freeman *et al.*, 1970; Sterky, 1970, Brenner *et al.*, 1976). Physical characteristics supplemented by estimates of neurological maturity afford an independent estimate of gestational maturity (Brazie and Lubchenko, 1974; Lubchenko, 1976; Usher *et al.*, 1966; Dubowitz *et al.*, 1970).

Studies of birth weight in relation to gestational age have been carried out in large population samples, usually on a retrospective basis. Tanner has emphasized that maternal height, weight gain to middle or end of pregnancy, parity, age, ethnic group, genetic factors, socioeconomic factors, nutrition before and during pregnancy, and smoking during pregnancy all may influence fetal growth. Using data reported for nearly the entire population of Aberdeen by Thomson *et al.* (1968), Tanner and Thomson have constructed centile charts for weight of newborn boys and for newborn girls which can be adjusted for parity (first born or later born), mother's height, and mid-pregnancy weight. These standards are applicable to the United Kingdom. Recently, Brenner *et al.* (1976) have proposed standards for fetal growth in the U.S.A., derived from 30,772 deliveries between 1962 and 1969 in Cleveland, Ohio, and 641 fetuses from 8 to 21 menstrual weeks gestation aborted with prostaglandins in Chapel Hill, North Carolina, from 1972 to 1975. The centile fetal weight curves are related to gestational age and adjusted for parity and race of the mother and sex of the baby. Their curves indicate that the U.S.A. 10 percentile standard observed weight at 40 weeks is 2750 g, rather than the 2500 g commonly used. The corrections for parity, race, and sex are appreciable (Figure 1). For example, to obtain the expected fetal weight for a singleton delivery to a white primagravida at 39 weeks gestation, to the median weight at that gestation (3170 g) is added the correction for primagravida (−65 g), race (white, +75 g), and sex (male, +75 g) = 3170 + 85 = 3255. A 2750-g baby born to such a mother would be almost 500 g underweight although the observed weight is at the 10th percentile. For a girl infant born to a multiparous black mother at 39 weeks the expected fetal weight would be 3170 + (multiparous +50), (black −80), (girl −70) = 3170 − 100 = 3070. At the 10th percentile also, a 2750-g baby born to this black mother would be only 320 g less than expected, relatively less underweight than the white baby. The report by Brenner *et al.* also provides very useful curves for fetal length and weight gain from 8 to 24 weeks of gestation and fetal "velocity" weight gain curves throughout pregnancy (Figure 2).

Fetal growth, like postnatal infant growth, undoubtedly is influenced by nutrient availability for mother and fetus and nutritional state of the mother before and during pregnancy. The nutritional state of the mother's mother, when the mother was a fetus, also may be a significant regulator of the growth of her fetus (Ounsted and Ounsted, 1966). A significant deviation below expected growth for gestational age is commonly attributed to fetal malnutrition.

*Body Composition.* Fetal body composition at different gestational ages indicates nutrient utilization for the synthesis and deposition of protein and fat and for the accumulation of water and ions. Detailed analyses of body composition are available (Widdowson and Spray, 1951; Widdowson and Dickerson, 1964).

The major components of fetal body composition are noted in Table VI,

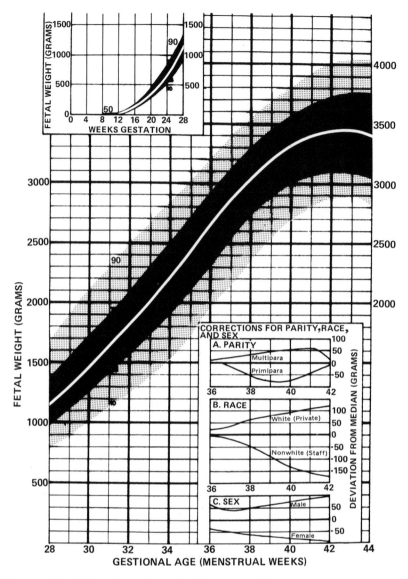

Fig. 1. Nomogram for adjusting observed weight of the newborn infant at a given gestational age for sex and for the parity and race of the mother. The observed weight at the indicated gestational age is compared with the expected weight. The expected weight is computed from the correction chart shown in the lower right-hand portion of the graph. For example, the expected baby weight for a multiparous white mother having a male infant at 40 weeks gestation would be the median at 40 weeks (approximately 3200 g) plus about 215 g +50 (=parity) +90 (=race) +75 (=sex), thus her infant should weigh about 3415 grams. This expected size should be compared with the observed weight of the baby. (From Brenner *et al.*, 1976, with permission.)

derived from Widdowson's data (1968). Shaw (1974) has calculated the rates of apposition of major nutrient components by human fetuses between 24 and 36 weeks, based on data published during the last 80 years (Figure 3). These rates may be the best estimate we have of the net flux of these substances across the placenta from mother to fetus. Dancis *et al.* (1973) ventured a calculation, based upon transport observations, that only about 20% of the fetal fat could be derived from free fatty acids provided by the mother and transported across the placenta. The

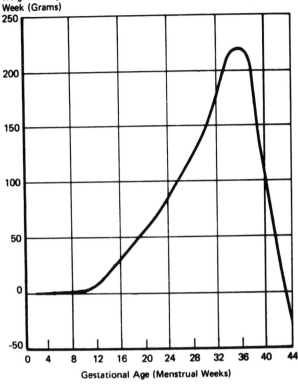

Fig. 2. The velocity of fetal weight gain per week throughout pregnancy. (From Brenner *et al.*, 1976, with permission.)

assumption has been made that most of the 350 g of body fat is deposited during the last month of pregnancy. Widdowson's data indicate that about 180 g of fat accumulate in the body between 35 and 38 weeks of gestation. At this rate, placental transport of fatty acids might account for as much as 40% of fetal fat.

Based upon 273 nitrogen balances summarized by Calloway (1974), it is possible to estimate the relationship between maternal nitrogen balance and fetal body composition. Between 26 and 38 weeks of gestation, the average maternal nitrogen balance of 92 g provides a margin of 42 g or 45% beyond that deposited as

*Table VI.  Fetal Body Composition[a]*

| | \multicolumn{8}{c}{Gestational age (weeks)} | | | | | | | |
| | 26 | % | 33 | % | 35 | % | 38 | % |
|---|---|---|---|---|---|---|---|---|
| Weight (g) | 1000 | | 2000 | | 2500 | | 3000 | |
| Water (g) | 860 | 86.0 | 1620 | 81.0 | 1940 | 77.6 | 2180 | 72.7 |
| Protein (g) | 87.5 | 8.8 | 231 | 11.6 | 306 | 12.2 | 344 | 11.5 |
| Fat (g) | 10 | 1.0 | 100 | 5.0 | 185 | 7.4 | 360 | 12.0 |
| Iron (mg) | 64 | | 64 | | 220 | | 260 | |
| Calcium (g) | 60 | | 60 | | 20.2 | | 25.0 | |
| Phosphorus (g) | 3.4 | | 3.4 | | 11.0 | | 14.0 | |
| Magnesium (mg) | 220 | | 460 | | 580 | | 700 | |

[a]From Widdowson, 1968.

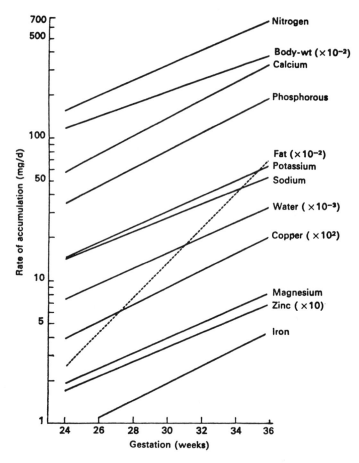

Fig. 3. Smoothed regression curves indicating rate of accumulation of minerals, water, nitrogen, and fat in the human fetus between 24 and 36 weeks of gestation. The data were computed from the available literature on body compositional analyses by Shaw (1974).

protein in the fetal and placental components (see Table I). Shaw has also computed the apposition of nitrogen content in 36 infants between 16 and 40 weeks of gestation, based on total body chemical analyses reported since 1894. As expected, the fetal accumulation of nitrogen, like other elements, is exponential: $y = y_0 e^{kt}$, where $y_0 = 0.346$ g and $k = 0.0186$/day, where $t$ is time in days.

### 3.2.2. Impaired Fetal Growth—Fetal Malnutrition

The gestation period is more than 37 weeks in a third to a half of low-birth-weight (less than 2500 g) babies and therefore cannot be ascribed to prematurity, but represents intrauterine growth impairment (Van den Berg and Yerushalmy, 1966). This has been attributed to maternal regulators of fetal growth which may be intrinsic characteristics of the mother (Ounsted and Ounsted, 1966), or due to hormonal, cardiovascular (Hellman and Pritchard, 1971) or, most commonly, nutritional events affecting the mother before or during pregnancy (Naeye et al., 1971).

About 5% of all live births in the U.S. (about 114,000 babies) may be compromised annually by fetal malnutrition. Thirty percent of fetal deaths may be attributable to fetal malnutrition (Usher, 1970), an additional 16,000–17,000 pregnancies (Metcoff, 1974a,b).

Recent estimates by WHO indicate that 15,000,000 babies will be small for gestational age when born this year (1977). The majority of of these babies will have fetal malnutrition, and many will be born in preindustrialized societies or in areas of poverty in industrialized societies. These babies will be in double jeopardy with risk of postnatal malnutrition. An indeterminate number will have impaired physical and mental development.

*Definition.* The prevalent view of fetal malnutrition holds that cell replication is impaired *in utero*. Cell replication depends on protein synthesis, which requires energy. Whether fetal malnutrition depends on maternal, placental, or fetal factors, or various combinations of these, is unknown. We suppose that idiopathic fetal malnutrition is essentially maternal in origin and is derived, ultimately, from nutritional causes (Metcoff, 1974a,b).

The synonyms intrauterine growth retardation (IUGR), small for dates, small for gestational age (SGA), and fetal malnutrition (FM), have been used to describe undersized babies born at or near term having characteristic clinical features and potential perinatal and postneonatal problems. We prefer the term fetal malnutrition (FM), since growth is not necessarily "retarded"; it is inadequate, and catch-up growth may not occur (Beck and Van den Berg, 1975); "dates" are subject to considerable error as previously noted; and size (e.g., small) at birth requires adjustment for more than gestational age. Currently, we define the fetally malnourished infant on the basis of weight and clinical features. The subcutaneous tissue is diminished and the loose skin of arms, legs, and interscapular regions appears several sizes too large for the baby. The expected weight is derived from a multiple-regression equation, which includes adjustments for maternal height, nonpregnant weight, age, parity, and race, and the baby's sex and gestational age. The "corrected" weight centiles of Brenner *et al.* (1976) or of Tanner and Thompson (1970) could be used for populations to which they apply.

The term "fetal malnutrition" was coined by Scott and Usher (1966) to describe babies who, at birth, were markedly underweight for their gestational age, with greatly reduced subcutaneous fat and a wrinkled skin with poor turgor. External features and gross neurologic responses usually are normal for gestational age. Except for lagging bone age, organs of FM babies are normally mature for the gestational period but have the morphologic characteristics of malnutrition which must have developed *in utero* (Naeye, 1965).

Fetal malnutrition may occur in four principal ways:

1. Decreased nutrient supply to the fetus resulting from reduced blood circulation and/or inadequate transport of nutrients via the placenta.

2. Decreased nutrient supply to the fetus resulting from some nutrient deficiency, excess, or imbalance in the mother. Placental circulation and transport may be normal, but placental metabolism and processing of nutrients also may be affected by the maternal state.

3. Decreased nutrient utilization and metabolism by the fetus.

4. Some combination of 1–3.

Since about 75–80% of babies with fetal malnutrition are born to mothers who do not have obvious evidences of malnutrition, and the placentas show little, if any, pathological changes, it appears that fetal malnutrition (specifically) most commonly derives from subtle maternal nutrient insufficiencies, excesses, or imbalances.

Clinical characteristics of the FM baby may depend on the timing and duration of the adverse factors affecting the mother during and/or before pregnancy. Three general types are commonly encountered (Metcoff, 1974a,b). For descriptive purposes, these have been called types I, II, and III, but they should be considered as representative of a family of possible impaired growth patterns. The type I fetus is uniformly small with respect to weight, length, and head circumference. The ponderal index (weight $\div$ length$^3$ $\times$ 100) may be normal. Other clinical features of FM usually are present. This type of FM most likely occurred during most of the period of high growth velocity, i.e., the last half of gestation. This type of fetal growth pattern was noted in babies born to low socioeconomic class, presumably malnourished, women in northern India (Ghosh *et al.*, 1971) and in Mexico (Urrusti *et al.*, 1972). In type II, weight is more affected than length or head circumference but both of these measures and the ponderal index are somewhat reduced. Clinical characteristics of FM are present. Marked deprivation of food for the pregnant women or limiting nutrient supply to the fetus during part or most of the last trimester seems the most likely cause. This pattern characterized babies born to mothers in the famine areas of Holland during the "starvation winter" of 1944–1945 (Stein and Susser, 1975). Some, perhaps one third, of the mothers with toxemia of pregnancy and possibly cigarette-smoking mothers may deliver this type of baby. In type III FM, the baby is markedly underweight, the skin seems several sizes too large, and subcutaneous fat is markedly diminished. Length and head circumference are not affected and are appropriate for gestational age. The ponderal index is markedly reduced. This type of fetus presumably suffers acute weight loss during the last weeks of gestation, after the period of rapid growth has ended. Although specific evidence is not available, one suspects that this is the type of newborn most likely to have exhausted hepatic glycogen stores (Shelley, 1961) and to be prone to hypoglycemia (Lubchenko and Bard, 1971). This type of FM may have occurred during the severe famine period between September 1941 and February 1942 in Leningrad (Bergner and Susser, 1970) or in March or April of 1945 in Holland. Mothers with severe toxemia may deliver type III FM babies.

From morphologic and demographic characteristics of stillborn and perinatal deaths with no abnormalities in the placenta or fetus, or evident diseases in the mother, Naeye *et al.* (1971) concluded that "maternal malnutrition during gestation provides the simplest explanation for the undernutrition found in newborn infants of the poor." Nonetheless many uncertainties about the relation between nutrition and fetal development persist, since most studies of these variables in human subjects, whether retrospective or prospective, have failed to account for the likely influence of multiple factors known to influence pregnancy outcome and fetal development: for example, genetic and ethnic characteristics of the mother, her age, height, prepregnancy weight, birth order, birth interval; age, height, and weight of the father, his ethnic and genetic influence. Environmental factors also are known to influence fetal growth; for example, women from lower socioeconomic classes in almost every country where data are available tend to have smaller babies. Whether this is the result of their adolescent growth, prepregnancy nutrition, cultural factors, health, or limited food intake during pregnancy is not known. In the British perinatal survey (Butler, 1974) and in a recent study by Miller and Hassanein (1974) smoking during pregnancy led to smaller babies, other factors being more or less constant. Babies born at higher altitudes are smaller than babies born at lower altitudes. Thus babies in Denver (7000 feet) might be smaller than babies in New York (sea level), although the nutrient status of Denver mothers might be at least as good as that of the general population of New York mothers.

Prepregnancy hypertension, congenital heart disease, and diabetes often exclude mothers with these problems from consideration in studies of FM, yet maternal nutrition rather than the primary disease may be responsible for impaired fetal growth. Viral or other infections before or during early pregnancy and the influence of serious acute diseases during childhood and adolescence upon the pregnancy and the development of the fetus have not been quantified. To date, few prospective studies employing multiple-regression analysis to adjust for the influence of many of these variables have been reported (cf. Lechtig *et al.*, 1975).

*Body Composition and Metabolism.* Apte and Iyengar (1972) analyzed the body composition of 41 fetuses of different gestational ages from a low socioeconomic group of mothers in Hyderabad, India (Table VII) and compared them with English fetuses described by Widdowson (1968). At 26–27 weeks of gestation the Indian fetuses weighed about 1000 g and contained about 86% water with amounts of calcium, phosphorus, and magnesium similiar to English fetuses of the same weight. The fat content of the Indian babies was more than twice that of the English fetuses (26 vs. 10 g), but they were beginning to lag in protein and iron deposition. At 40 weeks of gestation, a malnourished Indian fetus, weighing 2500 g, had body composition equivalent to a 5-weeks-younger English infant of 35 weeks gestation, with the same quantity of water, calcium, phosphorus, and magnesium; more fat; but less protein and iron (also see Table VI). A well-nourished 3000-g Indian fetus at 40 weeks gestation was equivalent to a 38-week English baby, but contained more water; less calcium, phosphorus, magnesium, iron, and protein; and about the same amount of fat.

Apte and Iyengar concluded that the chemical composition and nutrient stores of the developing fetus were influenced by the state of maternal nutrition. The average nitrogen balance of 48 g for Indian women afforded only a 6-g nitrogen margin or 12% over that deposited in the fetal body and placenta. Mothers of their study were considered malnourished, based on mean daily intakes of 1440 kcal (6.05 MJ) and protein. The intake of other nutrients also was considered to be grossly inadequate.

Although body composition was severely affected by maternal malnutrition, the liver of these Indian fetuses was not. The water and nitrogen contents of the liver were relatively constant from 28 to 40 weeks of gestation: 81–83 g and 1.8–1.9 g/100 g liver, respectively. The liver was about 4.2–4.7% of body weight. These values were not affected whether the fetal weight was more or less than 2250 g at 37–40 weeks of gestation. Total iron and copper contents of the liver were within the normal range (0.29–0.31 mg/g and 3.5–4.6 mg/100 g fresh liver, respectively).

Table VII.  *Body Composition with Fetal Malnutrition*

|  | FM[a] | % | Normal[a] | % |
|---|---|---|---|---|
| Gestational age (wks) | 40 |  | 40 |  |
| Weight (g) | 2500 |  | 3000 |  |
| Water (g) | 1935 | 77.4 | 2238 | 74.6 |
| Protein (g) | 242 | 9.7 | 291 | 9.7 |
| Fat (g) | 242 | 9.7 | 366 | 12.2 |
| Iron (mg) | 138 |  | 165 |  |
| Calcium (g) | 19.0 |  | 22.7 |  |
| Phosphorus (g) | 10.8 |  | 11.4 |  |
| Magnesium (g) | 540 |  | 655 |  |

[a]From Apte and Iyengar, 1972.

The ferritin content of the livers was lower in these babies than reported in other studies. Hepatic stores of vitamin A, $B_{12}$, and folate were reduced. Iyengar and Apte (1972) suggested that these values could reflect the poor maternal nutritional situation.

Analogously, among rats maintained for 12 generations on diets marginally deficient in protein, the proportion of small term offspring was 10 times higher than in a well-nourished control colony. Organs generally weighed less than controls of similar postnatal age but, except for the kidneys, were larger relative to body weight. On this basis, the brains were about 50% heavier than normal, although in absolute terms they were 5–5.5% lighter than those of controls (Stewart *et al.*, 1975).

While the mechanism is unclear, fetally malnourished babies have higher cord blood amino acid levels than normal term babies (Linblad *et al.*, 1969; Young and Prenton, 1969; Haymond *et al.*, 1974). The increased amino acid content of the maternal extracellular fluids and of infant cord blood suggests some kind of mass action effort to facilitate fetal growth by provision of amino acids and by gluconeogenesis. Haymond *et al.* (1974) suggest that impaired utilization of glucogenic amino acids may account for their elevation in cord blood.

Churchill *et al.* (1969) reported that mothers having lower levels of total amino nitrogen (4.0 mg/100 ml) had significantly smaller babies (3.06 vs. 3.43 kg). They found a significant difference in cranial volume: the mothers with lower amino acids had babies with smaller cranial volumes. The ratio of cranial volume to weight of the babies was not affected.

Mansani *et al.* (1973) described higher levels of total amino acids and glycine–valine ratios in cord blood of SGA than normal infants, and in both groups the values were higher than maternal levels. McClain *et al.* (1976) have demonstrated an association between mid-pregnancy maternal plasma free amino acid profiles and baby size at birth. Recently, they have shown a relationship between the maternal mid-pregnancy amino acid levels and those in cord blood at term. Apparently the free amino acid profiles from both sources are related to the baby's size at birth (McClain and Metcoff, 1977). Velazquez *et al.* (1976) consider that human fetal blood levels of amino acids at term normally are patterned after the placental rather than maternal ones. Further, the placenta acts as a sort of selective retention barrier to prevent some amino acids (arginine, glutamine, glutamic acid, proline, and alanine) from entering the mother's circulation from the fetal side. Similar studies have not been reported early in pregnancy or in FM infants.

The placentas of FM infants of nontoxemic mothers are significantly smaller than normal by size and weight. Morphometrically there appears to be a deficiency of villous tissue with occasional intervillous fibrin deposition (Laga *et al.*, 1972), but otherwise the placenta is histologically normal. While cell numbers (DNA) are reduced and the cells are hypertrophic (protein/DNA) (Winick, 1967; Laga *et al.*, 1972; Rosado *et al.*, 1972; Metcoff *et al.*, 1973), the levels of energy metabolites, enzyme activities (Rosado *et al.*, 1972), and RNA and protein synthesis (Laga *et al.*, 1972; Metcoff *et al.*, 1973) for the entire small, fetal malnutrition placenta are quantitatively equivalent to those of the larger, normal baby placenta. Significant positive correlations were found between birthweight ($y$) and placental rate of protein synthesis ($x$) of babies in Boston and Guatemala (Laga *et al.*, 1972):

$$\text{Boston:} \quad y = 1600(x) - (1400 \times 10^3), \quad r = 0.384$$
$$\text{Guatemala:} \quad y = 1908(x) - (2534 \times 10^3), \quad r = 0.583$$

Ribosomal protein synthesis by the placenta appears to decrease between 32 weeks of gestation and term (Metcoff *et al.*, 1973). This observation suggests that studies of term placenta may not accurately reflect fetoplacental events occurring during the high-velocity growth phase of the fetus earlier in pregnancy.

*Brain Development.* The effect of FM on brain development and function is controversial. In proportion to body weight or length, the head of the human FM infant is normal or large, although it may be small compared to the brain of a full-sized term baby of the same gestational age (McLean and Usher, 1970; Urrusti *et al.*, 1972). Cell numbers in the brain may be reduced (DNA content) but are normal per unit of brain mass (Winick, 1970; Dobbing and Sands, 1973). About a third to half of the total cortical DNA is present at birth in the human (Winick and Rosso, 1969); DNA content reflects numbers of both the supporting glial and the neuronal cells in the brain. Dobbing and Sands (1973) note that neuronal cell DNA in the cortex increases rapidly by mid-gestation and would be more likely affected by prenatal or early gestational malnutrition than would cerebellar DNA, which increases before birth but peaks postnatally. They also suggest that neuroblast formation in the human brain may be largely completed by mid-pregnancy. Prenatal or early gestational malnutrition may impair cortical neuroblast formation. Malnutrition later in pregnancy and in the early postnatal period may interfere with axon formation and limit dendritic arborization within the central nervous system. On the other hand, Drillien (1970) considers that the brains of small-for-date infants may be spared and the IQ unaffected, unless the infant is born to a low socioeconomic class mother. It has been noted, however, that the incidence of congenital anomalies was increased, and a small-for-date infant with congenital anomalies was more likely to have mental and neurological defects (e.g., Drillien, 1970); Neligan *et al.*, 1976). In the pregnant rhesus monkey, protein or protein + calorie deprivation did not appear to affect the chemical composition of the fetal brain. There were no significant differences in brain weight, DNA, RNA, protein, phospholipid, cholesterol, or water contents. The restricted diets contained 1.2 g/kg protein with 100 cal/kg, or 1.2 g/kg and 50 cal/kg daily, vs. the control of 4.2 g protein and 100 cal/kg. (Cheek *et al.*, 1976). While protein was restricted to comprise 2% of the diet, the quantities of protein and calories in the restricted primates were equivalent per kg body weight to those normally recommended for pregnant humans.

The animal investigations, however, indicate both biochemical and functional changes occur in the brain as a result of fetal nutrient deprivation. If postnatal malnutrition is superimposed upon fetal malnutrition, the ultimate formation of synaptosomal dendritic arborizations of cortical neurons may be impaired. The extent of the changes in myelination, dendritic arborization, cell numbers, etc., is related to the duration, timing, and intensity of the fetal deprivation. Intrauterine growth retardation in rats reduces synaptosomal number, RNA content, and sphingomyelin and cephalin concentrations, while increasing gamma-aminobutyric acid concentration (Bernal *et al.*, 1974). Subsequent adult behavior of fetally malnourished rats may be significantly impaired, without biochemical changes in DNA or phospholipid content in the brain although regional turnover (synthesis) of 5-hydroxytryptamine, a neurotransmitter, was increased in the hippocampus. As adults, these rats behaved less competently in response to stimuli (Smart *et al.*, 1976). Previously, altered activities of enzymes involved in metabolism of some neurotransmitters were found in brains of adult rats subjected to early malnutrition (Adlard and Dobbing, 1972; Adlard *et al.*, 1972). Perhaps neither cell number, size, morphology, nor DNA or RNA contents, etc., are the most important patterns relating to future development of subtle brain functions. Flexner (1954) found that activities of many enzymes in guinea pigs increased rapidly in the brain during the critical period of 41–45 days of the normal 67-day gestation. This coincided with the earliest evidence of cell function (appearance of electrical potentials and muscular response to cortical stimuli), although it may not be applicable to man. Drillien (1970) suggests that the late effect of undernutrition on brain function might relate more to disturbances in developing enzyme systems in the brain than to size, cell number, or myelination. The effect of FM on the ontogenesis and activities of important enzymes in fetal brain requires much more study.

By school age, children born with fetal malnutrition have an unusually high prevalence (30%) of neurological and/or intellectual sequelae. Late postnatal effects of FM may include impaired capacity for learning (Fitzhardinge and Steven, 1972) and difficulties with behavioral responses (Eaves *et al.*, 1970) and psychological performance. The prognosis for poor intellectual achievement is greater if the baby is born to low socioeconomic class parents. Presumably, catch-up brain cell differentiation and growth also are limited. In a follow-up study of 60 small-for-date babies, whose intrauterine growth was followed by ultrasound cephalometry, Fancourt *et al.* (1976) found slow growth and developmental performance at a mean age of 4 years. However, Winick (1976) and Cravioto (1976) suggest that the effects of undernutrition in early infancy may be "canceled" to a large extent by the stimulation provided by an "enriching" environment. Whether an enriched environment contributed to the normal mental performance at age 19 of cohort groups of Dutch males, born in Holland during the "famine winter" of 1944–1945 when they could have been exposed to prenatal undernutrition, is unknown (Stein *et al.*, 1972).

*Congenital Anomalies.* The incidence of congenital anomalies among babies who were born small-for-dates in New York City was about 8 times higher than that found in a population of appropriately sized infants born during the same interval (16 vs. 2%) (Van den Berg and Yerushalmy, 1966). Drillien (1970) also noted an increased incidence of congenital anomalies in small-for-date babies in Edinburgh, as did Mitchell (1966) in Aberdeen and Warkany *et al.* (1961) earlier. Presumably, FM can impair cell division and differentiation *in utero,* leading to teratogenic effects (see review by Hurley, 1976).

*Treatment.* Infants born with intrauterine growth retardation fail to catch up in weight or height by 10 years of age (Beck and Van den Berg, 1975) (see Chapter 16). Some possible effects on brain development and mental performance were described above. Treatment of fetal malnutrition must be directed toward preventing irreversible changes in cell functions leading to impaired mental and physical development. To be effective, treatments therefore must be started during or before critical periods of intrauterine development.

One way is to improve nutrition of all pregnant women at risk. Several intervention studies have provided either additional energy or protein supplements or both for poor pregnant women (see Jacobson, 1974). While the effect of these supplements upon fetal malnutrition has not yet been determined, in one study in rural Guatemala, mothers receiving more than 20,000-cal supplements, including protein, had larger babies and fewer with low birth weights (<2500 g) than those mothers receiving <20,000 supplemental calories (Lechtig *et al.*, 1976).

Another way is to attempt some specific treatments for those women likely to deliver FM babies. On the assumption that maternal levels of amino acids might be limiting fetal growth, Renaud *et al.* (1972), a few days before delivery, injected a mixture of amino acids into the amniotic sac of three pregnant women suspected of having FM babies. They found the amino acid levels in the placenta were increased and believed that head growth of one of the babies resumed. Mansani *et al.* (1974) infused amino acids daily, for at least a week in two and for 5–6 weeks before term in three mothers carrying undergrown babies. Four of the five mothers delivered, at term, normal-sized babies in good condition and with normal aminograms. The fifth was malnourished. They suggested that parenteral infusion of amino acids beginning at the 30–34 weeks of gestation might stimulate fetal growth. More studies with these and other nutrients associated with FM, e.g., trace minerals like zinc and chromium, are needed. Neither iron nor vitamin supplementation during pregnancy seem to affect fetal growth.

## 4. Correlations between Fetal Growth and Maternal Nutrition — Predicting Fetal Malnutrition

Fetal malnutrition is most likely to affect the fetus during the critical period of rapid cell replication and/or the period of maximal growth velocity. When cell numbers are reduced, postnatal treatments offer little hope for catch-up growth or for improved developmental performance. Prevention of fetal malnutrition appears to be the only effective known treatment at this time. This can be achieved by enhancing fetal growth *in utero*. Where some form of maternal malnutrition or nutrient imbalance is the etiologic factor, provision of the correct nutrients or correcting the imbalance would be effective treatments. To achieve this in a practical way it is necessary to identify those pregnant women carrying malnourished fetuses early enough in pregnancy to enhance fetal growth. Dietary supplementation with a significant number of calories may not help the fetus lacking specific amino acids, nor will provision of amino acids substitute for a deficit in zinc or chromium. For some pregnant women indiscriminant supplementation with calories and a variety of nutrients may prove harmful. Furthermore sociocultural and economic considerations might preclude mass supplementation of all pregnant women in an endeavor to prevent FM and improve the course of pregnancy. For identification of specific women whose fetuses are at risk for FM and identification of nutrient requirements, maternal measurements must be made no later than the second trimester of pregnancy and correlated with fetal growth.

Growth of the fetus is a complex process: it appears to be influenced by many variables, and it is unlikely to be regulated by a single maternal variable, nutrient or otherwise, under normal conditions. Therefore it is unlikely that simple product–moment correlations between birth-size measures and a single maternal variable will be reliable indicators of fetal growth. On the other hand, the interaction of several maternal variables may modulate fetal growth, for example, protein and calorie intakes, amino acid relationships, trace mineral availability, smoking during pregnancy, weight gain, etc. The influence of each of these variables can be related to the influence of the others with respect to birth size. Multiple-regression analysis provides but one method to assess this relationship.

### 4.1. Correlations

Human serum placental lactogen, urinary estriol excretion, and amniotic fluid components such as creatinine, osmolarity, and lecithin–sphingomyelin ratio have been recommended as measures of fetal maturity or well being. While these measures often are useful to predict fetal distress or gestational maturity, none have been satisfactory as predictors of FM (Metcoff, 1974*a*). Gestational age also may be assessed by radiological measure of the length of the fetal femur *in utero* with a reasonable degree of precision, with an error not exceeding 2 weeks (Martin and Higginbottom, 1971). Similarly, ultrasound biparietal measurements of fetal head size provide a range of values for each gestational week of pregnancy and correlate quite well with fetal age (Levi and Smets, 1973; Campbell and Newman, 1971). Repeated measurements of fetal biparietal diameters in small-for-dates babies can reveal a slowing of skull growth beginning around 34 weeks of gestation, 3–4 weeks before term (Fancourt *et al.*, 1976). Head circumference of infants with FM often is proportional to body length or weight and may be equivalent to head size of normal infants. Thus, ultrasound measures of biparietal diameters around mid-pregnancy are unlikely to identify the malnourished fetus.

Favier *et al.* (1972) found a correlation between the amniotic fluid levels of zinc, measured between the 29th and 42nd week of gestation, and birth weight of the baby:

(Birth weight) $y = 0.194 \times$ amniotic fluid zinc level, ($\mu$g/liter) $- 325.57$; $r = 0.769, P < 0.01$

In recent studies we have found correlations between adjusted birth-weight measures, as dependent variables, and mid-pregnancy measures of maternal variables, including some estimates of leukocyte metabolism and nutrient levels. Smoking mothers had significantly lower values for leukocyte RNA synthesis, plasma carotene, and plasma amino acids (Crosby *et al.*, 1977) (Table VIII). Length and head circumference also were correlated with various maternal measurements. The

Table VIII.  Differences between Smokers and Nonsmokers

| Variable | $n$ | Mean smokers | $n$ | Mean nonsmokers | $P$ |
|---|---|---|---|---|---|
| Adjusted birthweight | 59 | −94 | 99 | +56 | <0.05 |
| Adjusted baby length | 59 | −0.57 | 99 | +0.34 | <0.05 |
| Maternal height | 77 | 163 | 129 | 160 | <0.01 |
| PK | 47 | 11197 | 70 | 12414 | <0.05 |
| RNA synthesis | 43 | 18 | 67 | 23 | <0.01 |
| Total carotene | 42 | 96 | 67 | 115 | <0.01 |
| Amino acids | | | | | |
| Aspartic acid [b,c] | 13 | 42 | 39 | 57 | <0.01 |
| Threonine [a,b] | 13 | 166 | 39 | 211 | <0.001 |
| Serine [b] | 13 | 139 | 39 | 173 | <0.01 |
| Proline | 12 | 176 | 39 | 247 | <0.01 |
| Alanine | 12 | 310 | 39 | 388 | <0.01 |
| Valine [a] | 12 | 149 | 39 | 190 | <0.001 |
| Isoleucine [a] | 13 | 60 | 38 | 77 | <0.001 |
| Leucine [a,b] | 13 | 105 | 38 | 138 | <0.001 |
| Tyrosine [a,b] | 13 | 58 | 38 | 69 | <0.01 |
| Phenylalanine [a,b,c] | 13 | 64 | 38 | 80 | <0.001 |
| Lysine [a,b,c] | 13 | 143 | 38 | 191 | <0.001 |
| Histidine [a,b,c] | 13 | 90 | 39 | 102 | <0.05 |
| Arginine [b,c] | 13 | 132 | 39 | 164 | <0.01 |
| Ornithine [b] | 12 | 34 | 39 | 43 | <0.001 |

[a] Essential amino acid.
[b] Correlated with adjusted birth weight (entire group).
[c] Correlated with adjusted birth weight of nonsmokers.

maternal measurements were obtained once at mid-pregnancy. As described by others (Butler, 1974; Miller and Hassanein, 1974), mothers in our sample who smoked had significantly smaller babies with respect to weight and length. Of particular interest was the correlation between smokers and nonsmokers (Table IX).

Considering the complexities of the relation between maternal variables and fetal growth, a single maternal variable seems unlikely to provide a satisfactory predictor for birth size.

### 4.2. Toward Identifying Mothers Carrying Malnourished Fetuses

Pathak *et al.* (1974) proposed a prediction equation, using as variables three maternal serum enzyme levels measured at various intervals in 240 pregnancies, to identify mothers carrying small babies, which they ascribed to placental insufficiency. Of the three enzymes, heat-stable alkaline phosphatase, cystine aminopeptidase, and *n*-acetyl-beta-glucosaminidase, discriminant-function analyses indicated that the cystine aminopeptidase (oxytocinase) offered the most predictive power. The enzyme is synthesized in the placenta. They emphasized that a single test for placental function proved deceptive. Petrucco and associates (1973) observed that mothers, hypertensive or not, who delivered growth-retarded infants failed to show the progressive increase in serum oxytocinase, urinary estrogen, and serum heat-stable alkaline phosphatase seen in patients bearing normal-sized infants. The biochemical measurements were obtained in 12 women between 28 and 35.5 weeks and in 17 between 34 and 40 weeks. Discriminant functions were calculated and applied after the babies were born. The equations correctly identified 9 of the 12 and 13 of the 17 babies. Serum oxytocinase, combined with urine estrogen values, improved the prediction after 34 weeks of gestation.

Adequacy of fetal growth can only be determined by comparison of observed with expected size for the individual baby. Ideally, the maternal measurements should be obtained as early as possible in pregnancy. The standard growth chart (Figure 1) developed by Brenner and associates illustrates one method of comparing observed and expected weight of the baby after it is born.

We are in the process of developing equations from maternal measurements obtained at mid-pregnancy which will predict the difference between observed and expected weight, i.e., birth-weight residuals, and also length and head-circumference residuals. Progress is being made toward this goal. For example, preliminary prediction equations were derived from a linear combination of 30 independent variables measured once at an average $26 \pm 3$ weeks of pregnancy. The equations

Table IX. Relation of Smoking during Pregnancy to Adjusted[a] Size of the Baby at Birth

| Variable | Nonsmokers | | Smokers | | |
| --- | --- | --- | --- | --- | --- |
| | Mean | SD | Mean | SD | P |
| Weight residuals ($M \pm SD$) | 56.3 | 388.0 | −94.0 | 447.0 | 0.034 |
| Length residuals | 0.34 | 2.78 | −0.57 | 2.66 | 0.042 |
| Head circumference residuals | 0.15 | 1.66 | −0.25 | 1.65 | 0.157 |
| n | 99 | | 59 | | |

[a]Adjusted for gestational age, baby sex, race, mother's age, height, nonpregnant weight, weight/height ratio, and parity ($n = 158$).

then were used to compute the residuals for weight, length, and head circumference of babies born to those mothers. In Figure 4 the observed residuals for head circumference are compared with the head-circumference residuals computed from 14 of the 30 variables which entered the multiple-regression analysis and provided partial regression coefficients for the prediction equation. The multiple correlation was statistically significant. Similar comparisons were made for the adjusted weight and length of the baby for each of the 75 mothers having all needed data points, with similar results (Metcoff *et al.*, 1976). In similar fashion, plasma levels of 19 free amino acids, measured in 44 of these mothers at mid-pregnancy, were used to compute birth weights which compared favorably with the observed birth weights, adjusted for nonnutritional factors (Figure 5). Since cigarette smoking was found to affect maternal plasma amino acid levels, the multiple-regression analysis of the amino acids was controlled for cigarette smoking. Of the 19 free amino acids measured, proline, isoleucine, arginine, and ornithine contributed most heavily to the prediction equation (Metcoff *et al.*, 1976). As the number of mothers entering the ongoing study increases, the particular variables which will contribute most heavily to the prediction equations may differ from those observed to date. The process of prediction requires the development of equations from available data with continual testing and refining of the equations. Ultimately the refined equations

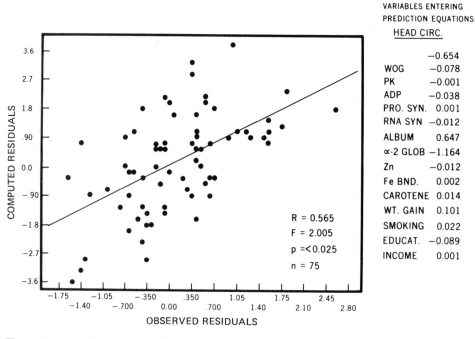

Fig. 4. The head circumference of the baby was computed by multiple regression analysis from 30 variables measured once in the mother at about mid-pregnancy. The variables which contributed most heavily to the prediction are listed, in their order of step wise appearance, with their partial correlation coefficients at the right of the chart. The measured head circumference of the baby at birth was adjusted for maternal age, parity, height, prepregnant weight, weight/height ratio, race, sex, and gestational age of the baby, as described in the text. The observed residual for adjusted head circumference of the baby then was compared with the residual computed from the maternal variables. The relation between the two is shown in this chart, which shows a significant correlation between the computed and observed residuals.

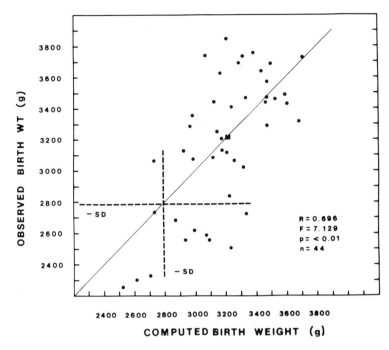

Fig. 5. Correlation of adjusted birthweight with maternal plasma amino acid levels and cigarette smoking at mid-pregnancy. Observed birthweight was adjusted for mother's age, height, non-pregnant weight, and wt./ht. ratio parity, and for baby's race, sex, and gestational age. Computed adjusted birth weight (from observations at 26 ± 3 weeks of pregnancy) = 112.91 − 22.06 (number of cigarettes/day − 3.94 (Proline) −17.99 (Ileu) +4.82 (Arg) +33.83 (Orn). (19 free AAs were measured by McClain.)

should permit reliable prediction of fetal growth from maternal measures made at a single time during the early velocity growth phase of the fetus. Of course the variables entering a reliable prediction equation will represent those having an association with fetal growth and may have therapeutic implications.

Prediction of fetal growth will be a powerful tool in the hands of the future physician. Thus mothers carrying malnourished fetuses should be identified early enough in pregnancy so that remedial action can be taken.

# 5. References

Ademowore, A. S., Courey, N. G., and Kime, J. S. 1972, Relationships of maternal nutrition and weight gain to new born birth weight, *Obstet. Gynecol.* **39**:460.

Adlard, B. P. F., and Dobbing, J., 1972, Vulnerability of developing brain. 8. Regional acetylcholinesterase activity in the brains of adult rats undernourished in early life. *Br. J. Nutr.* **28**:139.

Adlard, B. P. F., Dobbing, J., and Lynch A., 1972, Effect of undernutrition in early life on glutamate decarboxylase activity in the adult brain, *Biochem. J.* **130**:12.

Apte, S. V., and Iyengar, L., 1972, Composition of the human foetus, *Br. J. Nutr.* **27**:305.

Arroyave, G., 1975, Nutrition in pregnancy in Central America and Panama, *Am. J. Dis. Child.* **129**:427.

Balogh, M., Kahn, H. A., and Medalie, J. H., 1971, Random repeat 24 hour dietary recalls, *Am. J. Clin. Nutr.* **24**:304.

Beck, G. J., and Vandenberg, B. J., 1975, The relationship of the rate of intrauterine growth of low-birthweight infants to later growth, *J. Pediatr.* **86**:504–511.

Beilly, J. S., and Kurland, I. I., 1945, Relationship of maternal weight gain and weight of newborn infant, *Am. J. Obstet. Gynecol.* **50**:202.

Bergner, L., and Susser, M. W., 1970, Low birthweight and prenatal nutrition: An interpretative review, *Pediatrics* **46**:946.

Bernal A., Morales, M., Chew, S., and Rosado, A., 1974, Effect of intrauterine growth retardation on the biochemical maturation of brain synaptosomes in the rat, *J. Nutr.* **104**:1157.

Bernhardt, I. B., and Dorsey, D. J., 1974, Hypervitaminosis A and congenital renal anomalies in a human infant, *Obstet. Gynecol.* **43**:750.

Boyce, A., Mayaux, M. J., and Schwartz, D., 1976, Classical and "true" gestational postmaturity, *Amer. J. Obstet. Gynecol.* 911.

Brans, Y. W., Sumnars, J. E., Dweck, H. S., and Cassady, G., 1974, A noninvasive approach to body composition in the neonate: Dynamic skinfold measurements, *Pediatr. Res.* **8**:215.

Brazie, J. V., and Lubchenko, L. O., 1974, The newborn infant, in: *Current Pediatric Diagnosis and Treatment* (C. H. Kempe, H. K. Silver, and D. O'Brien, eds.), 3rd edition, p. 40, Lange, Los Altos.

Brenner, W. E., Edelman, D. A., and Hendricks, C. H., 1976, A standard of fetal growth for the U.S.A., *Am. J. Obstet. Gynecol.* **126**:555.

Brewer, T. H., 1966, *Metabolic Toxemia of Late Pregnancy: A Disease of Malnutrition,* Charles C. Thomas, Springfield, Illinois.

Brezezinski, A., Bomberg, Y. M., and Braun, K., 1952, Riboflavin excretion during pregnancy and early lactation, *J. Lab. Clin. Med.* **39**:84.

Brown, R. R., Thornton, M. J., and Price, J. M., 1961, The effect of vitamin supplementation on the urinary excretion metabolites by pregnant women, *J. Clin. Invest.* **40**:617.

Burke, B. S., Beal, V. A., Kirkwood, S. B., and Stuart, H. C., 1943, The influence of nutrition during pregnancy upon the condition of the infant at birth, *J. Nutr.* **26**:569.

Busch, H., Starbuch, W., Singh, E., and Sukro, J., 1964, Chromosomal proteins, in: *The Role of Chromosomes in Development,* (M. Loche, ed.), pp. 51–71, Academic Press, New York.

Butler, N. R., 1974, Late post-natal consequences of fetal malnutrition, in: *Nutrition and Fetal Development.* (M. P. Winich, ed.), p. 173, Wiley, New York.

Calloway, D. H., 1974, Nitrogen balance during pregnancy, in: *Nutrition and Fetal Development* (M. Winich, ed.), Vol. 2, p. 79, Wiley, New York.

Campbell, S., and Newman, G. B., 1971, Growth of the fetal biparietal diameter during normal pregnancy, *J. Obstet. Gynaecol. Br. Commonw.* **78**:513–519.

Canzler, E., Brosch, G., and Schlegel, C. H., 1972, Copper and ceruloplasmin content of the cord serum in relation to foetal age, *Zentralbl. Gynaekol.* **94**:646.

Chatterjee, M., Baliga, B. S., and Munro, H. N., 1976, Synthesis of human placental lactogen and human chorionic gonadotropin by polyribosomes and messenger RNA's from early and full term placentas, *J. Biol. Chem.* **251**(10):2945–2951.

Cheek, D. B., 1968, *Human Growth,* Lea Febiger, Philadelphia.

Cheek, D. B., Holt, A. B., Ellenberg, J. H., Hill, D. E., and Sever, J. L., 1976, Nutritional studies in the pregnant rhesus monkey. The effect of protein–calorie or protein deprivation on growth of the fetal brain, *Am. J. Clin. Nutr.* **29**:1149–1157.

Churchill, J. A., Moghissi, K. S., Evans, T. N., and Frohman, D., 1969, Relationships of maternal amino acid blood levels to fetal development, *Obstet. Gynecol.* **33**:493.

Clausen, S. W., and McCoord, A. B., 1938, The carotenoids and vitamin A of blood, *J. Pediat.* **16**:635.

Cleary, R. E., Lumeng, L., and Li, T. -K., 1975, Maternal and fetal levels of pyridoxal phosphate at term: Adequacy of vitamin $B_6$ supplementation during pregnancy, *Am. J. Obstet. Gynecol.* **121**:25.

Clements, F. W., 1960, Health significance of endemic goitre and related conditions, *WHO Monogr. Se.* **44**:245–260.

Cohenour, S. H., and Calloway, D. H., 1972, Blood, urine and dietary panto thenic acid levels of pregnant teenagers, *Am. J. Clin. Nutr.* **25**:512.

Conde, C., Martinez, M., and Ballaburga, A., 1974, Some chemical aspects of human brain development. I. Neutral glycosphingolipids, sulfatides, and sphingomyelin, *Pediatr. Res.* **8**:89–92.

Coursin, D. B., and Brown, V. C., 1961, Changes in vitamin $B_6$ during pregnancy, *Am. J. Obstet. Gynecol.* **82**:1307.

Cravioto, J., and DeLicardie, E.R., 1976, Microenvironmental factors in severe protein–calorie malnutrition, *Basic Life Science* **7**:25.

Crosby, W. M., Metcoff, J., Costilloe, P., *et al.,* 1977, Fetal malnutrition: An appraisal of correlated factors, *Am. J. Obstet. Gynecol.* **128**:22.

Dancis, J., Lind, J., Oratz, M., Smolens, J., and Vara, P., 1961, Placental transfer of proteins in human gestation, *Am. J. Obstet. Gynecol.* **82**:167.

Dancis, J., Jansen, V., Kayden, J. J., Schneider, H., and Levitz, M., 1973, Transfer errors in perfused human placenta. II. Free fatty acids, *Pediatr. Res.* **7**:192.

Darby, W. J., McGenety, W. J., Martin, M. P., Bridgforth, E., Densen, P. M., Kaser, M. M., Ogle, P. J., Newhill, J. A., Stockewell, A., Ferguson, M. E., Louster, O., McClellen, G. S., Williams, C., and Cannon, R. O., 1953, The Vanderbilt cooperative study of maternal and infant nutrition IV. Dietary, laboratory and physical findings in 2,129 delivered pregnancies, *J. Nutr.* **51**:565.

Davidson, I. W. F., and Burt, R. L., 1973, Physiologic changes in plasma chromium of normal and pregnant women: Effect of glucose load, *Am. J. Obstet. Gynecol.* **116**:601.

Dobbing, J., 1968, Vulnerable periods in developing brain, in: *Applied Neurochemistry* (A. N. Davison and J. Dobbing, eds.), p. 287, Blackwell, Oxford.

Dobbing, J., and Sands, J., 1973, Quantitative growth and development of human brain, *Arch. Dis. Child.* **48**:757.

Drillien, C. M., 1970, The small-for-date infant: Etiology and prognosis, *Pediatr. Clin. North Am.* **17**:9.

Dubowitz, L. M. S., Dubowitz, V., and Goldberg, C., 1970, Clinical assessment of gestational age in the newborn infant, *J. Pediatr.* **77**:1.

Duffus, G. M., MacGilliuray, I., and Dennis, K. J., 1971, The relationship between baby weight and changes in maternal weight, total body water, plasma volume, electrolytes and proteins and urinary oestriol excretion, *J. Obstet. Gynaecol. Br. Commonw.* **78**:97.

Eastman, N. J., and Jackson, E., 1968, Weight relationships in pregnancy: I. The bearing of maternal weight gain and pre-pregnancy weight on birth weight in full term pregnancies, *Obstet. Gynecol. Surv.* **23**:1003.

Eaves, L. E., Nuttal, J. C., Klonoff, H., and Dunn, H. G., 1970, Developmental and psychological test scores in children of low birth-weight, *Pediatrics* **45**:9.

Emerson, K., Jr., Saxena, B. N., and Poindexter, E. L., 1972, Caloric cost of normal pregnancy, *Obstet. Gynecol.* **40**:786.

Enesco, M., and Leblond, C. P., 1962, Increase in all numbers as a factor in the growth of the organs of the young rat, *J. Embryol. Exp. Morphol.* **10**:530.

Erway, L., and Mitchell, S. E., 1973, Prevention of otolith defect in pastel mink by manganese supplementation, *J. Hered.* **64**:110–119.

Erway, L., Fraser, A., and Hurley, L. S., 1971, Prevention of congenital otolith defect in pallid mutant mice by manganese supplementation, *Genetics* **67**:97–108.

Fancourt, R., Campbell, S., Harvey, D., and Norman, A. P., 1976, Follow-up study of small-for-dates babies, *Br. Med. J.* **1**:1435–1437.

Favier, M. Yacoub, M., Rocinet, C., Marka, C., Chabert, P., and Benbassa A., 1972, Lésions metalliques dans le liquide ammiotique au cours du troisième trimestre de la gestation, *Rev. Fr. Gynecol.* **67**:707–714.

Felig, P., and Lynch, V., 1970, Starvation in human pregnancy: Hypoglycemia, hypoinsulinemia and hyperkotonemia, *Science* **170**:990.

Felig, P. Marliss, E., Owen, O. E., and Cahill, G. F., Jr., 1969, Blood glucose and gluconeogenesis in fasting man, *Arch. Int. Med.* **123**:293.

Felig, P., Pozefsky, T., Marliss, E., and Cahill, G. F., Jr., 1970, Alanine: Key role in gluconeogenesis, *Science* **167**:1,003.

Felig, P., Kem, Y. J., Lynch, V., and Hendler, R., 1972, Amino acid metabolism during starvation in human pregnancy, *J. Clin. Invest.* **51**:1195.

Fitzhardinge, P. M., and Steven, E. N., 1972, The small-for-date infant. II. Neurological and intellectual sequelae. *Pediatrics* **50**:50.

Flexner, L. B., 1954, Enzymatic and functional patterns of the developing brain, in: *Biochemistry of the Developing Nervous System* (H. Waelsch, ed.), Academic Press, New York.

Fosmire, G. J., AL-Ubaidi, Y. Y., and Sandstead, H. H., 1975, Some effects of postnatal zinc deficiency on developing rat brain, *Pediatr. Res.* **9**:89–93.

Fraser, J. L., and Watt, H. J., 1964, Megaloblastic anemia in pregnancy and the puerperium, *Am. J. Obstet. Gynecol.* **89**:532.

Freeman, M. G., Graves, W. L., and Thompson, R. L., 1970, Indigent Negro and Caucasian birth weight–gestational age tables, *Pediatrics* **46**:9.

Friedman, W. F., and Mills, L. F., 1969, The relationship between vitamin D and the cranio facial and dental anomalies of the supravalvala aortic stenosis syndrome, *Pediatrics* **43**:12.

Garrow, J. S., and Hawes, S. F., 1971, The relationship of the size and composition of the human placenta to its functional capacity, *J. Obstet. Gynaecol. Br. Commonw.* **78**:22–28.

Gaull, G. E., Raiha, H. C. K., Saarikoski, S., and Sturman, J., 1973, Transfer of cyst(e)ine and methionine across the human placenta, *Pediatr. Res.* **7**:908.

Ghosh, S., Bhargava, S. K., Madhavan, S., Taskar, A. D., Bhargava, V., and Higam, S. K., 1971, Intrauterine growth of north Indian babies, *Pediatrics* **47**:826.

Giroud, A., 1970, *The Nutrition of the Embryo,* Charles C. Thomas, Springfield, Illinois, p. 121.

Gordon, J. E., 1975, Nutritional individuality, *Am. J. Dis. Child.* **129:**422–424.

Gruenwald, P., 1966, Growth of the human fetus. I. Normal growth and its variation, *Am. J. Obstet. Gynecol.* **94**(1):112.

Gurson, C. T., and Saner, G., 1971, Effect of chromuim on glucose utrilization in marasmic protein-calorie malnutrition, *Am. J. Clin. Nutr.* **24:**1313.

Habicht, J. P., Yarbrough, C., Lechtig, A., and Klein, R. E., 1974, Relation of maternal supplementary feeding during pregnancy to birth weight and other socio-biological factors, in: *Nutrition and Fetal Development* (M. Winick, ed.), pp. 127–145, Wiley, New York.

Hahn, P., and Skala, J., 1970, Some enzymes of glucose metabolism in the human fetus, *Biol. Neonate* **16:**362.

Halas, E. S., Hanlon, M. J., and Sandstead, H. H., 1975, Intrauterine nutrition and aggression, *Nature* **257:**222.

Hall, M. H., 1972, Folic acid deficiency and congenital malformation, *J. Obstet. Gynaecol.* **79:**159.

Hambridge, K. M., 1971, Chromium nutrition in the women and growing child, in: *Newer Trace Elements in Nutrition,* (W. Mertz and W. E. Cornatzer, eds.), p. 169, Dekker, New York.

Hambridge, K. M., and Droegemueller, W., 1974, Changes in plasma and hair concentrations of zinc, copper, chromium and manganese during pregnancy, *Obstet. Gynecol.* **44:**666.

Hamil, B. M., Munks, B., Moyer, E. Z., *et al.,* 1947, Vitamin C in the blood and urine of the newborn, and in cord and maternal blood, *Am. J. Dis. Child.* **74:**417.

Harding, P., Possmayer, F., and Seccombe, D., 1975, Lipid substrates and fetal development, in: *The Mammoliam Fetus. Comparative Biology and Methodology* (E. S. E. Hajez, ed.), pp. 85–103, Charles C. Thomas, Springfield, Illinois.

Harper, A. E., Benevenga, N. J., and Wolhueten, R. M., 1970, Effects of ingestion of disproportionate amounts of amino acids, *Physiol. Rev.* **50:**428.

Haymond, M. W., Karl, L. E., and Pagliara, A. S., 1974, Increased gluconeogenic sustrates in the small-for gestational-age infant, *N. Engl. J. Med.* **291:**322.

Heller, S., Salkeld, R. M., and Korner, W. F., 1974*a,* Vitamin $B_1$ status in pregnancy, *Am. J. Clin. Nutr.* **27:**1221.

Hellman, L. N., and Pritchard, J. A., 1971, *Williams Obstetrics,* 14th ed., Appleton-Century-Crofts, New York.

Henkin, R. I., Marshall, J. R., and Meret, S., 1971, Maternal–fetal metabolism of copper and zinc at term, *Am. J. Obstet. Gynecol.* **110:**131–134.

Hibbard, E. M., and Hibbard, E. D., 1966, Recurrance of defective folate metabolism in successive pregnancies, *J. Obstet. Gynaecol.* **73:**428.

Higgins, A., 1974, A preliminary report of nutrition study on public maternity patients, Report of a Workshop on Nutritional Supplementation and the Outcome of Pregnancy (H. N. Jacobson, ed.) National Academy Of Science National Research Council, Washington, D.C.

Hobbins, J., 1975, personal communication.

Hsu, J. M., Anthony, W. L., and Buchanan, B. J., 1969, Zinc deficiency and incorporation of $^{14}C$-labeled methionine into tissue proteins of rats, *J. Nutr.* **99:**425–432.

Hurley, L. S., 1976, Perinatal effects of trace element deficiencies, in: *Trace Elements in Human Health and Disease* (A. S. Prasad, ed.), Vol. 2, p. 301, Academic Press, New York.

Hurley, L. S., 1977, Nutritional deficiencies and excesses, in: *Handbook of Teratology* (J. G. Wilson and F. C. Fraser, eds.), Vol. 1, pp. 261–308, Plenum, New York.

Hurley, L. S., and Bell, L. T., 1974, Genetic influence on response to dietary manganese deficiency. *J. Nutr.* **104:**133–137.

Hurley, L. S., and Bell, L. T., 1975, Amelioration by copper supplementation of mutant gene effects in the crinkled mouse (38908), *Proc. Soc. Exp. Biol. Med.* **149:**830–834.

Hurley, L. S., Gowan, J., and Shroder, R., 1968, Genetic nutritional interaction in relation to manganese and calcification, in: *Les Tissues Calcifiés,* pp. 101–104, Fifth Symposium European Societe d' Enseignement Supérieur, Paris.

Hytten, F. E., and Leitch, I., 1971, *The Physiology of Human Pregnancy,* 2nd ed., Blackwell, Oxford.

Iyengar, L., and Apte, S. V. 1972, Nutrient stores in human foetal livers, *Br. J. Nutr.* **27:**313.

Jacobson, H. 1970, *Maternal Nutrition and the Course of Pregnancy,* National Academy of Science, Washington, D.C.

Jacobson, H. N., 1974, *Report of a Workshop on Nutritional Supplementation and the Outcome of Pregnancy,* National Academy of Science, National Research Council, Washington, D.C., in press.

Jameson, S., 1976, Effects of zinc deficiency in human reproduction, *Acta Med. Scand. Suppl.* **593:**1.

Kaminetsky, H. A., Langer, A., and Baker, H., 1973, The effect of nutrition in teenage gravidas on pregnancy and the status of the neonate. I. A nutritional profile, *Am. J. Obstet. Gynecol.* **115**:639.

Katz, M., Keusch, G. T., and Mata, L. (eds.), 1975, Malnutrition and infection during pregnancy: Determinants of growth and development of the child, *Am. J. Dis. Child.* **29**:419–463.

Kim, Y. J., and Felig, P., 1972, Maternal and amniotic fluid substrate levels during calorie deprivation in human pregnancy, *Metabolism* **21**(6):507.

Kirby, L., and Han, P., 1973, Enzyme induction in human fetal liver, *Pediatr. Res.* **7**:75.

Kitay, D. Z., 1969, Folic acid deficiency in pregnancy, *Am. J. Obstet. Gynecol.* **104**:1067.

Laga, E. M., Driscoll, S. G., and Munro, H. N., 1972*a*, Comparison of placentas from two socio-economic groups. I Morphometry, *Pediatrics* **50**:24.

Laga, E. M., Driscoll, S. G., and Munro, H. N., 1972*b*, Comparison of placentas from two socioeconomic groups. II. Biochemical characteristics, *Pediatrics* **50**:33.

Lechtig, A. Habicht, J-P., Delgado, H., Klein, R. E., Yarbrough, C., and Martorell, R., 1975, Effect of food supplementation during pregnancy on birth-weight. *Pediatrics* **56**:508.

Lechtig, A., Delgado, H., Lasky, R. E., *et al.*, 1975, Maternal nutrition and fetal growth in developing societies, *Am. J. Dis. Child.* **129**:434.

Levi, S., and Smets, P., 1973, Intra-uterine fetal growth studied by ultrasonic biparietal measurements. The percentiles of biparietal distribution, *Acta. Obstet. Gynecol. Scand.* **52**:193–198.

Linblad, B. S., Rahimtoola, R. J., Said, M. Haque, Q., and Khan, N., 1969, The venous plasma free amino acid levels of mother and child during delivery. III. In a lower socioeconomic group of a refugee area in Karachi, West Pakistan, with special reference to the 'small for dates' syndrome, *Acta Pediatr. Scand.* **58**:497.

Lind, T., Billewicz, W. Z., and Cheyne, G. A., 1971, Composition of amniotic fluid and maternal blood through pregnancy, *J. Obstet. Gynecol. Br. Commonw.* **78**:505–512.

Lindheimer, M. D., and Katz, A. I., 1973, Sodium and diuretics in pregnancy, *N. Engl. J. Med.* **288**:891.

Love, E. J., and Kinch, R. A., 1965, Factors influencing the birth weight, *Am. J. Obstet. Gynecol.* **91**:342.

Lubchenko, L. O., 1976, *The High Risk Infant,* Saunders, Philadelphia.

Lubchenko, L. O., and Bard, H., 1971, Incidence of hypoglycemia in newborn infants classified by birth weight and gestational age, *Pediatrics* **47**:831.

Lubchenko, L. O., Hansman, C., Dressler, M., and Boyd, E., 1963, Intrauterine growth as etimated from liveborn birthweight data at 24 to 42 weeks of gestation, *Pediatrics* **32**:793.

Mameesh, M., Metcoff, J., Costiloe, P., and Crosby, W., 1976, Kinetic properties of pyruvate kinase in human maternal leukocytes in fetal malnutrition, *Pediatr. Res.* **10**:561.

Mansani, F. E., Cavatorta, E., and Ceroti, M., 1973, Aminocidemia e malnutrizione fetale, *Ateneo Parmense, Acta Bio-Med.* **44**:211.

Mansani, F. E., Cavatprta, Ceruti, M., Coppola, F., and Vadora, E., 1974, Effetti della infusione di 1-eminoacidi naturali liberi alla madre mella terapia della insufficienza feto-placentare cronica. *Ateneo Parmense, Acta Bio-Med.* **45**:29–40.

Martin, M. P., Bridgforth, E., McGarrity, W. J., and Darby, W. J., 1957, The Vanderbilt cooperative study of maternal and infant nutrition. X Ascorbic Acid, *J. Nutr.* **62**:20.

Martin, R. H., and Higginbottom, J., 1971, A clinical and radiological assessment of fetal age, *J. Obstet. Gynaecol. Br. Commonw.* **78**:155–162.

Martinez, M., Conde, C., and Ballabuga, A., 1974, Some chemical aspects of human brain development. II. Phosphoglycende fatty acids, *Pediatr. Res.* **8**:93.

Mason, M., and Rivers, J. M., 1971, Plasma ascorbic acid levels in pregnancy, *Am. J. Obstet. Gynecol.* **109**:960.

McCartney, C. P., Pottinger, R. E., and Harrod, J. P., 1959, Alterations in body composition during pregnancy, *Am. J. Obstet. Gynecol.* **77**:1038.

McClain, P., and Metcoff, J., 1978, Relationship of maternal amino acid profiles at 24 weeks of gestation to fetal growth, *Am. J. Clin. Nutr.* in press.

McClain, P. E., Metcoff, J., Crosby, W. M., and Costiloe, J. P., 1976, Relationship between cord blood free amino acid profiles, maternal 25 week plasma profiles, and fetal growth, Presented at Fed. Amer. Soc. Exp. Biol., April 1977.

McGanity, W. J., Cannon, R. O., Bridgforth, E. B., Martin, D. P., Densen, P. M., Newbill, J. A., McClellan, G. S., Christie, A., Peterson, J. C., and Darby, W. J., 1954, The Vanderbilt cooperative study of maternal and infant nutrition. VI. Relationship of obstetric performance to nutrition, *Am. J. Obstet. Gynecol.* **67**:491, 501.

McLaren, D. S., 1968, To eat to see, *Nutr. Today* **3**:2.

McLean, F., and Usher, R., 1970, Measurements of liveborn fetal malnutrition infants compared with similar birth weight normal controls, *Biol. Neonate* **16:**215.

Mellanby, M., and Coumoulos, H., 1944, The improved dentition of 5-yr-old London school children, a comparison between 1943 and 1929, *Br. Med. J.* **1:**837.

Menkes, J., 1972, Kinky hair disease, *Pediatrics* **50:**181–183.

Mertz, W., 1976, Chromium and its relation to carbohydrate metabolism, *Med. Clin. North Am.* **60:**739.

Metcoff, J., 1974a, Maternal leukocyte metabolism in fetal malnutrition, in: *Advances in Experimental Medicine and Biology, Nutrition & Malnutrition, Identification and Measurement.* (A. F. Roche and F. Falkner, eds.), p. 49, Plenum, New York.

Metcoff, J., 1974b, Biochemical markers of intrauterine malnutrition, in: *Nutrition and Development* (M. Winick, ed.), Wiley, New York.

Metcoff, J., 1976, Maternal nutrition and fetal growth, in: *Textbook of Paediatric Nutrition* (D. S. McLaren and D. Burnan, eds.), pp. 19–45, Churchill Livingstone, London.

Metcoff, J., Wilkman-Coffelt, J., Yoshida, T., *et al.,* 1973, Energy metabolism and protein synthesis in human leukocytes during pregnancy and in placenta related to fetal growth, *Pediatrics* **51:**866.

Metcoff, J., Costiloe, P., Crosby, W., Sandstead, H., and McClain, P., 1976, Predicting fetal growth, (abstract) *Fed. Proc.* **36:**1109(abstract).

Miller, H. C., and Hassanein, K., 1974, Maternal smoking and fetal growth of full term infants, *Pediatr. Res.* **8:**960.

Miller, H. C., Hassanein, K., and Hensleigh, P. A., 1976, Fetal growth retardation in relation to maternal smoking and weight gain in pregnancy, *Am. J. Obstet. Gynecol.* **125:**55.

Mitchell, R. G., 1966, Nutritional influences in early life, *Proc. Roy. Soc. Med.* **59:**1073.

Mukherjee, A. B., Hastings, C., and Cohen, M. M., 1973, Nucleic acid synthesis in various organs of developing human fetuses, *Pediatr. Res.* **7:**696.

Naeye, R. L., 1965, Malnutrition: probable cause of fetal growth retardation, *Arch. Pathol.* **79:**284.

Naeye, R. L., Diener, M. M., Harcke, H. T. J., and Blanc, W. A., 1971, Relation of poverty and race to birthweight and organ and cell structure in the newborn, *Pediatr. Res.* **5:**17.

Nassi, L., Frangeni, V., Poggini, G., Pratesi, C., Vecchic, C., and Ratazzi, M., 1970, Plasmatic and erythrocytic copper in maternal, cord, and neonatal blood in the first week of life in relation to the quantity of haemoglobin. Determination of *in toto* placental copper (statistical analysis of results), *Pediatrics* **78:**818.

National Research Council, 1974, *Recommended Dietary Allowances,* National Academy of Sciences, Washington, D.C.

Neligan, G. A., McI. Scott, D., Kolvin, I., and Garside, R., 1976, Born too soon or born too small, *Clinics in Developmental Medicine,* No. 61, W. Heinemann, London.

Newberne, P. M., and Young, V. R., 1973, Marginal vitamin $B_{12}$ intake during gestation in the rat has long term effects on the offspring, *Nature* **242:**263.

Niswander, K. B., Singer, J., Westphal, M. J., and Weiss, W., 1969, Weight gain during pregnancy and prepregnancy weight, *Obstet. Gynecol.* **33:**482.

Ornov, A., Kaspi, T., and Nebel, L., 1972, Persistent defects of bone formation in young rats following hypervitaminosis $D_2$, *Isr. J. Med. Sci.* **7:**943.

Osofsky, H. J., 1975, Relationships between nutrition during pregnancy and subsequent infant and child development, *Obstet. Gynecol. Survey* **30:**227.

O'Sullivan, J. B., Gellis, S. S., and Tenny, B. O., 1965, Aspects of birth weight and its influencing variables, *Am. J. Obstet. Gynecol.* **92:**1023.

Ounsted, M., and Ounsted, C., 1966, Maternal regulation of intrauterine growth, *Nature* **212:**995.

Pathak, S., Himays, S., and Mosher, R., 1974, The small-for-dates syndrome: Some biochemical considerations in prenatal diagnosis, *Am. J. Obstet. Gynecol.* **120:**32.

Petrucco, O. M., Collier, K., and Fish Tall, A., 1973, Diagnosis of intrauterine fetal growth retardation by serial serum oxytocinase, urinary oestrogen & serum (HSAP) estimations in uncomplicated and hypertensive pregnancies, *J. Obstet. Gynaecol. Br. Commonw.* **80:**499.

Philipps, C., and Johnson, N. E., 1977, The impact of quality of diet and other factors on birth weight of infants, *Am. J. Clin. Nutr.* **30:**215.

Pitkin, R. M., 1975a, Risks related to nutritional problems in pregnancy, in: *Risks in the Practice of Modern Obstetrics* (S. Aladjem, ed.), pp. 165–181, C. V. Mosby, St. Louis.

Pitkin, R. M., 1975b, Vitamins and minerals in pregnancy, *Clin. Perinatol.* **2:**221.

Pitkin, R. M., 1976, Nutritional support in obstetrics and gynecology, *Clin. Obstet Gynecol.* **19:**19.

Pitkin, R. M., Kaminetzky, H. A., Newton, M., and Pritchard, J. A., 1972, Maternal nutrition. A selective review of clinical topics, *Obstet. Gynecol.* **40:**773.

Portliege, P. R., 1972, Hypervitaminosis $D_2$ in gravid rats, *Arch. Pathol.* **73:**29.

Raiha, N., 1958, On the placental transfer of vitamin C. An experimental study on guinea pigs and human subjects, *Acta Physiol. Scand.* **45**(Suppl. 155):5–53.

Renaud, R., Vincendon, G., Boog, G., Brettes, J. P., Schumacher, J. C., Koehl, C., Kirchstetter, L., and Gandar, R., 1972, Injections intra-aminio-tiques d'acides amines dans les cas de malnutrition foetale, *J. Gynecol. Obstet. Biol.* **1**:231–244.

Roopnarinesingh, S., Morris, D., and Chang, E., 1971, The underweight Jamaican parturient. *J. Obstet. Gynaecol. Br. Commonw.* **78**:379–382.

Rosado, A., Bernal, A., Sosa, A., Morales, M., Urrusti, J., Yoshida, P., Fresk, A., Velasco, L., Yoshida, T., and Metcoff, J., 1972, Human fetal growth retardation. III. Protein, DNA, RNA, adenosine nucleotides, and activities of the enzymes pyruvate and adenylate kinase in placenta, *Pediatric* **50**:568.

Rush, D., 1975, Maternal nutrition during pregnancy in industrialized societies, *Am. J. Dis. Child.* **129**:430.

Samborskaya, E. P., and Ferdman, T. D., 1966, The mechanism of termination of pregnancy by ascorbic acid, *Bull Exp. Biol. Med.* **62**:934 (cited by Hurley, 1977).

Sandstead, H. H., 1973, Zinc nutrition in the U.S., *Am. J. Clin. Nutr.* **26**:1251.

Sandstead, H. H., and Rinaldi, R. A., 1969, Impairment of deoxyribonucleic acid synthesis by dietary zinc deficiency in the rat, *J. Cell Physiol.* **73**:81–84.

Sandstead, H. H., Fosmire, G. J., Halas, E. S., Jacobs, R. A., Strobel, D. A., and Marks, E. D., 1977, Zinc deficiency: Effects on brain and behavior of rats and rhesus monkeys, *Teratology* in press.

Sandstead, H. H., Gillespie, D. D., and Brady, R. N., 1972, Zinc deficiency: Effect on brain of the suckling rat, *Pediatr. Res.* **6**:119–125.

Sarram, M., Younessi, M., Khowasky, P., Kjoury, G. A., and Reinhold, J. G., 1969, Zinc nutrition in human pregnancy in Fars Province, Iran. Significance of geographic and socioeconomic factors, *Am. J. Clin. Nutr.* **22**:726.

Schenker, J. G., Jungresis, E., and Polisheck, W. Z., 1972, Maternal and foetal serum copper levels at delivery, *Biol. Neonate* **20**:189.

Scott, K. K., and Usher, R. H., 1966, Fetal malnutrition: Its incidence, causes and effects, *Am. J. Obstet. Gynecol.* **94**:951.

Scott, D. E., Whalley, P. J., and Pritchard, J. A., 1970, Maternal folate deficiency and pregnancy wastage. II. Fetal Malformation, *Obstet. Gynecol.* **36**:26.

Sever, L. E., and Emanuel, I., 1972, Is there a connection between maternal zinc deficiency and congenital malformation of the central nervous system in men, *Teratology* **7**:117.

Shaw, J. C. L., 1974, Malnutrition in very low birth-weight, pre-term infants, *Proc. Nutr.* **33**:103.

Shelley, H. J., 1961, Glycogen reserves and their changes at birth and in anoxia, *Br. Med. Bull.* **17**:137.

Simpson, J. W., Lawless, R. W., and Mitchell, A. C., 1975, Responsibility of the obstetrician to the fetus II. Influence of pre-pregnancy weight & pregnancy weight gain on birthweight, *Obstet. Gynecol.* **45**:481.

Smart, J. L., Tricklebank, M. D., Adlard, B. P. F., and Dobbing, J., 1976, Nutritionally small-for-dates rats: Their subsequent growth, regional brain 5-hydroxytryptamine turnover, and behavior, *Pediatr. Res.* **10**:807–811.

Smith, C. A., 1947, Effects of maternal undernutrition upon the newborn infant in Holland (1944–45), *J. Pediatr.* **30**:229.

Smith, C. H., and Depper, R., 1974, Placental amino acid uptake II. Tissue preincubation, fluid distribution, and mechanisms of regulation, *Pediatr. Res.* **8**:697–703.

Smith, C. H., Adcock, E. W., III, Teasdale, F., *et al.,* 1973, Placental amino acid uptake. Tissue preparation, kinetics, and preincubation effect, *Am. J. Physiol.* **224**:558.

Somers, M., and Underwood, E. J., 1969, Studies of zinc deficiency in ram-lambs upon the digestability of the dry matter and the utilization of the nitrogen and sulphur of the diet, *Aust. J. Agric. Res.* **20**:899–903.

Stein, Z., and Susser, M., 1975a, The Dutch famine, 1944–1945, and the reproductive process. I. Effects on six indices at birth, *Pediatr. Res.* **9**:70.

Stein, Z., and Susser, M., 1975b, The Dutch famine, 1944–1945 and the reproductive process. II. Interrelations of calorie rations and six indices at birth, *Pediatr. Res.* **9**:76–83.

Stein, Z., Susser, M., Saenger, G., and Marolla, F., 1972, Nutrition and mental performance. Prenatal exposure to the Dutch famine of 1944–45 seems not related to mental performance at age 19, *Science* **178**:708–714.

Sterky, J., 1970, Swedish standard curves for intrauterine growth, *Pediatrics* **46**:7.

Stewart, R. J. C., Preece, R. F., and Sheppard, H. G., 1975, Twelve generations of marginal protein deficiency, *Br. J. Nutr.* **33**:233.

Taggart, N., Holliday, R. M., Billewicz, W. Z., Hytten, F. E., and Thomson, A. M., 1967, Changes in skinfolds during pregnancy, *Br. J. Nutr.* **21**:439.

Tanner, J. R., and Thomson, A. M., 1970, Standards for birthweight at gestation periods from 32 to 42 weeks, allowing for maternal height and weight, *Arch. Dis. Child.* **45**:566.

Terhune, M. W., and Sandstead, H. H., 1972, Decreased RNA polymerase activity in mammalian zinc deficiency, *Science* **177**:68–69.

Thomson, A. M., 1959, Diet in pregnancy, *Br. J. Nutr.* **13**:190.

Thomson, A. M., and Billewicz, W. Z., 1957, Clinical significance of weight trends during pregnancy, *Br. Med. J.* **1**:243.

Thomson, A. M., Billewicz, W. Z., and Hytten, F. E., 1968, The assessment of fetal growth, *J. Obstet. Gynaecol. Br.* **75**:903–916.

Tsang, R. C., Donovan, E. F., and Steichen, J. J., 1976, Calcium physiology and pathology in the neonate, *Ped. Clin. N. Amer.* **23**:611.

Underwood, E. J., 1971, *Trace Elements in Human and Animal Nutrition,* 3rd ed., pp. 83–87, Academic Press, New York.

Urrusti, J., Yoshida, P., Valasco, L., Frenks, S., Rosado, A., Sosa, A., Morales, M., Yoshida, T., and Metcoff, J., 1972, Human fetal growth retardation. I. Clinical features of sample with intrauterine growth retardation, *Pediatrics* **50**:547.

Usher, R. H., 1970, Clinical and therapeutic aspects of fetal malnutrition, *Pediatr. Clin. North Am.* **17**:169.

Usher, R., McLean, F., and Scott, K. E., 1966, Judgment of fetal age. II. Clinical significance of gestational age and an objective method for its assessment, *Ped. Clin. N. Amer.* **13**:835.

Utter, M. F., 1976, The biochemistry of manganese, *Med. Clin. North Am.* **60**:713.

Van den Berg, B. J., Yerushalmy, J., 1966, The relationship of the rate of intrauterine growth of infants of low birthweight to mortality, morbidity, congenital anomalies, *J. Pediatr.* **69**:531.

Velazquez, A., Rosado, A., Bernal, A., Noriega, L., and Arevalo, N., 1976, Amino acids pools in the feto-maternal system, *Biol. Neonate* **29**:28–40.

Wachstein, M., and Graffeo, L. W., 1956, Influence of vitamin $B_6$ on the incidence of preeclampsia, *Obstet. Gynecol.* **8**:177.

Warkany, J., Monroe, B. B., and Sutherland, B. S., 1961, Intrauterine growth retardation, *Am. J. Dis. Child.* **102**:127.

White, H. S. 1969, Inorganic elements in weighed diets of girls and young women, *J. Am. Diet. Assoc.* **55**:38.

Widdowson, E. M., 1968, Growth and composition of the fetus and newborn, in: *Biology of Gestation* (N. S. Assali, ed.), Vol. 2, pp. 1–49, Academic Press, New York.

Widdowson, E. M., 1974a, Nutrition, in: *Scientific Foundations of Paediatrics* (J. A. Davis and J. Dobbing, eds.) p. 44, Heineman Medical Books, London.

Widdowson, E. M., 1974b, Changes in body proportions and composition during growth, in: *Scientific Foundations of Paediatrics* (J. A. Davis and J. Dobbing, eds.), p. 153, Heinemann Medical Books, London.

Widdowson, E. M., and Dickerson, J. W. T., 1964, The chemical composition of the body, in: *Mineral Metabolism* (C. L. Comar and F. Bronner, eds.), Vol. 1A, pp. 1–246, Academic Press, New York.

Widdowson, E. M., and Spray, C. M., 1951, Chemical development *in utero, Arch. Dis. Child.* **26**:205.

Winick, M., 1967, Cellular growth of human placenta III. Intrauterine growth failure, *J. Pediatr.* **7**:390.

Winick, M., 1970, Nutrition and nerve cell growth, *Fed. Proc.* **29**:1510.

Winick, M., 1976, Early malnutrition and brain structure and function, *Nutr. Notes, Amer. Inst. Nutrition, p. 4.*

Winick, M., and Noble, A., 1965, Quantitative changes in DNA, RNA, and protein during prenatal and postnatal growth in the rat, *Dev. Biol.* **12**:451.

Woodard, J. C., and Newberne, P. M., 1967, The pathogenesis of hydrocephalus in newborn rats deficient in vitamin $B_{12}$, *J. Embryol. Exp. Morphol.* **17**:177.

Young, M., and Prenton, M. A., 1969, Maternal and fetal plasma amino acid concentrations during gestation and in retarded fetal growth, *J. Obstet. Gynaecol. Br. Commonw.* **76**:333.

Zamenhof, S., Van Marthens, E., and Gravel, L., 1971, DNA (cell number) in neonatal brain: second generation ($F_2$) alteration by maternal ($F_0$) dietary protein restriction, *Science* **172**:850.

# 18

# Carbohydrate, Fat, and Amino Acid Metabolism in the Pregnant Woman and Fetus

## PETER A. J. ADAM and PHILIP FELIG

## 1. Introduction

Throughout gestation, the human fetus accumulates the elements required for growth, oxidizes fuel transported from the mother, and excretes waste products transplacentally. In the last half of pregnancy, it also accumulates the energy stores which permit the independent maintenance of homeostasis during postnatal fasting. Thus, food ingested by the mother not only is consumed for her own needs but also is modified appropriately for uptake by the conceptus.

In this chapter, we shall examine the way in which the conceptus alters maternal metabolic responses to feeding and fasting. Three major metabolic compartments, involved in both the adaptation to pregnancy and to periodic maternal ingestion will be considered: those of the mother, the placenta, and the fetus.

First, maternal metabolism will be examined from the point of view of two apparently paradoxical events—the increased synthesis and storage of energy, and opposed to it, the more rapid catabolism and turnover of fuels during pregnancy. In mammalian models, pregnancy has been considered to be a phased event in which maternal synthesis is enhanced at mid-gestation, while catabolism is increased at term. In man, however, these events occur contiguously and are regulated by the periodic nature of maternal feeding. Gestational hormones, produced by the feto-placental unit, simultaneously enhance the capacity for synthesis of stores and the potential for diversion of maternal metabolism from carbohydrate to fat sources.

PETER A. J. ADAM • Case Western Reserve University School of Medicine, Cleveland Metropolitan General Hospital, Cleveland, Ohio. PHILIP FELIG • Yale University School of Medicine, New Haven, Connecticut.

PETER A. J. ADAM and
PHILIP FELIG

Meanwhile, glucose produced by gluconeogenesis and amino acids from maternal protein are diverted partially toward fetal requirements.

Second, the placental role of regulating substrate flow to the fetus will be examined. Finally, fetal consumption of maternally generated substrates will be evaluated in the context of fetal growth and development. Studies of mammals, the subhuman primate, and man will be examined to evaluate the potential or actual responses of the human fetus to maternal feeding, fasting, and labor.

In concluding, a hypothesis integrating our present knowledge of fetal fuel metabolism is presented. We propose that sequential hormonal events develop the fetal capacity for response to the altered intrauterine environment, but that the known fetal hormonal responses play a minor role in the fine regulation of fuel metabolism *in utero*. From mid-term onwards, the human fetus seems to have a highly developed capacity for both synthesis and mobilization of fuels which permits autonomous cycling of its responses according to maternal provisions. Thus an ordered response to maternal privation recruits first glucose and alternate fuels from the mother, then mobilizes fetal energy stores. These responses provide a continuous supply of the fuels required for the metabolism of fetal organs, including the central nervous system. Chronic disruption of the intrauterine environment by placental insufficiency or by maternal hyperglycemia distort the chemical and hormonal milieu, disrupt the balance between synthetic and lytic events in the fetus, and interfere with fetal growth and development.

## 2. Body Fuel Metabolism in Pregnancy

### 2.1. Metabolic Trends in Pregnancy

The pregnant state is associated with a variety of alterations in maternal fuel–hormone metabolism. A seeming paradox exists in which "accelerated starvation" and a tendency to fasting hypoglycemia (Freinkel, 1969; Felig and Lynch, 1970) are accompanied by a diabetogenic state characterized by insulin resistance, hyperinsulinemia, and a tendency to postprandial hyperglycemia (Burt, 1956; Baird, 1969). The major influences responsible for these disparate tendencies are: (1) continuous consumption of glucose and amino acids by the conceptus; (2) the production of hormones by the placenta which enter the maternal circulation and exert contrainsulin effects; and (3) the secretion of hormones by the fetoplacental unit which enhance maternal pancreatic insulin secretion. To understand the manner whereby these influences alter maternal metabolism, hormone–fuel relationships in fasting and fed adult man will be compared with those of the fasted and fed pregnant woman.

### 2.2. The Fasted State

#### 2.2.1. Fasting Metabolism in Adult Man

Maintenance of glucose homeostasis in the fasted, postabsorptive state depends on a balance between glucose production and utilization (Figure 1). Glucose uptake occurs primarily in the brain and to a much lesser extent in the formed elements of the blood and muscle tissue. In each of these sites, glucose

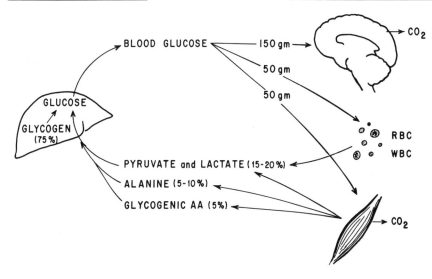

Fig. 1. Glucose balance in the postabsorptive state in the nonpregnant condition. Glucose utilization rates represent g/24 hr. The values in parentheses represent the proportion of total hepatic glucose production accounted for by the various precursors shown.

consumption is not insulin dependent. To meet the needs of these tissues, glucose is continuously released by the liver at rates of 2–3 mg/kg/min (150–250 g/day) (Felig, 1973a). The majority of the glucose released (75%) is derived from the breakdown of glycogen. The remaining 25% represents new glucose production (gluconeogenesis) from lactate, amino acids (principally alanine), and to a lesser extent, glycerol (Felig, 1973a).

The major hormonal signal regulating the production of glucose by the liver is the fall in circulating insulin levels from concentrations of 50–100 $\mu$U/ml in the fed state to 10–20 $\mu$U/ml in the fasted condition (Cahill *et al.,* 1966). The presence of basal concentrations of circulating glucagon (50–100 pg/ml) is also necessary for the maintenance of hepatic glucose production (Alford *et al.,* 1974). In addition to the proper hormonal milieu, it is evident that glucose precursors (such as lactate, alanine, and glycerol) must be delivered to the liver in adequate concentrations. This is particularly true in circumstances in which hepatic glycogen stores have been depleted (see below) and gluconeogenesis is responsible for an increased proportion of total hepatic glucose output.

As fasting extends beyond the postabsorptive state, the various metabolic processes are directed at maintaining glucose homeostasis as well as conserving protein stores. The over-all response may be characterized as biphasic, in which an initial and late phase may be distinguished (Felig *et al.,* 1969a). The initial response to fasting is concerned with the maintenance of glucose production to meet the needs of the brain (gluconeogenic phase). The late phase is directed at minimizing the rate of protein breakdown (protein conservation phase). Although it is difficult to place precise limits on the duration of each phase, the gluconeogenic phase prevails during the first 5–10 days of fasting and gradually is replaced by substitution of alternate fuels and increased protein conservation (Saudek and Felig, 1976).

PETER A. J. ADAM and
PHILIP FELIG

With regard to the first phase of fasting (Figure 2), carbohydrate stores are quite limited, so that maintenance of glucose homeostasis requires an increase in glucose production (gluconeogenesis) and a diminution in extracerebral glucose utilization. The main signal initiating the adaptive response to brief starvation is a small decline (10–15 mg/100 ml) in blood glucose concentration. The feedback sensitivity of the pancreatic islet cells is such that these small changes in blood glucose result in a decline in circulating insulin (Cahill *et al.,* 1966) and an increase in glucagon (Marliss *et al.,* 1970). As a result of the decline in plasma insulin, free fatty acids are released by lipolysis, oxidized, and provide the energy for gluconeogenesis and for muscle metabolism. The increased delivery of fatty acids to the liver and increased activity of intrahepatic enzymes engendered by the fall in plasma insulin result in a progressive rise of the blood ketones. Early in starvation the ketones contribute to muscle fuel requirements as an oxidizable substrate, while late in starvation (see below) ketones largely replace glucose as a fuel for the brain.

During early fasting, the liver is the sole site of synthesis and release of glucose into the bloodstream. Inasmuch as liver glycogen stores amount to no more than 70–90 g in the postabsorptive state and are dissipated within the first 1–3 days of fasting (Hultman and Nilsson, 1971), glucose production depends on intact gluconeogenic mechanisms; lactate and amino acids constitute the main gluconeogenic substrates. In this circumstance amino acids are of particular importance since they represent protein-derived glucose precursors.

Although all amino acids other than leucine have been recognized as potentially glucogenic, the classic "glucogenic/ketogenic" classification of amino acids fails to take into account the relative availability of individual amino acids as endogenous glucose precursors. When one examines amino acid balance across muscle tissue, the major reservoir of body protein stores, a specific pattern of amino acid exchange emerges. A net release is observed for virtually all amino acids. However, the output of alanine and glutamine exceeds that of all other amino acids (Felig *et al.,* 1970). Complementing this output of alanine from muscle is the pattern of amino acid uptake by the intestine and liver; the splanchnic uptake of alanine and glutamine exceeds that of all other amino acids (Felig *et al.,* 1969c). Whereas the gut is the site of splanchnic glutamine utilization (Felig *et al.,* 1973a,b), alanine is taken up solely by the liver. Evidence that alanine is used by the liver for gluconeogenesis is provided by studies with the perfused rat liver (Fisher and Kerly, 1964; Mallette *et al.,* 1969). In this sytem, the liver forms glucose from alanine more efficiently than from any other amino acid; furthermore, conversion of alanine to glucose is dependent primarily on the availability of alanine.

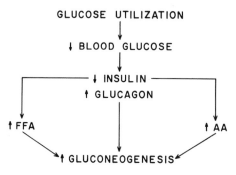

Fig. 2. Summary of the metabolic response to early starvation in the nonpregnant state. On-going glucose utilization leads to a small decline in blood glucose. The latter triggers a fall in insulin and rise in glucagon resulting in mobilization of fatty acids (lipolysis) and amino acids (proteolysis) and stimulation of gluconeogenesis.

**465**

*CARBOHYDRATE, FAT,
AND AMINO ACID
METABOLISM IN THE
PREGNANT WOMAN
AND FETUS*

During brief periods of fasting, muscle is thus in a catabolic state releasing large amounts of alanine for uptake by the liver. Over the initial 72 hr of starvation, the output of alanine increases progressively (Pozefsky *et al.*, 1974; Blackshear *et al.*, 1974). In addition, the fractional extraction of alanine by the liver at 72 hr of fasting is increased by 50% above the level observed in the postabsorptive (overnight fasted) condition. As a consequence of the increased muscle delivery and augmented hepatic extraction of gluconeogenic precursors, glucose production is able to keep pace with the rapid rates of glucose utilization in the brain.

Inasmuch as mammalian tissue lacks the enzymatic capacity to form glucose from fatty acids, it is clear that the early or gluconeogenic response to fasting is achieved at the expense of rapid proteolysis. As observed by Benedict many years ago, the rate of nitrogen loss in starvation does not persist unchanged (Benedict, 1915). As fasting extends beyond one week (Figure 3), nitrogen loss, particularly urea excretion, progressively declines reaching levels of 3–4 g/day after 4–6 weeks of starvation (Owen *et al.*, 1969). Concomitant with this reduction in protein catabolism, total glucose production declines from rates of 150–200 g/day in the postabsorptive state to less than 90 g/day after 3–6 weeks of starvation.

That the liver is not the prime regulatory site of the decrease in gluconeogenesis is suggested by the fact that the fractional extraction of alanine by the splanchnic bed is no less in prolonged fasting than in postabsorptive man (Felig *et al.*, 1969*c*). In contrast, the level of circulating alanine and the output of alanine from muscle are markedly reduced in prolonged starvation (Felig *et al.*, 1969*c*, 1970). Although similar declines are observed for a variety of amino acids, the magnitude of the fall in alanine level is greater than that of all other amino acids. Furthermore, when alanine levels are raised by infusion of this amino acid in subjects fasted for a prolonged period, a prompt rise in the blood glucose level is observed (Felig *et al.*, 1969*b*). The liver thus remains capable of rapidly converting alanine to glucose throughout starvation. The rate-limiting factor responsible for the reduction of hepatic glucose production as starvation progresses is the fall in circulating alanine levels (Felig *et al.*, 1970).

With respect to the mechanism of the fall in alanine, provision of alternate fuels for metabolism may be of major importance. Studies conducted by Sherwin *et al.* (1975) indicate that infusion of ketones results in a specific decline in plasma alanine. Thus, by inducing hyperketonemia, the hypoalaninemia of starvation can

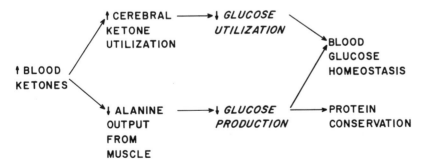

Fig. 3. Summary of the metabolic response to prolonged starvation. The elevation in blood ketones permits the latter to serve as substrate for the brain thereby decreasing glucose utilization. Simultaneously, hyperketonemia limits the outflow of alanine from muscle thereby reducing gluconeogenesis. In this manner glucose homeostasis and protein conservation are achieved.

be simulated. Furthermore, the alanine-lowering effect of an increase in blood ketones is demonstrable in prolonged fasted as well as in postabsorptive subjects. Finally, when hyperketonemia and hypoalaninemia are maintained by prolonged infusion of ketones, a concomitant reduction in urinary nitrogen excretion is observed. In considering the mechanism whereby ketones could influence protein catabolism and alanine production in muscle, it has been shown that physiologic increments in ketone acids can inhibit the oxidation of branched-chain amino acids in muscle (Buse *et al.*, 1972). As noted, the branched-chain amino acids are the major source of nitrogen for alanine synthesis in muscle (Odessey *et al.*, 1974; Wahren *et al.*, 1976).

These findings have been interpreted as indicating that ketones have a dual role in the late phase of starvation, serving as "substrate" as well as "signal" (Sherwin *et al.*, 1975). The ketones are the major energy-yielding substrate for the brain, thereby reducing the demand for glucose. Concomittantly they provide a signal to muscle, resulting in decreased amino acid catabolism and reduced output of alanine. The resulting hypoalaninemia in turn is responsible for the reduction in hepatic gluconeogenesis (Figure 3). In assigning a pivotal role to ketones, the postulated scheme provides a mechanism for the coordinated reduction of glucose production and utilization in prolonged fasting. It should be emphasized, however, that insulin undoubtedly contributes in a "permissive" manner to these effects of starvation ketonemia, inasmuch as the marked hyperketonemia of diabetic ketoacidosis is associated with rapid rates of protein breakdown (Benedict and Joslin, 1912).

### 2.2.2. Accelerated Fasting in Pregnancy

In the fasted condition, the effects of pregnancy on metabolism seem to result primarily from the continuous withdrawal of glucose and amino acids from the maternal to fetal circulation. Maternal blood glucose and plasma insulin decline more rapidly during fasting (Figures 4 and 5), resulting in an accelerated response to starvation. The continuous placental secretion of placental lactogen (HPL), also referred to as human chorionic somatomammotropin (HCS), may be important in providing a contrainsulin, lipolytic agent. This placental hormone promotes the diversion of maternal metabolism from the consumption of endogenously produced glucose to the mobilization and oxidation of FFA.

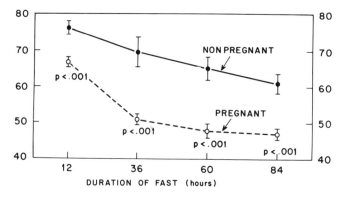

Fig. 4. Plasma glucose levels (shown in mg/100 ml) in fasted nonpregnant and pregnant subjects in the second trimester. (Based on the data of Felig and Lynch, 1970.)

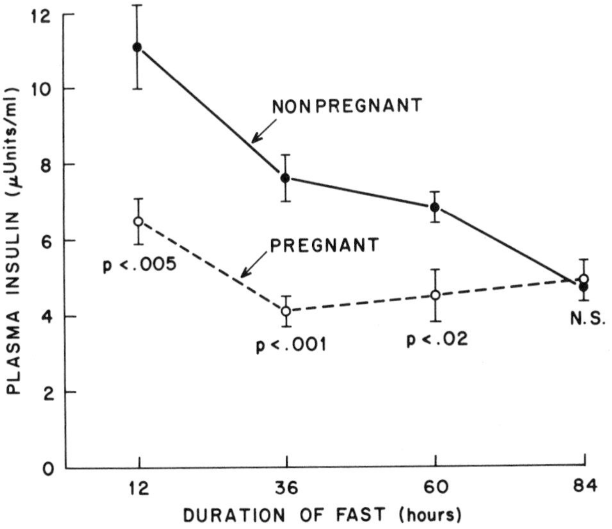

Fig. 5. Plasma insulin levels in fasted nonpregnant and pregnant subjects in the second trimester. (Based on the data of Felig and Lynch, 1970.)

The fuel requirements of the developing fetus are believed to be met mostly by the consumption of glucose (Dawes and Shelley, 1968). Glucose not only is utilized to provide energy necessary for protein synthesis but also constitutes the precursor for the synthesis of fat and for the formation of glycogen. The over-all level of glucose uptake required to meet these synthetic and oxidative needs has been estimated at 20 mg/min at term (Page, 1969); thus, the fetus alone consumes 12–20% of glucose being produced by the maternal liver and kidney, while depleting glycogen reserves more rapidly than in the nonpregnant state. Simultaneously, with this uptake of glucose, there is active transport of amino acids including alanine from mother to fetus. Largely as a result of this siphoning effect, plasma glucose levels after an overnight fast are significantly lower in pregnant than in nonpregnant women (Figure 5). That pregnancy accelerates the response to fasting is evident from the fact that the plasma glucose concentration has fallen to a plateau within 36 hr in pregnant women, but still is declining toward a similar level in nonpregnant subjects fasting for 84 hr (Felig, 1973b).

The effect of pregnancy on the response of plasma insulin to fasting is similar to that observed with glucose (Figure 5). After an overnight fast and for the first 60 hr of fasting, plasma insulin in pregnancy is reduced to approximately 50% of the concentration observed in the nonpregnant condition. Furthermore, whereas plasma insulin reaches its nadir in the pregnant group within 36 hr, a continuous decline is observed over 84 hr before comparable levels are reached in nonpregnant subjects. Throughout the fasting plasma insulin shows a significant direct linear correlation with plasma glucose. The data thus are consistent with the conclusion that the fasting hypoinsulinemia in pregnancy is a consequence of the hypoglycemia.

As a number of investigators have reported hyperinsulinemia in pregnancy (Spellacy and Goetz, 1963; Bleicher et al., 1964), the above data may seem at variance with established concepts. However, two aspects of the fasting data need

PETER A. J. ADAM and
PHILIP FELIG

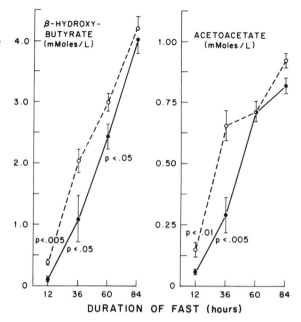

Fig. 6. Blood ketone acid levels in fasted nonpregnant and pregnant subjects in the second trimester. Ketone acids were significantly elevated in the pregnant group for the initial 36–60 hr of fasting (Felig and Lynch, 1970).

be emphasized. First, we are dealing with basal rather than glucose-stimulated or aminogenic-stimulated insulin concentrations. Thus, although the stimulative response of the beta cell increases with pregnancy, basal secretion falls in association with the fall in basal glucose levels (Tyson *et al.,* 1969). Second, as pregnancy progresses to the third trimester, a small rise of basal insulin levels is observed (Daniel *et al.,* 1974).

The influence of pregnancy on ketosis during fasting is shown in Figure 6. After an overnight fast and for the first 36–60 hr of starvation, blood $\beta$-hydroxybutyrate and acetoacetate are two- to threefold higher in the pregnant state. If fasting continues for 84 hr, however, blood ketone levels become virtually identical in the two groups of subjects in association with equalization of plasma insulin levels. Ketone acid concentration thus is higher in the pregnant group only as long as plasma insulin levels are significantly below those of nonpregnant controls. Thus, the data suggest that lowered plasma insulin levels are responsible for the heightened ketonemia of pregnancy. This effect of insulin lack is likely to be mediated via augmented adipose tissue lipolysis and hepatic ketogenesis. In summary, the maternal response to fasting with respect to circulating fuels and insulin may be characterized as an acceleration of the phenomena observed in the nongravid condition. Glucose concentration falls more rapidly and to a greater extent, thereby causing hypoinsulinemia which, in turn, precipitates an augmentation of ketosis during fasting.

Of particular interest is the mechanism responsible for the striking degree of fasting hypoglycemia observed in the pregnant group. It is clear that maternal gluconeogenic processes fail to keep pace with peripheral (maternal + fetal) glucose requirements as reflected by the fall in blood glucose. The question may be raised as to whether maternal hepatic gluconeogenesis is limited as a consequence of alterations in intrahepatic processes or alternatively as a result of changes in the supply of glucose precursors.

Concerning possible direct effects of gestation on the liver, studies in pregnant rats involving isotopically labeled glucose precursors have revealed an augmented rather than diminished capacity for conversion of exogenous substrate to glucose (Herrera *et al.*, 1969; Metzger *et al.*, 1971). However, similar data are neither available nor readily obtainable in humans. As to the role of endogenous substrate presentation, previous studies, as noted above, have identified alanine as a key gluconeogenic precursor (Felig *et al.*, 1970; Felig, 1973*a*). In prolonged fasting, alanine availability appears to be the rate-limiting factor in hepatic gluconeogenesis (Felig *et al.*, 1969*c*).

**469**

CARBOHYDRATE, FAT,
AND AMINO ACID
METABOLISM IN THE
PREGNANT WOMAN
AND FETUS

As shown in Figure 7, plasma alanine concentration is significantly lower in the pregnant state during the postabsorptive period (12-hr fast) and for the first 50 hr of fasting. Whereas a small but significant decline of plasma alanine occurs in the pregnant group between 12 and 60 hr of fasting, alanine levels fail to decline significantly in the nonpregnant group until 84 hr of fasting. These data indicate that pregnancy accelerates and exaggerates the hypoalaninemic response to fasting and suggest that lack of endogenous substrate contributes to gestational hypoglycemia by preventing a sufficient rise of hepatic glucose production during fasting. Further support of this conclusion is obtained from the glycemic response to infusion of alanine. When plasma alanine levels are increased in the pregnant group by intravenous infusion of this amino acid, a prompt increase of blood glucose is observed (Felig *et al.*, 1972). It should be noted, however, that urea and total nitrogen excretion in the fasting pregnant subject equal or exceed that in the nonpregnant state (Figure 8); therefore, it seems likely that proteolysis and hepatic gluconeogenesis are maintained at rates equaling those in the nonpregnant state (but insufficient to prevent the more rapid decline of blood glucose).

With respect to the mechanism of hypoalaninemia in pregnancy, significant urinary losses are unlikely to occur in early to mid-pregnancy (Christensen *et al.*, 1957). Alternatively, continuous fetal uptake of alanine, in the face of preferential utilization of this amino acid for maternal gluconeogenesis, may result in depletion of this circulating glucogenic substrate. A more likely explanation may relate to the hyperketonemia of pregnancy. As noted above in prolonged and briefly fasted individuals, infusion of β-hydroxybutyrate has been shown to reduce plasma alanine levels (Sherwin *et al.*, 1975). Thus the hyperketonemia of pregnancy may limit

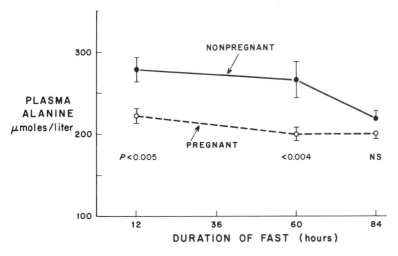

Fig. 7. Plasma alanine during starvation in nonpregnant and pregnant subjects during the second trimester. Plasma alanine was significantly reduced for the initial 60 hr of fasting. (Based on the data of Felig *et al.*, 1972).

PETER A. J. ADAM and
PHILIP FELIG

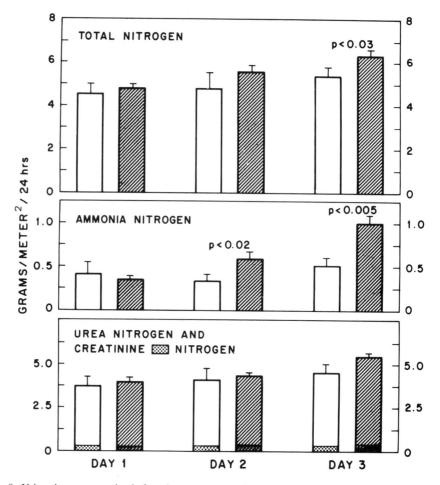

Fig. 8. Urine nitrogen excretion in fasted nonpregnant and pregnant subjects during the second trimester. In the pregnant group excretion of total nitrogen and ammonia was increased. Urea excretion was unchanged. (Based on the data of Felig and Lynch, 1970.)

alanine availability for gluconeogenesis by reducing the outflow of this amino acid from maternal protein reserves in muscle.

In view of the profound changes observed in maternal fuel metabolism during fasting, the question arises as to whether the secretion of placental hormones influences the fasting condition. Of particular interest is the role of human placental lactogen, a placental polypeptide hormone which has been postulated to accelerate maternal lipolysis and to enhance amino acid availability to the conceptus (Grumbach *et al.*, 1968). In view of the lipolytic and anabolic actions of HPL, placental secretion of this hormone may contribute to the augmented fasting ketosis of pregnancy and to the limitation of maternal hepatic gluconeogenesis.

### 2.3. The Fed State

#### 2.3.1. Responses to Nutrient Ingestion in Adult Man

Ingestion of nutrients in the diet is followed by their rapid assimilation and utilization either for energy purposes (oxidation to $CO_2$), storage as fuel (glycogen or triglyceride), or replacement of structural tissues (protein anabolism). The major nutrients in the average diet [carbohydrate (40%), fat (40%), and protein (20%)] are metabolized or stored in three tissues: liver, adipose tissue, and muscle. The primary hormonal signal regulating the metabolic response to feeding and the tissue utilization of nutrients is insulin. Following the administration of carbohydrate (as in a glucose-tolerance test) or the ingestion of mixed meals, plasma insulin levels increase two- to tenfold. As a consequence of this insulinemic response, glucose in the diet is taken up in a variety of tissues, but most importantly in the liver, where it may be stored as glycogen or converted to fatty acid and triglyceride. In this manner the blood glucose level during the course of the day is maintained within a very narrow range of 30–40 mg/100 ml despite intermittent administration of mixed meals (Victor, 1974). Hyperinsulinemia also facilitates the uptake of amino acids (particularly the branched-chain amino acids—valine, leucine, and isoleucine) by muscle tissue, where they are utilized for protein synthesis (Felig, 1975). It should be noted that in addition to an increment in insulin, plasma glucagon levels are suppressed by administration of glucose and stimulated by the ingestion of protein (Unger, 1974). The suppression of glucagon by carbohydrate is not, however, essential for the normal metabolism of this nutrient (Sherwin *et al.*, 1976). On the other hand, protein-stimulated glucagon secretion prevents the hypoglycemia that would otherwise accompany the hyperinsulinemia produced by a pure protein meal.

### 2.3.2. Enhanced Insulin Secretion and Contrainsulin Factors in Pregnancy

The metabolic response to feeding in pregnancy is characterized by hyperinsulinemia, hyperglycemia, hypertriglyceridemia (Freinkel and Metzger, 1975), and diminished sensitivity to insulin. An increase in plasma insulin levels in pregnancy was demonstrated soon after the availability of a radioimmunoassay for insulin (Spellacy and Goetz, 1963). This hyperinsulinemic effect is most marked in the third trimester and is demonstrable in response to administration of glucose or amino acids (Tyson and Merimee, 1970). The increase in insulin secretion is demonstrable even when blood glucose or amino acid levels are no higher than in control nonpregnant subjects (Freinkel and Metzger, 1975). Thus, a change in the responsiveness of the islets rather than an alteration in the circulating metabolic signal appears to be the responsible factor. Supporting this conclusion is the demonstration that pregnancy is characterized by hypertrophy and hyperplasia of the endocrine pancreas involving primarily the beta cells (Van Assche and Aerts, 1975).

Despite the hyperinsulinemia observed in pregnancy, the blood glucose response to an oral or intravenous carbohydrate load is higher than that observed in the nonpregnant state (O'Sullivan and Mahan, 1964). Studies in healthy humans have shown that the magnitude of elevation in blood glucose after carbohydrate feeding reflects the failure of glucose uptake by the liver and its escape into the systemic circulation (Felig *et al.*, 1975). In this context the augmented blood glucose response during gestation, in the face of hyperinsulinemia, seems to indicate that the liver is resistant to insulin. Diminished responsiveness to intravenously administered exogenous insulin suggests that peripheral tissues (adipose tissue, muscle) share in this resistance to insulin during pregnancy. Some quantification of this resistance may be obtained from studies in which plasma insulin is determined in response to a fixed degree of hyperglycemia maintained by a variable glucose infusion ("glucose clamp" technique). In such studies, the rate of glucose infusion divided by the plasma insulin level may provide an index of tissue sensitivity. By this technique recent reports indicate that tissue sensitivity seems to be reduced by as much as 80% in normal pregnancy (Sutherland *et al.,* 1975). Despite this diminution in tissue responsiveness to insulin, the excursions in blood glucose, during the course of a 24-hr day in which mixed meals are ingested, do not exceed 45 mg/100 ml in normal pregnancy and are less than in nonpregnant subjects (Victor, 1974). Maintenance of glucose homeostasis in this circumstance apparently is achieved by an increase of the plasma insulin response to the ingested nutrients.

The diminished tissue responsiveness to insulin of gravid subjects in the fed state coupled with the action of pregnancy in unmasking diabetes constitute the basis for characterizing the effects of pregnancy as "diabetogenic." Modifications of pancreatic glucagon responses, however, do not seem to contribute to the contrainsulin effects of pregnancy. In normal pregnancy fasting blood glucose is slightly lower and plasma glucagon levels are slightly higher than in the postpartum period; but suppression of glucagon by glucose administration is at least as great as in the nongravid state (Daniel *et al.,* 1974; Kuhl and Holst, 1976). Furthermore, the rise of plasma glucagon in response to protein feeding is not altered by pregnancy (Freinkel and Metzger, 1975). Thus, changes in pancreatic glucagon responses do not appear to play any role in the insulin resistance which characterizes normal pregnancy.

### 2.3.3. Effects of Gestational Hormones

*Human Placental Lactogen.* Human placental lactogen (HPL), also referred to as chorionic somatomammotropin (HCS), is a polypeptide hormone produced by the syncytiotrophblast (Josimovich and Brande, 1961). Chemically and immunologically it is similar to growth hormone (Josimovich and MacLaren, 1962). However, HPL circulates at term in a concentration 1000 times that of growth hormone, and in addition to having an anabolic effect on protein metabolism, it is both mammotropic and luteotropic (Josimovich and Brande, 1961; Grumbach *et al.,* 1968). A mild but definite impairment of glucose tolerance has been observed following the infusion of HPL in nonpregnant subjects over periods of 12 hr (Beck and Daughaday, 1967; Kahlkoff *et al.,* 1969). This is manifest as a small decrease in the rate of glucose utilization despite a parallel increment in circulating insulin levels (Beck and Daughaday, 1967; Kahlkoff *et al.,* 1969). HPL, like growth hormone, alters carbohydrate metabolism by diminishing the effectiveness of insulin. It also shares the lipolytic capabilities of growth hormone, permitting a marked increase in mobilization of free fatty acids from peripheral fat depots during fasting.

Although the total placental mass is the single most important factor determining HPL secretion, it has become apparent recently that nutrient availability also influences the maternal levels of this hormone. Thus, maternal starvation (Kim and Felig, 1971) and insulin-induced hypoglycemia have been demonstrated to cause a rise in HPL levels (Gaspard *et al.,* 1975). In contrast, a small but significant diminution in HPL levels is observed after intravenous administration of glucose (Freinkel *et al.,* 1974; Gaspard *et al.,* 1975); however, maternal HPL levels are not altered by physiologic fluctuations in blood glucose (Kuhl *et al.,* 1975; Soler *et al.,* 1974).

*Estrogen and Progesterone.* Following the luteal phase of pregnancy, the placental phase is characterized by increasing placental secretion of estrogen and progesterone, which readily enter the maternal circulation. Administration of physiologic quantities of purified natural estradiol-17β or progesterone for 3 weeks to virgin female rats enhances pancreatic islet secretion of insulin and plasma insulin

**473**

CARBOHYDRATE, FAT,
AND AMINO ACID
METABOLISM IN THE
PREGNANT WOMAN
AND FETUS

responses to glucose. This change is not produced acutely but is related to hypertrophy of the islets following chronic hormonal administration (Costrini and Kalkhoff, 1971). Similar changes occur in term, 3-week pregnant rats. Unlike pregnancy or progesterone treatment, estradiol administration alone, or with progesterone, enhances glucose tolerance. In the subhuman primate, however, 3 weeks of pharmacologic progesterone treatment enhances plasma insulin responses to glucose and reverses the glucose intolerance caused by cortisol administration to monkeys (Beck, 1969).

While progesterone alone does not enhance glucose tolerance, insulin-stimulated glucose uptake in isolated muscle tissue from rats pretreated with estradiol is augmented compared with untreated control animals (Shamoon and Felig, 1974). These findings are in agreement with the ameliorative effect of natural estrogen on diabetes in partially pancreatectomized animals (Houssay *et al.*, 1954). Thus, it is clear that estrogen and progesterone may contribute to hyperinsulinemia but are not major factors in the insulin resistance of pregnancy.

## 3. The Placenta

### 3.1. Transfer of Substrates

#### 3.1.1. Experimental Concepts

The placenta presents a complex interface between the fetus and mother which functions simultaneously to provide for active transfer (amino acids), carrier-mediated transport (glucose), diffusion (urea), and diffusion barriers (protein hormones) between mother and fetus. Through this interface a continuous exchange occurs by which substrates for fetal metabolism and growth flow toward the fetus, while waste products pass from the fetus to the mother for excretion. Our present knowledge of maternofetal transfer was acquired from several experimental approaches involving both the isolated placenta *in vitro,* and the mammalian fetus *in vivo;* however, net transfer must be demonstrated *in vivo* before quantitative importance can be attributed to a specific substance. The criteria which ideally should be fulfilled for the demonstration of net maternofetal transfer are the following:

1. Placental transfer of labeled substrate either *in vivo* or during perfusion of the maternal and fetal vessels of intact placenta or isolated cotyledon *in vitro.*

2. Appropriate maternal–fetal gradients and transfer kinetics *in vitro* for the proposed mechanism of transport. These experiments have used competitive inhibitors of carrier-mediated transport and toxins for energy metabolism to demonstrate active transport.

3. Maternofetal gradients *in vivo* appropriate to the mechanism, and a positive umbilical venous–fetal arterial gradient *in situ*. This is an essential criterion for net maternofetal transfer of a substrate.

4. Fetal excretion or placental clearance of the products from fetal metabolism of known substrates (for example, $CO_2$ and urea).

As will be evident, studies of human placental transfer during labor reflect a changing state, and net transfer *in vivo* has been quantified only in those animals where chronic catheterization of the umbilical vein and a fetal artery during gestation permits a steady state to be reestablished following surgery.

### 3.1.2. Glucose

PETER A. J. ADAM and
PHILIP FELIG

In the well-nourished pregnant woman transplacental glucose diffusion meets virtually the entire fetal substrate requirement for oxidative metabolism. The transfer of glucose occurs by a carrier-mediated process of facilitated diffusion (Hugget, 1961; Widdas, 1952) with the following characteristics: (1) The transfer occurs in the direction of a downhill gradient. (2) It is faster than would be expected on physiochemical grounds and exceeds the rate at which similar molecules such as fructose are transferred. (3) Competition by similar molecules, such as galactose, and saturation kinetics also can be demonstrated with the placenta *in situ*.

Within the physiological range, maternal and fetal blood glucose are closely related, so that fetal glucose rises in proportion to the maternal levels (Obenshain *et al.*, 1970; Stembera and Hodr, 1966; Morriss *et al.*, 1975).

### 3.1.3. Fats and By-Products

*Triglycerides, Free Fatty Acids, and Glycerol.* Even though maternal plasma triglyceride concentration and transport are increased during the last trimester of human gestation, no maternofetal transfer of triglycerides occurs as in other mammals, such as the guinea pig whose fetuses accumulate triglyceride stores near the end of pregnancy (Kayden *et al.*, 1969). Plasma triglyceride levels are very low in the human fetus at term (Robertson and Sprecher, 1968). When umbilical venous and arterial triglyceride levels have been measured in human subjects at delivery, the umbilical venous–arterial gradient for triglyceride tends to be negative (Sabata *et al.*, 1968*a*), suggesting that fetal production of esterified fatty acids occurs during labor and delivery. In any case, there is no evidence of net triglyceride transfer from the mother to fetus in any clinical or experimental circumstance.

Based on inferential data, free fatty acids are the major lipid moiety transferred from the maternal to the fetal circulation. Early in gestation, before the bulk of fetal adipose tissue triglycerides are synthesized from glucose, the distribution of esterified fatty acids extracted from fetal tissues resembles that of maternal plasma free fatty acids (Robertson and Sprecher, 1968). As term approaches, however, the transferred fatty acids become a diminishing proportion of total adipose tissue stores (Chen *et al.*, 1965; King *et al.*, 1971).

Szabo and Szabo (1974) have suggested that the high maternal and low fetal plasma FFA create a maternal arterial–fetal arterial gradient which would favor maternofetal transfer, but firm evidence of net transfer by demonstration of a positive umbilical venous–arterial gradient has been more difficult to obtain. Studies during labor are complicated by mobilization of free fatty acids from fetal adipose tissue so that both positive and negative umbilical venous–arterial FFA gradients have been measured (Sabata *et al.*, 1968*b*; Sheath *et al.*, 1972). Thus, net transfer has been inferred mainly from the compositional studies cited and from the marked downhill gradient from maternal to fetal plasma.

Although quantification of FFA transport by direct means is difficult, bidirectional transfer of palmitic and linoleic acids between the mother and fetus has been demonstrated in the pregnant monkey using dual-labeled fatty acids (Portman *et al.*, 1969); transfer in both directions also occurs in the isolated perfused human placenta (Szabo *et al.*, 1969) or cotyledon (Dancis *et al.*, 1973). The degree of aliphatic desaturation does not have a marked effect upon transfer rates across the human placenta, so that palmitic and linoleic acid cross the placenta at approximately equal rates *in vitro* (Dancis *et al.*, 1973).

Even though transfer is determined mainly by gradient, the mechanism appears to be more complex than simple diffusion. The long-chain fatty acids which comprise most of the free fatty acid transported during maternal fasting are insoluble in water, but are soluble in lipids and have a high degree of affinity for albumin (Dancis *et al.*, 1976). Protein binding permits the transport of large quantities of these FFA in plasma during fasting and consequently presents large amounts of FFA to the placenta in the fasted pregnant woman. When transfer of $^{14}C$-labeled $C_8$ to $C_{16}$ saturated fatty acids is studied in the isolated perfused cotyledon and conditions of albumin (1%) and free fatty acid (40 $\mu$M) concentrations are standardized, however, placental clearance of the FFA declines logarithmically with increasing chain length, lipid solubility, and affinity for protein (Dancis *et al.*, 1976). Nevertheless, the amount of palmitic acid transferred in this model falls within the physiological range for placental transfer when extrapolated to the entire cotyledonary mass; and raising the albumin concentration in the isolated fetal circulation enhances the FFA transfer. In summary, therefore, fatty acid transfer is a complex function of the amount of fatty acid presented to the placenta, lipid solubility, affinity for albumin in the maternal and fetal circulations, the plasma albumin concentration on both sides, and the FFA gradient between maternal and fetal arterial plasma. In normal pregnancy, the maternofetal FFA gradient is of the order of 500–1000 $\mu$M, and would favor FFA transfer to the fetus. When transfer by human placenta is quantified, however, it seems to account for a small proportion of the lipid stores accumulated in the last trimester of pregnancy.

The other product of lipolysis, glycerol, contributes in a minor way to placental transfer of carbon in man. The maternofetal gradient for glycerol (50–100 $\mu$M) is one tenth that of the free fatty acids, and umbilical venous–arterial gradient during labor either cannot be demonstrated (Sabata *et al.*, 1968*b*) or is negligible (Sheath *et al.*, 1972).

*Ketone Bodies.* At present, there are no studies which quantify maternofetal transfer of $\beta$-hydroxybutyrate or acetoacetate either *in vivo* or *in vitro*. Thus, our knowledge of ketone-body metabolism is inferred from studies of amniotic fluid in fasted women early in gestation and from measurements of the ketone bodies in the umbilical vessels during labor. If one may assume that direct exchange of ketone bodies between maternal plasma and amniotic fluid is restricted, while exchange between the fetus and amniotic fluid is rapid at mid-gestation, the levels of $\beta$-OH-butyrate and acetoacetate in the human fetus closely parallel those in the mother during fasting (Kim and Felig, 1972). In the immediate postabsorptive state, the total amniotic fluid ketone-body concentration of 170 $\mu$M is 37% of maternal levels. After 84 hr of fasting, the mean amniotic fluid levels of 3400 $\mu$M are 76% of maternal levels and correlate directly with peripheral venous plasma $\beta$-OH-butyrate and acetoacetate in the mother.

At term gestation, our concern for the newborn infant precludes rigorous examination of ketone-body metabolism during prolonged maternal fasting. During labor, however, maternal ketones rise progressively, and total ketones may be as high as 4 mM after prolonged labor (Paterson *et al.*, 1967). Paired umbilical venous–arterial differences have been measured, however, only in mothers whose total plasma ketones were 0.2–1.4 mM and in whom the fetal umbilical venous levels were less than 0.6 mM (Sabata *et al.*, 1968*b*). Under these conditions umbilical

PETER A. J. ADAM and
PHILIP FELIG

venous ketone bodies are approximately 45% of maternal levels; umbilical arterial concentrations are 70–85% of the umbilical venous levels. The average umbilical venous–arterial difference in these well-nourished mothers is 50 $\mu$M and can account for a significant but minor proportion of the carbon flux from mother to the fetus.

Thus, although the mechanism of ketone-body transfer across the human placenta has not been investigated thoroughly, fetal levels are closely related to those in maternal plasma during the last two thirds of pregnancy. The positive maternofetal gradient supports a variant of diffusion as the mechanism, and net fetal uptake is the rule, even in the well-nourished mother during labor.

### 3.1.4. Amino Acid Transfer

At term, the individual amino acids in fetal plasma are maintained at levels of 1.1–4.3 times those in maternal plasma, and the total amino acid concentration of the fetus is twice that in the mother (Lindblad and Baldesten, 1967; Ghadimi and Pecora, 1964). Even though there are data in the monkey (Kerr, 1968) and man (Ghadimi and Pecora, 1964) indicating a gradual decline of some plasma amino acids in the fetus, a comparison of the values between 15 and 20 weeks (Cockburn et al., 1970) with those at term (Lindblad and Zetterström, 1968) indicates that the changes are minor.

The greater concentration of amino acids observed in fetal plasma suggested their active transport by the placenta—an inference which was confirmed by Dancis and his co-workers (1968). When they infused pregnant guinea pigs with the nonmetabolizable amino acid, [$^{14}$C]$\alpha$-aminoisobutyric acid ([$^{14}$C]AIB), it was concentrated in the placenta. Following its accumulation the [$^{14}$C]AIB was secreted in both directions, but more rapidly into the fetus, so that its level was elevated in either fetal plasma or the perfusate of placentae in situ. Both human and guinea pig chorionic villi concentrated [$^{14}$C]AIB in vitro, while cyanide or 2,4-dinitrophenol inhibited this capacity for active concentration of the amino acid. Elaborating previous work by Page et al. (1957), Glendening and her co-workers (1961) also established the stereospecificity of human placental amino acid transport during infusion of excess amino acids to pregnant women by demonstrating more rapid appearance of L isomers in the fetus. Thus, active maternofetal transport of the L isomers against a gradient is supported by all the available evidence.

At present, however, net transfer of individual amino acids has been demonstrated only in the sheep (Lemons et al., 1976). When a steady state is reestablished after placement of catheters in a maternal artery, umbilical vein, and a fetal artery, measurement of the umbilical venous–fetal arterial difference for blood amino acid nitrogen reveals a gradient across the fetus approximating 450 $\mu$mol/liter. Essential amino acids account for 44% of the arteriovenous difference, while the rest of the gradient results from the nonessential acids. There is a positive umbilical venous–fetal arterial gradient for every essential amino acid and for the major nonessential ones, except for the dicarboxylic acids. Of the net amino acid nitrogen uptake, 22% results from the glutamine venous–arterial difference across the fetus, while glutamine, asparagine, alanine, glycine, and serine in combination account for more than 43% of the net amino acid uptake or 90% of the total nonessential amino acid nitrogen uptake. Thus glutamine may serve as a major transport form of carbon and nitrogen across the placenta. In contrast to the positive umbilical venous–fetal arterial nitrogen difference noted for most amino acids in the pregnant sheep, a

**477**

CARBOHYDRATE, FAT,
AND AMINO ACID
METABOLISM IN THE
PREGNANT WOMAN
AND FETUS

negative difference reflecting fetal excretion is observed with glutamate (Lemons *et al.*, 1976) and at least one product of amino acid oxidation: urea (Gresham *et al.*, 1972).

Even though umbilical venous–arterial uptake of amino acids has not been reported in the human or primate fetus, these observations apparently have general applicability to human fetal physiology. Battaglia and his associates (1968) have quantified placental clearance of urea by the subhuman primate *in vivo*. In the monkey, as in the sheep, placental urea clearance is limited primarily by permeability of the placental membrane. Since there is a positive fetal arterial–maternal arterial gradient, the net flow of urea would be from the fetus to the mother. Over a wide range of urea concentrations in the human, there is a rather constant maternal venous–fetal arterial urea difference of 2.5 mg/100 ml (Gresham *et al.*, 1971). If the human placenta resembles the monkey placenta with respect to urea clearance, then 0.5 g of urea nitrogen would be excreted transplacentally by the human fetus daily.

In summary, it appears that large quantities of L-amino acids are transported actively to the human fetus. Based on current evidence, the essential amino acid needs are met by net transfer from mother to fetus, while the nonessential amino acids may serve as an adaptable metabolic pool serving several functions. This aspect will be developed in Section 4.2 which discusses fetal adaptation to the maternal metabolic state.

### 3.2. Placental Production of Lactate

#### 3.2.1. Concepts from Mammalian Studies

Since pyruvate and lactate are produced within the three metabolic compartments being considered—maternal, placental and fetal—the intercompartmental exchange of carbon attributable to the products of glycolysis has been difficult to determine. While there is a direct correlation between ovine maternal and fetal levels of either blood lactate or pyruvate, fetal concentrations exceed those in the mother (Burd *et al.*, 1975; Char and Creasy, 1976). Furthermore, [l-$^{14}$C]lactate exchanges freely in both directions across the placenta of the pregnant monkey (Friedman *et al.*, 1960). Since the fetal arterial–maternal arterial gradient of lactate would favor net fetomaternal diffusion in all mammalian species studied, most investigators have assumed that the fetus produces quantities of this organic acid which exceed its uptake and that the excess diffuses to the mother. Based on ruminant studies, therefore, Ballard and his co-workers (1969) summarized the concepts of carbon flow which appeared to be likely for those species in which fetal and neonatal hepatic gluconeogenesis were possible:

> Although direct evidence is not available, it is assumed that the fetus produces lactate from glucose and that this is passed back to the mother for resynthesis. The extra glucose that is required during pregnancy can come from the diet in the monogastric animal, but the ruminant must increase its already high rate of gluconeogenesis to supply the additional glucose. It is therefore not surprising to find, in contrast to the fetal rat, a substantial gluconeogenic activity in both fetal cow and fetal sheep livers. This capacity to synthesize glucose allows the ruminant fetus to be less dependent on maternal glucose than fetuses from non-ruminants.

Although these workers recognized the possibility and potential importance of lactate uptake by the fetus, they assumed that the net flux would be from fetus to mother under most conditions. Two recent studies (Burd *et al.*, 1975; Char and

PETER A. J. ADAM and
PHILIP FELIG

Creasy, 1976) refute the concept that net fetal-to-maternal transfer of lactate occurs in the sheep. As indicated in Table I, ovine umbilical venous blood lactate exceeds the umbilical arterial levels, while uterine venous blood lactate exceeds the uterine arterial level in the chronically catheterized pregnant sheep. Thus, the sheep placenta *in vivo* produces lactate, presumably from maternal glucose, and the lactate is secreted in both directions. In the presence of the constant umbilical blood flow observed in these chronic preparations, there is a net uptake of lactate by the fetus, rather than a net flow from fetus to mother.

In summary, the ruminant placenta produces lactate which appears to serve as a fuel for fetal metabolism. In the perspective of present knowledge, the lactate probably originates from placental glycolysis but also could arise from amino acids.

### 3.2.2. Human Placental and Fetal Lactate Production

Studies of human fetal lactate metabolism *in vivo* have not been done in steady-state conditions; thus our knowledge of maternofetal relationships in man is restricted to information obtained at birth either following cesarian section or after labor and delivery. In the studies by Stembera and Hodr (1966) of healthy human fetuses at term, there was a direct correlation between maternal venous blood lactate and that in the umbilical vein at normal delivery. The mean umbilical venous blood lactate exceeded the maternal level slightly; the umbilical arterial lactate concentration exceeded that in the umbilical vein in 15 of the 24 umbilical vessel specimens. Thus, during labor and delivery, the data seem to indicate that net fetal production of lactate occurs in the majority and that the excess is transferred to the mother. Likewise, Hendricks (1957) detected higher blood levels of lactate in the umbilical artery than in the umbilical vein at caesarian section, suggesting that fetal lactate production occurred during the surgery. To date, inferential evidence for net uptake of lactate in some human fetuses is derived from the studies of Stembera and Hodr (1966) which are cited above and from those of Derom (1964) which were done when mean maternal lactate exceeded that in the fetus. Under this condition, a positive umbilical venous–umbilical arterial difference was observed in 10 of 21 fetuses studied.

To place these data in perspective, it appears that lactate and pyruvate are produced by the mother, placenta, and fetus. In a steady state, with adequate placental circulation, there appears to be rapid bidirectional exchange of lactate between the mother and the fetus. Prior to labor, fetal levels probably exceed those in the mother, but the net flux of lactate within the pregnant woman and her fetus is not known. Based on our knowledge of other species and man, lactate could be an

Table I. Whole Blood Lactate and Pyruvate Concentrations in the Pregnant Ewe[a] (mM; Mean ± SEM)

| | Lactate | Pyruvate | Lactate difference |
|---|---|---|---|
| Uterine | | | |
| Artery (A) | 0.86 ± 0.07 | 0.13 ± 0.02 | (V − A) |
| Vein (V) | 0.93 ± 0.07 | 0.14 ± 0.01 | 0.07 |
| Umbilical | | | |
| Vein (v) | 2.21 ± 0.15 | 0.25 ± 0.08 | (v − a) |
| Artery (a) | 2.05 ± 0.15 | 0.25 ± 0.08 | 0.16 |

[a]From Burd *et al.*, 1975.

adaptable substrate, produced by the fetus during labor or under duress, but taken up from the placenta in the more stable conditions which are expected throughout gestation. Because of the unstable nature of the umbilical circulation in man, we would not expect easy documentation or quantification of net lactate uptake by the human fetus.

## 4. Fetal Metabolism

### 4.1. Quantitative Relationships of Fetal Fuels

#### 4.1.1. Rationale and Methods of Mammalian Experimentation

In the past decade, our concepts of energy metabolism in the mammalian fetus have been modified by many data quantifying the uptake, oxidation, and accretion of food stuffs by the lamb *in utero*. Prior to this work, our assumption was that glucose provided virtually all the substrate required for oxidative metabolism *in utero*, particularly in man. The evidence for this concept is derived mostly from measurement of respiratory quotients in newborn infants (Cross *et al.*, 1957; Tunell *et al.*, 1976) and from mammalian studies of regional glucose uptake immediately after birth (Breuer *et al.*, 1967, 1968; Morriss *et al.*, 1973). In short, the respiratory quotient of the well newborn infant is 1.0 or higher, while glucose is the predominent fuel taken up by neonatal canine heart muscle or fetal ovine hind limb. Nevertheless, glucose uptake does not provide sufficient substrate for oxidative metabolism in the entire ovine fetus and accounts for less than half of the total carbon transferred from mother to fetus.

Three methods have been utilized to analyze data obtained from studies of the chronically catheterized fetus *in utero* as follows: (1) net transfer of substrates and waste products; (2) comparison of transfetal substrate gradients to the oxygen gradient; and (3) summation of fetal requirement for carbon or nitrogen accretion, oxidation, and excretion.

Net transfer is calculated either as the product of umbilical blood flow and the umbilical venous–arterial substrate concentration difference or as the product of the transplacental clearance of substrate and the maternal–fetal arterial concentration difference. In cases where the substrate or waste product is known, these measurements have the advantage of quantifying precisely the net flux of the substance between mother and fetus.

When investigators wish to infer the potential role of a substrate in oxidative metabolism, the second approach frequently is used. A nondimensional quotient is calculated from the molar quantity of $O_2$ required to fully oxidize one mole of substrate and the umbilical venous–arterial gradient for the substrate and oxygen. In the case of glucose (G), 6; lactate (L), 3; and $\beta$-hydroxybutyrate (B), 4.5 mol of $O_2$ are required to oxidize each substrate mole. The respective umbilical venous–arterial quotients, $6\Delta G/\Delta O_2$, $3\Delta L/\Delta O_2$, and $4.5\Delta B/\Delta O_2$, represent the fraction each substrate could contribute to oxygen consumption by the fetus if it were totally oxidized.

Finally, the third approach accounts for total fetal needs by summing the elements of its requirements. Fetal carbon and nitrogen accretion are estimated by carcass analysis at differing periods of gestation; excretion of carbon is quantified as the product of umbilical blood flow and the umbilical arterial–venous difference;

PETER A. J. ADAM and
PHILIP FELIG

Table II.   Carbon and Nitrogen Requirements of Fetal Sheep[a]

| Requirement | Carbon (g/day/ kg body weight) | Nitrogen (g/ day/kg body weight) |
|---|---|---|
| Accretion | 3.15 | 0.60 |
| Excretion | | |
| $CO_2$ | 4.38 | — |
| Urea | 0.16 | 0.36 |
| Total | 7.69 | 0.96 |

[a]From Cahill, 1972.

nitrogen excretion is assessed from the product of transplacental urea clearance and fetal–maternal arterial concentration differences. Although the sum of these does not account for all the excreted metabolites (e.g., creatinine and uric acid) or all the potential routes of excretion (e.g., transfer across the chorioamniotic membrane), it represents the best current estimate of fetal requirements.

In a recent editorial comment, Cahill (1972) estimated the daily carbon and nitrogen needs of the fetal sheep based on a summation of accretion and excretion rates (Table II).

If one compares these data with net transfer rates and the substrate/oxygen quotients in the ovine fetus, one can account for the major metabolic fates of three substrates: glucose, lactate, and amino acids (Table III).

Of course, this table represents an approximation based on average glucose/ oxygen, lactate/oxygen, and urea/oxygen quotients accounting for the relative proportions of oxidative metabolism fueled by glucose, lactate, and amino acids. Furthermore, carbon accreted cannot be construed as amino acid carbon alone. Throughout development, there is rapid exchange of carbon among the various carbohydrate, fat, and amino acid pools; but the table serves to give some idea of the quantitative relationships among substrates.

### 4.1.2. Uptake and Accretion of Carbon and Nitrogen by the Human Fetus

The principles developed by the investigators of fetal sheep metabolism may be applied to analysis of metabolism in the human fetus; however, the degree of

Table III.   Metabolic Fate of Carbon-Containing Substrates in Fetal Sheep

| Source of carbon | Net transfer (g carbon/day/kg body weight) | Oxidation (g carbon/day/kg body weight) | Accretion (g carbon/day/kg body weight) |
|---|---|---|---|
| Glucose | 1.8 | 1.8[a] | ? |
| Lactate | 1.4 | 1.4[a] | ? |
| Amino Acid | 3.9[b] | 0.9[c] | 3.0 |
| Total | 7.1 | | |

[a]Based on substrate/oxygen quotient (Lemons et al., 1976).
[b]Based on 0.364 g urea nitrogen/day.
[c]Based on comparison of fetal urea excretion and oxygen consumption.

certainty obtained from interpretation of the data is considerably less than that from ovine fetal studies. This uncertainty results from the instability of the human fetal umbilical circulation, the uncontrolled circumstances which cause spontaneous ejection of the fetus during the last trimester, the preceding clinical state of the fetus prior to stillbirth, and the necessity for extrapolation from subhuman primates in which chronic catheterization is possible.

In Table IV the influx and outflow of carbon from the human fetus during the last 60 days of gestation are calculated from the accretion rates of carbohydrate, fat, and nitrogen in the carcass (Widdowson, 1968), from minimal $CO_2$ excretion rates in the newborn (Tunell *et al.*, 1976; Jonxis *et al.*, 1967), and from urea clearance rates of the primate placenta (Gresham *et al.*, 1971). As indicated, the next influx of carbon near term would be approximately 6.7 g/day/kg body weight in the human fetus.

If one compares the known sources of carbon in the sheep and human (Table V), one may derive reasonable probabilities concerning the origin of the carbon flowing to the fetus of the well-nourished pregnant woman. In order to quantify influx the following assumptions were made:

1. Carbohydrate represented most of the oxidized carbon flowing to the human fetus. This assumption was based on fetal glucose/oxygen quotients close to 1.0 in carefully selected pregnant women undergoing cesarian section and on respiratory quotients which equal or exceed 1.0 immediately after birth.

2. Of the fatty acid in fetal stores, 6.75% is derived from maternal sources, and the fatty acids are minimally oxidized. This is based on a comparison between the fraction of linoleic acid in maternal venous plasma FFA at term and the fraction of fetal adipose tissue triglyceride FA represented by linoleic acid.

3. The small umbilical venous–arterial ketone-body gradient observed in the

*Table IV.  Accretion and Excretion of Carbon and Nitrogen in the Human Fetus during the Last 60 Days of Gestation*

| Parameter | Gestational age (weeks) 31 | Gestational age (weeks) 40 | Total increase | 60-day ΔC or ΔN (g/d) | ΔC at 2.5 kg (g/d/kg) |
|---|---|---|---|---|---|
| Accretion | | | | | |
| Body weight (kg) | 2.0 | 3.5 | 1.5 | | |
| Carbohydrate (g) | 9 | 34 | 25 | 0.2 ⎫ | 2.4 |
| Fat (g) | 100 | 560 | 460 | 5.8 ⎭ | |
| Protein (g) | 250[a] | 460[a] | 210 | 1.7 | 0.7 |
| Nitrogen (g) | 40 | 73.5 | 33.5 | 0.7 | |
| Total carbon accretion rate (g/day/kg) at 2.5 kg | | | | | 3.1 |
| Excretion | | | | | |
| $CO_2$ carbon (g) | | | | 8.7[b,c] | 3.5 |
| Urea carbon (g) | | | | 0.3[b,d] | 0.1 |
| Urea nitrogen (g) | | | | 0.6[b,d] | |
| Total carbon excretion rate (g/day/kg) | | | | | 3.6 |
| Total carbon influx (g/day/kg) | | | | | 6.7 |

[a]Based on average N content of 0.16 g N/g protein.
[b]Based on 2.5 kg body weight.
[c]Based on minimal $CO_2$ production of 4.5 ml/min/kg.
[d]Based on transplacental clearance of 15 ml/min/kg.

PETER A. J. ADAM and
PHILIP FELIG

Table V.  Comparison of Carbon Sources in Sheep and Human
Fetus near Term

| Parameter | Carbon (g/day/kg) | |
| --- | --- | --- |
| | Sheep | Human at 2.5 kg |
| Total accretion | 3.2 | 3.1 |
| Total excretion | 4.5 | 3.6 |
| Total Flux | 7.7 | 6.7 |
| Sources | | |
| Carbohydrate | 3.2[a] | 3.5[b] |
| Free fatty acids | ? | 0.2[c] |
| Ketone body | 0 | 0.2[d] |
| Amino acid | 3.9 | ? |
| Unknown | 0.6 | 2.8 |
| Total | 7.7 | 6.7 |

[a]Sum of ovine fetal glucose and lactate uptake.
[b]Based on human fetal G/O quotient of 1.0 and $O_2$ consumption of 4.5 ml/min/
kg.
[c]Based on fraction of linoleic acid in neonatal adipose tissue triglyceride
attributable to maternal free fatty acid (6.75%).
[d]Based on comparison of average glucose and ketone-body umbilical venous–
arterial differences during labor.

fetus of women in labor is representative of the fluctuating fetal responses in the
normally nourished mother.

Although such quantification is contingent upon extrapolation from several
sources, which are applied to a single point in gestation, there are sufficient data to
permit reasonable assumptions concerning the compartmentalization of human fetal
carbon sources in the well-nourished mother. Based on the evidence for active
maternofetal transport, one also might reasonably assume that amino acid accounts
for a large proportion of the "unknown" carbon flowing from the pregnant woman
to her fetus in the early postabsorptive state. Thus, maternally derived glucose
would seem to account for 50% or more of total fetal carbon uptake, while maternal
amino acids and placental lactate may account for about 40%. As indicated,
however, quantification of lactate contribution in human subjects will be difficult.
Free fatty acids and ketone bodies may account for a small proportion of the net
fetal carbon uptake, but the conclusion concerning fatty acids requires extended
inferences so that the exact proportion cannot be determined now.

Finally, we note again that these calculations are based upon averaged data and
that the information was collected mainly from "nourished" mothers after short
periods of fasting. Thus, we must restrict our present conclusions to the maternally
nourished state. In the following sections, we shall examine the adaptive mechanisms available to the human fetus during fluctuations in maternal nutrition and
during frank maternal privation.

### 4.2. Fetal Adaptation to the Maternal Nutritional State

### 4.2.1. Glucose Uptake

*Mammalian Models.* Investigators have explored fetal adaptation to changes
of maternal state in the aforementioned physiologic models. Ideally, they have
superimposed prolonged perturbation of maternal physiology upon the condition of
stable maternofetal exchange obtained with chronic catheterization. Based on such

**483**

*CARBOHYDRATE, FAT,
AND AMINO ACID
METABOLISM IN THE
PREGNANT WOMAN
AND FETUS*

an experimental model, our understanding of fetal adaptation *in vivo* is derived principally from the following manipulations of the maternal state: (1) comparing the maternally "fed" state with varying degrees of maternal fasting or starvation; (2) producing prolonged maternal hyperglycemia with continuous glucose infusion or drug-induced diabetes mellitus; (3) inducing maternal hypoglycemia with prolonged insulin infusion; or (4) maintaining maternal ketosis, without starvation, by means of a provocative ketogenic diet. In this section, we will evaluate the serial changes of fetal substrate uptake and oxidation which occur during the transition from maternal feeding to starvation. Most of this evidence was obtained from studies of the pregnant sheep.

If one examines the changes of ovine fetal $O_2$ uptake which occur with growth from 2 to 4 kg in single pregnancies, there is a progressive increase with a tendency to plateau at term (James *et al.*, 1972). Even though fetal glucose uptake accounts for only a fraction of the carbon accumulated and oxidized, the fetal uptake of glucose varies directly with the maternal glucose concentration. As maternal glucose concentration increases, the maternofetal arterial glucose concentration gradient increases and the umbilical vein–fetal arterial difference also increases. Thus, fetal glucose uptake, in the pregnant sheep fed *ad libitum*, is directly proportional to its availability.

In extensions of these studies, Boyd and co-workers (1973) quantified glucose uptake in fed and fasted pregnant sheep (Table VI).

As indicated, maternal ovine starvation for 3 days has the following effects:

1. Maternal and fetal arterial glucose levels decline to hypoglycemic levels.
2. Fetal $O_2$ consumption declines by $-0.81$ ml/min day rather than increasing normally by 0.70 ml/min day.
3. Although net fetal glucose uptake continues, it declines and accounts for a diminished proportion of fetal oxygen consumption.

Based on these and similar data, the ovine fetus adapts by several mechanisms to maternal starvation:

1. First, there is a reduction of metabolic rate.
2. Second, there is a reduction of glucose uptake from maternal sources which is subject to several interpretations:
   i. Alternate fuels may be substituted including other carbohydrates, or
   ii. The fetus may produce glucose by virtue of its gluconeogenic capacity.

*Table VI.* Effects of Maternal Starvation for 3 Days on Fetal Glucose Uptake and Oxidation in the Pregnant Sheep (Mean $\pm$ SEM)[a]

| Parameters | Units | Fed | Fasted |
|---|---|---|---|
| Maternal arterial glucose | mg/100 ml | $45.1 \pm 1.4$ | $26.6 \pm 1.7$ |
| Fetal arterial glucose | mg/100 ml | $20.0 \pm 0.7$ | $11.2 \pm 1.0$ |
| Glucose (UV − A)[b] | mg/100 ml | $2.31 \pm 0.15$ | $1.28 \pm 0.25$ |
| $O_2$ (UV − A) | mM | $1.53 \pm 0.06$ | $1.37 \pm 0.08$ |
| Umbilical blood flow | ml/min | $790.0 \pm 34.1$ | $739.0 \pm 45.7$ |
| Fetal glucose uptake | mg/min | $18.0 \pm 1.4$ | $9.7 \pm 1.7$ |
| Fetal $O_2$ uptake | ml/min | $26.7 \pm 1.3$ | $23.8 \pm 1.1$ |
| $6\Delta G/\Delta O_2$ ratio | | $0.52 \pm 0.04$ | $0.31 \pm 0.06$ |

[a]From James *et al.*, 1972.
[b]Umbilical venous minus fetal arterial level.

PETER A. J. ADAM and
PHILIP FELIG

Recent evidence (Schreiner *et al.*, 1976) demonstrates that ovine placental *lactate* production and fetal uptake of this carbohydrate are not increased by maternal starvation. Inasmuch as fetal urea excretion transplacentally is increased (Section 4.2.3, Fetal Oxidation of Amino Acid), fetal gluconeogenesis or direct oxidation of amino acids is increased. Furthermore, since ovine fetal brain and hind limb take up sufficient glucose to account for oxidative metabolism (Sections 4.2.2 and 4.2.3), fetal hepatic gluconeogenesis might be the means by which amino acids are converted to a major fuel for ovine fetal organ metabolism, i.e., glucose. The potential importance of amino acids in the sheep will be examined later (Section 4.2.3).

*Variations of Net Fetal Glucose Uptake in Man.* Although glucose uptake by the human fetus has not been quantified, there are sufficient data to demonstrate that similar physiological principles apply to glucose utilization *in utero*. Early in gestation, maternal glucose infusion raises umbilical venous and arterial blood glucose in proportion to maternal venous levels, and the maternal–fetal glucose difference is greater when the mother is hyperglycemic (Obenshain *et al.*, 1970; Adam, 1971). Similarly, prolonged maternal fasting early in gestation causes a decline of maternal venous and amniotic fluid glucose levels, while decreasing the maternal–amniotic glucose concentration difference (Kim and Felig, 1972). Although these data do not demonstrate variable fetal glucose uptake, they seem to indicate that higher maternal blood glucose favors more rapid maternofetal diffusion of glucose even early in gestation.

When the umbilical venous–arterial blood glucose difference has been examined at term during normal labor and delivery, the concentration difference across the fetus varies directly with the fetal blood glucose (Stembera and Hodr, 1966). Thus, the earliest data available favor the concept that human fetal glucose uptake throughout gestation varies according to the maternal supply.

More recently, Morriss and co-workers (1975) have evaluated transfetal blood glucose and oxygen gradients prior to labor under rigorously controlled conditions of fasting. The women were delivered at 38–41 weeks of gestation by cesarian section, after overnight fasting followed by intravenous infusion of Ringer's lactate instead of a glucose solution. None had begun to labor. Among the women studied, 11 had stable blood pressure throughout the procedure. In these, the fetal arterial blood glucose was directly related to the maternal venous blood glucose, and the maternal–fetal concentration difference rose with the increasing maternal blood glucose (Figure 9). Both the umbilical vessel glucose concentration difference and the umbilical glucose/oxygen quotient varied directly with the maternal blood glucose and the maternal–fetal glucose concentration difference (Figure 10). Thus, net glucose uptake and its contribution to oxidative metabolism in the human fetus apparently varies, as it does in the sheep, according to the maternal supply.

Unfortunately, the role of lactate was not evaluated in these studies. As indicated previously (Section 4.1.2), rapid exchange of lactate between the mother and fetus results in a close direct relationship between maternal and fetal blood lactate levels (Stembera and Hodr, 1966; Friedman *et al.*, 1960), and net fetal uptake may occur if maternal blood levels exceed those in the fetus (Derom, 1964). When the umbilical venous–arterial difference is evaluated after short spontaneous labor, approximately twice as many fetuses have higher arterial than umbilical venous levels (Stembera and Hodr, 1966). If fetal hypoxia occurs, the arterial

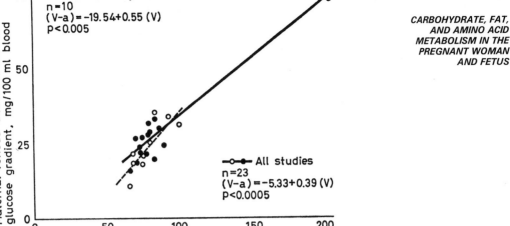

**485**

*CARBOHYDRATE, FAT,
AND AMINO ACID
METABOLISM IN THE
PREGNANT WOMAN
AND FETUS*

Fig. 9. Relationship of maternal venous–umbilical arterial blood glucose concentration gradient (V–a) to maternal venous blood glucose (V) in pregnant women (from Morriss *et al.*, 1975).

lactate levels always exceed those in umbilical vein. Thus, we may infer that the human fetus can produce or consume lactate, but the prevailing condition during a steady state can only be inferred. Based on the recent report of constant placental lactate production during maternal feeding or fasting in the sheep, however, it is unlikely that net fetal lactate uptake is a function of the mother's nutritional state (Schreiner *et al.*, 1976).

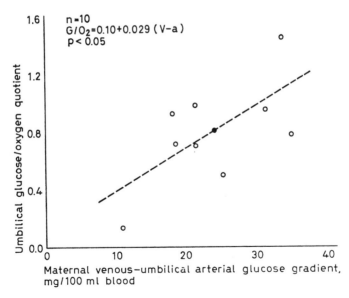

Fig. 10. Relationship of the human umbilical glucose/oxygen quotient (G/O$_2$) to maternal venous–umbilical arterial glucose gradient (from Morriss *et al.*, 1975).

**PETER A. J. ADAM and**
**PHILIP FELIG**

*Free Fatty Acids and Glycerol.* According to the principles elucidated previously (Section 3.1.3), we would expect free fatty acids to be transferred in proportion to their availability in plasma; however, transfer has not been quantified during maternal feeding and fasting so that our knowledge is restricted to inferences from indirect observations.

Early in gestation, prolonged fasting raises maternal plasma free fatty acids and blood glycerol (Table VII) (Kim and Felig, 1972), providing evidence that nutritional privation accelerates maternal lipolysis. If one examines the amniotic fluid levels of FFA and glycerol, however, only the glycerol concentration rises significantly.

These data from human gestation before the accumulation of fetal adipose tissue resemble those described in species such as the rat, in which little adipose tissue triglyceride accumulates *in utero* but fat reserves develop during suckling (Girard, 1975). In the rat, maternal fasting has no effect on fetal plasma FFA but causes a moderate rise of blood glycerol. The source of this glycerol apparently is maternal, since [$^{14}$C]glycerol crosses the placenta. Nevertheless, there are no data indicating that glycerol contributes a major portion of the maternofetal carbon transfer at any period of gestation, even during prolonged fasting.

In the last trimester of human pregnancy, maternally derived free fatty acids accumulate in fetal adipose tissue triglycerides; Szabo has hypothesized that the magnitude of this maternofetal transfer depends on the maternal plasma FFA levels (Szabo and Szabo, 1974). This concept may be inferred from the previously cited studies *in vitro* (Section 3.1.3 Triglycerides, FFA, and Glycerol) and from the transplacental and transfetal gradients which occur during labor and delivery. As demonstrated by Sabata and co-workers (1968*b*), plasma FFA rise during labor while umbilical venous plasma FFA correlate directly with maternal venous levels. In these studies, umbilical venous FFA exceeded umbilical arterial concentrations at high maternal venous levels, implying fetal FFA uptake. At lower maternal levels, the umbilical venous–arterial FFA gradient was reversed, implying fetal production. Thus, we might infer from such studies that labor is a state in which both mother and fetus produce FFA, but high maternal plasma FFA concentration may override net fetal production and permit some assimilation.

Studies of similar design, which confirm these conclusions, have not been reported elsewhere; but Sheath and her associates (1972) have examined the effect of rapid maternal glucose infusion, immediately before delivery on fetal FFA

Table VII. *Substrate Concentrations in Maternal Blood and Amniotic Fluid during the Second Trimester* [a]

|  | Maternal blood | | Amniotic fluid | |
|---|---|---|---|---|
|  | Fed | Fasted | Fed | Fasted |
| Glucose (mg/100 ml) | 75.1 ± 1.7 | 47.2 ± 2.0 | 32.7 ± 2.0 | 20.7 ± 1.1 |
| Free fatty acids (mM) | 0.96 ± 0.10 | 1.56 ± 0.09 | 0.08 ± 0.02 | 0.09 ± 0.02 |
| Glycerol (mM) | 0.17 ± 0.04 | 0.28 ± 0.04 | 0.13 ± 0.02 | 0.18 ± 0.04 |
| β-Hydroxybutyrate (mM) | 0.38 ± 0.04 | 3.79 ± 0.26 | 0.11 ± 0.02 | 2.78 ± 0.24 |
| Acetoacetate (mM) | 0.08 ± 0.01 | 0.66 ± 0.08 | 0.06 ± 0.01 | 0.61 ± 0.07 |

[a]Kim and Felig, 1972.

uptake. In 23 mothers who had received no glucose for 2–7 hr prior to delivery, the mean umbilical venous–arterial FFA gradient was 68 $\mu$M as would be expected with fetal uptake of the fatty acids. Intravenous injection of 13 women with 25 g of glucose 15–60 min before delivery was associated with a lower mean umbilical venous–arterial FFA gradient of 37 $\mu$M, while the gradient for glucose was increased. These trends, however, were not statistically significant. In summary, maternofetal transfer of FFA occurs, but its quantitative importance is not determined. Based on physicochemical principles and experimentation *in vitro*, we would expect effective placental transfer to depend mainly on the maternal levels of individual protein-bound free fatty acids. Evidence *in vivo* is fragmentary but supports some variation in net maternofetal FFA transfer based on the adaptive interrelationships between maternal carbohydrate and fat metabolism.

**487**

CARBOHYDRATE, FAT,
AND AMINO ACID
METABOLISM IN THE
PREGNANT WOMAN
AND FETUS

*Fetal Ketone-Body Metabolism.* As is evident from studies of pregnant mammals and man, preservation of fetal metabolism during maternal starvation may involve several mechanisms: (1) maternal production of an alternate fuel, (2) fetal production of the normally metabolized fuel from maternally derived substrates; or (3) oxidation of substrates which normally are assimilated by the fetus.

From our review of fetal metabolism in ruminants and man, it is apparent that the normal substrate, glucose, is not transferred from mother to fetus in sufficient quantities to account entirely for fetal oxidative metabolism, particularly during maternal fasting. Furthermore, although maternofetal free fatty acid exchange may undergo adjustments during fasting, the magnitude of transfer seems insufficient to account for the substitution required by a 50% or greater reduction of fetal glucose oxidation.

Among the physiologic responses proposed above, however, the first is preferable from the point of view of the fetus, since it involves neither fetal catabolism nor the consumption of otherwise valuable substrates. According to this concept, the mother would provide an alternate fuel derived from her ample stores which could perform the functions usually attributable to glucose. Although such an alternate has not been demonstrated in pregnant ruminants, $\beta$-hydroxybutyrate and acetoacetate produced by maternal liver from mobilized FFA diffuse freely from mother to fetus in both rat and man and provide ample quantities of fuel when the supply of glucose is diminished.

Girard (1975) has summarized the data of his associates which support this concept in the rat. After starvation for the 4 days prior to term, there is a marked decline of maternal blood glucose, lactate, and pyruvate, while blood $\beta$-hydroxybutyrate and acetoacetate rise to high levels. Concurrently, fetal blood glucose drops from a mean concentration of 3.7 to 2.3 mM, lactate declines from 7.8 to 4.0 mM, and ketone bodies rise in equilibrium with maternal levels from 0.1 to 9.5 mM. From these data, Girard infers reduced rates of maternal glucose production and transfer to fetus, decreased maternal and fetal glycolysis, and increased maternal production of ketone bodies with diffusion to the fetus. There are no murine studies *in vivo*, however, which demonstrate fetal ketone-body uptake.

Although studies of ketone-body metabolism in man have not been designed to demonstrate fetal uptake of this alternate fuel, analyzing available information by the methods of Battaglia and his co-workers provides substantial evidence that human fetal ketone-body consumption varies with the supply. If we hypothesize that the amniotic fluid early in gestation (before fetal renal function develops) is principally an extension of fetal extracellular fluid (Lind and Hytten, 1970), then the data of Kim and Felig (1972), obtained at mid-gestation, may be recalculated to determine the changes of maternofetal gradients for freely diffusible substances (Table VIII).

As indicated, the gradient between maternal and fetal fluids for glucose declines, while that for $\beta$-hydroxybutyrate increases markedly. Such an analysis provides, therefore, indirect evidence that ketone bodies produced early in gestation may be consumed during their passage through the fetus.

At term, ketone-body metabolism has been evaluated only during labor and delivery. With spontaneous labor, total ketones rise progressively in maternal plasma (Paterson *et al.*, 1967; Sabata *et al.*, 1968*b*). Umbilical venous plasma ketones increase in direct proportion to the maternal levels, but the increasing maternal–umbilical venous gradient is visually apparent in both studies. In addition, Sabata and associates (1968*b*) report 20 studies in which paired maternal–umbilical venous plasma ketone gradients are plotted and 14 such studies in which umbilical venous–arterial gradients are reported. Even in the narrow range of ketone-body concentration reported, they demonstrate increasing maternofetal and transfetal gradients. This becomes apparent when maternal venous and umbilical arterial concentrations of ketones are calculated from the reported regressions at each extreme of the umbilical venous levels observed (Table IX).

Finally, Paterson and his co-workers have demonstrated that administration of glucose to women with high plasma ketones during labor reduces maternal levels as one would expect; hence, the quantity of ketone bodies available to the fetus exemplifies a reciprocal relationship with the maternal supply of glucose.

In summary, the ketone bodies, particularly $\beta$-hydroxybutyrate, fulfill several of the criteria expected of a major alternate fuel replacing glucose *in vivo* and providing substrate for fetal energy metabolism during maternal fasting. These are as follow: (1) increased production by the mother as sources of carbohydrate are depleted; (2) transfer to the fetus; (3) an increasing gradient from maternal to fetal fluids as the supply increases; and (4) a rising umbilical venous–arterial ketone-body gradient as maternal levels rise, suggesting increased fetal uptake as the supply increases.

In order to place these data in perspective, we must acknowledge that rigorous studies during maternal fasting are not permissible in the third trimester; consequently, the human fetal capacity for ketone-body consumption *in vivo* probably will remain unquantified. Nevertheless, the physiologic advantage which accrues to the human fetus by utilizing maternal energy during starvation is evident.

### 4.2.3. Variations in Fetal Amino Acid Metabolism

*Maternofetal Transport.* We will comment on three lines of investigation which have elucidated the adjustments occurring in supply, uptake, and oxidation

Table VIII. Effects of Prolonged Fasting on Maternal Venous–Amniotic Fluid Gradients for Glucose and Ketone Bodies

|  | Fed | Fasted |
|---|---|---|
| Glucose (mg/100 ml) | 42.4 | 26.5 |
| $\beta$-Hydroxybutyrate (mM) | 0.29 | 1.49 |
| Acetoacetate (mM) | 0.02 | 0.05 |

**489**

CARBOHYDRATE, FAT,
AND AMINO ACID
METABOLISM IN THE
PREGNANT WOMAN
AND FETUS

Table IX.  Total Ketones in Maternal and Fetal Plasma from Relations Observed by Sabata (mM)

| Maternal vein (MV) | MV − UV | Umbilical vein[a] (UV) | UV − UA | Umbilical artery (UA) |
|---|---|---|---|---|
| 0.200 | 0.100 | 0.100 | 0.033 | 0.067 |
| 1.302 | 0.702 | 0.600 | 0.074 | 0.526 |

[a]Umbilical venous concentrations set arbitrarily within the range of values reported by Sabata. Other values calculated from the equations derived from the observed values: UV = 0.009 + 0.454 (MV); UA = 0.021 ± 0.976 (UV).

of amino acids by the mammalian fetus: (1) Adjustments of fetal supply have been evaluated by infusing or feeding excess quantities of individual amino acids to human or subhuman primates, while observing the maternofetal gradient. (2) The response of fetal uptake to maternal supply has been determined by measuring the umbilical venous–fetal arterial amino acid gradient in the pregnant ewe fed *ad libitum*. (3) Finally, the changes of fetal amino acid oxidation during maternal fasting have been evaluated by quantifying transplacental urea excretion in the starved pregnant sheep.

Based on studies of pregnant women receiving excess amino acid at term (Page *et al.*, 1957; Glendening *et al.*, 1961) or rhesus monkeys near term (Sturman *et al.*, 1973; Stegink *et al.*, 1975), three classes of amino acid with respect to transport characteristics may be distinguished:

1. Most amino acids maintain a higher blood level in the fetus than in the mother throughout gestation; and a fetal–maternal ratio greater than one is maintained or rapidly reestablished after administering a pulse to the mother. Such amino acids originate from maternal alimentation, are actively transported to the fetus, and respond to the maternal supply.

2. A second group of amino acids, exemplified by cystine and cysteine, maintain almost equal levels in the mother and fetus during maternal fasting, but are transferred slowly to the fetus after intravenous injection of the mother.

3. Finally, the dicarboxylic amino acids, glutamate and aspartate, are products of fetal intermediary metabolism and apparently maintain a higher level in the fetus because of fetal production. Intravenous infusions of glutamate to pregnant monkeys in the third trimester fail to raise fetal glutamate levels until maternal plasma levels are elevated 70-fold.

Based on these studies, therefore, one would expect the transport of essential amino acids and the nonspecific nitrogen source to vary according to the state of maternal alimentation. Adaptation of fetal uptake to this supply has been evaluated, however, only in the pregnant sheep.

*Effect of Maternal Supply on Fetal Uptake.* As was discussed previously, a positive umbilical venous–arterial gradient for all the amino acids measured (except glutamate, aspartate, taurine, and carnosine) was demonstrated in the pregnant ewe fed *ad libitum*. Essential amino acids accounted for 44% of the nitrogen uptake, while glycine, alanine, serine, asparagine, and glutamine equaled 43% of the average net transfetal nitrogen gradient. Among the amino acids studied, the fetal blood level was directly related to the maternal arterial concentration of glutamine, six essential amino acids (threonine, valine, leucine, isoleucine, phenylalanine, and lysine), 1- and 3-methylated histidine, and ornithine. Since glutamine accounts for 22% of the net alpha-amino nitrogen uptake by the fetus, it might provide a major adaptable component of nonspecific nitrogen uptake by the fetus; this concept remains unproven. Lemons and his co-workers (1976) have demonstrated, how-

PETER A. J. ADAM and
PHILIP FELIG

ever, that this relationship holds true for three amino acids to which 12% of the fetal uptake may be attributed (valine, citrulline, and phenylalanine). Based on this evidence, a general principal may be stated which is that fetal lamb nitrogen uptake varies directly as the maternal supply. We would expect periodic fetal assimilation of amino acids to attain maximal rates during maternal absorption of ingested food. As indicated by Simmons and co-workers (1974), however, fetal oxidation of the amino acids is not increased by greater maternal intake of protein.

*Fetal Oxidation of Amino Acid.* The fetal responses to the maternal nutritional state, which have been discussed previously, involve fetal assimilation of nutrients absorbed during maternal ingestion and progressive mobilization of maternal energy stores for fetal needs during fasting. Careful studies of fetal metabolism during starvation of pregnant ewes have elucidated an additional fetal response to the nutrient supply (Simmons *et al.*, 1974). In short, the ovine fetus accelerates the oxidation of amino acids normally required for fetal growth and development in the first week of maternal starvation. During the first two days of fasting, both maternal and fetal plasma glucose declined while plasma urea levels rise for three days. As indicated in Figure 11 (Simmons *et al.*, 1974), the maternal–fetal arterial glucose difference declines during the first two days of fasting, then stabilizes, while the fetal–maternal urea gradient increases for three days but declines to baseline levels in approximately one week. Since the transplacental urea clearance of fetal plasma does not decline for at least ten days of maternal starvation, fetal urea excretion

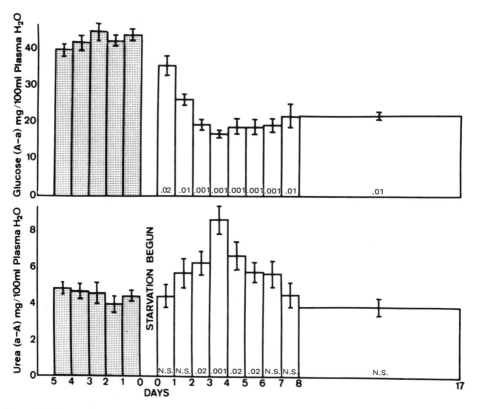

Fig. 11. Effect of maternal starvation on maternal arterial (A)–fetal arterial (a) differences of blood glucose and urea in the pregnant sheep (from Simmons *et al.*, 1974).

rises in proportion to the fetal–maternal arterial difference. Consequently, these data demonstrate conclusively that ovine fetal oxidation of amino acids is variable and that their oxidation increases temporarily when the supply of other fuels is diminished.

**491**

CARBOHYDRATE, FAT,
AND AMINO ACID
METABOLISM IN THE
PREGNANT WOMAN
AND FETUS

According to the calculated contributions of these substrates in the sheep, fetal oxidation of amino acids may account for 60% or more of the energy provided during maternal starvation (Table X). Although the human fetus ordinarily oxidizes amino acids (Gresham *et al.*, 1971), there is no evidence that the rate of oxidation may be increased by maternal starvation or that the provision of ketone bodies is inadequate. Furthermore, there is no direct evidence, at present, of active gluconeogenesis by the human fetus *in utero*. Thus, the role of amino acids in the adjustment of human fetus oxidative metabolism during maternal starvation has not been determined yet.

### 4.2.4. Fetal Growth Rates during Maternal Fasting

In the studies of ovine maternal starvation by Simmons and associates (1974), the fetuses were below predicted weight by an average of $0.61 \pm 0.33$ kg ($P < 0.001$). Thus, the cost of amino acid oxidation to the lamb fetus is apparent in either the reduced or arrested growth rates associated with maternal starvation.

### 4.2.5. Integrated Responses to Maternal Fasting

Our earlier consideration of fetal adaptation to cyclic maternal ingestion and fasting has been restricted to the variations of maternofetal flux and fetal oxidation of substrates provided by either maternal alimentation or depot catabolism. In well-nourished man, a hierarchy of events has developed to ensure a continuous supply of carbon-containing substrates to the fetus at rates maintaining homeostasis. Although periodic maternal ingestion of nutrients may result in episodes of accelerated depot storage by the fetus, the preponderance of evidence indicates that the human fetus receives sufficient substrate for maintenance of a normal metabolic rate during moderate maternal fasting. Even after overnight fasting, the carbohydrate derived from maternal sources provides most of the substrate required by the fetus. At this time, however, substitution is evident for approximately 20% of the glucose required to maintain oxidative metabolism.

In man, such fasting causes maternal production and fetal consumption of ketone bodies at rates which are sufficient to replace glucose as a fuel. Accordingly,

*Table X.  Calculated Average Contributions of Carbohydrate[a] and Amino Acids to Fetal Oxidative Metabolism of Two Substrates in the Fed, Fasted, and Starved Pregnant Ewe*

| | | | Starved | |
| --- | --- | --- | --- | --- |
| Fetal | Units | Fed | 3 days | 10 days |
| Carbohydrate uptake | mg/min/kg | 6.0[a] | 3.3[a] | 3.3[a] |
| Urea production | mg/min/kg | 0.7 | 1.3 | 0.7 |
| $O_2$ carbohydrate | ml/min/kg | 4.5 | 2.5 | 2.5 |
| $O_2$ urea | ml/min/kg | 2.0 | 4.0 | 2.0 |
| Carbohydrate | % | 69 | 38 | 56 |
| Amino acid | % | 31 | 62 | 44 |

[a]Includes glucose transferred from mother and fixed contribution of lactate produced by the placenta.

*PETER A. J. ADAM and PHILIP FELIG*

in the previously well-nourished pregnant woman we would expect greater substitution of the alternate fuels—rather than either a reduction of fetal metabolic rate or increased oxidation of amino acids—as the first line of defense against maternal fasting.

Human fetal metabolism has not been evaluated under conditions of chronic starvation, but prolonged maternal nutritional privation results in smaller newborn infants. At some point in maternal starvation, therefore, we would expect reduced rates of fetal accretion and metabolism and the recruitment of increased amino acid oxidation as demonstrated in the sheep. The consequences to fetal growth and development have been documented in mammalian experiments and clinical observation of man.

In summary, the responses to maternal fasting may be divided into two classes according to their consequences in the fetus (Table XI).

As indicated, the first order of responses in the hierarchy involve maternal catabolism and at least partial preservation of the fetal substrate supply. The second-order responses divert fetal substrates from accretion to oxidation, may catabolize fetal stores, and ultimately jeopardize fetal growth and development. In the sections which follow, we shall examine these inferences from the point of view of fetal organ metabolism and regional mobilization of fetal stores.

## 4.3. Consumption and Production of Fuels by Fetal Organs

### 4.3.1. Regional Considerations in Adaptation

In order to illustrate fetal tissue responses we have chosen for discussion four organs which epitomize the potential for adaptation *in utero*. These are as follow:

*The Brain.* From the point of view of fuel metabolism the central nervous system exemplifies an organ with limited endogenous provisions of energy—one which requires continuous substrate supplies from other sources. As the fetal organ with the greatest constraints on independent adaptation, therefore, brain is the tissue for which the integrated adaptive responses to maternal fasting are required to protect developmental potential in the fetus.

*Skeletal Muscle.* This metabolic mass is the model for a tissue which accumulates considerable endogenous energy stores in the form of glycogen and fat and which may respond to privation by consumption of its own depots. In addition,

*Table XI.  Ordered Responses to Maternal Fasting*

| Type | Site of response | Shift from anabolism to catabolism |
|------|-----------------|-----------------------------------|
| First order | Maternal glucose production | Maternal hepatic glycogenolysis<br>Maternal hepatic and renal gluconeogenesis |
| | Maternal ketone-body production | Maternal adipose tissue lipolysis<br>Maternal hepatic oxidation of free fatty acids |
| Second order | Reduced fetal consumption of primary substrates | Reduced fetal metabolic rate<br>Reduced fetal accretion and growth |
| | Altered metabolic fate of substrates derived from mother | Fetal oxidation of transported amino acids<br>Catabolism of fetal depots[a] |

[a]To be examined under regional responses.

skeletal muscle contains a potential pool of mobilizable energy providing substrate for hepatic and renal gluconeogenesis after birth.

*Adipose Tissue.* This depot, which accumulates in the third trimester, is the major energy store mobilized for maintenance of metabolic equilibrium during fasting in the newborn infant. No role in the fetus has been defined.

*Liver.* As the major organ regulating provision of fuels to the central nervous system postnatally, the liver illustrates a unique combination of phenomena, which potentially may apply to its role *in utero:* First, it accumulates a depot of glycogen early in gestation. Secondly, this depot may be mobilized rapidly for endogenous consumption and provision of substrate to other organs. Third, the liver produces both glucose from glucogenic amino acids, and also ketone bodies from transported FFA. Finally, the liver must be considered as a substrate consumer. In order to regulate intermediary metabolism, the liver is provided with a continuous supply of energy—either from its endogenous stores or by fuels transported from other sources.

The methods of study from which inferences concerning fetal organ metabolism have been drawn warrant documentation. These include the following: (1) assay of the enxymatic activities required for a specific metabolic pathway, for example, those required to metabolize glucose or ketone bodies; (2) demonstration, *in vitro,* that a labeled substrate is converted to the metabolic product under investigation, e.g., complete oxidation of $[U{-}^{14}C]$glucose to $^{14}CO_2$; (3) documentation of developing anatomical structures (e.g., dendritic arborization), functioning physiological units (e.g., muscle motor units) and accumulating depots (e.g., hepatic glycogen); (4) quantification of regional uptake *in vivo,* for example, glucose uptake by the hind limb, and (5) evaluation of compositional differences between body fluids *in vivo* and inferential analysis of each compartment represented (e.g., FFA composition in maternal and fetal plasma).

Finally, investigators have superimposed excessive or deficient provisions of specific substrates on the aforementioned experimental models, either *in vivo* or *in vitro,* in order to determine organ response to a surfeit or insufficiency of the fuel supply.

Ultimately our understanding of human fetal metabolism depends on inferences drawn from a multitude of mammalian experiments, including those in the subhuman primate, rodent, and ruminant. As an example of the method by which we construct a pattern of metabolic regulation, our knowledge of human fetal brain is restricted largely to information obtained from anatomical and histologic studies, compositional analysis, and assay of enzymatic activities at various points in gestation. Uptake of fuels by the fetal brain *in vivo* has been quantified only in the pregnant ruminant. Amino acid metabolism and substitution of alternate fuels *in vivo* have been correlated with structure and function in the suckling mouse and rat. Finally, these data from mammalian studies have been applied to previously cited structural and compositional studies, enzymatic activities, and metabolic profiles observed in man. Thus our understanding of organ metabolism in the human fetus is a collection of inferences and analogies drawn from a variety of sources.

### 4.3.2. Uptake of Substrates by Fetal Brain

*Metabolic Profile of the Mature Brain.* (a) Glucose metabolism: Glucose is the major substrate for brain metabolism and is taken up continuously in the well-nourished adult at a rate which supports oxidative metabolism. If the molecule is traced with $^{14}C$, it becomes apparent that glucose provides structural components

PETER A. J. ADAM and
PHILIP FELIG

and energy for several functions, including the following: (1) building and repairing complex molecules, (2) synthesizing neurotransmitters, and (3) maintaining the ionic gradient at synaptic junctions. This last function is associated with the major metabolic compartment of brain and consumes the bulk of substrate provided for oxidative metabolism (Siesjö *et al.,* 1974).

In the mature brain of well-nourished mammals, including man, glucose is consumed at a rate of approximately 30 $\mu$mol/min/100 g of brain. Virtually all of the glucose consumed eventually is converted to $CO_2$ and the remainder to lactate and pyruvate. Consequently, the molar quantity of $O_2$ extracted and $CO_2$ produced by brain in nourished mammals equals almost six times the number of moles of glucose consumed; the *RQ* is approximately 1.0. If [$^{14}$C]glucose is infused, however, the label distributes throughout several large metabolic pools, for example, among the glycolytic intermediates, the intermediary metabolites of the citric acid cycle, and ultimately, throughout the amino acid pool of brain.

(b) Alternate fuels during fasting: As demonstrated by Owen and co-workers (1967) $\beta$-hydroxybutyrate and acetoacetate are capable of replacing 50% or more of the cerebral glucose requirement during prolonged fasting in adult man. Furthermore, labeled ketone bodies enter the tricarboxylic acid cycle and perform most of the metabolic functions ordinarily reserved for glucose. These metabolic functions are manifest not only in the increased uptake of ketone bodies as blood levels rise, but also in the conversion of labeled $\beta$-hydroxybutyrate or acetoacetate to $^{14}CO_2$, $^{14}$C-labeled amino acids, and neurotransmitters by rat brain *in vivo* (Cremer, 1971) or *in vitro* (Itoh and Quastel, 1970). In addition, the metabolic compartmentation of the labeled ketone bodies, among amino acids in murine brain, resembles that of glucose *in vivo*. Thus, ketone bodies may replace a large fraction of the brain's requirement for glucose in those species which produce large quantities during fasting.

(c) Amino acid metabolism in mature brain: Lajtha and Toth (1961) and others have demonstrated in the mouse that the cerebral vasculature presents a barrier to free passage of amino acids from blood to the brain substance. This blood–brain barrier is exemplified by the inhibited rise of brain amino acids after elevation of their plasma levels by intraperitoneal injection or intravenous infusion. Although the mechanism is not completely understood (Meister, 1974), it is apparent that the penetration into brain of glutamate is strongly restricted (Schwerin *et al.,* 1950), that of lysine and proline is fairly strongly restricted (Dingman and Sporn, 1959), while that of glutamine and tyrosine is less restricted.

Nevertheless, the amino acid pool in brain can be maintained by fixation of nitrogen on carbon provided by glucose (Cooper *et al.,* 1974). As indicated in Figure 12, glucose can replenish, after its entry into the TCA cycle, the carbon skeleton for most of the major nonessential amino acid pools which characterize brain metabolism, for example, glutamate, glutamine, $\gamma$-aminobutyrate (GABA), and aspartate.

Finally, amino acid metabolism in the mature brain is characterized by compartmentalization into large and small glutamate pools. Patel and Balázs (1970) have presented a lucid discussion of the functional and anatomic correlates of these pools. In short, a large and a small glutamate pool may be inferred from the relative specific activities of brain glutamate and glutamine after the injection of $^{14}$C-labeled substrates *in vivo*. The large glutamate pool is part of the metabolically active pool associated with functioning nerve terminals and terminal dendrites. If a carbon tracer is introduced *in vivo* as [$^{14}$C]glucose or [$^{14}$C]$\beta$OH-butyrate, the specific

**495**

*CARBOHYDRATE, FAT,
AND AMINO ACID
METABOLISM IN THE
PREGNANT WOMAN
AND FETUS*

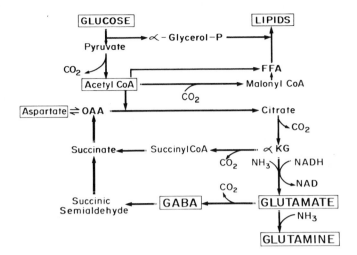

Fig. 12. Relationships between glucose, lipids, and nonessential amino acids in mature brain.

activity of brain glutamate exceeds that of glutamine so that the glutamate/glutamine specific activity ratio is greater than one (approximately 2.4 to 3).

The smaller glutamate pool, associated with glial and neuronal cell bodies, is labeled more rapidly by ketogenic amino acids oxidized to acetyl-CoA in brain than by glucose. Labeled carbon in leucine and phenylalanine enters the small pool more rapidly than glucose and results in a higher relative specific activity of glutamine than of glutamate (glutamate/glutamine relative specific activity ratio of approximately 0.3). This metabolic heterogeneity is characteristic of the brain only after it has progressed to moderate anatomical and functional maturity.

*Structural and Functional Considerations in Maturation.* Longitudinal studies relating fetal brain metabolism *in vivo* to structure have not been done in the pregnant primate. Thus, our knowledge of these relationships has been developed mainly from investigation of the suckling mouse and rat. Dobbing (1974) has presented the concepts succinctly and related them to human fetal growth and development. For reference, we have summarized in Table XII the early stages of brain development in the fetal and suckling rat and compared them with those in the human fetus.

If one examines the relationship between brain growth, glucose, and amino

*Table XII. Early Phases of Brain Growth in Rat and Man Following Neural Tube Formation[a]*

| | Developmental age | |
|---|---|---|
| Phase of growth | Rat | Man |
| Neuroblast multiplication | Last ⅓ pregnancy | 10–18 weeks gestation |
| Rapid brain growth | Birth–21 days | 18 weeks gestation–4th year |
|   Spongioblast multiplication and dendritic arborization | 0–9 days | 18 weeks gestation–2nd year |
|   Myelin synthesis and functional synaptic connection | 10–21 days | 32 weeks gestation–4th year |

[a]From Dobbing, 1974.

PETER A. J. ADAM and
PHILIP FELIG

acid metabolism in the rat, it is evident that rapid brain growth and functional maturity are closely related. In the newborn period, the blood–brain barrier restricts less than in the adult the uptake of amino acids such as glutamate (Himwich et al., 1957), lysine (Lajtha, 1958), and leucine (Lajtha and Toth, 1961). Furthermore, the brain is "metabolically homogeneous" in the early phase of growth between 0 and 9 days (Patel and Balázs, 1970). Brain protein, free glutamate, aspartate, and GABA increase rapidly throughout the period of rapid brain growth (Bayer and McMurray, 1967; Patel and Balázs, 1970); however, rapid development of the metabolic compartmentation does not occur until the second phase of rapid growth when myelin synthesis and functional synaptic connection occur. During this phase, glucose is incorporated into free amino acids at a progressively more rapid rate, and the relationships between glutamine and glutamate specific activities in the presence of $^{14}C$-labeled substrates develop the adult pattern (Patel and Balázs, 1970: Bayer and McMurray, 1967: Van den Berg, 1970). After 21 days of age, the glutamine/glutamate relative specific activity in the presence of the tracer substrates tends to plateau at a level resembling functional maturity. By this stage, the large glutamate pool is labeled predominently by glucose, while the small compartment is enriched after injection of phenylalanine (Van den Berg, 1970), or leucine (Patel and Balázs, 1970). Patel has suggested that functional maturity represented by such compartmentation results from maturation of glial–neuronal relations rather than from independent development of glia or of synaptic connections.

*Glucose Metabolism in Fetal Brain.* (a) Oxidative metabolism: As indicated previously the information concerning oxidative metabolism by the human fetal brain is derived from measurement of enzymatic activities and from studies of labeled substrate conversions *in vitro;* however, the data are fragmentary, and extrapolation should be restricted to inferences concerning potential pathways. When the maximal activities in brain of three glycolytic enzymes have been quantified (phosphofructokinase, glyceraldehyde phosphate dehydrogenase, and pyruvate kinase), they increase progressively during the period studied between 10 and 20 weeks of gestation (Hahn and Skala, 1970). At 10 weeks, brain glycogen is approximately 4.5 $\mu$mol/g; declines to 1.0 $\mu$mol/g at 20 weeks, and to 0.05 $\mu$mol/g at term (Villee, 1954). If glycogenolysis is evaluated at 20 weeks of gestation by dilution of [$^{14}C$]pyruvate *in vitro,* however, the quantity of unlabeled pyruvate formed from glycogen by brain slices is insignificant. Nevertheless, [$^{14}C$]glucose is converted to $^{14}CO_2$ (Villee, 1954), indicating that aerobic glycolysis and oxidative metabolism can occur.

The only regional studies *in vivo* of net cerebral glucose uptake by the fetus are those in pregnant ruminants during the last third of gestation (Makowski *et al.,* 1972; Tsoulos *et al.,* 1972; Jones *et al.,* 1976). Even though glucose accounts for only a fraction of net substrate uptake and oxidation by the ovine fetus *in vivo,* glucose and oxygen uptake by fetal sheep brain are consistent with the hypothesis that only glucose is oxidized. Despite the progressive decline of total fetal glucose/oxygen quotient as gestation progresses, the mean regional glucose/oxygen quotient of brain is 1.1 throughout gestation and does not drop significantly below 1.0 under various conditions of study. This is true whether the pregnant ewe is fed *ad libitum* (Tsoulos *et al.,* 1972) or fasted for 48–72 hr (Makowski *et al.,* 1972). The regional glucose uptake (30 $\mu$mol/min/100 g brain) and oxygen consumption (4 ml or 180 $\mu$mol/min/100 g brain) resemble the rates throughout life.

Furthermore, cerebral metabolism under steady-state conditions is entirely

**497**

*CARBOHYDRATE, FAT,
AND AMINO ACID
METABOLISM IN THE
PREGNANT WOMAN
AND FETUS*

*Table XIII. Mean Arterial Concentrations and Arterial–Sagittal Sinus Differences of Substrates for Oxidative Metabolism (mM)[a]*

|  | Arterial content | | | A–V difference | | |
|---|---|---|---|---|---|---|
|  | Adult | Lamb | Fetus | Adult | Lamb | Fetus |
| $O_2$ | 6.23 | 5.83 | 4.22 | 3.30 | 2.54 | 1.37 |
| Glucose | 2.81 | 5.39 | 0.99 | 0.57 | 0.42 | 0.22 |
| Lactate | 0.67 | 1.47 | 1.60 | −0.031[b] | −0.021 | 0.010 |
| Pyruvate | 0.15 | 0.14[c] | 0.31 | −0.013 | −0.053[c] | −0.020 |
| $6\Delta G/\Delta O_2$ |  |  |  | 1.03 | 0.99 | 0.98 |
| L/P | 4.37 |  | 5.21 |  |  |  |

[a] From Jones *et al.,* 1976.
[b] A negative value indicates production instead of uptake.
[c] Only 3 studies.

aerobic (Jones *et al.,* 1976). In Table XIII, various parameters of glucose metabolism are presented from comparative study of adult sheep and infant and fetal lamb. On the basis of these cerebral arteriovenous differences, glucose provides sufficient substrate for oxidative metabolism in the fetus, and negligible quantities are converted to lactate or pyruvate even though the arterial $O_2$ content is lower than in the lamb or adult sheep.

We may conclude from these studies that ovine fetal brain metabolism during the last third of pregnancy depends on a continuous supply of glucose as substrate and that there appears to be no alternate fuel for its oxidative metabolism in the fetal lamb.

*Relationship between Glucose Oxidation and Lipid Accretion.* (a) Synthesizing systems in the rat: Potentially, glucose could serve several major functions related to the accumulation of structural fat in the brain during development. These are summarized as follows: (1) glucose oxidation via the citric acid cycle generates energy, in the form of ATP, required for the synthesis of lipids; (2) partial oxidation of glucose by the pentose cycle reduces NADP to NADPH, providing reducing equivalents required for cytoplasmic synthesis of fatty acids; (3) glucose is the sole source of $\alpha$-glycerol phosphate—a necessary element of brain triglycerides and phospholipids; and (4) glucose might provide the acetyl-CoA ultimately synthesized into fatty acids. Examination of data leads one to conclude that most of the glucose taken up by brain is oxidized completely to $CO_2$, while the bulk of membrane and myelin lipid is derived from fatty acid carbon.

As indicated in the previous section, glucose can furnish most of the substrate required for oxidative metabolism by fetal brain; however, its role in supplying carbon for the skeleton of fatty acids has not been documented. Three intracellular systems for the synthesis of fat might incorporate glucose carbon into the fatty acids, and these have been investigated in the developing rat. Synthesis *de novo* occurs in the cytoplasm and is initiated by the production of malonyl-CoA from acetyl-CoA by the action of acetyl-CoA carboxylase. Subsequently, long-chain saturated fatty acids are manufactured by cytoplasmic fatty acid synthase and related enzymatic activities. Gross and Warshaw (1974) have done a developmental study of acetyl-CoA carboxylase in the rat, while Volpe and Kishimoto (1972) have reported a similar study of fatty acid synthase activity. In short, both activities are high in the rat during the phase of neuroblast multiplication and remain elevated throughout the early spurt of spongioblast proliferation. They start to decline at about 15 days of age during the peak of myelin synthesis but remain at substantial levels during the rest of infant and adult murine life. Throughout this period, aggregation of acetyl-CoA carboxylase by citrate, and allosteric inhibition by palmitoyl-CoA can be demonstrated. Furthermore, the enzymatic activities are highest during the period when [1-$^{14}$C]acetate is incorporated maximally into brain lipids *in vivo* (Dhopeshwarkar *et al.*, 1969). Thus, it appears that the potential for synthesis of lipids from glucose may exist in early development, although its actual contribution has not been quantified.

In order to evaluate the fatty acid elongating systems, Aeberhard and his associates (1969) studied isolated rat brain mitochondria and microsomes, with oxidation inhibited by anaerobic conditions *in vitro*. The mitochondrial system elongates saturated acids with greater lengths than C16, while the microsomal system is capable of desaturating, as well as elongating, these fatty acids. While the microsomal system requires substrate malonyl-CoA as does the cytoplasmic synthesizing system, the mitochondria add two carbon acetate radicals from acetyl-CoA to the long-chain fatty acids by reversal of $\beta$-oxidation. Thus, under anaerobic conditions *in vitro*, rat brain mitochondria reverse $\beta$-oxidation, while the microsomes incorporate [$^{14}$C]malonyl-CoA into fatty acids. Both these integrated systems have activities equivalent to those of adult animals during the early growth phase of spongioblast proliferation. The synthesizing capacity of both systems peaks at approximately 15 days and declines rapidly to adult levels in a manner similar to acetyl-CoA carboxylase activity. In short, the capacity to elongate fatty acids exists during all stages of development studied in the rat.

(b) Incorporation of substrate into brain lipids by man and the subhuman primate: Unfortunately, there are few data *in vivo* which might serve to elucidate the relationships between brain glucose and fat metabolism in the primate. Hahn and Skala (1970) have reported substantial levels of $\alpha$-glycerol phosphate dehydrogenase in human fetal brain from the 12th week of gestation onward, which permits us to infer that there is no limitation on the capacity of glucose to provide the 3-carbon skeleton of triglycerides and phospholipids. Furthermore, Roux and his associates have studied brain slices from the simian and previable human fetus under aerobic conditions (Roux and Myers, 1974; Yoshioka and Roux, 1972), and these studies may clarify the origin of brain lipids in the presence of physiological quantities of glucose and fat (Table XIV).

It is evident that human and simian fetal brains oxidize predominantly glucose *in vitro* under aerobic conditions and that most of the fatty acid is incorporated,

*Table XIV.*  *Substrate Incorporation into Lipids of Fetal Primate Brain in Vitro*[a]

| | [U-$^{14}$C] Glucose | [1-$^{14}$C] Palmitate | | |
| | Monkey | Human | Monkey | |
| Fetal age | 60–156 days | 12–13 weeks | 60–142 days | 156 days |
| Growth phase | All early phases | Neuroblast | Spongioblast[b] | Myelin[c] |
| Incorporation (%) | | | | |
| CO$_2$ | 98 | 3 | 1 | 4 |
| Total lipid[d] | 2 | 97 | 99 | 96 |
| Triglycerides | 0.5 | 52 | 34 | 31 |
| Phospholipids | 1.3 | 39 | 52 | 45 |
| Sterols | 0.4 | 7 | 13 | 20 |

[a]From Roux and Myers, 1974.
[b]These fetuses span neuroblast proliferation to early myelin synthesis.
[c]Includes formation of functional synapses.
[d]Data exclude free fatty acids which may include unaltered palmitic acid.

**499**

*CARBOHYDRATE, FAT,
AND AMINO ACID
METABOLISM IN THE
PREGNANT WOMAN
AND FETUS*

without being shortened, into triglycerides and phospholipids (Dhopeshwarkar and Mead, 1969). Nevertheless, a portion of the carboxyl end undergoes β-oxidation to acetyl-CoA, and is incorporated mainly into sterols such as cholesterol. Although glucose undoubtedly provides α-glycerol phosphate, only a minor portion of glucose carbons is consumed for this purpose in the compartment accreting fat, and the bulk of accumulating lipid is formed from the fatty acid provided.

During the phase of rapid myelin formation, cholesterol is a major component of the lipid accumulation. Our interest in cholesterol is derived from the elegant means by which Pitkin and his co-workers (1972) demonstrated the sites of cholesterol synthesis in fetal tissues. If tracer [4-$^{14}$C]- or [1-α-$^3$H]cholesterol (evenly distributed in dietary cholesterol) is fed to pregnant rhesus monkeys for periods up to 12 weeks, a steady-state specific activity of maternal plasma cholesterol is obtained between 4 and 7 weeks. At this time fetal serum cholesterol has a mean specific activity which is 43% of that in maternal serum. The cholesterol specific activity in the parenchyma of fetal tissues is 65–98% of that in fetal plasma; but brain is unique in having a $^{14}$C-specific activity that is only 5% of that in the plasma. These data resemble the pattern observed in suckling rats (Srere *et al.*, 1950; Morris and Chaikoff, 1961) where newborn brain actively synthesizes cholesterol. Thus most of the cholesterol is synthesized *de novo* in developing brain from an undefined source of acetyl-CoA. Based on present evidence, glucose is an unlikely source; β-oxidation of fatty acid carboxy-terminal carbon is possible; incorporation of amino acids oxidized in the small compartment into cholesterol has not been investigated extensively.

To summarize the relationships between glucose metabolism and fat accretion in brain, it appears that glucose provides a major component of the energy required, while much of the fat accumulates from fatty acids transported to the brain after maternofetal transfer. Most of the cholesterol is synthesized *de novo* from acetyl-CoA, but the source of this carbon has not been determined. Partial β-oxidation of long-chain fatty acids and oxidation of amino acids entering the compartment accreting lipid could provide carbon for lipid synthesis and modification *in vivo*.

*Alternate Fuel for Glucose.* (a) Activation and metabolism of ketone bodies: The physiologic criteria by which the ketone bodies have been judged alternate fuel for the fetus have been discussed in Section 4.2.2, Total Ketone-Body Metabolism. In addition to these, an alternative substrate necessarily would demonstrate specific characteristics of regional uptake and metabolism resembling those observed with the primary fuel. In order to be judged an appropriate substitute for glucose in brain, therefore, the ketone bodies must have some of the following characteristics of regional metabolism: (1) presence of enzymatic activities required to enter the metabolic pathway associated with the functions of the primary fuel; (2) replacement of glucose in regional uptake and adaptation of the uptake to the supply of glucose or the alternate fuel; (3) oxidation of the $^{14}$C-labeled substrate to $^{14}$CO$_2$; and (4) incorporation of the alternate fuel into amino acids and a pattern of compartmentalization resembling that characteristic of glucose.

In adult mammals and man, ketone bodies are formed by the liver and transported to other tissues for oxidation (Figure 13). During pregnancy, however, maternal liver produces the ketone bodies which are transported to and utilized by fetal as well as maternal tissues. While the liver produces ketone bodies, there is no pathway for their hepatic oxidation because of the absence of an activating enzyme linking ketone bodies to coenzyme A. As illustrated in Figure 13, the entry of

PETER A. J. ADAM and
PHILIP FELIG

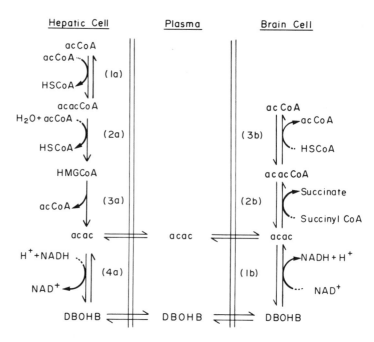

Fig. 13. Schematic representation of hepatic ketone-body production and cerebral ketone-body uptake. Liver is unable to oxidize ketone bodies because of the absence of 3-oxoacid acetyl-CoA transferase activity (2b). Glossary: 1a, 3b: acetoacetyl-CoA thiolase; 2a: hydroxymethylglutaryl-CoA synthase; 3a: hydroxymethylglutaryl-CoA lyase; 4a, 1b: 3-hydroxybutyrate dehydrogenase; 2b: 3-oxoacid acetyl-CoA transferase; acac: acetoacetate; acCoA: acetyl-CoA; acac CoA: acetoacetyl-CoA; HMGCoA: hydroxy-methylglutaryl-CoA; DBOHB: D-betahydroxybutyrate.

ketone bodies into the cycle of oxidative metabolism requires conversion of $\beta$-OH-butyrate to acetoacetate by $\beta$-OH-butyrate dehydrogenase, activation of acetoacetate by the 3-oxoacid CoA transferase exchanging CoA between succinyl-CoA and acetoacetate, and finally the formation of acetyl-CoA fragments in the presence of HSCoA and thiolase activity. The requisite enzymatic activities are present in the mitochondria of tissues oxidizing ketone bodies as an alternate fuel, while another mechanism of activation has been demonstrated in the cytoplasm of tissues utilizing ketone bodies as substrate for lipogenesis (Buckley and Williamson, 1973).

In the rat, the three enzymatic activities permitting entry of ketone bodies into the tricarboxylic acid cycle are present in the fetus (Dahlquist *et al.*, 1972; Dierks-Ventling, 1971). During the early period of postnatal life prior to weaning, in which rapid accretion of fat into the brain substance is occurring, $\beta$-hydroxybutyrate dehyrogenase and the 3-oxoacid CoA transferase activities rise rapidly from quite low levels to a sustained peak between 15 and 21 days, while the acetoacetyl-CoA thiolase activity is substantial throughout early life (Page *et al.*, 1971; Lockwood and Bailey, 1971; Klee and Sokoloff, 1967). Only one study of the activities in human fetal brain has been reported. In short, the three enzymatic activities cited were present in the cerebral cortex and cerebellum of one human fetus studied at 32 weeks of gestation (Page *et al.*, 1971).

(b) Ketone-body uptake by brain in early life and adaptation to the supply: In adult man and the rat, ketone-body uptake by brain *in vivo* is directly proportional to the plasma levels, whether they are raised by prolonged fasting or by infusion of acetoacetate (Hawkins *et al.*, 1971; Daniel *et al.*, 1971). Cerebral ketone-body uptake in the suckling rat also is directly proportional to the blood levels of D-$\beta$-

**501**

*CARBOHYDRATE, FAT,
AND AMINO ACID
METABOLISM IN THE
PREGNANT WOMAN
AND FETUS*

hydroxybutyrate and acetoacetate; however, the relationship of cerebral arteriove-nous differences to the arterial blood ketone-body levels is steeper in the suckling than in the adult rat. Thus a greater quantity of either ketone is taken up by suckling rat brain at every concentration studied. This greater uptake has been related to the higher cerebral activities of enzymes involved in metabolism of ketone bodies (Hawkins *et al.*, 1971), but it probably reflects other factors, such as diffusion barriers, as well. Such a hypothesis could be tested by measuring the influx rates of tracer [$^{14}$C]D-$\beta$-hydroxybutyrate or acetoacetate, as has been done in the adult rat (Daniel *et al.*, 1971).

In various experimental models, the uptake of ketone bodies by adult rat brain is proportional to the blood levels and is unaltered by prolonged starvation. Even though murine cerebral $\beta$-hydroxybutyrate dehydrogenase activity is increased by starvation for 48–72 hr (Smith *et al.*, 1969), the uptake of ketone bodies does not exceed the quantity expected on the basis of the blood level (Hawkins *et al.*, 1971).

During pregnancy, maternal starvation has minimal effects on fetal rat brain $\beta$-hydroxybutyrate dehydrogenase activity (Dahlquist *et al.*, 1972), while high fat intake in the mother causes a rise of cerebral thiolase and transferase activities (Dierks-Ventling, 1971). Such changes, however, may not affect ketone-body uptake by fetal brain—at least a direct relationship between the enzymatic activities and substrate utilization has not been shown to occur.

In early human life, ketone-body uptake by the brain has been studied only at two periods; those are in the mid-gestational fetus, and during early infancy. D-$\beta$-hydroxybutyrate is taken up by the isolated perfused fetal brain at a rate which would produce energy equaling that available from normal rates of glucose uptake (Adam *et al.*, 1975*b*); at 11 days to 12 years of age, there is a positive arterial–internal jugular venous gradient for both ketone bodies (Persson *et al.*, 1972; Settergren *et al.*, 1976). In studies up to 12 years of age the A:V difference for both acetoacetate and $\beta$-OH-butyrate also are directly related to the blood level, as in the rat. Furthermore, the ketone-body gradients in children exceed those reported in the adult at similar blood levels (Gottstein *et al.*, 1971). Thus, present evidence favors the concept that ketone bodies serve as alternate substrates for brain metabolism in the human fetus and infant, but no studies can be done during prolonged fasting to confirm this concept.

(c) Metabolic fate of ketone-bodies: Those species in which significant cerebral ketone-body uptake occurs *in vivo* also convert the $^{14}$C-labeled substrate to $^{14}$CO$_2$ *in vitro* (Itoh and Quastel, 1970; Cremer, 1971; Kahng *et al.*, 1974). In the adult rat, $\beta$-OH-butyrate and [$^{14}$C]acetoacetate are incorporated into metabolic products associated with passage through the tricarboxylic acid cycle in much the same manner as glucose. The labeled compounds recovered include $^{14}$CO$_2$, $^{14}$C-labeled amino acids, and the neurotransmitter acetylcholine.

Brain slices of the suckling rat, likewise, convert [3-$^{14}$C]acetoacetate to $^{14}$CO$_2$, glutamate, glutamine, $\gamma$-aminobutyrate, aspartate, and acetylcholine *in vitro* (Itoh and Quastel, 1970). Furthermore, Buckley and Williamson (1973) have demon-strated the presence of acetoacetyl-CoA synthase activity in the cytoplasm of developing rat brain, an enzyme which can activate acetyl-CoA and potentially initiate lipogenesis from ketone bodies. As indicated by Klee and Sokoloff (1967), the enzymatic activities which generate reducing equivalents for the synthesis of lipid (by permitting glucose entry into the pentose cycle) also rise in rat brain during early life. Thus ketone bodies may partially replace glucose as a fuel for oxidative metabolism, as a source of carbon replenishing the nonessential amino acid pool, and as a substrate for lipogenesis.

In the one study of human fetal cerebral ketone-body metabolism, $[3\text{-}^{14}C]\beta$-OH-butyrate also was taken up by isolated perfused brain and rapidly converted to $^{14}CO_2$ (Figure 14). Thus, the central nervous system in man has the potential for oxidizing the ketone-bodies as early as the first trimester *in utero*. Based on such evidence, ketone-body metabolism by human fetal brain would serve to ensure potential growth and development of the central nervous system, despite intermittent maternal caloric deprivation.

(d) Effects of ketone bodies on glucose metabolism: When the simultaneous metabolism of glucose and acetoacetate by infant or adult rat brain is examined *in vitro*, the ketone body inhibits oxidation of pyruvate, diverting it from entry into the tricarboxylic acid cycle toward its reduction to lactate (Itoh and Quastel, 1970). In the presence of maternal and intrauterine hyperketonemia, induced by feeding a 45% fat diet to pregnant rats for 7–8 days before term, the metabolic profile of the infant rat brain at birth resembles that of the puppies born to normally fed mothers (Wapnir *et al.*, 1973; Ozand *et al.*, 1975a,b). If their mothers are maintained on the previous diets, the puppies suckled by hyperketonemic mothers have higher plasma triglyceride and blood ketone levels. Brains in both groups grow normally, but cerebral intermediary metabolism differs *in vivo*. In the hyperketonemic newborns, brain levels of fructose diphosphate, glucose-6-phosphate, pyruvate, and glutamine are lower, while brain glutamate and lactate are higher than in the controls. Cerebral ATP, creatinine phosphate, and protein levels are unaltered by hyperketonemia. Such data may be interpreted to indicate attenuated entry of glucose into brain, diversion of aerobic glycolysis to lactate, oxidation of ketone bodies in the tricarboxylic acid cycle, and maintenance of the large glutamate pool by the provision of ketone bodies. Since growth and development are not jeopardized by a hyperketonemic regimen, it appears certain that ketone bodies may replace much of the cerebral glucose requirement during fetal life and during the period of rapid brain growth in the rat. This concept applies comfortably to our present knowledge of human fetal brain metabolism.

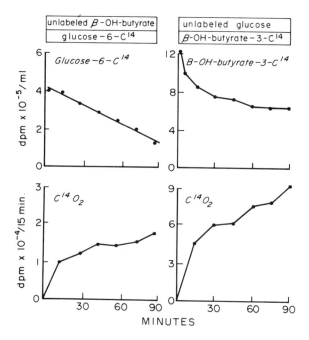

Fig. 14. Oxidative metabolism of glucose and $\beta$-OH-butyrate by isolated human fetal brain (from Adam *et al.*, 1975b).

*4.3.3. Muscle and Hind Limb* **503**

*CARBOHYDRATE, FAT,*
*AND AMINO ACID*
*METABOLISM IN THE*
*PREGNANT WOMAN*
*AND FETUS*

*Differentiation and Growth of Striated Muscle.* In a recent review, Mastaglia (1974) has presented a comprehensive analysis of the early development of muscle in the fetal trunk and limbs. Briefly, the musculature of the trunk arises from the primitive mesoderm, from which the full complement of 38 somites has formed by the end of the fifth week. The limb musculature is formed *in situ* from the mesoderm of the limb buds, and most of the individual muscles of the arm are recognizable by the end of the sixth week of gestation. Primitive uninucleate myogenic cells containing myosin and actin (myoblasts) divide repeatedly. They fuse to form a syncitium of myotubes with a peripheral rim of myofibrils and a central core of cytoplasm containing nuclei and other cellular organelles. At about the fifth week of gestation the first generation of myotubes appear; the central core of these cells is lost with the continuous formation of myofibrils, and few such cells remain after the 20th week of gestation. During this early phase of development, however, the muscle cell types characteristic of adult life cannot easily be distinguished. A common cell type with active oxidative and glycolytic systems seems to predominate between the 10th and 18th week of gestation, but type I and type II fibers differentiate thereafter. Between the 20th and 26th weeks, approximately 90% of the muscle consists of white glycolytic type II fibers; while the remainder may be distinguished by succinic dehydrogenase activity as the red, oxidative type I fibers. After the 26th week there is a progressive increase in the proportion of type I fibers until the 30th week, at which time the muscle has assumed a checkerboard pattern, similar to that of the mature adult, with approximately equal numbers of type I and II fibers.

*Accretion of Glycogen and Fat.* Even though muscle glycogen cannot be mobilized for other tissues, this depot is a major energy store, potentially available for fetal metabolism during maternal fasting. In the last half of gestation, striated muscle accounts for 20–25% of body weight, so that total muscle glycogen, at a lower relative concentration, equals the mobilizable pool present in liver. If skeletal muscle accounts for 25% of the metabolic rate after birth, muscle glycogen at any period in the last trimester would provide an ample energy depot for many hours of endogenous consumption (Table XV).

In the fetal rhesus monkey, the glycogen content of skeletal muscle rises progressively from the 50th day of gestation onwards (Bocek *et al.*, 1969; Kerr *et al.*, 1971), but the increases cannot be related directly to the activities *in vitro* of the enzymes controlling glycogen synthesis and degradation. Total activities of both glycogen synthase and phosphorylase increase throughout the period studied, but

*Table XV.  Glycogen Content of Human Striated Muscle[a]*

| Gestation (weeks) | Heart (mg/g) | Muscle (mg/g) | Skeletal muscle | | Provision (hr)[c] |
|---|---|---|---|---|---|
| | | | Total/kg body weight | | |
| | | | g/kg[b] | mM/kg | |
| 31 | 7.6 | 16.3 | 4.08 | 22.6 | 30 |
| 40 | 10.1 | 26.7 | 6.68 | 37.1 | 49 |

[a]Widdowson, 1968.
[b]250 g striated muscle/kg body weight.
[c]25% of energy consumption in a neutral thermal environment after birth.

the active component of each is a minor proportion of the total. Comparison of glycogen content/100 mg N with the enzymatic activities/100 mg N demonstrates that these parameters have no relationship to each other. Thus, Bocek et al. (1969) have concluded that accumulation and maintenance of glycogen in fetal rhesus muscle probably depends more on metabolite than on hormonal control. Furthermore, glycogen synthase D appears to be the predominant activity, suggesting that glucose-6-phosphate might be a major regulator. Although fetal muscle G6P in vivo is lower early in gestation than in the adult, the quantity of this synthase D activator at 150 days is twice that in the adult animals (Beatty et al., 1976). Furthermore, variation in G6P levels might result in variable net rates of glycogenolysis based on the regulation of glycogen synthetic rates. Based on this concept, G6P would decline if glucose entry into the hepatocyte were reduced or glycolysis were accelerated. Such an effect would permit glycogenolysis to prevail. Although this concept has not been investigated in primate muscle, such interactions of G6P with divalent cations, adenine nucleotides, and glycogen synthase activity have been described in human fetal liver (Section 4.3.5, Regulation of Enzymatic Activities and Metabolic Pathways).

*Muscle Lipids.* In order to enhance clarity, adipocytes intermingled with muscle are considered separately in Section 4.3.4. During development, then, the major lipid components of muscle cell are those incorporated into the sarcolemma and into the membranes of sarcoplasmic reticulum, the transverse tubular system, and the mitochondria. A small quantity of neutral lipids can be demonstrated histologically in the oxidative type I cells but their mobilization for muscle energy needs *in utero* has not been investigated.

*Capacity for Oxidative Metabolism.* (a) Oxidation of glucose: When regional consumption of various substrates has been examined during the late fetal or immediate neonatal period, it seems that glucose is the primary fuel for oxidative metabolism by fetal mammalian cardiac or skeletal muscle *in vivo*. Such data has encouraged studies *in vitro* which either measure the capacity for oxidation of alternate fuels or examine the biochemical regulation of fuel metabolism.

In detailed studies of metabolism by either homogenates of striated muscle or isolated muscle fibers from the fetal rhesus monkey, Beatty and her co-workers (1966, 1972) have demonstrated conclusively that the metabolic pathways for glycogenolysis, glycolysis, and oxidation of glucose are functional *in vitro*. Fetal rhesus muscle fibers or homogenates consume $[^{14}C]$glucose and oxygen, and produce $^{14}CO_2$ and $[^{14}C]$lactate. The labeled lactate is diluted by virtue of glycogenolysis and glycolysis occurring under aerobic conditions *in vitro*. When oxidation of $[1-^{14}C]$glucose is compared with that of $[6-^{14}C]$glucose, the specific yield of $^{14}CO_2$ from carbon-1 exceeds that from carbon-6, indicating that pentose cycle also may function. Quantification of the pathways *in vitro* is difficult, however, because glucose uptake exceeds the rate of oxidation, but only a fraction of the labeled carbon taken up is recovered in the $CO_2$ evolved. Thus, it appears that labeled carbon exchanges with large unlabeled metabolic pools in the fiber or the particulate homogenate under the conditions of study. Nevertheless, the metabolic pathways traced by the label appear to have been active *in utero*.

In one study of human fetal heart and diaphragm, tissue slices incorporated [U-$^{14}C$]glucose and $[2-^{14}C]$pyruvate into $^{14}CO_2$, indicating that glucose and pyruvate can be oxidized by human fetal striated muscle (Villee, 1954). The $^{14}CO_2$ recovered

*505*

*CARBOHYDRATE, FAT,
AND AMINO ACID
METABOLISM IN THE
PREGNANT WOMAN
AND FETUS*

had a lower specific activity than the substrates, presumably indicating oxidation of endogenous substrates or exchange of the label among unlabeled metabolic pools.

The biochemical regulation of aerobic glycolysis in the monkey fetus has been examined *in vivo* and *in vitro* at mid- and near-term gestation (Beatty *et al.*, 1976). When the glycolytic intermediates and cofactors were quantified in quick-frozen biopsies of fetal muscle, the calculated mass action ratio of hexokinase, phospho-fructokinase, and pyruvate kinase were similar in fetal and adult specimens. The values were exceeded more than 800-fold by the constants expected if these three reactions had proceeded to equilibrium, indicating probable regulation at these sites. Although the activation and inhibition of fetal phosphofructokinase activity occurred at different cofactor concentrations than in the adult, the general charac-teristics of allosteric regulation were qualitatively similar.

Furthermore, the Atkinson energy charge, (ATP + ½ ADP)/(ATP + ADP + AMP), was similar in the fetal and adult rhesus muscle, indicating oxidation of fuel was equally effective in the fetus and adult monkey. Thus, the biochemical regula-tion of glucose metabolism in the primate fetal muscle *in vivo* resembles that in the adult, and the provision of fuel for oxidative metabolism is adequate.

(b) Capacity for oxidation of fat: Based on the measurement of arterial–coronary sinus gradients, the heart of dogs 7–13 days old does not consume free fatty acids *in vivo* after a 6- to 8-hr fast (Breuer *et al.*, 1968). Although the mechanism restricting fatty acid uptake was not defined in these studies, a series of elegant experiments *in vitro* have elucidated some of the principals of control in the newborn rat and fetal calf. In Figure 15, the enzymatic activities required for the entry of long-chain fatty acids into mitochondria, for either oxidation or elongation, are illustrated (Tandler and Hoppel, 1972). As indicated, a long-chain fatty acid must be activated to an acyl-CoA, the CoA radical exchanged for carnitine by carnitine palmityltransferase A (CPT-A) activity, and the acyl-CoA regenerated during transport through the inner mitochondrial membrane by replacement of the carnitine radical with CoA (CPT-B activity).

Wittels and Bressler (1965) have compared the metabolism of [1-$^{14}$C]palmitate by coarse homogenates of neonatal rat heart with its oxidation by homogenates from adults. The homogenized heart of infant rats 24 hr old oxidizes [$^{14}$C]glucose at ten times the rate of homogenates from adult rats. Nevertheless, [1-$^{14}$C]palmitate is oxidized at 20% of the adult rate, even though there is no limitation on $\beta$-oxidation of hexanoate. In trying to localize the site of control, the investigators demonstrated that there are less fatty acyl thiokinase and carnitine palmitoyltransferase A activi-ties; however, these did not appear to be rate limiting since incorporation of [1-$^{14}$C]palmitate into triglycerides proceeded more rapidly in the newborn than in the adult heart homogenates. Furthermore, carnitine levels in the infant heart were 25% of those in the adult, and addition of this cofactor to the medium increased oxidation of palmitate to half the rate observed in the adult. Based on this evidence, Wittels and Bressler inferred that palmitate oxidation in the newborn rat heart was limited by lower CPT-A activity and carnitine cofactor concentration.

In contrast, Tomec and Hoppel (1975) have found no absolute limitation of palmitoyl-CoA oxidation by fetal bovine heart mitochondria *in vitro*. Palmitoyl-1-carnitine is oxidized at the same rate in mitochondria isolated from fetal and calf heart; but 40 $\mu$M palmitoyl-CoA is oxidized by fetal heart mitochondria at 15% of the rate in the newborn calf. Surprisingly, preincubation of the mitochondria for 10 min with palmitoyl-CoA and carnitine raises this rate of oxidation in fetal calf heart above that in the calf mitochondria which have not been preincubated. Carnitine

PETER A. J. ADAM and
PHILIP FELIG

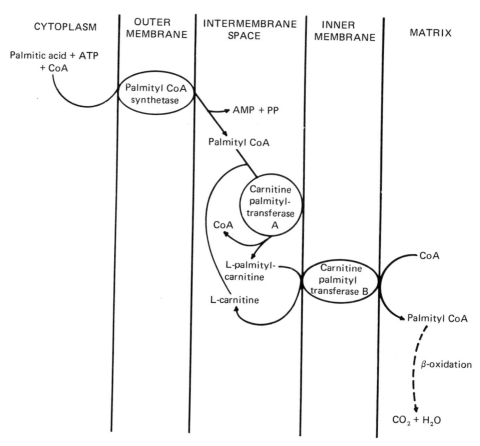

Fig. 15. Mitochondrial metabolism of palmitate to $CO_2$ and $H_2O$ (from Tandler and Hoppel, 1972, with permission of the authors and Academic Press).

palmitoyltransferase A and B activities are similar in the calf and fetal hearts, and the palmitoyl-CoA saturation curves are the same as in the older calf, indicating no overt deficiency of CPT-A. In summary, it appears that the limitation of long-chain fatty acid oxidation is environmentally determined and involves restrained transfer of palmitoyl-CoA into mitochondria for oxidation. The exact mechanism has not been determined.

Finally, the oxidation of [1-$^{14}$C]palmitate by primate muscle fibers has been evaluated in the presence of a physiologic glucose concentration (5.6 mM) *in vitro*

Table XVI.  Oxidation of [1-$^{14}$C] Palmitate by Fetal Rhesus Muscle Fibers in Vitro[a]

|  | $\mu$mol/g NCN/hr[b] | |
|---|---|---|
|  | 100 days | 155 days |
| Palmitate uptake | 50 | 27 |
| Lactate production | 1.1 | 0.6 |
| Palmitate→lipid | 36 | 21 |
| Palmitate→$CO_2$ | 2 | 2 |
| $O_2$ consumption | 870 | 635 |
| $CO_2$ production | 720 | 475 |

[a]From Beatty and Bocek, 1970.
[b]NCN = noncollagenous protein nitrogen.

(Beatty and Bocek, 1970). As indicated in Table XVI, the bulk of the long-chain fatty acid is incorporated into lipids, while only a small proportion undergoes β-oxidation. Thus, the long-chain fatty acid does not appear to be a major substrate for oxidative metabolism in fetal primate striated muscle. This conclusion is not firm, however, since fibers from adult monkeys produce $^{14}CO_2$ from [$^{14}C$]palmitate at similar rates, with a lesser uptake of the fatty acid, and fetal muscle FFA oxidation has not been evaluated *in vivo*.

(c) Potential for ketone-body utilization: The consumption of ketone bodies *in vivo* by fetal striated muscle has not been evaluated in animals such as the pregnant rat in which the fasting mother produces large quantities of ketone bodies in lieu of glucose. In the adult of species which augment hepatic ketone-body production during fasting, however, major organs such as heart, skeletal muscle, and kidney contain the enzymatic activities required for their activation and consumption (Williamson *et al.*, 1971). Furthermore, arteriovenous ketone-body differences across the human limbs have been demonstrated in adult man (Owen and Reichard, 1971).

Betahydroxybutyrate dehydrogenase, 3-oxoacid CoA transferase, and aceto-acetyl-CoA thiolase activities also appear in the heart and kidney of the rat before birth (Dierks-Ventling, 1971; Lockwood and Bailey,1971), but no study of these activities in fetal rat skeletal muscle has been reported. At 12 hr of age, slices of heart and kidney from the newborn pig oxidize $^{14}C$-labeled β-OH-butyrate, and the transferase and thiolase activities are present (Kahng *et al.*, 1974).

In short, the activities required for oxidation of ketone bodies apparently develop before birth in several organs of animals in which maternal production of β-hydroxybutryate and acetoacetate is a significant component of metabolic adaptation during pregnancy. The requisite activities develop in fetal heart but have not been evaluated in skeletal muscle before birth. It seems likely that ketone bodies provide a major alternate fuel for oxidative metabolism in human fetal striated muscle, but this hypothesis has not been tested adequately in the laboratory.

*Hormonal Regulation of Metabolism in Skeletal Muscle.* The response of isolated fetal rhesus muscle fibers to pharmacologic doses of insulin or epinephrine has been evaluated *in vitro* at 85–95 and 125 days of gestation (Bocek and Beatty, 1969; Bocek *et al.*, 1973). Insulin (10 mU/ml) enhances glucose uptake slightly and increases the exchange of labeled carbon from [6-$^{14}C$]glucose with glycogen, lactate, and $CO_2$; however, the tissue glycogen concentration is not increased. At present these results seem to indicate that insulin receptors are present but that the efficacy of insulin as a fine regulator of fetal muscle metabolism *in vivo* is in doubt.

Epinephrine ($6 \times 10^{-6}$ M) causes a reduction of hepatic glycogen content and glucose uptake, while enhancing lactate production by the muscle fiber. The incorporation of $^{14}C$ from [6-$^{14}C$]glucose into glycogen, lactate, and $CO_2$ is reduced by the pharmacologic concentration of epinephrine in the medium. Thus, the major effect of the catecholamine appears to be enhanced glycogenolysis.

Based on these results, the efficacy of catecholamines *in vivo* cannot be judged; Bocek and her co-workers (1973) have extended these investigations, however, to the evaluation of cyclic AMP responsiveness. In the isolated fetal muscle fiber, the incorporation of [8-$^{14}C$]adenine into [$^{14}C$]cyclic AMP is enhanced by a pharmacologic dose of epinephrine in the presence of the phosphodiesterase inhibitor, caffeine. When the phosphodiesterase activity in cell-free preparations of rhesus fetal striated muscle is quantified (Bocek and Beatty, 1976), both the high- and low-affinity forms of this activity are greater in the fetus than in the adult (Table XVII).

PETER A. J. ADAM and
PHILIP FELIG

Table XVII.  Total cAMP Phosphodiesterase Activity
(nmol cAMP Hydrolyzed/min/100 mg Protein)[a]

| Muscle | 100-day fetus | 155-day fetus | Adult |
|---|---|---|---|
| Skeletal | $272 \pm 25$ | $172 \pm 16$ | $44 \pm 4$ |
| Diaphragm | 266 | 193 | $49 \pm 5$ |
| Heart | 790 | 499 | $360 \pm 18$ |

[a]From Bocek and Beatty, 1976.

As indicated, there is a pattern of decline during fetal life, but the phosphodiesterase activity is four times the adult level, even as term approaches. Such high activities suggest that cyclic AMP turnover in the fetus would be rapid and that hormonal effects may be minimized, or of short duration, unless a prolonged stimulus occurs. In short, these data support Bocek's original hypothesis that hormonal regulation of fuel metabolism by fetal muscle may be less important to fine control than are adjustments of intracellular substrate and cofactor concentrations (Section 4.3.3. Accretion of Glycogen and Fat).

*Adaptation of Muscle to the Substrate Supply in Vivo.* Morriss and co-workers (1973) have demonstrated the responsiveness of fetal ovine hind limbs to variations in fetal arterial blood glucose concentrations. Utilizing the glucose/oxygen (G/O) quotients, they performed five studies of fetuses in ewes fed *ad libitum* and five in ewes fasted 2–3 days. As indicated in Table XVIII, fetal arterial glucose concentration was not affected significantly by maternal fasting in these studies, and the mean femoral arterial–venous quotient exceeded 1.0 in both groups. Thus glucose was the predominant fuel and could account for oxidative metabolism in most of the fetuses studied.

Nevertheless, the G/O quotient varied directly as the fetal glucose concentration. Three of the 10 fetuses studied had G/O quotients significantly less than 1.0, four were considerably higher than 1.0, while the remainder approximated 1.0. Thus the uptake of glucose by the hind limb adjusts to the ebb and flow of glucose as substrate. When there is an abundance, glucose is taken up at rates exceeding its oxidation, and presumably stored in striated muscle as glycogen. If the supply is diminished, alternate fuels are oxidized. Based on experiments *in vitro,* striated muscle glycogen stores can be mobilized, but the role of this depot during maternal fasting has not been documented. Alternatively, ketone bodies might be substituted for glucose in species other than the sheep, and the endogenous glycogen pool preserved. Although promising, neither of these hypotheses has been tested in the primate fetus.

Table XVIII.  Mean Fetal Blood Glucose and Femoral
Arterial–Venous G/O Quotient in Fed and Faster Ewe[a]

| | Mean $\pm$ SD | |
|---|---|---|
| Maternal feeding | Glucose (mM) | G/O quotient |
| Fed *ad libitum* | $0.82 \pm 32$ | $1.12 \pm 0.49$ |
| Fasted 2 to 3 days | $0.71 \pm 0.22$ | $1.01 \pm 0.53$ |

[a]From Morriss et al., 1973.

*4.3.4. Adipose Tissue*

**509**

*CARBOHYDRATE, FAT,
AND AMINO ACID
METABOLISM IN THE
PREGNANT WOMAN
AND FETUS*

*Growth and Development of Adipose Tissue.* While glucose serves as the major fuel for metabolism *in utero,* the largest energy depot at birth is contained in white adipose tissue. Development of this tissue commences in the third to fourth month of gestation at many sites around young capillary shoots lying beneath the dermis (Hull, 1974). At mid-gestation, the accumulation of triglycerides begins, but the major accumulation occurs in the last trimester (Widdowson, 1968, 1974) (Table XIX).

Our knowlege of white adipose tissue metabolism *in utero* is inferred largely from studies of tissue fragments or isolated adipocytes obtained from the gluteal region of newborn infants by biopsy. Even at birth, white adipose tissue differs structurally from that normally obtained in adults. Whereas all the adipocytes in the adult's tissue are unilocular, those isolated from the newborn infant present a mixture of multilocular and unilocular cells. The average calculated volume of infant cells (0.082 nl) is less than that of adult (0.53 nl), but the histological structure of the unilocular cells from both is similar. In short, the unilocular cells contain one large lipid droplet with a thin rim of cytoplasm and have an ovoid nucleus, small mitochrondria, and extensive smooth cysternae (Novak *et al.,* 1971). In contrast, the multilocular cells of the newborn infant contain multiple fat vacuoles giving cytoplasm a granular appearance, many glycogen granules arranged along the walls of the cysternae, and abundant large septate mitochondria. The copious mitochondria correlate with high activities of oxidative enzymes such as succinic dehydrogenase.

*Accumulation of Energy Depots.* (a) Glycogen: Although endogenous glycogen in adipose tissue is a minor energy store which can be utilized only within the adipocyte, it may have a considerable effect upon fat cell metabolism *in vivo* or *in vitro.* By extrapolation of adipose tissue glycogen content to the beginning of timed labor, its concentration *in utero* can be determined and is approximately 2 mg/g wet weight at term gestation (Novak and Monkus, 1972). If the glycogen provides $\alpha$-glycerol phosphate for fatty acid reesterification, the endogenous glycogen may attenuate the release of free fatty acids from adipose tissue for many hours even in the presence of low blood glucose levels. In any case, the glycogen seems to have such an effect upon isolated adipocytes *in vitro* obtained from the infant at birth (Novak and Monkus, 1972).

(b) Fat synthesis: The triglycerides of fetal adipose tissue might originate from glucose or from transported free fatty acids. As suggested earlier, fatty acids in fetal adipose tissue triglycerides also may be preserved by rapid reesterification exceed-

Table XIX.  *Accumulation of Lipid in the Human Fetus*[a]

| Gestation (weeks) | Body weight (kg) | Fat (%) | Fat (g) |
|---|---|---|---|
| 20 | | 0.5 | |
| 28 | 1.2 | 3.5 | 42 |
| 31 | 2.0 | 5.0 | 100 |
| 34 | 2.2 | 7.5 | 165 |
| 40 | 3.5 | 16.0 | 560 |

[a]From Widdowson, 1968.

PETER A. J. ADAM and
PHILIP FELIG

ing the rate of lipolysis; such a relationship between synthetic and lipolytic processes is illustrated in Figure 16.

Studies of enzymatic activities in adipose tissue and of metabolism by tissue fragments or isolated adipocytes seem to indicate that three mechanisms favoring accretion operate *in utero;* they are *de novo* synthesis, incorporation of circulating fatty acids, and fatty acid reesterification. Although the measurements are few, the enzymatic activities which provide substrates and reducing equivalents for cytoplasmic fatty acid synthesis *de novo* are substantial in the hours immediately after birth, tend to decline thereafter, but exceed adult levels (Novak *et al.,* 1973*a*). Alphaglycerol phosphate dehydrogenase, which provides the glycerol moeity of the triglycerides, also equals or exceeds adult activities.

When [U-$^{14}$C]glucose (Novak *et al.,* 1972) or [1-$^{14}$C]palmitate (Novak *et al.,* 1966) incorporation into triglycerides is evaluated *in vitro* during the early hours after birth, it is found that both substrates are incorporated after birth and the rates of incorporation decline in the hours which follow, even though the triglyceride pool becomes smaller. If these data are compared with both the low fetal plasma levels of FFA existing *in utero* and the evidence from compositional studies of a fetal contribution to its own circulatory FFA pool during labor, the biochemical "set" of the adipocyte *in vivo* is apparent. Evidently, the rates of triglyceride synthesis from glucose and circulating FFA by the human fetal adipocyte exceed the lipolytic rate and result in the rapid triglyceride accretion observed during the last trimester. Furthermore, a vigorous lipolytic stimulus such as labor apparently is countered by rapid reesterification of endogenous adipose tissue fatty acids. In light of this hypothesis, it seems unlikely that fetal adipose tissue would mobilize fuel for oxidative metabolism by other fetal organs if a reduction of the maternal substrate supply were to occur.

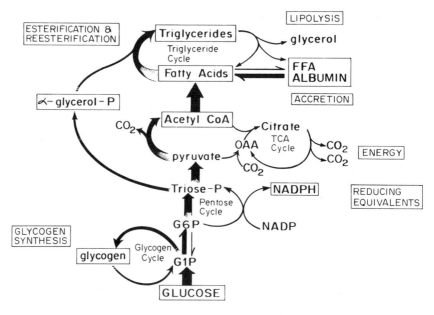

Fig. 16. Human fetal adipose tissue: relationships between synthesis, esterification, and lipolysis of triglycerides. Glossary: G6P: glucose-6-phosphate; G1P: glucose-1-phosphate; OAA: oxalacetate; FFA: free fatty acids.

*511*

CARBOHYDRATE, FAT,
AND AMINO ACID
METABOLISM IN THE
PREGNANT WOMAN
AND FETUS

*Oxidation of Circulating Substrates and Endogenous Depots.* Based on assays of maximal glycogen phosphorylase and phosphofructokinase activities shortly after birth, no absolute enzymatic limitation of glycogenolysis or glycolysis would be expected; [U-$^{14}$C]glucose is converted rapidly to $^{14}CO_2$ at this time (Novak *et al.*, 1972, 1973c). In contrast, oxidation of lipids apparently proceeds slowly *in utero*. In the newborn infant, 3–6 hr old, carnitine palmityltransferase A (CPT-A) activity is at higher levels than in the adult, but carnitine levels (30 $\mu$mol/g wet weight) are lower than expected in organs actively oxidizing long-chain fatty acids (Novak *et al.*, 1973b). At an unspecified time after birth, mitochondria isolated from neonatal white adipose tissue oxidize palmityl-CoA slowly (Novak *et al.*, 1975), but oxidize it in the presence of 1 mM carnitine. This indicates that CPT-A and CPT-B activities are present or develop shortly after birth. Conversion of [1-$^{14}$C]palmitate to $^{14}CO_2$ also can be demonstrated when fragments of adipose tissue obtained during the first day of life are incubated in the absence of glucose (Novak *et al.*, 1969); however, mitochondrial $\beta$-hydroxyacyl-CoA dehydrogenase activity is very low at birth, suggesting a specific enzymatic limitation of fatty acid oxidation in utero (Novak *et al.*, 1972). Although the evidence cited indicates that neonatal adipose tissue develops the capacity for fatty acid oxidation shortly after birth, the few data obtained immediately after birth suggest that this capacity is limited *in utero* and that the limitation is environmentally determined.

Finally, we should comment on the relationship between glucose oxidation and the accretion of triglycerides in human fetal adipose tissue. As indicated previously, the mitochondria and oxidative activities of the TCA cycle are abundant in this tissue. Furthermore, the ATP content of the adipose tissue is maintained at substantial levels until birth, and declines continuously during the first day as the endogenous glycogen is depleted (Novak *et al.*, 1973c). Based on these and the previously presented data, it seems most likely that glucose is oxidized at rates sufficient to maintain adipose tissue ATP content *in utero,* thus providing energy for the synthetic processes. To summarize some of the major roles of glucose in fetal adipose tissue, it performs the following functions: (1) provides substrate for fatty acid synthesis and elongation; (2) produces reducing equivalents via pentose cycle; (3) furnishes $\alpha$-glycerol phosphate for synthesis of triglycerides and reesterification of long-chain fatty acids; (4) generates energy for synthetic processes via aerobic glycolysis and oxidation; and (5) maintains an endogenous pool of glycogen which may replace the glucose requirement intermittently, thereby buffering adipose tissue metabolism from variations in the intrauterine environment. If true, this final function would serve to enhance the continuity of a synthetic process, thus preserving triglyceride accumulation for energy requirements during postnatal life.

*Lipolytic Capacity of Fetal Adipose Tissue.* (a) Mechanisms of control: Even though human fetal adipose tissue is environmentally "set" for accretion of triglycerides throughout the third trimester, it is equally apparent that a substantial potential for lipolysis is established simultaneously. Nevertheless, the lipolytic responses to hormonal and environmental modification differ from those which develop after birth. In the previous subsection, we have discussed the relationships of triglyceride synthesis and reesterification to lipolysis. Now we shall examine the lipolytic capacity of adipose tissue at birth and its hormonal regulation. Figure 17 illustrates several aspects of the hormonal regulation of lipolysis in adult man (Havel, 1972). As illustrated, sympathetic neural norepinephrine stimulates lipolysis by activating $\beta$-adrenergic adenylcyclase activity and hence the lipolytic cascade

PETER A. J. ADAM and
PHILIP FELIG

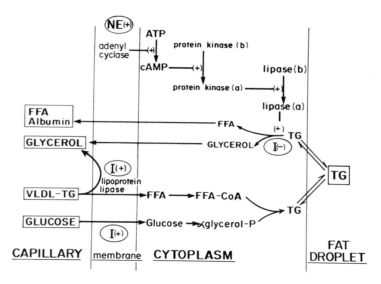

Fig. 17. Hormonal control of lipolysis and triglyceride synthesis in adipose tissue. Sympathetic neural catecholamine stimulates lipolysis as indicated. Insulin opposes lipolysis while enhancing glucose entry, triglyceride synthesis, and accretion of the fatty acid moeity of triglycerides carried to adipose tissue by the VLDL. (Derived from Havel, 1972.) Glossary: TG: triglyceride; FFA: free fatty acids; VLDL: very-low-density lipoproteins; I(+): enhanced by insulin; I(−): inhibited by insulin; NE(+): enhanced by sympathetic neural norepinephrine; (a): "active," (b): "inactive" enzymatic forms.

mediated by cAMP. In the adult man, insulin acts to reduce the efficacy of lipolysis by enhancing glucose entry into the adipocyte which in turn increases fatty acid reesterification, by activating capillary membrane lipoprotein lipase which promotes fatty acid entry from circulating triglycerides, and by inhibiting cyclic-AMP-mediated activation of the intracellular hormone-sensitive lipase. Review of the presently available evidence leads us to infer that lipolysis occurs continuously during labor and delivery but that triglyceride synthesis and fatty acid reesterification predominate. There are insufficient data to permit firm conclusions concerning the turnover of triglyceride fatty acid in the human fetus before the initiation of labor.

(b) Lipolysis *in vitro:* Suspensions of isolated adipocytes (Novak and Monkus, 1972) and fragments of adipose tissue (Novak *et al.,* 1968) from newborn infants release glycerol and FFA *in vitro*. Shortly after birth, however, the ratio of FFA/glycerol released from the adipocyte approximates 1.0, implying that ⅔ of the FFA are reesterified rapidly with α-glycerol phosphate derived from glucose or glycogen while the glycerol moiety of the triglycerides is released. Intact fragments of adipose tissue release a greater proportion of FFA relative to glycerol; however, less than ⅔ of that expected from complete lipolysis is released in the first 4 hr of life. If fluoride-stimulated adenylcyclase activity is quantified in order to estimate the potential for response to catecholamine, its activity rises progressively from the initiation of labor through the first 10 hr of life (Novak *et al.,* 1973*c*). Furthermore, pharmacologic quantities of epinephrine or norepinephrine ($5–6 \times 10^{-5}$ M) raise FFA and glycerol release by adipose tissue fragments *in vitro* from birth onwards (Novak *et al.,* 1968). Thus the potential for hormonal regulation of lipolysis exists at birth, but it may be modified by the endogenous store of carbohydrate.

**513**

*CARBOHYDRATE, FAT,
AND AMINO ACID
METABOLISM IN THE
PREGNANT WOMAN
AND FETUS*

(c) Lipolysis *in vivo:* From compositional differences between maternal and fetal plasma FFA at birth, one may calculate the approximate relative contributions of fetal adipose tissue and maternal plasma to circulating FFA *in utero* during labor. Based on the proportions of palmitic and linoleic acid (King *et al.,* 1971*a*), the fetal plasma FFA seem to be derived almost equally from maternal and fetal sources during labor. Thus, lipolysis in the fetus must be occurring continuously during labor and delivery. Nevertheless, fetal plasma FFA are low, implying that they are being deposited in adipose tissue as well as being exchanged continuously with the depot of triglyceride fatty acids. If this is so, then fetal metabolism, even during a stimulus to lipolysis, is adapted to the reassimilation of released fatty acids. Such control of the intrauterine adipose tissue metabolism would serve to preserve the major energy depot for postnatal life, rather than permitting its catabolism for fetal energy needs during transient maternal fasting or privation. There are, however, no data which address this concept directly.

Finally, we may examine the responses of premature infants to separation from the maternal substrate supply, in order to determine whether a developmental limitation exists which restricts the mobilization of fatty acids. From the era in which premature infants underwent prolonged fasting after birth, we may demonstrate that FFA and glycerol rise rapidly after birth in infants delivered at 32–36 weeks of gestation (Melichar and Wolf, 1967). Aside from a limitation imposed by the paucity of the triglyceride stores in early premature infants, there apparently is no restriction of lipolysis resulting from failure of neural, hormonal, or biochemical control. As inferred from studies of isolated adipocytes and mitochondria, therefore, accretion of adipose tissue triglyceride during the third trimester is determined mainly by the intrauterine environment after the appropriate hormonal mileu is established. During this period fine regulation of lipolysis by hormonal stimulation seems to be attenuated by the ample endogenous and circulating supplies of carbohydrates. There are no studies except those of the ovine hind limb which might elucidate the role of adipose tissue *in vivo* during maternal fasting. One would expect, however, that human adipose tissue responses would resemble the ovine hind limb and participate in the adaptation to reduced circulating glucose. Since provision of free fatty acids by fetal lipolysis seems to play a minor role in the adaptation, we expect that adipose tissue would merely vary the rate of triglyceride and glycogen accretion according to the glucose supply. This hypothesis requires further study.

### 4.3.5. Liver

*The Liver as a Regulatory Organ in Utero.* As was observed in the previous section, adipose tissue serves as the major caloric depot available for postnatal life. Although the hydrolytic capacity required to mobilize this fetal energy store exists throughout the last trimester, triglyceride accumulation results from the continued predominance of fatty acid synthesis and accretion. In short, synthetic capacity balances and exceeds lipolysis in adipose tissue so that fatty acids released *in utero* are redeposited and the energy store is preserved. Based on present evidence, one might infer that a similar situation prevails in the fetal liver. Immediately after birth, the neonatal liver serves as the major carbohydrate pool which can be mobilized for the provision of glucose to vital organs such as the brain. If the analogy with adipose tissue can be verified, the capacities for both glycogen synthesis and degradation

PETER A. J. ADAM and
PHILIP FELIG

develop in human fetal liver during early gestation; however, synthesis predominates under the influence of the intrauterine environment. The effects of varied
maternal nutrition are buffered *in utero,* and the hepatic glycogen store is mobilized
only under extreme conditions. The purpose of the current discussion is to examine
this postualte, i.e., the hypothesis that glycogen synthesis prevails because the
conditions obtaining in the intrauterine environment enhance the synthetic arm of
the glycogen cycle. Although extreme nutritional privation may deplete this depot,
the battery of maternal responses may act to preserve this fetal glycogen store.

In our attempt to verify these concepts, we shall draw on evidence from several
sources in mammals and man. These will include: (1) the enzymatic activities of the
glycogen cycle and gluconeogenesis, which define the limits imposed on metabolic
responses; (2) the incorporation of substrates into glucose *in vitro* which confirm
the early establishment of metabolic pathways required for glucose synthesis, albeit
at a low level; (3) the responses of the isolated perfused liver to modification of the
medium glucose and substrate supply, which verify the adaptability of the fetal
hepatic glycogen cycle to the nutrient supply; (4) the variation of fetal glucose
production caused by an excess or insufficiency of maternal provisions *in vivo,*
which corroborates fetal hepatic responsiveness to the maternal nutritional state,
and finally (5) the magnitude of systemic glucose production in prematurely born
infants which reveals that whatever constraints to fetal hepatic glucose secretion
exist during the last trimester are determined by the intrauterine environment.

*Energy Depots.* (a) Glycogen: Human fetal accumulation of a hepatic glycogen
store commences at approximately the 13th week of gestation (Čapkova and
Jirasek, 1968) and progresses until term (Shelley, 1969; Widdowson, 1968). In Table
XX, we have expressed hepatic glycogen content as the concentration in liver, the
quantity available per kg body weight, and the number of hours during which
hepatic glycogen could provide approximately 50% of the postnatal energy needs.
This last assumption is based on the measurement of fasting systemic glucose
production (Kalhan *et al.,* 1976) and the respiratory quotient in a neutral thermal
environment (Tunell *et al.,* 1976) at 2 hr of age.

As is shown, hepatic glycogen could provide a considerable proportion of fetal
caloric needs, if the maternal energy supply were diminished.

(b) Lipid stores and their significance: During fetal life there is no accretion of a
triglyceride depot in the liver which resembles that of glycogen. Based on chemical

*Table XX. Human Fetal Hepatic Glycogen Content*

| Gestational age (weeks) | Liver glycogen | | Potential glucose[b] at 50% of energy (hr) |
|---|---|---|---|
| | Liver ($\mu$mol/g) | Body weight (mmol/kg)[a] | |
| 8 | 19.0 | 0.9 | 0.6 |
| 12 | 20.9 | 1.0 | 0.7 |
| 14 | 35.8 | 1.8 | 1.2 |
| 16 | 54.4 | 2.7 | 1.8 |
| 18 | 136.8 | 6.8 | 4.6 |
| 21 | 146.6 | 7.3 | 5.0 |
| 40 | 217.8 | 10.9 | 7.4 |

[a]Liver 4–5% of body weight.
[b]Glucose production at 2 hr of age = 24.7 $\mu$mol/min/kg body weight.

**515**

*CARBOHYDRATE, FAT,
AND AMINO ACID
METABOLISM IN THE
PREGNANT WOMAN
AND FETUS*

estimates in the human and subhuman primate, the average hepatic lipid concentration declines with progression of gestation (Widdowson, 1974; Kerr *et al.*, 1971). [1-$^{14}$C]Palmitic acid and cholesterol are incorporated into the triglyceride, phospholipid, and sterol components of liver (Roux and Myers, 1974; Yoshioka and Roux, 1972; Pitkin *et al.*, 1972), presumably reflecting membrane synthesis and turnover. Based on this evidence, the hepatic lipids perform a major structural role *in utero* and play a minor role in energy metabolism and transport. Furthermore, the plasma levels of very-low-density lipoproteins (VLDL) are negligible *in utero,* indicating that hepatic triglyceride synthesis coupled with transport to distal sites is negligible in the human fetus.

*Potential Metabolic Pathways.* (a) The glycogen cycle: In Figure 18 we have summarized the relationships between major potential substrates for hepatic glycogen, glucose, and lipids. Rate and directional control reside at irreversible enzymatically regulated steps which oppose one another, so that synthesis and hydrolysis can be enlisted separately, but their rates are coordinated by reciprocal activation and inhibition. Synthesis of glycogen requires either glucose or three carbon substrates, an energy supply, the enzymatic activities permitting the entry of glucose into the intermediate pool, and the specific enzymes promoting glycogen

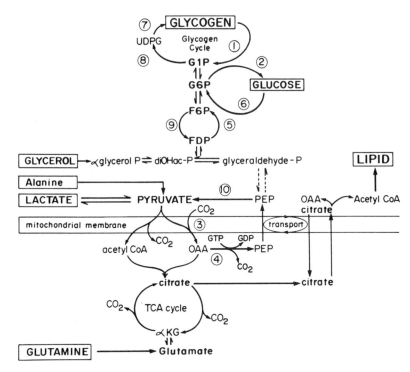

Fig. 18. Relationships between potential substrates for glucose, glycogen, and lipids in the human liver. Glossary: G6P: glucose-6-phosphate; G1P: glucose-1-phosphate; F6P: fructose-6-phosphate; FDP: fructose-1,6-diphosphate; PEP: phosphoenolpyruvate; OAA: oxalacetate; KG: alphaketoglutarate. Enzymatic activities exhibiting unidirectional control of glucose metabolism: Glucose production; (1) glycogen phosphorylase; (2) glucose-6-phosphatase; (3) pyruvate carboxylase; (4) phosphoenolpyruvate carboxykinase; (5) fructose diphosphatase. Glucose consumption and assimilation: (6) hexokinase and glucokinase; (7) glycogen synthase; (8) glucose-1-phosphate uridyltransferase; (9) phosphofructokinase; (10) pyruvate kinase.

PETER A. J. ADAM and
PHILIP FELIG

synthesis. The potential for control of synthesis, either by activation or allosteric regulation, dwells mainly in glycogen synthase activity.

Accumulation of glycogen in the human fetal liver coincides with a progressive increase of fetal glycogen synthase activity; however, the more active glycogen synthase (I form) remains at low levels throughout the early period of rapid glycogen accumulation. Similar observations in monkey skeletal muscle have led Bocek and her associates (1969) to postulate that the less active phosphorylated form, glycogen synthase D, may serve as the ''active'' synthetic enzyme *in utero*. In a recent investigation, Schwartz and Rall (1975*b*) have examined this hypothesis extensively by duplicating the conditions of substrate and cofactor concentrations *in vitro* which exist in the human fetal liver during intrauterine life. At mid-gestation, total glycogen synthase activity under ideal conditions approximates 2 $\mu$mol/min/g liver, while the activity when conditions *in utero* are duplicated approximates 0.2 $\mu$mol/min/g. Such levels are more than sufficient to account for the average daily hepatic glycogen accumulation rates in the human fetus, provided glycogenolysis is adequately inhibited.

Although there are no similar investigations of the human fetus at term, Glinsmann and his co-workers have evaluated fetal hepatic glycogen synthase activity in the rhesus monkey near term gestation (Glinsmann *et al.*, 1975; Sparks *et al.*, 1976). As it does in the human, the D form of glycogen synthase activity predominates. Similarly, the total activity is approximately 2 $\mu$mol/min/g of liver. Thus, it seems that the human and subhuman primate accumulate liver glycogen and maintain the depot *in utero* by similar mechanisms. The role that allosteric regulation of synthase activity plays will be discussed on pp. 527–529.

(b) Glycogenolysis: The rate-controlling activity of glycogenolysis, glycogen phosphorylase, has been quantified in human fetal liver obtained at abdominal hysterotomy, in the isolated perfused liver, and organ explant (Gensser *et al.*, 1971; Schwartz *et al.*, 1975; Adam *et al.*, 1972). When measured in the direction of glycogen synthesis instead of degradation, maximal phosphorylase activity exceeds that of glycogen synthase and is similar in the three reports. Only one of these investigations has quantified maximal fetal hepatic phosphorylase activity in the hydrolytic direction according to its action *in vivo* (Schwartz *et al.*, 1975). In this study, the total activity was approximately 3 $\mu$mol/min/g liver. Thus, total glycogen synthase and phosphorylase activities are of similar magnitude early in gestation, and maximal rates of glycogenolysis could override synthesis if it were not for inhibition of the hydrolytic activity.

Hepatic glycogen phosphorylase activity has not been quantified in the third trimester in the human fetus. In the rhesus monkey fetus near term, however, glycogen phosphorylase activity measured in the synthetic direction resembles that of the early gestational human fetus and bears a similar relationship to maximal glycogen synthase activities (Sparks *et al.*, 1976). Based on this resemblance, the rhesus fetal liver has served as a model for hepatic glycogen metabolism and its regulation.

(c) Hepatic glucose release: In the human, hepatic glucose-6-phosphatase activity may be measured in fetal specimens obtained during the first trimester (Schwartz *et al.*, 1974). Although lower than after birth, the mean activity early in gestation ranges from 1 to 3 $\mu$mol/min/g liver and would not be expected to restrain hepatic glucose release from glycogen (Aurrichio and Rigillo, 1960; Gensser *et al.*, 1971; Schwartz *et al.*, 1974). The potential opposing activities, which permit glucose uptake, have not been studied in the human fetus; however, hepatic glucose uptake and consumption always occur when glucose is provided *in vitro* (Adam *et al.*, 1972).

(d) Gluconeogenic potential: The major substrates for gluconeogenesis enter the pathway reversing glycolysis in the form of three carbon intermediates. As illustrated in Figure 18, lactate, pyruvate, and amino acids entering intermediary metabolism through pyruvate are shuttled past pyruvate kinase activity by the combined actions of pyruvate carboxylase and phosphoenolpyruvate carboxykinase (PEP-CK). Energy for initiation of gluconeogenesis enters as ATP, GTP, or ITP, and the phosphoenolpyruvate generated can be shuttled from mitochondrion to cytoplasm in exchange for citrate (Robinson, 1973). Then, the reversible activities of glycolysis catalyze formation of fructose diphosphate, at which point another regulated shuttle is required. Fructose diphosphatase activity reverses the glycolytic sequence and generates fructose-6-P, which is isomerized to glucose-6-P, and converted to glucose by the action of glucose-6-phosphatase in liver or kidney.

In one report of the three major activities controlling gluconeogenesis, all are present from early gestation but at lower levels than after birth (Räihä and Lindros, 1969). As indicated in Table XXI, pyruvate carboxylase activity is substantial at midgestation and sufficient to permit anaplerotic enrichment of the oxalacetate (OAA) pool.

Although quantified at $1/10$ the activities of adult life, maximal PEP-CK likewise seems to be sufficient for the transport of carbon entering the intramitochondrial oxalacetate pool from pyruvate to phosphoenolpyruvate. In the early human fetus, this activity is distributed mainly in mitochondria but also is present in the cytoplasm (Kirby and Hahn, 1973), as it is in the other mammals such as the fetal guinea pig which can initiate gluconeogenesis immediately after birth (Robinson, 1976). Furthermore, maximal fructose diphosphatase activity is of similar magnitude and should not place an absolute restriction on hepatic gluconeogenesis.

Although all the enzymatic activities required for reversal of glycolysis are present, attempts to quantify pyruvate, lactate, or glycerol incorporation into glucose result in low estimates of gluconeogenesis *in vitro*. Either prior conditions *in utero* or the conditions of experimentation may limit such estimates. Alternatively, $^{14}C$ introduced within substrates for glucose may exchange rapidly with large unlabeled intracellular carbon pools, even though a net flux in the direction of glucose is occurring.

In any case, liver slices from the early human fetus convert [2-$^{14}C$]pyruvate to small quantities of [$^{14}C$]glucose (Villee, 1954). When dual labeling is used to quantify glucose production by the isolated perfused human liver, however, glucose production proceeds at a physiologic rate by postnatal standards, while lactate

*Table XXI.  Enzymatic Activities Reversing Glycolysis [a] ($\mu mol/min/g$ liver)*

| Subjects | Gestational or postnatal age | Pyruvate carboxylase | PEP-CK[b] | Fructose diphosphatase |
|---|---|---|---|---|
| Fetal | 2–3 mo | $2.0 \pm 1.6$[c] (4)[d] | $0.7 \pm 0.5$ (5) | $0.5 \pm 0.1$ (4) |
| | 3–4 mo | $2.3 \pm 1.4$ (4) | $0.8 \pm 0.5$ (4) | $1.0 \pm 0.2$ (3) |
| | 4 mo | $2.0 \pm 1.3$ (3) | 1.0 (2) | 1.4 (2) |
| Adult | 20 yr | $3.0 \pm 1.2$ (4) | $9.1 \pm 3.7$ (4) | 2.6 (2) |

[a]Räihä and Lindros, 1969.
[b]Phosphoenolpyruvate carboxykinase activity.
[c]Mean ± SE.
[d]Number of subjects.

incorporation accounts for only $\frac{1}{35}$ to $\frac{1}{12}$ of the total glucose production rate (Adam et al., 1972).

The uptake and incorporation of amino acids into glucose has been evaluated only in explants of human fetal liver after depletion of hepatic glycogen (Schwartz, 1974a,b). Briefly, the nonmetabolizable amino acid, [$^{14}$C]α-aminoisobutyrate ([$^{14}$C]AIB), is concentrated by human fetal liver in the first half of gestation; uptake of this amino acid is competitively inhibited by alanine; and [U-$^{14}$C]alanine is incorporated into glucose linearly for 48 hr of study (Schwartz and Rall, 1975a). The rates of incorporation, even after glycogen depletion, are somewhat less than in the isolated perfused liver, and they do not exceed 0.05 μmol/min/g liver, even when stimulated maximally with triamcinalone and dibutyrl cyclic AMP.

To summarize human fetal hepatic glucogenic potential, the capacity for gluconeogenesis at low rates exists from early gestation. Although there is no absolute limitation preventing incorporation of three carbon intermediates into glucose, gluconeogenesis apparently plays a minor role in normal fetal adaptation to the usual maternal nutritional state.

*Alternate Energy Sources for Fetal Liver.* (a) Carbohydrate: Glycogen: At present, no regional study elucidating fetal hepatic oxidative metabolism in man has been reported from a stable surgical preparation *in vivo*. Our inferences are drawn from experiments *in vitro* which serve to illustrate potential substrate sources and tissue responses. Although hepatic glycogen synthesis *in vitro* has been demonstrated in the isolated perfused adult rat liver, human or primate fetal livers have not been studied under conditions which demonstrate such synthesis conclusively. In every study of isolated livers, slices, or explants, glycogen concentration declines spontaneously unless very high doses of insulin (0.1–1.0 U/ml) are added to the medium after glycogen depletion (Adam *et al.*, 1972; Villee, 1954; Sparks *et al.*, 1976; Schwartz *et al.*, 1975). If the liver is studied in the presence of a recirculating, initially aglycemic medium, glucose is released until its concentration stabilizes in a physiologic range, but the rate of glycogen loss from the liver exceeds the net rate of glucose release. (Figure 19). Nevertheless, continuous dilution of [$^3$H]- or [$^{14}$C]glucose occurs even after the medium glucose concentration stabilizes, at a rate which equals the decline of hepatic glycogen content (Adam *et al.*, 1972). Data such as these indicate that glycogenolysis proceeds *in vitro* at rates which would be considered physiologic postnatally; that the glycogen is consumed to provide the energy requirements of the liver *in vitro;* and finally, that glucose already released from the liver exchanges rapidly with the unlabeled glucose-6-phosphate pool formed from glycogen. Thus, hepatic glycogen potentially may serve as a source of energy for other organs or for the liver. Based on currently available evidence it is unlikely that human fetal hepatic glycogen serves as a net energy source for the liver *in vivo* except under hypoxic conditions (Villee *et al.*, 1958).

Glucose: Early in gestation the human fetal liver furnished with no alternate substrate *in vitro* is glucoregulatory (Adam *et al.*, 1972). In the absence of added substrate for oxidative metabolism, net glucose release occurs continuously until the medium glucose concentration reaches a physiologic range. Glucose, 2.8 mM (50 mg/100 ml), suppresses net glucose release completely unless an additional caloric source is provided (Figure 19). After prolonged perfusion until glycogen depletion occurs, there is a net uptake of medium glucose *in vitro*. As would be expected, there are no studies of the perfused human liver at term; however, the

**519**

*CARBOHYDRATE, FAT,
AND AMINO ACID
METABOLISM IN THE
PREGNANT WOMAN
AND FETUS*

perfused fetal rhesus monkey liver exhibits net glucose release with an aglycemic medium, and a relatively stable glucose concentration when perfused with media containing more than 6 mM glucose (Sparks *et al.*, 1976).

When the oxidation of glucose is examined in studies of liver slices from the early human (Villee, 1954) and the term monkey fetus (Roux and Myers, 1974), uniformly labeled glucose is converted to $^{14}CO_2$. The monkey fetal slices incorporated almost 90% of the glucose uptake into $^{14}CO_2$ *in vitro*. Thus glucose potentially could provide the bulk of hepatic energy requirements in the fetus of a well-nourished mother; this concept remains unproven.

Role of lactate and pyruvate: According to the metabolic pattern observed in studies of the ovine conceptus, lactate is produced continuously by the placenta and assimilated by the fetus (Burd *et al.*, 1975; Char and Creasy, 1976). Furthermore, striated muscle, such as the heart, produces lactate in early life (Breuer *et al.*, 1967), but fetal skeletal muscle has not been evaluated *in vivo* from this point of view. Based on the sheep studies, therefore, it seems that most of the lactate produced by elements of the conceptus are retained or consumed by the fetus, rather than being transported back to the mother for metabolic conversions.

Although three carbon compounds have been considered as substrates for hepatic gluconeogenesis, their role in providing energy for the fetal liver has not been emphasized recently. When slices of fetal liver are incubated *in vitro* [$^{14}C$]pyruvate is incorporated into $^{14}CO_2$ to a greater extent than is [$^{14}C$]glucose

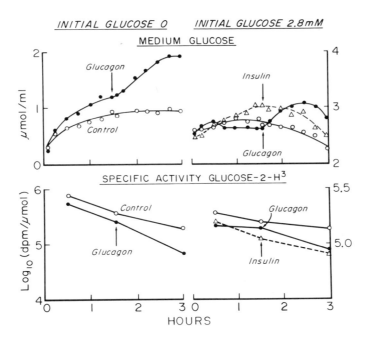

Fig. 19. Human fetal hepatic glucose production by the isolated liver perfused with no exogenous substrate except glucose (2.8 mM) where indicated. Hormonal additions commenced at the arrows and were maintained by continuous infusions into the recirculating media. [2-$^3$H]Glucose tracer in the media was diluted continuously as illustrated in the lower panels. Glucagon increased the hepatic output of unlabeled glucose, while insulin effects could not be demonstrated. Glucose (2.8 mM) enhanced glucose uptake by the liver, thereby inhibiting net hepatic glucose output (Adam *et al.*, 1978). o——o control; •——• glucagon ($8.6 \times 10^{-7}$ M) from 90–180 min; △----△ insulin (1400 -mgU/ml) from 90–180 min.

PETER A. J. ADAM and
PHILIP FELIG

(Villee, 1954). Lactate also is taken up by the isolated perfused human fetal liver (Adam *et al.*, 1972) and enhances hepatic glucose production despite a low rate of incorporation into glucose (Figure 20). Based on these data, lactate can furnish substrate for at least two of the metabolic pools involved in oxidative metabolism by the fetal liver after being oxidized to pyruvate: (1) Lactate can provide carbon for the anaplerotic enrichment of intramitochondrial oxalacetate, thus maintaining sufficiency of the intermediates required for turnover of oxidative metabolism. (2) The pyruvate formed also can be decarboxylated and activated, thus entering the tricarboxylic acid cycle as substrate for oxidative metabolism.

These functions are included in Figure 21, which illustrates the relationships expected between energy sources, oxidative metabolism, and the hepatic storage depot of the human fetus.

(b) Role of amino acids: Since Battaglia and his co-workers have demonstrated that variable oxidation of amino acids *in vitro* is a major component of ovine fetal adaptation to maternal fasting, we may wonder where such oxidation occurs *in vivo*. [$^{14}$C]Alanine is converted to $^{14}CO_2$ by slices of human fetal liver, but accounts for only a small proportion of the hepatic $CO_2$ formation *in vitro* (Villee, 1954). Labeled glycine is incorporated into $^{14}CO_2$ to a greater extent; but has not been shown to have a major role *in vivo*. Finally, glutamine is the major constituent of the amino acid uptake by the sheep fetus and potentially could enter the TCA cycle following its deamination to glutamate and subsequent oxidation to $\alpha$-ketoglutarate. Since glutamine serves as the major substrate for renal gluconeogenesis in adult life,

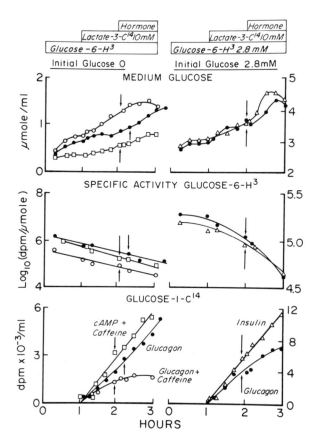

Fig. 20. Effect of lactate (10 mM) on human fetal hepatic glucose metabolism by the isolated perfused liver. Hepatic glucose production was quantified by the dilution of [6-$^3$H]glucose while lactate incorporation into glucose was evaluated by the addition of [3-$^{14}$C]lactate to the recirculating medium. As indicated, glucagon or insulin seemed to enhance hepatic glucose production in the presence of lactate (10 mM) plus glucose (2.8 mM), but none of the stimulating agents affected acutely lactate incorporation into glucose (Adam *et al.*, 1972): ●——● glucagon (8.7 × 10$^{-7}$ M) from 135 to 195 or 120 to 180 min; ○——○ glucagon plus caffeine (0.5 mM) from 120 to 180 min; □——□ cyclic AMP (0.5 mM) plus caffeine from 120 to 180 min; △——△ insulin (1400 $\mu$U/ml) from 120 to 180 min.

**521**

*CARBOHYDRATE, FAT,
AND AMINO ACID
METABOLISM IN THE
PREGNANT WOMAN
AND FETUS*

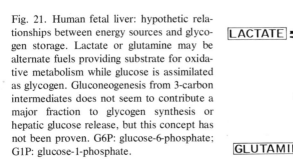

Fig. 21. Human fetal liver: hypothetic relationships between energy sources and glycogen storage. Lactate or glutamine may be alternate fuels providing substrate for oxidative metabolism while glucose is assimilated as glycogen. Gluconeogenesis from 3-carbon intermediates does not seem to contribute a major fraction to glycogen synthesis or hepatic glucose release, but this concept has not been proven. G6P: glucose-6-phosphate; G1P: glucose-1-phosphate.

we might expect renal uptake; but this does not seem to occur in the newborn baboon even when renal glucose production is occurring (Levitsky *et al.*, 1976). At present, however, hepatic gluconeogenesis from glutamine or its direct oxidation has not been quantified in fetal materials from the human or subhuman primate. Thus, the role of this potential substrate is undefined.

(c) Fats and the role of ketone bodies: In fasting adult man, free fatty acids transported from adipose tissue after lipolysis are oxidized and provide the bulk of energy for synthetic processes (Havel *et al.*, 1970). After glycogen depletion, the energy required for hepatic gluconeogenesis is generated by the oxidation of fatty acids. During maternal fasting, however, there is no evidence to indicate that oxidation of fat *in utero* provides the fuel for fetal hepatic oxidative metabolism.

Although ketone bodies cannot be oxidized by liver, the fetus of ketogenic species can accumulate glycogen during maternal starvation. This is true in the rat, even when such starvation precedes the period of rapid glycogen accumulation. Even though hepatic glycogen content is somewhat lower in the fetus of starved rats (Girard, 1975), it has been demonstrated that substantial glycogen accretion occurs (Dahlquist *et al.*, 1972). Furthermore, maternal starvation increases cytosolic phosphoenolpyruvate carboxykinase activity fourfold, indicating that the glucogenic potential of fetal rat liver can be induced prenatally by maternal starvation (Girard, 1975). Thus the carbon sources for such glycogen remain in doubt.

At present, we may analyze the rat as a potential model for human fetal metabolism and propose the following construct:

By substituting for glucose in other maternal and fetal organs, ketone bodies seem to spare glucose and permit its accretion in energy depots such as hepatic glycogen. In addition, the ketone bodies inhibit the oxidation of pyruvate in fetal and maternal organs which utilize the ketones, thus sparing carbohydrate consumption by these organs. Diversion of glucose metabolism from complete oxidation to glycolytic production of lactate can provide greater quantities of this three-carbon compound for oxidative metabolism in the liver. Alternatively, lactate and amino acids oxidized in the liver also could provide carbon incorporated into hepatic glycogen, while glutamine enriches the intermediate pool of the TCA cycle and is oxidized. These hypotheses concerning the response to maternal starvation are summarized in Section 4.4.3, but require further testing by regional study *in vivo*

PETER A. J. ADAM and
PHILIP FELIG

before they can be confirmed. In any case, recruitment of increased amino acid oxidation and gluconeogenesis apparently are unnecessary for providing adequate nutrition to the human fetus during the normal fluctuation occurring with maternal ingestion and fasting.

*Regulation of Enzymatic Activities and Metabolic Pathways.* (a) Principles of fetal hormonal regulation: Investigation of the endocrinologic control of fetal fuel metabolism has been aimed at several problems: The first of these concerns induction of a capacity to metabolize substrates or to synthesize and accrete the energy depots. Secondly, and related to the first, is the maintenance of such a capacity. After the metabolic potential has developed, such maintenance by hormonal action can be considered "permissive." Finally, when the capacity is established and maintained, fine control is permitted. Such regulation may involve rapid activation of enzymatic activities already present or allosteric modulation of activity by substrates, cofactors, and the products of metabolism.

Each of these actions must be judged in the context of present understanding of fetal hormonal responses to their physiologic stimuli. As was discussed earlier, several principals govern the responses of fetal hormones which are known to regulate fuel metabolism. These are as follows:

1. Although the placenta is not an absolute barrier to all the hormones, in man it effectively divides the maternal from the fetal compartment with respect to the hormones controlling fuel metabolism.

2. For the most part, fetal hormonal regulation is autonomous, responding to the intrauterine environment and fetal tropic hormones, even when the substrate required for their production originates from the mother.

3. Even though fetal hormonal responses to prolonged stimuli originating in the mother can be demonstrated, the magnitude of the response frequently is attenuated. For example, fetal pancreatic insulin is secreted in response to prolonged maternal hyperglycemia (Obenshain *et al.,* 1970) but very slowly in response to acute elevation of maternal glucose (Milner and Hales, 1965; Adam *et al.,* 1969). Furthermore, no fetal response to maternally administered arginine can be demonstrated (King *et al.,* 1971*b*). Such failure to respond reflects either diminished responsiveness of the endocrine organ, as with a glucose stimulus (Milner *et al.,* 1972; Milner, 1970), or restricted maternofetal transport of the agent, as with arginine (King *et al.,* 1971*b*). Nevertheless, the human fetal pancreas is capable of responding to pharmacologic stimuli *in vitro* from early fetal life, either with increased insulin (Milner *et al.,* 1971) or glucagon release (Assan and Girard, 1976). Since the plasma pancreatic hormonal levels are modulated effectively by many physiologic stimuli soon after premature term birth, the attenuated hormonal responses observed *in utero* are governed mainly by the prevailing intrauterine environment, rather than by any absolute restriction upon the potential for fetal endocrinologic adaptation.

(b) Sequence of hormonal induction in rodents: Methods: The discussion which follows will focus on the developmental sequence required for induction of a capacity to regulate fetal carbohydrate metabolism. The principals involved have been elucidated by studies of animals with short periods of gestation, such as the rat and rabbit, in which the events are accelerated and amenable to controlled investigation. After a fetal hormone has been demonstrated by direct measurement or inferred from histologic studies, several approaches have been utilized to demonstrate an inductive role. The methods involve either ablation of the hormonal source, followed by hormonal replacement, or introduction of exogenous hormone into the developmental sequence before its normal appearance. We shall refer to four such types of experiment.

1. Ablation of a tropic hormonal source in the fetus with or without replacement of the tropic or target organ hormones. This has been done by fetal encephalotomy and hypophosectomy of the rat or rabbit fetus. Replacement has involved injection of the fetus with large quantities of prolactin, growth hormone, ACTH, or cortisol.

2. Extirpation of the target endocrine organs such as the adrenal or thyroid gland, with or without appropriate hormonal replacement.

3. Premature treatment of the fetus with massive doses of corticosteroids, glucagon, or thyroid hormone before endogenous secretion of each commences.

4. Isolation of tissue explants at various stages of development, and supplementation of the incubation medium with pharmacologic quantities of the hormone under investigation.

This last investigative tool is particularly powerful, since fetal tissues remain viable *in vitro* for periods of several days, the sequence of hormonal exposure may be carefully controlled, and responses prevented with pharmacologic inhibitors or toxins applied only to the tissue under investigation.

Fetal hepatic glycogen metabolism: The experimental models just described have elucidated several aspects of fetal glycogen metabolism including (1) the conditions required for development of glycogen synthase activity, (2) an additional condition required for the onset of glycogen accretion, and (3) the responsiveness of glycogenolysis to glucagon or adrenergic stimuli. If the fetal rabbit is decapitated and hypophysectomized prior to 23 days of gestation, hepatic glycogen accumulation cannot be induced by hormonal replacement (Jost and Picon, 1970). After fetal encephalotomy at 25 days of gestation, however, administration of prolactin plus cortisol to the fetus will permit hepatic glycogen accumulation, while only cortisol replacement is required when the pituitary is ablated after the 26 days. Growth hormone can substitute for prolactin in this experimental model; thyroidectomy has no effect.

In the rat fetus, cortisol levels increase spontaneously at 17–18 days of gestation. Decapitation at this time prevents the increase of the hepatic glycogen synthase activity and glycogen content occurring normally at the end of gestation; early treatment with a cortisol excess induces glycogen synthase and glycogen accretion *in vivo*. Such premature induction requires RNA and protein synthesis (Greengard, 1973).

Finally, from studies of the rat fetal liver in organ culture (Eisen *et al.*, 1973), we learn that glycogen synthase activity appears appropriately in explants obtained from the fetus at 16 days of gestation if cortisol is added to the medium, but glycogen does not accumulate unless insulin as well as cortisol is present during incubation. From experiments such as these our appreciation of sequenced conditions required for glycogen synthesis emerges. Although the initiating endocrine message has not been identified, the liver must be exposed to several hormonal factors before the hepatic glycogen depot can accrete. In the order of requirement these are a pituitary factor such as prolactin (or growth hormone), cortisol, and insulin. As has been demonstrated in human fetuses (Gitlin and Biasucci, 1969; Kaplan *et al.*, 1972; Aubert *et al.*, 1975; Shepard, 1967; Milner *et al.*, 1971; Assan

PETER A. J. ADAM and
PHILIP FELIG

and Boillot, 1971), these hormones are present from early gestation, and the synthetic arm of the glycogen cycle commences its effective activity at 10–11 weeks of gestation.

Hepatic glycogenolysis and glucose release: The events required for early establishment of a glycogenolytic capacity are not as clear as those for glycogen synthesis; however, the activation of glycogenolysis has been examined extensively. In the context of the experimental models, activation of phosphorylase activity and glycogenolysis coincides with deactivation of glycogen synthase I, and neither involves protein synthesis. Briefly, the rat fetal liver near term gestation responds to premature birth, glucagon, and adrenergic catecholamines. Receptors for glucagon and $\beta$-adrenergic activation of adenylcyclase can be demonstrated (Sherline et al., 1974); but the $\alpha$-adrenergic cAMP-independent receptor, which provides an alternate mechanism for activation of phosphorylase in the isolated perfused adult rat liver (Sherline et al., 1972) cannot be demonstrated in the rat fetal organ explants.

Although maintenance of blood glucose in the newborn rat does not require postnatal induction of glucose-6-phosphatase, adequate levels of this activity are established just before birth. Decapitation of the rat fetus prevents this increase (Jacquot and Kretchmer, 1964), while administration of thyroid hormone after the 20th day of gestation causes a premature increase in its activity (Greengard, 1973). As indicated by Greengard, cyclic AMP can induce the rise of hepatic glucose-6-phosphatase activity at day 18, but a glucagon receptor cannot be demonstrated until the 20th day of gestation. Induction of this activity in either instance requires protein synthesis.

In summary, the fetal capacity for glycogen accretion and glucose production develop at approximately the same stage of gestation. Although the hormonal events are not completely understood, there is sufficient evidence to demonstrate that development of a fetal capacity for autonomous regulation of the glycogen cycle requires an ordered sequence of endocrinologic events.

Fetal capacity for gluconeogenesis: A similar requirement applies to the capacity for gluconeogenesis in the rat fetus (Hanson et al., 1975; Chang, 1976). Development of such a capacity requires the induction of mitochondrial pyruvate carboxylase and cytoplasmic PEP-CK activities; the increase in activity of each coincides with the appearance of the specific enzyme protein. At birth cytoplasmic PEP-CK activity in the newborn rat is rate limiting and insufficient for the maintenance of glucose production during fasting (Yeung and Oliver, 1967); the activity is induced rapidly by either term or premature delivery. During the last day of gestation, fetal hepatic PEP-CK can be induced prematurely by pharmacologic treatment in utero with glucagon or dibutyryl cyclic AMP, and such induction can be blocked by high doses of insulin.

In the guinea pig, glucose-6-phosphatase, fructose diphosphatase, mitochondrial PEP-CK and pyruvate carboxylase activities are sufficient to support gluconeogenesis at birth, and gluconeogenesis can be demonstrated in the newborn guinea pig (Robinson, 1976). Although cytoplasmic PEP-CK (analogous to rat PEP-CK) has attained only a very low activity in the fetus, its role in species where mitochondrial PEP-CK is sufficient for gluconeogenesis has not been defined. Furthermore, the mitochondrial activity remains constant after birth, while the cytoplasmic activity is induced. Since the newborn infant appears to have sufficient levels of mitochondrial PEP-CK (Räihä and Lindros, 1969), experiments concerning premature induction of gluconeogenesis in the rat fetus and newborn may not apply to the control of human fetal glucose production.

(c) Capacity for response to hormonal stimuli in the human and subhuman primate: Pancreatic control of glycogen metabolism: Our understanding of the fetal hepatic responsiveness early in human gestation is derived mainly from studies *in vitro* of liver slice, organ explant, and isolated perfused liver. As indicated earlier, the isolated liver *in vitro* releases and consumes glucose continuously as hepatic glycogen declines. During perfusion, glucose release from hepatic glycogen can be accelerated by treatment with large doses of glucagon (Adam *et al.,* 1972). In addition, if one waits until hepatic glycogen in explants decreases to a constant level, treatment with pharmacologic quantities of glucagon, cyclic AMP, or dibutyryl cyclic AMP causes a further decline of the glycogen content (Schwartz *et al.,* 1975).

When cyclic AMP production is estimated by its release into the perfusion medium, it occurs continuously *in vitro* but at higher rates per gram of tissue from the smallest fetuses (Adam *et al.,* 1972). Nevertheless glucagon-mediated cAMP production appears to be attenuated. During perfusion, cyclic AMP release by glucagon-treated fetal liver is indistinguishable from that in controls, but amplified in the presence of a phosphodiesterase inhibitor (Adam *et al.,* 1972). Pharmacologic treatment of fetal hepatic tissue explants with glucagon ($2 \times 10^{-6}$ M) for 6 hr also increases cAMP release into the incubation medium, an effect which is enhanced greatly by phosphodiesterase inhibitors (Schwartz *et al.,* 1975).

After glycogen depletion in the explant, insulin at very high doses (1 U/ml) will cause partial reaccumulation of glycogen if glucose is provided and will oppose the glycogenolytic effect of glucagon (Schwartz *et al.,* 1975). The inhibitory effect appears to occur after cyclic AMP production and represents opposition to its action rather than to its formation. Finally, conversion of glycogen synthase from the I to D form can be demonstrated in the presence of dibutyryl cAMP.

In summary, there is no point after seven weeks of gestation at which pharmacologic regulation of the glycogen cycle by pancreatic hormone cannot be demonstrated *in vitro*. The responses of the isolated perfused fetal liver or explant appear to be attenuated, perhaps because cAMP degradation by phosphodiesterase activity is rapid.

Although there are no data from the term human fetus, a series of elegant experiments employing the isolated perfused rhesus fetal liver near term elucidate hepatic carbohydrate metabolism *In vitro,* the glycemic responses resemble those of the early human fetal liver (Glinsmann *et al.,* 1975). As was observed, an aglycemic medium also causes net glucose release by the isolated perfused monkey fetal liver, but a medium glucose concentration set at normoglycemic or hyperglycemic levels prevents the net release. Glucagon $10^{-8}$ M or cyclic AMP $10^{-3}$ M accelerates net glucose release briskly and overrides the inhibition by glucose.

If an aglycemic medium is used, insulin (0.1 U/ml) inhibits the initial net glucose release; no effect of insulin on glucose uptake can be demonstrated in the presence of a hyperglycemic perfusion medium. Prolonged exposure of the liver to insulin *in vitro* or *in vivo*, however, does increase hepatic glycogen synthase I; glucose and insulin in concert depress hepatic phosphorylase activity.

To summarize, the monkey fetal liver at term demonstrates physiologic responses similar to those of the early human fetal liver. Although glucose "cycling" has not been evaluated in the term monkey liver, the autoregulatory

PETER A. J. ADAM and
PHILIP FELIG

responses and hormonal effects upon net glucose release are similar. During fetal life glycogen accumulation occurs in both species; but net degradation is the rule under the conditions of study *in vitro*.

Control of glucose release: The hormonal induction of glucose-6-phosphatase activity has been evaluated in explants of early human fetal liver (Schwartz *et al.*, 1974) but not in the term fetal monkey. As was observed with the rat fetus *in vivo*, induction of glucose-6-phosphatase activity is mediated by cyclic AMP and requires protein synthesis. In explants of the early human fetal liver, dibutyryl cyclic AMP increases glucose-6-phosphatase activity four- to eightfold, a response which is abolished by cycloheximide or reversed by actinomycin D. Insulin (1 U/ml) also reduces glucose-6-phosphatase activity and reverses an increase induced by the cyclic AMP analog. Since mean hepatic glucose-6-phosphatase activity throughout early gestation resembles the maximal activities attained by enzymes controlling the glycogen cycle at mid-gestation, however, one would not expect a limitation exerted upon fetal hepatic glucose production at its point of release.

Adrenergic effects on glucose release from hepatic glycogen: At present, reports concerning adrenergic control of fetal primate hepatic glycogenolysis are scanty, so our understanding of its regulation depends to a large extent on information obtained in nonprimate newborn mammals. As indicated earlier, Sherline and his co-workers (1974) have demonstrated $\beta$-adrenergic adenylcyclase receptors in the fetal rat liver explant. In newborn calves, Edwards and Silver (1970) have shown that the glycemic response to splanchnic nerve stimulation *in vivo* is $\frac{1}{5}$ that ultimately attained and that a progressive increase occurs between the first and 33rd day of life. Furthermore, this glycemic response to sympathetic neural stimulation occurs after pancreatectomy or adrenalectomy indicating that the glycemic response is not entirely mediated by glucagon or cortisol. Finally, when the isolated canine liver is perfused with maximally stimulating doses of norepinephrine ($10^{-6}$ M), glucose production by the term fetal and 3-hr neonatal liver is less than that at one day of age (Chlebowski and Adam, 1975). Collectively these and similar data indicate that the hepatic glycogenolytic response to direct sympathetic neural stimulation depends on at least two factors—one being the progressive sympathetic innervation of the liver, the other being the environmentally controlled change of responsiveness which occurs around the time of birth.

In this context, we may examine the few data available from studying liver explants of two previable human fetuses (Schwartz, 1974a). Pharmacologic doses of epinephrine or norepinephrine ($10^{-4}$ M) cause a slight increase of explant cyclic AMP output—an increase which is less than that occurring when the phosphodiesterase inhibitor alone (theophylline $5 \times 10^{-4}$ M) is added to the incubation medium. A moderate glycogenolytic effect also occurs with both catecholamines. Early in human fetal life, therefore, the catecholamines apparently fit into a construct similar to that described for responses to pancreatic glucagon. Although a pharmacologic receptor for catecholamines can be demonstrated, its physiologic utility apparently is attenuated and its metabolic role early in human fetal life is doubtful. At present, there are insufficient data to evaluate the role of alpha- or beta-adrenergic receptors in controlling hepatic glycogenolysis by the fetal primate; there are no data which would elucidate the opposing effects of parasympathetic regulation by acetylcholine.

Pharmacologic regulation of hepatic gluconeogenesis: Amino acid uptake and alanine incorporation into glucose by the explanted human fetal liver have been modified by incubation with glucagon, insulin, dibutyryl cyclic AMP, and triamcina-

**527**

*CARBOHYDRATE, FAT,
AND AMINO ACID
METABOLISM IN THE
PREGNANT WOMAN
AND FETUS*

lone (Schwartz, 1974*a,b*; Schwartz and Rall, 1975*a*). *In vitro,* pharmacologic quantities of insulin, glucagon, or dibutyryl cyclic AMP enhance fetal hepatic concentration of the nonmetabolizable amino acid, $\alpha$-aminoisobutyric acid. Such an effect requires 5–7 hr, apparently involves protein synthesis, and can be prevented by cyclohexamide. At least two stimulating mechanisms are involved since glucagon and dibutyryl cyclic AMP effects are nonadditive, while the response to insulin adds quantitatively to the glucagon or dibutyryl cyclic AMP effect.

Although insulin enhances glucagon-induced amino acid uptake, it redirects amino acid metabolism and has the opposite effect on exchange of [$^{14}$C]alanine with glucose in the medium. While glucagon, dibutyryl cAMP, or triamcinalone increase the incorporation of $^{14}$C-labeled alanine into the medium glucose, insulin reduces the exchange of label and opposes the increase caused by glucagon on the cyclic AMP analog. Finally, triamcinalone and dibutyryl cyclic AMP enhance the incorporation of labeled alanine into glucose by separate mechanisms and are additive at maximally effective doses.

Although these observations have important implications concerning the long-range pharmacologic modification of the fetal metabolism, their relevance to fetal nutritional responses remain in doubt. As indicated previously, no net glucose production has been demonstrated in this model; furthermore the rate at which alanine label exchanges with the medium glucose, even after maximal stimulation, would contribute only marginally to total body glucose metabolism or hepatic glucose release.

In summary, the fetal hormonal milieu is of major importance in the chronologic development of glycogen stores and the capacity for glycogenolysis or gluconeogenesis. The potential for fine regulation of hepatic glucose production and release, however, does not seem to reside only in the known hormonal adjustments.

(d) Substrate and cofactor regulation: Glycogen cycle: Although hormonal effect seems to be attenuated in the fetus, the preponderance of evidence favors effective regulation of the hepatic glycogen cycle by substrate entry into the fetal liver. Apparently, accumulation of glycogen after induction of synthase D activity results from allosteric modulation of the synthetic activity, while glycogenolysis is inhibited by net hexose entry into the liver. Conversely, reduced entry of substrate results in reduced glycogen synthase activity and increased phosphorylase activity. These concepts have been reduced and summarized in Figure 22. For a more comprehensive overview, the authors direct the reader to recent work by Glinsmann and his co-workers, in which they have examined the regulation of the glycogen phosphorylase and synthase by several kinase and phosphoprotein phosphatase activities (Glinsmann *et al.,* 1970; Huang *et al.,* 1975, 1976; Zieve and Glinsmann, 1973; Huang and Glinsmann, 1975, 1976). Briefly, such work has required the use of natural and synthetic substrates with differing affinities for the enzymes under study, and this body of work has distinguished independent regulatory mechanisms for at least two kinase and two phosphoprotein phosphatase activities. Although Figure 22 emphasizes the cAMP-dependent mechanism, independent activities also can be demonstrated.

In studies of specimens obtained from the previable human fetus immediately after abdominal hysterotomy (Schwartz *et al.,* 1975), it is apparent that glycogen accumulation and total glycogen synthase and phosphorylase activities increase simultaneously early in gestation. Concurrently, intrahepatic glucose-6-phosphate and 5'-AMP concentrations increase (Schwartz and Rall, 1975*b*), while that of inorganic phosphate declines (Figure 23). For perspective one might judge these

PETER A. J. ADAM and
PHILIP FELIG

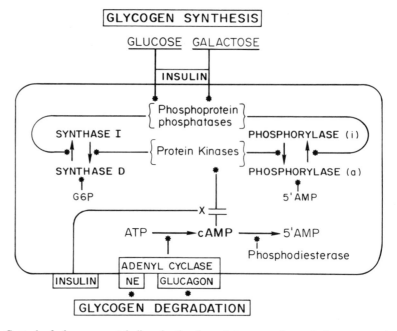

Fig. 22. Control of glycogen metabolism in the liver. Interconversions of glycogen synthase and phosphorylase activities. Asterisks denote activation, whereas X represents a blocking effect. As illustrated, hexose entry into the liver enhances conversion of glycogen synthase to the "active" independent form (I), while inactivating phosphorylase (a) by dephosphorylation to produce the "inactive" (i) form. Activation of hepatocyte adenyl cyclase by glucagon or sympathetic neural norepinephrine (NE) enhances glycogenolysis by conversion of glycogen synthase to the less active glucose-6-phosphate-dependent (D) form, while stimulating activation of phosphorylase. For a complete discussion of the protein kinases and phosphoprotein phosphatase, refer to the work of Glinsmann and co-workers cited in the text.

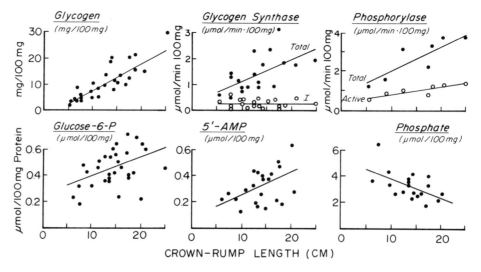

Fig. 23. Control of glycogen accumulation in the human fetal liver (Schwartz *et al.*, 1975; Schwartz and Rall, 1975*b*).

data in the context of Mersmann's investigations of hepatic glycogen synthase activity in supernatants of homogenates obtained from fetal and newborn pig (Mersmann *et al.*, 1972). In order to evaluate the effects of glycogen synthase D to I interconversion, Mersmann and his co-workers "activated" glycogen synthase prior to assay by incubating the homogenate supernatants for 90 min at room temperature. In examining the affinity of glycogen synthase for substrate, these investigators demonstrate a tenfold increase of enzymatic affinity for UDPG by the combined effects of enzymatic activation and optimal glucose-6-phosphate (G6P) concentrations (Table XXII). In addition, they showed that enzymatic conversion to the I form increased the efficacy of the glucose-6-phosphate activation (Table XXII).

Based on the measurements of substrate and cofactor concentration obtaining early in human gestation (Schwartz and Rall, 1975*b*), we would expect glycogen synthase activity to reflect mainly the D form and that substrate and cofactor regulation of both forms (I + D) could occur. In support of this concept intrahepatic G6P and UDPG concentrations are in the expected range of their activation and affinity constants during early gestation. Furthermore, in studies of human fetal liver homogenates, Schwartz demonstrated the potential for regulation of synthase activity by G6P, 5'-AMP, Pi, and divalent cations under conditions *in vitro* resembling those *in utero*. In short, G6P, 5'-AMP, $Ca^{2+}$ and $Mg^{2+}$ enhanced homogenate glycogen synthase activity under the conditions expected *in utero,* while inorganic phosphate would inhibit the activity. Thus, the progressive increase of intrahepatic G6P and 5'-AMP concentrations during early gestation, combined with the decline of inorganic phosphate, would favor glycogen synthesis.

Balance between glycogen synthesis and degradation: In Table XXIII the activities of hepatic glycogen synthase and phosphorylase in the human and rhesus monkey fetus are compared at three stages of gestation. As indicated, phosphorylase activity is expressed in the lytic direction either by direct measurement or by a calculated approximation where estimated originally by the rate of glycogen synthesis from glucose-1-phosphate *in vitro*.

Although hormone-induced conversion of phosphorylase from active to the inactive dephosphorylated form cannot be demonstrated in the human fetal organ explant, phosphorylase phosphatase activity is present during incubation of a homogenate *in vitro* (Schwartz *et al.*, 1975). Fetal hepatic phosphorylase activity in the synthetic direction can be modulated by 5'-AMP, UDPG, and glycogen concentrations *in vitro* (Gensser *et al.*, 1971), but the role of these *in vivo* has not been investigated.

When the human fetal or monkey liver has been studied *in vitro*, however, net glycogenolysis has always occurred. In explants of human fetal liver intrahepatic

Table XXII.  Effect of Glycogen Synthase (GS) Conversion to the I Form on Enzymatic Affinity for UDPG and Activation by Glucose-6-Phosphate (G6P)[a]

| State of GS | UDPG (mM) | $K_a$ G6P (mM) | G6P (mM) | $K_m$ UDPG (mM) |
|---|---|---|---|---|
| Unactivated | 5.34 | ~0.7 | 0 | ~2.5 |
|  |  |  | 5.35 | 0.56 |
| Activated | 5.34 | ~0.1 | 0 | ~0.7 |
|  |  |  | 5.35 | 0.22 |

[a]From Mersmann *et al.*, 1972.

PETER A. J. ADAM and
PHILIP FELIG

Table XXIII.  Maximal Hepatic Glycogen Synthase and Phosphorylase
Activities in Utero (µmol/min/100 mg Protein)

| Activity | Human fetus | | Normoglycemic monkey term fetus |
|---|---|---|---|
| | 8 cm | 20 cm | |
| Glycogen synthase | | | |
|    Total (I + D) | 1.0 | 1.9 | 1.9 |
|    Active (I) | 0.2 | 0.2 | 0.1 |
| Phosphorylase | | | |
|    Total | 1.7 | 3.1 | 5.8[a] |
|    Active (a) | 0.6 | 1.1 | 1.3 |

[a]Estimated by exposing isolated liver to hypoxia for 45 sec (Sparks *et al.*, 1976) and calculated to be ⅓ the activity in the synthetic direction (Gennser *et al.*, 1971; Maddaiah and Madsen, 1966*a,b*).

glucose-6-phosphate falls to ⅛ and UDPG declines to ⅓ the level prevailing *in utero* (Schwartz and Rall, 1975*b*). As was demonstrated by Buschiazzo and co-workers (1970), intrahepatic G6P can be maintained by perfusion of the isolated rat liver with 11 mM glucose. An aglycemic medium results in a decline of the G6P concentration and has no significant effect on the UDPG level. Glycogen reaccumulates in the perfused livers of fasted rats when the hepatic G6P concentration is maintained. These, and data *in vivo* which indicate that intrahepatic G6P declines after birth in the rat (Dawkins, 1963), suggest that G6P levels *in vivo* may be critical in maintaining the synthetic arm of the glycogen cycle. Glycogenolysis would predominate if the G6P level declines.

The foregoing discussion has focused on the direct effects of substrate and cofactor concentrations upon the activities of glycogen synthase D and phosphorylase. Another dimension of metabolic regulation by substrates has become apparent from the elegant investigations of Sparks *et al.* (1976), who have employed the isolated perfused fetal monkey liver. As demonstrated by their investigations, exposure of the liver to high glucose concentrations (15–22 mM for 30 min) reduces active hepatic phosphorylase but has no effect on glycogen synthase I activity Figure 24. Although galactose is not a substrate *in utero,* addition of this hexose to the medium results in a doubling of glycogen synthase I activity, even in the absence of exogenous hormone. Insulin, $10^{-8}$ M, enhances the effect of galactose and results in glycogen synthase I activities of approximately 0.75 µmol/min/100 mg of liver protein, which equal 40% of the total. Under the conditions of study, glucose is released by all the livers, but galactose uptake exceeds glucose release and causes a net uptake of hexose.

Although the intrahepatic concentrations of G6P, UDPG, nucleotides, and other regulatory metabolites have not been quantified in these studies, the independent conversion of glycogen synthase D to I and of phosphorylase *a* to *i* activities does not require the presence of hormone. Maintenance of an environment with high hexose concentrations is sufficient to affect the activity of individual phosphoprotein phosphatases, either inactivating phosphorylase or activating glycogen synthase. Insulin amplifies the effects, perhaps by inhibition of protein kinase activities or by redirecting the metabolism of hexoses entering the liver; however, these hypotheses have not been tested in this experimental model.

If we are to encapsulate the regulation of the hepatic glycogen cycle *in utero,* the net synthesis which prevails depends on the provision of a glucose supply and

**531**

*CARBOHYDRATE, FAT,
AND AMINO ACID
METABOLISM IN THE
PREGNANT WOMAN
AND FETUS*

its uptake by the liver. After induction of the capacity to synthesize and degrade glycogen, the internal environment in the fetus maintains a high intrahepatic level of G6P which is sufficient to render synthase D the effective synthetic activity. Active phosphorylase is insufficient to override the net glycogen synthesis, but the reasons for this are not apparent.

While cofactor regulation plays a major role in ensuring synthesis of depot glycogen, the primate fetal liver at term responds briskly to pharmacologic glucagon stimulation, suggesting that it may be sensitive to physiologic amounts immediately after birth (Glinsmann *et al.*, 1975). Insulin, *in vitro*, has been relatively ineffective in modifying fetal hepatic glucose uptake. Thus, one might conclude that maintenance of a hepatic glycogen depot *in utero* requires suppression of pancreatic glucagon secretion. This concept has not been proven and could be contested by virtue of data from the rat (Girard *et al.*, 1974). When the pregnant rat is fasted for four days at the end of gestation, fetal plasma insulin declines, while pancreatic glucagon levels rise. Nevertheless, glycogen accumulates in fetal liver, although at a reduced rate (Dalhquist *et al.*, 1972). As indicated previously, alternate fuels and hepatic gluconeogenesis might play a role under conditions of extreme starvation (Girard *et al.*, 1973), but this concept, likewise, has not been confirmed.

Gluconeogenesis: Regulation of hepatic gluconeogenesis by substrate concentration has not been evaluated in the primate fetus. In the guinea pig, hepatic mitochondrial phosphoenolpyruvate carboxykinase (PEP-CK) is sufficient, as it appears to be in man, for the maintenance of hepatic gluconeogenesis without postnatal enzymatic induction (Robinson, 1976). The affinity of PEP-CK for oxalacetate is in the physiologic concentration range ($K_m = 6 \times 10^{-6}$ M), and the enzyme is activated by $Mn^{2+}$ ($K_a = 10^{-4}$ M) or $Mg^{2+}$ ($K_a = 10^{-2}$ M). There are, however, no studies which support the regulation of gluconeogenesis by substrate provision in the newborn guinea pig. After depletion of hepatic glycogen, medium lactate concentrations of 1, 2, and 10 mM in a nonrecirculating perfusate result in indistin-

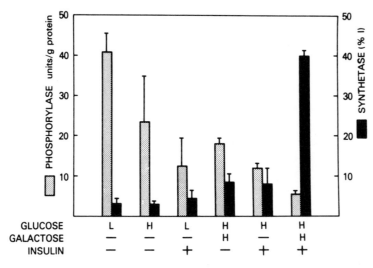

Fig. 24. Glycogen synthase and phosphorylase activities (mean ± SEM) in the isolated near-term monkey fetal liver perfused for 30 min with various combinations of glucose, galactose, and insulin (from Sparks *et al.*, 1976). Glucose or galactose, low (L) = 100–250 mg/100 ml; high (H) = 275–400 mg/100 ml; Insulin (+) = $10^{-7}$ M.

guishable glucose release rates approximating 0.5 $\mu$mol/min/g of liver (Arinze, 1975). Although lactate enhances net hepatic glucose release in the undepleted isolated perfused human fetal liver, this effect could be accounted for by oxidation of circulating lactate in lieu of hepatic glycogen and medium glucose (Adam *et al.*, 1972). Thus, there are no data in man supporting substrate regulation of gluconeogenesis *in utero*. Although the potential for hepatic gluconeogenesis seems to exist at term gestation, its role in maintaining fetal homeostasis remains in doubt.

*Integrated Responses of the Liver to Glucose Privation and Surfeit.* (a) Fetal glucose production: Rat fetal glucose production *in vivo*: Three lines of evidence indicate that the fetal liver responds to reduced substrate supply by initiating glucose production. The first of these is experimentation in which the maternal blood glucose is maintained at low levels and the fetal contribution to fetal plasma glucose estimated. The second line involves estimation of fetal hepatic glucose production *in vitro*. Finally, quantification of hepatic glucose production in the premature infant serves to determine whether the fetal capacity for response to cessation of substrate supply is developmentally limited in the last trimester.

Goodner and Thompson (1967; Goodner *et al.*, 1969) have examined the relationships between maternal and fetal glucose metabolism in pregnant rats that were hyperglycemic, fasted, or hypoglycemic. In the hyperglycemic pregnant rat infused continuously with glucose and tracer [U-$^{14}$C]glucose, the specific activity of the tracer in fetal blood stabilized at about 85% of the level in the mother. Fasting the pregnant rats lowered the fetal specific activity to 65% of that in maternal plasma glucose. These data indicate that the fetus either can produce glucose at a variable rate or that the labeled carbon exchanges rapidly with large unlabeled fetal pools; however, induction of maternal hypoglycemia by continuous insulin infusion results in fetal blood glucose levels exceeding those in the mother. Thus, fetal glucose production is initiated in the rat when glucose provisions are insufficient.

Most investigators have assumed that the glucose produced originated from fetal hepatic glycogen because cytoplasmic phosphoenolpyruvate carboxykinase activity was known to be low in the normal fetus. Bossi and Greenberg (1972) have confirmed that fasting the pregnant rat for 16 hr on the 21st day of gestation results in a fetal blood glucose concentration which is higher than in the mother and that fetal dilution of labeled [2-$^{3}$H]- and [$^{14}$C]glucose progresses with gestation between 18 and 21 days. Although fetal consumption of glycogen could not have provided all the caloric needs of the fetus during the entire period of fasting, the depot could have provided 50% of the needs for 6 hr. In looking for an alternate source of glucose, the investigators were unable to demonstrate fetal gluconeogenesis from pyruvate by the rat fetus *in utero*, the 45-min-old prematurely born rat, or liver slices from fetuses. These studies of gluconeogenesis were not performed after fasting, however, as were the studies of maternal starvation which demonstrated induction of fetal hepatic PEP-CK (Girard *et al.*, 1973). Thus, the role of fetal hepatic gluconeogenesis in the preservation of homeostasis in the rat fetus remains unresolved.

(b) Early human fetal liver *in vitro*: The glycemic responses of the isolated perfused human fetal liver from early gestation are quantified in Table XXIV. As indicated, net glucose release occurs in the absence of exogenous substrate but is suppressed by a glucose concentration of 50 mg/100 ml. This occurs because glucose production is balanced or exceeded by hepatic glucose uptake at physio-

**533**

CARBOHYDRATE, FAT,
AND AMINO ACID
METABOLISM IN THE
PREGNANT WOMAN
AND FETUS

Table XXIV.  *Mean Glucose Production and Uptake by Isolated Perfused Human Fetal Liver in the Absence of 3-Carbon Substrate (μmol/min/g Liver)*

| | Medium glucose (0) | | | Medium glucose (2.8 mM) | | |
|---|---|---|---|---|---|---|
| | 5 min | 15–90 min[a] | 90–180 min | 5 min | 15–90 min | 90–180 min |
| Glucose (mM) | 0.31 | 1.03[a] | 1.35 | 2.52 | 2.84 | 2.47 |
| Production[b] | — | 0.23 | 0.35 | — | 0.24 | 0.23 |
| Uptake | — | 0.13 | 0.30 | — | 0.26 | 0.38 |
| Net release | 1.22 | 0.10 | 0.05 | −1.41 | −0.01 | −0.16 |

[a]Medium glucose at end of period indicated.
[b]Quantified by dilution of [2-³H]glucose in the recirculating medium.

logic glucose concentrations. In addition, all the glucose produced by the isolated livers can be accounted for by the decline of glycogen concentration.

Provision of lactate for oxidative metabolism alters hepatic glucose production rates when the medium glucose concentration is in the physiologic range (Table XXV). As demonstrated, lactate, 10 mM, enhances hepatic glucose production at physiologic glucose concentrations but production is partially balanced by uptake. Addition of glucagon to the medium enhances glucose production, but again, it is largely balanced by hepatic glucose uptake.

Besides demonstrating that the human liver responds to lack of glucose by releasing it into the medium, these results show that there is no limitation of hepatic glucose uptake in the range of physiologic glucose concentrations. Furthermore, the early human fetal liver is capable of producing glucose at physiologic rates for several hours, if there is an adequate provision of substrate for oxidative metabolism. Although incorporation of label into glucose is limited, lactate could provide substrate for oxalacetate and acetyl-CoA formation in the fetal liver. The significance of lactate as a potential alternate fuel has been discussed on p. 519.

Although the balance between glucose production and uptake has not been evaluated in the isolated monkey fetal livers at term, the responses of net glucose release resemble those of the human fetal livers (Sparks *et al.,* 1976). The role of glucogenic substrates has not been reported in this model, but glycogen degradation accounts for more than 80% of the glucose produced by the isolated perfused canine liver at term gestation (Chlebowski and Adam, 1975). In the presence of 10 mM lactate, glucose uptake by the liver from newborn dogs is insignificant until glucose in the perfusate exceeds 6 mM (100 mg/100 ml). Thus, the regulation of hepatic

Table XXV.  *Glucose Production and Uptake by Isolated Human Fetal Liver Provided with Lactate, 10 mM (μmol/min/g Liver)*

| | Medium glucose (0) | | | Medium glucose (2.8 mM) | | |
|---|---|---|---|---|---|---|
| | No lactate 60 min | Lactate 10 mM 120 min | Lactate + glucagon[a] 180 min | No lactate 60 min | Lactate 10 mM 120 min | Lactate + glucagon 180 min |
| Glucose (mM) | 0.61 | 0.98 | 1.36 | 2.92 | 3.34 | 3.98 |
| Production | 0.16 | 0.25 | 0.35 | 0.19 | 0.48 | 1.06 |
| Uptake | 0.05 | 0.14 | 0.27 | 0.09 | 0.39 | 0.91 |
| Net release | 0.11 | 0.11 | 0.08 | 0.11 | 0.09 | 0.16 |

[a]Glucagon 8.6 × 10⁻⁷ M.

glucose uptake in the dog fetus at term may differ from that of the human early in gestation, but this inference may reflect species differences as well as stage of gestation and should be examined in the primate at term.

(c) The human fetus and premature infant: The evidence cited earlier (Section 4.2.1, Variations of Net Fetal Glucose Uptake in Man) demonstrates that net glucose uptake by the human fetus *in vivo* is variable and directly proportional to the maternal blood glucose concentration. Such data were considered from the point of view of net glucose accretion and oxidation by the entire fetus; however, several additional lines of evidence indicate that the variation of transplacental and transfetal glucose gradients results from modulation of net fetal hepatic glucose output.

First, the potential for hepatic glucose production at physiologic rates has been demonstrated in the isolated perfused human fetal liver (Adam *et al.*, 1972). Secondly, the $^{13}C$ abundance of glucose carbon in the normoglycemic human fetus occasionally exceeds that in the mother, suggesting that some of the glucose in the fetus originates from fetal rather than maternal sources. When hyperglycemia is induced in the normal mother by glucose infusion, or in the diabetic mother by less stringent control of endogenous glucose production with insulin therapy, the differences of naturally occurring $^{13}C$ abundance in the mother and fetus are virtually obliterated (Adam *et al.*, 1976). Thus, fetal hepatic glucose production may commence before birth and may account for a greater proportion of the fetal glucose pool during maternal fasting than it does when the maternal blood glucose is raised.

Finally, evaluation of the neonatal blood glucose and quantification of glucose production by infusion of stable isotopic tracers indicates that fetal hepatic glucose production is well established at birth (Kalhan *et al.*, 1976). If the normal pregnant woman is infused with isotonic saline solution during labor and delivery, blood glucose in her newborn stabilizes above 50 mg/100 ml without declining into the hypoglycemic range. Between 2 and 4 hr of age, half the metabolic requirement is met by the consumption of endogenously produced glucose (Tunell *et al.*, 1976) and the systemic glucose production rate in the newborn is 4.5 mg/min/kg body weight

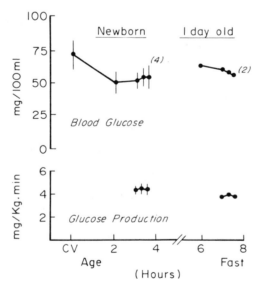

Fig. 25. Glucose production in newborn infants (from Kalhan *et al.*, 1976).

**535**

*CARBOHYDRATE, FAT,*
*AND AMINO ACID*
*METABOLISM IN THE*
*PREGNANT WOMAN*
*AND FETUS*

*Table XXVI. Systemic Glucose Production Rates in Newborn Infants (mg/min/kg Body Weight; Mean ± SD)*

| | | Basal | | 1-hr alanine (2.8 $\mu$mol/min/kg) |
|---|---|---|---|---|
| | Age: | 3 hr | 3.5 hr | 4.5 hr |
| Normal (4) | | 4.36 ± 0.52 | 4.57 ± 0.39 | 3.79 ± 0.59 |
| IDM (3) | | 2.96 ± 0.24 | 2.73 ± 0.34 | 3.01 ± 0.19 |

(Figure 25). Furthermore, no absolute developmental constraint exists, since the early premature infant weighing as little as 650 g produces glucose at physiologic rates (Bier *et al.*, 1976).

The sources of glucose in the human fetus at term have been inferred from postnatal studies. Hepatic glycogenolysis can account for virtually all the glucose produced immediately after birth in the dog (Chlebowski and Adam, 1975), while circulating alanine seems to account for a negligible quantity. In the newborn infant, bolus infusions of this amino acid do not raise blood glucose at 1 hr of age (Lowry and Adam, 1975; Adam *et al.*, 1974), while pharmacologic doses of intravenous glucagon during the first day of life lower blood glutamine but have no effect on the circulating alanine concentration (Reisner *et al.*, 1973). In a more physiologic study, continuous infusion of alanine at rates of 2.8 $\mu$mol/min/kg body weight does not raise the hepatic glucose production rate in normal infants or IDMs (Table XXVI).

As demonstrated in the newborn dog and isolated canine liver, however, gluconeogenesis from lactate and lactate recycling become significant components of hepatic glucose production during the first day of life in that species (Adam *et al.*, 1975*a*). In addition, the kidney of the newborn baboon seems to produce significant quantities of glucose from lactate but not from glutamine (Levitsky *et al.*, 1976).

*Overview of Fetal Liver.* Like adipose tissue, the human fetal liver has the potential for recycling a major substrate stored in its energy depot. In adipose tissue the synthetic arm of triglyceride synthesis predominates and fatty acid mobilization for oxidative metabolism is unimportant *in utero*. In contrast, the fetal liver from early gestation releases glucose in response to reduction of the maternal supply. Although hormonal regulation may be recruited by maternal starvation, the autonomous responses which occur during the nutrient fluctuation depend largely on allosteric regulation of the glycogen cycle. Finally, although fetal autonomy ordinarily reflects allosteric regulation, prolonged modification of the intrauterine environment may cause distorted hormonal controls to override the prevailing set. Such distortions become manifest in the starved pregnant rat in which fetal pancreatic glucagon and other hormones may induce a capacity for gluconeogenesis (Girard *et al.*, 1973). In contrast, prolonged hyperglycemia *in utero* may enhance the synthetic arm of the glycogen cycle and prevent adequate glucose production rates in the newborn period (Kalhan *et al.*, 1976). This concept has been discussed in detail elsewhere (Adam *et al.*, 1976).

## 4.4. Regulation of Fetal Metabolism by Maternal Substrates

### 4.4.1. Hypothesis of High-Flux Cycles in Utero

In this paper we have emphasized the cyclic nature of fetal metabolism. As independent endocrine development permits, the human fetus induces the synthetic

PETER A. J. ADAM and
PHILIP FELIG

and lytic arms of the glycogen cycle in liver and of the triglyceride depot in adipose tissue. Normally, the mother exchanges substrates with the fetus at a higher rate than is required by its basal metabolic needs. Since the flux of substrate through each depot exceeds the requirement, a degree of inefficiency is tolerated; however, the autonomy of fetal metabolic control resides largely in the cyclic nature and allosteric regulation of depot metabolism. Such inefficiency of fetal metabolism is easily afforded in the well-nourished woman because of the ample provision in her depots. Thus, maternal catabolism acts as the first order of reponse in the postabsorptive state and fetal depot synthesis may continue. As has been discussed earlier, synthesis predominates throughout gestation, and the human fetus is well protected by the mother's ability to provide alternate substrates and fuels. Consequently, the human fetus is less vulnerable to maternal fasting or starvation than are the fetuses of some other species.

### 4.4.2. Conditions of Predominant Hormonal Effect

During prolonged disturbances of the maternal nutrition, the endogenous hormonal milieu of the fetus is modified and overrides the normal cyclic fetal responses to periodic maternal ingestion. For example, a prolonged surfeit of glucose associated with maternal diabetes causes fetal hyperinsulinemia, enhances hepatic glycogen synthesis *in utero,* inhibits the initiation of hepatic glucose production during labor and delivery, and causes neonatal hypoglycemia with insufficient mobilization of alternate fuels (Adam *et al.,* 1974, 1976). Conversely, maternal starvation probably raises pancreatic glucagon levels in fetal plasma and may induce hepatic gluconeogenesis (Girard *et al.,* 1973). Ultimately, significant fetal catabolism is possible.

### 4.4.3. Integrated Responses to Moderate Maternal Fasting: The Role of Substrate Provision

Although endogenous hormonal effects can override the usual cyclic autonomy of the fetus, the major known regulatory factors evident during moderate maternal fasting are a reduced glucose supply to the fetus, and the provision of alternate fuels from maternal sources. Hormonal regulation of the responses seems to reside mainly in the maternal compartment, while allosteric modulation of fetal metabolism governs the fetal response to the substrate supply.

Figure 26 summarizes our concept of the fetal response to a moderate maternal fast. As indicated, maternal hepatic energy metabolism depends mainly on the oxidation of fatty acids, while glucose is provided from glucogenic amino acids. Ketone bodies are generated from the fatty acids oxidized by the maternal liver and diffuse to the fetus where they substitute for glucose in fetal tissues such as brain and striated muscle. Partial diversion of oxidative metabolism from glucose to ketone-body consumption redirects oxidation of glucose to glycolysis and may result in lactate production by these tissues. Then the continuous contribution of lactate from placenta is added to that from fetal glycolysis, and both are oxidized by the fetal liver. Similarly, glutamine from maternal muscle may provide fuel for hepatic oxidative metabolism, but this has not been confirmed.

**537**

*CARBOHYDRATE, FAT,
AND AMINO ACID
METABOLISM IN THE
PREGNANT WOMAN
AND FETUS*

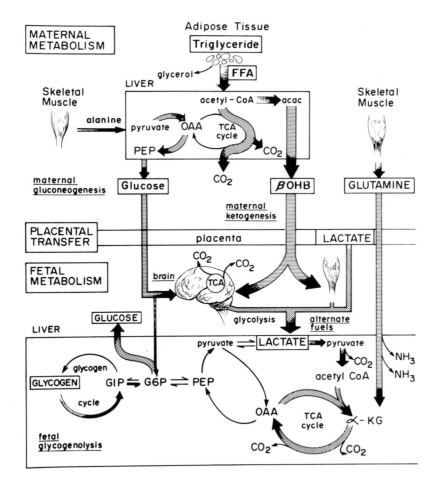

Fig. 26. Integrated hypothesis for the responses of human fetal and maternal metabolism to moderate maternal fasting. In the mother, fat provides the major fuel for hepatic oxidative metabolism, while maternal gluconeogenesis is enhanced. Nevertheless, maternal and fetal blood glucose decline, and ketone bodies produced by maternal liver become the major alternate fuel for oxidative metabolism in the fetus. Note that ketone bodies seem to divert aerobic glycolysis, in tissues where they are utilized, and cause lactate production from the reduced glucose supply. *In utero* lactate derived from placental and fetal metabolism continues to act as a substrate for fetal hepatic oxidative metabolism. Glutamine and other amino acids mobilized from maternal striated muscle may serve a similar function in fetal liver. Ultimately, net glucose production from glycogen might occur, after exceptional privation, but this would ordinarily be balanced by fetal hepatic glucose uptake in the previously well-nourished fetus.

If maternal fasting is sufficiently prolonged, the glycogen cycle may be modified so that hepatic glycogenolysis predominates and glucose is provided from this source. Since the carbon of glucose formed in the fetus has a $^{13}C$ natural abundance exceeding that in the mother, exchange of fetal amino acid with glucose carbon may be occurring, but this has not been shown to be important in the human fetus or newborn infant at the moment of birth.

Finally, note that fatty acids have not been shown to contribute in a major way to human fetal oxidative metabolism; instead, the fetal triglyceride store ordinarily is preserved by reesterification when a lipolytic stimulus occurs, and fetal adipose tissue conserves the major energy depot required in the fasting newborn infant.

PETER A. J. ADAM and
PHILIP FELIG

Adam, P. A. J., 1971, Control of glucose metabolism in the human fetus and newborn infant, *Adv. Metab. Disord.* **5:**183.

Adam, P. A. J., Teramo, K., Räihä, N., Gitlin, D., and Schwartz, R., 1969, Human fetal insulin metabolism early in gestation. Response to acute elevation of the fetal glucose concentration and placental transfer of human insulin-$^{131}$I, *Diabetes* **18:**409.

Adam, P. A. J., Kekomäki, M., Rahiala, E.-L., and Schwartz, A. L., 1972, Autoregulation of glucose production by the isolated perfused human fetal liver, *Pediatr. Res.* **6:**396 (abstract).

Adam, P. A. J., Chlebowski, R., and Lowry, M., 1974, Fuel metabolism in the human fetus and newborn infant, *Perinat. Med.* **5:**321.

Adam, P. A. J., Glazer, G., and Rogoff, F., 1975*a*, Glucose production in the newborn dog. I. Effects of glucagon *in vivo*, *Pediatr. Res.* **9:**816.

Adam, P. A. J., Räihä, N., Rahiala, E.-L., and Kekomäki, M., 1975*b*, Oxidation of glucose and D-β-OH-butyrate by the early human fetal brain, *Acta Paediatr. Scand.* **64:**17.

Adam, P. A. J., Kalhan, S. C., and Savin, S. M., 1976, Fuel metabolism in the infant of the diabetic mother: Attenuated mobilization of alternate fuels, in: *Symposium on Diabetes and other Endocrine Disorders during Pregnancy and in the Newborn* (M. New and R. Fiser, eds.) pp. 51–67, Alan R. Liss Inc., New York.

Adam, P. A. J., Schwartz, A. L., Rahiala, E.-L., and Kekomäki, M., 1978, Glucose production in midterm human fetus. I. Autoregulation of glucose uptake, *Am. J. Physiol.* **234** (in press).

Aeberhard, E., Grippo, J., and Menkes, J. H., 1969, Fatty acid synthesis in the developing brain, *Pediatr. Res.* **3:**590.

Alford, F. P., Bloom, S. R., Nabarro, J. D. N., Hall, R., Besser, G. M., Coy, D. H., Kastin, A. J., and Schally, A. V., 1974, Glucagon control of fasting glucose in man, *Lancet* **2:**974.

Arinze, I. J., 1975, On the development of phosphoenolpyruvate carboxykinase and gluconeogenesis in guinea pig liver, *Biochem. Biophys. Res. Commun.* **65:**184.

Assan, R., and Boillot, J., 1971, Pancreatic glucagon and glucagon-like material in tissues and plasmas from human foetuses 6–12 weeks old, in: *Nutricia Symposium—Metabolic Processes in the Foetus and Newborn Infant,* (J. H. P. Jonxis, H. K. A. Visser, and J. A. Troelstra, eds.), pp. 210–219, Williams and Wilkins Co., Baltimore; H. E. Stenfert Kroese N.V.-Leiden.

Assan, R., and Girard, J. R., 1976, Glucagon in the human fetal pancreas, in press.

Aubert, M. L., Grumbach, M. M., and Kaplan, S. L., 1975, The ontogenesis of human fetal hormones. III. Prolactin, *J. Clin. Invest.* **56:**155.

Aurrichio, S., and Rigillo, N., 1960, Glucose-6-phosphatase activity of the human fetal liver, *Biol. Neonat.* **2:**146.

Baird, J. D., 1969, Some aspects of carbohydrate metabolism in pregnancy with special reference to the energy metabolism and hormonal status of the infant of the diabetic woman and the diabetogenic effect of pregnancy, *J. Endocrinol.* **44:**139.

Ballard, F. J., Hanson, R. W., and Kronfeld, D. S., 1969, Gluconeogenesis and lipogenesis in tissue from ruminant and nonruminant animals, *Fed. Proc.* **28:**218.

Battaglia, F. C., Behrman, R. E., Meschia, G., Seeds, A. E., and Bruns, P. D., 1968, Clearance of inert molecules, Na and Cl irons across the primate placenta, *Am. J. Obstet. Gynecol.* **102:**1135.

Bayer, S. M., and McMurray, W. C., 1967, *J. Neurochem.* **14:**695.

Beatty, C. H., and Bocek, R. M., 1970, Metabolism of palmitate by fetal, neonatal, and adult muscle of the rhesus monkey, *Am. J. Physiol.* **219:**1311.

Beatty, C. H., Basinger, G. M., and Bocek, R. M., 1966, Pentose cycle activity in muscle from fetal, neonatal and infant rhesus monkeys, *Arch. Biochem. Biophys.* **117:**275.

Beatty, C. H., Young, M. K., Dwyer, D., and Bocek, R. M., 1972, Glucose utilization of cardiac and skeletal muscle homogenates from fetal and adult rhesus monkeys, *Pediatr. Res.* **6:**813.

Beatty, C. H., Young, M. K., and Bocek, R. M., 1976, Control of glycolysis in skeletal muscle from fetal rhesus monkeys, *Pediatr. Res.* **10:**149.

Beck, P., 1969, Progestin enhancement of the plasma insulin response to glucose in the rhesus monkey, *Diabetes* **18:**146.

Beck, P., and Daughaday, W., 1967, Human placental lactogen: Studies of its acute metabolic effects and disposition in man, *J. Clin. Invest.* **46:**103.

Benedict, F. G., 1915, A study of prolonged fasting, *Carnegie Inst. Washington Publ.* **203.**

Benedict, F. G., and Joslin, E. P., 1912, A study of metabolism in severe diabetes, *Carnegie Inst. Washington Publ.* **176.**

**539**

*CARBOHYDRATE, FAT,
AND AMINO ACID
METABOLISM IN THE
PREGNANT WOMAN
AND FETUS*

Bier, D. M., Leake, R. D., Arnold, K. J., Haymond, M., Gruenke, L. D., Sperling, M. A., and Kipnis, D. M., 1976, Glucose production rates in infancy and childhood, *Pediatr. Res.* **10**:405 (abstract).

Blackshear, P. J., Holloway, P. A. H., and Alberti, K. G. M. M., 1974, The effects of starvation and insulin on the release of gluconeogenic substrates from the extrasplanchnic tissues *in vivo*, *FEBS Lett.* **48**:310.

Bleicher, S. J., O'Sullivan, J. B., and Freinkel, N., 1964, Carbohydrate metabolism in pregnancy. V. The interrelationships among glucose, insulin, and free fatty acids in late pregnancy, *N. Engl. J. Med.* **271**:866.

Bocek, R. M., and Beatty, C. H., 1969, Effect of insulin on carbohydrate metabolism of fetal rhesus monkey muscle, *Endocrinology* **85**:615.

Bocek, R. M., and Beatty, C. H., 1976, Cyclic AMP phosphodiesterase activity in fetal and adult muscle, *Develop. Biol.* **48**:382.

Bocek, R. M., Basinger, G. M., and Beatty, C. H., 1969, Glycogen synthase, phosphorylase and glycogen content of developing rhesus muscle, *Pediatr. Res.* **3**:525.

Bocek, R. M., Young, M. K., and Beatty, C. H., 1973, Effect of insulin and epinephrine on the carbohydrate metabolism and adenylate cyclase activity of rhesus fetal muscle, *Pediatr. Res.* **7**:787.

Bossi, E., and Greenberg, R. E., 1972, Sources of blood glucose in the rat fetus, *Pediatr. Res.* **6**:765.

Boyd, R. D., Morris, F. H., Meschia, G., Makowski, E. L., and Battaglia, F. C., 1973, Growth of glucose and oxygen uptakes by fetuses of fed and starved ewes, *Am. J. Physiol.* **225**:897.

Breuer, E., Barta, E., Pappova, E., and Zlatos, L., 1967, Developmental changes of myocardial metabolism. I. Peculiarities of cardiac carbohydrate metabolism in the early postnatal period in dogs, *Biol. Neonat.* **11**:367.

Breuer, E., Barta, E., Zlatos, L., and Pappova, E., 1968, Developmental changes of myocardial metabolism. II. Myocardial metabolism of fatty acids in the early postnatal period in dogs, *Biol. Neonat.* **12**:54.

Buckley, B. M., and Williamson, D. H., 1973, Acetoacetate and brain lipogenesis. Developmental pattern of acetoacetyl-coenzyme A synthetase in the soluble fraction of rat brain, *Bioch. J.* **132**:653.

Burd, L. I., Jones, M. D., Simmons, M. A., Makowski, E. L., Meschia, G., and Battaglia, F. C., 1975, Placental production and fetal utilisation lactate and pyruvate, *Nature* **254**:710.

Burt, R. L., 1956, Peripheral utilization of glucose in pregnancy: Insulin tolerance, *Obstet. Gynecol.* **7**:658.

Buschiazzo, H., Exton, J. H., and Park, C. R., 1970, Effects of glucose on glycogen synthetase, phosphorylase, and glycogen deposition in the perfused rat liver, *Proc. Natl. Acad. Sci. U.S.A.* **65**:383.

Buse, M. G., Biggers, J. F., Frederici, K. H., and Buse, J. F., 1972, Oxidation of branched chain amino acids by isolated hearts and diaphragms of the rat, *J. Biol. Chem.* **247**:8085.

Cahill, G. F., Jr., 1972, Prenatal nutrition of lambs, bears—and babies? *Pediatrics* **50**:357.

Cahill, G. F., Jr., Herrera, M. G., Morgan, A. P., Soeldner, J. S., Steinke, J., Levy, P. L., Reichard, G. A., Jr., and Kipnis, D. M., 1966, Hormone-fuel interrelationships during fasting, *J. Clin. Invest.* **45**:1751.

Čapkova, A., and Jirasek, J. E., 1968, Glycogen reserves in organs of human foetuses in the first half of pregnancy, *Biol. Neonat.* **13**:129.

Chang, L. O., 1977, The development of pyruvate carboxylase in rat liver mitochondria, *Pediatr. Res.* **11**:6.

Char, V. C., and Creasy, R. K., 1976, Lactate and pyruvate as fetal metabolic substrates, *Pediatr. Res.* **10**:231.

Chen, C. H., Adam, P. A. J., Laskowski, D. E., McCann, M. L., and Schwartz, R., 1965, The plasma free fatty acid composition and blood glucose of normal and diabetic pregnant women and of their newborns, *Pediatrics* **36**:843.

Chlebowski, R. T., and Adam, P. A. J., 1975, Glucose production in the newborn dog. II. Evaluation of autonomic and enzymatic control in isolated perfused canine liver, *Pediatr. Res.* **9**:821.

Christensen, P. J., Date, J. W., Schonheyder, F., and Volqvartz, K., 1957, Amino acids in blood plasma and urine during pregnancy, *Scand. J. Clin. Lab. Invest.* **9**:54.

Cockburn, F., Robins, S. P., and Forfar, J. O., 1970, Free amino acid concentrations in fetal fluids, *Br. Med. J.* **3**:747.

Cooper, J. R., Bloom, F. E., and Roth, R. H., 1974, *The Biochemical Basis of Neuropharmacology*, Oxford University Press, London, pp. 202–233.

Costrini, N. V., and Kalkhoff, R. K., 1971, Relative effects of pregnancy, estradiol, and progesterone on plasma insulin and pancreatic islet insulin secretion, *J. Clin. Invest.* **50**:992.

*PETER A. J. ADAM and*
*PHILIP FELIG*

Cremer, J., 1971, Incorporation of label from D-β-hydroxy[14C]butyrate and [3-14C]acetoacetate into amino acids in rat brain *in vivo, Bioch. J.* **122**:135.

Cross, K. W., Tizard, J. P. M., and Trythall, D. A. R., 1957, The gaseous metabolism of the newborn infant, *Acta Paediatr.* **46**:265.

Dahlquist, G., Persson, U., and Persson, B., 1972, The activity of D-β-hydroxybutyrate dehydrogenase in fetal, infant and adult rat brain, and the influence of starvation, *Biol. Neonat.* **20**:40.

Dancis, J., Money, W. L., Springer, D., and Levitz, M., 1968, Transport of amino acids by placenta, *Am. J. Obstet. Gynecol.* **101**:820.

Dancis, J., Jansen, V., Kayden, H. J., Schneider, H., and Levitz, M., 1973, Transfer across perfused human placenta. II. Free fatty acids, *Pediatr. Res.* **7**:192.

Dancis, J., Jansen, V., and Levitz, 1976, Transfer across perfused human placenta. IV. Effect of protein binding on free fatty acids, *Pediatr. Res.* **10**:5.

Daniel, P. M., Love, E. R., Moorehouse, S. R., Pratt, O. E., and Wilson, P., 1971, Factors influencing utilisation of ketone-bodies by brain in normal rats and rats with ketoacidosis, *Lancet* **2**:637.

Daniel, R. R., Metzger, B. F., Freinkel, N., Faloona, G. R., Unger, R. H., and Nitzan, M., 1974, Carbohydrate metabolism in pregnancy. XI. Response of plasma glucagon to overnight fast and oral glucose during normal pregnancy and in gestational diabetes, *Diabetes* **23**:771.

Dawes, G. S., and Shelley, H. J., 1968, Physiological aspects of carbohydrate metabolism in the fetus and newborn, in: *Carbohydrate Metabolism and Its Disorders* (F. Dickens, R. J., Randle, and W. J. Whelan, eds.), Vol. 2, Academic Press, New York.

Dawkins, M. J. R., 1963, Glycogen synthesis and breakdown in rat liver at birth, *Q. J. Exp. Physiol.* **48**:265.

Derom, R., 1964, Anaerobic metabolism in the human fetus: I. Normal delivery, *Am. J. Obstet. Gynecol.* **89**:241.

Dhopeshwarkar, G. A., and Mead, J. F., 1969, Fatty acid uptake by the brain. II. Incorporation of [1-14C]palmitic acid into the adult rat brain, *Biochim. Biophys. Acta* **187**:461.

Dhopeshwarkar, G. A., Maier, R., and Mead, J. F., 1969, Incorporation of [1-14C]acetate into the fatty acids of developing brain, *Biochim. Biophys. Acta* **187**:6.

Dierks-Ventling, C., 1971, Prenatal induction of ketone-body enzymes in the rat, *Biol. Neonat.* **19**:426.

Dingman, W., and Sporn, M. B., 1959, The penetration of proline and proline derivatives into brain, *J. Neurochem.* **4**:148.

Dobbing, J., 1974, The later development of the brain and its vulnerability, in: *Scientific Foundations of Paediatrics* (J. A. Davis and J. Dobbing, eds.), pp. 565–577, W. B. Saunders, Philadelphia.

Edwards, A. V., and Silver, M., 1970, The glycogenolytic response to stimulation of the splanchnic nerves in adrenolectomized calves, *J. Physiol.* **211**:109.

Eisen, H. J., Goldfine, I. D., and Glinsmann, W. H., 1973, Regulation of hepatic glycogen synthesis during fetal development: Roles of hydrocortisone, insulin, and insulin receptors, *Proc. Natl. Acad. Sci. U.S.A.* **70**:3454.

Felig, P., 1973a, The glucose–alanine cycle, *Metabolism* **22**:179.

Felig, P., 1973b, Maternal and fetal fuel homeostasis in human pregnancy, *Am. J. Clin. Nutr.* **26**:998.

Felig, P., 1975, Amino acid metabolism in man, *Annu. Rev. Biochem.* **44**:933.

Felig, P., and Lynch, V., 1970, Starvation in human pregnancy: Hypoglycemia, hypoinsulinemia, and hyperketonaemia, *Science* **170**:990.

Felig, P., Marliss, E., Owen, O. E., and Cahill, G. F., Jr., 1969a, Blood glucose and gluconeogenesis in fasting man, *Arch. Intern. Med.* **123**:293.

Felig, P., Marliss, E., Owen, O. E., and Cahill, G. F., Jr., 1969b, Role of substrate in the regulation of hepatic gluconeogenesis in man, *Adv. Enzyme Regul.* **7**:41.

Felig, P., Owen, O. E., Wahren, J., and Cahill, G. F., Jr., 1969c, Amino acid metabolism in prolonged starvation, *J. Clin. Invest.* **48**:584.

Felig, P., Pozefsky, T., Marliss, E., and Cahill, G. F., Jr., 1970, Alanine: Key role in gluconeogenesis, *Science* **167**:1003.

Felig, P., Kim, Y. J., Lynch, V., and Hendler, R., 1972, Amino acid metabolism during starvation in human pregnancy, *J. Clin. Invest.* **51**:1195.

Felig, P., Wahren, J., Karl, I., Cerasi, E., Luft, R., and Kipnis, D. M., 1973a, Glutamine and glutamate metabolism in normal and diabetic subjects, *Diabetes* **22**:573.

Felig, P., Wahren, J., and Raf, L., 1973b, Evidence of interorgan amino acid transport by blood cells, *Proc. Natl. Acad. Sci. U.S.A.* **70**:1775.

Felig, P., Wahren, J., and Hendler, R., 1975, Influence of oral glucose ingestion on splanchnic glucose and gluconeogenic substrate metabolism, *Diabetes* **24**:468.

Fisher, M. M., and Kerly, M., 1964, Amino acid metabolism in perfused liver, *J. Physiol.* **174**:273.

Freinkel, N., 1969, *Homeostatic Factors in Fetal Carbohydrate Metabolism,* Vol. 4, Appleton-Century-Crofts, New York.

Frienkel, N., and Metzger, B., 1975, Some considerations of fuel economy in the fed state during late human pregnancy, in: *Early Diabetes in Early Life* (R. A. Camerini-Davalos and H. S. Cole, eds.), Academic Press, New York.

Freinkel, N., Metzger, B., Nitzan, N., Daniel, R., Surviaczynska, B. Z., and Nagel, T. C., 1974, Facilitated anabolism in late pregnancy: Some novel material compensations for accelerated starvation, in: *Proceedings, VIII Congress of the International Diabetes Federation,* Excerpta Medica.

Friedman, E. A., Gray, M. J., Grynfogel, M., Hutchinson, D. L., Kelly, W. T., and Plentl, A. A., 1960, Distribution of $C^{14}$-labeled lactic acid and bicarbonate in pregnant primates, *J. Clin. Invest.* **39:** 227.

Gaspard, V. J., Sandront, H. M. Luyckx, A. S., and Lefebvre, P. J., 1975, The control of human placental lactogen (HPL) secretion and its interrelation with glucose and lipid metabolism in pregnancy, in: *Early Diabetes in Early Life* (R. A. Camerini-Davalos and H. S. Cole, eds.), Academic Press, New York.

Gennser, G., Lundquist, I., and Nilsson, E., 1971, Glycogenolytic activity in the liver of the human foetus, *Biol. Neonat.* **19:**1.

Ghadimi, H., and Pecora, P. D., 1964, Free amino acids of cord plasma as compared with maternal plasma during pregnancy, *Pediatrics* **33:**500.

Girard, J. R., 1975, Metabolic fuels of the fetus, *Isr. J. Med. Sci.* **11:**591.

Gitlin, D., and Biasucci, A., 1969, Ontogenesis of immunoreactive growth hormone, follicle-stimulating hormone, thyroid-stimulating hormone, luteinizing hormone, chorionic prolactin and chorionic gonadotropin in the human conceptus, *J. Clin. Endocrinol. Metab.* **29:**926.

Girard, J. R., Caquet, D., Bal, D., and Guillet, I., 1973, Control of rat liver phosphorylase, glucose-6-phosphatase, and phosphoenopyruvate carboxykinase activities by insulin and glucagon during the perinatal period, *Enzyme* **15:**272.

Girard, J. R., Kevran, A., Soufflet, M. S., and Assan, R., 1974, Factors affecting the secretion of insulin and glucagon by the rat fetus, *Diabetes* **23:**310.

Glendening, M. D., Margolis, A. J., and Page, E. W., 1961, Amino acid concentrations in fetal and maternal plasma, *Am. J. Obstet. Gynecol.* **81:**591.

Glinsmann, W., Pauk, G., and Hern, E., 1970, Control of rat liver glycogen synthetase and phosphorylase activities by glucose, *Biochem. Biophys. Res. Commun.* **39:**774.

Glinsmann, W. H., Eisen, H. J., Lynch, A., and Chez, R. A., 1975, Glucose regulation by isolated near term fetal monkey liver, *Pediatr. Res.* **9:**600.

Goodner, C. J., and Thompson, D. J., 1967, Glucose metabolism in the fetus *in utero:* The effect of maternal fasting and glucose loading in the rat, *Pediatr. Res.* **1:**443.

Goodner, C. J., Conway, M. J., and Werrbach, J. H., 1969, Relation between plasma glucose levels of mother and fetus during maternal hyperglycemia, hypoglycemia, and fasting in the rat, *Pediatr. Res.* **3:**121.

Gottstein, U., Muller, W., Berghoff, U., Gartner, H., and Held, K., 1971, Zur Utilization von nichtveresterten Fettsauren und Ketonkorpen im Gehirn des Menschen, *Klin. Wochenschr.* **49:**406.

Greengard, O., 1973, Effects of hormones on development of fetal enzymes, *Clin. Pharmacol. Ther.* **14:**721.

Gresham, E. L., Simons, P. S., and Battaglia, F. C., 1971, Maternal–fetal urea concentration difference in man: Metabolic significance, *J. Pediatr.* **79:**809.

Gresham, E. L., James, E. J., Raye, J. R., Battaglia, F. C., Makowski, E. L., and Meschia, G., 1972, Production and excretion of urea by the fetal lamb, *Pediatrics* **50:**372.

Gross, I., and Warshaw, J. B., 1974, Fatty acid synthesis in developing brain: Acetyl-CoA carboxylase activity, *Biol. Neonat.* **25:**365.

Grumbach, M. M., Kaplan, S. L., Sciarra, J. J., and Burr, I. M., 1968, Chorionic growth hormone-prolactin (CGP): secretion, disposition, biological activity in man, and postulated function as the "growth hormone" of the second half of pregnancy, *Ann. N.Y. Acad. Sci.* **148:**501.

Hahn, P., and Skala, J., 1970, Some enzymes of glucose metabolism in the human fetus, *Biol. Neonat.* **16:**362.

Hanson, R. W., Reshef, L., and Ballard, J., 1975, Hormonal regulation of hepatic *P*-enolpyruvate carboxykinase (GTP) during development, *Fed. Proc.* **34:**166.

Havel, R. J., 1972, Caloric homeostasis and disorders of fuel transport, *N. Engl. J. Med.* **287:**1186.

Havel. R. J., Kane, J. P., Balasse, E. O., Segel, N., and Basso, L. V., 1970, Splanchnic metabolism of free fatty acids and production of triglycerides of very low density lipoproteins in normotriglyceridemic and hypertriglyceridemic humans, *J. Clin. Invest.* **49:**2017.

Hawkins, R. A., Williamson, D. H., and Krebs, H. A., 1971, Ketone-body utilization by adult and suckling rat brain *in vivo, Biochem. J.* **122**:13.

Hendricks, C. H., 1957, Studies on lactic acid metabolism in pregnancy and labor, *Am. J. Obstet. Gynecol.* **73**:492.

Herrera, E., Knopp, R. H., and Freinkel, N., 1969, Carbohydrate metabolism in pregnancy. VI. Plasma fuels, insulin, liver composition, gluconeogenesis, and nitrogen metabolism during late gestation in the fed and fasted rat, *J. Clin. Invest.* **48**:2260.

Himwich, W. A., Petersen, J. C., and Allen, M. L., 1957, Hematoencephalic exchange as a function of age, *Neurology* **7**:705.

Houssay, B. A., Foglia, V. G., and Rodrigues, R. R., 1954, Production or prevention of some types of experimental diabetes by oestrogen or corticosteroids, *Acta Endocrinol.* **17**:146.

Huang, F. L., and Glinsmann, W. H., 1975, Inactivation of rabbit muscle phosphorylase phosphatase by cyclic AMP-dependent kinase, *Proc. Natl. Acad. Sci. U.S.A.* **72**:3004.

Huang, F. L., and Glinsmann, W. H., 1976, A second heat-stable protein inhibitor of phosphorylase phosphatase from rabbit muscle, *FEBS Lett.* **62**:326.

Huang, K.-P., Huang, F. L., Glinsmann, W. H., and Robinson, J. C., 1975, Regulation of glycogen synthetase activity by two kinases, *Biochem. Biophys. Res. Commun.* **65**:1163.

Huang, K.-P., Huang, F. L., Glinsmann, W. H., and Robinson, J. C., 1976, Effect of limited proteolysis on activity and phosphorylation of rabbit muscle glycogen synthetase, *Arch Biochem. Biophys.* **173**:162.

Hugget, A. St. G., 1961, Carbohydrate metabolism in placenta and foetus, *Br. Med. Bull.* **17**:122.

Hull, D., 1974, The function and development of adipose tissue, in: *Scientific Foundations of Paediatrics* (J. A. Davis and J. Dobbing, eds.), pp. 440–455, W. B. Saunders, Philadelphia.

Hultman, E., and Nilsson, L. H., 1971, Liver glycogen in man. Effect of different diets and muscular exercise, *Adv. Exp. Med. Biol.* **11**:143.

Itoh, T., and Quastel, J. H., 1970, Acetoacetate metabolism in infant and adult rat brain *in vitro, Bioch. J.* **116**:641.

Jacquot, R., and Kretschmer, N., 1964, Effect of fetal decapitation on enzymes of glycogen metabolism, *J. Biol. Chem.* **239**:130l.

James, E. J., Raye, J. R., Gresham, E. L., Makowski, E. L., Meschia, G., and Battaglia, F. C., 1972, Fetal oxygen consumption, carbon dioxide production, and glucose uptake in a chronic sheep preparation, *Pediatrics* **50**:361.

Jones, M. D., Jr., Burd, L. I., Makowski, G., and Battaglia, F. C., 1976, Cerebral metabolism in sheep: a comparative study of the adult, the lamb, and the fetus, *Am. J. Physiol.* **229**:235.

Jonxis, J. H. P., Van der Vengt, J. J., DeGroot, C. J., Boersma, E. R., and Meijers, E. D. K., 1967, The metabolic rate in praemature, dysmature, and sick infants in relation to environmental temperature, in: *Aspects of Praematurity and Dysmaturity (Nutricia Symposium)* (J. H. P. Jonxis, H. K. A. Visser, and J. A. Troelstra, eds.), pp. 201–209, Charles C. Thomas, Springfield, Illinois.

Josimovich, J. B., and Brande, B. I., 1961, Chemical properties and biologic effects of human placental lactogen (HPL), *Trans. N.Y. Acad. Sci.* **27**:161.

Josimovich, J. B., and MacLaren, J. A., 1962, Presence in the human placenta and term serum of a highly lactogenic substance immunologically related to pituitary growth hormone, *Endocrinology* **71**:209.

Jost, A., and Picon, L., 1970, Hormonal control of fetal development and metabolism, *Adv. Metab. Disord.* **4**:123.

Kahng, M. W., Sevdalian, D. A., and Tildon, J. T., 1974, Substrate oxidation and enzyme activities of ketone-body metabolism in the developing pig, *Biol. Neonat.* **24**:187.

Kalhan, S. C., Savin, S. M., and Adam, P. A. J., 1976a, Measurement of glucose turnover in the human newborn with glucose-1-$^{13}$C, *J. Clin. Endocrinol. Metab.* **43**:704.

Kalhan, S. C., Savin, S. M., Uga, N., and Adam, P. A. J., 1976b, Quantification of glucose turnover with glucose-1-$^{13}$C tracer: Attenuated glucose production in newborn infants of diabetic mothers, *Pediatr. Res.* **10**:411 (abstract).

Kalkhoff, R. R., Richardson, B. I., and Beck, P., 1969, Relative effects of pregnancy, human placental lactogen and prednisone on carbohydrate tolerance in normal and subclinical diabetic subjects, *Diabetes* **18**:153.

Kaplan, S. L., Grumbach, M. M., and Shepard, T. H., 1972, The ontogenesis of human fetal hormones. I. Growth hormone and insulin, *J. Clin. Invest.* **51**:3080.

Kayden, H. J., Dancis, J., and Money, W. L., 1969, Transfer of lipids across the guinea pig placenta, *Am. J. Obstet. Gynecol.* **104**:564.

543

CARBOHYDRATE, FAT,
AND AMINO ACID
METABOLISM IN THE
PREGNANT WOMAN
AND FETUS

Kerr, G. R., 1968, The free amino acids of serum during development of *Macaca mullatta*. II. During pregnancy and fetal life, *Pediatr. Res.* **2**:493.

Kerr, G. R., Campbell, J. A., Helmuth, A. C., and Waisman, H. A., 1971, Growth and development of the fetal rhesus monkey *(Macaca mulatta)*. II. Total nitrogen, protein, lipid, glycogen and water composition of major organs, *Pediatr. Res.* **5**:151.

Kim, Y. J., and Felig, P., 1971, Plasma chorionic somatomammotropin levels during starvation in midpregnancy, *J. Clin. Endocrinol.* **32**:864.

Kim, Y. K. K., and Felig, P., 1972, Maternal and amniotic fluid substrate levels during calorie deprivation in human pregnancy, *Metabolism* **21**:507.

King, K. C., Adam, P. A. J., Laskowski, D. E., and Schwartz, R., 1971*a*, Sources of fatty acids in the newborn, *Pediatrics* **47**:192.

King, K. C., Butt, J., Raivio, K., Räihä, N., Roux, J., Teramo, K., Yamaguchi, K., and Schwartz, R., 1971*b*, Human maternal and fetal insulin response to arginine, *N. Engl. J. Med.* **285**:607.

Kirby, L., and Hahn, P., 1973, Enzyme induction in human fetal liver, *Pediatr. Res.* **7**:75.

Klee, C. B., and Sokoloff, L., 1967, Changes in D (−)-β-hydroxybutyric dehydrogenase activity during brain maturation in the rat, *J. Biol. Chem.* **242**:3880.

Kuhl, C., and Holst, J. J., 1976, Plasma glucagon and the insulin–glucagon ratio in gestational diabetes, *Diabetes* **25**:16.

Kuhl, C., Gaede, P., Klebe, J. G., and Pederson, J. G., 1975, Human placental lactogen concentration during physiological fluctuations of serum glucose in normal pregnant and gestational diabetic women, *Acta Endocrinol.* **80**:365.

Lajtha, A., 1958, Amino acid and protein metabolism of the brain. II. The uptake of L-lysine by brain and other organs of the mouse at different ages, *J. Neurochem.* **2**:209.

Lajtha, A., and Toth, J., 1961, The brain barrier system. II. Uptake and transport of amino acids by the brain, *J. Neurochem.* **8**:216.

Lemons, J. A., Adcock, E. W., III, Jones, M. D., Jr., Naughton, M. A., Meschia, G., and Battaglia, F. A., 1976, Umbilical uptake of amino acids in the unstressed fetal lamb, *J. Clin. Invest.* **58**:1428.

Levitsky, L. L., Paton, J. B., Fisher, D. E., and Delannoy, C. W., 1976, Blood levels of gluconeogenic precursors and renal gluconeogenesis in the fasting baboon infant, *Pediatr. Res.* **10**:412 (abstract).

Lind, T., and Hytten, F. E., 1970, Relation of amniotic fluid volume to fetal weight in the first half of pregnancy, *Lancet* **1**:1147.

Lindblad, B. S., and Baldesten, A., 1967, The normal venous plasma free amino acid levels of non-pregnant women and of mother and child during delivery, *Acta Paediatr. Scand.* **56**:37.

Lindblad, B. S., and Zetterström, R., 1968, The venous plasma free amino acid levels of mother and child during delivery. II. After short gestation and gestation complicated by hypertension with special reference to the "small-for-dates" syndrome, *Acta Paediatr. Scand.* **57**:195.

Lockwood, E. A., and Bailey, F., 1971, The course of ketosis, and the activity of key enzymes of ketogenesis and ketone-body utilization during development of the postnatal rat, *Biochem. J.* **124**:249.

Lowry, M. F., and Adam, P. A. J., 1975, Lack of gluconeogenesis from alanine at birth, *Pediatr. Res* **9**:353 (abstract).

Maddaiah, V. T., and Madsen, N. B., 1966*a*, Kinetics of purified liver phosphorylase, *J. Biol. Chem.* **241**:3873.

Maddaiah, V. T., and Madsen, N. B., 1966*b*, Studies on the biological control of glycogen metabolism in liver. I. State and activity pattern of glycogen phosphorylase, *Biochim. Biophys. Acta* **121**:261.

Makowski, E. L., Schneider, J. M., Tsoulos, N. G., Colwill, J. R., Battaglia, F. C., and Meschia, G., 1972, Cerebral blood flow, oxygen consumption, and glucose utilization of fetal lambs *in utero, Am. J. Obstet. Gynecol.* **114**:292.

Mallette, L. E., Exton, J. H., and Park, C. R., 1969, Control of gluconeogenesis from amino acids in the perfused rat liver, *J. Biol. Chem.* **244**:5713.

Marliss, E. B., Aoki, T. T., Unger, R. H., Soeldner, J. S., and Cahill, G. F., Jr., 1970, Glucagon levels and metabolism effects in fasting, *J. Clin. Invest.* **49**:2256.

Mastaglia, F. L., 1974, The growth and development of skeletal muscles, in: *Scientific Foundations of Paediatrics* (J. A. Davis and J. Dobbing, eds.), pp. 348–375, W. B. Saunders, Philadelphia.

Meister, A., 1974, An enzymatic basis for the blood brain barrier? The γ-glutamyl cycle—background and considerations relating to amino acid transport in brain, in: *Brain Dysfunction in Metabolic Disorders,* Vol. 53 (F. Plum, ed.), pp. 273–291, Raven Press, New York.

Melichar, V., and Wolf, H., 1967, Postnatal changes in the blood serum content of glycerol and free fatty acids in premature infants. Influence of hypothermia and respiratory distress, *Biol. Neonat.* **11**:50.

Mersmann, H. J., Phinney, G., Mueller, R. L., and Stanton, H. C., 1972, Glycogen metabolism in pre- and postnatal pigs, *Am. J. Physiol.* **222**:1620.

Metzger, B. E., Agnoli, F. S., and Freinkel, N., 1971, Effect of sex and pregnancy on formation of urea and ammonia during gluconeogenesis in the perfused rat liver, *Horm. Metab. Res.* **2**:367.

Milner, R. D. G., 1970, The development of insulin secretion in man, in: *Nutricia Symposium— Metabolic Processes in the Foetus and Newborn Infant* (J. H. P. Jonxis, H. K. A. Visser, and J. A. Troelstra, eds.), pp. 192–207, Williams and Wilkins, Baltimore; H. E. Stenfert Kroese N.V.-Leiden.

Milner, R. D. G., and Hales, C. N., 1965, Effect of intravenous glucose on concentration of insulin in maternal and umbilical-cord plasma, *Br. Med. J.* **1**:284.

Milner, R. D. G., Barson, A. J., and Ashworth, M. A., 1971, Human foetal pancreatic insulin secretion in response to ionic and other stimuli, *J. Endocrinol.* **51**:323.

Milner, R. D. G., Ashworth, M. A., and Barson, A. J., 1972, Insulin release from human foetal pancreas in response to glucose, leucine, and arginine, *J. Endocrinol.* **52**:497.

Mitchell, F. L., 1967, Steroid metabolism in the fetoplacental unit and in early childhood, *Vitam. Horm.* **25**:191.

Morris, M. D., and Chaikoff, I. L., 1961, Concerning incorporation of labeled cholesterol, fed to mothers, into brain cholesterol of 20-day-old suckling rats, *J. Neurochem.* **8**:226.

Morriss, F. H., Boyd, R. D. H., Makowski, E. L., Meschia, G., and Battaglia, F. C., 1973, Glucose/oxygen quotients across the hindlimbs of fetal lambs, *Pediatr. Res.* **7**:794.

Morriss, F. H., Makowski, E. L., Meschia, G., and Battaglia, F. C., 1975, The glucose/oxygen quotient of the term human fetus, *Biol. Neonate* **25**:44.

Novak, M., and Monkus, E., 1972, Metabolism of subcutaneous adipose tissue in the immediate postnatal period of human newborns. I. Developmental changes in lipolysis and glycogen content, *Pediatr. Res.* **6**:73.

Novak, M., Melichar, V., and Hahn, P., 1966, Lipid metabolism in adipose tissue from human infants, *Biol. Neonate* **9**:105.

Novak, M., Melichar, V., and Hahn, P., 1968, Changes in the reactivity *in vitro* to epinephrine and norepinephrine during postnatal development, *Biol. Neonate* **13**:175.

Novak, M., Hahn, P., and Melichar, V., 1969, Postnatal development of human adipose tissue, oxygen consumption, and oxidation of fatty acids, *Biol. Neonate* **14**:203.

Novak, M., Monkus, E., and Pardo, V., 1971, Human neonatal subcutaneous adipose tissue. Function and ultrastructure, *Biol. Neonate* **19**:306.

Novak, M., Monkus, E., Wolf, H., and Stave, U., 1972, The metabolism of subcutaneous adipose tissue in the immediate postnatal period of human newborns. II. Developmental changes in metabolism of $^{14}$C-(U)-Glucose and in enzyme activities of phosphofructokinase (PFK; EC 2.7.1.11) and $\beta$-hydroxyacyl-CoA dehydrogenase (HAD; EC 1.1.1.35), *Pediatr. Res.* **6**:211.

Novak, M., Hahn, P., Penn, D., Monkus, E., and Kirby, L., 1973a, Metabolism of subcutaneous adipose tissue in the immediate postnatal period of human newborns. Developmental changes in some cytoplasmic enzymes, *Biol. Neonate* **23**:19.

Novak, M., Hahn, P., Penn, D., Monkus, E., and Skala, J., 1973b, The role of carnitine in subcutaneous white adipose tissue from newborn infants, *Biol. Neonate* **23**:11.

Novak, M., Monkus, E., and Wolf, H., 1973c, The metabolism of subcutaneous adipose tissue in the immediate postnatal period of human neonates. III. Role of fetal glycogen in lipolysis and fatty acid esterification in the first hours of life, *Pediatr. Res.* **7**:769.

Novak, M., Penn Walker, D., and Monkus, E. F., 1975, Oxidation of fatty acids by mitochondria obtained from newborn subcutaneous (white) adipose tissue, *Biol. Neonate* **25**:95.

Obenshain, S. S., Adam, P. A. J., King, K. C., Teramo, K., Raivio, K. O., Räihä, N., and Schwartz, R., 1970, Human fetal insulin response to sustained maternal hyperglycemia, *N. Engl. J. Med.* **283**:566.

Odessey, R., Khairlah, A., and Goldbert, A. L., 1974, Origin and possible significance of alanine production by skeletal muscle, *J. Biol. Chem.* **249**:7623.

O'Sullivan, J. B., and Mahan, C. M., 1964, Criteria for the oral glucose tolerance test in pregnancy, *Diabetes* **13**:278.

Owen, O. E., and Reichard, G. A., Jr., 1971, Human forearm metabolism during progressive starvation, *J. Clin. Invest.* **50**:1536.

Owen, O. E., Morgan, A. P., Kemp, H. G., Sullivan, J. M., Herrera, M. G., and Cahill, G. F., 1967, Brain metabolism during fasting, *J. Clin. Invest.* **46**:1589.

Owen, O. E., Felig, P., Morgan, A. P., Wahren, J., and Cahill, G. F., Jr., 1969, Liver and kidney metabolism during prolonged starvation, *J. Clin. Invest.* **48**:574.

Ozand, P. T., Stevenson, J. H., Tildon, J. T., and Cornblath, M., 1975a, The effects of hyperketonemia on glycolytic intermediates in the developing rat brain, *J. Neurochem.* **25**:61.

**545**

*CARBOHYDRATE, FAT,
AND AMINO ACID
METABOLISM IN THE
PREGNANT WOMAN
AND FETUS*

Ozand, P. T., Stevenson, J. H., Tildon, J. T., and Cornblath, M., 1975*b,* The effects of hyperketonemia on glutamate and glutamine metabolism in developing rat brain, *J. Neurochem.* **25**:67.

Page, E. W., 1969, Human fetal nutrition and growth, *Am. J. Obstet. Gynecol.* **104**:378.

Page, E. W., Glendening, M. B., Margolis, A. J., and Harper, H. A., 1957, Transfer of D- and L-histidine across the human placenta, *Am. J. Obstet. Gynecol.* **73**:589.

Page, M. A., Krebs, H. A., and Williamson, D. H., 1971, Activities of enzymes of ketone-body utilization in brain and other tissues of suckling rat, *Biochem. J.* **121**:49.

Patel, A. J., and Balázs, R., 1970, Manifestation of metabolic compartmentation during the maturation of the rat brain, *J. Neurochem.* **17**:955.

Paterson, P., Sheath, J., Taft, P., and Wood, C., 1967, Maternal and fetal ketone concentrations in plasma and urine, *Lancet* **1**:862.

Persson, B., Settergren, G., and Dahlquist, G., 1972, Cerebral arteriovenous difference of acetoacetate and D-$\beta$-hydroxybutyrate in children, *Acta Paediatr. Scand.* **61**:273.

Pitkin, R. M., Connor, W. E., and Lin, D. S., 1972, Cholesterol metabolism and placental transfer in the pregnant rhesus monkey, *J. Clin. Invest.* **51**:2584.

Portman, O. W., Behrman, R. E., and Soltys, P., 1969, Transfer of fatty acids across the primate placenta, *Am. J. Physiol.* **216**:143.

Pozefsky, T., Tancredi, R. G., Moxley, R. T., Dupre, J., and Tobin, J., 1974, Forearm tissue metabolism in postabsorptive and 60-hr fasted man: Studies with glucagon, *J. Clin. Invest.* **53**:61a (abstract).

Räihä, N. C. R., and Lindros, K. O., 1969, Development of some enzymes involved in gluconeogenesis in human liver, *Ann. Med. Exp. Fenn.* **47**:146.

Reisner, S. H., Aranda, J. V., Colle, E., Papageorgiou, A., Schiff, D., Scriver, C. R., and Stern, L., 1973, The effect of intravenous glucagon on plasma amino acids in the newborn, *Pediatr. Res.* **7**:184.

Robertson, A. F., and Sprecher, H., 1968, A review of human placental lipid metabolism and transport, *Acta Paediatr. Suppl.* **183**:1.

Robinson, B. H., 1973, The role of mitochondrial tricarboxylate anion transport in metabolism, *Symp. Soc. Exp. Biol.* **27**:195.

Robinson, B. H., 1976, Development of gluconeogenic enzymes in the newborn guinea pig, *Biol. Neonate* **29**:48.

Roux, J. F., and Myers, R. E., 1974, *In vitro* metabolism of palmitic acid and glucose in the developing tissue of the rhesus monkey, *Am. J. Obstet. Gynecol.* **118**:385.

Sabata, V., Stembera, Z. K. S., and Novak, M., 1968*a,* Levels of unesterified and esterified fatty acids in umbilical blood of hypoxic fetuses, *Biol. Neonate* **12**:194.

Sabata, V., Wolf, H., Lausmann, S., 1968*b,* The role of free fatty acids, glycerol, ketone-bodies and glucose in the energy metabolism of the mother and fetus during delivery, *Biol. Neonate* **13**:7.

Saudek, C. D., and Felig, P., 1976, The metabolic events of starvation, *Am. J. Med.* **60**:117.

Schreiner, R. L., Burd, L. I., Douglas Jones, M., Jr., Lemons, J. A., Sheldon, R. E., Simmons, M. A., Battaglia, F. C., and Meschia, G., 1976, Fetal metabolism in fasting sheep, *Pediatr. Res.* **10**:325 (abstract).

Schwartz, A. L., 1974*a,* Hormonal regulation of glucose production in human fetal liver, Ph.D. thesis, Case Western Reserve University.

Schwartz, A. L., 1974*b,* Hormonal regulation of amino acid accumulation in human fetal liver explants, *Biochim. Biophys. Acta* **362**:276.

Schwartz, A. L., and Rall, T. W., 1975*a,* Hormonal regulation of incorporation of alanine-U-$^{14}$C into glucose in human fetal liver explants—effects of dibutyryl cyclic AMP, glucagon, insulin, and triamcinolone, *Diabetes* **24**:650.

Schwartz, A. L., and Rall, T. W., 1975*b,* Hormonal regulation of glycogen metabolism in human fetal liver. II. Regulation of glycogen synthase activity, *Diabetes* **24**:1113.

Schwartz, A. L., Räihä, N. C. R., and Rall, T. W., 1974, Effect of dibutyrl cyclic AMP on glucose-6-phosphatase activity in human fetal liver explants, *Biochim. Biophys. Acta* **343**:500.

Schwartz, A. L., Räihä, N. C. R., and Rall, T. W., 1975, Hormonal regulation of glycogen metabolism in human fetal liver. I. Normal development and effects of dibutyryl cyclic AMP, glucagon, and insulin in liver explants, *Diabetes* **24**:1101.

Schwerin, P., Bessman, S. P., and Waelsh, H., 1950, The uptake of glutamic acid and glutamine by brain and other tissues of the rat and mouse, *J. Biol. Chem.* **184**:37.

Settergren, G., Lindblad, B. S., and Persson, B., 1976, Cerebral blood flow and exchange of oxygen, glucose, ketone-bodies, lactate, pyruvate and amino acids in infants, *Acta Paediatr. Scand.* **65**:343.

Shamoon, H., and Felig, P., 1974, Effects of estrogen on glucose uptake by rat diaphragm, *Yale J. Biol. Med.* **47**:227.

Sheath, J., Grimwade, J., Waldron, K., Bickley, M., Taft, P., and Wood, C., 1972, Arteriovenous monesterified fatty acids and glycerol differences in the umbilical cord at term and their relationship to fetal metabolism, *Am. J. Obstet. Gynecol.* **113**:358.

Shelley, H. J., 1969, Carbohydrate metabolism in the foetus and the newly born, *Proc. Nutr. Soc.* **28**:42.

Shepard, T. H., 1967, Onset of function in the human fetal thyroid: Biochemical and radioautographic studies from organ culture, *J. Clin. Endocrinol. Metab.* **27**:945.

Sherline, P., Lynch, A., and Glinsmann, W., 1972, Cyclic AMP and adrenergic receptor control of rat liver glycogen metabolism, *Endocrinology* **91**:680.

Sherline, P., Eisen, H., and Glinsmann, W., 1974, Acute hormonal regulation of cyclic AMP content and glycogen phosphorylase in fetal liver in organ culture, *Endocrinology* **94**:935.

Sherwin, R. S., Hendler, R. G., and Felig, P., 1975, Effect of ketone infusion on amino acid and nitrogen metabolism in man, *J. Clin. Invest.* **55**:1382.

Sherwin, R. S., Fisher, M., Hendler, R., and Felig, P., 1976, Hyperglucagonemia and blood glucose regulation in normal, obese, and diabetic subjects, *N. Engl. J. Med.* **294**:455.

Siesjo, B. K., Johannson, H., Ljunggren, B., and Norberg, K., 1974, Brain dysfunction in cerebral hypoxia and ischemia, in: *Brain Dysfunction in Metabolic Disorders* (F. Plum, ed.), Vol. 53, pp. 75–112, Research Publications, Association for Research in Nervous and Mental Disease, Raven Press, New York.

Simmons, M. A., Meschia, G., Makowski, E. L., and Battaglia, F. C., 1974, Fetal metabolic response to maternal starvation, *Pediatr. Res.* **8**:830.

Smith, A. L., Satherthwaite, H. S., and Sokoloff, L., 1969, Induction of brain D(−)-β-hydroxybutyrate dehydrogenase activity by fasting, *Science* **163**:79.

Soler, N. G., Nicholson, H. O., and Malins, J. M., 1974, Serial determinations of human placental lactogen in the last half of normal and complicated pregnancies, *Am. J. Obstet. Gynecol.* **120**:214.

Sparks, J. W., Lynch, A., Chez, R. A., and Glinsmann, W. H., 1976, Glycogen regulation in isolated perfused near term monkey liver, *Pediatr. Res.* **10**:51.

Spellacy, W. N., and Goetz, F. C., 1963, Plasma insulin in normal late pregnancy, *N. Engl. J. Med.* **268**:988.

Srere, P. A., Chaikoff, I. L., Treitman, S. S., and Burstein, L. S., 1950, The extrahepatic synthesis of cholesterol, *J. Biol. Chem.* **182**:629.

SteginK, L. D., Pitkin, R. M., Reynolds, W. A., Filer, L. J., Jr., Boaz, D. P., and Brummel, M. C., 1975, Placental transfer of glutamate and its metabolites in the primate, *Am. J. Obstet. Gynecol.* **122**:70.

Stembera, Z. K., and Hodr, J., 1966, I. The relationship between blood levels of glucose, lactic acid, and pyruvic acid in the mother, and in both umbilical vessels of the healthy fetus, *Biol. Neonate* **10**:227.

Sturman, J. A., Nieman, W. H., and Gaull, G. E., 1973, Metabolism of $^{35}$S-methionine and $^{35}$S-cystine in the pregnant rhesus monkey, *Biol. Neonate* **22**:16.

Sutherland, H. W., Fisher, P. M., and Stowers, J. M., 1975, Evaluation of maternal carbohydrate metabolism by the intravenous glucose tolerance test, in: *Early Diabetes in Early Life* (R. A. Camerini-Davalos and H. S. Cole, eds.), Academic Press, New York.

Szabo, A. J., and Szabo, O., 1974, Placental free-fatty-acid transfer and fetal adipose-tissue development: An explanation of fetal adiposity in infants of diabetic mothers, *Lancet* **2**:498.

Szabo, A. J., Grimaldi, R. D., and Jung, W. F., 1969, Palmitate transport across perfused human placenta, *Metabolism* **18**:406.

Tandler, B., and Hoppel, C. L., 1972, *Mitochondria,* Academic Press, New York, p. 16.

Tomec, R. J., and Hoppel, C. L., 1975, Carnitine palmitoyl transferase in bovine fetal heart mitochondria, *Arch. Biochem. Biophys.* **170**:716.

Tsoulos, N. G., Schneider, J. M., Colwill, J. R., Meschia, G., Makowski, E. L., and Battaglia, F. C., 1972, Cerebral glucose utilization during aerobic metabolism in fetal sheep, *Pediatr. Res.* **6**:182.

Tunell, R., Copher, D., and Persson, B., 1976, The pulmonary gas exchange and blood gas changes in connection with birth, in: *Neonatal Intensive Care* (P. R. Swyer and J. B. Stetson, eds.), pp. 89–109, Warren H. Green, St. Louis, Missouri.

Tyson, J. E., and Merimee, T. J., 1970, Some physiologic effects of protein ingestion in pregnancy, *Am. J. Obstet. Gynecol.* **107**:797.

Tyson, J. E., Rabinowitz, D., Merimee, T. J., and Friesen, H., 1969, Response of plasma insulin in human growth hormone to arginine in pregnant and postpartum females, *Am. J. Obstet. Gynecol.* **103**:313.

Unger, R. H., 1974, Alpha and beta cell interrelationships in health and disease, *Metabolism* **23**:581.

Van Aasche, F. A., and Aerts, L., 1975, Morphologic and ultrastructure modifications in the endocrine pancreas in pregnant rats, in: *Early Diabetes in Early Life* (R. A. Camerini-Davalos and H. S. Cole, eds.), Academic Press, New York.

Van den Berg, C. J., 1970, Compartmentation of glutamate metabolism in the developing brain: Experiments with labeled glucose, acetate, phenylalanine, tyrosine, and proline, *J. Neurochem.* **17**:973.

Victor, A., 1974, Normal blood sugar variation during pregnancy, *Acta Obstet. Gynecol. Scand.* **53**:37.

Villee, C. A., 1954, The intermediary metabolism of human fetal tissues, *Cold Spring Harbor Symp. Quant. Biol.* **19**:186.

Villee, C. A., Hagerman, D. D., Holmberg, N., Lind, J., and Villee, D. B., 1958, The effects of anoxia on the metabolism of human fetal tissues, *Pediatrics* **22**:953.

Volpe, J. J., and Kishimoto, Y., 1972, Fatty acid synthetase of brain: Development, influence of nutritional and hormonal factors, and comparison with liver enzyme, *J. Neurochem.* **19**:737.

Wahren, J., Felig, P., and Hagenfeldt, J., 1976, Effect of protein ingestion on splanchnic and leg metabolism in normal man and diabetes mellitus, *J. Clin. Invest.* **57**:987.

Wapnir, R. A., Tildon, J. T., and Cornblath, M., 1973, Metabolic differences in offspring of rats fed high fat and control diets, *Am. J. Physiol.* **224**:596.

Widdas, W. F., 1952, Inability of diffusion to account for placental glucose transfer in the sheep and consideration of the kinetics of a possible carrier transfer, *J. Physiol. (London)* **118**:23.

Widdowson, E. M., 1968, Growth and composition of the fetus and newborn, in: *Biology of Gestation* (N. S. Assali, ed.), Vol. 2, p. 1. Academic Press, New York.

Widdowson, E. M., 1974, Changes in body proportions and composition during growth, in: *Scientific Foundations of Paediatrics* (J. A. Davis and J. Dobbing, eds.), pp. 440–455, W. B. Saunders, Philadelphia.

Williamson, D. H., Bates, M. W., Page, M. A., and Krebs, H. A., 1971, Activities of enzymes involved in acetoacetate utilization in adult mammalian tissues, *Biochem. J.* **121**:41.

Wittels, B., and Bressler, R., 1965, Lipid metabolism in the newborn heart, *J. Clin. Invest.* **44**:1965.

Yeung, D., and Oliver, I. T., 1967, Development of gluconeogenesis in neonatal rat liver. Effect of premature delivery, *Biochem. J.* **105**:1229.

Yoshioka, T., and Roux, J. F., 1972, *In vitro* metabolism of palmitic acid in human fetal tissue, *Pediatr. Res.* **6**:675.

Zieve, F. J., and Glinsmann, W. H., 1973, Activation of glycogen synthetase and inactivation of phosphorylase kinase by the same phosphoprotein phosphatase, *Biochem. Biophys. Res. Commun.* **50**:872.

# 19

# Pre- and Perinatal Endocrinology

## PIERRE C. SIZONENKO and
## MICHEL L. AUBERT

## 1. Introduction

The understanding of prenatal and perinatal endocrinology in man still remains in a primitive stage compared to the accumulating knowledge obtained in animal studies, particularly in the rat, rabbit, sheep, and more recently in the monkey (Jost and Picon, 1970). Obvious ethical limitations restrict the investigation of perinatal endocrinology in man, even though the relevance of such studies is critical to the understanding of endocrine development after birth. However, the increasing availability of specific assays, using very small samples of biologic fluids, has accelerated the investigation of endocrine factors that represent important markers for maturation and behavior in the perinatal period.

A precise knowledge of embryology (morphogenesis, differentiation) and of the chronology of fetal growth and maturation (enzyme activation, parturition) is essential for the understanding of the ontogenesis of endocrine secretions. The concept of growth should be associated with the maturation of the organism, and more particularly the maturation of each organ, and in this context, that of endocrine glands. Each gland or organ has developmental patterns, sharply scheduled to occur at preset times. This maturation very often depends on many hormonal factors, either of maternal or of fetal origin. Frequently, it has been impossible to trace the precise origin of these factors, due to the fact that many hormones or factors can pass through the placenta or even undergo transformation within the placenta. This role of the placenta has rendered the studies of prenatal endocrinology particularly difficult in the human. Mathematical models have been proposed (Di Stephano *et al.*, 1973) which integrate the combined effects of placental transfer of hormones and synthesis and secretion of new substances or metabolites by the placenta. Many review articles have already been published and have analyzed the

*PIERRE C. SIZONENKO and MICHEL L. AUBERT* • Clinique Universitaire de Pédiatrie, Université de Genève, Genève, Switzerland.

development of endocrine function in the human placenta and fetus (Klevit, 1966; Liu, 1966; Villee, 1969; Cleveland, 1970; Diczfaluzy, 1974). The specific endocrine secretions of the placenta are discussed in Chapter 14.

Essentially, data and concepts on the endocrinology of the human fetus and the newborn infant up to one year of age will be presented; animal experiments will only by quoted occasionally, in order to provide further information when data are unavailable in man. Obstetrical, nutritional, and toxic factors will not be developed in the present chapter.

Fetal development is essentially described as a function of gestational age. In most embryologic studies reviewed here, gestational age is meant to start at conception. Some endocrinological studies, however, follow the obstetrical dating system consisting of using the 1st day of the last menstrual period (LMP) as time zero for "gestational age." In the first case, the gestation lasts 266 days, in the second, 280 days or 40 weeks. In this chapter, unless indicated, gestational age is counted from conception.

## 2. Hypothalamus and Posterior Hypophysis

### 2.1. Embryology

The brain development of the embryo begins very early in human gestation. The forebrain can be identified by 22 days of gestational age; differentiation of the telencephalon and diencephalon occurs by 34 days. The anterior pituitary appears between 28 and 35 days from the epithelium of Rathke's pouch and attaches itself to the diencephalon. The primordium of the posterior lobe appears by 49 days (Falin, 1961).

The capillaries of the human hypophyseal portal system develop between 60 and 100 days. The transmission of neurosecretory material into the hypophyseal portal system starts between 14 and 18 weeks of gestation (Raiha and Hjelt, 1957; Rinne, 1962) with the concurrent development of the hypothalamus. At this period, the hypothalamic nuclei and fibers of the supraoptic tract appear with further differentiation of pars tuberalis and the median eminence.

Three types of hypothalamic neurohormones or factors have been identified: (1) The most important ones are the hypophysiotropic-releasing and release-inhibiting factors or hormones which are peptides most likely secreted by the hypothalamus and transported to the pituitary by the hypophyseal portal system. The history and physiology of these factors has been reviewed extensively (Harris, 1955; Blackwell and Guillemin, 1973; Schally et al., 1973). (2) Aminergic neurotransmittor substances include dopamine, norepinephrine, and serotonin. (3) Oxytocin and vasopressin nonapeptides are synthetized in the paraventricular and supraoptic nuclei of the hypothalamus and transported by neurons to the posterior pituitary or neurohypophysis.

The appearance of monoamine fluorescence in the hypothalamus at 10 weeks of fetal life and in the median eminence at the 13th week probably are signs of secretory activity (Hyyppa, 1972). However, nothing is known of the metabolism of monoamines in the human fetal hypothalamus. The mean concentration of dopamine in the hypothalamus of the fetus of 11–15 weeks is above that of the hypothalamus in adults, the mean concentration of norepinephrine and serotonin are low. Nothing to date is known about melatonin concentrations in the fetus and the newborn. The relationship of catecholamines and indolamines in the maturation of the neurosecretory function of the fetal hypothalamus is not known. By the 19th–21st week, the continuity of the primary and secondary plexus of the portal system is complete. By mid-gestation, the anatomical development of the hypothalamic–pituitary complex is well advanced (Niemineva, 1950; Jost *et al.*, 1970).

## 2.2. Hypothalamic-Releasing Hormone

### 2.2.1. Thyrotropin-Releasing Hormone (TRH)

This was found in the brain of the human fetus as early as 4.5 weeks after conception (Winters *et al.*, 1974*a*). Although the bulk of TRH was found in the hypothalamus (51 ± 19 pg/mg of tissue), significant concentrations were measured in the cerebellum (8 ± 2 pg/mg) and in the cerebrum (0.9 ± 0.2). TRH is present in the amniotic fluid and in the placenta (Gibbons *et al.*, 1975). The possible role of this placental TRH on the fetal pituitary thyroid axis is as yet unknown. The content of TRH in the hypothalamus varied from 0.65 to 184 ng between 10 and 22 weeks of gestation, whereas concentrations varied from 0.2 to 218 pg/mg (Kaplan *et al.*, 1976). Similar concentrations of TRH were found in the cerebral cortex. Only 20–30% of the TRH present in the CNS of mammals is located within the hypothalamus and only 5% in the pituitary gland (Jackson and Reichlin, 1974). Preliminary data demonstrate that TRH is present in maternal (15 pg/ml) and cord blood (41.8 pg/ml) at a much higher level (Czernichow *et al.*, 1975). Twenty and 40 min after birth, plasma TRH in the newborn remains high (39.6 and 44 pg/ml, respectively). The high levels of plasma TRH are probably in relation to the sudden activation of the pituitary thyroid axis which occurs immediately after birth.

### 2.2.2. Luteinizing-Hormone-Releasing Hormone (LHRH)

This has been shown to appear early during fetal life, both by immunofluorescence (Barry and Dubois, 1974) and by direct radioimmunochemical measurement (Winters *et al.*, 1974*a*). In the human fetus, LHRH was observed in the whole brain of a 4.5-week-old fetus. The hypothalamic concentrations of LHRH varied from 4 to 65 pg/mg between 8 and 24 weeks of gestation (Winters *et al.*, 1974*a*). Little or no LHRH was detectable in extrahypothalamic areas. No significant correlation of the concentration of LHRH was observed either with sex or with gestational age between the 10th and 22nd week (Kaplan *et al.*, 1976).

### 2.2.3. Growth-Hormone-Inhibiting Hormone or Somatostatin (GIH)

This has been found widely distributed throughout the CNS and also has been detected in the pancreas and in the stomach. In the hypothalamus of human fetuses aged 10–14 weeks, the mean concentration was 10.2 pg of GIH per mg of hypothalamus and increased to 28.5 pg/mg by mid-gestation (Kaplan *et al.*, 1976). The content rose from 7.8 ng to 36.6 ng. There was a positive correlation between somatostatin concentration in the hypothalamus and gestational age. In the cerebral cortex, an appreciable amount of GIH was demonstrable (1/3–1/5 of the content present in the hypothalamus).

### 2.2.4. Other Hypothalamic Factors

The existence of other releasing factors, such as corticotrophin-releasing factor (CRF), growth-hormone-releasing factor (GRF), or inhibiting factors such as pro-

PIERRE C. SIZONENKO
and MICHEL L. AUBERT

lactin-inhibiting factor (PIF), has been either postulated or demonstrated, but no data are available in humans, probably due to the fact that these factors have not yet been isolated in their definitive form.

## 2.3. Oxytocin

The presence of secretory granules in the ganglia cells, demonstrating the secretion of neurohormones, has been shown in the hypothalamus by 19 weeks and in the posterior pituitary by 23 weeks (Benirschke and MacKay, 1953; Rinne *et al.*, 1962). The fetus has been shown to release massive amounts of oxytocin during labor and at the time of delivery (Seppälä *et al.*, 1972). The role of oxytocin at labor remains unclear. Suggestion of an action of oxytocin on the onset of labor has been inferred from the findings that anencephalic fetuses without hypothalamus usually have prolonged pregnancy (Malpas, 1933). Experimentally, the administration of oxytocin to the fetus can stimulate uterine contraction; whether the endogenous release of oxytocin is the cause or the consequence of labor remains to be proven (Honnebier and Swaab, 1973). It has been suggested that fetal corticosteroids may also play a role in the initiation of parturition. Most data were obtained in the sheep (Liggins, 1968; Anderson *et al.*, 1969).

## 2.4. Vasopressin

Vasopressin, like oxytocin, was demonstrable at mid-gestation in the hypothalamus and in the posterior pituitary of human fetuses by 19–23 weeks (Rodeck and Caesar, 1956; Raiha and Hjelt, 1957). From animal experiments Pearson *et al.*, (1975) have suggested that synthesis of vasopressin and neurophysins was maximum between the 40th and 55th day of gestation in the guinea pig. They have described a hypothalamic factor which appears on the 35th day of gestation, reaches its maximum activity by 50 days gestation, and then disappears. This factor was shown to stimulate vasopressin synthesis. At birth the vasopressin content of the pituitary gland in humans ranged from 350 to 400 $\mu$U (Heller and Zaimis, 1949). The mean plasma level of arginine-vasopressin was 21 $\mu$U/ml in newborns delivered by cesarian section, and 80 $\mu$U/ml in newborns delivered vaginally (Hoppenstein *et al.*, 1968). Stress and the relative hypoxia resulting from vaginal delivery or decreased placental circulation during labor may stimulate the release of arginine-vasopressin (Chard *et al.*, 1971). Levels of vasopressin in the blood of infants are not known. However, the ability of the newborn infant to appropriately respond to a water load or to an infusion of either isotonic dextran or hypertonic saline provides evidence of a functional capacity of the volume and osmoreceptor systems at birth. Immature renal function, rather than blunted responsiveness to vasopressin or decreased secretion of vasopressin, may be presumed responsible for the decreased concentration ability of the newborn infant (Fisher *et al.*, 1963).

## 3. Anterior Pituitary

The development of the anterior pituitary gland was first studied by histological methods. The pituitary gland of the human fetus has the capacity to synthesize and store protein hormones in the first trimester. Acidophilic cells have been demonstrated as early as the 8th week of gestation by histochemical methods (Pearse,

1953; Falin, 1961). Electron microscopic studies (Dubois, 1968) and immunofluorescent techniques (Ellis *et al.*, 1966; Pasteels *et al.*, 1972) have shown growth-hormone-secreting cells by the 9th week of fetal life. *In vitro* cultures of fetal pituitary cells have shown secretion of growth hormone as early as the 8th week of gestation (Pasteels *et al.*, 1963; Gitlin and Biasucci, 1969a; Pierson *et al.*, 1973).

## 3.1. Growth Hormone

### 3.1.1. Growth Hormone in Fetal Life

The content of fetal anterior pituitary glands and changes in serum growth hormone were studied throughout gestation either by bioassay (Parlow, 1974) or by radioimmunoassay (Franchimont *et al.*, 1970; Matsuzaki *et al.*, 1971; Kaplan *et al.*, 1972; Grumbach and Kaplan, 1973). At 68 days of gestation, content of GH was 0.04 $\mu$g/pituitary gland; it increased to 225.9 $\pm$ 40.5 $\mu$g at 25–29 weeks of gestation and further rose to 675.2 $\pm$ 112.3 $\mu$g at 35–40 weeks (Figure 1). Growth hormone isolated from fetal pituitary gland or from *in vitro* culture of fetal pituitary cells is biochemically similar to GH extracted from adult pituitaries (Kaplan *et al.*, 1972). In serum, fetal GH concentration rises steadily from 14.5 ng/ml at 10 weeks of gestational age to peak values of 131.9 $\pm$ 21.9 ng/ml between the 20th and 24th week of gestation. Serum GH then decreases markedly, ranges about 53.5 $\pm$ 10.8 ng/ml between the 25th and 29th weeks, and later reaches, at term, a mean concentration of 33.5 $\pm$ 4.2 ng/ml (Kaplan *et al.*, 1976). Fetal GH secretion can be stimulated *in vivo* by stress, anoxia, acidosis following hysterotomy at mid-gestation (Turner *et al.*, 1971) or at delivery (Aubert *et al.*, 1971; Turner *et al.*, 1973). Maternal or fetal growth hormone does not pass the placenta (Gitlin *et al.*, 1965a; King *et al.*, 1971b).

Available evidence suggests that neither maternal or fetal human growth hormone is essential. The birth length of decapitated fetuses (Jost, 1954), apituitary fetuses, anencephalic fetuses, or children with idiopathic hypopituitarism is usually within the normal range (Blizzard and Alberts, 1956; Brewer, 1957; Reid, 1960). Children born to women with isolated growth-hormone deficiency (Tyson *et al.*, 1970) or to mothers who underwent hypophysectomy during gestation (Little *et al.*, 1958), do not show evidence of growth retardation. Similar data were obtained in the rhesus monkey (Smith, 1954; Chez *et al.*, 1970). The role of other factors such as insulin or somatomedin on fetal skeletal growth remains speculative. Somatomedin is probably present in the fetus. Levels in cord blood at birth were reported to be low (Tato *et al.*, 1975), although they were higher than those in the maternal circulation.

### 3.1.2. Growth Hormone in Newborn Infants

At birth the plasma concentration of growth hormone is higher in the fetal than in the maternal circulation. Levels range from 10 to 40 ng/ml in the cord blood (Cornblath *et al.*, 1965; Kaplan and Grumbach, 1965; Laron and Pertzelan, 1969; Yen *et al.*, 1967). After an initial increase with peak values at 48 hr of life, plasma levels of GH decrease progressively in normal full-term infants during the first week of life and reach adult levels at one month. The pattern is different in premature infants in whom plasma growth hormone remains high much longer during the first 2–8 weeks of life (Cornblath *et al.*, 1965).

PIERRE C. SIZONENKO
and MICHEL L. AUBERT

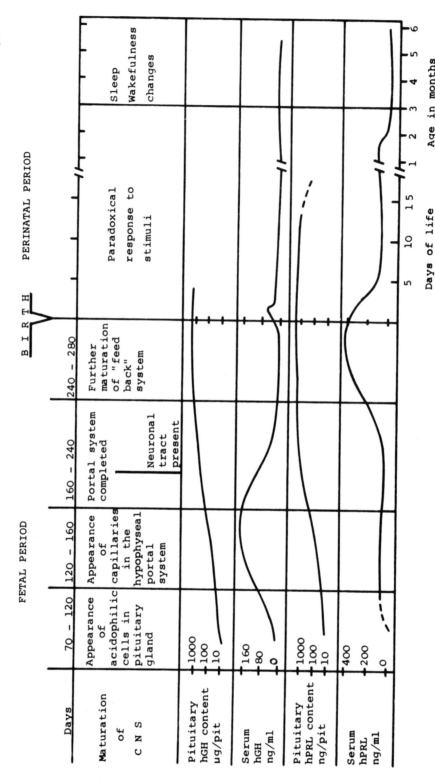

Fig. 1. Pituitary content and serum concentration of growth hormone (hGH) and prolactin (hPRL) in relation to the maturation of the central nervous system (CNS) during fetal and perinatal life. (Adapted from Kaplan *et al.*, 1972; Aubert *et al.*, 1975.)

Evidence of a negative correlation between plasma growth-hormone levels and either birth weight or fetal age, has been reported (Cornblath *et al.,* 1965; Cramer *et al.,* 1971; Poonai *et al.,* 1975), although the occurrence of such a relationship has been questioned (Humbert and Gotlin, 1971). Fetal hyperglycemia, resulting from glucose infusion to the mother during labor, induces a paradoxical increase in neonatal plasma growth-hormone levels (Wolf *et al.,* 1970).

Stimulation of the release of growth hormone by the pituitary gland has been studied in relation to fetal and postnatal age. A paradoxical rise of growth hormone in normal term infants has been obtained after infusion or oral administration of glucose (Cornblath *et al.,* 1965; Milner and Wright, 1966; Pildes *et al.,* 1969; Cole *et al.,* 1970; Falorni *et al.,* 1972). Similar rises of growth hormone were observed in premature infants (Milner *et al.,* 1972*b*; Ponte *et al.,* 1972*b*) or in newborns small for gestational age (Milner *et al.,* 1972*c*). A paradoxical (slight) decrease of plasma growth hormone in response to stress (heel pricks) was observed during the first days of life (Milner and Wright, 1967; Stubbe and Wolf, 1971). Glucagon administration induced a very marked rise of plasma growth hormone during the newborn period (Figure 2) when compared to normal children (Milner and Wright, 1967; Sizonenko *et al.,* 1972). Normal responses of growth-hormone secretion to insulin-induced hypoglycemia (Cornblath *et al.,* 1965) and to arginine (Ponte *et al.,* 1972*a*) were observed in full-term and premature infants. It is probable that during the first postnatal weeks, a further maturation of the regulatory mechanisms of growth-hormone secretion occurs, with disappearance of paradoxical responses to stimuli. By 3 months of age, there is an initiation of sleep-associated growth-hormone release, which is absent in very young infants (Finkelstein *et al.,* 1971; Shaywitz *et al.,* 1971). Secretion of growth hormone is affected by a variety of factors, such as metabolic substrates, glucose, amino acids, and free fatty acids. Alpha- and beta-adrenegic receptors, as well as dopaminergic and serotoninergic mechanisms, are involved in the modulation of the secretion of growth hormone. Differential maturation of neurotransmitter mechanisms, in addition to other maturation mechanisms, are likely to play an important role in the regulation of growth-hormone secretion during pre- and perinatal life (Grumbach and Kaplan, 1974).

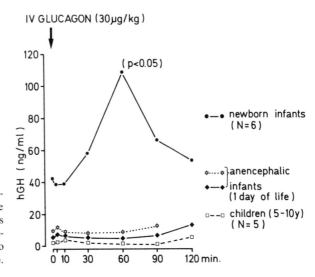

Fig. 2. Effect of intravenous glucagon on serum growth hormone (hGH) in newborn infants (less than 72 hr of life), and anencephalic infants in comparison to normal children (aged 5/10 years).

## 3.2. Prolactin

*PIERRE C. SIZONENKO
and MICHEL L. AUBERT*

Biologically active prolactin (PRL) has been demonstrated in the fetal pituitary gland at mid-gestation by *in vitro* culture techniques (Pasteels *et al.*, 1963) and by direct extraction of pituitary glands (Levina, 1968). Recently, Siler-Khodr *et al.*, (1974) demonstrated the presence of PRL in the primordium of the pituitary gland as early as 5 weeks. Using immunofluorescence techniques, Pasteels *et al.*, (1974) detected lactotrope cells in the pituitary gland of a 4-month-old fetus. By radioimmunochemical assay, human prolactin was detected by 68 days of gestation (Aubert *et al.*, 1975). Prolactin content of the anterior pituitary gland was above 2.0 ng between 10 and 15 weeks gestation. The mean content of prolactin increased sharply from $14.8 \pm 4.6$ ng at 15–19 weeks to $405 \pm 142$ ng at 20–24 weeks and $542 \pm 204$ ng at 25–29 weeks of gestation. At term, the mean content was $2039 \pm 459$ ng and the mean concentration $15.9 \pm 2.4$ ng/mg. In fetal serum, the mean concentration of prolactin between 12 and 24 weeks was $19.5 \pm 2.5$ ng/ml. By 26 weeks, fetal serum prolactin increased sharply and reached a mean level of 268.3 ng/ml at late gestation. At delivery, the mean plasma concentration of prolactin was $167.8 \pm 14.2$ ng/ml in the umbilical vein and $111.8 \pm 12.3$ ng/ml in the maternal peripheral vein. Similar levels were reported at birth (Tyson *et al.*, 1972; Guyda and Friesen, 1973). During the perinatal period, mean prolactin content of pituitary glands aged 1–2 months was $5429 \pm 2275$ ng (Aubert *et al.*, 1975). Serum levels decreased rapidly during the first five days of life but then still remained above normal levels until about 6 weeks of life (Guyda and Friesen, 1973). The patterns of secretion of prolactin during pregnancy are similar in the fetus and the mother with maximum increase during late gestation and at term (Figure 1). No placental transfer of prolactin has been reported and evidence against such a transfer in the rhesus monkey has been documented (Josimovich *et al.*, 1974). Levels of prolactin comparable to those of normal newborn infants have been reported in the anencephalic infant (Hayek *et al.*, 1973; Aubert *et al.*, 1975), suggesting that a specific hypothalamic releasing factor is not necessary for synthesis and release of prolactin. It has been postulated that the steady increase of prolactin secretion in both maternal and fetal circulation is in direct correlation with the increase of circulating estrogens during gestation (Tyson *et al.*, 1972; Aubert *et al.*, 1975).

### 3.3. ACTH

Adrenocorticotrophic hormone (ACTH) has been demonstrated by immunofluorescence techniques in human pituitaries during the 2nd trimester of gestation, before the 14th week (Dubois *et al.*, 1973). Transplacental passage of ACTH is likely to be minimal (Miyakawa *et al.*, 1974). Fetal plasma ACTH levels have been detected by 12 weeks of gestational age (Allen *et al.*, 1973; Winters *et al.*, 1974*b*). Elevated fetal plasma levels of ACTH ($249 \pm 65.7$ pg/ml) are present by 12–14 weeks, and there is a decrease in the concentration of ACTH by 35–40 weeks. In cord blood, plasma ACTH level is $143 \pm 7.0$ pg/ml at term. The same ACTH plasma concentrations were observed in neonates born vaginally after spontaneous or oxytocin-induced labor, or after cesarian section before or after spontaneous labor. After birth, plasma ACTH remains elevated during the first week ($120 \pm 8$ pg/ml) compared to normal older children ($43 \pm 3.7$ pg/ml). These results suggest that plasma ACTH is elevated during the critical period of perinatal life, very likely before normal nycthemeral cortisol cycles appear. Maternal cortisol may in part modulate fetal ACTH secretion, as the administration of ACTH to the mother decreased fetal ACTH significantly. This is a consequence of the transplacental passage of the corticosteroids produced by the maternal adrenal glands (Miyakawa et al., 1974). Studies with human anencephalic fetuses (Benirschke, 1956; Mosier, 1956; Allen et al., 1974) and experimental data (Jost, 1966a) have provided evidence for the critical role of fetal ACTH in the normal growth of the adrenal cortex, both of its fetal zone and neocortex. However, the mechanism of the growth of the fetal zone during prenatal life and its involution during the perinatal period still remain unexplained. Additional trophic factor or factors are very likely necessary. There is some evidence that human chorionic gonadotropin (hCG) may stimulate the development of the fetal adrenal cortex in the presence of fetal ACTH (Johannisson, 1968, Araï et al., 1972). The existence of a human chorionic corticotropin (hCC), which has been debated for many years, has not clearly been established. Recently, Genazzani et al., (1975) found some evidence of the presence of hCC in the human placenta and reinforced the belief of an active role of the placenta in stimulating the fetal adrenal zone.

Biologically active $\beta$-melanocyte-stimulating hormone (MSH) has been reported in the pituitary gland as early as the 10th–11th week of gestation (Levina, 1968; Kastin et al., 1968; Dubois et al., 1975). There was no significant rise in pituitary concentration of bioactive $\beta$-MSH through the mid-gestational period. Recently, an elevation of $\beta$-MSH was found in samples of amniotic fluid throughout gestation (Ances and Pomerantz, 1974).

### 3.4. Thyrotropin (TSH)

#### 3.4.1. TSH During Fetal Life

TSH activity has been demonstrated in fetuses from the age of 12 weeks after conception (Levina, 1968; Gitlin and Biasucci, 1969b; Fisher et al., 1970; Fukuchi et al., 1970; Greenberg et al., 1970). Plasma TSH has been detected as early as 11 weeks of gestation, coincidental with the onset of iodine uptake by the fetal thyroid gland and the synthesis of iodothyronines. Plasma TSH levels increased from about 4.0 $\mu$U/ml at 18–20 weeks to about 10 $\mu$U/ml at term. A good correlation was found between plasma TSH and plasma free and total thyroxine (Fisher et al., 1970, Greenberg et al., 1970). However, a relative triiodothyronine deficiency is present at birth and will be discussed later (Fisher et al., 1973). In addition to fetal pituitary TSH, two placental glycoprotein hormones with TSH biological activity have been described, human chorionic thyrotropin (Hennen et al., 1969; Herschman and Starnes 1969) and human chorionic gonadotropin (Nisula and Ketelslegers, 1974). Pituitary TSH does not cross the placenta (Dussault et al., 1972).

#### 3.4.2. TSH in Newborn Infants

At birth, plasma TSH in cord blood is more elevated (9.5 $\mu$U/ml) than in maternal sera (3.9 $\mu$U/ml) and increases very rapidly at 10 min (60 $\mu$U/ml) and at 30 min (86 $\mu$U/ml). Plasma TSH plateaus during 3–4 hr and decreases to 13 $\mu$U/ml at 48 hr of life (Fisher and Odell, 1969; Czernichow et al., 1971). This rise of plasma TSH is accompanied by a state of chemical T3 thyrotoxicosis (Figure 3). The

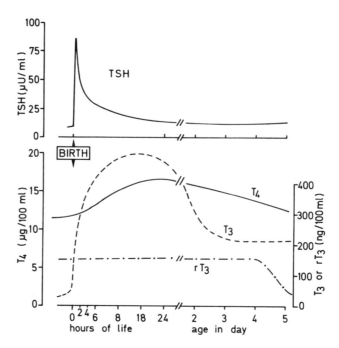

Fig. 3. Plasma concentration of thyrotropin (TSH), thyroxine (T4), triiodothyronine (T3) and reverse triiodothyronine (rT3) in newborn infants during the first five days of life. (Adapted from Fisher and Odell, 1969, Erenberg *et al.*, 1974; Chopra *et al.*, 1975.)

mechanism of the neonatal TSH surge is not entirely clear. Cooling of the newborn infant augments serum TSH; however, prevention of the cooling at delivery does not prevent the rise of TSH (Fekete *et al.*, 1972; Fisher, 1975).

### 3.5. Gonadotropins

Both, follicle-stimulating hormone (FSH) and luteinizing hormone (LH) are synthetized by the fetal pituitary under the influence of luteinizing-hormone-releasing hormone (LHRH). Histochemically, stains of FSH and LH have been detected in pituitary cells as early as 10 weeks of gestation (Pearse, 1953; Falin, 1961; Mitskevich and Levina, 1965; Dubois and Dubois, 1974). Biological activities were found at the same period of gestation (Levina, 1970; Parlow, 1974). Furthermore, tissue culture of 10- to 20-week-old fetal pituitary cells resulted in a release of gonadotropins which could be modulated either by LHRH or by estradiol added to the medium (Gitlin and Biasucci, 1969a; Gailaini *et al.*, 1970; Groom *et al.*, 1971; Siler *et al.*, 1972; Groom and Boyns, 1973). Finally, radioimmunochemical methods have confirmed the presence in the fetus of gonadotropic materials both in the pituitary gland and in the serum (Levina, 1972; Grumbach and Kaplan, 1973).

### 3.5.1. Prenatal Period

Immunoreactive LH was detected in pituitary glands as early as 10 weeks of gestation. In both sexes, the pituitary content of LH showed a sharp rise between 10–14 and 15–19 weeks of gestation. After 20 weeks, there was little change. Pituitary LH concentrations peaked between 15 and 24 weeks, earlier in female than male fetuses, and then decreased from 25 weeks to term (Figures 4 and 5). LH concentration was significantly higher in the female than male pituitary gland between 125 and 160 days of gestation (Grumbach and Kaplan, 1974). Immunoreactive FSH was also present in the pituitary gland by 10 weeks. There was a striking increase in the pituitary content of FSH between 10–14 weeks and 25–29 weeks in fetuses of both sexes. From 29 weeks until term, FSH content of the pituitary gland remained constant. In both sexes, the FSH pituitary concentration increased from about 10 weeks on, peaked at around 24 weeks of gestation, and then decreased significantly until term (Figures 4 and 5). In the female, pituitary content and concentration of FSH was strikingly greater than in the male throughout gestation. In the fetal serum, FSH was detected by 11 weeks (Grumbach and Kaplan, 1974). From a mean plasma concentration of 6.6 ng/ml at 12–14 weeks of gestation, there was a rise to mean peak FSH levels of 16–18 ng/ml at 15–24 weeks, followed by a sharp decline to 1.7–2.0 ng/ml from 25 weeks to term. In the umbilical cord serum, the mean concentration was $1.8 \pm 0.2$ ng/ml and in maternal serum at delivery, $1.2 \pm 0.1$ ng/ml. The concentration of serum FSH was generally higher in the female than in the male at mid-gestation (Levina, 1972; Winter and Faiman, 1972; Reyes et al., 1974; Grumbach and Kaplan, 1974).

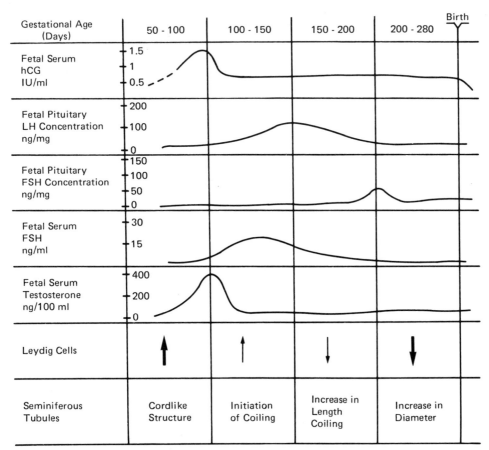

Fig. 4. Pituitary and serum concentrations of human chorionic gonadotropin (hCG), luteinizing hormone (LH), follicle-stimulating hormone (FSH), and testosterone in relation to the development of Leydig cells and seminiferous tubules during fetal life. (Adapted from Grumbach and Kaplan, 1973; Kaplan et al., 1976.)

PIERRE C. SIZONENKO
and MICHEL L. AUBERT

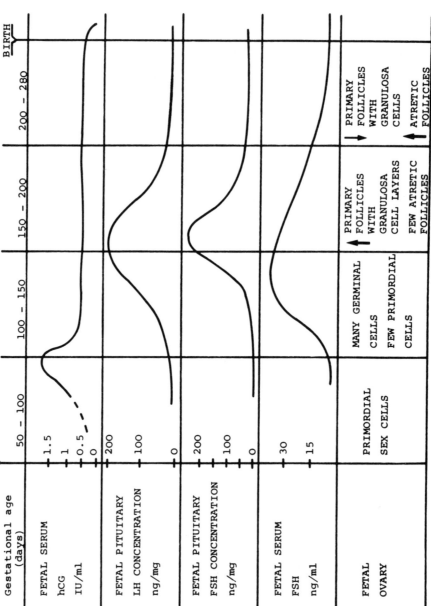

Fig. 5. Pituitary and serum concentrations of human chorionic gonadotropin (hCG), luteinizing hormone (LH), and follicle-stimulating hormone (FSH) in relation to the development of the ovary during fetal life. (Adapted from Grumbach and Kaplan, 1973; Kaplan *et al.*, 1976.)

The study of serum LH during gestation has been difficult because until recently radioimmunoassays could not discriminate between LH and hCG. Since the development of specific radioimmunoassays for LH and hCG using antisera generated against the respective $\beta$ subunit, specific determinations of LH or hCG have been possible. In addition, a qualitative analysis of gonadotropin by separation on gel chromatography of the different forms ($\alpha$ and $\beta$ subunits or intact hormones) present in biological specimens has provided a better understanding of gonadotropin secretion during gestation (Hagen and McNeilly, 1975a; Kaplan et al., 1976). Serum hCG in fetal and maternal circulation follows similar secretory patterns with the important difference that maternal levels are approximately 10–20 times higher throughout gestation. Following a peak by 90–120 days, fetal hCG decreases to a mean concentration of $268.5 \pm 16.1$ mU/ml. The pattern of change of hCG in the fetal circulation is related to that reported for serum testosterone in the male fetus. Abramovich and Rowe (1973), Reyes et al., (1974), and Diez d'Aux and Murphy (1974) have documented elevated levels of testosterone between 11 and 18 weeks in the male fetus with a decrease between 17 and 24 weeks of gestation (Figure 4).

Upon gel filtration, large amounts of free $\alpha$ subunits of glycoprotein hormones have been found in fetal plasma or pituitary extracts. Fetal pituitary glands studied between 17 and 31 weeks contain an excess of $\alpha$-glycoprotein subunit in comparison to intact hormones. Intact LH is present in a significantly higher concentration in the female than male pituitary gland. Similarly, intact FSH is higher in the female than in the male fetus. In serum, a similar pattern is observed, again with predominance of $\alpha$ subunits which represent a mixture of variable proportions of $\alpha$-LH, $\alpha$-hCG, $\alpha$-FSH, or $\alpha$-TSH subunits. Intact LH and FSH are in higher concentration in female than male serum.

At term, $\alpha$-glycoprotein subunits predominate in cord sera with some intact hCG, low intact LH, and some intact FSH demonstrable. In maternal sera at delivery, $\alpha$-glycoprotein subunits are high as well, with the presence of a substantial concentration of intact hCG. In the umbilical cord, no arteriovenous difference in any of these hormones or subunits were observed, and no correlation could be made between levels of hormones in maternal and fetal circulation (Hagen and McNeilly, 1975b).

The predominance in the serum and pituitary gland of the human fetus of the free $\alpha$ subunit component of glycoprotein hormones is in agreement with *"in vitro"* studies in which $\alpha$ subunits were shown to be released predominantly by fetal pituitary cells in tissue culture with little of the $\beta$ subunit or intact hormone secreted in absence of hypothalamic stimulation (Franchimont and Pasteels, 1972).

### 3.5.2. Perinatal Period

In the newborn infant, no sex difference has been noted in the plasma concentration of hCG by some authors (Faiman et al., 1968; Geiger, 1973); others, measuring both hCG and LH, have found a higher mean concentration of LH–hCG in the cord serum of male infants (Brody and Carlström, 1965; Penny et al., 1974).

At birth, plasma FSH is higher in female than in male infants, and this sex difference remains until the end of the second year of life (Faiman and Winter, 1971, 1974; Forest et al., 1974b; Ryle et al., 1975). Serum LH is found to be low during the first days of life and increases during the first month, more in the male than in the female (Faiman et al., 1974; Forest et al., 1974b). Further decreases of LH occur during the first two years of life. During the first months, serum LH and FSH are also above prepubertal levels in male infants.

PIERRE C. SIZONENKO
and MICHEL L. AUBERT

Kaplan *et al.* (1976) interpreted the high serum values observed in the fetal circulation at mid-gestation to be due either to autonomous secretion of FSH and LH or to the relatively unrestrained secretion of LHRH into the hypothalamic pituitary portal system. As fetal development advances, the inhibitory feedback mechanism matures and the hypothalamus secretes less LHRH, which results in the decreased secretion of fetal FSH and LH. Progressive acquisition of increasing sensitivity of the hypothalamus to sex steroids is very likely in the fetus. However, the hypothalamic regulatory mechanism may not be fully developed at birth. Recent data (Forest *et al.*, 1974*a*; Forest and Cathiard, 1975) would sustain the notion that in early infancy the regulatory mechanism of gonadotropin secretion has not yet attained the high degree of sensitivity to sex steroid feedback that is reached after two years of life during the prepubertal period (Grumbach *et al.*, 1974).

The relation of fetal secretion of pituitary gonadotropins and fetal sex differentiation has not been established in man, as compared to animal models (Jost, 1965). Anencephalic and apituitary male fetuses have hypoplastic male external genitalia and hypoplastic undescended testes (Blizzard and Alberts, 1956; Reid, 1960; Zondek and Zondek, 1974). It is likely that the hCG secreted by the placenta and present in the fetal serum is sufficient to stimulate testosterone secretion by the fetal testis and to induce male sex differentiation.

## 4. Thyroid Gland

### 4.1. Ontogenesis of the Thyroid Gland

The mammalian thyroid gland evolves from an outpouching of the ventrobuccal endoderm, visible by 16–17 days of gestation, which forms the midline part. The ultimobranchial portions of the fourth pharyngeal pouches contribute also to the midline main gland. The gland progresses from a bilobed hollow vesicle (precolloid phase) to a solid mass of tissue which descends along the thyroglossal tract to its definitive location in the anterior lower neck (Fisher and Dussault, 1974). Between 73 and 80 days, intracellular colloid appears (early colloid phase) with iodide uptake, oxidation of iodide, its incorporation into tyrosine, and synthesis of thyroid hormones (Shepard, 1968). From 80 days to term, follicular spaces appear and development of capacity of the thyroid gland to concentrate iodine and synthesize hormone progressively increases (Shepard, 1967). Growth of the thyroid gland does not seem to be dependent on fetal TSH. The gland will develop, as observed histologically, and store colloid as well as small amounts of hormone in the absence of the pituitary gland. Thyroglobulin synthesis also occurs in the thyroid gland by the 29th day of gestation (Gitlin and Biasucci, 1969*b*) even in the absence of fetal pituitary TSH (Jost, 1966*b*). Jost has shown in animal experiments that, in contrast to man, concentration of radioiodine and synthesis and release of hormone by the fetal thyroid gland are TSH dependent.

In the human fetus, thyroxine was observed as early as 78 days (Greenberg *et al.*, 1970). Serum total and free thyroxine increase rapidly from the 13th–24th weeks to the 25th–34th weeks, with a further progressive rise from the 34th week to term (Fisher *et al.*, 1969, 1970, 1973; Greenberg *et al.*, 1970; Fisher, 1975). A similar rise of total T3 and free T3 occurs progressively. However, levels of T3 and free T3 are lower in late gestation fetuses than in normal children. In the human fetus there is a relative T3 deficiency (Fisher *et al.*, 1973). The concentrations of thyroid-binding globulin (TBG) and thyroid-binding prealbumin in fetal serum increase from the 14th to the 24th week of gestation and later plateau until term (Greenberg *et al.*, 1970). In amniotic fluid, T4 is present as early as the 15th week and seems to increase until term with a wide scatter (Sack *et al.*, 1975*a*). At term the concentration of total T4 in the amniotic fluid is less than in fetal or maternal serum. Most of the T4 in amniotic fluid is protein bound, presumably to TBG, but the mean free T4 concentration in amniotic fluid at term is significantly greater than the levels in fetal or maternal serum. T3 was not measurable in amniotic fluid. There is no correlation between amniotic fluid T4 level and fetal serum T4 concentration.

## 4.2. Placental Transfer of Thyroid Hormones

A number of studies in man indicate that placental transfer of thyroid hormone is limited (Grumbach and Werner, 1956; Myant, 1958, Kearns and Hutson, 1963; Raiti *et al.*, 1967; Dussault *et al.*, 1969). Less than 1% of a large dose of T4 given to mothers in labor is transferred to the fetal circulation. Large doses of T3 (50–300 $\mu$g daily) given to women near term slightly increase fetal serum T3 and provide very minimal decrease of fetal serum T4 concentration. There is no correlation between maternal and fetal serum concentrations of total and free T3 or T4 or TSH. There is a marked maternal-to-fetal gradient of T3 and free T3 throughout gestation (Erenberg *et al.*, 1974). Prior to 20 weeks, there is a maternal-to-fetal gradient of T4 and free T4. Near term, the gradient of T4 and free T4 tends to favor the fetus (Greenberg *et al.*, 1970; Erenberg *et al.*, 1974). The placenta is able to concentrate iodine in the fetal compartment. Administration of iodine at high doses or radioactive iodine is a danger for the very active fetal thyroid gland. Similarly, the long-acting thyroid stimulator (LATS) which is present in some cases of maternal thyrotoxicosis crosses the placenta (McKenzie, 1964) and may cause neonatal Grave's disease. Antithyroid drug administration for maternal thyrotoxicosis can cross the placenta and cause a goiter in the fetus.

## 4.3. Perinatal Secretion of Thyroid Hormones

At birth, as already mentioned, cord blood T4 and free T4 are similar or slightly lower than maternal concentrations. T3 and free T3 are lower in the cord blood than in the maternal serum (Robin *et al.*, 1970; Lieblich and Utiger, 1973; Montalvo *et al.*, 1973). Reverse T3 (rT3) is much higher in cord blood serum than in normal adult serum (Chopra *et al.*, 1975). At birth, the newborn infant switches from a situation of chemical T3 deficiency to a transient state of chemical T3 thyrotoxicosis (Figure 3). Immediately after delivery, in response to the postnatal TSH surge, mean concentrations of total T4, free T4, total T3, and free T3 increase dramatically and remain elevated during the first 24–28 hr after birth (Czernichow *et al.*, 1971; Abuid *et al.*, 1973; Montalvo *et al.*, 1973; Erenberg *et al.*, 1974; Similä *et al.*, 1975). TBG remains unchanged during this period, with levels of 5.0 mg/100 ml. This period of neonatal thyrotoxicosis is transient, with its maximum at 24–72 hr, and disappears by 3–4 weeks. The mechanism of the neonatal TSH surge and its thyroidal response is not well understood. Reverse T3 levels, initially high in the cord blood, remain elevated for 3–5 days and then gradually fall by 10–14 days to levels comparable to those measured in adults. The conversion of T4 to rT3 and T3 is not TSH dependent. The patterns observed suggest that peripheral enzymatic pathways for thyroxine metabolism rapidly mature in the newborn infant, probably via autonomic activation of the thyroxine hydroxylase activity (Fisher, 1975; Sack *et al.*, 1975*b*). By one to two weeks, the pattern of serum thyroid hormones is similar to that of adults.

Thyroid hormones very likely play an important role in the development of brain enzymatic systems and myelinization. There is a critical period of thyroid hormone dependency for the development of the CNS in the human. Therefore, a screening program for early diagnosis of congenital hypothyroidism is of interest (Klein *et al.*, 1974; Dussault *et al.*, 1975).

## 5. Adrenal Cortex

The fetal adrenal cortex plays a very important role not only for its direct action on the fetus but also by supplying metabolites which will be further utilized by the placenta, thus constituting a very important element of the fetoplacental unit (Cathro, 1969; Villee, 1969, Diczfaluzy, 1969, 1974; Siiteri, 1974).

### 5.1. Fetal Adrenal Cortex

#### 5.1.1. Embryology

The fetal adrenal cortex is of mesodermal origin. At the junction of the mesonephros and dorsal mesentery, proliferation of cortical cells form invasive cords which separate from the coelomic epithelium and form two large masses on either side of the aorta. Adjacent to these cortical cells are the medullary crest cells which migrated from the neural crest. By 6–7 weeks of gestation the human fetal adrenal cortex is divided into a thin outer zone and a large inner (fetal) zone (Johannisson, 1968). The outer zone is thought to originate from a second downgrowth of cells from the coelomic epithelium which envelops the original cortical cells and their invading medullary cells (Crowder, 1957). The fetal zone of the adrenal cortex represents 80% of the gland and is very active in steroid metabolism. During the first trimester of gestation, cells of the fetal zone become increasingly complex and presumably functionally mature.

#### 5.1.2. Fetal Zone

The fetal or transitional zone possesses little or no $\Delta$5-3$\beta$-hydroxysteroid dehydrogenase activity (Figure 6) and thus is virtually incapable of converting 5-3$\beta$-hydroxysteroids such as cholesterol or $\Delta$5-pregnenolone into $\Delta$4-3-ketosteroids (Goldman *et al.*, 1966; Villee, 1972). These findings are in contradiction with Niemi and Baillie's data (1965). The fetal zone is very sensitive to ACTH and sensitive to hCG only at high doses (Johannisson, 1968). It has been suggested that the increased levels of hCG present during the first trimester of pregnancy stimulate adrenal growth. Maintenance of the adrenal cortex after the fourth to fifth months of gestation (when hCG production decreases) would depend on other trophic factors of hypothalamo–pituitary origin such as fetal pituitary ACTH (Stark *et al.*, 1965). The adrenal glands of anencephalic fetuses appear to develop normally during the first trimester of gestation, but after the 20th week, the fetal zone undergoes involution (Benirschke, 1956; Tuchman-Duplessis and Mercier-Parrot, 1963). From many *in vitro* culture studies of human fetal adrenal, it may be concluded that only

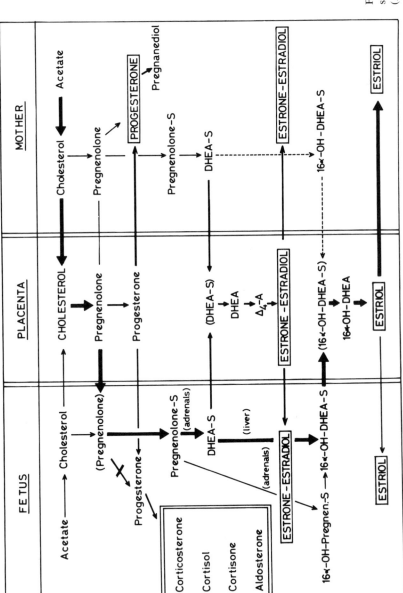

Fig. 6. Steroid pathways and interrelations of steroid metabolism in the feto-placental unit. (Adapted from Diczfaluzy, 1974; Siiteri, 1974.)

cells of the adult zone of the fetal adrenal cortex are able to synthesize glucocorti-coids (Villee, 1972). The inner fetal zone produces primarily dehydroepiandroster-one (DHEA) and its sulfate (Villee and Loring, 1969). From cholesterol and Δ5-pregnenolone, human fetal adrenal tissue is able to make DHEA, DHEA sulfate, Δ5-pregnenolone sulfate, and 17-hydroxy-Δ5-pregnenolone sulfate. Sulfokinase activity is higher in the fetal adrenal gland than in the fetal liver (Solomon *et al.*, 1967). Although the fetal adrenal cortex is relatively deficient in Δ5-3β-hydroxyste-roid dehydrogenase, the fetal adrenal cortex is able to form cortisol from progester-one which is abundant. The 17-, 21-, and 11β-hydroxylases are present and active *in vitro* after 10 weeks of gestation (Villee *et al.*, 1961). In addition, the fetal adrenal cortex has a very active 16α-hydroxylase activity. Progesterone and Δ5-pregneno-lone are very actively transformed into 16α-hydroxylated compounds (Villee *et al.*, 1961). Similarly the fetal liver is capable of converting DHEA-S into 16α-hydroxy-DHEA-S (Diczfaluzy, 1969), which are precursors of estriol (Figure 6).

### 5.1.3. Outer Zone

The outer zone is responsible for the synthesis of cortisol which is active after 8–10 weeks of gestation. During gestation cortisol content of the adrenal gland increases and this steroid is probably mostly synthesized by transformation of placental progesterone (Villee, 1969).

### 5.1.4. Placental Transfer of Cortisol

In man, placental transfer of cortisol from mother to fetus has been demon-strated. In cases of excessive maternal cortisol production, levels of fetal ACTH are decreased and may cause adrenal insufficiency in the newborn infant (Kreines and De Vaux, 1971). Similarly, with the administration of cortisol or ACTH to women before delivery, a decrease of 16α-hydroxy-DHEA and its sulfate in the cord plasma is observed, suggesting placental transfer of maternal cortisol and decrease of metabolite production from the fetal adrenal cortex (Simmer *et al.*, 1975). Plasma concentrations of cortisol in both fetal and maternal circulations are dependent on cortisol-binding globulins (CBG). Plasma levels and binding capacities of CBG are markedly increased during pregnancy on the maternal side (Migeon *et al.*, 1956). The importance of fetal cortisol for the development of surfactant factor in the newborn infant and for maturation of pulmonary alveolar lining has been suggested (Spellacy *et al.*, 1973). In the amniotic fluid, cortisol concentration increases regularly with gestational age (Figure 7) as well as 16α-OH-DHEA in its free or sulfated form (Murphy *et al.*, 1975; Lacourt *et al.*, 1975; Schindler and Siiteri, 1968).

### 5.2. Fetoplacental Unit

The concept of a fetoplacental unit responsible for the secretion of steroids during pregnancy was suggested a decade ago when it became apparent that both the placenta and fetus lack certain enzymatic functions which are essential for steroidogenesis. It then became apparent that the enzymes lacking in the placenta are present in the fetus and *vice versa*, and that, by integration of fetal and placental functions, the so-called *fetoplacental unit* can elaborate most, if not all, active

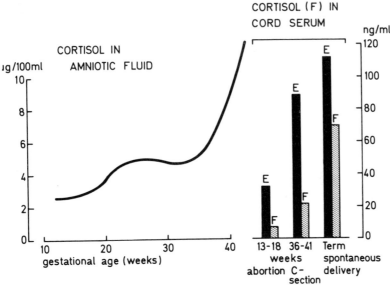

Fig. 7. Cortisol concentration in amniotic fluid; on the right, cortisone and cortisol concentrations in cord blood of aborted fetuses, and of newborn infants after caesarian section or spontaneous delivery. (Adapted from Murphy and Diez d'Aux, 1972; Murphy *et al.*, 1975; Lacourt *et al.*, 1975.)

steroid hormones (Diczfaluzy, 1969). Many review articles and books on the fetoplacental unit have already been mentioned, and the reader should refer to Chapter 14.

A schematic diagram of the fetoplacental unit is presented in Figure 6. The fetal liver, adrenals, and testes were shown to be able to incorporate acetate and synthesize cholesterol. However, the placenta cannot produce cholesterol from acetate; therefore, all steroidogenic reactions which occur in the placenta involve circulating cholesterol mainly of maternal but also of fetal origin. The human placenta is able to convert cholesterol to pregnenolone and then progesterone, whereas little, if any, pregnenolone is converted into progesterone by the human fetus. The fetal adrenal cortex is able to convert placental progesterone to a variety of hydroxylated compounds in positions 21, 11$\beta$, 17$\alpha$, and 18. Thus, the fetal adrenal gland can synthesize all biologically important corticosteroids, including deoxycorticosterone, corticosterone, cortisol, aldosterone, 16$\alpha$-hydroxyprogesterone and 11$\beta$-hydroxy-$\Delta$4-androstenedione (Diczfaluzy, 1969). However, the main function of the fetal adrenal cortex is devoted to the synthesis from pregnenolone sulfate, via 17-hydroxypregnenolone sulfate, of DHEA sulfate and 16$\alpha$-hydroxy-DHEA sulfate. The major portion of DHEA sulfate and 16$\alpha$-OH-DHEA sulfate moves to the placenta, in which aromatization occurs with the formation of 17$\beta$-estradiol and estriol, respectively (Ryan, 1959; Magendantz and Ryan, 1964; Siiteri and MacDonald, 1966). DHEA sulfate (half maternal and half fetal in origin) is the most important precursor of placental estrone and 17$\beta$-estradiol (Bolté *et al.,* 1964; MacDonald and Siiteri, 1965). Hence estriol synthesis is a complex process which involves functions of both fetal and maternal adrenal cortices, fetal and maternal liver, and sulfate cleavage and aromatization by the placenta. The assessment of

PIERRE C. SIZONENKO
and MICHEL L. AUBERT

fetal viability by urinary estriol (Greene and Toughstone, 1963) will depend on these different functions, mainly secretions of DHEA sulfate and 16α-hydroxylation (Diczfaluzy, 1974).

### 5.3. Fetal Cortisol

A rise of fetal corticosteroids has been implicated in the fetal maturation and onset of labor (Smith and Shearman, 1972; Murphy and Diez d'Aux, 1972; Murphy, 1973a). Clinical experience has suggested that fetal adrenal insufficiency, as observed in anencephalic fetuses, may be the cause of prolonged pregnancy (Anderson et al., 1969; Roberts and Cawdery, 1970; Nwosu et al., 1975). Conversely, hyperplasia of the fetal adrenal in the human (Anderson et al., 1971) or infusion of ACTH or cortisol into fetal lamb (Liggins, 1968) may cause premature parturition.

### 5.4. Perinatal Regulation of Corticosteroid Secretion

In the newborn, secretion of corticosteroids by the fetal zone decreases progressively during the first few weeks. DHEA, 16α-hydroxy-DHEA, and estriol are high in urine and in plasma of newborn infants and decrease rapidly (Migeon et al., 1955; Bongiovanni, 1962; Reynolds, 1964; France, 1971; Fenner et al., 1974). High rates of urinary excretion of 6β-hydroxy metabolites of cortisol have also been observed (Reynolds et al., 1962). 15α-Hydroxylated compounds of DHEA have been found in the urine of newborn infants and are thought to be active markers of the fetal compartment (Anderson et al., 1974). All the urinary metabolites are excreted in conjugated forms, either as glucuronide or sulfate (Anderson et al., 1974). The decrease in metabolites with Δ5- and 16-hydroxylation reflects the involution of the fetal zone.

Plasma cortisol in cord blood is 2–5 times lower than in maternal blood (62 μg/ 100 ml) (Aarskog, 1965; Murphy, 1973b). Fetal corticosteroid levels are higher after normal delivery when compared with levels in umbilical cord plasma after elective caesarian section, suggesting that the hypothalamo–pituitary–adrenal axis is active at birth (Smith and Shearman, 1974; Murphy and Diez d'Aux, 1972). In infants, circadian rhythms of cortisol secretion are not apparent in the infant until a couple of months of age (Franks, 1967), although the young infants are responsive to stress (Anders et al., 1970).

In the first few days of life, the preponderance of cortisone over cortisol in the plasma or in the urine is marked (Eberlein, 1965; Klein et al., 1969; Murphy and Diez d'Aux, 1972). The half-life of cortisol is longer in the newborn (Bongiovanni et al., 1958; Bertrand et al., 1963). The secretion rate of cortisol (22 mg/m²/24 hr) during the first few days of life is 1.5–2 times greater than that of the adult (Migeon, 1959; Bertrand et al., 1963; Kenny et al., 1966). It decreases to an average of about 16.4 mg/m²/24 hr during the first year of life.

### 5.5. Aldosterone Secretion and Its Regulation

The fetal adrenal gland is able to synthesize aldosterone from progesterone and corticosterone (Dufau and Villee, 1969) as early as the 15th week of gestation, in in vitro experiments. At term, a significant amount of the aldosterone is secreted by the fetus, and only a small portion of the total fetal aldosterone is of maternal origin

(Bayard *et al.*, 1970). Secretion rates of aldosterone in newborn infants from one to eight days of life were found to be low when compared to older infants or children (Weldon *et al.*, 1967). Increases of secretion rates and plasma levels of aldosterone following low-sodium diet administered to the mother before delivery were observed (Weldon *et al.*, 1967; Beitins *et al.*, 1972; Siegel *et al.*, 1974). In cord blood, plasma aldosterone increased from 22.9 ± 17.3 ng/100 ml (normal sodium diet) to 75.2 ± 51.6 ng/100 ml (low-sodium diet). Similarly, plasma aldosterone remained high during the first three days of life with a low-sodium diet (Beitins *et al.*, 1972). It has been suggested that the high levels of progesterone found in the circulation of newborn infants might aggravate the sodium loss observed at this period of time, concomitantly with the water loss which occurs during the first 3–5 days of life (Cheek *et al.*, 1961; Conly *et al.*, 1970).

### 5.5.1 Renin–Angiotensin–Aldosterone System

The renin–angiotensin–aldosterone system is fully active in newborn infants. Renin and its substrate do not cross the placenta (Symonds and Furler, 1973). Fetal kidney has been shown to contain renin and could therefore be the source of some of this hormone (Ljungqvist and Wagermark, 1966). Chorionic laevi cells are capable of renin production, and this could explain the high levels of renin in the amniotic fluid (Brown *et al.*, 1964; Symonds *et al.*, 1968). Plasma renin activity and plasma levels of angiotensin II are higher in the cord blood than in the maternal blood. Concentration of plasma renin substrate is lower in the fetal blood at term when compared to that of the mother (Godard *et al.*, 1975).

## 6. The Gonads

### 6.1. Prenatal Period

#### 6.1.1. Sexual Differentiation

The gonads play an active part during fetal life in sexual differentiation in mammals (Jost, 1953), probably including man. Sexual differentiation begins with the fertilization of the ovum by the spermatozoon. The nature of the sex chromosome of the male gamete determines the genetic sex of the embryo and initiates a chain of events which result in the development of male or female gonads from undifferentiated gonadal primordium and in the differentiation of the genital tract and external genitalia. Male differentiation of the primordial genital tract requires the presence of a functioning testis, as there are several tissues responsive to the action of fetal testicular secretions. In agonadal fetuses the somatic sex structures develop a female pattern irrespective of chromosomal sex structure (Jost and Picon, 1970; Jost *et al.*, 1973). The ovary does not play a critical role in human sex differentiation.

The gonadal primordium which is common to both sexes develops on the ventral surface of the mesonephros where primordial germ cells migrate and settle into the coelomic epithelium. As the mesenchymal layer between the germinal epithelium and the mesonephric tubules increases, the gonadal ridge becomes visible. By the 7th week of gestation (14–16 mm in crown–rump length), male gonadal differentiation begins in the XY embryo (Gillman, 1948). Leydig cells in the testes and evidence of steroidogenesis appear at 30 mm crown–rump length in the human fetus. Ovarian differentiation is essentially a passive procedure: the ovary is still undifferentiated at 10 weeks of gestation. The young ovary has a lobulated aspect due to the fact that blood vessels and connective tissue radiate toward the center of the gonad (Van Wagenen and Simpson, 1965; Falin, 1969). True ovarian organogenesis appears at the 13th week and primordial follicles occur at the 17th week. According to Witschi (1967), the undifferentiated gonad is potentially bisexual, with two zones: an outer zone, the cortex, which would give the ovarian organogenesis in the female, and an inner zone, the medulla, which in the male would give the testicular organogenesis. More recently, Jost (1970) has submitted the working hypothesis that only male differentiation is genetically induced in the young gonad. In the absence of a genetic stimulus, the young gonad develops as an ovary (Jost *et al.*, 1973).

The young human embryo has a double set of genital ducts, the Wolffian ducts and the Mullerian ducts. Both sets of ducts reach the posterior wall of the anogenital sinus below the neck of the bladder. Up to 32 mm in crown–rump length, both sets of ducts are present and the external genitalia are also undifferentiated. At this time of development, male sexual differentiation of internal ducts begins with the regression of the Mullerian ducts and the development of the Wolffian ducts into vasa deferentia, seminal vesicles and prostatic buds. In the meanwhile, the genital folds fuse, leading to the formation of a phallic and perineal urethra, after the 50-mm stage. In the female, the Mullerian ducts are retained and develop into the uterus and fallopian tubes. The Wolffian ducts progressively degenerate. The lower part of the urogenital sinus develops as the lower part of the vagina. The genital folds do not fuse, and instead form the labia after the 70-mm stage.

### 6.1.2. Fetal Gonadal Secretions

Sexual differentiation of the genital tract and of the external genital organs is dependent, in the male, on the secretion of the fetal testis (Jost, 1970). In the absence of a male gonad, genital duct and external genitalia on the same side would develop along female lines. Grafting fetal testicular tissue has the opposite effect. The fetal testis secretes two types of substances, one which inhibits the development of the Mullerian ducts and one which stimulates the Wolffian structures.

The Mullerian-duct-inhibiting factor is present only during a short time of human fetal life, around 25 mm in crown–rump length and not after 33 mm (Josso, 1972a; Josso, 1975). This factor is a nonsteroidal non-species specific macromolecule (Josso, 1972b) and is probably synthesized by the Sertoli cells (Blanchard and Josso, 1974).

The nature of the fetal testicular secretion responsible for the masculinization of the genital tract is not yet completely elucidated. Testosterone and many derivates are active in the fetuses, and antiandrogens like cyproterone acetate inhibit the normal development of the Wolffian ducts and of the male external genitalia. Testosterone is present in the fetal testes as early as the 5-mm stage. Leydig cells appear in the fetal testis at 60 days of gestation and greatly increase between the 10th and the 18th week. After the 20th week, there is a rapid decline in size and number. In human fetuses of 5.0–15.3 mm crown–rump, fetal testes are sensitive to LH and to hCG (Villee, 1974). Testosterone levels rise markedly at 10–20 weeks of gestation (Abramovitch and Rowe, 1973; Reyes *et al.*, 1973, 1974; Diez d'Aux and Murphy, 1974). This is in agreement with *in vitro* studies in which fetal testes were able to secrete testosterone after incubation with $\Delta 5$-pregnenolone and progesterone (Siiteri and Wilson, 1974; Payne and Jaffe, 1975). The capacity to synthesize testosterone is absent before 8–10 weeks and reaches a peak between 12 and 17 weeks, after which time a sharp decline in testosterone formation is observed. Testosterone *per se* and not dihydrotestosterone (DHT) is probably the hormone which acts on the Wolffian ducts as these ducts lack the $5\alpha$-reductase system present in many other target tissues. In contrast, DHT is very likely the active mediator of male differentiation of the urogenital sinus and the external genitalia (Imperato-McGinley *et al.*, 1974). The changes in testosterone synthesis and secretion coincide with the high concentration of fetal serum hCG early in gestation and later with the peak concentration of fetal serum LH (Figure 4).

In the female fetus, testosterone secretion is very low (Abramovitch and Rowe, 1973). Important morphological changes and enzymatic maturation take place in the ovary and the uterus of the female fetus, newborn, and infant (Pryse-Davies and Dewhurst, 1971). In anencephalic fetuses, ovaries are usually hypoplastic and exhibit a decreased number of primordial and primary follicles (Ross, 1974). Estrogen secretion has been widely studied during pregnancy, but very little data is available on the secretion of estrogens by the fetal ovaries. It is evident that the fetal circulating estrogen levels are part of the placental production (Maner *et al.*, 1963; Laatikainen and Peltonen, 1974). In pregnant women, estrone sulfate, estradiol and estriol plasma levels are in better correlation with fetal weight than with placental weight (Abdul-Karim *et al.*, 1971; Loriaux *et al.*, 1972). Mean total estrogens in the human fetal plasma increase from 22.7 $\mu$g/ml at 17–20 weeks, to 108.9 $\mu$g/ml at term (Shutt *et al.*, 1974). The human fetal ovary is active as early as the 12th week and can synthesize DHEA but not testosterone or estrogens, as demonstrated in *in vitro* studies (Payne and Jaffe, 1974).

Two main types of enzymatic abnormalities of testosterone secretion are known (Grumbach and Van Wyk, 1974): (1) excessive androgen secretion by the adrenal gland which will cause virilization of female fetuses, as in congenital adrenal hyperplasia with female pseudohermaphroditism; and (2) enzymatic biosynthetic defects affecting either testosterone formation or the end-organ sensitivity to testosterone, which will cause a defect in the development of the male ducts and external genitalia leading to male pseudohermaphroditism.

### 6.2. Perinatal Period

During the perinatal period, testosterone and estrogen secretion have been widely studied. In mixed cord blood, testosterone levels are slightly higher in male (35.7 $\pm$ 10.5 ng/100 ml) than in female newborn infants (25.9 $\pm$ 8.0 ng/100 ml) (Forest *et al.*, 1973, 1974*a*). Recently, Forest and Cathiard (1975) have documented a striking difference at birth between peripheral and umbilical venous concentrations of plasma testosterone and $\Delta 4$-androstenedione. In mixed cord blood, which represents mostly venous blood returning from the placenta, a mean concentration of 38.3 $\pm$ 10.5 ng/100 ml plasma testosterone was measured in 25 male fetuses, whereas the mean value of 19 specimens obtained either from heel stab or peripheral venipuncture during the first day of life was 227.7 $\pm$ 128.7 ng/100 ml. A similar concentration gradient was found in female infants: 29.2 $\pm$ 13.9 (cord) vs. 46.3 $\pm$ 13.9 (periphery). These findings suggest that fetal testosterone and $\Delta 4$-androstenedione are metabolized in the placenta, probably by aromatization; they also suggest a more active testicular function in late gestation than the levels in cord blood had previously indicated. In male newborn infants, plasma testosterone levels in peripheral blood decreased dramatically to $32.6 \pm 5.0$ ng/100 ml at 5–6 days of life, then markedly increased from the 10th day of life on to levels above 200 ng/100 ml with a peak of $247.7 \pm 16.4$ ng/100 ml in the second month of life (Forest and Cathiard, 1975). There is then a decline between 60 and 210 days of life (Figure 8). This increase of plasma testosterone at 60 days is observed both in normal term and in premature babies (Forest *et al.*, 1974*a*). In female newborn infants, plasma testosterone in peripheral blood decreases from 46.3 ng/100 ml to $13.1 \pm 9.9$ at 6 days, and $8 \pm 4.7$ ng/100 ml at 60 days (Figure 9). $\Delta 4$-Androstenedione concentrations in cord plasma are $92.7 \pm 37.9$ ng/100 ml in the female and $86.7 \pm 30.2$ ng/100 ml in the male. In the peripheral blood at the first day of life, $\Delta 4$-androstenedione levels are $173.3 \pm 75.4$ in the female newborn and $197 \pm 92.2$ in the male infant. Levels of $\Delta 4$-androstenedione decreased to $33.2 \pm 17$ in the female and $34.9 \pm 17.3$ in the male at 6 days and to $8 \pm 4.7$ ng/100 ml and $4.3 \pm 1.3$ ng/100 ml, respectively, between 30 and 60 days of life (Forest and Cathiard, 1975). Plasma levels of DHEA and DHEA-S, which are elevated at birth in the cord blood, progressively decrease during the first days of life (Sizonenko, 1976).

In the female newborn infant, the ovary presents numerous graafian follicles with luteinization of the ovarian theca (Kraus and Neubecker, 1962, Valdes-Dapena, 1967). In cord blood, similar high concentrations of estrone and estradiol were observed for both sexes (Bidlingmaier *et al.*, 1974). During the first 72–96 hr of life, in both sexes, an initial rapid decline of estradiol and estrone is observed (Anderson *et al.*, 1965; Kenny *et al.*, 1973; Bidlingmaier *et al.*, 1974).

Very little data are available on the secretion of progesterone and $17\alpha$-OH-progesterone by the fetus and the newborn infant. There is a progressive decrease of both steroids during the first days of life in both sexes. From high levels in cord blood (413 ng/ml) progesterone decreases after the first day of life to 11.6 ng/ml and at the 6th day declines to 0.29 ng/ml. Cord blood concentration of $17\alpha$-hydroxypro-

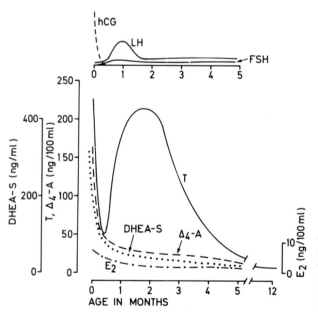

Fig. 8. Serum concentrations of human chorionic gonadotropin (hCG), luteinizing hormone (LH), follicle-stimulating hormone (FSH), testosterone (T), $\Delta 4$-androstenedione ($\Delta 4$-A), dehydroepiandrosterone sulfate (DHEA-S), and estradiol ($E_2$) in male infants in the first months of life. (Adapted from Forest *et al.*, 1974*b*; Forest and Cathiard, 1975; and personal data.)

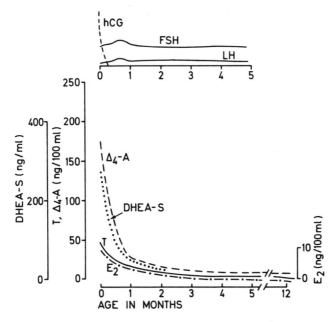

Fig. 9. Serum concentrations of human chorionic gonadotropin (hCG), luteinizing hormone (LH), follicle-stimulating hormone (FSH), testoterone (T), Δ4-androstenedione (Δ4-A), dehydroepiandrosterone sulfate (DHEA-S), and estradiol (E₂) in female infants in the first months of life. (Adapted from Bidlingmaier *et al.*, 1974; Forest *et al.*, 1974*a,b*, and personal data.)

gesterone is 45.9 ng/ml. Plasma levels thereafter decrease to 3.6 ng/ml at 24 hr, to 1.3 ng/ml at one month, and to 0.2 ng/ml at 6 months (Toublanc *et al.*, 1975).

## 7. Calcium, Parathyroid Glands, and Calcitonin

Calcium needs are extremely important during intrauterine life and after birth during the first year of life (see Vol. 2, Chapter 5). At birth, the newborn infant has a calcium body mass of about 10 g, two thirds of which have been passed to the fetus by the mother during the last trimester of gestation. At one year, the calcium mass of the infant body is 100 g. At birth, total calcium (11.7 mg/100 ml) in newborn plasma is slightly higher than in maternal plasma (10.4 mg/100 ml) in vaginal births. After cesarian section plasma total calcium is 10.9 mg/100 ml in the newborn infant and 9.3 mg/100 ml in the mother. Similarly plasma ultrafiltrable calcium and plasma phosphorus in the fetus (8.0 mg/100 ml and 5.5 mg/100 ml, respectively) are higher than in the mother (5.9 and 2.6 mg/100 ml, respectively). Both total and ultrafiltrable calcium in plasma decrease during the first 48 hr (Bergman, 1974). Phosphorus levels increase during this period (Anast, 1969; David and Anast, 1974).

The parathyroid glands derive from the endoderm of the pharyngeal pouches. The superior pair arises from the 4th branchial pouch in relation to the thyroid gland with which they descend; the inferior pair arises from the 3rd branchial cleft and descends with the thymus. Many animal experiments have shown that the parathyroid glands are active during fetal life (Norris, 1946; Garel, 1971; Pic, 1973). In the thyroid gland, calcitonin-secreting cells have been found, suggesting the secretion of calcitonin during fetal life (Pearse and Carvalheira, 1967).

PIERRE C. SIZONENKO
and MICHEL L. AUBERT

Placental transfer of parathormone and calcitonin very likely does not occur in the human, as it is suggested by animal experiments (Garel *et al.*, 1969; Garel, 1972). Depending on the antiserum and standard used for the radioimmunoassay of parathormone, values obtained are different from one group of investigators to another (Fischer *et al.*, 1975).

In cord blood immunoreactive parathormone levels are undetectable or low while calcium levels are high (David and Anast, 1974). During the first 48 hr (Figure 10) serum parathormone and calcium increase in parallel (David and Anast, 1974; Milner and Woodhead, 1975). Parathormone secretion is low in newborn infants, particularly in premature infants. It has been suggested that a functional hypoparathyroidism might be the cause of transient neonatal tetany. This is particularly observed in infants of shortened gestational age and in infants during the first two days of life (Tsang *et al.*, 1973). The removal of placental passage of calcium at delivery may affect calcium levels after birth. Maternal hyperparathyroidism has been described as inducing neonatal transitory hypoparathyroidism (Arsdel, 1955; Walton, 1954). Since it is likely that PTH does not cross the placenta, it appears that fetal parathyroid glands are suppressed by high extracellular calcium concentration. In the cord, calcitonin level is higher in arterial blood (7.3 ± 2.2 ng/ml) than in venous blood (4.1 ± 1.2 ng/ml). Plasma calcitonin progressively decreases from the first day of life to 1 month and to five years of age (Samaan *et al.*, 1975). The hypersecretion of calcitonin during the first few days of life may play a role in blood calcium regulation.

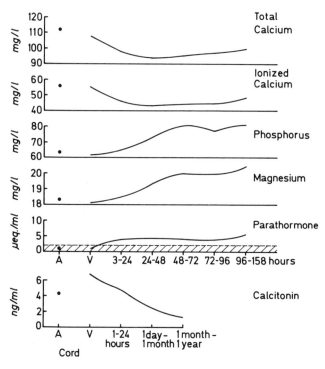

Fig. 10. Serum total and ionized calcium, phosphorus, magnesium, parathormone, and calcitonin in cord blood (A = arterial, V = venous) and during perinatal life. (Adapted from David and Anast, 1974; Samaan *et al.*, 1975.)

The possibility that kidney tubule responsiveness to parathormone is impaired in newborn infants has been suggested. A tubular maturation is observed during the first days and weeks of life: renal phosphorus clearance and urinary cyclic AMP excretion after parathormone increased considerably between the first and the fifth day of life (Linarelli, 1972).

## 8. Pancreas

The fetal pancreas comes from an outgrowth of the duodenal endoderm. Dorsal and ventral embryonic buds join by the 7th week of gestation. Endocrine cells of the pancreatic islets differentiate from the digestive epithelium. Pancreatic islets are first observed at 54 mm crown–rump length. Differentiation of $\alpha$ and $\beta$ cells occurs at stage 130 mm.

### 8.1. Insulin

Secretion grains of insulin at the capillary pole of $\beta$ cells are first seen by the 12th week of gestation (Steinke and Driscoll, 1965; Grillo and Shima, 1966). Insulin content of the human fetal pancreas is in direct correlation with the number of islet cells (Gasparo et al., 1969; Espinosa et al., 1970; Rastogi et al., 1970; Wellman et al., 1971; Ashworth et al., 1973). In the fetus, serum insulin was detected as early as 84 days and remained from 1 to 30 $\mu$U/ml (mean 8.0 $\pm$ 1.5 $\mu$U/ml) until term (Kaplan et al., 1972). In vitro incubation of human fetal pancreas has shown that while glucose is not a good stimulating agent of insulin release, arginine is a very potent factor for insulin secretion in pancreas from fetuses weighing more than 200 g. Leucine had a very variable effect on the insulin release of these pancreas. The results obtained by different groups, particularly in in vitro preparations of animal pancreas (Lambert et al., 1972), illustrate the development of different mechanisms for the release of insulin from the human fetal $\beta$ cell (Milner et al., 1972a; Leach et al., 1973). In vivo infusion of glucose or arginine into fetuses delivered by hysterotomy at 15–20 weeks gestational age failed to cause a rise in fetal plasma insulin levels (Adam et al., 1969; Thorell, 1970; King et al., 1971c). Fetal serum insulin levels do not change at term when fetal hyperglycemia is obtained by injecting glucose to the mother (Tobin et al., 1969; Obenshain et al., 1970). Fetal insulin is radioimmunologically identical to adult insulin. Only a very minimal fraction of plasma insulin passes the placenta (Gitlin et al., 1965b).

After birth a progressive maturation of the mechanisms of insulin release is very likely in the human, as in the subhuman primate (Mintz et al., 1969). In cord blood, insulin levels range from 6 to 16 $\mu$U/ml. At birth there is a positive correlation between umbilical or peripheral insulin concentrations and birth weight (Shima et al., 1966). Infusion of glucose induces a sluggish rise of insulin in the newborn infant (Isles et al., 1968; Pildes et al., 1969; Le Dune, 1972a; Falorni et al., 1974; Cser and Milner, 1975). The response of insulin is greater in newborn infants with birth weights above 3500 g. (Mølsted-Pedersen and Jørgensen, 1972). Response to arginine in newborn infants has been shown to be much smaller than in normal children (Grasso et al., 1968; Reitano et al., 1971; King et al., 1971c, 1974; Ponte et al., 1972a; Falorni et al., 1975a,b). Similar sluggish responses of insulin to intravenous glucagon were observed in newborns (Milner and Wright, 1967; Le

Dune, 1972b; Hunter and Isles, 1972; Sizonenko et al., 1972). With age, maturation of the mechanisms of secretion of pancreatic insulin occurs: $\beta$ cells become more and more sensitive to usual stimuli. Basal plasma insulin levels remain identical during the first days of life (Sperling et al., 1974). Fasting plasma levels of insulin increase from 3 months of age to the first year (Grant, 1967).

### 8.2. Glucagon

Glucagon is present in the human fetus as early as the 10th–12th week of gestation. Fetal and adult glucagon behave identically in radioimmunochemical determinations. Alanine stimulates glucagon secretion in the fetus (Wise et al., 1973). Glucagon does not cross the placenta in the human (Adam et al., 1972; Johnston et al., 1972; Moore et al., 1974).

The characteristic fall in blood glucose, reaching a nadir within hours of birth, is associated with a significant increase in glucagon concentration (Figure 11). Despite the persistence of low blood glucose levels, glucagon does not change appreciably between 2 and 24 hr of life (Luyckx et al., 1972; Fekete et al., 1972; Sperling et al., 1974). A further significant rise of glucagon levels occurred from day 1 to day 3 and was associated with a rise of blood glucose to normal levels. Arginine or alanine infusion stimulate plasma glucagon during the first days of life (Sperling et al., 1974; Falorni et al., 1975a,b; Fiser et al., 1975). Glucose alone poorly suppresses portal glucagon in the newborn infant (Luyckx et al., 1972). However, the administration of insulin in addition to glucose results in a significant decrease in portal plasma glucagon (Massi-Benedetti et al., 1974).

## 9. Hormonal Control of Carbohydrate and Lipid Metabolism

During intrauterine life and in the perinatal period many hormones, in particular pancreatic hormones, induce metabolic changes and maintain homeostasis. Hormonal secretions influence developmental processes by stimulating or inhibiting

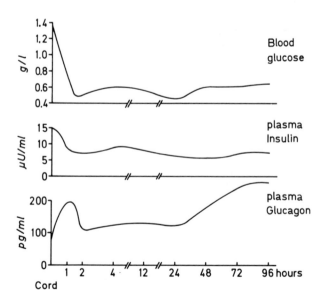

Fig. 11. Blood glucose, plasma insulin, and glucagon in normal newborn infants during the first 96 hr of life. (Adapted from Sperling et al., 1974.)

the transformation of tissues and organs. The role of pituitary secretions on growth and maturation of peripheral endocrine glands and enzymatic activities has been well investigated by using decapitation of the fetus (Jost and Picon, 1970).

### 9.1. Liver Glycogen

Glycogen appears in the fetal liver at a particular step of its development. Glycogen is found as early as the 18th day of gestation in the rat fetus, 25th day in the rabbit, and 9th week in the rhesus monkey. In the human, glycogen is present for the first time between the 15th and 25th week of gestation and increases markedly during the last month of pregnancy (Villee, 1953, 1961). Liver glycogen decreases very abruptly after birth and remains low during the first days (Girard *et al.*, 1973; Snell and Walker, 1973). Most of the liver enzymes necessary for glycogenesis and gluconeogenesis are present at birth (Dawkins, 1963; Haworth, 1969).

Gluconeogenic enzyme activities in the fetal and newborn liver are increased by administration of glucagon, cortisol, epinephrine, and norepinephrine (Jost and Picon, 1970). In infants, small for gestational age, a delayed maturation of these hepatic enzymes with ineffective gluconeogenesis has been suggested, following the observation of high levels of glucogenic amino acids, in particular alanine, in presence of low blood glucose (Haymond *et al.*, 1974; Mestyan *et al.*, 1975; Williams *et al.*, 1975). Glucagon responses to alanine have been shown to be normal in these small-for-gestational-age babies. However, in premature babies and in a group of small-for-gestational-age babies, low alanine levels were found (Sizonenko *et al.*, 1974), suggesting that some heterogenity may exist in the small-for-gestational-age babies.

It is very likely that other hormones like human chorionic somatotropin (hCS) play a role in the supply of maternal glucose to the fetus (Grumbach *et al.*, 1968). Fetal growth hormone and prolactin may also play a role in the production of glycogen. Knowledge of the regulatory factors that govern this storage process is fragmentary, but an intact pituitary–adrenal axis seems to be essential.

### 9.2. Fat Formation and Utilization

The normal human fetus accumulates fat rapidly during the last three months of gestation (Widdowson and Spray, 1951). Apparently, the fat is primarily synthesized *in utero* from glucose (Hirsch *et al.*, 1960) and is of different composition than that of the adults (King *et al.*, 1971*a*). The source of fatty acids in the human fetus is not known, nor is the differential role of hormones like thyroxine and triiodothyronine, growth hormone, glucagon, insulin, and cortisol on lipid metabolism in the fetus. After birth, the newborn infant presents a rapid rise of plasma free fatty acids, a portion of which is converted to ketone bodies, concurrent with the fall of blood glucose (Persson and Gentz, 1966) and the increase of glucagon (Sperling *et al.*, 1974).

## 10. Adrenal Medulla

The adrenal medullary gland derives from chromaffin cells of the neuroectodermal tissue. These cells have been seen as early as 27 mm crown–rump length and in

PIERRE C. SIZONENKO
and MICHEL L. AUBERT

all subsequent stages of development. The intraadrenal chromaffin cells develop independently and more slowly than the chromaffin tissue of the aortic region. Actually, during fetal life, the bulk of chromaffin tissue lies outside the adrenal gland, whereas intraadrenal chromaffin tissue remains underdeveloped. The suggestion has been made that these tissues achieve maturity during the latter half of fetal life. Pressor amines have been found in the adrenal medulla of 95-mm fetuses and older specimens (Coupland, 1952). Nothing is known of the secretion of catecholamines during human fetal life. The fetus seems to be protected against maternal catecholamines, as suggested in cases of pheochromocytoma, by the low levels of catecholamines in cord blood compared to the high levels observed in the mother (Barzel *et al.*, 1964). This might be due to the fact that the placenta possesses a catechol-*o*-methyl transferase which might inactivate catecholamines by methylation.

At birth, the chromaffin cell bodies attain their maximum size. After birth, most of the cells located in the extraadrenal bodies (mainly paraaortic) undergo atrophy. In contrast to the degeneration of the fetal zone of the adrenal cortex, the adrenal medulla develops rapidly and attains maturity during the first 3 years of postnatal life (West *et al.*, 1951; Coupland, 1965). Norepinephrine is the principal, if not the exclusive, catecholamine secreted by fetal chromaffin tissue. Secretion of epinephrine occurs also in the human newborn infant, although it is much lower than that of norepinephrine. Little is known about the relation of the adrenal cortex and the medulla (Jost and Picon, 1970). The secretion of epinephrine is probably influenced by ACTH. The phenylethanolamine-*N*-methyl transferase activity responsible for the methylation of norepinephrine to epinephrine is dependent on the amount of cortisol secreted by the adrenal cortex and present in the medulla (Wurtman and Axelrod, 1956). During the first three years of life, the adrenal medulla is able to secrete increasing amounts of epinephrine as observed from plasma levels or urinary excretions of metabolites such as metanephrine, normetanephrine, and vanillylmandelic acid (VMA). Urinary VMA excretion increased with age and body weight until 3 years of age (Greenberg and Gardner, 1960; McKendrick and Edwards, 1965). However, the newborn is able to increase the secretion of epinephrine or norepinephrine in relation to stress, hypoxia, or hypoglycemia (Greenberg *et al.*, 1960; Cheek *et al.*, 1963; Kaye *et al.*, 1970; Anagnostakis and Lardinois, 1971).

## 11. General Conclusions

During intrauterine life, the endocrine systems of the fetus develop and mature progressively in relation with the maternal and placental endocrine systems. Knowledge of the exact physiological events is very scarce. Timing of maturation of the different endocrine systems is not yet well known. At birth, very important changes occur. Removed from a constant infusion of glucose, oxygen, free fatty acids, and other essential nutrients, the newborn infant must readjust his fetal circulation to one compatible with pulmonary oxygen, maintain his temperature and metabolic needs, and coordinate his energy expenditure in relation to his intake, utilization, storage, and waste of substrates (Cornblath and Schwartz, 1966). The interrelationships among the different regulations of carbohydrate, fat, and protein metabolism and the endocrine systems is very important.

There are good correlations between unusual clinical findings and abnormal endocrine developments. A typical example is the hypoglycemia in anencephalics

due to lack of growth hormone or ACTH resulting from the absence of hypotha-
lamic-releasing factors. A further example is the finding of normal prolactin levels in
these patients which confirms that prolactin secretion is not controlled by a hypo-
thalamic-releasing factor (Aubert *et al.*, 1975). The relationship of poor insulin
secretion with growth-hormone deficiency present in anencephalics has been noted
too. Finally the absence of ambiguous genitalia in male anencephalics, although
they are small and hypoplastic, suggests that pituitary gonadotropins are not
essential, or are secreted in a sufficient amount for normal sex differentiation at the
critical period of the first trimester of pregnancy, or that hCG present in the fetal
circulation is sufficient for this purpose (Kaplan *et al.*, 1976). The low cortisol
secretion of these patients is one cause of low glycogen deposition in the liver (Allen
*et al.*, 1974).

Many neonatal endocrine diseases are explained by the knowledge of placental
transfer of steroid hormones, pathologic factors such as LATS and medication such
as antithyroid drugs. Pathological conditions such as diabetes during pregnancy
change the fetal environment and cause diseases in the newborn (Fletcher, 1975;
Stern, 1975). Abnormal hormone synthesis in the fetus, such as congenital adrenal
virilizing hyperplasia, induces important developmental changes. Abnormal administration of sex steroids to the mother may cause sexual ambiguity. Hormones
administered prenatally may affect the organization and differentiation of certain
patterns of human cognition and intellectual functioning (Reinisch, 1974). The effect
of these hormones may influence the acquisition of sexually dimorphic behavior and
interest in man. The effects of fetal hormones on the development of the central
nervous system have been shown to exert their influence up to adulthood both in
humans and animals, suggesting that these effects on brain organization and differentiation are permanent (Money and Ehrhardt, 1972). The use for diagnostic
purposes of measurements of hormone concentrations in amniotic fluid represents a
new approach for the prenatal detection of abnormal conditions such as congenital
adrenal hyperplasia (Jeffcoate *et al.*, 1965; Merkatz *et al.*, 1969; Nichols, 1970),
congenital hypothyroidism (Sack *et al.*, 1975a), or neonatal Grave's disease (Hol-
lingswirth and Austin, 1971), as well as detection of chromosomal abberations or
metabolic defects. However, this approach in many cases is limited or has not
always been reliable when attempted in early or mid-gestation (Merkatz *et al.*,
1969). In cases of early diagnosis of hypothyroidism (Klein *et al.*, 1974; Dussault,
1975; Fisher and Sack, 1975), mass screening of newborn infants might be more
contributory.

ACKNOWLEDGMENT

Current research referred to in this paper was supported by Swiss National
Science Foundation Grants 3.091-0.73, 3.496-0.75, and 3.747-0.76.

## 12. References

Aarskog, D., 1965, Cortisol in the newborn infant, *Acta Paediatr. Scand.* Suppl. 158.
Abdul-Karim, R. W., Nesbitt, R. E. L., Drucker, M. H., and Rizk, P. T., 1971, The regulatory effects of
    oestrogens on fetal growth. 1. Placental and fetal body weights, *Am. J. Obstet. Gynecol.* **109:**656.
Abramovitch, D. R., and Rowe, P., 1973, Foetal plasma testosterone levels at mid-pregnancy and at
    term: Relationship to foetal sex. *J. Endocrinol.* **56:**621.
Abuid, J., Stinson, D. A., and Larsen, P. R., 1973, Serum triiodothyromine and thyroxine in the neonate
    and the acute increases in these hormones following delivery, *J. Clin. Invest.* **52:**1195.

*PIERRE C. SIZONENKO*
*and MICHEL L. AUBERT*

Adam, P. A. J., Teramo, K., Räihä, N., Gitlin, D., and Schwartz, R., 1969, Human fetal insulin metabolism early in gestation. Response to acute elevation of the fetal glucose concentration and placental transfer of human insulin I-131, *Diabetes* **18**:409.

Adam, P. A. J., King, K. C., Schwartz, R., and Teramo, K., 1972, Human placental barrier to $^{125}$I-glucagon early in gestation, *J. Clin. Endocrinol. Metab.* **34**:772.

Allen, J. P., Cook, D. M., Kendall, J. W., and McGilvra, R., 1973, Maternal–fetal ACTH relationship in man, *J. Clin. Endocrinol. Metab.* **37**:230.

Allen, J. P., Greer, M. A., McGilvra, R., Castro, A., and Fisher, D. A., 1974, Endocrine function in an anencephalic infant, *J. Clin. Endocrinol. Metab.* **38**:94.

Anagnostakis, D. E., and Lardinois, R., 1971, Urinary catecholamine excretion and plasma NEFA concentration in small-for-date infants, *Pediatrics* **47**:1000.

Anast, C. S., 1969, Tetany of the newborn, in: *Endocrine and Genetic Diseases of Childhood* (L. I. Gardner, ed.), pp. 352–364, W. B. Saunders, Philadelphia.

Ances, I. G., and Pomerantz, S. H., 1974, Serum concentrations of $\beta$-melanocyte-stimulating hormone in human pregnancy, *Am. J. Obstet. Gynecol.* **119**:1062.

Anders, T. F., Sachar, E. J., Kream, J., Roffwarg, A. P., and Hellman, L., 1970, Behavioral state and plasma cortisol response in the human newborn, *Pediatrics* **46**:532.

Anderson, A. B., Laurence, K. M., Turnbull, A. C., 1969, The relationship of anencephaly between the size of the adrenal cortex and the length of gestation, *J. Obstet. Gynaecol. Br. Commonw.* **76**:196.

Anderson, A. B. M., Laurence, K. M., Davies, K., Campbell, H., and Turnbull, A. C., 1971, Fetal adrenal weight and the cause of premature delivery in human pregnancy, *J. Obstet. Gynaecol. Br. Commonw.* **78**:481.

Anderson, H. A., Bojesen, E., Jensen, P. K., and Sorensen, B., 1965, Concentrations of free oestradiol, oestrone and oestradiol: Precursors in human cord plasma and foetal systemic venous plasma at term, *Acta Endocrinol.* **48**:114.

Anderson, R. A., Chambaz, E. M., Madani, C., and Defaye, G., 1974, Etudes des stéroïdes urinaires chez le nouveau-né humain, in: *International Symposium on Sexual Endocrinology of the Perinatal Period* **32**:267–280, INSERM, Paris.

Araï, K., Kuwabara, Y., and Okinaga, S., 1972, Effect of adrenocorticotropic hormone and dexamethesone, administered to the fetus *in utero,* upon maternal and fetal estrogens, *Am. J. Obstet. Gynecol.* **113**:316.

Arsdel, P. P. Van, 1955, Maternal hyperparathyroidism as a cause of neonatal tetany, *J. Clin. Endocrinol. Metab.* **15**:680.

Ashworth, M. A., Leach, F. N., and Milner, R. D. G., 1973, Development of insulin secretion in the human fetus, *Arch. Dis. Child.* **48**:151.

Aubert, M. L., Sistek, J., Chabot, V., and Bossart, H., 1971, L'hormone de croissance hypophysaire foetale pendant l'accouchement, *Schweiz. Med. Wochenschr.* **101**:1102.

Aubert, M. L., Grumbach, M. M., and Kaplan, S. L., 1975, The ontogenesis of human fetal hormones. III. Prolactin, *J. Clin. Invest.* **56**:155.

Barry, J., and Dubois, M. P., 1974, Differenciation perinatale de l'hypothalamus gonadotrope chez les mammifères, in: *International Symposium on Sexual Endocrinology of the Perinatal Period* **32**:63–78, INSERM, Paris.

Barzel, U. S., Barilan, Z., Rumney, G., Lazebnïk, Y., Eckerling, B., and Devries, A., 1964, Pheochromocytoma and pregnancy, *Am. J. Obstet. Gynecol.* **89**:519.

Bayard, F., Ances, I. G., Tapper, A. J., Weldon, V. V., Kowarski, A., and Migeon, C. J., 1970, Transplacental passage and fetal secretion of aldosterone, *J. Clin. Invest.* **49**:1389.

Beitins, I. Z., Bayard, F., Levitsky, L., Ances, I. G., Kowarski, A., and Migeon, C. J., 1972, Plasma aldosterone concentration at delivery and during the newborn period, *J. Clin. Invest.* **51**:386.

Benirschke, K., 1956, Adrenals in anencephaly and hydrocephaly, *Obstet. Gynecol.* **8**:412.

Benirschke, K., and MacKay, D. B., 1953, The antidiuretic hormone in fetus and infant, *Obstet. Gynecol.* **1**:638.

Bergman, L., 1974, Studies on early neonatal hypocalcemia, *Acta Endocrinol. Suppl.* **248**:1–25.

Bertrand, J., Loras, B., Gilly, R., and Cautenet, B., 1963, Contribution à l'étude de la sécrétion et du métabolisme du cortisol chez le nouveau-né et le nourrisson de moins de trois mois, *Pathol. Biol.* **11**:997.

Bidlingmaier, F., Versmold, H., and Knorr, D., 1974, Plasma estrogens in newborns and infants, in: *International Symposium on Sexual Endocrinology of the Perinatal Period* **32**:299–314, INSERM, Paris.

Blackwell, R. E., and Guillemin, R., 1973, Hypothalamic control of adrenohypophysial secretions, *Annu. Rev. Physiol.* **35**:357.

Blanchard, M. G., and Josso, N., 1974, Sources of the anti-Müllerian hormone synthesized by the fetal testis: Müllerian-inhibiting activity of fetal bovine sertoli cells in tissue culture, *Pediatr. Res.* **8:** 968.

Blizzard, R. M., and Alberts, M., 1956, Hypopituitarism, hypoadrenalism and hypogonadism in the newborn infant, *J. Pediatr.* **48:**782.

Bolté, E., Mancuso, S., Eriksson, G., Wiquist, N., and Diczfalusy, 1964, Studies on the aromatisation of neutral steroids in pregnant women. 3. Overall aromatisation of dehydroisoandrosterone sulfate circulating in the foetal and maternal compartments. *Acta Endocrinol.* **45:**576.

Bongiovanni, A. M., 1962, The adrenogenital syndrome with deficiency of 3β-hydroxysteroid dehydrogenase, *J. Clin. Invest.* **41:**2086.

Bongiovanni, A. M., Eberlein, W. R., Westphal, M., and Boggs, T., 1958, Prolonged turnover rate of hydrocortisone in the newborn infant, *J. Clin. Endocrinol. Metab.* **18:**1127.

Brewer, D. B., 1957, Congenital absence of the pituitary gland and its consequences, *J. Pathol. Bacteriol.* **73:**59.

Brody, S., and Carlström, G., 1965, Human chorionic gonadotropin pattern in serum and its relation to the sex of the fetus, *J. Clin. Endocrinol. Metab.* **25:**792.

Brown, J. J., Davies, D. L., Doak, P. B., Lever, A. F., and Robertson, J. I. S., 1964, The presence of renin in human amniotic fluid, *Lancet* **2:**64.

Cathro, D. M., 1969, Adrenal cortex and medulla in: *Paediatric Endocrinology* (D. Hubble, ed.), pp. 187–327, Blackwell Scientific Publications, Oxford.

Chard, T., Hudson, C. N., Edwards, C. R. W., and Boyd, N. R. H., 1971, The release of oxytocin and vasopressin by the human fetus during labor, *Nature* **234:**352.

Cheek, D. B., Maddison, T. G., Malinek, M., and Coldbeck, J. H., 1961, Further observations on the corrected bromide space of the neonate and investigation of water and electrolyte status in infants born of diabetic mothers, *Pediatrics* **28:**861.

Cheek, D. B., Malinek, M., and Fraillon, J. M., 1963, Plasma adrenaline and noradrenaline in the neonatal period, and in infants with respiratory distress syndrome and placental insufficiency, *Pediatrics* **31:**374.

Chez, R. A., Hutchinson, D. L., Salazar, H., and Mintz, D. H., 1970, Some effects of fetal and maternal hypophysectomy in pregnancy, *Am. J. Obstet. Gynecol.* **108:**643.

Chopra, I. J., Sack, J., and Fisher, D. A., 1975, Circulating 3,3′,5′-triiodothyromine (reverse $T_3$) in the human newborn, *J. Clin. Invest.* **55:**1137.

Cleveland, W. W., 1970, Maternal–fetal hormone relationships, *Pediatr. Clin. North Am.* **17:**273.

Cole, H. S., Bilder, J. H., Camerini-Davalos, R. A., and Grimaldi, R. D., 1970, Glucose tolerance, insulin and growth hormone in infants of gestational diabetic mothers, *Pediatrics* **45:**394.

Conly, P. W., Morrison, T., Sandberg, D. H., and Cleveland, W. W., 1970, Concentrations of progesterone in the plasma of mothers and infants at the time of birth, *Pediatr. Res.* **4:**76.

Cornblath, M., and Schwartz, R., 1966, *Disorders of Carbohydrate Metabolism in Infancy,* W. B. Saunders, Philadelphia.

Cornblath, M., Parker, M. L., Reisner, S. H., Forbes, A. E., and Daughaday, W. H., 1965, Secretion and metabolism of growth hormone in premature and full term infants, *J. Clin. Endocrinol. Metab.* **25:**209.

Coupland, R. E., 1952, Prenatal development of abdominal para-aortic bodies in man, *J. Anat.* **86:**357.

Coupland, R. E., 1965, *The Natural History of the Chromaffin Cell,* Longmans, Green, London.

Cramer, D. W., Beck, P., and Makowski, E. L., 1971, Correlation of gestational age with maternal human chorionic somatomammotropin and maternal and fetal growth hormone plasma concentrations during labor, *Am. J. Obstet. Gynecol.* **109:**649.

Crowder, R. E., 1957, Development of the adrenal gland in man, with special reference to origin and ultimate location of cell types and evidence in favor of all "migration" theory, *Contr. Embryol. (Carnegie Inst.)* **36:**195.

Cser, A., and Milner, R. D. G., 1975, Glucose tolerance and insulin secretion in very small babies, *Acta Paediatr. Scand.* **64:**457.

Czernichow, P., Greenberg, A. H., Tyson, J., and Blizzard, R. M., 1971, Thyroid function studied in paired maternal-cord sera and sequential observations of thyrotropic hormone release during the first 72 hours of life, *Pediatr. Res.* **5:**53.

Czernichow, P., Oliver, C. H., and Friedman, R., 1975, TRH and prolactin (HPr) in plasma of newborns during the first hours of life, *Pediatr. Res.* **9:**668.

David, L., and Anast, C. S., 1974, Calcium metabolism in newborn infants. The interrelationship of parathyroid function and calcium, magnesium and phosphorus metabolism in normal "sick" and hypocalcemic newborns, *J. Clin. Invest.* **54:**287.

*PIERRE C. SIZONENKO
and MICHEL L. AUBERT*

Dawkins, M. J. R., 1963, Glycogen synthesis and breakdown in fetal and newborn rat liver, *Ann. N.Y. Acad. Sci.* **111:**203.

Diczfalusy, E., 1969, Steroid metabolism in the foeto-placental unit, in: *The Foeto-Placental Unit* (A. Pecile and C. Finzi, eds.) Vol. 183, p. 65, Excerpta Medica Foundation, Amsterdam.

Diczfalusy, E., 1974, Endocrine functions of the human fetus and placenta, *Am. J. Obstet. Gynecol.* **119:**419.

Diez d'Aux, R. C., and Murphy, B. E. P., 1974, Androgens in the human fetus, *J. Steroid Biochem.* **5:**207.

Di Stephano, J. J., III, Durando, A. R., Jang, M., Jenkins, D., Johnson, D. J., Mak, P., Marshall, T., Mons, B., Warsavsky, A., and Fisher, D. A., 1973, Estimates and estimation errors of hormone secretion, transport and disposal rates in the maternal–fetal system, *Endocrinology* **93:**324.

Dubois, P., 1968, Données ultrastructurales sur l'antehypophyse d'un embryon humain à la huitième semaine de son développement, *C. R. Soc. Biol.* **162:**689.

Dubois, P. M., and Dubois, M. P., 1974, Mise en évidence par immunofluorescence de l'activité gonadotrope LH dans l'antehypophyse foetale humaine, in: *International Symposium on Sexual Endocrinology of the Perinatal Period* **32:**37, INSERM, Paris.

Dubois, P. M., Vargues-Regairaz, H., Dubois, M. P., 1973, Human foetal anterior pituitary: Immunofluorescent evidence for corticotropin and melanotropin activities, *Z. Zellforsch.* **145:**131.

Dubois, P. M., Bethenod, M., Gilly, R., and Dubois, M. P., 1975, Etude des activités cortico-melanotropes dans l'antehypophyse d'un anencéphale et d'un nouveau-né normal, *Arch. Fr. Pediatr.* **32:**647.

Dufau, M. L., and Villee, D. B., 1969, Aldosterone biosynthesis by human fetal adrenal *in vitro, Biochim. Biophys. Acta* **176:**637.

Dussault, J., Row, V. U., Lickrish, G., and Volpé, R., 1969, Studies of serum triiodothyronine concentration in maternal and cord blood: Transfer of triiodothyronine across the human placenta, *J. Clin. Endocrinol. Metab.* **29:**595.

Dussault, J. H., Hobel, C. J., Distefano, J. J., III, Erenberg, A., and Fisher, D. A., 1972, Triiodothyronine turnover in maternal and fetal sheep, *Endocrinology* **90:**1301.

Dussault, J. H., Coulombe, P., Laberg, C., Letarte, J., Guyda, H., and Khoury, K., 1975, Preliminary report on mass screening program for neonatal hypothyroidism, *J. Pediatr.* **86:**670.

Eberlein, W. R., 1965, Steroids and sterols in umbilical cord blood, *J. Clin. Endocrinol. Metab.* **25:**1101.

Ellis, S. T., Beck, J. S., and Currie, A. R., 1966, The cellular localisation of growth hormone in the human foetal adenohypophysis, *J. Pathol. Bacteriol.* **92:**179.

Erenberg, A., Dale, L., Phelps, D. L., Lam, R., and Fisher, D. A., 1974, Total and free thyroid hormone concentrations in the neonatal period, *Pediatrics* **53:**211.

Espinosa, A., De los Monteros, M., Driscoll, S. G., and Steinke, J., 1970, Insulin release from isolated human fetal pancreatic islets, *Science* **168:**1111.

Faiman, C., and Winter, J. S. D., 1971, Sex differences in gonadotropin concentration in infancy, *Nature* **232:**130.

Faiman, C., and Winter, J. S. D., 1974, Gonadotropins and sex hormone patterns in puberty, clinical data, in: *Control of Onset of Puberty* (M. M. Grumbach, G. D. Grave, and F. E. Mayer, eds.), pp. 32–55, Wiley, New York.

Faiman, C., Ryan, R. J., Zwirek, S. J., and Rubin, M. E., 1968, Serum FSH and HCG during human pregnancy and puerperium, *J. Clin. Endocrinol. Metab.* **28:**1323.

Faiman, C., Reyes, F. I., and Winter, J. S. D., 1974, Serum gonadotropin patterns during the neonatal period in man and in the chimpanzee, in: *International Symposium on Sexual Endocrinology of the Perinatal Period* **32:**281–298, INSERM, Paris.

Falin, L. I., 1961, The developmental of human hypophysis and differentiation of cells of its anterior lobe during embryonic life, *Acta Anat.* **44:**188.

Falin, L. I., 1969, The development of genital glands and the origin of germ cells in human embryogenesis, *Acta Anat.* **72:**195.

Falorni, A., Fracassini, F., Massi-Benedetti, F., and Amici, A., 1972, Glucose metabolism, plasma insulin, and growth hormone secretion in newborn infants with erythroblastosis fetalis compared with normal newborns and those born to diabetic mothers, *Pediatrics* **49:**682.

Falorni, A., Fracassini, F., Massi-Benedetti, F., and Maffei, S., 1974, Glucose metabolism and insulin secretion in the newborn infant. Comparisons between the responses observed the first and the seventh day of life to intravenous and oral glucose tolerance tests, *Diabetes* **23:**172.

Falorni, A., Massi-Benedetti, F., Gallo, S., and Romizi, S., 1975*a,* Levels of glucose in blood and insulin in plasma and glucagon response to arginine infusion in low birth weight infants, *Pediatr. Res.* **9:**55.

Falorni, A., Massi-Benedetti, F., Gallo, G., and Trabalza, N. 1975*b,* Blood glucose serum insulin and glucagon response to arginine in premature infants, *Biol. Neonate* **27:**271.

Fekete, M., Milner, R. D. G., Soltez, G. Y., Assan, R., and Mestyan, J., 1972, Plasma glucagon, thyrotropin, growth hormone and insulin response to cold exposure in the human newborn, *Acta Paediatr. Scand.* **61**:435.

Fenner, A., Lange, G. U., Moenkemeier, D., and Ohlenroth, G., 1974, Estriol excretion in the first voided urine of male newborns, *Biol. Neonate* **25**:267.

Finkelstein, J. W., Anders, T. R., Sachar, E. J., Roffwarg, H. P., and Hellman, L. D., 1971, Behavioral state, sleep stage and growth hormone levels in human infants, *J. Clin. Endocrinol. Metab.* **32**:368.

Fischer, J. A., Blum, J. W., Hunziker, W., and Binswanger, U., 1975, Regulation of circulating parathyroid hormone levels: Normal physiology and consequences in disorders of mineral metabolism, *Klin. Wochenschr.* **53**:939.

Fiser, R. H., Jr., Williams, P. R., Fisher, D. A., Delamater, P. V., Sperling, M. A., and Oh, W., 1975, The effect of oral alanine on blood glucose and glucagon in the human newborn infant, *Pediatrics* **56**:78.

Fisher, D. A., 1975, Thyroid function in fetus, in: *Perinatal Thyroid Physiology and Disease* (D. A. Fisher and G. N. Burron, eds.), pp. 21–32, Raven Press, New York.

Fisher, D. A., and Dussault, J. H., 1974, Development of the mammalian thyroid gland, in: *Handbook of Physiology–Endocrinology, Vol. III, The Thyroid* (M. A. Greer, and D. H. Salomon, eds.), pp. 21–38, American Physiological Society, Washington, D.C.

Fisher, D. A., and Odell, W. D., 1969, Acute release of thyrotropin in the newborn, *J. Clin. Invest.* **48**:1670.

Fisher, D. A., and Sack, J., 1975, Thyroid function in neonate and possible approaches to newborn screening for hypothyroidism, in: *Perinatal Thyroid Physiology and Disease* (D. A. Fisher and G. N. Burrow, eds.), pp. 197–210, Raven Press, New York.

Fisher, D. A., Pyle, H. R., Porter, J. C., Beard, A. G., and Panos, T. C., 1963, Control of water balance in the newborn, *Am. J. Dis. Child.* **106**:137.

Fisher, D. A., Odell, W. D., Hobel, C. J., and Garza, R., 1969, Thyroid function in the term fetus, *Pediatrics* **44**:526.

Fisher, D. A., Hobel, C. J., Garza, R., and Pierce, C. A., 1970, Thyroid function in the preterm fetus, *Pediatrics.* **46**:208.

Fisher, D. A., Dussault, J. H., Hobel, C. J., and Lam, R., 1973, Serum and thyroid gland triiodothyronine in the human fetus, *J. Clin. Endocrinol. Metab.* **36**:397.

Fletcher, A. B., 1975, The infant of diabetic mother, in: *Neonatology* (G. B. Avery ed.), pp. 203–215, J. B. Lippincott, Philadelphia.

Forest, M. G., and Cathiard, A. M., 1975, Pattern of plasma testosterone and Δ4-androstenedione in normal newborns: Evidence for testicular activity at birth, *J. Clin. Endocrinol. Metab.* **41**:977.

Forest, M. G., Cathiard, A. M., and Bertrand, J., 1973, Total and unbound testosterone levels in newborns and normal and hypogonadal children: Use of a sensitive radioimmunoassay for testosterone, *J. Clin. Endocrinol. Metab.* **36**:1132.

Forest, M. G., Cathiard, A. M., Bourgeois, J., and Genoud, J., 1974a, Androgènes plasmatiques chez le nourrisson normal et prématuré. Relation avec la maturation de l'axe hypothalamo-hypophyso-gonadique, in: *International Symposium on Sexual Endocrinology of the Perinatal Period* (M. G. Forest and J. Bertrand, eds.), pp. 315–336, INSERM, Paris.

Forest, M. G., Sizonenko, P. C., Cathiard, A. M., and Bertrand, J., 1974b, The hypophyso-gonadal functions in humans during the first year of life. I. Evidence for testicular activity and feed back mechanism in early infancy, *J. Clin. Invest.* **53**:819.

France, J. T., 1971, Levels of 16α-hydroxy-dehydroepiandrosterone, dehydroepiandrosterone and pregnenolone in cord plasma of human normal and anencephalic fetuses, *Steroids* **17**:697.

Franchimont, P., and Pasteels, J. L., 1972, Sécrétion indépendante des hormones gonadotropes et de leurs sous-unités, *C. R. Acad. Sci. Paris Ser. D.* **275**:1799.

Franchimont, P., Legros, J. J., Deconinck, B., Demeyts, P., Goulard, M., Ketelslegers, J. M., and Schaub, C., 1970, Anterior pituitary function in human fetal life, *Symp. Deut. Ges. Endokrinol.* **16**:47.

Franks, R. C., 1967, Diurnal variation of plasma 17-hydroxysteroids in children, *J. Clin. Endocrinol. Metab.* **27**:75.

Fukuchi, M., Inoue, T., Abe, H., and Kumahara, Y., 1970, Thyrotropin in human fetal pituitaries, *J. Clin. Endocrinol. Metab.* **31**:565.

Gailaini, S. D., Nussbaum, A., McDougall, W. J., and McLimans, W. F., 1970, Studies on hormone production by human fetal pituitary cell cultures, *Proc. Soc. Exp. Biol. Med.* **134**:27.

Garel, J. M., 1971, Fetal calcemia and fetal parathyroids, *Isr. J. Med. Sci.* **7**:349.

Garel, J. M., 1972, Distribution of labeled parathyroid hormone in the rat fetus, *Horm. Metab. Res.* **4**:131.

Garel, J. M., Milhaud, G., and Sizonenko, P., 1969, Thyrocalcitonine et barrière placentaire chez le rat, *C.R. Soc. Biol.* **269:**1785.

Gasparo, de M., Van Assche, A., Gepts, W., and Hoet, J. J., 1969, The histology of the endocrine pancreas and the insulin content of the microdissected islets of foetal pancreas, *Rev. Fr. Etud. Clin. Biol.* **14:**904.

Geiger, W., 1973, Radioimmunological determination of human chorionic gonadotropin, human placental lactogen, growth hormone and thyrotropin in the serum of mother and child during the early puerperium, *Horm. Metab. Res.* **5:**342.

Genazzani, A. R., Fraiolo, F., Hurlimann, J., Fioretti, P., and Felber, J. P., 1975, Immunoreactive ACTH and cortisol plasma levels during pregnancy. Detection and partial purification of corticotrophin-like placental hormones. The human chorionic corticotrophin (HCC), *Clin. Endocrinol.* **4:**1.

Gibbons, J. M., Jr., Mitnick, M., and Chieffo, V., 1975, *In vitro* biosynthesis of TSH- and LH-releasing factor by the human placenta, *Am. J. Obstet. Gynecol.* **121:**127.

Gillman, J., 1948, Development of the gonads in man with a consideration of the role of fetal endocrines and the histogenesis of ovarian tumors, *Contrib. Embryol. (Carnegie Inst.)* **32:**81.

Girard, J. R., Cuendet, G. S., Marliss, E. B., Kervran, A., Rieutort, M., and Assan, R., 1973, Fuels, hormones and liver metabolism at term and during early postnatal period in the rat, *J. Clin. Invest.* **52:**3190.

Gitlin, D., and Biasucci, A., 1969*a,* Ontogenesis of immunoreactive growth hormone, follicle-stimulating hormone, thyroid-stimulating hormone, luteinizing hormone, chorionic prolactin and chorionic gonadotropin in the human conceptus, *J. Clin. Endocrinol. Metab.* **29:**926.

Gitlin, D., and Biasucci, A., 1969*b,* Ontogenesis of immunoreactive thyroglobulin in the human conceptus, *J. Clin. Endocrinol. Metab.* **28:**849.

Gitlin, D., Kumate, J., and Morales, C., 1965*a,* Metabolism and materno-fetal transfer of human growth hormone in the pregnant woman at term, *J. Clin. Endocrinol. Metab.* **25:**1599.

Gitlin, D., Kumate, J., and Morales, C., 1965*b,* On the transport of insulin across the human placenta, *Pediatrics* **35:**65.

Godard, C., Hufschmid, U., Gaillard, R., and Vallotton, M. B., 1975, Le Système rénine-angiotensine dans la période périnatale, *Helv. Paediatr. Acta Suppl.* **35:**20.

Goldman, A. S., Yakovac, W. C., and Bongiovanni, A. M., 1966, Development of activity of 3β-hydroxy-steroid dehydrogenase in human fetal tissues and in two anencephalic newborns, *J. Clin. Endocrinol. Metab.* **26:**14.

Grant, D. B., 1967, Fasting serum insulin levels in childhood, *Arch. Dis. Child.* **42:**375.

Grasso, S., Messina, A., Saporita, N., and Reitano, G., 1968, Serum insulin response to glucose and aminoacids in the premature infant, *Lancet* **2:**755.

Greenberg, R. E., and Gardner, L. I., 1960, The excretion of free catecholamines by newborn infants, *J. Clin. Endocrinol. Metab.* **20:**1207.

Greenberg, R. E., Lind, J., and von Euler, U. S., 1960, Effect of posture and insulin hypoglycemia on catecholamine excretion in the newborn, *Acta Paediatr. Scand.* **49:**780.

Greenberg, A. H., Czernichow, P., Reba, R. C., Tyson, J., and Blizzard, R. M., 1970, Observations on the maturation of thyroid function in early fetal life, *J. Clin. Invest.* **49:**1790.

Greene, J. W., and Toughstone, J. C., 1963, Urinary estriol as an index of placental function. A study of 274 cases, *Am. J. Obstet. Gynecol.* **85:**1.

Grillo, T. A. I., and Shima, K., 1966, Insulin content and enzyme histochemistry of the human foetal pancreatic islet, *J. Encodrinol.* **36:**151.

Groom, G. V., and Boyns, A. R., 1973, Effect of hypothalamic releasing factors and steroids on release of gonadotropins by organ cultures of human foetal pituitaries, *J. Endocrinol.* **59:**511.

Groom, G. V., Groom, M. A., Cooke, I. D., and Boyns, A. R., 1971, The secretion of immuno-reactive luteinizing hormone and follicle stimulating hormone by the human foetal pituitary in organ culture, *J. Endocrinol.* **49:**335.

Grumbach, M. M., and Kaplan, S. L., 1973, Ontogenesis of growth hormone, insulin prolactin and gonadotropin secretion in the human fetus, in: *Foetal and Neonatal Physiology,* pp. 462–487, Cambridge University Press, Cambridge.

Grumbach, M. M., and Kaplan, S. L., 1974, Human fetal pituitary hormones and the maturation of central nervous system regulation of anterior pituitary function, in: *Modern Perinatal Medicine* (L. Glück, ed.), pp. 247–271, Year Book Medical Publishers, Chicago.

Grumbach, M. M., and Van Wyk, J., 1974, Disorders of sex differentiation, in: *Textbook of Endocrinology* (R. H. Williams, ed.), pp. 423–501, W. B. Saunders, Philadelphia.

Grumbach, M. M., and Werner, S. C., 1956, Transfer of thyroid hormone across the humaa placenta at term, *J. Clin. Endocrinol. Metab.* **16:**1392.

Grumbach, M. M., Kaplan, S. L., Sciarra, J. J., and Burr, I. M., 1968, Chorionic growth hormone prolactin (CGP): Secretion, disposition, biologic activity in man, and postulated function as the "growth hormone" of the second half of pregnancy. *Ann. N.Y. Acad. Sci.* **148**:501.

Grumbach, M. M., Roth, J. C., Kaplan, S. L., and Kelch, R. P., 1974, Hypothalamic pituitary regulation of puberty: Evidence and concepts derived from clinical research, in: *The Control of the Onset of Puberty* (M. M. Grumbach, G. D. Grave, and F. Mayer, eds.), pp. 115–166, Wiley, New York.

Guyda, H. J., and Friesen, H. G., 1973, Serum prolactin levels in humans from birth to adult life, *Pediatr. Res.* **7**:534.

Hagen, C., and McNeilly, A. S., 1975a, Identification of human luteinizing hormone, follicle-stimulating hormone, luteinizing β-subunit and gonadotrophin α-subunit in foetal and adult pituitary glands, *J. Endocrinol.* **67**:49.

Hagen, C., and McNeilly, A. S., 1975b, The gonadotropic hormone and their subunits in human maternal and fetal circulation at delivery, *Am. J. Obstet. Gynecol.* **121**:926.

Harris, G. W., 1955, *Neural Control of the Pituitary Gland,* Edward Arnold, London.

Haworth, J. C., 1969, Carbohydrate metabolism in the fetus and the newborn, in: *Endocrine and Genetic Disorders of Childhood* (L. I. Gardner, ed.), pp. 788–798, W. B. Saunders, Philadelphia.

Hayek, A., Driscoll, S. G., and Warshaw, J. B., 1973, Endocrine studies in anencephaly, *J. Clin. Invest.* **52**:1636.

Haymond, M. W., Karl, I. E., and Pagliara, A. S., 1974, Increased gluconeogenic substrates in the small-for-gestational age infant, *N. Engl. J. Med.* **291**:322.

Heller, H., and Zaimis, E. J., 1949, The antidiuretic and oxytoxic hormones in the posterior pituitary glands of newborn infants and adults, *J. Physiol.* **109**:162.

Hennen, G., Pierce, J. G., and Freychet, P., 1969, Human chorionic thyrotropin: Further characterization and study of its secretion during pregnancy, *J. Clin. Endocrinol. Metab.* **29**:581.

Hershman, J. M., and Starnes, W. R., 1969, Extraction and characterization of a thyrotropic material from the human placenta, *J. Clin. Invest.* **48**:923.

Hirsch, J., Farquhar, J. W., Ahrens, E. H., Peterson, M. L., and Stoffel, W., 1960, Studies of adipose tissue in man, *Am. J. Clin. Nutr.* **8**:499.

Hollingsworth, D. R., and Austin, E., 1971, Thyroxine derivatives in amniotic fluid, *J. Pediatr.* **79**:923.

Honnebier, W. J., and Swaab, D. F., 1973, The influence of anencephaly upon intrauterine growth of fetus and placenta and upon gestation length, *J. Obstet. Gynaecol. Br. Commonw.* **80**:577.

Hoppenstein, J. M., Miltenberger, F. W., and Moran, W. H., 1968, The increase in blood levels of vasopressin in infants during birth and surgical procedures, *Surg. Gynecol. Obstet.* **127**:966.

Humbert, J. R., and Gotlin, R. W., 1971, Growth hormone levels in normoglycemic and hypoglycemic infants born small for gestational age, *Pediatrics* **48**:190.

Hunter, D. J. S., and Isles, T. E., 1972, The insulinogenic effect of glucagon in the newborn, *Biol. Neonate* **20**:74.

Hyyppa, M., 1972, Hypothalamic monoamines in human foetuses, *Neuroendocrinology* **9**:257.

Imperato-McGinley, J., Guerrero, L., Gautier, T., and Peterson, R. E., 1974, Steroid 5α-reductase deficiency in man: An inherited form of male pseudohermaphroditism *Science* **186**:1213.

Isles, T. E., Dickson, M., and Farquhar, J. W., 1968, Glucose tolerance and plasma insulin in newborn infants of normal and diabetic mothers, *Pediatr. Res.* **2**:198.

Jackson, I. M. D., and Reichlin, S., 1974, Thyrotropin-releasing hormone (TRH): Distribution in hypothalamic and extra-hypothalamic brain tissues of mammalian and submammalian chordates, *Endocrinology* **95**:854.

Jeffcoate, T. N. A., Fliegner, J. R. H., Russell, S. H., Davis, J. C., and Wade, A. P., 1965, Diagnosis of the adreno-genital syndrome before birth, *Lancet* **2**:553.

Johannisson, E., 1968, Foetal adrenal cortex in human: Its ultrastructure at different stages of development and in different functional states, *Acta. Endocrinol.* **58**(Suppl. 130):7.

Johnston, D. I., Bloom, S. R., Greene, K. R., and Beard, R. W., 1972, Failure of human placenta to transfer pancreatic glucagon, *Biol. Neonate* **21**:375.

Josimovich, J. B., Weiss, G., and Hutchinson, D. L., 1974, Sources and disposition of pituitary prolactin in maternal circulation, amniotic fluid, fetus and placenta in the pregnant rhesus monkey, *Endocrinology* **94**:1364.

Josso, N., 1972a, Evolution of the müllerian-inhibiting activity of the human testis. Effect of fetal, perinatal, and post-natal human testicular tissue on the müllerian duct of the fetal rat in organ culture, *Biol. Neonat.* **20**:368.

Josso, N., 1972b, Permeability of membranes to the müllerian inhibiting substance synthetized by the human fetal testis *in vitro:* A clue to its biochemical nature. *J. Clin. Endocrinol. Metab.* **34**:265.

Josso, N., 1975, L'hormone anti-müllerienne: Une foeto-proteine? *Arch. Fr. Pediatr.* **32**:109.

Jost, A., 1953, Problems of fetal endocrinology: The gonadal and hypophyseal hormone, *Rec. Prog. Horm. Res.* **8**:379.

Jost, A., 1954, Hormonal factors in the development of the fetus, *Cold Spring Harbor Symp. Quant. Biol.* **19**:167.

Jost, A., 1965, Gonadal hormones, in: *Sex Differentiation of the Mammalian Fetus in Organogenesis* (P. L. De Hann, and G. H. Vesprung, eds.), p. 611, Holt, Rinehart and Winston, New York.

Jost, A., 1966*a*, Problems of foetal endocrinology: The adrenal glands, *Rec. Prog. Horm. Res.* **22**:541.

Jost, A., 1966*b*, Anterior pituitary function in foetal life, in: *The Pituitary Gland* (G. W. Harris, and B. T. Donovan, eds.), Vol. 2, pp. 299–323, Butterworths, London.

Jost, A., 1970, Hormonal factors in the sex differentiation of the mammalian foetus, *Phil. Trans. R. Soc. London.* **259**:119.

Jost, A., and Picon, L., 1970, Hormonal control of fetal development and metabolism, *Adv. Metab. Disord.* **4**:123.

Jost, A., Dubony, J. P., and Geloso-Mayer, A., 1970, Hypothalamohypophyseal relationship in the fetus, in: *The Hypothalamus* (L. Martini, M. Motta, and F. Fraschini, eds.), p. 605, Academic Press, New York.

Jost, A., Vigier, B., Prepin, J., and Perchellet, J. P., 1973, Studies on sex differentiation in mammals, *Rec. Prog. Horm. Res.* **29**:1.

Kaplan, S. L., and Grumbach, M. M., 1965, Serum chorionic "growth hormone-prolactin" and serum pituitary growth hormone in mother and fetus at term, *J. Clin. Endocrinol. Metab.* **25**:1370.

Kaplan, S. L., Grumbach, M. M., and Shepard, T. H., 1972, The ontogenesis of human fetal hormones. I. Growth hormone and insulin, *J. Clin. Invest.* **51**:3080.

Kaplan, S. L., Grumbach, M. M., and Aubert, M. L., 1976, The ontogenesis of pituitary hormones and hypothalamic factors in the human fetus: Maturation of central nervous system. Regulation of anterior pituitary function. *Rec. Prog. Horm. Res.* **32**:161.

Kastin, A. J., Gennser, G., Arimura, A., Miller, M. C., III, and Schally, A. V., 1968, Melanocyte-stimulating and corticotrophic activities in human foetal pituitary glands, *Acta Endocrinol.* **58**:6.

Kaye, R., Baker, L., Kunzman, E. E., Prasad, A. L. N., and Davidson, M. H., 1970, Catecholamine excretion in spontaneously occurring asymptomatic neonatal hypoglycemia, *Pediatr. Res.* **4**:295.

Kearns, J. E., and Hutson, W., 1963, Tagged isomers and analogues of thyroxine: Their transmission across the human placenta and other studies, *J. Nucl. Med.* **4**:543.

Kenny, F. M., Preeyasombat, C., and Migeon, J., 1966, Cortisol production rate. II. Normal infants, children and adults, *Pediatrics* **37**:34.

Kenny, F. M., Angsusingha, K., Stinson, D., and Hotchkiss, J., 1973, Unconjugated estrogens in the perinatal period, *Pediatrics* **7**:826.

King, K. C., Adam, P. A. J., Laskowski, D. E., and Schwartz, R., 1971*a*, Sources of fatty acids in the newborn, *Pediatrics* **47**:192.

King, K. C., Adam, P. A. J., Schwartz, R., and Teramo, K., 1971*b*, Human placental transfer of human growth hormone I$^{125}$, *Pediatrics* **48**:534.

King, K. C., Butt, J., Raivio, K., Räihä, N., Roux, J., Teramo, K., Yamaguchi, K., and Schwartz, R. 1971*c*, Human maternal and fetal insulin response to arginine, *N. Engl. J. Med.* **285**:603.

King, K. C., Adam, P. A. J., Yamaguchi, K., and Schwartz, R., 1974, Insulin response to arginine in normal newborn infants and infants of diabetic mothers, *Diabetes* **23**:816.

Klein, A. H., Agustin, A. V., and Foley, T. P., 1974, Successful laboratory screening for congenital hypothyroidism, *Lancet* **2**:77.

Klein, G. P., Chan, S. K., and Giroud, C. J. P., 1969, Urinary excretion of 17-hydroxy- and 17-deoxysteroids of the pregn-4-ene series by the human newborn, *J. Clin. Endocrinol. Metab.* **29**:1448.

Klevit, H. D., 1966, Fetal–placental–maternal interrelations involving steroid hormones, *Pediatr. Clin. North Am.* **13**:59.

Kraus, F. F., and Neubecker, R. D., 1962, Luteinization of the ovarian thera in infants and children, *Am. J. Clin. Pathol.* **37**:389.

Kreines, K., and De Vaux, W. D., 1971, Neonatal adrenal insufficiency associated with maternal Cushing's syndrome, *Pediatrics* **47**:516.

Laatikainen, T., and Peltonen, J., 1974, Levels of estriol, estriol sulfate, progesterone and ventral steroid mono- and disulfates in umbilical cord arterial and venous plasma, in: *International Symposium on Sexual Endocrinology of the Perinatal Period,* **32**:225–266, INSERM, Paris.

Lacourt, G., Sizonenko, P. C., Engelhorn, A., Berli, C., Arendt, J., Paunier, L., and Beguin, F., 1975, Détermination du cortisol dans le liquide amniotique. Indice de maturation foetale? *Helv. Paediatr. Acta Suppl.* **35**:13.

Lambert, A. E., Bzondel, B., Kanazawa, Y., Orci, L., and Renold, A. E., 1972, Monolayer cell culture of neonatal rat pancreas: Light microscopy and evidence for immunoreactive insulin synthesis and release, *Endocrinology* **90**:239.

Laron, Z., and Pertzelan, A., 1969, Somatotrophin in antenatal and perinatal growth and development, *Lancet* **1**:680.

Leach, F. N., Ashworth, M. A., Barson, A. J., and Milner, R. D. G., 1973, Insulin release from human foetal pancreas in tissue culture, *J. Endocrinol.* **59**:65.

Le Dune, M. A., 1972a, Intravenous glucose tolerance and plasma insulin studies in small-for-dates infants, *Arch. Dis. Child.* **47**:111.

Le Dune, M. A., 1972b, Response to glucagon in small-for-dates hypoglycaemic and non-hypoglycaemic newborn infants, *Arch. Dis. Child.* **47**:754.

Levina, S. E., 1968, Endocrine features in development of human hypothalamus, hypophysis, and placenta, *Gen. Comp. Endocrinol.* **11**:151.

Levina, S. E., 1970, Regulation of secretion of hypophyseal gonadotropins in human embryogenesis, *Probl. Endocrinol.* **16**:50.

Levina, S. E., 1972, Times of appearance of LH and FSH activities in human foetal circulation, *Gen. Comp. Endocrinol.* **19**:242.

Lieblich, J. M., and Utiger, R. D., 1973, Triiodothyronine in cord serum, *J. Pediatr.* **82**:290.

Liggins, G. C., 1968, Premature parturition after infusion of corticotrophin or cortisol into foetal lambs, *J. Endocrinol.* **42**:323.

Linarelli, L. G., 1972, Newborn urinary cyclic AMP and developmental renal responsiveness to parathyroid hormone, *Pediatrics* **49**:14.

Little, B. O., Smith, O. W., Jessiman, A. G., Selenkow, H. A., Van't Hoff, W., Eglin, J. M., and Moore, F. D., 1958, Hypophysectomy during pregnancy in a patient with cancer of the breast: Case report with hormone studies, *J. Clin. Endocrinol. Metab.* **18**:425.

Liu, N., 1966, Some aspects of endocrinology in the fetus and in the newborn, *Pediatr. Clin. North Am.* **13**:1047.

Ljungqvist, A., and Wagermark, J., 1966, Renal juxta-glomerular granulation in the human foetus and infant, *Acta Pathol. Microbiol. Scand.* **67**:257.

Loriaux, D. L., Ruder, H. J., Knab, D. R., and Lipsett, M. B., 1972, Estrone sulfate, estrone, estradiol and estriol plasma levels in human pregnancy, *J. Clin. Endocrinol. Metab.* **35**:887.

Luyckx, A. S., Massi-Benedetti, F., Falorni, A., and Lefebvre, P., 1972, Presence of pancreatic glucagon in the portal plasma of human neonates: Differences in the insulin and glucagon responses to glucose between normal infants and infants from diabetic mothers, *Diabetologia* **8**:296.

MacDonald, P. C., and Siiteri, P. K., 1965, Origin of estrogen in women pregnant with an anencephalic fetus, *J. Clin. Invest.* **44**:465.

Magendantz, H. G., and Ryan, K. J., 1964, Isolation of a new estriol precursor, *Fed. Proc.* **23**:275.

Malpas, P., 1933, Post maturity and malformation of foetus, *J. Obstet. Gynaecol. Br. Commonw.* **40**:1046.

Maner, F. D., Saffan, B. D., Wiggins, R. A., Thompson, J. D., and Preedy, J. R. K., 1963, Interrelationship of estrogen concentrations in the maternal circulation, fetal circulation and maternal urine in late pregnancy, *J. Clin. Endocrinol. Metab.* **23**:445.

Massi-Benedetti, F., Falorni, A., Luyckx, A., and Lefebvre, P., 1974, Inhibition of glucagon secretion in the human newborn by simultaneous administration of glucose and insulin, *Horm. Metab. Res.* **6**:392.

Matsuzaki, F., Irie, M., and Shizume, K., 1971, Growth hormone in human fetal pituitary glands and cord blood, *J. Clin. Endocrinol. Metab.* **33**:908.

McKendrick, T., and Edwards, R. W. H., 1965, The excretion of 4-hydroxy-3-methoxy mandelic acid by children, *Arch. Dis. Child.* **40**:418.

McKenzie, J. M., 1964, Neonatal Graves' disease, *J. Clin. Endocrinol. Metab.* **24**:660.

Merkatz, I. R., New, M. I., Peterson, R. E., and Seaman, M. P., 1969, Prenatal diagnosis of adrenogenital syndrome by amniocentesis, *J. Pediatr.* **75**:977.

Mestyan, J., Soltész, Gy., Schultz, K., and Horvath, M., 1975, Hyperaminoacidemia due to the accumulation of glucogenic amino acid precursors in hypoglycemic small-for-gestational-age infants, *J. Pediatr.* **87**:409.

Migeon, C. J., 1959, Cortisol production and metabolism in the neonate, *J. Pediatr.* **55**:280.

Migeon, C. J., Keller, A. R., and Holmstrom, E. G., 1955, Dehydroisoandrosterone, androsterone and 17-hydroxycorticosteroid levels in maternal and cord plasma in cases of vaginal delivery, *Bull. Johns Hopkins Hosp.* **97**:415.

*PIERRE C. SIZONENKO*
*and MICHEL L. AUBERT*

Migeon, C. J., Prystowsky, H., Grumbach, M. M., and Byron, M. C., 1956, Placental passage of 17-hydroxycorticosteroids: Comparison of the levels in maternal and fetal plasma and effect of ACTH and hydrocortisone administration, *J. Clin. Invest.* **35**:488.

Milner, R. D. G., and Woodhead, J. S., 1975, Parathyroid hormone secretion during exchange transfusion, *Arch. Dis. Child.* **50**:298.

Milner, R. D. G., and Wright, A. D., 1966, Blood glucose, plasma insulin and growth hormone response to hyperglycaemia in the newborn, *Clin. Sci.* **31**:309.

Milner, R. D. G., and Wright, A. D., 1967, Plasma glucose, nonesterified fatty acid, insulin and growth hormone response to glucagon in the newborn, *Clin. Sci.* **32**:249.

Milner, R. D. G., Ashworth, M. A., and Barson, A. J., 1972*a* Insulin release from human foetal pancreas in response to glucose, leucine and arginine, *J. Endocrinol.* **52**:497.

Milner, R. D. G., Fekete, M., and Assan, R., 1972*b*, Glucagon, insulin, and growth hormone response to exchange transfusion in premature and term infants, *Arch. Dis. Child.* **47**:186.

Milner, R. D. G., Fekete, M., Assan, R., and Hodge, J. S., 1972*c*, Effect of glucose in plasma glucagon, growth hormone, and insulin in exchange transfusion, *Arch. Dis. Child.* **47**:179.

Mintz, D. H., Chez, R. A., and Horger, E. O. III., 1969, Fetal insulin and growth hormone metabolism in the subhuman primate, *J. Clin. Invest.* **48**:176.

Mitskevich, M. A., and Levina, S. E., 1965, Investigation on the structure and gonadotropic activity of the anterior pituitary in human embryogenesis, *Arch. Anat. Microsc. Morphol. Exp.* **54**:129.

Miyakawa, I., Ikeda, I., and Sheyama, M., 1974, Transport of ACTH across human placenta, *J. Clin. Endocrinol. Metab.* **39**:440.

Mølsted-Pedersen, L., and Jørgensen, K. R., 1972, Aspects of carbohydrate metabolism in newborn infants of diabetic mothers. III. Plasma insulin during intravenous glucose tolerance test, *Acta Endocrinol.* **71**:115.

Money, J., and Ehrhardt, A. A., 1972, *Man and Woman, Boy and Girl: Differentiation and Dimorphism of Gender Identity from Conception to Maturity,* Johns Hopkins University Press, Baltimore.

Montalvo, J. M., Wahner, H. W., Mayberry, W. E., and Lum, R. K., 1973, Serum triiodothyronine, total thyroxine, and thyroxine to triiodothyronine ratios in paired maternal-cord sera and at one week and one month of age, *Pediatr. Res.* **7**:706.

Moore, W. M. O., Ward, B. S., and Gordon, C., 1974, Human placental transfer of glucagon, *Clin. Sci. Mol. Med.* **46**:125.

Mosier, H. D., 1956, Hypoplasia of the pituitary and adrenal cortex, *J. Pediatr.* **48**:63.

Murphy, B. E. P., 1973*a*, Does the human fetal adrenal play a role in parturition? *Am. J. Obstet. Gynecol.* **115**:521.

Murphy, B. E. P., 1973*b*, Steroid arteriovenous differences in umbilical cord plasma: Evidence of cortisol production by the human fetus in early gestation, *J. Clin. Endocrinol. Metab.* **36**:1037.

Murphy, B. E. P., and Diez d'Aux, R. C., 1972, Steroid levels in the human fetus: Cortisol and cortisone, *J. Clin. Endocrinol. Metab.* **35**:678.

Murphy, B. E. P., Patrick, J., and Denton, R. L., 1975, Cortisol in amniotic fluid during human gestation, *J. Clin. Endocrinol. Metab.* **40**:164.

Myant, N. B., 1958, Passage of thyroxine and triiodothyronine from mother to fetus in pregnant women, *Clin. Sci.* **17**:75.

Nichols, J., 1970, Antenatal diagnosis and treatment of the adenogenital syndrome, *Lancet* **1**:83.

Niemi, M., and Baillie, A. H., 1965, 3-$\beta$-hydroxysteroid deshydrogenase activity in the human foetal adrenal cortex, *Acta Endocrinol.* **48**:423.

Niemineva, K., 1950, Observations on the development of the hypophyseal-portal system, *Acta Paediatr.* **39**:366.

Nisula, B. C., and Ketelslegers, J. M., 1974, Thyroid-stimulating activity and chorionic gonadotropin, *J. Clin. Invest.* **54**:494.

Norris, E. H., 1946, Anatomical evidence of prenatal function of the human parathyroid glands, *Anat. Rec.* **96**:129.

Nwosu, U., Wallach, E. E., Boggs, T. R., Nemiroff, R. L., and Bongiovanni, A. M., 1975, Possible role of the fetal adrenal glands in the etiology of post-maturity, *Am. J. Obstet. Gynecol.* **121**:366.

Obenshain, S. S., Adam, P. A. J., King, K. C., Teramo, K., Raivio, K. O., Räihä, N., and Schwartz, R., 1970, Human fetal insulin response to sustained maternal hyperglycemia, *N. Engl. J. Med.* **283**:566.

Parlow, A. F., 1974, Human pituitary growth hormone, adrenocorticotropic hormone, follicle stimulating hormone and luteinizing hormone concentrations in relation to age and sex, as revealed by bioassay, in: *Advances in Human Growth Hormone Research, Symposium in Baltimore* (S. Raiti, ed.), pp. 658–666, DHEW Publication No. (NIH) 74-612, Washington, D.C.

Pasteels, J. L., Brauman, H., and Brauman, J., 1963, Etude comparée de la sécrétion d'hormone somatotrope par l'hypophyse humaine *in vitro* et de son activité lactogénique, *C. R. Acad. Sci.* **256**:2031.

Pasteels, J. L., Gausset, P., Danguy, A., Ectors, F., Nicoll, C. S., and Varavudhi, P., 1972, Morphology of the lactotropes and somatotropes of man and rhesus monkey, *J. Clin. Endocrinol. Metab.* **34**:959.

Pasteels, J. L., Gausset, P., Danguy, A., and Ectors, F., 1974, Gonadotropin secretion by human foetal and infant pituitaries, in: *International Symposium on Sexual Endocrinology of the Perinatal Period,* **32**:13–35, INSERM, Paris.

Payne, A. H., and Jaffe, R. B., 1974, Androgen formation from pregnenolone sulfate by the human fetal ovary, *J. Clin. Endocrinol. Metab.* **39**:300.

Payne, A. H., and Jaffe, R. B., 1975, Androgen formation from pregnenolone sulfate by fetal, neonatal, prepubertal and adult human testes, *J. Clin. Endocrinol. Metab.* **40**:102.

Pearse, A. G. E., 1953, Cytological and cytochemical investigations on the foetal and adult hypophysis in various physiological and pathological states, *J. Pathol. Bacteriol.* **65**:355.

Pearse, A. G. E., and Carvalheira, A. F., 1967, Cytochemical evidence for an ultimobronchial origin of rodent thyroid C cells, *Nature* **214**:929.

Pearson, D. B., Goodman, R., and Sachs, H., 1975, Stimulated vasopressin synthesis by a fetal hypothalamic factor, *Science* **187**:1081.

Penny, R., Olambiwonnu, N. O., and Frasier, S. D., 1974, Follicle-stimulating hormone (FSH) and luteinizing hormone, human chorionic gonadotropin (LH-HCG) concentrations in paired maternal and cord sera, *Pediatrics* **53**:41.

Persson, B., and Gentz, J., 1966, The pattern of blood lipids glycerol and ketone bodies during the neonatal period, infancy and childhood, *Acta. Paediatr. Scand.* **55**:353.

Pic, P., 1973, Rôle des parathyroïdes foetales dans la régulation de la calcémie et de la phosphorémie du foetus du rat, *Ann. Endocrinol.* **34**:621.

Pierson, M., Malaprade, D., Grignon, G., Harteman, P., Bellevelle, F., Lemoine, D., and Nabet, P., 1973, Etude de la sécrétion hypophysaire du foetus humain: Correlation entre morphologie et activité sécrétoire, *Ann. Endocrinol.* **34**:418.

Pildes, R. S., Hart, R. J., Warrner, R., and Cornblath, M., 1969, Plasma insulin response during oral glucose tolerance tests in newborns of normal and gestational diabetic mothers, *Pediatrics* **44**:76.

Ponte, C., Gaudier, B., Deconinck, B., and Fourlinnie, J. C., 1972*a,* Blood glucose, serum insulin and growth hormone response to intravenous administration of arginine in premature infants, *Biol. Neonate* **20**:262.

Ponte, C., Gaudier, B., Franchimont, P., Deconinck, B., Fourlinnie, J. C., and Delabre, M., 1972*b,* Etude des sécrétions de l'hormone de croissance et de l'insuline chez le nouveau-né. *Arch. Fr. Pediatr.* **92**:801.

Poonai, A. P. V., Tang, K., and Poonai, P. V., 1975, Relation between glucose, insulin and growth hormone in the fetus during labor and at delivery, *Am. J. Obstet. Gynecol.* **45**:155.

Pryse-Davies, J., and Dewhurst, C. J., 1971, The development of the ovary and uterus in the foetus, newborn and infant: A morphological and enzyme histochemical study, *J. Pathol.* **103**:5.

Räihä, N., and Hjelt, L., 1957, The correlation between the development of the hypophysial portal system and the onset of neurosecretory activity in the human fetus and infant, *Acta Paediatr.* **46**:610.

Raiti, S., Holzman, G. B., Scott, R. L., and Blizzard, R. M., 1967, Evidence for the placental transfer of triiodothyronine in human beings, *N. Engl. J. Med.* **277**:456.

Rastogi, G. I., Letarte, J., and Fraser, T. R., 1970, Immunoreactive insulin content of 203 pancreases from foetuses of healthy mothers, *Diabetologia* **6**:445.

Reid, J. R., 1960, Congenital absence of the pituitary gland, *J. Pediatr.* **56**:658.

Reinisch, J. M., 1974, Fetal hormones, the brain and human sex differences: A heuristic integrative review of the recent literature, *Arch. Sex. Behav.* **3**:51.

Reitano, G., Grasso, S., Di Stefano, G., and Messina, A., 1971, The serum insulin and growth hormone response to arginine and to arginine with glucose in the premature infant, *J. Clin. Endocrinol. Metab.* **33**:924.

Reyes, F. I., Winter, J. S. D., and Faiman, C., 1973, Studies on human sexual development. I. Fetal gonadal and adrenal sex steroids, *J. Clin. Endocrinol. Metab.* **37**:74.

Reyes, F. I., Boroditsky, R. S., Winter, J. S. D., and Faiman, C., 1974, Studies on human sexual development. II. Fetal and maternal serum gonadotropin and sex steroid concentrations, *J. Clin. Endocrinol. Metab.* **38**:612.

Reynolds, J. W., 1964, The isolation of 16 keto-androstenediol ($3\beta$, $17\beta$-dihydroxyandrost-5-en-16-one) from the urine of a newborn infant, *Steroids* **3**:77.

Reynolds, J. W., Colle, E., and Ulstrom, R. A., 1962, Adrenocortical steroid metabolism in newborn infants. V. Physiologic disposition of exogenous cortisol loads in the early neonatal period, *J. Clin. Endocrinol. Metab.* **22**:245.

Rinne, U. K., 1962, Neurosecretory material passing into the hypophysial portal system in the human infundibulum and its foetal development, *Acta Neuroveg.* **25**:310.

Rinne, U. K., Kivalo, E., and Talanti, S., 1962, Maturation of human hypothalamic neurosecretion, *Biol. Neonat.* **4**:351.

Roberts, G., and Cawdery, J. E., 1970, Congenital adrenal hypoplasia, *J. Obstet. Gynaecol. Br. Commonw.* **77**:654.

Robin, N. I., Refetoff, S., Gleason, R. E., and Selenkow, H. A., 1970, Thyroid hormone relationships between maternal and fetal circulations in human pregnancy at term: A study in patients with normal and abnormal thyroid function, *Am. J. Obstet. Gynecol.* **108**:1269.

Rodeck, H., and Caesar, R., 1956, Zur Entwicklung des neurosekretorischen Systems bei Säugern und Mensch und der Regulationsmechanismen des Wasserhaushaltes, *Z. Zellforsch.* **44**:666.

Ross, G. T., 1974, Gonadotropins and preantral follicular maturation in women, *Fertil. Steril.* **25**:522.

Ryan, K. J., 1959, Aromatization of steroids, *J. Biol. Chem.* **234**:268.

Ryle, M., Stephenson, J., Williams, J., and Stuart, J., 1975, Serum gonadotrophins in young children, *Clin. Endocrinol.* **4**:413.

Sack, J., Fisher, D. A., Hobel, C. J., and Lam, R., 1975a, Thyroxine in human amniotic fluid, *J. Pediatr.* **87**:364.

Sack, J., Beaudry, M., De La Mater, P., Oh, W., and Fisher, D., 1975b, The mechanism of the $T_3$ response to parturition, *Pediatr. Res.* **9**:682.

Samaan, N. A., Anderson, G. D., and Adam-Mayne, M. E., 1975, Immunoreactive calcitonin in the mother, neonate, child and adult, *Am. J. Obstet. Gynecol.* **121**:622.

Schally, A. V., Arimura, A., and Kastin, A. J., 1973, Hypothalamic regulatory hormones. At least nine substances from the hypothalamus control the secretion of pituitary hormones, *Science* **179**:341.

Schindler, A. E., and Siiteri, P. K., 1968, Isolation and quantitation of steroids from human amniotic fluid, *J. Clin. Endocrinol. Metab.* **28**:1189.

Seppälä, M., Aho, I., Tissari, A., and Ruoslahti, E., 1972, Radioimmunoassay of oxytocin in amniotic fluid, fetal urine and meconium during late pregnancy and delivery, *Am. J. Obstet. Gynecol.* **114**:788.

Shaywitz, B. A., Finkelstein, J., Hellman, L., and Weitzman, E. D., 1971, Growth hormone in newborn infants during sleep-wake periods, *Pediatrics* **48**:103.

Shepard, T. H., 1967, Onset of function in the human fetal thyroid: Biochemical and radioautographic studies from organ culture, *J. Clin. Endocrinol. Metab.* **27**:945.

Shepard, T. H., 1968, Development of the human fetal thyroid, *Gen. Comp. Endocrinol.* **10**:174.

Shima, K., Price, S., and Foa, P. P., 1966, Serum insulin concentration and birth weight in human infants, *Proc. Soc. Exp. Biol. Med.* **121**:55.

Shutt, D. A., Smith, I. D., and Shearman, R. P., 1974, Oestrone, oestradiol-17β and oestriol levels in human foetal plasma, during gestation and at term, *J. Endocrinol.* **60**:333.

Siegel, S. R., Fisher, D. A., and Oh, W., 1974, Serum aldosterone concentrations related to sodium balance in the newborn infant, *Pediatrics* **53**:410.

Siiteri, P. K., 1974, Steroid hormones in pregnancy, in: *Modern Perinatal Medicine* (L. Gluck ed.), pp. 231–241, Year Book Medical Publishers, Chicago.

Siiteri, P. K., and MacDonald, P. C., 1966, Placental estrogen biosynthesis during human pregnancy, *J. Clin. Endocrinol. Metab.* **26**:751.

Siiteri, P. K., and Wilson, J. D., 1974, Testosterone formation and metabolism during male sexual differentiation in the human embryo, *J. Clin. Endocrinol. Metab.* **38**:113.

Siler, T. M., Morgenstern, L. L., and Greenwood, F. C., 1972, The release of prolactin and other peptide hormones from human anterior pituitary tissue cultures, in: *Lactogenic Hormones, Ciba Foundation Symposium* (G. E. W. Wolstenholme and J. Knight, eds.), pp. 1071–1073, Churchill and Livingstone, London.

Siler-Khodr, T. M., Morgenstern, L. L., and Greenwood, F. C., 1974, Hormone synthesis and release from human fetal adenohypophyses *in vitro, J. Clin. Endocrinol. Metab.* **39**:891.

Similä, S., Koivisto, M., Ranta, T., Leppäluoto, J., Reinalä, M., and Haapalahti, J., 1975, Serum triiodothyronine, thyroxine, and thyrotrophin concentrations in newborns during the first two days of life, *Arch. Dis. Child.* **50**:565.

Simmer, H. H., Frankland, M., and Greipel, M., 1975, On the regulation of fetal and maternal 16α-hydroxy dehydroepiandrosterone and its sulfate by cortisol and ACTH in human pregnancy at term, *Am. J. Obstet. Gynecol.* **121**:646.

Sizonenko, P. C., 1976, Plasma concentrations of DHEA, and DHEA-sulfate, in the newborn during the first ten days of life, unpublished data.

Sizonenko, P. C., Zahnd, G., Paunier, L., Lacourt, G., and Kohlberg, I. J., 1972, Plasma glucose, insulin, cortisol, and growth hormone to glucagon stimulation in newborns, *Pediatr. Res.* **6**:63.

Sizonenko, P. C., Paunier, L., Zahnd, G., Lacourt, G., and Bieler, E., 1974, Alanine, insulin and glucose levels in newborns, *Pediatr. Res.* **8**:918.

Smith, I. D., and Shearman, R. P., 1972, The relationship of human umbilical arterial and venous plasma levels of corticosteroids to gestational age, *J. Endocrinol.* **55**:211.

Smith, I. D., and Shearman, R. P., 1974, Fetal plasma steroids in relation to parturition. II. The effect of gestational age upon umbilical plasma corticosteroids following hysterotomy and caesarian section, *J. Obstet. Gynaecol. Br. Commonw.* **81**:16.

Smith, P. E., 1954, Continuation of pregnancy in rhesus monkeys *(Macaca mulatta)* following hypophysectomy, *Endocrinology* **55**:655.

Snell, K., and Walker, D. G., 1973, Glyconeogenesis in the newborn rat: The substrates and their quantitative significance, *Enzyme* **15**:40.

Solomon, S., Bird, C. E., Lind, W., Iwamoya, M., and Young, P. C. M., 1967, Formation and metabolism of steroids in the fetus and placenta, *Rec. Prog. Horm. Res.* **23**:297.

Spellacy, W. N., Buhi, W. C., Riggall, F. C., and Holsinger, K. L., 1973, Human amniotic fluid lecithin/sphingemyelin ratio changes with estrogen or glucocorticoid treatment, *Am. J. Obstet. Gynecol.* **115**:216.

Sperling, M. A., Delamater, P. V., Phelps, D., Fiser, R. H., Oh, W., and Fisher, D. A., 1974, Spontaneous and amino-acid-stimulated glucagon secretion in the immediate postnatal period. Relation to glucose and insulin, *J. Clin. Invest.* **53**:1159.

Stark, E., Gyévai, A., Szalay, K., and Acz, Z. S., 1965, Hypophyseal–adrenal activity in combined human foetal tissue cultures, *Can. J. Physiol. Pharmacol.* **43**:1.

Steinke, J., and Driscoll, S. G., 1965, The extractable insulin content of pancreas from fetus and infants of diabetic and control mothers, *Diabetes* **14**:573.

Stern, L., 1975, Disturbances in glucose, calcium, and magnesium homeostasis, in: *Neonatalogy* (G. B., Avery ed.), pp. 423–435, J. B. Lippincott, Philadelphia.

Stubbe, P., and Wolf, H., 1971, The effect of stress on growth hormone, glucose, and glycerol levels in newborn infants, *Horm. Metab. Res.* **3**:175.

Symonds, E. M., and Furler, I., 1973, Plasma renin levels in the normal and anephric fetus at birth, *Biol. Neonate* **23**:133.

Symonds, E. M., Stanley, M. A., and Skinner, S. L., 1968, Production of renin by *in vitro* cultures of human chorion and uterine muscle, *Nature* **217**:1152.

Tato, L., Du Caju, M. V. L., Prévot, C., and Rappaport, R., 1975, Early variations of plasma somatomedin in the newborn, *J. Clin. Endocrinol. Metab.* **40**:534.

Thorell, J. I., 1970, Plasma insulin levels in normal human foetuses, *Acta. Endocrinol.* **63**:134.

Tobin, J. D., Roux, J. F., and Soeldner, J. S., 1969, Human fetal insulin response after acute maternal glucose administration during labor, *Pediatrics* **44**:668.

Toublanc, J. E., Tea, N. T., Roger, M., and Joab, N., 1975, Plasma 17-hydroxyprogesterone and progesterone in normal newborns and infants, *Pediatr. Res.* **9**:675.

Tsang, R. C., Chen, I. W., Friedman, M. A., and Chen, I., 1973, Neonatal parathyroid function: Role of gestational age and post-natal age, *J. Pediatr.* **83**:728.

Tuchmann-Duplessis, H., and Mercier-Parrot, L., 1963, Etude comparative de la structure de l'hypophyse et de la surrénale des anencéphales et des hydrocéphales humains, *C. R. Soc. Biol.* **157**:977.

Turner, R. C., Schneeloch, B., and Paterson, P., 1971, Changes in plasma growth hormone and insulin of the human foetus following hysterotomy, *Acta Endocrinol.* **66**:577.

Turner, R. C., Oakely, N. W., and Beard, R. W., 1973, Human foetal plasma growth hormone prior to the onset of labour, *Biol. Neonate* **22**:169.

Tyson, J. E. A., Barnes, A. C., Merimee, T. J., and McKusik, V. A., 1970, Isolated growth hormone deficiency: Studies in pregnancy, *J. Clin. Endocrinol. Metab.* **31**:147.

Tyson, J. E., Hwang, P., Guyda, H., and Friesen, H. G., 1972, Studies of prolactin secretion in human pregnancy, *Am. J. Obstet. Gynecol.* **113**:14.

Valdes-Dapena, M. A., 1967, The normal ovary of childhood, *Ann. N.Y. Acad. Sci.* **142**:597.

Van Wagenen, G., and Simpson, M. E., 1965, *Embryology of the Ovary and Testis,* Yale University Press, New Haven.

Villee, C. A., 1953, Regulation of blood glucose in the human fetus, *J. Appl. Physiol.* **5**:437.

Villee, C. A., 1961, Enzymes in the development of homeostatic mechanisms, in: *Somatic Stability in the Newly Born, Ciba Foundation Symposium* (G. E. W. Wolstenholme and M. O'Connor, eds.), p. 246, Churchill, London.

Villee, C. A., and Loring, J. M., 1969, The synthesis and cleavage of steroid sulfates in the human fetus and placenta, in: *The Foeto-Placental Unit* (A. Pecile and C. Finzi, eds.), p. 182, Excerpta Medica Foundation, Amsterdam.

Villee, D. B., 1969, Development of endocrine function in human placenta and fetus, *N. Engl. J. Med.* **281**:473.

Villee, D. B., 1972, The development of steroidogenesis, *Am. J. Med.* **53**:533.

Villee, D. B., 1974, The control of androgen synthesis in human fetal testicular cells in culture, *International Symposium on Sexual Endocrinology of the Perinatal Period,* **32**:247–254, INSERM, Paris.

Villee, D. B., Engel, L. L., Loring, J. M., and Villee, C. A., 1961, Steroid hydroxylation in human fetal adrenals. II. Formation of 16α-hydroxyprogesterone, 17-hydroxyprogesterone and deoxycorticosterone, *Endocrinology* **69**:354.

Walton, R. L., 1954, Neonatal tetany in two siblings: Effect of maternal hyperparathyroidism, *Pediatrics* **13**:227.

Weldon, V. V., Kowarski, A., and Migeon, C. J., 1967, Aldosterone secretion rates in normal subjects from infancy to adulthood, *Pediatrics* **39**:713.

Wellman, K. F., Volk, B. W., and Brancato, P., 1971, Ultrastructure and insulin content of the endocrine pancreas in the human fetus, *Lab. Invest.* **25**:97.

West, G. B., Shepherd, D. M., and Hunter, R. B., 1951, Adrenaline and noradrenaline concentrations in adrenal glands at different ages and in some diseases, *Lancet* **2**:966.

Widdowdson, E. M., and Spray, C. M., 1951, Chemical development *in utero, Arch. Dis. Child.* **26**:205.

Williams, P. R., Fiser, R. H., Sperling, M. A., and Oh, W., 1975, Effects of oral alanine feeding blood glucose, plasma glucagon and insulin concentrations in small-for-gestational-age infants, *N. Engl. J. Med.* **292**:612.

Winter, J. S. D., and Faiman, C., 1972, Serum gonadotropin concentrations in agonadal children and adults, *J. Clin. Endocrinol. Metab.* **35**:561.

Winters, A. J., Eskay, R. L., and Porter, J. C., 1974*a,* Concentration and distribution of TRH and LRH in the human fetal brain, *J. Clin. Endocrinol. Metab.* **39**:960.

Winters, A. J., Oliver, C., Colston, C., MacDonald, P. C., and Porter, J. C., 1974*b,* Plasma ACTH levels in the human fetus and neonate as related to age and parturition, *J. Clin. Endocrinol. Metab.* **39**:269.

Wise, J. K., Lyall, S. S., Hendler, R., and Felig, P., 1973, Evidence of stimulation of glucagon secretion by alanine in the human fetus, *J. Clin. Endocrinol. Metab.* **37**:345.

Witschi, E., 1967, Biochemistry of sex differentiation in vertebrate embryos, in: *The Biochemistry of Animal Development* (R. Weber, ed.), Vol. 2, p. 193, Academic Press, New York.

Wolf, H., Stubbe, P., and Sabata, V., 1970, The influence of maternal glucose infusions on fetal growth hormone levels, *Pediatrics* **45**:36.

Wurtman, R. J., and Axelrod, J., 1956, Adrenaline synthesis. Control by the pituitary gland and adrenal glucocorticoids, *Science* **150**:1464.

Yen, S. S. C., Samaan, N., and Pearson, O. H., 1967, Growth hormone levels in pregnancy, *J. Clin. Endocrinol. Metab.* **27**:1341.

Zondek, L. H., and Zondek, T., 1974, The influence of complications of pregnancy and of some congenital malformations on the reproductive organs of the male foetus and neonate, in: *International Symposium on Sexual Endocrinology of the Perinatal Period,* Vol. 32, pp. 79–96, INSERM, Paris.

# 20

# Development of Immune Responsiveness

## A. R. HAYWARD

## 1. Introduction

The immunologist's view of the ontogeny of the immune response has changed over the past few years from the concept of a period of total unresponsiveness followed by the chaos of random somatic mutation to a concept of a genetically determined sequence of maturation occuring in an orderly predetermined manner. Although the numerous mechanisms which must be involved in this process are largely unknown, it has become clear that factors other than those which might be considered strictly immunological are important. These include the adverse consequences of intrauterine malnutrition or lack of thyroid or pituitary hormones. Rather than review all the information related to the development of immune responses, I have concentrated principally on the following topics because of their biological interest:

1. Evidence that human fetuses develop the capacity for immunity responses during intrauterine life
2. Observations linking the development of lymphocytes and their subpopulations to responsiveness
3. The relationship to ontogeny of the generation of diversity of antigen specificities
4. The relationship of tolerance to ontogeny

## 2. Specific Immune Responses

The mechanisms of specific immunity, which form the basis for most of this chapter, are summarized briefly below. Information which directly relates growth to immunity development is sparse and is therefore discussed last.

*A. R. HAYWARD* • Department of Pediatrics, and the Comprehensive Cancer Center, University of Alabama, Birmingham, Alabama. On leave from Department of Immunology, Institute of Child Health, University of London.

Specific immune responses depend on the recognition of a foreign structure (an antigen) by appropriate clones of lymphocytes with complementary receptors for that structure. Such recognition requires contact between the lymphocyte and the antigen, which is often most effectively mediated by macrophages. Stimulated lymphocytes generally proliferate, and they and their progeny mediate the two major effector limbs of the immune response. One effector mechanism depends on the secretion of antibody by plasma cells, which are the descendants of a class of lymphocytes called B lymphocytes. This class is characterized by the presence of small amounts of immunoglobulin (which is their receptor for antigen) on the cell surface. Antibody activity is restricted to the group of serum proteins called immunoglobulins, of which there are five classes (or isotypes) called IgG, IgA, IgM, IgD, and IgE. The specificity of an antibody is determined by the amino acid sequences of the variable regions of the heavy and light chains, which together constitute a complete immunoglobulin molecule (Figure 1). The structure of the constant regions of the heavy chains determines the biological properties of the immunoglobulin isotype, such as opsonization, complement fixation and, in the case of IgE, binding to mast cells. Both complement and phagocytes are essential for humoral immunity, as illustrated by the susceptibility to infection of patients who lack either, but nevertheless make good antibody responses.

The other effector mechanism of the specific immune response is cell-mediated immunity, and this depends on a population of thymus derived (or T) lymphocytes. Activated T lymphocytes kill cellular targets directly, and they also recruit other cell types, particularly macrophages, into inflammatory responses by releasing a

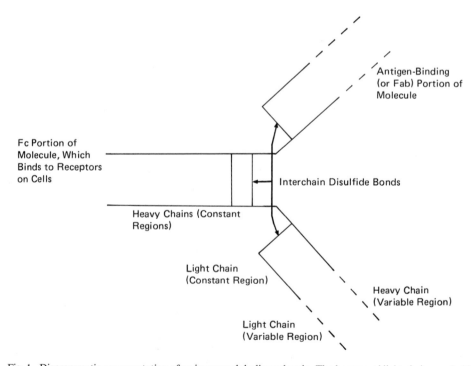

Fig. 1. Diagrammatic representation of an immunoglobulin molecule. The heavy and light chains are held together by disulfide bonds and each chain has a series of loops formed through intrachain bonds (which are not illustrated). The antigen-binding cleft is composed of the variable regions of the heavy and light chains.

number of mediators called lymphokines. Human T lymphocytes are identified by their ability to rosette with sheep erythocytes and also by staining with specific anti-T antisera raised in other animals. When exposed to antigens such as lymphocytes bearing foreign tissue-type (or histocompatibility) antigens, T lymphocytes proliferate: this is called the mixed lymphocyte reaction. T lymphocytes also proliferate following stimulation *in vitro* by certain mitogenic plant extracts such as phytohemagglutinin. This latter response is considered to be nonspecific since most T cells are stimulated; in contrast, stimulation by antigen elicits a response only by those cells with appropriate receptors for the antigen tested.

The T lymphocytes provide essential help to B cells which enables the latter to respond to antigen by differentiation into plasma cells, but T cells themselves never make antibody. B lymphocytes cultured *in vitro* with pokeweed mitogen respond by differentiating into plasma cells, provided that T cells are present to provide help. This laboratory model provides information on the maturity of the responding lymphocytes and their ability to interact. A subpopulation of T lymphocytes, different from those which help, mediates suppression, so that antibody formation is subject to both positive and negative control.

The immune system discriminates between a large number ($10^5$ or more) of antigens, and it is generally accepted that this diversity of response depends on the existence of an equivalent number of lymphocyte clones, each with a single antigen-binding specificity. Two of the central questions in contemporary immunology are: (1) What is the nature and specificity of the T lymphocyte receptor for antigen? (2) How is the diversity of receptor specificities (on T or B cells) generated? Developmental studies have contributed little to the former, but they are pertinent to the latter since diversity appears during ontogeny.

## 2.1. Immune Responsiveness in the Human Fetus

Healthy babies are born with low serum levels of IgM and IgA and their lymph nodes are small, with few if any plasma cells and no germinal centers. In contrast, prenatally infected infants, who have been subject to antigen stimulation *in utero,* have both morphological and functional evidence of active immune responses. These observations provide secure evidence for the immunological potential of the fetus. The morphological evidence is based on the finding of plasma cells in the tissues of syphilitic fetuses. These were recognized as long ago as 1904 and, more recently, Silverstein (1962) established that 20 weeks was the earliest gestational age at which they appeared. This age may well be the earliest at which the fetus can respond to the spirochete since organisms, but not plasma cells, were seen in those few younger infected fetuses which were examined. Plasma cells were also found in fetuses aborted with congenital toxoplasmosis, and in this infection, as in congenital syphilis, it is likely that these cells are engaged in making antibody to the infecting organism. Direct evidence, however, is lacking until birth and even then is clouded by the presence of maternal IgG antibody. In the case of syphilis it is likely that the fetus contributes to the Wasserman reactive antibody since the titer was found to rise following delivery, instead of falling as would be expected if maternal immunoglobulin was solely responsible. Infants with congenital toxoplasmosis have raised IgM levels in cord serum, and this material has antibody activity in the Sabin Feldman test. Since maternal IgM normally does not cross the placenta, any found in cord serum is assumed to be of fetal origin. Evidence of fetal production of antibody is better established in the case of congenital rubella and cytomegalovirus infections. Both these viruses can cross the placenta to produce a range of birth defects, the severity of which are, in part, related to the gestational age at the time of infection. Since the defects are milder as gestational age increases, it is possible that the fetuses' own immune responses help to limit infection, although less mature fetuses may also be more vulnerable because the virus can interfere with earlier stages of differentiation. Newborn infants with congenital rubella have raised cord serum levels of IgM, and frequently of IgA, each with specific antibody activity. The immunopathology of congenital rubella is complicated by the interference of this virus with growth and secondary effects on the immune system. A few infants with congenital rubella have IgA deficiency or, more rarely, IgG deficiency. Such immunoglobulin class restriction seems unlikely to result from direct damage to plasma cell precursors by the virus, and an alternative possibility, that the virus interferes with thymus-dependent lymphocytes which control immune responses, is discussed later.

The survival of infants given intrauterine blood transfusions for the treatment of hemolytic disease of the newborn due to rhesus incompatibility provides indirect evidence for another type of immune response by fetuses. Such transfusions contain mature lymphocytes from the donor which, if given to an immunologically unresponsive individual, would be expected to cause a fatal graft-versus-host disease. The fact that, with certain rare exceptions, the transfused cells have not been harmful suggests that the donor's lymphocytes were killed by the fetus. Fetal lymphocytes can kill target cells (in tissue culture) from about 15 weeks of gestation, so this effector mechanism of the immune response appears some 10 weeks before intrauterine transfusion is usually attempted.

The mechanisms of immunological responsiveness of the healthy human fetus are largely unknown and, because of their inaccessibility, are likely to remain so. Specific immune responses depend on lymphocytes, so morphological and functional studies of the appearance and distribution of these cells give some information on the potential of the developing fetus.

## 3. Lymphocytes

### 3.1. Stem Cells and the Development of Lymphocytes

In mice, hemopoietic and leukopoietic cells have a common precursor which is derived from the yolk sac and subsequently migrates to hemopoietic organs, including the fetal liver, and later the bone marrow; this sequence is likely also to be true of other mammals. During adult life some precursor cells persist in the marrow as judged by the ability of transferred marrow cells to repopulate irradiated recipients. Colonies derived from a single stem cell, grown either in agar or isolated from the spleens of the irradiated animals, have few cells with obvious lymphoid morphology but contain cells which can differentiate into lymphocytes, megakaryocytes, and red cells in further transfers. Recent evidence in man, based on differences in the G6PD isoenzymes of different cell lines from patients with polycythemia vera, indicates that later stages of stem cells may differentiate to provide separate precursors for lymphocytes and for the erythroid and myeloid series. This division is consistent with the findings in some rare immunodeficiency diseases in which there may be a combined failure of both lymphocytes and granulocytes or a failure of lymphocyte differentiation alone.

Lymphocytes are heterogeneous in origin and function, and conventional morphological criteria do not distinguish between them. For simplicity the histological observations on human fetuses are described first and the differences in lymphocyte function and their development second. In fact the site at which the first lymphocytes appear in man is disputed; most early workers have found them in the thymus first, between 7 and 9 weeks of gestation. Others have found lymphocytes almost simultaneously in the thymus, blood, and mesenteric lymph nodes. The actual number of lymphocytes in the thymus increases at an exponential rate from about 10 weeks of gestation. Corticomedullary differentiation becomes apparent at about 10 weeks, and Hassals corpuscles appear after about 12 weeks. The thymus grows rapidly throughout the fetal period and reaches its maximum size in relation to body weight at birth. Postnatal growth is slower and continues until puberty, after which much of its substance is replaced by fatty tissue. The reticular, vascular, and macrophage components of lymph nodes mature early but, as stated previously, signs of secondary lymphoid activity only develop after antigen exposure; this occurs at birth for healthy fetuses, but earlier for those with congenital infections.

## 3.2. Development of T Lymphocytes

The first T lymphocytes, as judged by rosette formation with sheep erythrocytes or *in vitro* proliferative responses, appear in the human fetal thymus between 10 and 12 weeks of gestation. Experiments in animals suggest that the precursors of these fetal T cells arise in the fetal liver and, later, in the bone marrow. They reach the thymus via the blood stream, and their localization in the thymus is dependent on the activity of thymic epithelial cells. The differentiation of these precursors into T cells is also dependent on thymic epithelial cells and is mediated by humoral factors. Maturation is accompanied by the acquisition of new surface antigens and receptors. In mice both TL and Θ antigens are acquired, and in man the capacity to make rosettes is acquired. Comparable appearance of new surface antigens can be induced in subpopulations of precursor lymphoid cells, isolated from marrow or spleen, by incubating them with thymus extracts or with cyclic AMP or its inducers. However, such induced cells still localize to the thymus when injected into suitable recipients so they do not behave like fully mature T cells. One thymic humoral factor which is active *in vitro* has been isolated and characterized as a 49 amino acid polypeptide; a synthetic pentapeptide subunit was also effective at inducing T-cell antigens on spleen cells from athymic mice.

Human fetal thymus cells proliferate more in response to PHA than do adult thymus cells. This probably means that the fetal thymus contains more mature lymphocytes than adult thymus, perhaps because such cells are held in the fetal thymus before joining the recirculating pool. Mature lymphocytes in adult animal thymuses do not disappear during cortisone-induced involution, and they are mostly located in the medulla. Fetal thymus has a relatively larger proportion of medullary lymphocytes which presumably include those responsive to PHA. The ability of fetal thymus lymphocytes to respond with immunological specificity indicates that maturation can take place completely within the thymus, while in adults some steps in the maturation may be extrathymic. Dissemination of human T lymphocytes to the blood and to peripheral lymphoid tissues starts at about 14 weeks of gestation as judged by the appearance of T cells in the fetal spleen and liver (Figure 2).

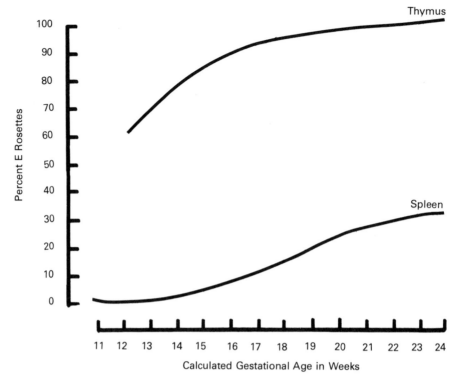

Fig. 2. Percentage of T lymphocytes (identified by E rosetting) in human fetal thymus and spleen.

### 3.3. The Thymus and Generation of Diversity

The wide range of antigens which can be distinguished by the immune system is thought to reflect an equal diversity in the specificity of individual lymphocytes. It has become clear over the past few years that at least part of the specificity repertoire is encoded in the genome, but whether it all is, or whether a significant expansion of diversity takes place by somatic mutation during ontogeny, remains controversial. In any event, a large number of lymphocytes must be produced in the developing animal, and the thymus, because of its high rate of antigen-independent lymphopoiesis, is an obvious site for this production. Since not all lymphocytes that arise in the thymus leave it (the precise fraction is unclear), it has been suggested that those cells that are destroyed within it are cells reactive to self-antigens. Consistent with this hypothesis is the observation that congenitally thymus-deficient mice grafted with an allogeneic thymus become at least partially tolerant of the thymus donor's histocompatibility antigens.

The early appearance and high proportion of antigen-binding cells in the human fetal thymus is hard to interpret. Many of the cells which bind one antigen can also bind another, so if these cells are T lymphocyte precursors, they do not appear to be restricted in their immunological reactivity. Recent evidence suggesting that T cells use the same heavy-chain variable regions to bind antigens as do B cells should provide the means to resolve this dilemma.

### 3.4.1. Pre-B Cells

The development of the B lymphocyte series is much more amenable to study than that of T cells because the cells involved can be securely identified by their synthesis of immunoglobulin. In mammals the most primitive B cell yet identified by direct immunofluorescence is the pre-B cell. This is a mononuclear cell which contains small amounts of IgM in its cytoplasm but has no detectable surface immunoglobulin. In man the first such cells appear in the fetal liver at about 8 weeks of gestation, and they are found from 13 weeks onwards in the fetal bone marrow. In adults they are found in the bone marrow only. The pre-B cells of fetuses and adults divide rapidly; this behavior in even young fetuses suggests that division is not dependent on antigen stimulation and reflects some as yet undiscovered mechanisms which maintain B cell homeostasis. Pre-B cells mature into B lymphocytes as shown by the appearance of cells bearing surface IgM in cultures of rabbit or mouse liver cells obtained from fetuses at a time before the first B cells develop *in vivo*. In addition to the absence of cell surface immunoglobulin, pre-B cells lack other cell-surface characteristics of B lymphocytes such as receptors for C3 or the Fc region of IgG and, in mice, cell surface antigens coded for in the Ia region of the genome. Experimental animals treated with anti-immunoglobulin antisera (anti-$\mu$ in mice and anti-allotype in rabbits) lose their B cells, but their pre-B cells survive. This is further evidence that pre-B cells lack functional surface receptors for antigen, and it suggests that these precursor cells could survive in antigenic environments which would induce tolerance in more mature cells. The tendency for experimentally induced tolerance to wane spontaneously supports the notion that precursor cells are not permanently inactivated.

Pre-B cells contain both the heavy- and light-chain components of IgM molecules so, if these were associated within the cell to form a complete molecule, it is likely that they would have antigen-binding activity. By producing antibodies specific for the antigen-binding site, it may soon become possible to identify the pre-B precursors of myeloma cells. Such a finding would imply restriction of diversity at a pre-B cell stage and so dissociate the generation of diversity, in the B series at least, from the response to antigen.

### 3.4.2. B Lymphocytes

In man the first B lymphocytes appear in the fetal liver at about 10 weeks of gestation, some 2 weeks after appearance of pre-B cells. Their development in relation to T cells is summarized in Table I. In fetal rabbits and mice the interval between the development of these two cell types is only about 3–4 days. In all three species the earliest B cells lack Fc and C3 receptors, and in mice they lack Ia determinants for about the first week of life (Table II). Immature mouse and human B cells lack cell-surface IgD; this immunoglobulin class has not been identified in rabbits. It is possible that the lack of Ia antigens from the surface of immature B cells contributes to the ease with which these cells become specifically tolerant to multivalent antigens to which they have been exposed in the absence of specific helper T cells. The latter observation applies also to the immature B cells obtained from the bone marrow of adult animals, so this possible mechanism of tolerance

Table I.   Summary of Lymphocyte Development in Human Fetuses

| T cells | Postmenstrual age (weeks) | B cells |
|---|:---:|---|
| Epithelial thymus | 5 | |
| | 6 | |
| | 7 | Pre-B cells in liver |
| Thymus becomes lymphoid | 8 | |
| | 9 | Surface IgM cells in liver |
| Cortico medullary differentiation | 10 | Surface IgG cells in liver |
| | 11 | Surface IgA cells in liver |
| | 12 | Most sIgM cells have sIgD |
| | 13 | |
| Hassals corpuscles appear | 14 | B cells in blood and tissues |
| | 15 | |
| T cells in spleen | 16 | |

induction could apply throughout an animals' life. Curiously the development of C3 receptors and IgD on the surface of mouse B cells lags 1–2 weeks behind the appearance of surface Ia. Little is known of the functional significance of these surface receptors.

In the human fetus, B cells are found in the blood and spleen in the weeks immediately following their appearance in fetal liver. The earliest B cells with surface IgG appear at about 11 weeks, and those with surface IgA follow at about 12 weeks of gestation. From about 14 weeks onwards the percentages of B cells bearing surface IgM, IgG, IgA, or IgD in blood or spleen is close to that found in adults. Since the antigen exposure of human fetuses is likely to be small, it seems probable that the production and dissemination of B lymphocytes with the various classes of surface immunoglobulin is not dependent on antigen stimulus. The normal development of the B lymphocyte population in inbred germ-free mice confirms this. A major difference between mature and immature B lymphocytes is the presence of multiple immunoglobulin heavy-chain classes on the surface of the latter. This may in part reflect the much higher proportion of memory B cells (which would be expected to be restricted to a single heavy-chain class) in mature animals. The mechanisms by which immature cells can simultaneously synthesize molecules with different heavy chains but with, presumably, the same combining specificity are obscure. At present the most likely explanation appears to be that the developing B cell makes one or more copies of the selected variable region gene and links these with the appropriate genes for the heavy chains. An alternative theory is that individual cells are preprogrammed to switch from surface IgM expression to a single other class but that there is persistence of messenger RNA for the surface IgM for perhaps one or more cycles of cell division after the switch is made.

Table II.   Special Characteristics of Immature B Lymphocytes

1. Have IgM as only class of cell surface immunoglobulin
2. Expression of surface immunoglobulin readily inhibited by anti-immunoglobulin antibodies
3. Lack IgD
4. Lack C3 receptor
5. Lack surface Ia antigens (in mice)

Plasma cells appear spontaneously in human fetal spleen and liver at 18–20 weeks of gestation. The earliest cells stain for IgM; about 2 weeks later others stain for IgG. Few or no IgA plasma cells have been seen during fetal life. The maturation of lymphocyte interactions, on which the induction of plasma cells depends, is discussed later.

## 4. Development of Immunoglobulins in Human Fetuses

In general, the observations of immunoglobulin synthesis by human fetal tissues have complemented the results of the cellular studies so far described. Ten weeks is the earliest age at which IgM biosynthesis has been described using radiolabeled amino acids and fetal spleen lymphocytes. IgM in fetal serum becomes detectable by radial immunodiffusion around 20 weeks of gestation, and levels rise to 5–20% of adult values by birth. This rise results from fetal synthesis since IgM does not cross the placenta.

The first IgG is found in fetal sera as early as 38 days of gestation, but it is probably derived from the mother. Its concentration remains at 5–8% of maternal concentration until after 17–22 weeks when it rises to between 10 and 20% and after 26 weeks it equals maternal concentration. These changes result from altera-tions in placental permeability and the maturation of an IgG-specific carrier mecha-nism. Fetal tissues incorporate amino acids into IgG at about 11 weeks of gestation, close to the time at which the first IgG B lymphocytes appear. Synthesis of IgE by liver and lung was detected in one report as early as 11 weeks, but none was made by spleen cells until 21 weeks. The positive results with 11-week fetal lung are remarkable and require confirmation. Unfortunately it has been difficult in the past to verify the specificity of anti-IgE antisera. The total amounts of IgE made by the fetus are likely to be small since IgE is not normally found in cord blood serum and newborns do not react to skin tests with anti-IgE. They do react in the Prausnitz-Kustner test, so the effector limb of the immediate hypersensitivity response matures early. IgA is present in low concentration in homogenized preparations of fetal gut at 13 weeks and in trace amounts in cord serum; however, biosynthesis in tissue culture has not been demonstrated. The IgA found in the gut may have been derived from swallowed maternal amniotic fluid.

## 5. Immune Responses

### 5.1. Animal Fetuses

This section is concerned predominantly with experiments on fetal sheep because the remarkable tolerance of the mother and fetus to surgery facilitates experiments in this species. Ovine gestation takes 150 days, and the thymus is the first tissue to become lymphoid, between 41 and 43 days. Lymphocytes accumulate in nodes after about 45 days; the spleen and gut become lymphoid later, after 58 and 75 days, respectively. Compared with these times, the appearance of antibody on day 41 to bacteriophage $\phi$X 174 which had been injected on day 37 is remarkably early. With increasing gestational age the fetuses make antibody to a range of antigens in a reproducible sequence (Table III).

Other animal species respond to antigens in a different sequence, but within the

Table III. Gestational Age at
Which Fetal Sheep Make Immune
Responses[a]

| Antigen | Age |
|---|---|
| $\phi$X 174 bacteriophage | 41 |
| Ferritin | 56 |
| Q fever vaccine | <65 |
| Skin graft rejection | 75 |
| Haemocyanin | 80 |
| SV 40 virus | 90 |
| DNP | <110 |
| Arsanilate | <110 |
| Ovalbumin | 120 |
| Blue-tongue virus | 122 |
| LCM virus | 140 |

[a]Based on information presented by A. M. Silverstein at the 2nd International Congress of Immunology, Brighton, 1974.

species the order is constant. Fetal lambs below 84 days of gestation made antibody mainly of the IgM class, but apart from this the quality of the fetal immune response as assessed by antigen elimination was similar to that of adults. Fetuses injected with antigen before the time at which they became responsive to it did not become tolerant, even if the antigen was made to persist by incorporating it in adjuvant. These results suggest that the development of capacity to respond to antigen is genetically determined, and experiments to determine the cellular basis of the control are discussed next.

### 5.1.1. T Lymphocyte Responses

Fetal T lymphocytes provided help in antibody responses to ovalbumin at 89 days, which is about 30 days before the fetuses makes antibody to this antigen. It is therefore unlikely that the helper function of T lymphocytes determines the sequential appearance of antigen reactivity. Another T-lymphocyte-dependent response, the mixed lymphocyte reaction, is detectable early in gestation in several species, including rats.

### 5.1.2. Antigen-Binding Cells

The immune response in mature animals is initiated by the binding of antigen to receptors on the surface of lymphocytes. Using appropriately labeled antigens, such binding can be visualized directly, without relying on the effector phase of the immune response. Most tests for antigen binding detect B lymphocytes, although the tissue distribution of responsive cells suggests that T lymphocytes can also react. Antigen-binding cells appear early in gestation in mice and rabbits and other species including man, providing evidence for antigen recognition by the reactive cells. Ovalbumin-binding cells were present in the 58-day fetal sheep spleen, which is long before the fetuses make antibody to this antigen. Binding of a range of antigens developed simultaneously in mice, so it seems unlikely that this controls the sequential appearance of antibody. It is, however, uncertain that antigen binding by fetal lymphocytes necessarily indicates a capacity to react because the percentage of binding cells for single antigens is high and, in the thymus, tends to fall with increasing gestational age.

### 5.1.3. Other Control Mechanisms

Antibody responses by neonatal mice to sheep erythrocytes and neonatal rats to a range of antigens, including *Listeria* monocytogenes, were increased when injections of adult monocytes were added, suggesting that immaturity of these cells was a possible limiting factor in immunity responses by immature animals. This interpretation was open to the objection that some adult lymphocytes had been transferred along with the monocytes, but other *in vivo* experiments suggested that the neonatal mononuclear phagocyte system was more susceptible to blockage than that of adults. More recently Mosier and Johnson (1975) found that spleen cells of young mice responded well to a T-independent antigen, DNP-ficoll, especially if the T lymphocytes were eliminated. The response to a macrophage and T-cell-dependent antigen was found only if the autologous T lymphocytes were replaced by T cells from adult animals. Since the antibody response by mature cells was inhibited by neonatal T lymphocytes, it seemed likely that antibody responses by immature animals might be modulated by suppressor T cells.

Another, but not necessarily exclusive, possibility is that the delayed appearance of lymphocyte-cooperation-determining factors on the surface of fetal B lymphocytes determines the sequential appearance of antibody responses. Ia antigens, which are expressed after SmIg, are obvious candidates since they may be antigen specific.

### 5.2. Lymphocyte Interactions in the Development of Human Fetal Responses

As indicated above, the response of B lymphocytes to most antigens is dependent on appropriate interactions with T lymphocytes and macrophages. It seems likely that the development of antibody responses in human fetuses is regulated more by such cooperating factors than by the appearance of B lymphocytes. For example, B lymphocytes are present in the fetal spleen by 15 weeks of gestation, but they do not differentiate into IgM-containing plasma cells in pokeweed mitogen (PWM) stimulated tissue cultures unless adult T lymphocytes are added. By 18 weeks and beyond there are sufficient T cells in the fetal spleen to permit an IgM plasma cell response to PWM, but the addition of adult T cells increases their number and enables a few IgG-containing cells to appear. The plasma cell response of PWM-stimulated newborn blood lymphocytes is very low compared with that of the adult unless the newborn B lymphocytes are separated out and cultured with adult T cells. It therefore appears that newborn T lymphocytes are relatively inefficient helpers for the B-cell response to PWM and they also help adult B cells poorly. Newborn infants' T lymphocytes tend to suppress the response of adult blood lymphocytes to PWM, although their activity in this respect is somewhat variable. In addition, T cells of the newborn suppress adult T lymphocyte responses to PHA and to a foreign histocompatibility stimulus in mixed lymphocyte culture. These various examples of suppression by newborn T cells could be of biological importance in maternofetal relationships, since the fetus has to escape rejection as an allograft by the mother.

A. R. HAYWARD

In general, healthy individuals do not make antibodies to their own tissues or serum proteins. The fact that they do so in a variety of diseases indicates that B cells with specificity for self can exist, but they must normally be kept in check. It is clear that at some stage in the development of the immune system a distinction has to be made between what is self and what is foreign. The instructions for making this decision might be carried in the genome or they might be acquired during the period of embryonic life which precedes immunological competence. These are not mutually exclusive alternatives, and there is some evidence that both may operate. There are numerous examples in laboratory animals of a genetically determined failure to respond to certain simple synthetic antigens. As might be expected, the capacity to respond is inherited as a dominant. The complexity of most naturally occurring antigens makes comparable studies in man difficult, but the segregation within families of susceptibility to some infections makes it likely that such "germ-line" tolerance could exist in humans too. Interestingly, genetically unresponsive animals possess B lymphocytes which can bind the tolerated antigen, and their failure to respond is determined by T lymphocytes.

Acquired tolerance can be induced in lymphocytes of the T or B series. Interference with the response of either cell line is sufficient to prevent an immune response to most antigens because of the requirement for lymphocyte cooperation. Repeated injections of small amounts of antigens into mature animals induces a state called "low-zone tolerance" in which the T cells become unresponsive; in some situations this results from the production of specific suppressor T cells. It is conceivable that new potential antigens, as they arise during ontogeny, might initially be present in low enough concentrations to induce low-zone tolerance, but there is no experimental evidence relating to this.

B lymphocytes of newborn mice are particularly sensitive to tolerance induction. The reasons for this susceptibility are not clear; it may result from intrinsic characteristics of immature B cells (see above), or it may result from excessive antigen exposure in the absence of T cells or macrophages. Mouse splenic B cells lose much of their susceptibility to tolerance induction by the age of 2 weeks, and prior to this they are more resistant to tolerance if they are exposed to antigen in the presence of specific helper T cells.

Whatever mechanisms are involved, it has generally been easier to induce tolerance in young animals than old, and this leads to the tempting speculation that the events which facilitate experimental tolerance induction resemble those which result in tolerance of self. If this is so, then the period of self-tolerance induction during embryonic life must be very short since antigen (in the form of congenital infection) experienced in the first trimester induces antibody and not tolerance.

## 7. Development of Nonspecific Effector Mechanisms

Antigen elimination is a requirement for the normal maturation of immune responses. Complement, macrophages, and polymorphs all contribute to the clearance of opsonized antigen, and immaturity of any of these could impair the immune responses of the young animal. The functional immaturity of newborn animals' macrophages was mentioned above but there is no comparable data for humans. Newborn infants' neutrophils migrate in response to chemotactic stimuli less well than adult cells, although their random mobility is similar. They are actively phagocytic and kill ingested bacteria well, except in the first few hours following birth. The principal difference between the over-all phagocytic activity of cord and adult blood lies in the relatively poor opsonic capacity of cord serum. Much of this deficiency is due to lack of IgM antibodies, but some of it is also due to the reduced susceptibility of cord serum complement to activation by the alternative pathway. Most of the individual complement components which have been studied are synthesized from an early stage of fetal development, but the over-all activity of the complement sequence is low at birth, with hemolytic titers about half of adult values.

## 8. Immunity at Birth

Several maternal factors contribute to the infant's immunity potential at the time of birth. The best characterized is maternal IgG which has been acquired across the placenta. Maternal antibody has clearly protective effects, as seen in the reduced incidence of tetanus infections in infants born to mothers with high titres of antitetanus antibody. In experimental animals, passively administered IgG-specific antibody usually reduces the recipient's own immune response to the corresponding antigen, largely because the antibody hastens antigen elimination. In man, this effect is apparent in infants immunized early while they still have significant amounts of maternal antibody, since their antibody responses are low compared with infants who do not have maternal antibody. Maternal IgA in breast milk is also protective (mortality from gastroenteritis is apparently lower in breast-fed infants) and interferes with oral immunization by attenuated viruses such as those of poliomyelitis. In these examples protection from infection and interference with immunization are different consequences of passively conferred immunity.

Some studies have indicated that the newborns' lymphocytes proliferate in tissue culture when stimulated by antigens to which their mothers' lymphocytes respond. Since the fetus is not likely to have been exposed to the environmental antigens tested during intrauterine life, these observations suggest that antigen-specific factors capable of priming for cell-mediated immune responses are transferred across the placenta. Maternal lymphocytes could do this, but several karyotype studies of male infants' blood lymphocytes have failed to disclose significant numbers of XX cells. This may not be surprising since fetal lymphocytes would be expected to be primed against maternal histocompatibility antigens, so that any cells which did enter the fetus would be killed. Maternal lymphocytes could be acquired from breast milk, as well as possible placental transfer of cells, since this contains large numbers of T and B lymphocytes. Conceivably, lymphocytes acquired by any means might liberate transfer factor and confer specific responsiveness on the newborn infant.

## 9. Growth Failure and Immunodeficiency

### 9.1. Animal Models

Congenital hypopituitarism, with autosomal recessive inheritance, is found in two unrelated inbred mouse strains, known as Snell-Bragg and Ames. Affected animals appear normal at birth but after weaning they become lymphopenic, and the thymus and thymus-dependent areas of spleen and lymph node atrophy. They reject skin grafts poorly and make feeble antibody responses unless repeatedly immunized, although their serum immunoglobulin concentrations remain normal. The immunological effects are interpretable as a lack of T lymphocyte function and the endocrinological effects are predominantly those of thyroxine and growth-hormone deficiency. The life-span of these mice is considerably reduced with some features of premature senescence. The most important observation is that their immune responsiveness and their normal life-span are restored by injections of growth hormone and thyroxine. Immunological reconstitution by these hormones is abrogated by prior thymectomy, presumably because maturation of precursor cells requires an intact thymic microenvironment. The immunodeficiency may be mediated through a primary effect of hormonal lack on thymic epithelial cells, or it may operate at a stem cell level. If it affects the latter, then only lymphoid precursor is involved since colony-forming units differentiated normally.

Ames and Snell-Bragg mice frequently develop a chronic runting disease similar to that seen in homozygous nude mice. These hairless variants have severe hypoplasia of the thymus and the T-dependent lymphocyte series. However, in germ-free environments they grow normally so the runting is secondary to infection.

### 9.2. Human Correlations

Most patients with severe congenital immunodeficiencies fail to thrive unless they are maintained in germ-free environments. Their growth failure is presumably secondary to the consequences of the immunodeficiency, as in nude mice. In a small number of patients, such as those with adenosine deaminase deficiency, the development of both T lymphocytes and bone cells is adversely affected by a primary enzyme deficiency. In the case of lymphocytes, the interference with development appears to result from excessive intracellular levels of adenosine and ATP, which render the cell less susceptible to triggering. Insensitivity of bone cells to their hormonal influences could presumably be mediated through the same pathway. Other rare, genetically determined, disorders in which various degrees of immunodeficiency are associated with various degrees of developmental retardation include ataxia telagiectasia, cartilage-hair hypoplasia, and a syndrome incorporating antibody deficiency, ectodermal dysplasia, and short-limbed dwarfism. However, the adverse effects of immunodeficiency on growth outlined above make these associations hard to interpret, particularly in view of their phenotypic variations. Patients with primary hypopituitarism or isolated growth-hormone deficiency generally have normal lymphocyte counts and serum immunoglobulins. Apart from genetic differences, the normal results in humans compared with the immunodeficiency found in hypopituitary mice may reflect the protection afforded to human fetuses during their long intrauterine gestation, permitting adequate time for the immune system to develop before birth. Intrauterine malnutrition has contrasting effects, producing both growth retardation and congenital (but usually transient) immunodeficiency.

### 10. Conclusions

Lymphocytes which can recognize antigen appear by the third month of gestation in human fetuses. Multiple factors appear to influence the responsiveness of these cells, particularly enhancing their tendency to make IgM antibody responses or to become tolerant. These include specific control by fetal T lymphocytes, deficiency in function or number of macrophages, and possibly delay in the full maturation of B lymphocytes. Presumably these control mechanisms contribute to the regulation of maternofetal immune responses, but they do not abolish the capacity of the developing fetus to respond to strong antigen stimuli such as congenital infections.

ACKNOWLEDGMENTS

I am grateful to Professors J. F. Soothill and M. D. Cooper, Drs. A. R. Lawton and J. F. Kearney, and to Bill Gathings for helpful discussions and access to unpublished data. Drs. E. Pearl and L. Vogler kindly criticized the manuscript, which was typed by Mrs. Nancy Perry.

Current research referred to in this chapter was supported by Grants 1-354, awarded by The National Foundation–March of Dimes, and in part by AI 11502, awarded by the NIAID, USPHS, and CA 16673, awarded by the National Cancer Institute, DHEW.

## 11. Bibliography

Adinofi, M. 1969, *Immunology and Development*, Spastic International Medical Publishers, London.

Blaese, R. M., and Lawrence, E. C., 1977, Development of macrophage function and the expression of immunocompetence, in: *Development of Host Defenses* (M. D. Cooper and D. H. Dayton, eds.), Raven Press, New York.

Hayward, A. R., and Soothill, J. F., 1972, Reaction to antigen by human fetal thymus lymphocytes, in: *Ontogeny of Acquired Immunity*, (R. Porter and J. Knight, eds.), pp. 261–268, Ciba Foundation Symposium.

Hess, M. W., 1968, *Experimental Thymectomy. Possibilities and Limitations*, Springer-Verlag, Berlin.

Lawton, A. R., and Cooper, M. D., 1978, Ontogeny of immunity, in: *Immunologic Disorders of Infants and Children* (R. Stiehm and V. Fulginiti, eds.), Saunders, Philadelphia.

Mosier, D. E., and Johnston, B. M., 1975, Ontogeny of mouse lymphocyte function. II. Development of the ability to produce antibody is modulated by T lymphocytes, *J. Exp. Med.* **141**:216.

Silverstein, A. M., 1962, Congenital syphilis and the timing of immunogensis in the human fetus, *Nature (London)* **194**:196.

Solomon, J. B., 1971, *Foetal and Neonatal Immunology*, North-Holland, Amsterdam.

# 21

# *Fetal Growth: Obstetric Implications*

## *KARLIS ADAMSONS*

## *1. Introduction*

The objective of this chapter is to provide the reader with a review of certain aspects of fetal growth as they relate to obstetric management. The term "obstetric" has been used in a narrow sense, which has limited the scope of discussion to clinical issues. Emphasis has been placed upon the gross morphologies of the fetus as they relate to obstetric management because functional development has already been covered in detail by other contributors to this volume.

## *2. Growth of the Placenta*

### *2.1. Normal Placenta*

#### *2.1.1. Normal Implantation*

The implantation of the placenta in the anterior aspect of the uterus is occasionally viewed by the obstetrician as less favorable than that occurring in the posterior wall of the uterus. This is due to the fact that amniocentesis, if required for diagnosis of the condition of the fetus, is more readily performed when the placenta does not occupy a major portion of the anterior aspect of the uterus. With the widespread use of ultrasonic localization of the placenta prior to amniocentesis, this factor is of less consequence today. In addition, the most favorable site for amniocentesis is in the midline in the lower uterine segment, which is rarely occupied by the normally implanted placenta.

---

***KARLIS ADAMSONS*** • Brown University Program in Medicine, Providence, Rhode Island.

The anteriorly implanted placenta may present difficulties at the time of cesarean section, particularly when a fundal incision is performed. This may lead to substantial fetal hemorrhage and to an increase in fetal blood transmitted to the mother by absorption from the mother's peritoneal cavity. The anterior location is obviously an asset when aspiration of placental tissue or fetal red cells is needed for diagnostic purposes. The same also pertains for detailed morphometric measurements by ultrasound.

### 2.1.2. Abnormal Implantation

Implantation of the placenta in the lower uterine segment with or without complete obstruction of the endocervical canal occurs in approximately 0.25% of all pregnancies. The relative frequency is increased in multiparous patients and in patients with history of dilatation and curettage. The obstetric implications depend upon the degree of obstruction of the endocervical canal. Total placenta previa, which occurs in 20–25% of such patients, obviously necessitates termination of pregnancy by hysterotomy. In patients with partial placenta previa, normal delivery can be anticipated in a substantial proportion as long as maternal or fetal bleeding does not require prompt termination of labor.

Fetal growth is likely to be affected by the abnormally implanted placenta. In a series of 150 cases of uncomplicated placenta previa followed to term, nearly 25% of fetuses had head sizes below the 10th percentile. Fetal–maternal bleeding is more common in patients with placenta previa, which is of particular significance in the Rh-negative mother.

Placenta accreta denotes a condition in which there is either a partial or a total absence of the decidua basalis, and the villi are in direct contact with the myometrium. Depending on the degree of penetration of the uterine wall by the chorionic tissue, one refers to placenta accreta, placenta increta, or placenta percreta. This abnormal attachment of the placenta may involve only a few cotyledons or the entire organ. The incidence in the reported series ranges between 0.1 and 0.003%. The morphologic details of this condition and the conditions which predispose the patient to placenta previa have been covered elsewhere. Suffice it to say that there is no cogent evidence that trauma to the decidua, such as may be inflicted at the time of curettage, increases the frequency of this condition significantly. The relative frequency of placenta accreta is markedly increased among patients with previous history of manual removal of the placenta for placental retention. As expected, there is also an association between placenta previa and the placenta accreta. It has been estimated that about 20% of placenta accreta is placenta previa.

The clinical management of patients with placenta accreta depends on the extent of the abnormal attachment, the timing of diagnosis, and the projected procreational expectations of the patient. The treatment of choice is abdominal hysterectomy as soon as the diagnosis is unequivocally established. The maternal mortality remains high, nevertheless.

The so-called "conservative" management of a patient with placenta accreta denotes a management in which the uterus is preserved. Because there is no natural cleavage plane between the placenta and the uterine wall, removal of the placenta inevitably leads to extensive trauma to the uterine wall and the attendant complications such as hemorrhage and infection. Because the diagnosis of placenta accreta is rarely, if ever, made before efforts to remove the placenta manually have failed, little data are available on the clinical course of those patients in whom the placenta was left *in situ* undisturbed. There are, however, a few reports in which the placenta accreta was not only spontaneously eliminated but in which subsequent pregnancies were entirely normal.

The risks to the mother with the "conservative" management of placenta accreta are judged considerably higher than those following hysterectomy. In larger series of cases maternal mortality rate has exceeded 40%. Because of the low relative frequency of placenta accreta, the statistical material is often gathered over many years, during which substantial changes in management of patients may have occurred. Thus, the maternal mortality rate of 40% may be misleadingly high under the standards and supportive measurements available today.

In contrast to the high risk for the mother, fetal outcome seems to be unaffected by placenta accreta, as measured by perinatal mortality rates and malformation rates for corresponding populations.

## 2.2. Abnormal Placenta

### 2.2.1. Acquired Placental Abnormalities

There are a variety of acquired conditions in which the alterations in placental structure have considerable bearing upon obstetric management. Such abnormalities may affect fetal tolerance to labor and also affect certain diagnostic and therapeutic measures.

Maternal anemia is known to be associated with an increase in placental mass, and, for still unknown reasons, with an increased incidence of premature separation of the placenta. Because maternal anemia is often associated with other adverse social and medical variables, the specific contribution of a lowering in oxygen-carrying capacity of maternal blood remains to be elucidated. It can be proposed that the lower oxygen tension in the blood of the intervillous space promotes growth and spread of the young villi.

The large, edematous placenta of the fetus suffering from erythroblastosis fetalis and other conditions leading to hemolysis of fetal red cells not infrequently creates problems for amniocentesis and performance of intraperitoneal transfusions of the fetus. The transfusions are particularly affected by the large, anteriorly implanted placenta because the site of penetration of the uterine wall is dictated by the location of the abdominal cavity of the fetus. This occasionally will necessitate that a rather large-bore needle be intentionally passed through the placenta, with its attendant risks of fetal hemorrhage. Conversely, the large placenta is an easy target for placental biopsy by aspiration.

Progressive edema of the villi, which is at least partly due to the relatively high pressure in the umbilical vein of the erythroblastotic fetus, reduces the volume of the intervillous space in spite of the large overall mass of the placenta. This, in turn, will reduce oxygen tension in the fetus and will decrease the fetal tolerance to uterine contractions. Similar considerations apply also to patients with diabetes mellitus without vascular involvement and those characterized as gestational diabetics.

The small placenta of the fetus of the mother with hypertension or other vascular disorders is associated with a significantly increased risk for both mother and fetus. The risk to the fetus is due to increased probability of premature separation of the placenta and to a decrease in villous surface and conductance through the intervillous space. Depending on the degree of reduction of blood flow through the intervillous space, there will be a reduction in fetal body mass and a reduction of fetal tolerance to labor. Fetal maturity, on the other hand, is greater than that of the normally supplied fetus.

The risks to the mother are secondary to the complications resulting from premature separation of the placenta. They include hemorrhage, infarction of the myometrium, intravascular coagulation, and shock. The risks to the fetus are also increased due to the abnormal uterine irritability often seen in patients with partial separation of the placenta. In recognition of this double jeopardy, the contemporary obstetric management of the patient with partial separation of the placenta is prompt delivery by hysterotomy unless adequate fetal oxygenation has been demonstrated by analysis of fetal blood.

The relative accuracy with which placental volume can be calculated from placental morphometry using ultrasound makes such measurement of considerable prognostic significance for both maternal and perinatal outcomes.

### 2.2.2. Developmental Abnormalities

Velamentous insertion of the umbilical cord, which is characterized by the separation of the umbilical vessels prior to their reaching the surface of the placenta, occurs in approximately 1% of all pregnancies. It is of relatively little consequence unless the placenta is implanted low in the uterine cavity and the separated vessels cross the amnion in the region near the internal os of the cervical canal. The latter condition is referred to as vasa previa and is of considerable obstetric significance. The clinical findings are painless vaginal bleeding of moderate amount, occasionally coupled with fetal heart rate changes suggestive of fetal hypovolemia or asphyxia. The sources of bleeding are ruptured fetal vessels. The diagnosis can be readily established by examining a sample of the blood present in the vagina for its resistance to denaturation by sodium hydroxide or the presence of nucleated erythrocytes. Because fetal bleeding can lead to acute fetal hypovolemia and death, even when the external blood loss appears to be only moderate, it is recommended that tests for fetal hemoglobin be done in all cases of painless vaginal bleeding. Vasa previa can also be diagnosed prior to their rupture by amnioscopy. Velamentous insertion of the umbilical cord also increases the likelihood of compression of fetal vessels by the presenting part.

Placenta membranacea, which is characterized by a retention of the functioning villi over the entire surface of the chorion, is a rare abnormality. The clinical picture is similar to that seen in patients with placenta previa. It is characterized by vaginal bleeding, increasing progressively through the second and third trimesters. Occasionally it becomes necessary to terminate pregnancy by hysterotomy. As in all other cases of painless vaginal bleeding, it is highly desirable to examine the blood for presence of fetal red cells. Poor separation of the placenta after delivery of the fetus is a known complication of placenta membranacea.

Placenta circumvallate, which is a variety of extrachorial placenta, is usually diagnosed only postpartum. With increasing use of sonography, however, it is likely that this form of placenta can be at least suspected prior to birth. Placenta circumvallate is associated with an increased rate of antepartum hemorrhage and prematurity.

## 3.1. Normal Fetus

Morphometry of fetal head by ultrasonography has added considerably to the diagnostic accuracy of various fetal conditions during gestation and in the management of the patient during labor. The biparietal diameter of the normally growing fetus appears to correlate better with fetal age than any other morphometric or biochemical indicator. Between the 12th and 20th week of gestation, the increase in biparietal diameter is approximately 0.3 cm/week. During this period fetal age in weeks can be rather accurately calculated by the following equation:

$$\text{Age (weeks)} = \frac{\text{biparietal diameter (cm)} + 1.65}{0.3}$$

Between the 20th week of gestation and term, the mean slope is approximately 0.21 cm, being greater than that initially and less toward the end of gestation, particularly during the last 5 weeks. Because of the greater variability of biparietal diameters toward the end of gestation, the predictive value of this measurement is less during that period.

The other clinical usefulness of sonographic cephalometry is to identify fetuses who are not growing normally *in utero*. Because fetal head growth is least affected by intrauterine malnutrition, irrespective of its cause, the reduced rate in head growth as ascertained by serial measurements of the biparietal diameter correlates well with the relative frequency of subnormal birth weight. Fetuses whose head growths are below the fifth percentile have about 80% probability of being either borderline or distinctly small for gestational age at delivery.

When sonographic measurements are depended upon in making irreversible clinical decisions, it is important to recognize that, even in the hands of experienced personnel, measurement errors do occur. This factor, in addition to the well-appreciated variability of fetal head size, particularly in the last 5 weeks of gestation (the mean at 35 weeks is within −2 standard deviations of the mean at 40 weeks), emphasizes the importance of including either biochemical or clinical indicators in the overall assessment of the fetus.

The assessment of fetal head size, in particular that of the biparietal diameter, is also important in the management of patients in labor with suspected disproportion between the fetus and the osseous structures of the pelvis, particularly when the relevant pelvic measurements have been obtained by X-ray examination. Without such information adequate clearance for a normal-sized pelvis is expected when the biparietal diameter does not exceed 9.6 cm. The decreasing use of X-ray pelvimetry, which is partly due to the growing concern for exposure of the fetus *in utero* and to the recognition that in the majority of cases the clinical progress of labor is more influential in determining the management than the numerical values obtained at cephalometry and pelvimetry, has somewhat reduced the role of accurate assessment of the size of the normal fetal head during labor.

## 3.2. Abnormal Fetus

Internal hydrocephalus, which complicates approximately one out of 2000 pregnancies, is perhaps the most significant cephalic abnormality from the obstetric point of view. Because of the progressive expansion of the cranium, which in some instances has been as much as 5 liters, there is disproportion between the fetal head and the bony pelvis. If the fetus is in breech presentation, which is often the case with hydrocephalic fetuses, the diagnosis may not be made until the delivery of the trunk. Under such circumstances, the head can only be delivered after partial decompression by aspirating the intraventricular fluid.

Because of the high relative frequency of other severe malformations, some of which are not compatible with extrauterine survival, there has been considerable reluctance to recommend the delivery of all fetuses with hydrocephaly by hysterotomy. Recently attention has been focused on diagnosing hydrocephalus early in gestation before extensive destruction of the cerebral hemispheres has taken place. The intent has been to perform a ventriculo amniotic shunt as a temporary measure until the delivery of the child.

There are no specific guidelines as to what biparietal diameter constitutes an unequivocal case of hydrocephaly. To eliminate serious errors in diagnosis, it is suggested that serial measurements be performed in all cases in which the biparietal diameter of the fetus near term is more than 11 cm. In those cases in which the age of the fetus is known with precision, more rigid criteria can be applied. For example, a biparietal diameter of 8.5 cm at 28 weeks, which is about 4 standard deviations above the mean, should provide an exceedingly high probability that one is dealing with a hydrocephalic fetus.

Anencephaly is a morphologic abnormality of the fetus which has several important obstetric implications. Presumably because of the low oxygen consumption of the fetus due to a greatly reduced brain mass, the fetus can be maintained *in utero* for a considerably longer period than the normal fetus. The average duration of a pregnancy complicated by anencephaly is about 310 days, and there is a well documented case in which the pregnancy continued with the live fetus for one year and 24 days. The total body mass of the anencephalic fetus, unlike that of the normal fetus, continues to increase past the 42nd week of gestation. Prior to that the anencephalic fetus weighs less than the normal fetus, the difference near term being approximately 500 g.

Because of the shape and size of the head structures and the often larger-than-normal circumference of the arms and thorax, the anencephalic fetus postterm can present formidable obstacles for vaginal delivery. In order to minimize the risk to the soft tissues of the mother, it may be necessary to deliver the fetus by hysterotomy.

Anencephaly is often suspected from the abnormal increase in the volume of amniotic fluid. It is confirmed by sonography or by X-ray examination of the uterine contents. Elevated concentration of alpha-fetoprotein in amniotic fluid is not specifically diagnostic of anencephaly since this protein is present in increased concentrations in amniotic fluid with other forms of neural tube defects such as meningomyelocele and rachischisis.

Microcephaly of the fetus is a rare disorder and usually coexists with generalized growth retardation of the fetus. The diagnosis is usually made by sonography. The neurologic outcome for the child, if it survives, is usually unfavorable.

Hydranencephaly cannot be diagnosed *in utero* by morphometric techniques. It can be suspected if, during labor, prolonged uterine contractions fail to exhibit heart rate patterns characteristic of transient hypoxia. Otherwise the condition has no specific obstetric considerations.

Because the volume of the fetal head is greater in proportion to its body earlier in pregnancy the premature fetus in breech presentation constitutes a special obstetric problem. Breech presentation is about ten times more common at 28 weeks of gestation than at term, when the incidence is about 3%. The principal risk for the premature fetus in breech presentation is due to the fact that the dilatation of the cervix and the soft tissues of the maternal birth canal by the trunk of the fetus is less than that required for the safe delivery of the head. This is further compounded by the lesser physical strength of the younger fetal tissues, including the blood vessels of the central nervous system. Due to the above and the fact that fetal oxygenation is often impaired in breech presentations, there has been an increasing trend to deliver all premature breeches by hysterotomy, as long as they are candidates for extra-uterine survival.

The obstetric considerations for the fetus classified as small for gestational age are similar to those for the young fetus, with the qualification that the disproportion between the circumference of the trunk and the head may be even greater than for the young fetus and that the tolerance of the fetus to oxygen deprivation is distinctly less than that of the young fetus.

The macrosomic fetus of diabetic mothers without vascular disease or mothers classified as gestational diabetics often presents an obstetric problem known as shoulder distocia. This term denotes the difficulty or inability to deliver the shoulders and the thorax of the fetus after the delivery of the head. This is chiefly due to the fact that the biparietal diameter of the fetus exposed *in utero* to hyperglycemia and hyperinsulinemia is not larger than that of the normal fetus, whereas the circumference at the level of the shoulders is. The risks to the fetus may range from a transient Erb's-type palsy, resulting from forceful traction of the arm, to death from asphyxia.

The increased awareness that certain groups of pregnant patients may exhibit significant hyperglycemia without other clinical or laboratory evidence of abnormal carbohydrate metabolism, and hence have hypersomia of the fetus, has placed the obstetrician on alert for this complication. Sonographic evaluation of the fetus toward the end of gestation has been particularly valuable. Attention is being paid to the diameter of the thorax, limbs, and the thickness of the placenta; all are significantly increased if the fetus is exposed to hyperglycemia and hyperinsulinemia. Unless the pelvic measurements are substantially larger than average, the delivery of the hypersomic fetus is best accomplished by hysterotomy.

Extreme distention of the fetal abdomen by ascites, tumors, or distended urinary bladder, may occasionally present obstetric complications and account for a relative fetopelvic disproportion. In most instances the underlying condition, as in the case of erythroblastosis fetalis, has provided the obstetrician with the needed morphometric or biochemical data for optimal management.

## 5. Tolerance of Fetus to Labor as a Function of Fetal Age and Size
### 5.1. The Prematurely Born Fetus

The obstetric considerations for a fetus prior to term are chiefly directed toward delaying delivery until sufficient maturation of the respiratory system is attained to ensure adequate oxygenation in the neonatal period. The therapy has been directed toward suppression of premature labor and acceleration of maturation of the fetus. Because subnormal oxygenation of the fetus (which often results in the presence of conditions either leading to or associated with premature labor) is known to accelerate rapidly the maturation of the lung, delay of delivery by as little as 24–36 hr has become of considerable clinical significance. A variety of pharmacologic agents, notably diazoxide and synthetic beta-mimetic sympathomimetic agents, have been found particularly efficacious tocolytics. Acceleration of fetal maturity has relied upon the biologic effect of glucocorticoids, which are readily transmitted across the human placenta. There have been isolated efforts to achieve the same objective by intraamniotic administration of thyroxin.

Oxygenation of the premature fetus and its monitoring by prevailing techniques are not substantially different than those pertaining to the fetus near term. Although it is correct to infer that the rate of change and the composition of fetal tissues for any given duration of interruption in oxygenation will be less in the immature fetus in contrast to that at term, it is incorrect to assume that certain clinical signs such as specific changes in fetal heart rate have basically different diagnostic significance. Specifically, the loss of variability of fetal heart rate or the appearance of uterine-contraction-induced hypoxic bradycardia ("late deceleration") must be viewed as unequivocal signs of marked impairment in fetal oxygenation.

### 5.2. The Normal Fetus Postterm

It is recognized that perinatal mortality rates are significantly affected by the duration of gestation, except over the interval of 38 to about 41 weeks. A variety of factors contribute to this phenomenon, with immaturity of the ventilatory system being the most important. The reasons for the increased mortality among fetuses older than 42 weeks (calculated from the last normal menstrual period) are less clear. Reduced tolerance to labor and the physical trauma of the delivery of the larger fetus are thought to be the major contributors. Impaired oxygenation of the older fetus during labor is due to the fact that fetal oxygen consumption increases exponentially as gestation progresses, whereas there is little change in the perfusion of the intervillous space, and hence in the delivery of oxygen to the fetus, during the last months of pregnancy. The statement that fetal oxygen consumption increases exponentially during gestation is based on direct measurements of oxygen consumption in a neutral thermal environment of newborn babies delivered between the 25th and 42nd weeks. According to these data, the doubling time of fetal oxygen consumption is about 40 days, reaching about 15 ml/min for the average-sized fetus at term. These measurements obviously do not include the oxygen consumption of the placenta, which must be taken into account in calculating oxygen availability to the fetus. Besides an over-all increase in fetal oxygen requirements, one has to consider the specific tolerance of the central nervous system to acute oxygen deprivation. Experiments with rhesus monkeys have demonstrated that the period of total asphyxia required to produce morphologically demonstrable CNS injury is three times as long at mid-gestation than at term. Recent data indicate that this reduction in tolerance to oxygen lack may be linked to the rate of accumulation of lactic acid in brain tissue, which in turn is contingent upon the carbohydrate stores of the fetus.

In considering the impact of intermittent reduction in intervillous space perfusion, as brought about by uterine contractions, upon oxygen availability to the fetus, the age of the fetus is of greater consequence than its total mass. This is due to the high relative proportion of total body oxygen consumption allocated to the fetal brain. The progressive diminution in the relative size of the extracellular volume of the fetus, which does not measurably participate in energy transformation, further dictates that total body oxygen consumption of the fetus will relate better to gestational age than to the total body mass.

It is customary in clinical practice to use the term "postmature" to denote a fetus older than 280 days from the time of presumed fertilization or 294 days from the time of last normal menstrual period. The perinatal mortality rate of the so-called postterm fetus has been found in large series to be about 2.2%, which is approximately three times that observed in term and near-term fetuses. Other indicators of fetal tolerance to labor, such as presence or absence of meconium in the amniotic fluid and fetal heart rate patterns indicative of hypoxia, are all supportive of the contention that the postmature fetus is less tolerant to oxygen deprivation.

Applying the changes in fetal oxygen consumption referred to above, the fetus at 42 weeks will consume approximately 30% more oxygen than the fetus at 40 weeks. This will necessitate an increased extraction of oxygen from the blood in the intervillous space and some lowering of fetal $Po_2$. Interruption in the perfusion of intervillous space will thus place the postmature fetus in double jeopardy due to smaller oxygen stores and an accelerated rate of their depletion.

Because labor is often induced in the pregnant patient postterm, which in itself may produce a less favorable pattern of uterine activity, the above-mentioned problems in fetal oxygenation are likely to be aggravated. Indeed, it has been found that the perinatal mortality rate is higher among patients in whom labor is induced than among those in whom labor begins spontaneously. In recognition of this fact, close surveillance of the fetus by electronic monitoring of fetal heart rate and periodic sampling of fetal blood is indicated in this high-risk group of patients.

### 5.3. The Abnormal Fetus

Reduction in metabolically active tissue results in an increased tolerance by the fetus to transient oxygen deprivation. This is best illustrated in cases of anencephaly, hydranencephaly, and hydrocephaly. During labor such fetuses characteristically exhibit a fixed heart rate and fail to develop bradycardia of the "late-deceleration type" even after an abnormally long uterine contraction. Such tolerance is anticipated in view of the high proportion of oxygen consumed by the central nervous system of the fetus. As mentioned elsewhere, the low total oxygen consumption of the fetus is thought to explain the fact that anencephalic fetuses can continue to grow *in utero* and remain well nourished up to ten weeks beyond the normal duration of gestation.

The converse is true for fetuses traditionally referred to as small for gestational age. Although satisfactory statistics are not available for this group, it is generally appreciated that intrapartum asphyxia occurs more frequently among such patients than in a normal population. The present explanation, which pertains particularly to infants whose mothers are hypertensive, is that the total oxygen consumption of the fetus is only slightly less than that of the normal fetus, whereas the placental mass, or more specifically the surface of the villi, is substantially less. Detailed morphometric studies have revealed that the mean villous surface of the placenta at term of fetuses whose mothers have hypertension may be less than 60% of norm.

# 6. Bibliography

KARLIS ADAMSONS

Brent, R. L., and Harris, M. I., eds., 1976, *Prevention of Embryonic, Fetal, and Perinatal Disease,* DHEW Publication No. 76-853, U.S. Government Printing Office, Washington, D.C.

Gluck, L., ed., 1977, *Intrauterine Asphyxia and the Development of Fetal Brain,* Year Book Publishing Company, New York.

Gruenwald, P., ed., 1975, *The Placenta and Its Maternal Supply Line,* University Park Press, Baltimore.

Notake, Y., and Suzuki, S., eds., 1977, *Biological and Clinical Aspects of the Fetus,* University Park Press, Baltimore.

Pritchard, J. A., and MacDonald, P. C., eds., 1976, *Williams Obstetrics,* Appleton-Century-Crofts, New York.

# *Index*